SPECIAL OPERATIONS FORCES
MEDICAL HANDBOOK

Special Operations Forces Medical Handbook
Editorial Staff, Reviewers, Authors and Contributors

Editorial Staff
COL Steve Yevich, MC, US Army, Sponsoring Editor
COL Warren Whitlock, MC, US Army, Chief Editor
LTC Richard Broadhurst, MC, ARNG, Senior Medical Editor
Gay Dews Thompson, RN, MPH, CHES, Managing Editor
Pete Redmond, Style Editor
CSM (Ret) Ron Packard, US Army, Production Manager

Joint Editorial Review Board
LTC Arthur Baker, MC, US Army
LTC Francis Balog, MS, US Army Reserve
LTC Charles Cannon, MS, US Army
Robert T. Clayton, SF, US Army (Ret)
SSgt Steven Cum, SOF Medic, US Air Force
CPT Leonard Gruppo, SP, PA, US Army
SFC Richard Hodges, SOF Medic, US Army
MSG Russell Justice, SOF Medic, US Army
MAJ Dennis Kilian, MS, US Army
HM1 Lynn Larraway, SOF Medic, US Navy
LTC Mark Lovell, MC, US Army
MAJ Robert Lutz, MC, US Army
Lt Col John McAtee, US Air Force, BSC, PA-C
COL George McMillian, DC, US Army
MSG Peter Pease, SOF Medic, US Army
MAJ Andre Pennardt, MC, US Army
HMC Dennis Polli, SOF Medic, US Navy
HM1 Alan Saviano, SOF Medic, US Navy
Maj Frederick Shuler, US Air Force, MC
LTC (Ret) Jon Zotter, VC, US Army

Authors and Contributors
LTC Eugene Acosta, AN, US Army
Lt. Col Robert Allen, US Air Force, MC
COL Naomi Aronson, MC, US Army
1LT Harold Becker, SP, PA, US Army Reserve
CPT William Bosworth, VC, US Army
Col Stephen Brietzke, US Air Force, MC
LTC Richard Broadhurst, MC, ARNG
COL David Burris, MC, US Army
LTC Howard Burtnett, AN, US Army
CAPT Frank Butler, MC, USN
Mark Calkins, MD
LTC Brian Campbell, MC, US Army
LTC Lee Cancio, MC, US Army
COL Ted Cieslak, MC, US Army
COL Clifford Cloonan, MC, US Army
SFC Jeff Crainich, SOF Medic, US Army
SSgt Steven Cum, SOF Medic, US Air Force
Maj Michael Curriston, US Air Force, MC
MAJ Michael Doyle, MC, US Army
LTC David DuBois, DC, US Army
COL Edward Eitzen, MC, US Army
CPT Joseph Fasano, AN, US Army
CDR Scott Flinn, MC, US Navy
MAJ Ann Friedmann, MC, US Army
CDR Raymond A. Fritz, MSC, US Navy Reserve
COL (Ret) Joel C. Gaydos, MC, US Army
SFC Dominique Greydanus, US Army
CPT Leonard Gruppo, SP, PA, US Army
Murray Hamlet, DVM
MAJ Karla Hansen, MC, US Army
Lt Col Aimee Hawley, US Air Force, MC
MAJ Steven Hendrix, AN, US Army
MAJ John P. Hlavnicka, AN, US Army
SFC Richard Hodges, SOF Medic, US Army
LTC John Holcomb, MC, US Army
CAPT Elwood Hopkins, MC, US Navy
LTC Duane Hospenthal, MC, US Army
CAPT Michael Hughey, MC, US Navy Reserve
LCDR Gary Ivey, MSC, US Navy
LCDR Christopher Jankosky, MC, US Navy
CAPT Robert Johnson, MC, US Navy
LTC Niranjan Kanesa-thasan, MC, US Army
COL Patrick Kelley, MC, US Army
LTC Richard Kramp, MC, US Army
CAPT Leo Kusuda, MC, US Navy
COL James Leech, MC, US Army
MAJ Mark A. Leszczynski, AN, US Army
CDR D. Mark Llewellyn, MC, US Navy
MAJ Thomas Lovas, MC, US Army
CPT Robert Mabry, MC, US Army
CPT Paul J. Maholtz III, AN, US Army
LTC Michael Matthews, MC, US Army
Lt Col John McAtee, US Air Force, BSC, PA-C
COL (Ret) Peter McNally, MC, US Army
COL (Ret) Alan Mease, MC, US Army
LCDR (Sel) Edward Moldenhauer, MSC, US Navy
CPT Brooks Morelock, MC, US Army
CPT Jeffrey Morgan, MC, US Army
CPT Donald L. Nance, AN, US Army
MSG Peter Pease, SOF Medic, US Army
MAJ Andre Pennardt, MC, US Army
Lt Col Gerald Peters, US Air Force, MC

Point Loma Industries, Inc., San Diego, CA
COL Glenn J. Reside, DC, US Army
Lt Col (Sel) Kevin Riley, US Air Force, MSC
COL Paul Rock, MC, US Army
MAJ Michael Roy, MC, US Army
HM1 Alan Saviano, SOF Medic, US Navy
MAJ Daniel Schissel, MC, US Army
Daryl Scurry, MD
Maj Frederick Shuler, US Air Force, MC
CPT Michael Smith, MSC, US Army
MSG (Ret) Albert Stallings, SOF Medic, US Army
MAJ Seth Stankus, MC, US Army
MAJ Christoth Stouder, AN, USA
CAPT Kurt Strosahl, MC, US Navy
COL Richard Tenglin, MC, US Army
The Geneva Foundation, Tacoma, WA
Technical And Management Services Corporation, Calverton, MD
LTC Richard Trotta, MC, US Army
CDR Robert Wall, MC, US Navy
LTC Winston Warme, MC, US Army
COL Roland J. Weisser, Jr., MC, US Army
COL Warren Whitlock, MC, US Army
Lt Col John M Wightman, US Air Force, MC
MAJ Joseph Wilde, MC, US Army
MAJ Marvin Williams, MC, US Army
MAJ Joseph Williamson, VC, US Army
LTC Glenn Wortmann, MC, US Army
COL Steve Yevich, MC, US Army
MAJ Victor Yu, SP, US Army

Production Staff
Jeanette Rasche, MS, Multimedia Director
Brad Sullivan, MS, Medical Illustrator
Barry Keel, BA, Audio/Video Engineer
Matt Morrey, MS, Medical Illustrator
David Dust, BA, Graphic Artist
Knox Hubard, MS, Medical Illustrator

Additional information on the Editors and Contributors is available on the Special Operations Forces Medical CD-ROM.

Preface

Special Operations Forces Medical Handbook, 2001, is the first edition of a comprehensive medical reference resource designed for Special Operations Forces (SOF) medics. This "single-source" reference provides many revolutionary approaches to accessing medical information, such as a treatment hierarchy based on available medical resources and mission circumstances commonly facing the SOF Medic. The Special Operations Forces Medical Handbook is an innovative achievement in military medical knowledge, with contributions by over 80 medical specialists organized into a problem-oriented template for reference to diagnoses and treatments.

As an Infantry Special Forces Officer in Vietnam and later along the Thai-Laotian border, I lived with valiant medics who did their utmost to care for our wounded and sick. These medics, called "Doc" by the men they served, primarily relied on the only medical resources at their disposal- what they brought to the war, ingenuity and courage. To those SOF medic warriors and all those who have served so ably and well since, this handbook is given in remembrance of noble deeds and honorable intentions, so that those SOF medics who follow can serve at an even higher level of medical skill and knowledge.

I dedicate this work to all combat medics, especially those of the Special Operations Forces, who have made and continue to make unsung sacrifices to accomplish their mission: providing exceptional care to their comrades. If this innovative medical knowledge base makes a difference in the life of one soldier, it is well worth all the time and effort it took to make it a reality. If this handbook makes that difference, then we have accomplished our mission.

Brigadier General (P) Darrel R. Porr, MD
Commanding General
Southeast Regional Medical Command (SERMC)
 and Eisenhower Army Medical Center

Acknowledgements

The editorial staff and contributors wish to acknowledge their gratitude to Congressman Randy Cunningham for his support of Special Operations Forces (SOF) and the Clinical Assessment Recording Environment (C.A.R.E.) program through which this medical handbook is co-developed. We further wish to acknowledge the important contributions of the United States Special Operations Command, Office of the Command Surgeon, for inspiring this work, and most importantly for allowing their medics to participate in the production and editing of this publication from start to finish.

We are profoundly grateful for the time and effort put forth by the authors of the Special Operations Forces Medical Handbook. We recognize with heartfelt appreciation the time sacrificed from busy schedules, families and friends in order to meet this requirement. Though the rules of engagement were stringent, and the template a struggle, the perseverance demonstrated is embodied within these pages.

A special debt of gratitude is extended to the United States Army Medical Research and Material Command (USAMRMC) and the Telemedicine and Advanced Technology Research Center (TATRC) for providing the opportunity and challenge to develop a comprehensive military medical work to meet the needs of an elite group, the SOF Medics. We look forward to the future challenges of integrating this work with the Systematized Nomenclature of Medicine (SNOMED) architecture.

We appreciate Technical and Management Services Corporation (TAMSCO) for providing excellent program management support and Point Loma Industries, Inc. for pioneering the medical system software interface for the digital version of this work.

A host of other individuals have our deepest gratitude, but we need to acknowledge the Geneva Foundation for providing a non-profit organization and cooperation of the publisher, Teton New Media, for their assistance in building a renewable resource that will hopefully continue providing the most effective and comprehensive medical information for the SOF medical community.

Colonel Warren L. Whitlock
Chief Editor
Center for Total Access
Southeast Regional Medical Command (SERMC)

Introduction

Twenty years have passed since the forerunner of this book, the last edition of The U.S. Army Special Forces Medical Handbook, was published. The world has changed dramatically in those twenty years, with new weapons, threats and diseases emerging. Advancing technology and an explosive proliferation of medical information on the Internet have revolutionized medical practice. However, the mission of the Special Operations Forces (SOF) medic has not changed: unaided, the medic provides care with limited resources in austere, hostile, stressful, and isolated environments, without the capability to evacuate the patient for up to seventy-two hours. The scope and standards of the SOF medical mission are radically different from those found in a fixed-facility, fully equipped and staffed hospital in a peaceful setting in the United States. Few in the civilian (or even military) medical profession ever are challenged with the conditions in which SOF medics practice medicine daily. The SOF medic still has the ultimate medical mission. Despite the proliferation of medical information, no single reference source has emerged addressing the varied and complicated needs of the SOF medic. As a result, conflicting information and even misinformation have created confusion over the most basic medical questions, possibly endangering the lives of those we are committed to helping. The SOF medical community had to remedy this dangerous situation by creating a new SOF Medical Handbook, one that provides guidance to medics in our special environment, answering the hard diagnostic and treatment questions as best as possible. These answers are based on the best possible knowledge and tailored to the austere mission — in plain, straightforward language, without excuses, conditions or academic musings.

The Handbook is written for SOF medical personnel performing the mission with the understanding that medicine is not a sacred subject practiced only by physicians, but rather skills and knowledge that can be learned and used to save lives. There is advice in this handbook that will be viewed as outrageous in traditional, conservative, hospital-based medical settings. Only someone struggling with life and death decisions in the difficult environment of a SOF medic can appreciate the need for this advice. In other contexts it could be viewed as inappropriate, possibly even bordering on malpractice. **Be advised that the Handbook has limited application outside of the SOF context and is not intended for anything other than use by highly trained SOF medics.**

This Handbook is part of an evolving system, one that takes advantage of new information technology. The printed version is limited by size constraints, so the system includes a CD that covers topics in more exhaustive detail. On the CD you will find hotlinks in the text to even more information on the Web. Furthermore, we will no longer stand for an obsolete text — the Handbook will be revised annually, with improvements in format and updated information reflecting the rapid advances in medicine and information technology. We knew this project would be immense and that the first edition would be incomplete. It is wiser to get this system in the hands of SOF medics and start the dynamic process of evolution now than to wait years until the development process was completed.

All the authors have experience with SOF and its environment. All were challenged with the same question: "How would you diagnose and treat this patient if it was your wife, child or parent and you were alone, with no assistance, evacuation or consultation, in an isolated environment, armed only with the most basic of medical tools?" These authors struggled answering this difficult question, knowing they had to break with the conservative paradigms of medicine, possibly facing the censure of their peers in doing so. The SOF community and I salute your pioneering efforts. This Handbook will be instrumental in saving lives. We will maintain the course — continually improving and updating the Handbook, helping medics make the right medical decisions in the field, helping them save lives.

"Unconventional Warfare - Unconventional Medicine!"

Colonel Steve Yevich
USSOCOM Surgeon

Special Operations Forces Medical Handbook
First Edition 1 June 2001

Table of Contents

Editorial Staff and Contributors	i
Preface	iii
Acknowledgements	iv
Introduction	v

PART 1: OPERATIONAL ISSUES

Care Under Fire	1-1
MedCAP (H/CA) Guide	1-2
Hospital Survey Form	1-4
General Medical Site Survey Checklist	1-8
Site Survey, Veterinary Annex	1-10
Pararescue Primary Medical Kit Packing List	1-12
USAF SOF Trauma Ruck Pack List	1-15
USAF SOF Trauma Vest Pack List	1-17
Suggested M5 Packing List	1-18
Naval Special Warfare Combat Trauma AMAL	1-20
Patient Considerations	1-22
9 Line Medevac Request	1-23
Helicopter Landing Sites	1-25
Semi-Fixed Base Operations (Day)	1-27
Semi-Fixed Base Operations (Night)	1-28
Field Expedient Landing Zone (Day)	1-29
Field Expedient (Y) LZ (Night)	1-30
CASEVAC with Fixed Wing Aircraft	1-31
Marking and Lighting of Airplane LZ (Day)	1-31
Marking and Lighting of Airplane LZ (Night)	1-32
Air Evacuation Telephone List	1-33
Aircraft Patient Loads	1-33

PART 2: CLINICAL PROCESS

Medical History and Physical Examination	2-1

PART 3: GENERAL SYMPTOMS

Acute Abdominal Pain	3-1
Anxiety	3-2
Back Pain, Low	3-6
Breast Problems	
Mastitis in a Lactating Mother	3-7
Procedure - I & D of Breast Abscess	3-9
Chest Pain	3-10
Constipation	3-13
Cough	3-13
Depression and Mania	3-15
Diarrhea, Acute	3-18
Dizziness	3-20
ENT Problems	3-22
Eye Problems	
Acute Vision Loss without Trauma	3-22

Acute Red Eye without Trauma	3-24
Orbital or Periorbital Inflammation	3-26
Eye Injury	3-27
Fatigue	3-29
Fever	3-31
Gynecologic Problems	
Pelvic Examination	3-37
Abnormal Uterine Bleeding	3-39
Pelvic Pain, Acute	3-41
Pelvic Pain, Chronic	3-43
Vaginitis	3-47
Vaginitis Chart	3-48
Bacterial Vaginosis	3-47
Candida Vaginitis/Vulvitis	3-49
Pelvic Inflammatory Disease	3-50
Bartholin's Duct Cyst/Abscess and I & D Procedure	3-52
Headache	3-55
Jaundice	3-57
Joint Pain	3-59
Joint Dislocations	3-64
Shoulder Pain	3-70
Hip Pain	3-72
Knee Pain	3-74
Ankle Pain	3-76
Male Genital Problems	
Male Genital Inflammation	3-77
Testis/Scrotal Mass	3-79
Prostatitis	3-80
Testis Torsion	3-82
Epididymitis	3-84
Memory Problems	3-85
Obstetric Problems	
Pregnancy	3-87
Vaginal Delivery	3-87
Preterm Labor	3-93
Procedure: Shoulder Dystocia	3-95
Procedure: Breech Birth	3-96
Procedure: C-Section	3-100
Procedure: Episiotomy and Laceration Repair	3-102
Pre-eclampsia & Eclampsia	3-104
Palpitations	3-110
Rash with Fever	3-112
Itching	3-113
Shortness of Breath (Dyspnea)	3-115
Syncope (Fainting)	3-117

PART 4: ORGAN SYSTEMS
Chapter 1: Cardiac/Circulatory

Acute MI	4-1
Congestive Heart Failure (Pulmonary Edema)	4-3
Hypertensive Emergency	4-5
Pericarditis	4-6
Cardiac Resuscitation	4-7

Chapter 2: Blood	**4-8**
Chapter 3: Respiratory	
Common Cold and Flu	4-10
Pneumonia	4-12
Pleural Effusion	4-14
Thoracentesis	4-15
Empyema	4-17
Allergic Pneumonitis	4-18
Asthma	4-19
COPD	4-21
Pulmonary Embolism	4-23
Acute Respiratory Distress Syndrome	4-24
Apnea	4-25
Chapter 4: Endocrine	
Adrenal Insufficiency	4-27
Diabetes Mellitus	4-28
Hypoglycemia	4-30
Thyroid Disorders	4-31
Chapter 5: Neurologic	
Seizure Disorders and Epilepsy	4-34
Meningitis	4-35
Bell's Palsy	4-37
Chapter 6: Skin	
Introduction and Clinical Approach to Dermatology	4-38
Bacterial Skin Infections	
Disseminated Gonococcal Infection	4-40
Meningococcemia	4-41
Erysipelas	4-41
Staphylococcal Scalded Skin Syndrome	4-42
Impetigo Contagiosa	4-43
Cutaneous Tuberculosis	4-44
Leprosy	4-45
Viral Skin Infections	
Ecthyma Contagiosum: Milker's Nodules (MN)/Human Orf (HO)	4-45
Herpes Zoster (Shingles)	4-47
Molluscum Contagiosum	4-47
Warts	4-48
Superficial Fungal Infections	
Dermatophyte (Tinea) Infections	4-49
Pityriasis (Tinea) Versicolor	4-51
Parasitic Infections	
Loiasis (loa loa)	4-52
Myiasis	4-52
Onchocerciasis	4-53
Swimming Dermatitis - Seabather's eruption (salt water), Seaweed dermatitis (salt water), Swimmer's itch	4-54
Bug Bites and Stings	
Bed Bug	4-55
Centipede	4-55
Millipede Exposure	4-56
Hymenoptera (Bee, Fire Ant, Hornet, Wasp)	4-57
Mites	4-59
Spiders (Black Widow, Brown Recluse)	4-60
Scabies	4-61

Lice (Pediculosis)	4-62
Spirochetal Diseases	
Yaws	4-63
Pinta	4-64
Skin Disorders	
Psoriasis	4-64
Pseudofolliculitis Barbae	4-65
Skin Cancer (Basal Cell, Squamous Cell, Malignant Melanoma)	4-66
Seborrheic Keratosis	4-68
Contact Dermatitis	4-69
Chapter 7: Gastrointestinal	
Appendicitis	4-70
Appendectomy	4-72
Acute Cholecystitis	4-78
Food Poisoning	4-79
Acute Gastritis	4-80
Pancreatitis	4-82
Acute Peritonitis	4-83
Acute Peptic Ulcer	4-84
Acute Organic Intestinal Obstruction	4-86
Chapter 8: Genitourinary	
Urinary Tract Problems	4-87
Urinary Incontinence	4-90
Urolithiasis (Urinary Stones)	4-91
Urinary Tract Infection	4-93

PART 5: SPECIALTY AREAS

Chapter 9: Podiatry	
Heel Spur Syndrome	5-1
Ingrown Toenail	5-2
Plantar Warts	5-3
Bunion	5-4
Corns/Calluses	5-5
Stress Fractures	5-6
Friction Blisters	5-7
Chapter 10: Dentistry	
General Information	5-9
Dental Caries	5-12
Tooth/Crown Fractures	5-12
Acute Periapical Abscess	5-13
Untreated Acute Periapical Abscess	5-13
Luxated (Dislocated) Tooth	5-14
Avulsed (Completely Removed/Loss) Tooth	5-14
Periodontal Abscess	5-15
Acute Necrotizing Ulcerative Gingivitis	5-15
Herpetic Lesions	5-16
Aphthous Ulcers	5-16
Pericoronitis	5-17
Localized Osteitis (Dry Socket)	5-17
Dislocation of the Temporomandibular Joint(s)	5-18
Dental Antibiotics	5-19
Thermal Test for Caries	5-20
Dental Anesthesia	5-20
Temporary Restorations	5-24

Tooth Extraction	5-25
Draining Abscess	5-26
Preserving/Transporting Avulsed Tooth	5-26

Chapter 11: Sexually Transmitted Diseases

Urethral Discharge	5-26
Genital Ulcers	5-28
Vaginal Trichomonas	5-30

Chapter 12: Zoonotic Diseases Chart — **5-31**

Chapter 13: Infectious Diseases

Parasitic Infections

Amebiasis	5-33
Ascariasis (Roundworm)	5-34
Babesiosis	5-34
Clonorchiasis	5-35
Cyclosporiasis	5-36
Enterobiasis (Pinworms)	5-37
Fasciolopsiasis	5-38
Fascioliasis	5-38
Filariasis	5-39
Giardiasis	5-40
Hookworm and Cutaneous Larvae Migrans	5-41
Leishmaniasis	5-42
Malaria	5-44
Paragonimiasis	5-45
Schistosomiasis	5-46
Strongyloidiasis (Cutaneous Larva Currens)	5-47
Taeniasis (Tapeworm)	5-48
Trichinellosis (Trichinosis)	5-50
Trichuriasis (Whipworm)	5-51
Trypanosomiasis, African (Sleeping Sickness)	5-52
Trypanosomiasis, American (Chagas' Disease)	5-53

Mycobacterial Infections

Tuberculosis	5-54
Non-tuberculosis Mycobacterial Disease	5-56

Fungal Infections

Introduction	5-57
Candidiasis	5-57
Blastomycosis (North American Blastomycosis, Gilchrist Disease)	5-59
Coccidioidomycosis (Valley Fever)	5-60
Histoplasmosis (Darling's Disease)	5-61
Paracoccidioidomycosis (South American Blastomycosis)	5-62

Viral Infections

Introduction	5-63
Adenoviruses	5-64
Dengue Fever	5-65
Arboviral Encephalitis (TBE, JE, WN, St Louis)	5-66
Hantavirus	5-68
Yellow Fever	5-69
Hepatitis A	5-71
Hepatitis B & D	5-72
Hepatitis C	5-73
Hepatitis E	5-75
Hepatitis Chart	5-74
Human Immunodeficiency Virus	5-77

Infectious Mononucleosis	5-78
Poliovirus	5-80
Rabies	5-81
Rickettsial Infections	5-83
Q Fever	5-85
Spirochetal Infections	
Leptospirosis	5-86
Lyme Disease	5-88
Relapsing Fever	5-89
Bacterial Infections	
Anthrax	5-90
Bartonellosis	5-92
Brucellosis	5-94
Ehrlichiosis	5-95
Plague	5-95
Rat-bite Fever	5-97
Acute Rheumatic Fever	5-98
Streptococcal Infections	5-99
Tetanus	5-100
Tularemia	5-102
Typhoid Fever	5-104

Chapter 14: Preventive Medicine

Introduction	5-105
Immunizations	5-105
Surveillance for Illness and Injury	5-107
Field Sanitation	5-108
Waste Disposal	5-111
Field Water Purification	5-118
Malaria Prevention and Control	5-120
Pest Control	5-120
Rabies Prevention and Control	5-121
Landfill Management	5-121

Chapter 15: Veterinary Medicine

Antemortem Exam	5-123
Humane Slaughter & Field Dressing	5-123
Postmortem Exam	5-125
Food Storage & Preservation	5-126
Physical Exam & Restraint	5-128
Large Animal OB	5-130
Administer IV Infusion to an Animal	5-131
Animal Diseases	
Bloat in Bovine	5-132
Colic in Equine	5-132
Foot Rot in Caprine	5-133
Equine Lameness	5-134
Mastitis	5-135
Diarrhea in Porcine	5-135
Conjunctivitis	5-136

Chapter 16: Nutritional Deficiencies

Vitamin and Mineral Chart	5-137

Chapter 17: Toxicology

Poisoning: General	5-140
Venomous Snake Bites	5-142

Chapter 18: Mental Health
Operational Stress — 5-147
Suicide Prevention — 5-149
Substance Abuse — 5-150
Psychosis vrs. Delirium — 5-151
Recovery of Human Remains — 5-153

Chapter 19: Anesthesia
Total Intravenous Anesthesia — 5-155
Local/Regional Anesthesia — 5-159

PART 6: OPERATIONAL ENVIRONMENTS

Chapter 20: Dive Medicine
- Ears — 6-1
- Other (Dental, Sinus, GI, Skin, And Face Mask) — 6-3

Decompression Injuries
- Decompression Sickness — 6-4
- POIS (AGE, Pneumothorax, etc.) — 6-7

Gas Problems
- Hypoxia — 6-8
- Oxygen Toxicity — 6-9
- Carbon Monoxide Poisoning — 6-11
- Carbon Dioxide Poisoning — 6-12

Marine Hazards
- Dangerous Marine Lifez: Venomous — 6-13
- Dangerous Marine Life: Biting Animals (Trauma) — 6-16
- Blast/Explosion, Sound — 6-18
- Caustic Cocktail Chemical Burn — 6-20
- Disqualifying Conditions for Military Diving — 6-21

Dive Treatment Tables — 6-23

Chapter 21: Aerospace Medicine
Hypoxia — 6-31
Barodontalgia — 6-32
Barosinusitis — 6-33
Barotitis — 6-34
Decompression Sickness — 6-35

Chapter 22: High Altitude Illnesses
Acute Mountain Sickness — 6-37
High Altitude Cerebral Edema — 6-38
High Altitude Pulmonary Edema — 6-40

Chapter 23: Cold Illnesses and Injuries
Frostbite — 6-42
Hypothermia and Chart — 6-43
Non-freezing Cold Injury (Trenchfoot and Immersion Foot) — 6-44

Chapter 24: Heat-Related Illnesses
Introduction — 6-47
Heat Cramps — 6-48
Heat Exhaustion — 6-49
Heat Stroke — 6-50

Chapter 25: Chemical
Nerve Agents — 6-52
Blood Agents — 6-52
Blister Agents — 6-53
Set Up a Casualty Decontamination Station — 6-54

Chapter 26: Biological

Introduction/Suspected Biological Exposure	6-56
Inhalational Anthrax	6-56
Botulism	6-57
Pneumonic Plague	6-58
Smallpox	6-59
Tularemia	6-60
Viral Hemorrhagic Fevers	6-60

Chapter 27: Radiation
Radiation Injuries	6-61

PART 7: TRAUMA
Chapter 28: Trauma Assessment
Primary and Secondary Survey	7-1
Primary and Secondary Survey Checklists	7-5

Chapter 29: Human and Animal Bites — 7-7

Chapter 30: Shock
Introduction	7-10
Anaphylactic Shock	7-10
Hypovolemic Shock	7-11
Fluid Resuscitation Tables	7-14
Routes of Fluid Administration Table	7-16

Chapter 31: Burns, Blast, Lightning, & Electrical Injuries
Burns	7-17
Blast	7-23
Lightning & Electrical	7-26

Chapter 32: Non-Lethal Weapons Injuries
Laser	7-28

PART 8: PROCEDURES
Chapter 33: Basic Medical Skills
Airway Management	8-1
Intubation	8-4
Cricothyroidotomy, Needle and Surgical	8-5
Thoracostomy, Needle and Chest Tube	8-7
Pulse Oximetry Monitoring	8-9
3 Lead EKG	8-10
Pericardiocentesis	8-12
Apply & Remove Pneumatic Anti Shock Garment	8-13
Assess and Perform Blood Transfusion	8-14
Field Blood Transfusion	8-18
Blood Transfusion Reactions	8-19
Intraosseous Infusion	8-21
Suturing	8-22
Wound Debridement	8-25
Surgical Treatment of Skin Masses	8-27
Joint Aspiration	8-29
Compartment Syndrome Management	8-30
Splint Application	8-32
Common External Traction Devices	8-33
Straight Urinary Bladder Catheterization	8-33
Suprapubic Bladder Aspiration (Tap)	8-35
Portable Pressure Chamber	8-36
Pain Assessment and Control	8-37

Chapter 34: Lab Procedures

Specimen Transportation 8-40
Urinalysis 8-41
Wet Mount and KOH Prep 8-43
Gram's Stain 8-44
Brucellosis Test 8-45
Wright's Stain using Cameco Quick Stain 8-46
Ziehl-Neilson Stain 8-47
Giemsa Stain for the Presence of Blood Parasites 8-47
Tzanck Stain 8-48
Culture Interpretation 8-49
Macroscopic Exam of Feces 8-50
Feces for Ova & Parasites 8-51
Concentration Techniques for Ova & Parasites 8-52
Microhematocrit Determination 8-53
RBC Count/Morphology 8-54
WBC Count on Whole Blood 8-56
WBC Differential Count 8-58
ABO Grouping and Confirmation Tests 8-59
Rh Typing 8-59
Crossmatch Procedure 8-60

APPENDICES
Anatomical Plates ..... A-1 to A-20
Color Plates ..... A-21 to A-26
Identification of Cellular Blood Components A-27, A-28
Antibiotic Chart ..... A-29 to A-31
Photosensitivity Drugs A-31
IV Drip Rates A-32
Glasgow Coma Scale A-33
Mini Mental Status A-33
Neurological Examination Checklist A-34
Dermatomes of Cutaneous Innervation A-36, A-37
Wind Chill Chart A-38
Temperature Guidelines for Physical Activity A-39
Fluid Replacement Guidelines for Warm Weather A-39
WBGT A-40
Do Not Resuscitate Guidelines A-41
Lab Values A-42
Abbreviations ..... A-44 to A-49

PART 1: OPERATIONAL ISSUES

Operational Issues: Care Under Fire
Lt Col John Wightman, USAF, MC

Introduction: The primary types of injuries caused by weapons are penetrating, blast, and thermal trauma. Penetrating injuries from bullets, fragments, shrapnel, and secondary debris are the most common and easily identified injuries. Thermal injuries are likewise easily noticed. Blunt injury can be caused by blast winds propelling a casualty, causing them to tumble or hit objects. Injuries from blunt and blast trauma may not be immediately apparent. Attending to or retrieving the casualty is important, but may not be worth drawing fire or exposing others to risk. Predetermined hand signals should be used to communicate with conscious casualties. Binoculars may help assess unconscious casualties from a concealed site. "The best medicine on the battlefield is fire superiority."

Subjective: Symptoms
Focused History (if attending casualty): *Where does it hurt?* (may help identify location of wounds with potentially exsanguinating hemorrhage) *Can you breathe OK?* (may help decide urgency of movement to cover) *Can you shoot back or make it to cover?* (determines whether the casualty will be operationally useful, or be able to assist in his own rescue)

Objective: Signs
Using Basic Tools: General: Altered mental status (AMS) may range from confusion to coma. Seizures may occur. However, airway problems are rare on the battlefield and intervention, beyond placing casualty in coma position if the casualty is unable to move, is not worth the risk while under fire. Altered mental status is most likely due to penetrating or blunt head trauma or shock from bleeding, but two unique features of blast injury are less common causes: blast overpressure on lungs can cause vasovagal syncope with bradycardia and hypotension, lasting minutes to hours even with conventional treatment; and stress-induced tears in lung tissue allowing air into pulmonary veins, which can then be ejected to brain (stroke) or heart (heart attack). Inspection: Identify sites of life-threatening external hemorrhage first. The volume of bleeding is the critical parameter— exsanguinating hemorrhage from penetrated extremities is the #1 cause of preventable death on battlefield. Traumatic amputation, ranging from tips of digits to entire limbs, and penetrating vascular injury are common in casualties close to explosions.
Auscultation: Not necessary while under fire.
Palpation: Rapidly touching all body surfaces may help identify wounds with significant bleeding. Rapid palpation of spine or extremities may be appropriate to decide if casualty can move under own power.

Assessment:
Make Decision Rapidly: Significant external hemorrhage and respiratory distress are the only medical reasons for attending a casualty under fire. But the benefit of rescue must outweigh the cost to the mission from losing more personnel in the rescue. The casualty and potential rescuers may continue to be targets due to exposure and movement. Unconsciousness alone is not reason to expose additional personnel to danger. Without respiratory distress or arrest, the casualty's airway can be considered intact. Blast-induced vasovagal syncope will resolve on its own. Penetrating head and torso trauma, arterial gas embolism (AGE), and seizures cannot be managed under fire.
Differential Diagnosis
Loss of consciousness or seizures manifesting after detonation may indicate release of chemical nerve agent or cyanide.

Plan:
Treatment
1. Have casualty return fire (if capable) as directed or required, take cover or otherwise prevent additional

injury and don chemical-biological-radiological (CBR) protection, if appropriate.
2. If conscious but unable to assist in firefight, direct casualty to move to cover. If unable to move, direct casualty to lay motionless in order to avoid drawing hostile fire. Make tactical decision whether or not to have other personnel attempt rescue.
3. Stop exsanguinating external hemorrhage with tourniquet on any bleeding extremity. Ignore non-life-threatening hemorrhage.
4. Move casualty to cover so direct pressure can be applied to bleeding wounds in other locations. Potential hazards of time and exposure do not warrant immobilization of cervical spine before movement.
If airway can be managed by gravity, or AGE is suspected, place casualty in coma position: left side down (halfway between left-lateral decubitus and prone) and head at same level as heart (Figure 1-1).

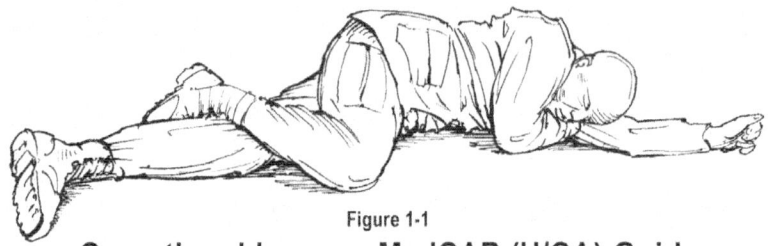

Figure 1-1

Operational Issues: MedCAP (H/CA) Guide
CPT Leonard Gruppo, SP, PA, USA

The entire 86 page MedCAP guide, including all 24 appendices, can be accessed on the SOFMH CD-ROM. Below is a summary of MedCAP planning procedures, the Hospital Survey checklist and the Village Survey checklist. These are provided as a quick reference for field use. This guide was originally written for Army SOF use but is easily adapted by sister services.

MedCAP Planning Checklist
1. **Preparation.** Get passports and international drivers license; computers: programs, reports, peripheral equipment, backup data.
2. **Receive the order.** Verify timeline, analyze order, analyze initial budget request/amount, request information, request country clearances, manage information through S-3; research using: people who have been there, unit records, higher HQ medical sections, Armed Forces Medical Intelligence Center, medical capabilities studies, U.S. embassy, host nation personnel, World Health Organization, Centers for Disease Control, internet; outline the mission, what is the required clothing
3. **Initial planning conference (IPC).** Confirm your duties, arrange initial meetings, verify VetCAPs/DentCAPs, set time for IPC; prepare for IPC: how much money, dental and veterinary assets, additional expertise; meet with counterparts: what type of MedCAP is requested, where and when will the MedCAP take place, what personnel are needed, how many interpreters, what diseases are endemic, what immunizations are requested, what equipment is desired/needed; VetCAP specifics: # and type of animals, location of herds; what not to discuss, committing, get temporary license to practice, verify weather considerations for mission, prepare tentative schedule, write memorandum of agreement or understanding, get price lists; get MedEvac information: hospitals, agencies; identify computer services, prepare reports, list points of contact, verify mid planning conference, get organized
4. **Post IPC.** Review IPC, make informal coordinations; organize team composition: choose personnel, cover your unit med section; request personnel, prepare the medical supply order, consider the patient populations, consider the types of medications/supplies needed, attempt to maintain standards of care, consider preventive medicine issues, order Lexington Bluegrass Army Depot (LBAD) equipment, request civilian clothing funds if required; prepare a tentative mission budget: keep 10% reserve funds, track expenditures, request subsistence support, reserve $$ for medical resupply; issue warning order
5. **Mid Planning Conference (MPC).** Review IPC, review logistics procedures, find host nation (HN) suppliers, reserve area for storage of supplies, reserve rental vehicles, detail food and water sources, prepare email account for MWR, confirm next meeting at Predeployment Site Survey (PDSS), reserve

supplies
6. **Post MPC.** Review MPC; place medical supply orders, CONUS; OCONUS: USAMMCE (USA Med Material Com, Europe), HN; Let your higher HQ medical supply section play their role, time of supply deliveries, receive personnel requests, submit temporary license requests, submit country clearance requests, issue the operations order, prepare the mission brief, request MIPRs (Military Interdepartmental Purchase Request), order non-medical supplies, order certificates of appreciation
7. **Predeployment Site Survey (PDSS).** Make general preparations, confirm previous arrangements; visit treatment areas: village chief and shaman; make strip maps, visit patient treatment structures, visit sleeping quarters, make force protection plan
8. **Post PDSS.** Review plan, make coordinations, prepare for FPC
9. **Final Planning Conference (FPC).** Required? If so where, make MOU addendums if necessary
10. **Advanced Party (ADVON).** Send minimum number of personnel, confirm all previous arrangements, confirm MIPR received, complete DA form 3953 (Purchase Request and Commitment) and DD form 1155 (Order for Supplies or Services), receive supplies, perform site surveys, finalize schedule, receive the main body
11. **Main Body.** Pack and deploy per unit SOP.
12. **Mission.** Review MOU wisdom, maintain HN relations, keep fuel logs, put maps and first aid kits in vehicles, secure medical supplies each night, inventory sensitive items daily, take pictures, send situation reports daily, prepare certificates of appreciation, leave unused supplies but get a hand receipt from HN first.
13. **Stay Behind.** Keep best man behind, make travel reservations
14. **Post Mission.** Tie up loose ends, submitting awards and pay requests, submit reports
15. **Stand-Alone MedCAPs.** Realize special problems, use "A-Team" assistance (i.e., non-medical SOF personnel), verify funding amounts, schedule HN conferences
16. **Force Protection.** Very important, follow guidance, make a plan; weapons: plan for transportation, plan for securing; how many HN Guards needed, watch for agitators in the crowds
17. **Non-Governmental Organizations (NGOs).** Know the types, contact them, see them for free vaccines
18. **Civil Affairs.** Know their capabilities, they use primarily their own funds, request best suited personnel for the mission
19. **Medical Issues.** Have a MedEvac, send emergency medical funds MIPR, keep first aid kits and emergency medical supplies in vehicles, obtain required immunizations, prophylaxis as necessary
20. **Appendices:** 1) H/CA Funding Legal Arguments; 2) Title 10, section 401 U.S.C.; 3) Title 10, section 2551 U.S.C.; 4) MIPR examples; 5) Schedule example; 6) (Memorandum of Understanding (MOU); 7) Hospital Survey; 8) Personnel Requests; 9) Example Medical supply order; 10) Example Veterinary supply order; 11) Example Preventive Medicine supply order; 12) Clothing Allowance Request; 13) Budget examples; 14) Subsistence Support Request; 15) Strip Map; 16) DA form 3953 (Purchase Request and Commitment); 17) Fuel Log; 18) Humanitarian Service Award request.; 19) Force Protection Plan; 20) Site Survey Checklist, General; 21) Site Survey Checklist, Medical; 22) Site Survey Checklist, Preventive Medicine; 23) Site Survey Checklist, Veterinarian; 24) Site Survey Checklist, Dental
21. **Hospital Survey Checklist.** See Below.
22. **Site Survey Checklist.** See Below.
23. **Veterinary Site Survey Checklist.** See below.

Operational Issues: Hospital Survey

1. City/Country:_____
2. Hospital name:_____
3. Trauma level:_____
4. Address:_____
5. Hospital operator telephone #:_____
6. Date of assessment:_____
7. Assessor:_____
8. Primary Point of Contact (POC)
    a. Name, title and position:_____
    b. Office telephone:_____
    c. Home telephone:_____
    d. Cell phone:_____
    e. Pager:_____
    f. Fax:_____
    g. E-mail:_____
9. Patient admissions/information POC and telephone #:_____
10. Security POC and telephone #:_____
11. Emergency department
    a. Location (floor, wing):_____
    b. Number of trauma beds:_____
    c. Trauma capacity (case load at one time):_____
    d. POC:_____
    e. 24 hour desk phone #
        1) Primary:_____
        2) Secondary:_____
        3) Alternate:_____
    f. Radio frequencies
        1) Primary:_____
        2) Secondary:_____

3) Alternate: _____

12. Helipad/Landing Zone (LZ)

    a. General description of surface material and location: _____
    _____

    b. Grid coordinates: _____

    c. Cardinal direction and distance from emergency department: _____
    _____

    d. Elevation:

        1) Above ground level: _____

        2) Above sea level: _____

    e. Max capacity (largest aircraft): _____

    f. Weight limit: _____

    g. Dimensions: _____

    h. Air POC:

        1) Name: _____

        2) Pager: _____

        3) Telephone: _____

        4) Mobile phone: _____

13. Helipad drawing or pictures (attach to survey):

14. Recommended ambulance services:

	Name	Phone #	Freq.	Call sign
a.				
b.				
c.				

d. _____

15. Operating room
    a. Location (floor, wing): _____
    b. Number of suites: _____

16. Intensive care unit
    a. Location (floor, wing): _____
    b. Number of beds: _____
    c. Number of ventilators: _____

17. Indicate in-hospital availability of service or specialist, and response times for service or specialist if/when not in hospital
    1) Emergency dept: _____
    2) CT scan/technician: _____
    3) MRI/technician: _____
    4) X-ray/technician: _____
    5) Radiologist: _____
    6) Clinical lab: _____
    7) Whole blood: _____
    8) Pathologist: _____
    9) Internist: _____
    10) Cardiologist: _____
    11) Coronary care unit: _____
    12) Anesthesiologist: _____
    13) General surgeon: _____
    14) Thoracic surgeon: _____
    15) Orthopedic surgeon: _____
    16) Neurosurgeon: _____
    17) Operation team: _____
    18) Intensive car unit: _____
    19) Hazardous material casualty care: _____
    20) Air EVAC: _____
    21) Ground EVAC: _____

18. Distances and times to mission/training sites:

Site	Kilometers	Drive Time	Fly Time

a._____

b._____

c._____

19. Type of cases referred to other hospitals:

a._____

b._____

20. Name of other referral hospitals:

Name	Location	Phone #	POC

a._____

b._____

c._____

Operational Issues: General Medical Site Survey Checklist

Date:_____

Village:_____

Grid:_____

Map sheet:

1. Village leader's name:_____

2. Population information

    a. Number of infants:_____

    b. Number of children:_____

    c. Number of adults:_____

    d. Total:_____

3. General information.

    a. Description of living conditions

(1) Economy/commerce: _____

(2) Type of housing: _____

(3) Sanitation (method of handling garbage, human and animal refuse): _____

_____

(4) Main food supply, typical diet: _____

(5) Water supply (location in village, height of water table and type and number of wells): _____

_____

(6) Electricity (source and type): _____

(7) Personal hygiene, clothing, shoes: _____

b. School

   (1) School teacher(s): _____

      (a) Level of training: _____

      (b) Language spoken: _____

      (c) Number of students: _____

      (d) Age and grade distribution: _____

   (2) Facilities

      (a) Dimensions of schoolhouse: _____

      (b) Condition of building

Inside: _____

Outside: _____

      (c) Nearby terrain feature (obstacles for LZ setup of identifying area): _____

_____

      (d) Nearest large town and distance: _____

      (e) Local modes of communication, number and effectiveness: _____

_____

4. Potential landing zones

   a. Grid coordinates: _____

   b. LZ capabilities (type and number of aircraft): _____

_____

c. Obstacles and security considerations on LZ:_____
_____

5. Sketch a diagram of village and area for the mission (include LZ, schools, water supplies, and potential medical areas).

6. Comments:_____
_____
_____
_____
_____

7. Members of site survey team (Include name, rank, MOS or title):_____
_____
_____
_____

Operational Issues: Site Survey, Veterinary Annex

Date:_____

Village:_____

1. Animal population

    a. Cats:_____

    b. Dogs:_____

    c. Horses/mules:_____

    d. Sheep/goats:_____

    e. Pigs:_____

    f. Cows:_____

    g. Fowl:_____

2. What is the common usage of animals in the village:_____
_____

3. What problems have been experienced with the animals of the village within the past year?

    a. Cats:_____

    b. Dogs:_____

    c. Horses/mules:_____

    d. Sheep/goats:_____

    e. Pigs:_____

    f. Cows:_____

    g. Fowl:_____

4. What type of animal/veterinary facilities are available in the village (housing, stalls, working chutes)?
_____
_____

5. Describe any on-going agriculture improvement project (US, religious, host national):_____
_____
_____

6. Date of last veterinary assistance visit (work done, by whom, and scope of work):_____

_____

_____

7. Name of nearest veterinary service:_____

    a. Point of contact:_____

    b. Location:_____

    c. Size of location:_____

8. List any specific requests for assistance by villagers or local government:_____

_____

9. Points of contact:

    a. Country team USDA member:_____

    b. Ministry of Agriculture or Natural Resources (address and phone number):_____

_____

_____

    c. Local representative:_____

_____

    d. Local agriculture related organizations (i.e., Association of Ganaderos):_____

_____

_____

_____

10. Do animals roam free in the village, are they penned or corralled? (Sketch where pens of corrals are in relation to the village. Specify what type animals are in what pens.):_____

_____

_____

Operational Issues: Pararescue Primary Medical Kit Packing List

FIELD PACK W/FRAME, ALCE, LARGE 8465-01-286-5356
  (All quantities are minimums) (Suitable substitutes may be used upon request)
UPPER LEFT SMALL PKT
Packing Note: Scissors stored loose in pocket.

Item	NSN	Qty
Band-Aid	6510-00-913-7909	1pg
First aid kit, eye dressing	6545-00-853-6309	1pg
4x4 post op sponge	6510-00-148-9770	2ea
Chapstick	6508-01-436-0607	2ea
Bactrocan	6505-01-375-5686	1tu
Ear plugs	6515-00-137-6345	2pr
Bandage scissors, large	6515-00-935-7138	1ea

UPPER CENTER SMALL PKT
Packing Note: Steri-Strips 6510-00-054-7256 will need to be added to Surgical kit (not a component of kit).

Item	NSN	Qty
Space blanket	7210-00-935-6666	2ea
Tape, surgical, 1" waterproof	6510-00-879-2258	1ea
Surgical kit:	**6545-00-957-7650**	**1ea**

(Following are the components)

NSN	Item	Qty
6545-00-913-6525	Case, Minor Surgery, Surgical Instrument Set	1ea
6515-00-660-0011	Blade, Surgial knife, Detachable, CS no. 10, 6S	2pg
6515-00-660-0010	Blade, Surgial knife, Detachable, CS no. 11, 6S	2pg
6515-00-299-8736	Holder, Suture, Needle, Hegar-Mayo, 6 inch	1ea
6515-00-333-3600	Forceps, Dressing, Straight, 5 ½ inch	1ea
6515-00-334-6800	Forceps, Hemostatic, Straight, Kelly 5 1/3 inch	2ea
6515-00-344-7800	Handle, Surgical Knife, Detachable Blade	1ea
6515-00-352-4500	Needle, Suture, Surg, Reg, Size 12 3/8 Circle 6S	1pg
6515-00-352-4540	Needle, Suture, Surg, Reg, Size 16 3/8 Circle 6S	1pg
6515-01-119-0018	Probe, General Operating, Straight, 5inch	1ea
6515-00-967-6983	Suture, Nonab, Surg, Silk, Braided, Size 0 12S	6/12pg
6515-00-754-2812	Suture, Nonab, Surg, Silk, Braided, Size 00 12S	6/12pg
6515-00-365-1820	Scissors, Straight 5-1/2 inch	1ea
6510-00-054-7256	Steri-strip 1/8 inch	2ea

LOWER CENTER LRG PKT (BANDAGE POCKET)
NOTE: Bandage scissors stored loose in pocket.

Item	NSN	Qty
Bandage scissors, large	6515-00-935-7138	1ea
Povidine/Iodine 4" applicator	6510-01-371-9636	6ea
11" battle dressing	6510-00-201-7425	1ea
Battle packs		
(consisting of)		
Tourniquet	6515-00-383-0565	1ea
Battle dressing small	6510-00-201-7430	1ea
Ace wrap	6510-00-935-5822	1ea
Petrolatum gauze	6510-00-201-0800	2ea
And or		
Sodium Chloride gauze	6510-01-342-5935	2ea
Muslin bandage	6510-00-201-1755	1ea
Kerlix (in mfg wrapper)	6510-00-582-7992	1ea
4x4 gauze sponges	6510-00-148-9770	1ea

8x8 ziplock bag	8105-00-837-7755	1ea
Gloves, High Risk Lg	6515-01-342-3002	1pr

LOWER LEFT LRG PKT (AIRWAY POCKET)
Packing Note: Either type laryngoscope blade may be used.

Handle, laryngoscope	6515-01-153-5295	1ea
Blade, laryngoscope Miller #2	6515-01-307-7474	1ea
Blade, laryngoscope, Macintosh #3	6515-00-955-8865	1ea
Tape, surgical, 1"waterproof	6510-00-879-2258	1ro
Endotracheal tube 7.5	6515-01-036-9034	3ea
Stylet, ET	6515-01-276-6850	3ea
Berman airway adult	6515-00-687-8052	3ea
18 Ga. cath	6515-01-337-3681	1ea
Alcohol pads	6510-00-786-3736	2pg
Kelly hemo 5 ½ inch	6515-00-334-6800	1ea
Heimlich valve	6515-00-926-9150	1ea
Nasopharyngeal airway (trumpet)	6515-01-230-9953	2ea
Surgilube packets	6515-00-111-7829	2ea
Knife, Gen. surg #10	6515-01-149-8097	1ea
Finger cot	6515-00-935-1193	4ea
Pocket mask	6515-01-276-1417	1ea
Syringe hypo 10cc	6515-00-754-0412	1ea
12x12 ziplock bag	8105-00-837-7757	1ea

LOWER RIGHT LRG PKT (FLUID POCKET)
NOTE: 1. Fluid bag will be unwrapped and stored in the infusor cuff.
2. Remaining IV items will be packed in an appropriate sized ziplock bag and stored between the fluid bag and the infusor cuff or taped directly to IV/infusor cuff.

IV infusor kit		**1ea**
(consisting of)		
Sodium Chloride 1000ML	6505-00-083-6544	1ea
Infusor cuff	6515-01-280-8163	1ea
IV admin set	6515-00-115-0032	1ea
Alcohol pad	6510-00-786-3736	3ea
80lb test line, 36"	8305-00-264-2088	1ea
Penrose drain	6515-01-385-1697	1ea
Or Penrose drain	6515-01-385-2013	1ea
18 ga cath	6515-01-337-3681	1ea
20 ga cath	6515-01-337-3682	1ea
14 ga cath	6515-01-340-5429	1ea
8x8 ziplock bag	8105-00-837-7755	1ea

BOTTOM ACCESSORY POUCH

Poleless litter	6530-00-783-7510	1ea
Cervical collar, reg.	6515-01-235-2648	1ea

RADIO PKT INSIDE PACK (FLUID/DIAGNOSTIC POCKET)
NOTE: Battle packs are packed in individual bags.
IV Infusor Kit
2ea

Battle Packs		2ea
Needle, 21 ga	6515-01-274-4690	2ea
Normal Saline 1000cc	6505-00-083-6544	1ea

RADIO PKT INSIDE PACK (FLUID/DIAGNOSTIC POCKET)

Diagnostic Kit		1ea
(consisting of)		
BP cuff	6515-01-280-8163	1ea
Stethoscope	6515-01-361-8596	1ea
Penlight	6230-00-125-5528	2ea
Subnormal thermometer	6515-01-375-3244	1ea
Foley catheter	6515-01-098-3623	1ea
Surgilube packets	6505-00-111-7829	3ea
Rectal thermometer	6515-00-149-1407	1ea
12x12 ziplock bag	8105-00-837-7757	1ea

INSIDE MAIN POCKET
NOTE:
1. Battle packs will be in individual bags.
2. Splints may be wire ladder, SAM, or both.
3. Kerlex and 4x4 sponges may be packed together in one bag or as separate bags.

Battle packs		3ea
Ice pack, (as required)	6530-01-444-5476	2ea
Heat pack, (as required)	6530-01-317-1131	2ea
Flexible splint, padded		
Wire ladder splint	6515-00-373-2100	
SAM splint	6515-01-225-4681	
Kerlex	6510-00-058-3047	6ea
4x4 sponges, post op	6510-00-148-9770	6ea
Kendrick Traction Device (KTD)	6515-01-346-9186	1ea
12x12 ziplock bag	8105-00-837-7757	1ea
V-Vac Suction	6515-01-364-1047	1ea

UPPER FLAP POCKET		
Non-medical items		A/R
Batteries, AAA	6135-00-826-4798	A/R
Batteries, AA	6135-00-985-7845	A/R
Medication and Procedure Handbook		1ea
Patient Treatment Cards		3ea
High Risk gloves Med	6515-01-375-4105	4ea
Or		
High Risk gloves Large	6515-01-342-3002	4ea

Operational Issues: USAF SOF Trauma Ruck Pack List

NSN	Nomenclature	LevelQty/UI
4240012954305	Carabiner	2ea
6135008264798	Battery nonrechar AAA	1pg
6135009857846	Battery nonrecharge12	1pg
6240005529672	Lamp incandescent 2.5 v	1pg
6240007970420	Lamp incandescent .280 amps	1ea
6260010744229	Light. chemiluminescent	1bx
6260011785559	Light chemiluminescent	1bx
6260011785560	Light chemiluminescent	1bx
6260011960637	Shield light chemilumenescent	2ea
6260012094434	Light chemiluminescent	1bx
6505000797867	Naloxone HCL inj 10s	1bx
6505001117829	Lubricant surg 5gm 144S	1bx
6505001334449	Epinephrine inj1ml10s	1pg
6505001838820	Diphenhydramine HCL10	1pg
6505005986116	Lidocaine 1% 50 ml	1bx
6505012629508	Cefazolin sod f/inj1gm25s	1pg
6505013306269	Sodium chl inj 1000ml 12s	1pg
6505014517338	Ketorolac tromethamin	10ea
6510001594883	Dressing first aid	2ea
6510002011755	Bandage 37x37x52 in	12ea
6510002017425	Dress fld 11-3/4 in	2ea
6510002017430	Dressing first aid 7.75in	4ea
6510007219808	Sponge surg 4x4in 1200s	1pg
6510007241017	Bandage gauze 6x180in 48s	1pg
6510007863736	Pad isopropyl alcohol	1pg
6510009268882	Adh tape surg wht 1in 12S	1pg
6510009268884	Adhesive tape surg 3"	1pg
6510009355823	Band elas 4.5yds x 6in12s	1pg
6510010087917	Appli povidone-iodine150S	1pg
6510011126414	Gauze petro3x36in 12s	1pg
6510011940252	Sheet burn-traum66x99	2ea
6510012104453	Dressing occlusive adh500	1pg
6515001050720	Tube trach disp 28cm 10s	1pg
6515001050744	Tube trach 7mm id 10s	1pg
6515001050759	Tube trach 32cm lg 10s	1pg
6515001179021	IV inj 78in lg 48s	1pg
6515002259719	Tu stomach surg 16fr 50S	1pg
6515002998712	Stylet cath-tu copper	1ea
6515003323300	Forceps trach tu adl	1ea
6515003349500	Forceps hemo 8.75-9.25 in	2ea
6515003634100	Saw finger ring 6"lg	1ea
6515003830565	Tourniquet nonpneu 41.5IN	2ea
6515004054007	Sut nonabs surg sz0 36s	1dz
6515005842893	Infusor bld col-dispn	2ea
6515006555751	Needle hypo 25ga 100s	1pg
6515006878052	Airway pharyn 100mm	1bx
6515007344342	Suture nonabs 4-0 36s	1pg
6515007542834	Needle hypo 18ga 100s	1pg
6515007542838	Needle hypo 21ga 100s	1pg

NSN	Description	Qty
6515008669073	Tube drain 36fr 10s	1pg
6515009134607	Blade laryngosc 158mm	1ea
6515009269150	Valve surg drain 4.5in10s	1pg
6515009354088	Stethoscope comb adult sz	1ea
6515009357138	Scissors bandage7.25"	2pg
6515009558865	Blade laryngosc 130mm	1ea
6515009608192	Suture nonabs 3-0 36s	1pg
6515010394884	Sphygmomanometer	1ea
6515010615374	Suture nonabs 5-0 36s	1pg
6515010694405	Syringe hypo 60cc 20s	1pg
6515011250121	Airway nasopharyngeal 12S	1pg
6515011256615	Blade laryn adult sz	1ea
6515011295439	Airway nasopharyn 32fr10S	1pg
6515011398387	Catheter introducer	1pg
6515011467794	Tourniquet adult14x1"	2ea
6515011498097	Knife gen sz10 100s	1pg
6515011498841	Gloves surg sz 7-1/2 50s	1pg
6515011676637	Airway nasopharyngeal	1pg
6515011961748	Handle laryngoscope	1ea
6515012090699	Resuscitator hand-pow	1ea
6515012254681	Splint univ 36x4.5"12	1pg
6515012309953	Airway nasoph 26fr 12	1pg
6515013180463	Mask/rebreath bag disp50S	1pg
6515013365874	Catheter/ndl 16ga 200	1pg
6515013373681	Catheter/ndl 18ga 200	1pg
6515013405429	Catheter & ndl 200s	1pg
6515013448487	Inj tube plast reusable	1ea
6515013469186	Traction apparatus	1ea
6515013568511	Syringe hypo10ml luer	1pg
6515013641047	Suction unit airway	1ea
6515013682874	Detector end-tidal 6s	1pg
6515013852013	Tube drainage 200s	1pg
6515013882484	Cricothyroidotomy kit	1ea
6515014524435	Support cervical	2ea
6515014527697	Oximeter pulse110/220	1ea
6515014686154	Container disposal	1pg
6530007837510	Litter poleless nylon	1ea
6530010780365	Bag sterilization200s	1pg
6530014221267	Pack medical trauma	1ea
6530014685819	Bag oxygen manageme	1ea
6540012901157	Goggles protective100	1pg
6545009577650	Surg instr se mnr sur	1se
6545012811237	Snake bite kit	2ea
6550001656538	Test kt occult bld100	1ea
6640013918391	Glove laboratory 500s	1pg
6680012346789	Regulator oxygen pres	1ea
7210009356666	Blanket casu od sil3oz	4ea
7230002523394	Hook shower curtain susp	1bx
7510000744961	Tape press sensitive	1ro
8105008377757	Bag plas 4a flat 12in	1bx
8120012731465	Cylinder compressed	1ea

Operational Issues: USAF SOF Trauma Vest Pack List

NSN	Nomenclature	Level Qty/UI
5330NCM990007	Black butt pack	1ea
6135008357210	Battery nonrec 1.5v 12s	1pg
6230012917531	Flashlight 3v dc black	1ea
6260010744229	Light,chemiluminesc	1bx
6260011785559	Light chemiluminesc	1bx
6260011785560	Light chemiluminesc	1bx
6510001594883	Dressing first aid	6ea
6510002011755	Bandage 37x37x52 in	4ea
6510002017425	Dress fld 11-3/4 in	1ea
6510002017430	Dressing first aid 7.75iN	4ea
6510007219808	Sponge surg 4x4in 1200S	1pg
6510007241017	Bandage gauze 6x180in 48S	1pg
6510009268882	Adh tape surg wht 1in 12S	1pg
6510009268884	Adhesive tape surg 3"	1pg
6510009355822	Band elas 4in x 4.5yd 10S	1pg
6510011126414	Gauze petro 3x36in 12s	1pg
6515003830565	Tourniquet nonpneu 41.5IN	2ea
6515006878052	Airway pharyn 100mm	1bx
6515009357138	Scissors bandage 7.25"	1pg
6515011250121	Airway nasopharyngeal 12S	1pg
6515011295439	Airway nasopharyn 32FR10S	1pg
6515011535373	Catheter-ndl 14ga 50s	1bx
6640013918391	Glove laboratory 500s	1pg
7210009356666	Blanket casu OD sil 3oz	1ea
7510NCM990008	Clips all purpose	1pg
8415014228753	Vest medical trauma	1ea
8465013221966	Belt individual equ	1ea

Operational Issues: Suggested M5 Packing List
General Purpose, USA, SOF

Airway/Breathing Management

	Item	Quantity
1.	Airway, nasopharyngeal	1
2.	Airway, oropharyngeal, lg	1
3.	Airway, oropharyngeal, med	1
4.	Ambu bag	1 ea
5.	Asherman Device	2 ea
6.	Endotracheal Tube, 7 Fr	1 ea
7.	Endotracheal Tube, 8 Fr	1 ea
8.	Gauze, Vaseline	4 ea
9.	Heimlich Valve	1 ea
10.	Laryngoscope blade, Macintosh, #4	1 ea
11.	Laryngoscope blade, Miller, #3	1 ea
12.	Laryngoscope Handle, pediatric	1 ea
13.	Pocket Mask	1 ea
14.	Stethoscope	1 ea
15.	Stopcock, 3 way	2 ea
16.	Syringe, 50-60cc	1 ea
17.	Thoracotomy tube, 32 fr.	2 ea
18.	Tracheostomy Tube, Cuffless	1 ea

Circulation Management

19.	Bandage, Dyna-Flex, Cohesive Compression 2" x 5 yd	1 ea
20.	Bandage, Kerlex Gauze, 4" roll	4 ea
21.	Bandage, Muslin	8 ea
22.	Blade, surgical, #10	4 ea
23.	Blade, surgical, #11	2 ea
24.	Case Medical Inst & Supply Set, complete w/instruments	1 ea
25.	Catheter, 14ga	2 ea
26.	Catheter, 16ga	4 ea
27.	Catheter, 18ga	6 ea

Circulation Management

28.	Drain, Penrose, 1", sterile	3 ea
29.	Dressing, field 11 3/4" x 11 3/4"	2 ea
30.	Dressing, field 4" x 7"	8 ea
31.	Dressing, field 7 1/2" x 8"	2 ea
32.	Infusion Set, IV 10gtts/ml	4 ea
33.	Normal Saline, 1L	2
34.	Pads, Nonadherent, 4" x 4"	5 ea
35.	Tape, Microfoam, 3"	1 ea
36.	Tape, nylon, 1"	2 ea

Disability/Exposure/Vitals Asses/Mgt

37.	Oto/Ophthalmoscope set, Heine-mini	1 ea
38.	Scissors Bandage	1 pr
39.	Sphygmomanometer	1 ea
40.	Thermometer, Oral	1 ea
41.	Thermometer, Rectal	1 ea

Surgery
42.	Gauze Sponge, 2" x 2"	10ea
43.	Gauze Sponge, 4" x 4"	10ea
44.	Gloves, sterile	2 pr
45.	Marcaine, 0.5%, 50ml	1 vl
46.	Betadine, 2oz	1 bt
47.	Betadine Swab Sticks, 3/pk	3 pk
48.	Sponge w/Brush, Surgical, Betadine	1 ea
49.	Steri-strips, 1/2"	2pk
50.	Suture Nylon, 3-0 w/ needle	2 pk
51.	Suture Nylon, 4-0 w/ needle	2 pk
52.	Suture Silk, size 0 w/o needle	2pk
53.	Suture Vicryl, 3-0 w/ needle	2 pk
54.	Suture Vicryl, 4-0 w/ needle	2 pk
55.	Suture Silk, 0-0, w/ needle	2 pk

Emergency Medications
56.	Benadryl, 50mg/ml, 1ml vial	2ea
57.	Epinephrine, 1:1000, 10ml	1 vial
58.	Morphine Sulfate, 15mg/ml, 30ml	1 bt
59.	Naproxen, 500 mg tablets	60 ea
60.	Narcan, 0.4mg/ml, 1ml vial	4 ea
61.	Needle, 18ga, 1.5"	4 ea
62.	Needle, 25ga, 1.5"	4 ea
63.	Phenergan, 25mg/ml	6 vl
64.	Syringe, 10cc	4 ea
65.	Pads, Alcohol	20ea
66.	Pads, Betadine	20ea

Antibiotics/Antifungals
67.	Augmentin 875mg tablets	20 ea
68.	Nitroglycerin Tablets, 0.4mg, 25/bt	1 bt
69.	Erythromycin 500mg tablets	40 ea
70.	Fluconazole, 150 mg tablets	12 ea
71.	Levaquin, 500mg tablets	20 ea
72.	Metronidazole 500mg tablets	40 ea
73.	Unasyn, 3 gm, powder	1 pk

General Medications
74.	Bags, drug dispensing	12 ea
75.	Imodium, 2mg capsules	24 ea
76.	Mupirocin ointment, 15gm	1 tube
77.	Phenergan, 25 mg tablets	24 ea
78.	Psuedo-Gest, 60mg tablets	50 ea
79.	Triamcinolone cream, 15 gm	1 tb
80.	Valium, 10 tablets	20 ea
81.	Vicodin, tablets	20 ea
82.	Zantac, 150mg tablets	50 ea
83.	Zyrtec, 10mg tablets	50 ea

Miscellaneous Items

84.	Batteries, AA	6 ea
85.	Blade, tongue	5 ea
86.	Eye patch kit	1 ea
87.	Field Medical Cards	1 book
88.	Flashlight, Mini-Mag	1 ea
89.	Gloves, exam	6pr
90.	Jackstrap, Mini-Mag Head strap	1 ea
91.	Moleskin, 20" square	1 ea
92.	Splint, flexible, "Sam"	2
93.	Snake bite kit, Sawyer, "Extractor"	1 ea
94.	Surgilube, 5 gm packet	4 ea
95.	Swab, Benzoin Tincture	4pk
96.	Bandage elastic, 4" x 4.5 yds	2 ea
97.	Bandaids 3/4" x 3"	10 ea

Operational Issues: Naval Special Warfare Combat Trauma AMAL

ITEMS	QTY
CASE	
LBT 1468A-Advanced Life Support Combat Medical Bag	1
PATIENT EXAM/TREATMENT	
Sets of exam gloves	2
Stethoscope	1
Skin marker	1
Exam penlight	1
Syringe, sterile 10cc	3
Syringe, sterile 5cc	2
Needle, sterile disposable 18g	3
Needle, sterile disposable 21g	3
Safety Pins, Large	3
PHARMACEUTICAL	
Betadine solution, .5oz bottle	2
Morphine 10mg/ml, 60mg injectable	2
Tubex injector tube	1
Nalbuphine (Nubain) 20mg/ml, 2ml inj.	4
Epinephrine 1mg/ml (1:1,000) 1ml	5
Diphenhydramine 50mg/ml, 1ml	2
Water for injection, 5ml	5
Ceftriaxone (Rocephine) 1g vial	2
Clindamycin 150mg/ml, 6ml vial	2
Cefoxitin (Mefoxin) 1g vial	2
Saline/Heparin lock flush sets	2*
Naloxone (Narcan) .4mg/ml, 1ml	6
Promethazine injection 25mg/ml, 10ml vial	1
INTRAVENOUS AND DRUG THERAPY	
Hetastarch (Hespan) 500ml	1
Lactated Ringers 500ml (1000ml LR optional)	1

Item	Qty
Macro drip IV administration sets	2
IV Catheter 18g 1 1/4"	4
IV Catheter 16g 1 1/4"	2
IV Catheter 14g 1 1/4"	1
Blood collection tourniquet or Penrose drain	1
Pressure infuser IV disposable	1
Luer Lock	2
IV Vein-a-guard tape/opsite	2
IV Stopcock, 3 way	1
Velcro IV tourniquet	1*

AIRWAY

Item	Qty
Oral Pharyngeal airway #4	1
Oral Pharyngeal airway #5	1
Oral Pharyngeal airway #6	1
Nasopharyngeal airway # 32 adult	1
Xylocaine 2% Jelly, 5ml tube	2
Laryngoscope handle (RUSH)	1
Laryngoscope blade (RUSH) Macintosh #3	1
Laryngoscope blade (RUSH) Macintosh #4	1
Laryngoscope blade (RUSH) Miller #3	1
Intubation stylet (14f)	1*
Endotracheal tube 7.5mm	2
Endotracheal tube 6.0mm (Cricothyrotomy)	1*
Tongue blades	2

TRAUMA

Item	Qty
Vaseline gauze 4 x 4"	4
Asherman Chest Dressing	2
Bandage, cotton, elastic, wrap/Ace	2
Bandage 6 ply x 3yds Kerlix	4
Dressing field 4" x 7"	2
Dressing field 7 1/2" x 8"	1
Bandage 37" x 37" x 52" (Cravat)	3
Surgical blade #10	2
Surgical blade #11	1*
Gauze sponges 2" x 2"	2
Gauze sponges 4" x 4"	4
Adhesive tape 1"	1
Adhesive tape 2"	1
Suture 2-0 Nylon armed with cutting needle	3
Alcohol pads	8
Hemostats, curved Kelly	2*
30cc syringe	1
Trauma scissors	1

WATERPROOF BAG

Item	Qty
LBT1468B-WATERPROOF BAG	1

COMBAT/JUNGLE LITTER

Item	Qty
LBT 1681 COMBAT/JUNGLE LITTER	1

(*) Indicates a recommended, but not required to carry item
Note: this is the min. Combat and or Combat training loadout requirements.

Operational Issues: Patient Considerations

The following are patient considerations and classifications that should be used to inform the Patient Movement Control Center (PMCC) of your patient(s') status.

Patient Classifications/Category: 1 – Psychiatric, 2 – Inpatient Litter, 3 – Ambulatory Inpatient, 4 – Infant, 5 – Outpatient, 6 - Attendant

NOTE: When contacting the PMCC, use the numerical description. Additional patient description/categorization should not be necessary.

Air Force Patient Movement Precedence: (Joint Pub 4-02.2)

Urgent: Immediate movement to save life, limb or eyesight
Priority: Patients requiring prompt medical care not available locally, used when the medical condition could deteriorate and the patient cannot wait for routine evacuation, (movement within 24 hours)
Routine: Patient requires medical evacuation, but their condition is not expected to deteriorate significantly (movement within 72 hours)

NOTE: These differ from the Army and Navy patient movement precedences. Ensure that the PMCC understands you are with a SOF unit and your request is not an everyday request. Relay any unusual circumstances or need to send this patient to a particular destination. They will respond accordingly.

Patient Information: You should be prepared to provide the follow information to the PMCC:

Name, Rank, SSN, Organization, Nationality, Date of Departure, Destination, Diagnosis, Special equipment needed (including oxygen), Special medical considerations, Patient classification

NOTE: The PMCC can assist you in finding an accepting physician if needed.

Patient Preparation/Documentation: Document on one of the following forms, if available:

DD Form 1380 (US Medical Card), DD Form 600 (Chronological Record of Medical Care), DD Form 602 (Patient Evacuation Tag) or AF Form 3899 (AE Patient Record). Any other available clinical documentation format

It should be noted that there is no intent here to tell you how to take care of your patients. These are simply some considerations for you. When communicating with the PMCC, have as much of the patient information readily available as possible. Be brief:

Why the patient is being aeromedically evacuated, i.e., what is clinically/medically wrong with the patient? Brief synopsis of current history, if known, past significant medical history, including allergies if none, so state, Current knowledge of patient medications, if known, send documentation, if none, so state.

Operational Issues: 9 Line MEDEVAC Request

LINE	ITEM	EXPLANATION	REASON
1	Location of Pickup Site	Encrypt the grid coordinates of the pickup site. When using the DRYAD Numeral Cipher the same "set" line will be used to encrypt the grid zone letters and coordinates. To preclude misunderstanding, a statement is made that grid zone letters are included in the message unless unit SOP specifies its use at all times.	Required so evac vehicle knows where to pickup patient. Also, so the unit coordinating the evacuation mission can plan the route for the route for the vehicle or if vehicle must pick up from more than one location.
2	Radio frequency Call sign and Suffix	Encrypt the frequency of the radio at the pickup site, not a relay frequency, The call sign (and suffix if used) of person to be contacted at the pickup site may be transmitted in the clear.	Required so evac vehicle can contact requesting unit en route to obtain additional information or change in situation or directions.
3	Number of patients by precedence	Report only applicable information and encrypt the brevity codes. A-URGENT, B-URGENT-SURG, C-PRIORITY, D-ROUTINE, E-CONVENIENCE If 2 or more categories must be reported in the same request, insert the word "BREAK" between each category.	Required by unit controlling the evac vehicles to assist in prioritizing missions.
4	Special equipment required	Encrypt the applicable brevity codes: A-None, B-Hoist, C-Extraction equipment, D-Ventilator	Required so that the equipment can be placed on board the evac vehicle prior to the start of the mission.
5	Number of patients by type	Report only applicable information and encrypt the brevity code. If requesting MEDEVAC for both types, insert the word "BREAK" between litter entry and ambulatory entry. L+# of pts – Litter A+# of pts – Ambulatory (sitting)	Required so that the appropriate number evac vehicles may be dispatched to the pickup site. They should be configured to carry the patients requiring evac.
6	Security of pickup site (Wartime)	N- No enemy troops in area P- Possible enemy troops in area (approach with caution) E- Enemy troops in the area (approach with caution) X- Enemy troops in area (armed escort required)	Required to assist the evac crew in assessing the situation and determining if assistance is required. More definitive guidance can be provided to the evac vehicle while en route (specific location of enemy to assist aircraft in planning approach).

6	Number and type of wound, injury or illness (Peacetime)	Specific information regarding patient wounds by type. Report serious bleeding with patient blood type (if known)	Required to assist evac personnel in treatment and special equipment needed.
7	Method of marking pickup site	Encrypt the brevity codes. A- Panels B- Pyrotechnic signal C- Smoke D-None E-Other	Required to assist crew in identifying the pickup site. Note that the color of panels or smoke should not be transmitted until the evac vehicle contacts the unit just prior to its arrival. For security the crew should identify the color and the unit confirm.
8	Patient nationality and status	The number patients in each category need not be transmitted. Encrypt only the applicable brevity codes. A- US military, B- US civilian, C- Non-US military, D- Non-US civilian, E- EPW	Required to assist in planning for destination facilities and need for guards. Unit requesting support should ensure that there is English speaking representative at the pickup site.
9	NBC contamination (Wartime)	Include this line only when applicable. Encrypt the applicable brevity codes. N- Nuclear, B- Biological, C- Chemical	Required to assist in planning for the mission (determines which evac vehicle will accomplish the mission and when it will be accomplished)
9	Terrain Description (Peacetime)	Include details of terrain features in and around proposed LZ. If possible, describe relationship of site to prominent terrain features.	Required to allow evac personnel to assess approach into area.

Operational Issues: Helicopter Landing Sites

Responsibility: The unit requesting aeromedical evacuation support is responsible for selecting and properly marking the helicopter LZs.

Criteria for Landing Sites: The helicopter LZ and the approach zones to the area should be free of obstructions. Sufficient space must be provided for the hovering and maneuvering of the helicopter during landing and takeoff. The approach zones should permit the helicopter to land and take off into the prevailing wind whenever possible. It is desirable that landing sites afford helicopter pilots the opportunity to make shallow approaches. Definite measurements for LZs cannot be prescribed since they vary with temperature, altitude, wind, terrain, loading conditions, and individual helicopter characteristics. The minimum requirement for light helicopters is a cleared area of 30 meters in diameter with an approach and departure zone clear of obstructions.

Removing or Marking Obstructions: Any object (paper, cartons, ponchos, blankets, tentage, or parachutes) likely to be blown about by the wind from the rotor should be removed from the landing area. Obstacles, such as cables, wires, or antennas at or near LZs, which cannot be removed and may not be readily seen by a pilot, must be clearly marked. Red lights are normally used at night to mark all obstacles that cannot be easily eliminated within a LZ. In most combat situations, it is impractical for security reasons to mark the tops of obstacles at the approach and departure end of a LZ. If obstacles or other hazards cannot be marked, pilots should be advised of existing conditions by radio.

Identifying the Landing Site: When the tactical situation permits, a landing site should be marked with the letter "H" or "Y", using identification panels or other appropriate marking material. Special care must be taken to secure panels to the ground to prevent them from being blown about by the rotor wash. Firmly driven stakes will secure the panels tautly; rocks piled on the corners are not adequate.

If the tactical situation permits, the wind direction may be indicated by a:
- Small windsock or rag tied to the end of a stick at the edge the LZ.
- Man standing at the upwind edge of the site with his back to the wind and his arm extended forward.
- Smoke grenades, which emit colored smoke as soon as the helicopter is sighted. Smoke color should be identified by the aircrew and confirmed by ground personnel.

In night operations, the following factors should be considered: One of the many ways to mark a landing site is to place a light, such as a chemical light, at each of the four corners of the usable LZ. These lights should be colored to distinguish them from other lights that may appear in the vicinity. A particular color can also serve as one element in identifying the LZ. Flare pots or other types of open lights should only be used as a last resort as they are usually blown out by the rotor downwash. Further, they often create a hazardous glare or reflection on the aircraft's windshield. The site can be further identified using a coded signal flash to the pilot from a ground operator. This signal can be given with the directed beam of a signal lamp, flashlight, vehicle lights, or other means. When using open flames, ground personnel should advise the pilot before he lands. Burning material must be secured in such a way that it will not blow over and start a fire in the LZ. Precautions should be taken to ensure that open flames are not placed in a position where the pilot must hover over or be within 3 meters of them. The coded signal is continuously flashed to the pilot until recognition is assured. After recognition, the signal operator, from his position on the upwind side of the LZ, directs the beam of light downwind along the ground to bisect the landing area. The pilot makes his approach for landing in the line with the beam of light and toward its source, landing at the center of the marked area. All lights are displayed for only a minimum time before arrival of the helicopter. The lights are turned off immediately after the aircraft lands.

When standard lighting methods are not possible, pocket-sized white (for day) or blue (for night) strobe lights are excellent means to aid the pilot in identifying the LZ.

During takeoff, only those lights requested by the pilot are displayed; they are turned off immediately after the aircraft's departure.

When the helicopter approaches the LZ, the ground contact team can ask the pilot to turn on his rotating beacon briefly. This enables the ground personnel to identify the aircraft and confirm its position in relation to the, LZ (north, south, east, or west). The rotating beacon can be turned off as soon as the ground contact team has located and identified the aircraft. The ground contact team helps the pilot by informing him of his location in relation to the LZ, observing the aircraft's silhouette, and guiding the aircraft toward the LZ. While the aircraft is maneuvering toward the LZ, two-way radio contact is maintained and the type of lighting or signal being displayed is described by the pilot and verified by ground personnel via radio. The signal should be continued until the aircraft touches down in the LZ.

The use of FM homing procedures can prove to be a valuable asset, especially to troops in the field under adverse conditions. Through the use of FM homing, the pilot can more accurately locate the ground personnel. The success of a homing operation depends upon the actions of the ground personnel. First, ground personnel must be operating an FM radio, which is capable of transmitting within the frequency range of 30.0 to 69.95 megahertz; then they must be able to gain maximum performance from the radio (refer to appropriate technical manual for procedure). The range of FM radio communications is limited to line of sight; therefore, personnel should remain as clear as possible of obstructions and obstacles that could interfere with or totally block the radio signals. Ground personnel must have knowledge of the FM homing procedures. For example, when the pilot asks the radio operator to "key the microphone," he is simply asking that the transmit button be depressed for a period of 10 to 15 seconds. This gives the pilot an opportunity to determine the direction to the person using the radio.

NOTE: When using FM homing electronic countermeasures, the possible site detection of LZs by means of electronic triangulation presents a serious threat and must be considered.

Figure 1-2: Semi-Fixed Base Operations (Day)

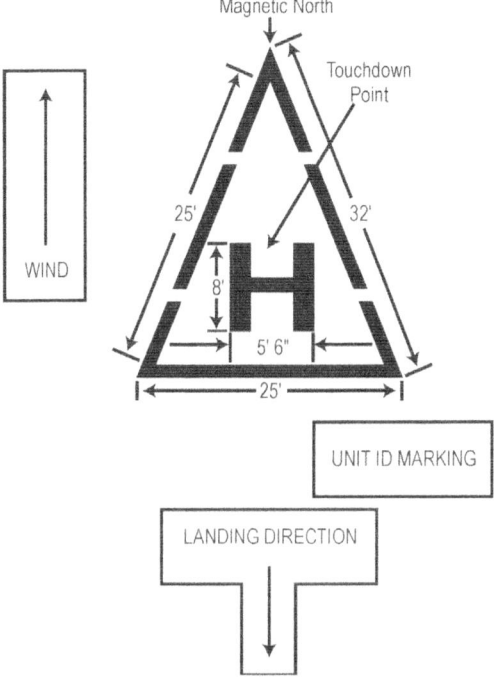

Figure 1-3: Semi-Fixed Base Operations (Night)

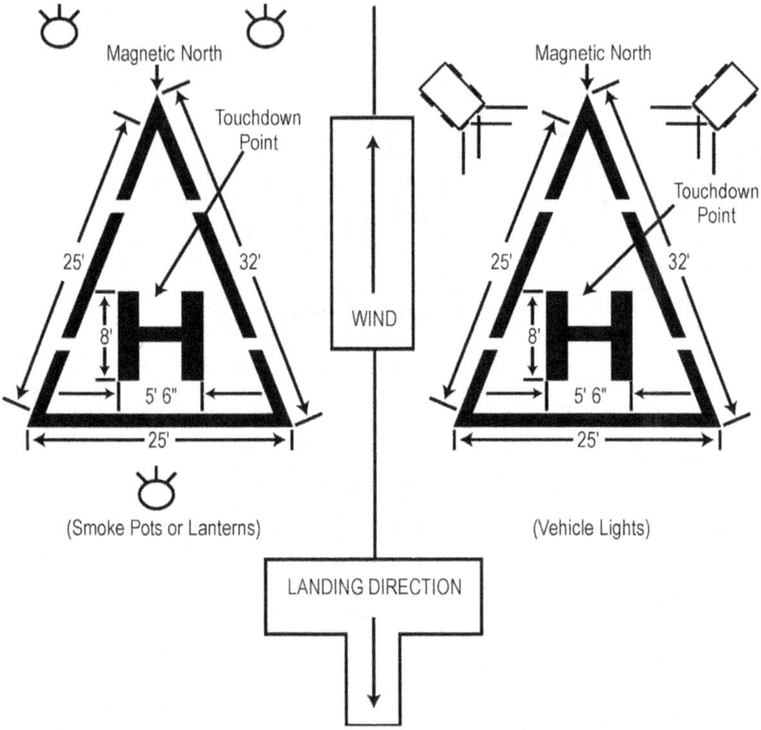

Figure 1-4: Field Expedient Landing Zone (Day)

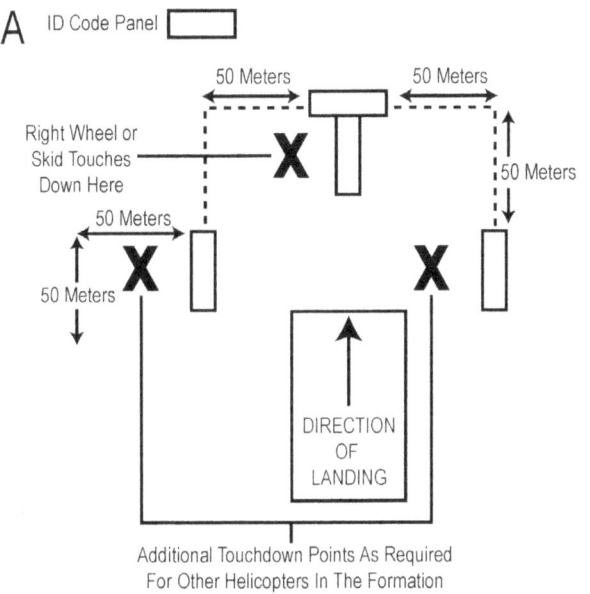

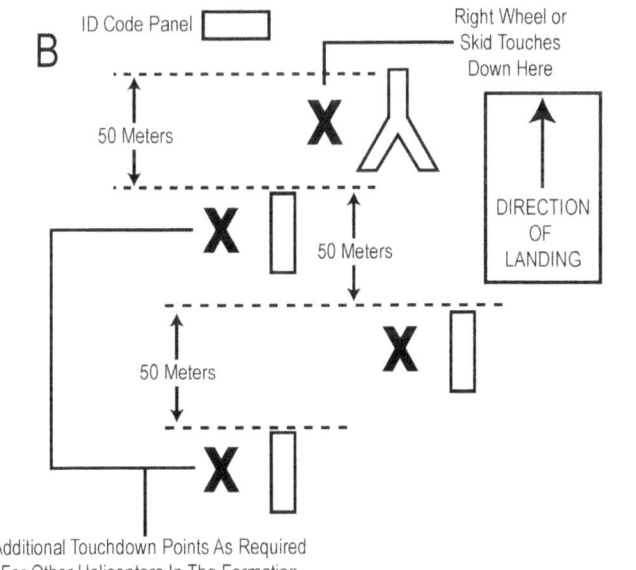

Figure 1-5: Field Expedient (Y) LZ (Night)

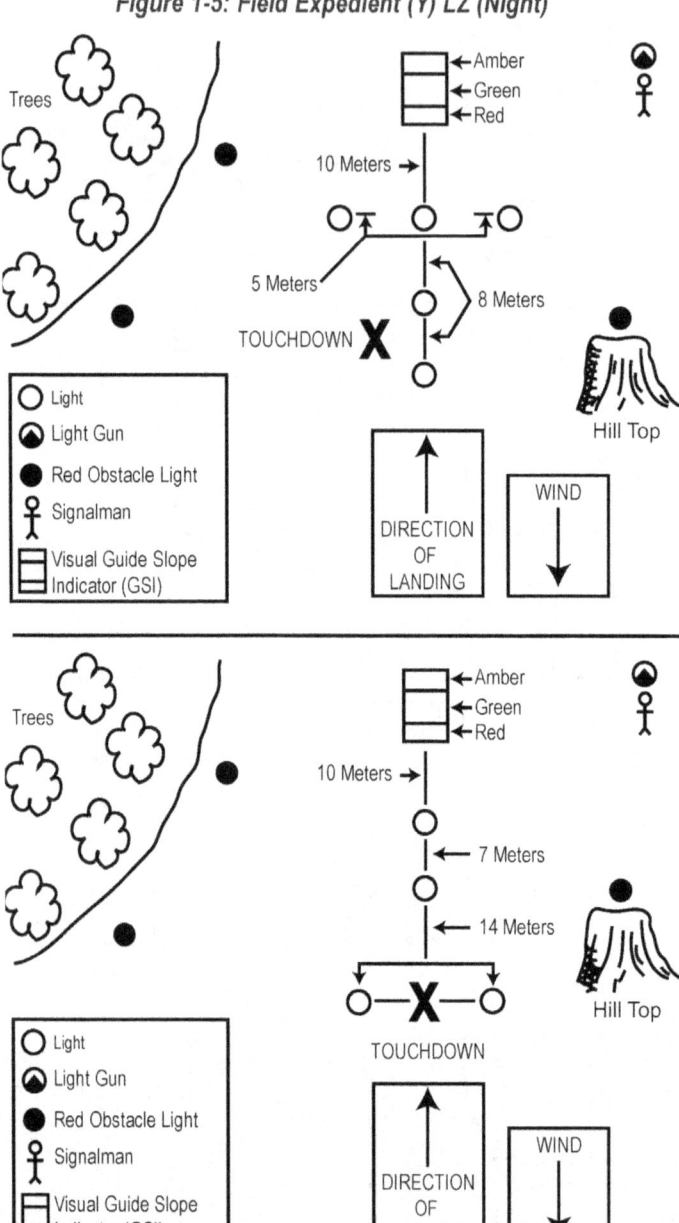

Figure 1-6: Field Expedient (T) LZ (Night)

Operational Issues: CASEVAC with Fixed Winged Aircraft
Lt Col (sel) Kevin Riley, USAF, MSC

The capability of fixed-wing aircraft to land or take off from selected small, unprepared areas permit the evacuation of patients from AOs which would be inaccessible to larger aircraft. These aircraft can fly slowly and maintain a high degree of maneuverability. This capability further enhances their value in forward areas under combat conditions. Small fixed-wing aircraft are limited in speed and range as compared with larger transport-type aircraft. When adequate airfields are available, fixed-wing aircraft may be used in forward areas for patient evacuation. This is a secondary mission for these aircraft, which will be used only to augment dedicated air ambulance capabilities.

Figure 1-7: Marking and Lighting of Airplane LZ (Day)

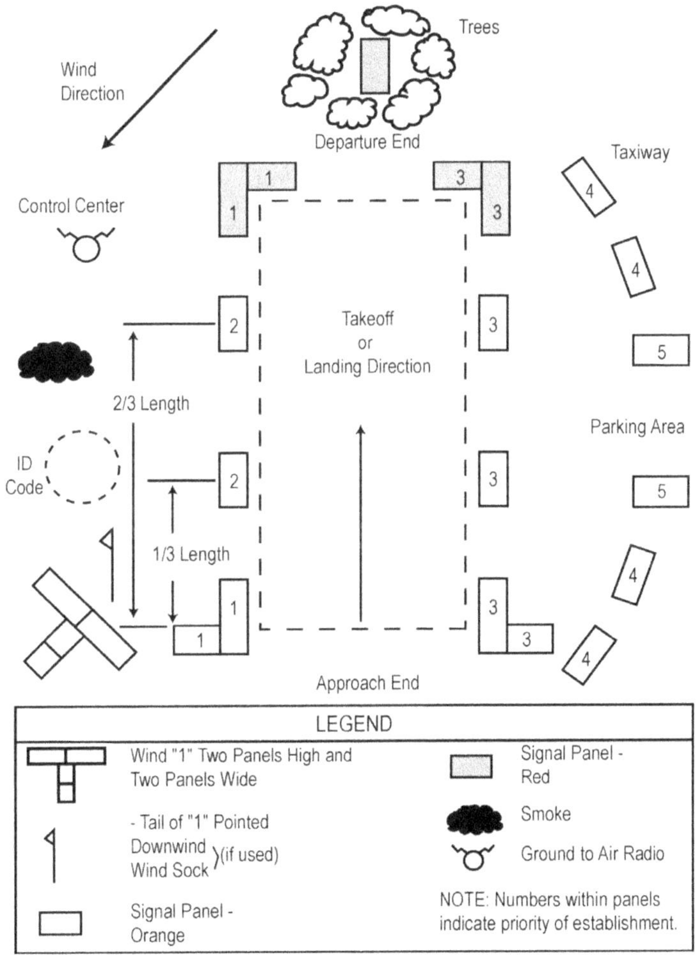

Figure 1-8: Marking and Lighting of Airplane LZ (Night)

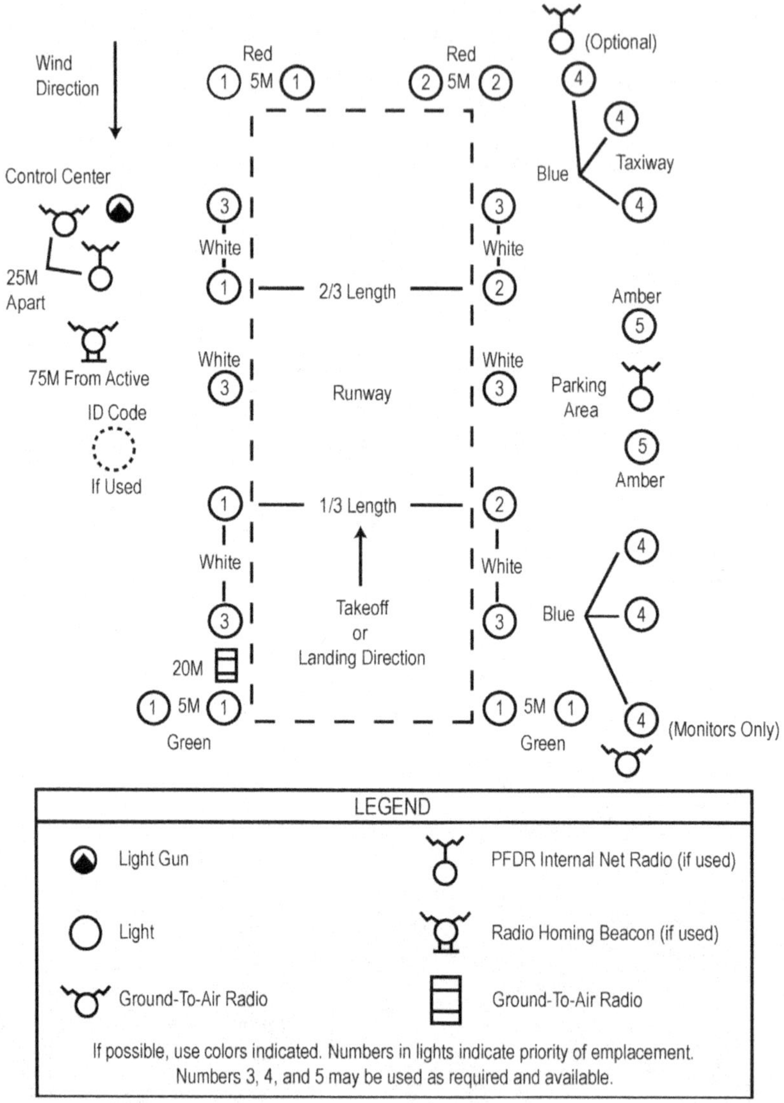

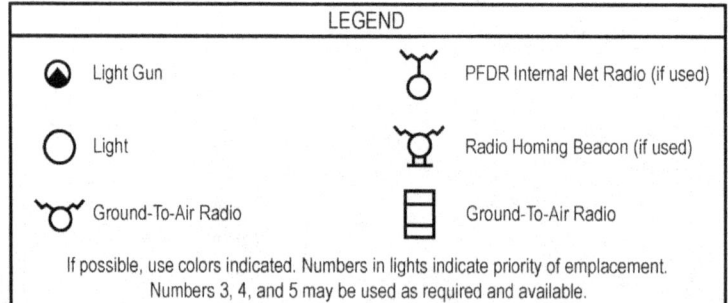

Operational Issues: Air Evacuation Phone List

Global Patient Movement Requirement Center (GPMRC) Intertheater and CONUS AE.	1-800-874-8966	DSN 779-7155
AE Support EUCOM	011-49-6371-47-2264	DSN 314-480-2264
AE Support PACOM	011-81-3117-55-4700	DSN 315-225-4700
AE Support SWA		DSN 318-434-7826
Joint Recovery Coordination Center (JRCC)		
CONUS Langley AFB, VA	804-764-8112	DSN 574-8112
PACOM Hickam AFB, HI	808-531-1112	DSN 315-448-6665
EUCOM Ramstein AB, GE		DSN 314-480-2692
SWA Al Karj, SA		DSN 318-435-7811
North Atlantic Keflavik, Iceland		DSN 450-5007
South and Central America Key West, Fl	305-295-5415	DSN 483-5835
Gulf of Mexico / Caribbean San Juan, PR		DSN 894-1490

Aircraft Patient Loads

Aircraft	Max Litter	Max Amb
C-130	70	92
C-141	103	168
C-9	40	40
MH-53	14	20
CH-47	24	33-44
UH-1	6	12
MH-60	4	11
C-17	36	54

PART 2: CLINICAL PROCESS

Clinical Process: Medical History and Physical Examination
MAJ Andre Pennardt, MC, USA & COL Warren Whitlock, MC, USA

Problem-based learning had its origin in medical education in the 1960s at McMaster University in Canada. A method to communicate understandable, organized, and essential medical information was developed, which later became the S-O-A-P note format of the modern problem-oriented approach to patient care[1]. This system has been so successful that it is used in all other healthcare disciplines (nursing, dentistry, pharmacy, veterinary medicine, public health) and is taught at every medical university in the US.

The value of the SOAP format is in standardizing the collection of symptoms, organization of signs, generation of a clinical diagnosis, and formulation of a multidimensional plan. In this book, this powerful system has been utilized for possibly the first time to organize medical information about diseases in the order in which medics collect and report it. The goal is to make complete medical information readily available concerning any disease or injury in the same SOAP format used by physicians, PAs, nurses and SOF medics. However, this may the first time that specialists have organized diseases and injuries into the format that is used to collect information, and form diagnoses and treatment plans.
- Subjective complaints
- Objective findings
- Assessment of information and development of diagnosis
- Plan of management and treatment

S-O-A-P Format CONDENSED

Subjective
Symptoms:_____ Acute_____ Sub acute_____ Chronic_____
Constitutional_____
CNS_____
Skin_____
HEENT_____
Chest_____
Back_____
GU_____
Abdominal_____
Extremity_____
Hand_____
Foot_____
Other_____
Focused History_____
Objective
Signs:_____ Acute_____ Sub acute_____ Chronic_____
   Inspection (see)_____
   Palpation (feel)_____
   Percussion (tap)_____
   Auscultation (listen)_____
Diagnostic Tests and Procedures (Essential, Recommended)_____

Assessment
Differential Diagnoses:_____
   1._____ 2._____ 3._____ etc._____
Plan
Treatment:_____
   Primary_____ Alternative_____
   Primitive_____ Empiric_____

Patient Education: _____
   General_____ Diet_____
   Activity_____ Medications_____
   Prevention and Hygiene_____ No Improvement/deterioration_____
   Wound Care_____
Follow-up Actions:_____
   Return evaluation_____
   Evacuation/Consultation Criteria_____

Medical Examination:

The comprehensive medical examination is an organized hierarchy for all medical history and physical findings, which constitute the major portions of the Subjective and Objective portions of the SOAP format. These sections of the SOAP note contain all findings, including tests and laboratory studies, to provide the basis upon which all "clinical" diagnosis is made. The final assessment of information and plan to confirm the diagnosis of an illness or injury, begin treatment, educate the patient, and form any follow up plans is contained in the Assessment and Plan portion of the SOAP format. Despite being simple in concept, the SOAP problem-oriented approach is a universal standard in medical education, recording medical information, and communicating medical information to other health care providers.

A necessary part of all medical histories is patient identifying data, including name, rank, social security number, unit, sex, and date of birth. The medical history may be divided into four parts - the chief complaint (CC), history of present illness (HPI), review of systems (ROS), and past, family, and or social history (PFSH).

Chief Complaint:
This consists of a concise statement describing the symptom, problem, condition, diagnosis, or other factor that is the reason the patient is seeking treatment. It is usually stated in the patient's own words.

History of Present Illness:
This consists of a chronological description of the patient's illness or injury. There are 8 elements associated with the HPI.
   LOCATION: Specific area of body involved; radiation; bilateral, anterior, distal, etc.
   QUALITY: Specific patterns and descriptions: dull, sharp, throbbing, stabbing, constant, intermittent, worsening, etc.
   SEVERITY: Degree of severity or intensity (scale of 1-10): "feels like when...", severe, mild, etc.
   DURATION: Onset of problem or symptom: started 3 days ago, 1 hour ago; since yesterday; until this morning; for about 2 months, etc.
   TIMING: Indicates frequency and progression, how long it lasts, how often it occurs, etc.
   CONTEXT: Setting in which it occurs: what was patient doing when signs/symptoms started; occurs after meals, etc.
   MODIFYING FACTORS: What has patient done to relieve signs/symptoms: type of medications taken, how it relieved or made worse; rest makes it better; movement makes it worse, etc.
   ASSOCIATED SIGNS AND SYMPTOMS: Other signs and symptoms patient has experienced or has at presentation: Medic should ask direct questions (e.g., nausea/vomiting, blurred vision, change in bowel habits, etc.).

Review of Systems:
This is an inventory of body systems obtained through a series of questions seeking to identify any signs/symptoms the patient may be experiencing. There are 14 elements to the ROS.

CONSTITUTIONAL:	Weight changes, fever, fatigue, weakness, etc.
EYES:	Pain, redness, blurring, photophobia, decreased visual acuity, diplopia, etc.
ENT:	Decreased hearing, pain, vertigo, tinnitus, epistaxis, sore throat, etc.
CARDIOVASCULAR:	Palpitations, fainting, tachycardia, orthopnea, EKG results, etc.
RESPIRATORY:	Wheezing, cough, sputum (color and quantity), dyspnea, pleuritic pain, etc.
GASTROINTESTINAL:	Nausea, vomiting, diarrhea, bloody stools, constipation, abdominal pain, anorexia, etc.
GENITOURINARY:	Frequency, painful urination, hematuria, testicular pain, incontinence, penile discharge, last menstrual period, etc.
MUSCULOSKELETAL:	Joint pain or stiffness, muscle pain, decreased range of motion, swelling, etc.
SKIN/BREAST:	Dryness, rashes, itching, jaundice, lumps, sores, changes in hair or skin, nipple discharge, etc.
NEUROLOGICAL:	Seizures, focal weakness, slurred speech, tremors, difficulty walking, paralysis, numbness, headaches, etc.
PSYCHIATRIC:	Nervousness, depression, mood changes, insomnia, etc.
ENDOCRINE:	Diabetes, excessive thirst, cold intolerance, sweating, etc.
HEMATOLOGIC/LYMPHATIC:	Anemia, easy bruising or bleeding, adenopathy, past transfusions, etc.
ALLERGIC/IMMUNOLOGIC:	Allergies, prior immunizations, HIV test results, etc.

Past, Family, and Social History:

PAST MEDICAL HISTORY:	This includes the patient's significant prior medical problems, past surgeries, current medications, medication allergies, and immunization history.
FAMILY HISTORY:	This consists of a review of medical events in the patient's family that may be hereditary, or place patient at increased risk (e.g., colon cancer, father had heart attack at age 40, etc.).
SOCIAL HISTORY:	This is an age-appropriate review of past and current activities that may affect the illness/injury. It includes employment, marital status, alcohol, drug and tobacco use, living arrangements, etc.

Physical Examination:
Constitutional:

VITAL SIGNS:	Height, weight, blood pressure, pulse rate, respiratory rate, and temperature.
GENERAL APPEARANCE:	Development, nutrition, growth, body habitus, attention to grooming, etc.

Ear, Nose, Mouth, and Throat:

EAR:	Overall appearance of auricles, auditory canal (swelling, drainage, etc.), tympanic membrane (erythema, blood, mobility, etc.), and assessment of hearing acuity.
NOSE:	Examination of external nose, nasal mucosa, sinuses, septum, and turbinates for swelling, redness, polyps, blood, rhinorrhea, deviation, perforation, etc.
MOUTH:	Examination of lips, teeth, tongue, gums, etc. for dental caries, gingivitis, periodontal disease, tooth loss, cyanosis, etc.
THROAT:	Examination of oropharynx for lesions, symmetry, erythema, tonsillitis, etc.

Respiratory:
Examination should include inspection of chest (shape, symmetry, expansion, use of accessory muscles, and intercostal retractions), percussion of chest (dullness, hyperresonance), palpation of chest (tenderness, masses, tactile fremitus), and auscultation of lungs (equality of breath sounds, rubs, rales, rhonchi, and wheezes).

Cardiovascular:
Examination should include palpation of heart (location, forcefulness of the point of maximal impact, thrills, etc.), auscultation of heart (murmurs, abnormal sounds), assessment of pulse amplitude and presence of bruits in various arteries (carotid, femoral, popliteal, etc.), assessment of jugular veins (distention, A, V, or

cannon A waves), palpation of pedal and other pulses, and measurement of ankle-brachial index.

Breasts (Chest):
The breasts may be inspected for contour, symmetry, nipple discharge, gynecomastia, and palpated for lumps/masses and tenderness.

Gastrointestinal:
The abdomen should be inspected for obesity, distention, and scarring, followed by auscultation in all four quadrants for bowel sounds. Percussion may be performed to detect abdominal tenderness, ascites, or tympani. Palpation is performed to assess for tenderness, guarding, rebound, and other signs of peritoneal irritation, enlargement of the spleen or liver, masses, and pulsatile enlargement of the aorta (abdominal aortic aneurysm). Digital examination of the rectum may be performed to detect hemorrhoids or rectal masses, assess rectal tone, obtain stool for Hemocult determination, and examine the prostate.

Genitourinary (Male):
External examination of the penis may be performed to detect lesions or discharge. The testicles may be examined for symmetry, tenderness, masses, hydrocele, or varicocele. The prostate may be assessed for enlargement, tenderness, or masses during digital rectal examination. The bladder may be palpated to assess for distention or tenderness.

Genitourinary (Female):
The external genitalia and vagina may be examined for general appearance, estrogen effect, lesions, or discharge. The cervix may be inspected for lesions or discharge using a speculum, at which time specimens may be obtained for microscopy and culture. Bimanual examination of the internal GU organs is performed to detect cervical motion tenderness as well as uterine and/or ovarian enlargement, tenderness, or masses. The bladder may be palpated to assess for distention or tenderness.

Musculoskeletal:
The spine may be examined for tenderness, range of motion, step-offs, scoliosis, or other deformity. Muscles may be examined for strength, tenderness, swelling, or spasm. Joints may be examined for range of motion, tenderness, warmth, discoloration (erythema/ecchymosis), swelling, and instability. Other bones should be palpated for tenderness, deformity, and crepitus as appropriate.

Neurologic:
The patient should have his mental status assessed (mini-mental status exam) for higher cognitive function (including level of consciousness). Glasgow Coma Scale is a useful adjunct to assess the current mental status and progression of trauma victims. Cranial nerves II through XII are routinely assessed as part of the neurologic examination.

CN II	Visual acuity and fields
CN III, IV, and VI	Extraocular movements, pupillary reflex (III)
CN V	Facial sensation and corneal reflex
CN VII	Facial symmetry and strength
CN VIII	Hearing
CN IX	Gag reflex and reflex palatal movement
CN X	Voluntary movement of soft palate or vocal cord function
CN XI	Shoulder shrug strength
CN XII	Tongue protrusion (midline)

CN I (olfactory) is difficult to assess in the field since this requires testing of smell. Cerebellar function is tested by having the patient perform actions requiring coordination such as finger/nose or rapid alternating movements. Motor assessment includes strength and symmetry of major muscle groups. Standard deep tendon reflexes tested include the knee, ankle, and biceps. Sensation testing may include light touch, pinprick, vibration, and proprioception. Gait testing is another good measure of central nervous system function.

Psychiatric:
If psychiatric examination is indicated, it should include a number of elements, including a description

of speech (rate, volume, pressured, etc.), assessment of thought process (rate and content), association (loose, tangential, intact, etc.), abnormal or psychotic thoughts (hallucinations, delusions, suicidal or homicidal ideation, etc.), and mood/affect (depression, anxiety, agitation, etc.). Other psychiatric components that may be assessed as part of overall examination include orientation, memory, concentration, and attention span.

Lymphatic:
Evaluation may include palpation for enlarged nodes in the neck, axillae, and groin.

Integumentary:
Examination may include quantity, texture, and distribution of hair, as well as assessment of skin for rashes, lesions, moles, birthmarks, and hyperhidrosis etc.

1. Weed LL. Medical records that guide and teach. N Engl J Med. 1968;278:652-7.

PART 3: GENERAL SYMPTOMS

Symptom: Acute Abdominal Pain
COL (Ret) Peter McNally, MC, USA

Introduction: Acute abdominal pain is an internal response to a mechanical or chemical stimulus. The pain can be separated into three categories: visceral (dull and poorly characterized), somatoparietal (more intense and precisely localized) and referred (pain felt remote from the origin). The most important elements in the evaluation of acute abdominal pain are the history and physical examination. Attention to the chronology and description of the pain can often suggest the origin of acute abdominal pain. Acute abdominal pain caused by blunt or penetrating trauma is covered on the SOF Medical CD-ROM.

Subjective: Symptoms
Listed on Table 3-1 are some of the most common causes of acute abdominal pain and their associated symptoms. Some patients will voluntarily provide a typical description of the details about the onset, location, and character of the pain. For others, the medic will have to ask pertinent questions (e.g., *Where does it hurt? How would you describe the pain?*, etc.) to obtain the necessary information. Integrate past medical and surgical history, family history and medications into the search for the origin of acute abdominal pain. If the patient also has jaundice, constipation, diarrhea and vomiting, see the appropriate symptom section. GU and GYN illnesses may present as abdominal pain, although they typically present as flank pain or pelvic pain respectively. See the respective chapters for additional information.

Objective: Signs
Using Basic Tools: Temperature: Fever suggests infection or inflammation, i.e., appendicitis, cholecystitis, pancreatitis, diverticulitis, gastroenteritis or pelvic inflammatory disease.
BP and Pulse: Pain typically causes a reflex increase in heart rate and BP. If the BP < 90, consider causes of blood or vascular fluid loss, i.e., bleeding ulcer, pancreatitis (fluid third spacing), gastroenteritis (diarrheal losses). Signs of shock suggest rapid loss of blood, i.e., ruptured ectopic pregnancy, hemorrhaging ulcer or ruptured abdominal aneurysm.
Inspection: Surgical scars may suggest small bowel obstruction. Assumption of the fetal or knee-chest position by the patient may suggest pancreatitis or sickle cell crisis.
Palpation: The abdominal examination should start gently away from the site of discomfort. Localization of pain in the RLQ suggests appendicitis or pelvic inflammatory disease. Sudden inspiratory arrest during steady palpation of the RUQ (Murphy's Sign) suggests cholecystitis. Rebound tenderness and involuntary guarding highly suggest peritonitis from bowel perforation.
Pelvic Examination: Severe cervical motion tenderness or a tender adnexal mass, coupled with fever, suggests pelvic inflammatory disease.
Extremities: Loss of lower extremity pulse(s) suggests abdominal aneurysm.
Rectal: Tarry, sticky, foul-smelling stool (melena) suggests bleeding ulcer. Bright red blood on rectal exam can indicate torrential ulcer bleeding or ischemic colitis.
For further objective signs, see Table 3-2.
Using Advanced Tools: Lab: CBC for infection and anemia, and UA infection, stones, etc.

Assessment:
Differential Diagnosis: Self-limiting causes of abdominal pain are usually milder in severity and remit either spontaneously within 24 hrs, or after administration of antacids, H-2 blockers, laxatives, etc. Examples of common self-limiting causes of abdominal pain would include gastroesophageal reflux, gastritis, intestinal gas, constipation, etc. See discussion and Table 3-1 for other diagnoses to consider. Also include OB (labor), GYN, GU causes of abdominal pain (see respective chapters in this book).

Plan:
Treatment
Goals for Field Management: Eliminate pain, maintain intravascular volume and treat infection.
1. Place 1 large IV (≥18 gauge). If hypotensive or bleeding then place 2 IVs.
2. Use D5Lactated Ringer's or normal saline at 100 cc/hr or boluses of 500 cc to normalize blood pressure and resuscitate.
3. Insert NG tube for gastric decompression for significant abdominal distention or vomiting.
4. Use pain control medications (see Procedure: Pain Assessment and Control).
5. Use antiemetic of choice (e.g., **Compazine** 5-10 mg IM q 3-4 hr, max 40mg/day).
6. Acid suppression: **Pepcid** 20mg per NG tube q 6 hrs, or **Pepcid** 20mg (10mg/ml) IV over 2 minutes q 12 hrs.
7. If fever and/or peritoneal signs present, then initiate antibiotics:
Single Agents: **Cefoxitin** 2 gm IV q 8 hr, **cefotetan** 2 gm IV q 12 hr, **cefmetazole** 2 gm IV q 8-12 hr
Combination Agents: **Aztreonam** 2 gm q 8 hr plus **metronidazole** 500 mg IV q 8 hr
8. Evacuate for potential surgery if any of the following: persistent or worsening abdominal pain with duration >4 hours, associated fever, signs of hypovolemia, intestinal bleeding, shock or peritonitis.

Patient Education
General: Maintain healthy diet with high fiber, low fat content. Exercise daily.

Follow-up Actions
Evacuation/Consultant Criteria: Evacuate urgently for continuing pain or unstable condition. Consult general surgery early and other appropriate specialties as needed.

Symptom: Anxiety
MAJ Michael Doyle, MC, USA

Introduction: Anxiety is a vague feeling of apprehension due to the anticipation of danger. It is a common, normal reaction to any internal or external threat, is usually transient and does not tend to recur frequently. Some situations—like jumping out of an airplane—are inherently anxiety provoking. When the symptoms of anxiety begin to interfere with duty or with social/occupational functioning, the medic may need to intervene. Anxiety, as a symptom, is often associated with most mental disorders and Combat and Operational Stress Reactions. This section identifies those specific conditions in which anxiety is the disorder and not just a symptom of a condition.

Subjective: Symptoms
Free-floating anxiety not attached to any particular idea or notion, fear, agitation, tension, panic. Patients may then complain of sleep, appetite or activity disturbances.
Focused History: *When did you start feeling this way? Have you ever felt this way before?* (identify precipitating events) *Are these feelings constant or do they come and go? How long do the spells last?* (Panic attacks come and go and are usually brief; anxiety due to an underlying medical condition or from post traumatic stress disorder (PTSD) or other chronic anxiety conditions usually is present always.) *Does the anxiety keep you from sleeping or wake you up? How is your appetite?* (If there are significant appetite and sleep problems, then a mood disorder may be the culprit.) *Can you do your job?* (Occupational impairment is important to document and monitor.) *Do you have thoughts of hurting yourself or anyone else?* (always consider safety) *What helps you feel better?* (incorporate the patient in treatment plans)

Objective: Signs
Using Basic Tools: Distracted, jittery, skittish, easily startled and often confused.
Mental Status Exam:
    Alert and oriented in all spheres; may appear easily distracted or startled
    Activity—restless, hypervigilant, easily startled
    Speech—may be rapid, breathless, but also can be slowed with hesitancy or stutterin

Table 3-1 Common causes of acute abdominal pain and their associated symptoms.

Descriptor	Onset/Intensity	Location/Radiation	Character	Key History	Etiology
Burning, Knifelike	Sudden/severe	Epigastric/back	Localized early, diffuse late	Aspirin or NSAID* ingestion	Perforated ulcer
Agonizing	Sudden/severe	Periumbilical/none	Diffuse	ASHD*, DM*, high cholesterol, smoker	Intestinal ischemia/infarction
Tearing	Sudden/severe	Abdomen/back & flank	Diffuse	Pulsatile abdomen	Ruptured aneurysm
Excruciating	Sudden/severe	Periumbilical	Diffuse	African or Mediterranean heritage	Sickle cell crisis
Orthostatic	Sudden/severe	Either adnexal (pelvic) area/none	Localized	Missed menses, nipple discharge, & AM nausea	Ruptured ectopic pregnancy
Constricting	Rapid/moderate	RUQ*/scapula	Localized	Family history of gallstones, rapid weight reduction	Cholecystitis
Boring/Drilling	Rapid/severe	Epigastric/back	Localized	History of gallstones, high triglycerides, abdominal trauma.	Pancreatitis
Ache	Slow/moderate	Periumbilical- early RLQ* - late/none	Diffuse early, localized late	Unless prior appendectomy, it is always a potential cause.	Appendicitis
Ache	Slow/mild to moderate	LLQ*/none	Localized	Often onset after straining at defecation	Diverticulitis
Crampy	Slow/moderate	Periumbilical/none	Diffuse	Previous abdominal surgery & vomiting	Small bowel obstruction
Spastic	Slow/mild to moderate	Periumbilical	Diffuse	Nausea, vomiting and diarrhea	Gastroenteritis
Ache	Slow/moderate	Either adnexal (pelvic) area/none	Localized	Sexual promiscuity; vaginal discharge	Pelvic inflammatory disease

*nonsteroidal anti-inflammatory drugs (NSAIDs), arteriosclerotic heart disease (ASHD), diabetes mellitus (DM), right upper quadrant (RUQ), right lower quadrant (RLQ), left lower quadrant (LLQ)

Table 3-2 Abdominal Pain Objective Signs.

Basic Tools	Clinical Findings	Interpretations
Vital Signs	• Pulse > 100 beats per minute • Systolic BP < 90 • Orthostatic change in VS (systolic BP drop of 20 mm Hg or pulse rise 20 beats per minute) • Temperature (above 101.5 F)	• Probable hypovolemia • Probable hypovolemia • Significant hypovolemia • Suggests infection or inflammation
Appearance	• Pallor of anemia, diaphoresis	• Suggests significant blood loss
Gastric Contents [examine if (+) history of melena or hematemesis]	NG aspirate • Bile; no blood or coffee grounds • Coffee ground • Bright red blood	• No active bleeding • Recent bleeding • Active bleeding
Abdomen	• Rigid abdomen with guarding • Abdominal distention without bowel sounds • Abdominal distention with high pitched and tinkling bowel sounds • Disproportionate pain to abdominal exam	• Peritonitis • Ileus • Small bowel obstruction • Intestinal ischemia, infarction, sickle crisis, or abdominal aneurysm
Rectal Examination	• Melena (black, sticky, and tar-like stool)	• Recent UGI bleeding

Thought content—not delusional, may have hopelessness and dread that gives rise to suicidal ideation. Obsessions (recurring irresistible thoughts or feelings that cannot be eliminated by logical effort) may be present.
Thought processes—usually logical, linear and goal directed; may perseverate (go over and over) on one idea or theme
Mood—generally miserable, worried, or sad
Affect—often anxious, but if describing panic attacks, may appear normal.
Cognition—intact, though present anxiety may slow cognition or responsiveness
Insight—variable; may be poor to good

Assessment:
Differential Diagnosis - Always maintain a high index of suspicion for a physical or CNS injury!
Occult injury→ a hypotensive or hypoxic service member will appear anxious! DO NOT MISS THESE!
Substance Withdrawal - patients in early alcohol withdrawal look anxious. (see Mental Health chapter)
Hyperthyroidism - see Endocrine Chapter.
Combat or Operational Stress Reaction - see Mental Health chapter
Battle Fatigue - see Mental Health: Operational Stress
Mental Disorders associated with anxiety are:
    Panic Disorder - discrete recurring episodes of sudden onset panic attacks
    Phobias - specific fears, triggered by environmental stimuli, that are unreasonable under the circumstances
    Generalized Anxiety Disorder - a pervasive, nearly constant and impairing sense of free-floating anxiety
    Acute Stress Disorder - circumscribed period lasting 2+ days of anxious symptoms and unpleasant, intrusive recollections of a recent unusual or traumatic event; occurring within 4 weeks of the event and resolving within 4 weeks of onset.
    Post Traumatic Stress Disorder - chronic symptoms of anxiety with recurring, unpleasant, intrusive recollections of a past unusual or traumatic event, beginning anywhere from immediately following the event to years later.

Plan:
Primary Treatment - Basic
Symptomatic relief through rest, reassurance.
Benzodiazepines (**lorazepam** mg po q 6-8 hours or **diazepam** 2-5 mg po q 8-12 hours as needed)
Relaxation exercise:
1. Slow deep breathing—use a paper bag or simply work with patient to take slow deep breaths.
2. Progressive muscle relaxation—focus on separate muscle groups (such as the balls of the feet) contract them then relax slowly on the count of 5, move on to next muscle group
3. Visualization—encourage patient to visualize a relaxing setting like sitting on a beach or fishing by a cool stream.

Primary Treatment - Advanced
When available, consider initiation of definitive treatment with a Selective Serotonin Reuptake Inhibitor (SSRI), starting with a low dose (1/2 therapeutic dose). (See Symptom: Depression)

Patient Education
General: Reassure patient that this condition is not life threatening, and he is not going crazy.
Activity: Normal. Try to keep on duty.
Diet: Avoid caffeine or other stimulants.
Prevention and Hygiene: Sleep, relaxation, stress management

Follow-up Actions
Return evaluation: Frequent, scheduled follow-ups as opposed to "come in as needed", support and assist patients with management of their anxiety.
Evacuation/Consultation Criteria: Most anxiety disorders do not need to be evacuated. Consult when there is evidence of mild impairment in function that has not been responsive to rest and reassurance.

Symptom: Back Pain, Low
CDR Scott Flinn, MC, USN

Introduction: Low back pain is an extremely common affliction. Most low back pain results from strain or mechanical stress, is self-limited and resolves in 4-6 weeks. Identification of worrisome signs or symptoms (e.g., pain over 6-8 weeks, night pain, weight loss, neurological injury including loss of bowel and bladder control) will determine which patients require additional testing or treatment. Evaluate trauma causing low back pain for the presence of a fracture. Immobilize properly if possible. Although very common in adults, low back pain is unusual in children and adolescents and warrants investigation.

Subjective: Symptoms
Constitutional: Worrisome symptoms include persistent fever, night pain, weight loss and progressive neurological symptoms such as progressing weakness or saddle anesthesia. Loss of bowel or bladder control in a non-trauma patient suggests cauda equina syndrome, a rare condition that is a surgical emergency to prevent chronic neurologic damage.
Location: Low back pain may be midline, one-sided, radiate into the hip or buttock. Numbness or tingling radiating past the knee, and/or lower extremity weakness suggests a herniated disc pushing on a nerve.
Focused History: *Was there any trauma?* (suspect a fracture). *Are there neurological symptoms such as numbness, tingling or weakness?* (if acute from trauma, suspect fracture, otherwise suspect a nerve impingement from a herniated disc or other cause). *Are there warning signs that pain is due to a serious condition?* (night pain and unexplained weight loss of a large amount [e.g., 20 pounds] suggest a cancerous cause) *Is the pain chronic?* (if greater than 2-3 months in duration, may need evaluation for worrisome cause like herniated disc, tumor) *Is there a persistent fever (greater than 2 weeks) and fatigue or malaise?* (suggests infection or other cause)

Objective: Signs
Using Basic Tools: Acute traumatic low back pain – screen for signs of fracture.
Inspection: Obvious deformities –acute trauma (think fracture) or chronic pain (look for scoliosis). Any mass – tumor. Skin erythema – infection or tumor.
Palpation: Step-off on spinous processes –sign of fracture. Palpable spasm—sign of trauma. Palpable mass--tumor. Abnormal neuro exam including motor function (extensor hallucis longus [great toe pulled up], peroneals [feet held up and out/inverted], and quadriceps extension), sensation (first metatarsal to anus), or deep tendon reflexes (Achilles and patellar tendon) indicates a possible CNS lesion or trauma. If there is loss of sensation in the anal area, check the anal sphincter tone. Loss of sphincter tone and sensation about the anus suggests neurologic damage to the sacral nerves, such as in cauda equina syndrome or serious damage to the spinal cord. Unless other red flags are present, initial evaluation of low back pain does not require X-rays. Manual muscle test scale is 0-5 with 0 being absent and 5 normal. Deep tendon reflexes are 0-4 scale, with 0 being absent, 2 normal, and 4 being hyperactive with clonus.

Assessment:
Differential Diagnosis
The differential diagnosis of low back pain is extensive and includes mechanical low back pain, sciatica, herniated disc with or without nerve impingement, spondylolysis with or without spondylolisthesis, scoliosis, sacroiliac joint dysfunction, infection, ankylosing spondylitis, spinal stenosis, abdominal aortic aneurysm in elderly patients, various benign and malignant tumors, fracture, and cauda equina syndrome. See Symptom: Joint Pain and other related topics in this book. Urological conditions such as stone disease and pyelonephritis may present as back pain (see GU chapter). Other problems may be referred to the back from the abdomen, including labor (see Symptom: OB Problems) and pancreatitis (see GI: Pancreatitis).

Plan:
Treatment
Primary: Usual treatment of mechanical low back pain includes ice, anti-inflammatories such as **ibuprofen**

(800 mg tid with food) and progressive range of motion exercises and trunk strengthening. Bed rest is not indicated unless absolutely essential, as it merely causes deconditioning. Epidural steroids are sometimes used; oral steroids are not recommended. Cauda equina syndrome, a rare complication where there is compression of the cauda equina in the spinal column causing neurological impairment, may become permanent if not surgically repaired in 12-24 hours. Suspected fractures should be immobilized on a spine board or the nearest field equivalent and evacuated to the nearest appropriate facility that can perform appropriate radiological studies and surgery if necessary.

Patient Education
General: Most low back pain is self-limited and will resolve in 4-6 weeks in most people.
Activity: Gradually resume activity. Avoid bedrest if possible - it only weakens the back muscles.
Diet: Normal
Medications: Anti-inflammatory medicine may cause bleeding ulcers, kidney and liver problems with chronic use.
Prevention and Hygiene: Use proper mechanics when lifting – bend at the knees not the waist
No Improvement/Deterioration: Loss of bowel or bladder control warrants immediate referral.

Follow-up Actions
Return evaluation: Worrisome signs for further referral for imaging (such as x-ray and MRI) and evaluation include fever, night pain, unexplained weight loss, persistent pain (greater than 4-6 weeks)
Evacuation/Consultation Criteria: Loss of bowel or bladder control, and/or urinary retention warrants immediate evacuation. Obtain delayed evacuation for imaging and evaluation, if fever, night pain, unexplained weight loss, persistent pain (greater than 4-6 weeks)

Symptom: Breast Problems: Mastitis
MAJ Ann Friedmann, MC, USA

Introduction: Mastitis is inflammation of the breast most commonly presenting as a cellulitis of the subcutaneous tissues in a lactating breast (1-3% of breastfeeding women). The causative organisms, *S. Aureus, E. Coli* and *streptococcus* (rarely), are easily treated with antibiotics. Tuberculosis mastitis is very rare (1% of cases) even where TB is endemic. One of the risk factors for mastitis is plugging or obstruction of one of the milk ducts which drain to the nipple. Obstruction can be secondary to delayed infant feedings, which can lead to engorgement, and tight clothing (poorly fitting brassieres and underwires that dig in). Other risk factors include cracked nipples, maternal stress and fatigue. **Do not let mastitis interrupt breastfeeding.** The infected breast will worsen if the baby does not empty it, and the infection cannot be transmitted to the infant through the milk. Untreated or delayed and inappropriate treatment can lead to breast abscesses and stop lactation in the affected breast, which deprives the infant of its food source. This may be devastating, particularly in developing countries.

Subjective: Symptoms
Localized pain, redness, swelling, warmth in one breast; fever; chills; body aches; fatigue; headache; occasionally nausea and vomiting.

Objective: Signs
Using Basic Tools: Fever - often greater than 101°F
Inspection: Pink, wedge-shaped area on the breast. Patient appears in mild to moderate distress.
Palpation: Tender, occasionally indurated, warm area. There SHOULD NOT be a palpable, fluctuant mass - that is a sign of abscess.

Assessment:
Differential Diagnosis:
Plugged Duct - Tender lump in the breast of a mother who is otherwise well. Caused by partial obstruction

of a duct. Infection is not present but MAY RESULT if the duct remains blocked. See end of this section for treatment.

Breast Engorgement - Gradual onset in the immediate postpartum period (peak on days 2-4) of bilateral breast swelling and warmth. Pain is generalized. Fever may occur but is rarely over 101°F. The breasts feel better after they are emptied. Caused by inadequate emptying of the breasts. A risk factor for mastitis. See end of this section for treatment.

Breast Abscess - Painful, fluctuant mass. 10-15% of women who delay treatment of mastitis will develop a breast abscess. Should be suspected if a patient on antibiotics for mastitis does not improve after 72 hours of antibiotic therapy. See Breast Abscess Incision and Drainage Procedure description in following section.

Breast Cancer - Very rare. Unilateral, unchanging lump or mass that persists despite treatment for engorgement. Plugged duct or persistent mastitis must be evaluated by appropriate radiological and surgical approaches, if possible.

Plan:
Treatment for Mastitis
Primary:
1. Ensure infant nurses on both breasts, starting on the unaffected side. Even after feeding the affected breast may need to be more thoroughly emptied by manual expression or pumping.
2. Put mother on bedrest. Maternal fatigue and stress are risk factors for recurrent mastitis. The baby should be right next to the mother either in the same bed or readily available in a nearby crib to facilitate frequent emptying of the breast.
3. Ensure mother completes full course of therapy:
   a. **Dicloxacillin** or **cephalexin** 500 mg po q 6 hrs x 10 days
   b. Patients allergic to **penicillin**: **clindamycin** 300 mg po q 6 hours x 10 days or **erythromycin** 500 mg po q 6 hours x 10 days
   c. If patient does not improve after 48 hours of rest and therapy, switch to **Augmentin (amoxicillin/ clavulanate)** bid if 875/125 mg or tid if 500/125 mg.
4. Apply ice packs or warm packs to the breast (whichever the mother prefers). Hot packs provide drainage and pain relief.
5. Ensure mother drinks plenty of fluids.
6. Give **ibuprofen** or **acetaminophen** for pain relief.
7. Advise mother to wear a support bra or other supportive clothing that does not cause painful pressure on the breast.

Alternate: Patients who are intolerant of oral medications may need IV therapy. You may use **nafcillin** or **oxacillin** 2.0 gm q 4 hrs IV or **cefazolin** 1.0 gm IV q 8 hours. Penicillin allergic patients can be given **clindamycin** 300 mg IV q 6 hrs.

Primitive: If antibiotics are not available, initiate rest, hydration and MOST IMPORTANTLY, drainage (nursing) of the affected breast. Mastitis will recur in at least 50% of women not treated with antibiotics, and the breast abscess rate will be high.

Patient Education
General: Counsel family to assist the mother.
Activity: Bedrest. If patient is improving she may gradually increase her activity level after 72 hours of strict rest.
Diet: Fluids and well-balanced diet
No Improvement/Deterioration: Improvement in systemic symptoms is expected in 24-48 hours. The focal breast tenderness should be significantly improved in 72 hrs then gradually resolve in 7-10 days.

Follow-up Actions
Return evaluation: 48-72 hrs. Reassess vital signs and perform breast examination for mass or fluctuance (evaluate for abscess). Reemphasize compliance with antibiotic therapy and rest. Assess breastfeeding frequency and nutritional status of the infant. Be sure the mother is emptying the infected breast. If the patient is not improving, consider changing antibiotics to **Augmentin**.

If the patient remains febrile after 72 hours:
1. If possible, transfer the patient immediately. If not, then see below.
2. If non-compliant, administer **ceftriaxone** 1 gm IM. See if the patient would also take one gram of oral **azithromycin**. Follow-up in 24 hrs and repeat the **ceftriaxone**.
3. If compliant add **clindamycin** IV as above.
4. Find an experienced person in the community who can stay with the patient and assist with breastfeeding and manual expression.

Evacuation/Consultation Criteria: Suspicion of breast abscess, continued fever after 72 hours of antibiotics. Breast abscess requires incision and drainage, which should be done by a trained physician if possible.

Treatment for Plugged Duct: Massage the area, gently pressing toward the nipple. Warm compresses help. The most important intervention is frequent feeding on the affected side. Consecutive feedings should be started on the affected side to facilitate flow from the obstruction. Because different lobes of the breast are drained better with different nursing positions, place the infant with its chin pointing toward the blocked duct. If the mother is separated from her infant for any reason the breast should be emptied by hand-expression or by using an effective breast pump. Be sure to follow the mother closely as mastitis can occur. **NOTE:** Plugged ducts will last only for short periods of time (a few days). Any lump that persists for many days must be evaluated for malignancy.

Treatment for Engorgement: Frequent breastfeeding is the most effective treatment. If the nipples are engorged it may be difficult for the baby to latch on. Relieve nipple engorgement by applying warm compresses before the feeding, gently express some milk to soften the breast, or lean the breasts into a large bowl of warm water (or take a warm shower) just before the feeding to facilitate milk release and soften the nipples. After the feeding, apply cold compresses, or cool cabbage leaves to the breast for 20 minutes leaving the nipple exposed. Use standard doses of **acetaminophen** and/or **ibuprofen** for pain relief. Use mild narcotics in severe cases (**Tylenol #3**). Support the breasts with a good-fitting brassiere. Do not bind the breasts as this will increase the engorgement.

Symptom: Breast Problems: Breast Abscess Incision and Drainage Procedure
MAJ Ann Friedmann, MC, USA

When: A breast abscess is causing systemic symptoms that are unresponsive to less invasive therapy. The abscess will be a fluctuant breast mass related to non-resolving or worsening mastitis (see Mastitis section). The condition is rare, except when antibiotic treatment has been delayed or discontinued too early. Needle aspiration of a recurrent abscess should be attempted twice before incision and drainage is required. Do not attempt this procedure unless evacuation to a physician is unavailable.

What You Need: Sterile prep and drape, 18 and 24-26 gauge needles, 5 cc and 10cc syringes, alcohol prep pads, local anesthetic agent such as 1% **lidocaine with epinephrine**, 2x2 and 4x4 dressings, scalpel with #15 blade (but any blade will work), sterile irrigation if available, gloves, and a small Penrose drain

What To Do:
Needle Aspiration: Anesthetize skin and subcutaneous tissues over fluctuant area using 5cc syringe and 24-26 gauge needle. Insert 18 gauge needle attached to 10cc syringe into the abscess, aspirating as you advance the needle. Drain as much pus as possible once the cavity is entered. Then remove needle and cover puncture site with a small dressing.

Incision and Drainage of Abscess: Anesthetize the skin and subcutaneous tissues over the fluctuant mass. If possible, choose your incision point close to but not in the areola (allow room for an infant to nurse without contacting the incision). Make the incision parallel to the edge of the areola and over the fluctuant area. Try not to make transverse incisions- they leave an unacceptable scar. Circumareolar incisions will heal with a better cosmetic result. Keep the incision small, only large enough to allow entrance of your 5th digit. Do not cut deeply into the breast tissue, but start superficially and advance carefully. Make your incision deeper until the cavity is just reached and pus begins to drain. Insert your 5th digit into the wound to break up any loculations and to ensure complete drainage. Irrigate the wound, and if possible place a Penrose drain, pack with 2x2 dressings and cover with 4x4 dressings.

Follow-up care: Remain at bedrest, continue warm soaks and hydrate well. Continue antibiotic therapy with **Augmentin (amoxicillin/clavulanate)** po bid if 875/125 mg or tid if 500/125 mg for 10 days after resolution of the abscess. Engorgement of the breast will interfere with healing. Feed infant or empty the affected breast on a regular basis, every 2-2 ½ hours. The milk is clean as long as the abscess drains to the outside (through the skin). Nursing can continue when the abscess is surgically drained as long as the incision and drainage tube are far enough from the areola. If the drain is too near the areola for the infant to nurse, the breast still must be emptied by manual expression or pumping. Milk may drain from the incision during breastfeeding. Apply pressure over the incision while breastfeeding to minimize leakage. If the patient cannot do this herself, her spouse or another person may hold light pressure over the incision.

Follow-up Exam: 24 hours and as needed thereafter. Change the dressings and advance the drain at each visit. The breast should heal well from the inside out within 3-5 days.
Referral Criteria: Persistent fever despite treatment; worsening pain; increased size of abscess.

What Not To Do:
Do not make incision any deeper than necessary.
Do not make the incision too close to the areola to avoid compromising breastfeeding.
Do not allow the skin to close over the incision until the abscess has healed from the inside out to the surface.

Symptom: Chest Pain
CAPT Kurt Strosahl, MC, USN & COL Warren Whitlock, MC, USA

Introduction: Chest pain can arise from any of the structures within the chest or be referred from outside of the chest (from the cervical spine or abdominal organs). Immediate, life-threatening causes of chest pain include: acute coronary ischemic syndromes, pericardial tamponade, pneumothorax, arterial gas embolism, pulmonary embolism, aortic dissection, esophageal rupture, and perforated ulcer. A high level of suspicion for these diagnoses is necessary to institute potentially life saving therapy. The history and physical examination and basic tests may be insufficient to exclude these diagnoses in the field, so many patients may be treated "unnecessarily." A prior history of any of these diseases should lead to a presumptive diagnosis in favor of recurrence. Other causes of chest pain may not be life threatening if they are misdiagnosed. These include musculoskeletal chest wall pain, decompression sickness, hiatal hernia with reflux disease or dyspepsia, herpes zoster (shingles), bronchitis or pneumonia, gall stones, mitral valve prolapse, subacromial bursitis, mastitis, pancreatitis. (See appropriate portion of the Handbook for detailed discussion of these conditions). Heart attacks occur in healthy twenty year olds. Most heart attack victims die in the first 1-4 hours due to ventricular fibrillation, asystole, myocardial rupture or cardiogenic shock. There is very little to be done in the field to prevent this! Heparin helps acute coronary ischemia and pulmonary emboli but harms esophageal rupture, pneumothorax, tamponade, and aortic dissection.

Subjective: Symptoms
Focused History: Quality: *Is the pain sharp?* (Musculoskeletal pain is sharp, localized to a finger point area, and lasts seconds to minutes. Cardiac pain is better described as a discomfort, localized to the inside of the chest, builds in intensity over several minutes and lasting 5 minutes or more. Pneumothorax pain is sharp, localized to the inside but can be stopped by holding the breath. GI pain is sharp or a discomfort but is relieved by antacids, swallowing liquids, and located centrally under the breastbone or in the upper abdomen. Aortic dissection is a severe, tearing pain that starts suddenly, builds rapidly, moves into the back and causes

the member to not want to move for fear of their life. Tamponade pain is sharp or discomfort associated with fullness in the neck. Shingles pain is burning and localized to a dermatome, not crossing the midline and confirmed by the development of blistering lesions.)

Duration: *When did it start and how long did it last? Does it come and go?* (Angina is usually relieved over 5 minutes while myocardial infarction pain may last for 30 minutes to an hour. Musculoskeletal pain may come and go. Pneumothorax pain lasts until relieved by an intervention. Aortic dissection pain continues for hours. Dyspepsia resolves in minutes with treatment.)

Alleviating or Aggravating Factors: *What makes it better or worse?* (Angina is brought on by exertion and relieved by rest. Dyspepsia comes on after eating, is worsened by lying down and relieved by antacids. Nothing relieves the pain of aortic dissection, perforated peptic ulcer or pneumothorax except narcotics. Musculoskeletal pain is brought on by movement of the arms or chest and alleviated by not moving or aspirin/NSAIDs. Gallstone pain is brought on by fatty foods. Movement of neck aggravates cervical disk pain. Shingles pain is burning and worsened by clothing touching the area. Esophageal rupture is worsened by eating or drinking and relieved only by narcotics. Eating or antacids relieve peptic ulcer disease pain. Drinking cold liquids can bring on esophageal spasm. Pulmonary embolism pain is worsened by deep breaths and partially relieved by oxygen.)

Associated Symptoms: *Does the pain move anywhere?* (Pain moving to the jaw or left arm suggests a cardiac cause. Pain boring through to the back may be a peptic ulcer or aortic dissection. Pain moving to the right shoulder blade is likely gall bladder. Pain moving to the top of the shoulder is likely to be a pneumothorax or subacromial bursitis. Splenic pain may be referred to the left shoulder.) *Are there other symptoms?* (Sweating, nausea, vomiting, shortness of breath are non-specific. Palpitations may accompany heart disease but are also present whenever fear is present. An acid taste in the mouth [water brash] suggests esophageal reflux. Pneumonia is associated with fever and a productive cough.)

Coronary Risk Factors: The presence of 3 or more coronary risk factors increases the likelihood of the disease presence: smoking, hypertension, family history of heart attack before age 55, diabetes, high total cholesterol, low HDL (good cholesterol), obesity, peripheral vascular disease.

Pearls:
1. Pain that lasts seconds is not serious.
2. Pain that moves from the chest into the arms and then the legs is not coronary disease.
3. Pain relieved by **nitroglycerin** sublingual is smooth muscle: coronary or lower esophageal sphincter or gall bladder or intestinal angina.
4. Chest pain resulting in collapse and shock is due to one of the life threatening causes.
5. Anxiety and psychogenic chest pain can be sharp and last for seconds or be dull and last for days. While it may be unresponsive to all empiric therapy, psychogenic chest pain remains a diagnosis of exclusion.

Objective: Signs

Using Basic Tools: General: Pale, sweaty, cool clammy skin suggests decreased cardiac output; cyanosis suggests PE. Vital Signs: Fever suggests infectious cause such as pneumonia or bronchitis. Irregular pulse suggests cardiac cause. Absent pulses in the left arm or a blood pressure >10 mm lower than the right suggest an aortic dissection. BP <100 systolic with heart rate >120 suggests decreased cardiac output (e.g., tamponade, myocardial infarction and others). Respirations >30 suggest decreased cardiac output or hypoxemia. Oxygen saturation <85 suggests pulmonary embolism but may be present with pneumothorax or MI with pulmonary edema.

Neck Veins: Elevated suggests tamponade.

Chest Wall: Point tenderness over the costochondral junction or the intercostal muscles suggests chest wall pain (inflammation of the costochondral junction, muscle strain). Numbness in a subcostal nerve distribution suggests nerve injury from trauma. Lateral compression of the chest cage will accentuate pain from a rib fracture. Tenderness and warmth of the breast suggests mastitis.

Lungs: Absence of breath sounds or increased resonance to percussion suggests pneumothorax. A shift in the trachea supports a tension pneumothorax. Fine-crackling sounds of fluid (rales) in the alveoli suggests cardiac cause. Coarse crackling sounds (rhonchi) suggest pneumonia or bronchitis. A finding of consolidation (E->A change) suggests pneumonia.

Heart: Irregular beating suggests cardiac cause. Muffled sounds suggest tamponade or pneumothorax.

Galloping heart sounds (like hoof beats) suggest cardiac or pulmonary embolism. Velcro rubbing sounds with each heartbeat suggests pericarditis.
Abdomen: Right upper quadrant tenderness suggests gall bladder. Peritoneal findings suggest ruptured peptic ulcer or pancreatitis
Extremities: Pulse deficit suggests aortic dissection. Swelling and tenderness of the calf or thigh suggests a source for pulmonary embolism.
Using Advanced Tools:
WBC for infection (pneumonia). EKG: ST elevation (1mm or more) is supportive of an acute myocardial infarction. ST depression of 1mm suggests ischemia. ST elevation, along with PR depression suggests pericarditis that can lead to tamponade.

Plan:
Primary Treatment – Basic
Rest to decrease oxygen consumption. Oxygen to bring the oxygen saturation above 95%
Sips of water only, until stable
Aspirin 325 mg chewed (A single aspirin reduces the risk of angina going on to myocardial infarction or death by 50%.)
Intravenous line for IV drug access and Normal Saline Solution at 100cc/hr (or bolus 1000cc to bring the systolic BP over 100)
Insert 16 gauge, 6 inch IV needle over the 2nd rib on the side of the chest with decreased breath sounds if tension pneumothorax is suspected
Endotracheal intubation if in shock with oxygen saturation under 80%
Morphine sulfate 2mg IV repeated q 5 minutes until relief of pain or sedation
Use **Narcan** if over-sedation occurs
Repeat vital sign determinations (q 15 minutes) until stable

Primary Treatment – Advanced
Suspect Coronary Cause: **ASA** 325 mg po daily. Low molecular weight **heparin** 1mg/kg SC q 12 hrs.
Propranolol 40-80 mg po qid (to keep HR < 80) (also treats MVP). **Verapamil** 80 mg po qid if history of asthma precludes use of propranolol. **Nitroglycerin** 0.4mg SL q 5 minutes X 3 doses or pain relief
Diazepam 5mg po qid for 48 hours for sedation. **Furosemide** 40mg po if rales present. **Atropine** 1mg IV (repeat X1 in 5 minutes) if HR <50. **Lidocaine** 100mg IM if pulse is irregular (or PVCs present)
Suspect Pulmonary Embolism: **Heparin** 10,000 units IV followed by 12,500 units SC q 12 hrs. Treatment of pulmonary embolism is directed solely at the prevention of further emboli. (see Respiratory: PE)
Suspect Aortic Dissection: **Propranolol** 40-80 mg po qid. **Diazepam** 5mg po qid. NO HEPARIN!
Suspect Pneumothorax: Oxygen to >95% saturation; consider needle decompression (see Procedures: Thoracostomy, Needle)
Suspect GI Cause (reflux, dyspepsia): Liquid antacid, 1 tablespoon q 2-4 hrs. **Cimetidine** 400mg po tid or **famotidine** 20mg po bid.
Suspect Rupture of the Esophagus or Stomach: NPO and refer to GI Chapter for treatment
Suspect Pericarditis: **ASA** 650 mg po q 4 hrs or **ibuprofen** 800 mg tid. If pain not relieved in 12 hours, **prednisone** 60mg po qd
Suspect Tamponade: If BP is falling, perform pericardiocentesis (see Procedures)
Suspect Shingles: **Codeine** 60mg po q 4-6 hrs (see Dermatology: Herpes Zoster)

Patient Education
General: Healthy lifestyle.
Activity: Do not restrict physical exertion if medical condition not life threatening.
Diet: High fiber, low cholesterol, low fat
Medications: Take exactly as prescribed. Beta-blockers cause tiredness and slow HR
Prevention: Discontinue smoking; obtain treatment for hypertension and hyperlipidemia

Follow-up Actions
Wound Care: Keep puncture sites clean

Return evaluation: Repeat EKG and other pertinent exams.
Evacuation/Consultation Criteria: All life threatening causes should be evacuated at the first window of opportunity for further evaluation and treatment. Recurrent chest pain without objective findings may be treated with **aspirin** and **diazepam** but should be further evaluated upon completion of the mission.

Symptom: Constipation
COL (Ret) Peter McNally, MC, USA

Introduction: Constipation is a conscious, unpleasant, and subjective sensation of deviating from the "normal" defecation pattern. Most will feel uncomfortably distended and "backed up." Healthy and active persons may become acutely constipated after a change in lifestyle, decrease in dietary bulk, dehydration or inactivity. When the urge to defecate is repeatedly repressed or ignored, constipation may arise.

Subjective: Symptoms
The definition of constipation varies person to person but a reasonable definition is as follows: Two or fewer bowel movements per week, straining > 25% of the time, hard stools > 25% of the time, incomplete evacuation > 25% of the time. Constipation is much more common among women than men, and the young and aged persons are especially prone. Common causes of constipation include inadequate fiber & food intake, repression or ignoring the urge to defecate, and immobility. Medications such as opiates, anticholinergics and antidepressants can slow intestinal transit and promote constipation. A preceding history of prolonged confinement in a vehicle, airplane or ship with inactivity and decreased intake is typical.

Objective: Signs
Uncomfortable and restless; normal vital signs; distended abdomen; stool-filled loops may be palpable but abdominal tenderness is uncommon. Always look for signs of hypothyroidism: fatigue, feeling cold, loss of hair.

Assessment:
For acute constipation temporally associated with change in diet and activity, no testing is necessary. For chronic constipation, tests to exclude structural and systemic disease are necessary.

Plan:
Treatment
Primary: Laxative: **senna bisacodyl**, single to few doses (onset <24 hr).
Alternate: **Magnesium citrate** 12 oz po (effective in 6-8 hrs); **psyllium** or **methylcellulose**, daily dosing with increased fluid intake (effect within a week).
Primitive: Perform a digital rectal examination and remove fecal impaction if present. If hard stool is present, **glycerin** or **bisacodyl** suppositories may be helpful. Position the suppository against the rectal wall.

Patient Education
General: Promote healthy, high fiber diet, increased fluid consumption and daily exercise. Answer the urge to defecate. Develop a regular bowel habit.

Follow-up Actions
Evacuation/Consultation Criteria: Evacuation is not usually necessary. Consult as needed.

Symptom: Cough
COL Warren Whitlock, MC, USA

Introduction: Cough is a normal physiologic function of the respiratory system, and is the only mechanism that clears secretions from the lung. Conditions that cause coughing include respiratory obstruction from increased secretions or aspirating foreign material, irritation (infectious, chemical or thermal injury), chronic lung disease, or a non-pulmonary cause (congestive heart failure) or an idiosyncratic effect (medications like an ACE inhibitor). Most acute coughs, divided equally between upper airway (ENT) and lower airway (lung) causes, are related to infections (viral upper respiratory tract infections, bronchitis, or laryngitis), allergies (often seasonal) and postnasal drip. Most chronic coughs are due to underlying lung disease such as emphysema, chronic bronchitis (especially in smokers) or asthma. Environmental irritants, perennial allergies, and gastroesophageal (GE) reflux/aspiration can also cause chronic cough. Cough due to heart failure, tuberculosis or lung cancer may be more likely, depending on patient history.

Subjective: Symptoms

Focused History: Quality: *Does anything come up when you cough?* (indicates secretions and inflammation in the airway, which are common in infections) *What color is the stuff you cough up?* (green sputum: associated with bacterial infection; blood: generally associated with infection or underlying lung disease; clear or white sputum or non-productive cough: asthma or pneumonia from mycoplasma) *Do cough at night?* (CHF, asthma, GE reflux)...*while lying flat?* (GE reflux) ...*after exercise?* (asthma) *Have you had a cough that changed? How?* (COPD may have a chronic cough that increases and becomes productive with an infection).
Duration: *How long have you been coughing?* (viral URTI or viral bronchitis usually lasts 7-10 days; if > 14 days, consider underlying lung disease or a more serious type of infection, such as atypical pneumonia).
Alleviating or Aggravating Factors: *What makes the cough better, and what makes it worse?* (cough that worsens with talking is usually due to infection or allergy; cough that improves by sitting up, gets worse after eating or when lying down suggests GE reflux; cough that only occurs after exercise, and is worse in cold weather [below 32°F the airway is devoid of moisture] is highly suggestive of asthma; cough that improves with cold medications suggests allergy and post-nasal drip.) *Do you smoke?* (Morning cough is common, due to chronic bronchitis.) *Did you stop smoking recently?* (After stoping, the clearance mechanisms of the lung begin to recover and mobilize secretions from the lower airways. This type of cough is beneficial to the lungs and improves over several months if they do not resume smoking).

Sputum	Acute cough (< 24 hrs)	Subacute (> 24 < 96 hrs)	Chronic (Over 5 days)
Non-productive	Viral	or Mycoplasma	or Asthma/COPD
Productive-Clear	Viral or Asthma	Viral or Asthma	Asthma and/or Allergic
Purulent	Bacterial	Bacterial	and/or Chronic Bronchitis
Thick/dark	Bacteria	and/or Chronic Bronchitis	and/or underlying lung disease
Bloody	Bacterial	Bacterial	Bacterial/Tuberculosis/Cancer

Pearls:
1. Cough associated with eating suggests a mechanical swallowing problem causing aspiration, or a tracheoesophageal fistula (connection between trachea and esophagus), or gastroesophageal reflux (associated with heartburn or a sour taste).
2. Persistent morning cough that improves after expectorating sputum is typical of chronic bronchitis.
3. Nighttime cough associated with SOB may suggest congestive heart failure (especially in elderly patients) or asthma (if wheezing). Nighttime cough without SOB indicates a sinus infection or an allergic origin such as post-nasal drip.

Objective: Signs

Using Basic Tools: Vital Signs: Low-grade fever (99–100.5°F): viral or mycoplasma infection; high-grade fever (>102°F): bacterial infection; <14 breath/min: allergies, post-nasal drip, bronchitis, GE reflux; >14 breaths/min: asthma, pneumonia, acute exacerbation of chronic bronchitis
Lungs: Resonant: normal – think upper airway cause (sinusitis, etc.); wheeze: asthma or emphysema (rarely foreign body aspiration); rhonchi: secretions in the airway - bronchitis, pneumonia; rales: inflammation or fluid in the alveoli – pneumonia; dull: parapneumonic effusion, pneumonia or collapsed lung; barrel chest: COPD/chronic bronchitis; splinting respiration: pleurisy or pneumonia.
Using Advanced Tools Labs: Gram stain and culture of sputum: >15 white cells/high power field indicates bacterial infection

CXR: clear: asthma or upper airway cause, infiltrate: pneumonia, cavity: lung cancer or tuberculosis

Assessment:
Differential Diagnosis
Infections - bronchitis (acute or chronic), pneumonia (viral, bacterial, fungal, mycoplasma including TB), sinusitis, pharyngitis, laryngitis
Aspiration - foreign body, gastric contents.
Allergic or sensitization response - asthma, pneumonitis (chemical, biological), allergic rhinitis
Chronic lung disease - emphysema, COPD, chronic bronchitis, smoking, cancer.
Other lung disease - pulmonary embolism, pneumothorax, pleurisy
Other non-pulmonary disease - congestive heart failure, irritant rhinitis, gastroesophageal (GE) reflux, medication effect

Plan:
Treatment
1. See separate Respiratory sections for treatment for asthma, pneumonia, pleurisy, pneumothorax, allergic pneumonitis, emphysema, COPD, pulmonary embolism.
2. Antibiotics are only indicated in patients with evidence of a mycoplasma or bacterial infection, or at high risk due to a chronic underlying pulmonary disease (empirically treat for both Gram positive and negative). Antibiotics are not generally needed for acute bronchitis.
3. Treat symptomatically when the findings on history and physical examination do not warrant antibiotics. Do not suppress a productive cough unless it interferes with obtaining adequate rest/sleep or jeopardizes concealment. **Dextromethorphan** in tablet or liquid form (children over age 2: 2.5 mg, to age 12: 10 mg, Adults: 30-60 mg, q hs or q 4 hrs) or combined with expectorants like **guaifenesin** are often used. Expectorants have not been proven to be effective. **Codeine** (children over age 6: 5 mg; age 12, 10 mg; adults: 30-60 mg, po q hs or q 4 hrs for severe cough. QHS is preferred for severe cough that interferes with the ability to rest (use no more than 3 nights) and will not impair the ability to clear secretions during the daytime.
4. If you cannot determine a clear etiology for the cough, treat empirically for allergy, since this is one of the most common causes in otherwise healthy individuals and treatment is well tolerated.

Patient Education
Follow-Up: Return if cough persists for more than 2 weeks or worsens.

NOTES: Cough can be associated with psychological symptoms. However, "psychogenic cough" is a diagnosis of exclusion, but can be associated with severe anxiety or becomes part of a conversion reaction.

Symptom: Depression and Mania
MAJ Michael Doyle, MC, USA

Introduction: Everyone experiences happiness and sadness. However, command and medical personnel should be concerned when service members' variations in mood begin to impair duty performance. 20% of the general population will at some point experience depression outside of the level of sadness expected in daily life. Another 2-5% will experience sustained mood elevation, called mania or hypomania. Both depression and mania appear in all cultures. SOF medics may be called upon to evaluate host nation civilian and military personnel.

Subjective: Symptoms

Depression: Loss of pleasure in activities, social isolation and withdrawal, subjective cognitive impairments, anxiety, worry, excessive guilt, preoccupation with thoughts of death or suicide, insomnia or hypersomnia, changes in appetite and weight, loss of energy and feelings of helplessness, hopelessness, and worthlessness.

Mania: Mood elevation, grandiosity, increased seeking of pleasurable stimuli (hyper-sexuality, spending money, etc.), intrusiveness, belief in special powers, skills or relationships.

Focused History: *When did you start feeling this way, and have you ever felt this way before?* (Many people with depression have had it before.) *Are these feelings constant or do they come and go?* (Severe depression is usually ever-present; mood swings are related to personality; mania rarely switches rapidly enough to seem to "come and go.") *How is your sleep?* (Depressed patients complain of poor sleep or too much sleep, but never feeling rested; manic patients insist they feel fine on very little sleep.) *How is your appetite?* (often decreased in depression, but some patients complain of cravings or eating too much in order to feel better) *Can you do your job?* (Hypo-manic patients appear to do their job very well [and try to do everyone else's]; manic and depressed patients have obvious impairment in work functioning.) *Do you have thoughts of hurting yourself or anyone else?* (Always, always ask!) *What helps you feel better?* (Manic patients feel even better with sex, alcohol or spending money; depressed patients find that little interests them or helps them feel better.)

Objective: Signs

Let a patient talk for a few minutes uninterrupted and try to sit quietly and follow where the train of thought goes. *What themes are present? Depressed themes? Grand themes? Are the thoughts connected? Are there delusions? How severe is the depression—on a scale of 1 to 10.*

Using Basic Tools: Normal vitals; manic patients may have slightly elevated pulse and BP.
Mental Status Exam:

Depressed persons appear sad or sometimes worried. They move slowly but may also be agitated, unable to sit still. Look for hand wringing. A depressed person will often avoid eye contact, preferring to gaze downward. Depressed persons may have difficulty concentrating or completing thoughts. Mental processes are generally slowed.

Manic persons are happy or irritable. They talk profusely—that is, with pressured speech. Interrupting them is like trying to stop a freight train. Manic persons can be grandiose—believing they are God, or simply have grand ideas about how to solve the world's problems. They are energetic and intrusive. Sometimes ideas are very loosely connected or hard to follow.

Always ask about suicidal and homicidal ideations, intents or plans.

Both depressed and manic patients may develop fixed false beliefs called delusions. This represents more severe illness.

Assessment:
Differential Diagnosis

Condition	Manic Symptoms	Depressive Symptoms
Substance Intoxication	PCP, LSD, amphetamines, cocaine	Barbiturates, benzodiazepines, alcohol, marijuana
Trauma or Mass lesion	Head injury, tumor	Head injury, tumor
Endocrine	Hyperthyroidism	Hypothyroidism
Mental Disorder	Bipolar Disorder—a severe illness	Major depressive disorder—a severe

wherein there is a prominent and distinct period of significant mood elevation that impairs the affected individual's social and occupational functioning. Cyclothymia—periods of low mood alternating with periods of elevated mood: distinct from "mood swings" in response to social cues or irritability. Low or elevated mood or the swings between interfere with social or occupational functioning. illness of 2 or more weeks duration with most of the symptoms and signs described. Dysthmia–a longstanding pattern of low mood more days than not, for more than 2 years, but not as severe as Major Depression. Adjustment Disorders—as the name indicates, there is some difficulty adjusting to a new stressor that results in functional impairment. Of these listed, adjustment disorders are the most common conditions causing depressive symptoms.

Always maintain a high index of suspicion for a physical or CNS injury to explain a change in mood!

Plan:
Treatment
Primary
1. Ensure safety of the patient (suicide risk) through physical or chemical restraint, 1 on 1 watch or merely increased supervision (as the situation dictates) until definitive care is available.
2. **Mania**: Manage the behavioral disturbance through benzodiazepines or neuroleptics.
   a. Benzodiazepines (**diazepam** 5-10 mg po or IM or **lorazepam** 2 mg po, IM or IV)
   b. Neuroleptics (**haloperidol** 2-5 mg IM or po or **chlorpromazine** 50-100 mg po or IM). Consider **diphenhydramine**, 25-50 mg po or IM coincident with the neuroleptic.
   c. If these fail to settle down an agitated patient, see the Psychiatric Restraint Procedure on CD-ROM. If leather restraints are unavailable, consider physical restraint with sheets wrapped around patient on litter.
3. **Depression**: Selective Serotonin Re-uptake Inhibitors (SSRIs) such as **Prozac**, **Zoloft**, and **Paxil**, are mainstays of therapy, but take 1-2 weeks before becoming effective. In an uncomplicated patient without other medical problems, starting an anti-depressant medicine like an SSRI sooner (rather than later) can be very helpful. Starting dose is ½ the therapeutic dose:

	STARTING DOSE	THERAPEUTIC DOSE
Prozac (fluoxetine)	10 mg	20 mg
Zoloft (sertraline)	50 mg	100 mg
Paxil (paroxetine)	10 mg	20 mg

4. Refer to local civilian or military health authorities as soon as possible. (This may not be immediate.)
5. Treat side effects that may emerge with use of SSRIs (nausea, loose stools and headache) symptomatically. Evaluate other physical symptoms (severe headache, diarrhea) as a separate condition and rule out distinct causes.
6. Implement weapons access restrictions.
7. Educate commanders to provide time, opportunity, and conditions for sleep.

Patient Education
General: Get adequate rest. Go to bed and arise on same schedule daily. Avoid tobacco, alcohol, caffeine in the evenings.
Medications: Take SSRIs with food. They may cause headache, diarrhea and nausea.

Follow-up Actions
Return evaluation: Follow frequently with scheduled and prn visits.
Evacuation/Consultation Criteria: Evacuate urgently, particularly if functionally impaired, suicidal or a

danger to others.

Symptom: Acute Diarrhea
COL (Ret) Peter McNally, MC, USA

Introduction: Diarrhea is a change in bowel habits marked by numerous, watery stools. Diarrhea episodes lasting longer than 30 days are chronic and may require sophisticated evaluation and management. Acute diarrhea is common, with Americans suffering 1-2 episodes each year. Fortunately, most episodes (>90%) of acute diarrhea are mild and self-limited. Proper management of acute severe diarrhea is vital as this illness is the second leading cause of death in developing countries where SOF teams are frequently deployed. Consult medical intelligence sources before deployment for major causes of diarrhea in each area of operations. There are numerous causes of diarrhea: parasites, bacteria, viruses, food poisoning, chemical agent exposure, chronic disease, malabsorption and many others.

Subjective: Symptoms
Diarrhea (with or without blood, mucus, pus); fever often above 102°F; orthostatic dizziness, mental status changes; abdominal pain (may be severe and unremitting). The basic clinical presentations can be divided as follows:
Bloody diarrhea (dysentery): e.g., *Shigella*, some *E. coli*
Purging with voluminous diarrhea (>1 liter of stool/hr): e.g., cholera, antibiotic related pseudomembranous colitis (*C. difficile*)
Vomiting with minimal diarrhea: e.g., food poisoning (toxins), some viruses
Chronic diarrhea: e.g., *Giardia*, malabsorption of food
Gross bleeding per rectum: e.g., GI bleed: perforated hollow viscus (intestine, etc.) or hemorrhoid; hemorrhagic colitis (*E. coli*)
Simple diarrhea without a specific cause: usually viral, inflammatory bowel disease
Focused History: *Have you ever had this diarrhea before?* (Recurrent diarrhea may be due to chronic disease.) *Have you had blood or pus in your diarrhea?* (typical of dysentery) *Have you had diarrhea that was all just blood?* (typical of GI bleed) *How long have you had the diarrhea?* (Chronic diarrhea does not resolve in 4 weeks; cholera may cause extreme diarrhea and death within days.) *Are you vomiting also?* (typical of toxin ingestion) *Have you been taking antibiotics?* (typically within the past month for pseudomembranous colitis)

Objective: Signs
Using Basic Tools:
Vital Signs: Fever, hypotension, tachycardia, orthostatic changes in heart rate and blood pressure.
Inspection: May appear dehydrated, weak and disoriented.
Auscultation: Usually rapid bowel sounds are evident. Absent bowel sounds with abdominal distention and tympani suggest megacolon or perforation.
Percussion: Tympani and distention suggest dilated bowel loops.
Palpation: Mild diffuse tenderness is usual. Spleen enlargement may be seen in typhoid fever.
Using Advanced Tools: Lab: CBC with differential for infection, anemia; urinalysis to avoid renal failure from dehydration; stool sample for O & P, blood and leukocytes

Assessment:
Differential Diagnosis
Shigellosis dysentery - fever, abdominal pain, urgency (tenesmus), diarrhea contains blood (Hemocult positive), mucus and pus.
Salmonellosis (typhoid fever) - minimal diarrhea, prolonged high fever, delirium, bacteremia, enlargement of the spleen and abdominal pain. Without prompt diagnosis and treatment, this can be a severe 4-week illness.
Amebic dysentery - bloody diarrhea, toxic megacolon (extreme dilation of colon), liver abscess and death.
Cholera - rapid progression, voluminous diarrhea can lead to hypovolemic shock and **DEATH** within 1-2 days.
Pseudomembranous (*C. difficile*) colitis - watery diarrhea may be voluminous (Hemocult positive), and associated with fever and acute megacolon.

Inflammatory bowel disease - chronic, abdominal pain, urgency and diarrhea; may be associated with anxiety.
GI Bleeding - bright red blood per rectum (lower GI bleed); black (Hemocult positive), tarry stool (upper GI bleed); may be accompanied by vomiting blood if perforated peptic ulcer, ruptured esophageal veins; anemia.
Giardiasis - chronic diarrhea, cramping, weight loss, pale and greasy stools.
Viral Gastroenteritis (stomach flu) - various viruses can cause diarrhea, which may be associated with outbreaks in daycare or other institutional facilities (Rotavirus), epidemics in military personnel (e.g., Norwalk virus), and food-borne and waterborne outbreaks.
Food poisoning - usually with vomiting; (see GI: Acute Bacterial Food Poisoning); includes toxic *E. coli*; some viruses can also be spread by food (e.g., hepatitis A)
Traveler's diarrhea - non-specific term; diarrhea of bacterial or viral origin originating in conjunction with foreign travel
A multitude of other infectious agents including parasitic worms, malaria, and many other tropical illnesses.
Other chronic diseases including malabsorption syndromes.

Plan:
Treatment
1. Fluid Resuscitation: Either intravenous or oral rehydration solution (ORS).
   World Health Organization ORS: 1 liter of purified water + 20 gm glucose + 3.5 gm salt + 5 gm sodium bicarbonate + 1.5 gm potassium chloride
2. Food: Continue nutrition intake, but avoid lactose and caffeine.
3. Antibiotics (see GI: Acute Bacterial Food Poisoning also)

Primary/Empiric: Ciprofloxacin 500 mg po
Alternate: Norfloxacin 400 mg q 12 hr for 5 days
   Specific Organism:
      Shigella - **Ampicillin** 500 mg po qid or **Bactrim** 1 po bid x 5 days
      Clostridium difficile - **Metronidazole** 250-500 mg po tid x 10 days
      Salmonella, hemorrhagic *E. coli* - **Ampicillin** 50-100 mg/kg/day in four doses x 10-14 days
      Amebiasis - **Metronidazole** 750 mg tid x 10 days then **iodoquinol** 650 po tid x 20 days
      Giardia lamblia - **Metronidazole** 250 mg tid X 5 days
      E. coli - **Bactrim** 1 po bid x 5 days
4. Antimotility drugs: **Codeine, paregoric, tincture of opium, loperamide**. Use only in cases of simple or minimal diarrhea, or if necessary to enable operator to complete the mission. There is a significant risk of toxic megacolon, sepsis and perforation.
5. Blood. Consider transfusion for excessive GI bleeding, depending on stability of patient.
6. Treat other chronic illnesses if possible.

Patient Education
General: Most patients recover from acute diarrhea without sequelae. Fluid resuscitation with a glucose, bicarbonate and potassium containing liquid is the essential method to avoid dehydration. Gatorade is an excellent commercial drink.
Diet: Liquids first, then bland solids as tolerated. Bananas can improve consistency of stool.
Prevention and Hygiene: Wash hands. Consume only approved water and food (see Preventive Medicine chapter).

Follow-up Actions
Return evaluation: Review history and consider alternate treatment or evacuation.
Evacuation/Consultation Criteria: Evacuate all with severe diarrhea, especially if associated with change in mental status, sepsis. Most cases of simple or minimal diarrhea do not require evacuation. Consult infectious disease specialist for severe or chronic diarrhea.

Symptom: Dizziness
CPT Brooks Morelock, MC, USA

Introduction: The complaint of dizziness is extremely vague and must be clarified. When a patient presents with dizziness, the examiner must ascertain whether the person is describing an alteration of consciousness (see Symptome: Syncope), an alteration of balance, a sensation of motion, or a feeling of lightheadedness that accompanies standing up. It should become readily apparent that the etiology of dizziness may involve in the inner ear, the central nervous system or a systemic disorder.

Subjective: Symptoms
Focused History: Does the patient have a prior history that can account for recurrent dizziness such as Meniere's disease or vertigo? **Duration** How long has the patient had symptoms? (acute symptoms likely a self-limited illness such as otitis media or labyrinthitis. Chronic symptoms suggest either anatomic abnormalities, such as acoustic neuroma, or chronic illness such as Meniere's) **Illness** Has the patient been ill, especially any upper respiratory illnesses? (recent URI can lead to vertigo through otitis media, either serous or purulent; labyrinthitis or benign paroxysmal vertigo.) **Fullness** Has the patient been experiencing ear pain / fullness? (can be associated with otitis media.) **Trauma** Has the patient been exposed to any direct trauma to the ear or barotrauma? (Can result in serous otitis media.) Has the patient been flying or diving recently? (possible decompression sickness–see Diving Medicine) **Hearing/Ringing** Does the patient have a persistent ringing (tinnitus) in their ears? If so, do they also have a hearing loss? (the combination makes Meniere's more likely). **Spinning** Does the patient have a sensation of motion/spinning? If so, does head movement bring it on? (classic symptoms for benign paroxysmal vertigo, which may be accompanied by vomiting.) **Walking** Does the patient have difficulty walking? If so, do they feel dizzy when this happens? (abnormal gait without dizziness is most likely ataxia (difficulty walking), a motor control problem) **Falling** Does the patient constantly fall toward the same direction? (An anatomic abnormality [i.e., tumor] in the middle ear will classically cause the patient to fall toward the affected side.)

Objective: Signs
Using Basic Tools: Vital signs: Low blood pressure (or a change with standing of >20mm Hg systolic) suggests possible dehydration causing pre-syncope due to volume depletion. fever: possible inner ear infection.
Ear Exam (otoscope): Tympanic membrane (TM) normal: typical finding that only rules out vertigo caused by infection. TM injected, loss of light reflex, +/- purulent fluid in middle ear: purulent otitis media can cause vertigo by stimulating of the vestibular apparatus. TM bulging, clear fluid in middle ear: suggests serous otitis media.
Neurologic: Dix-Hallpike Maneuver*- positive symptom reproduction and rotatory nystagmus – vertigo. Abnormal neuro exam: Possible CNS dysfunction- consider evacuation.
* Dix-Hallpike Maneuver: Have the patient sit, so that when lying supine, the head extends over the end of the table. Instruct the patient to keep their eyes open and to stare at the examiner's nose during the test. To test the left posterior canal, have the patient turn his head 45° to the left. Keeping the head in this position, lie the patient down rapidly until the head is dependent and extended below the table. In each position, observe the eyes closely for up to 40 seconds for development of nystagmus. Return the patient to the upright position. To test the right posterior canal, repeat maneuver with the head turned 45° to the right side.

Assessment:
Differential Diagnosis
Meniere's Disease - a chronic disorder resulting in decreased hearing acuity over long duration, accompanied by multiple exacerbations of vertigo and tinnitus.
Labyrinthitis - causes dizziness and a decrease in hearing acuity. This can be due to bacterial or viral infection.
Benign Positional Vertigo (BPV) / Benign Paroxysmal Positional Vertigo (BPPV) - an acute spinning sensation brought on with head movement, and associated with a rotatory nystagmus on physical exam. There is no change in the patient's hearing. The vertigo and nystagmus can be elicited by Dix-Hallpike Maneuvers*. The

cause is a displaced otolith (debris) in a vestibular organ. Symptoms will usually respond to antihistamines. Returning the head to neutral position may be instantly curative.
If the patient has dizziness only when walking or standing, he does not have true vertigo.

Plan:
Treatment
BPV/BPPV:
Basic/Primitive: Curative maneuvers (Valsalva, Epley maneuver)
Epley maneuver: The goal is to herd debris in the vestibular canals away from the hair cells that directly influence balance.
1. Rotate posterior canal backward close to its planar orientation. This directs foreign material out of the canal.
2. Change the angular displacement of the head by about 90° with each position change.
3. Rapidly perform the changes in head positions and maintain each position until nystagmus has disappeared, indicating cessation of endolymph flow. If no nystagmus is visible, the latency and duration of nystagmus observed during Dix-Hallpike testing may serve as a guideline.
4. Guide head movements from behind and execute each change in position within one second; maintain each position for at least 30 seconds.
5. If vertigo is severe, pre-medicate patient with a vestibular sedative, such as **prochlorperazine** or **dimenhydrinate**, 30-60 minutes before performing the maneuver.
Perform the Epley maneuver:
   a. Have the patient sit upright with head turned 45° to the affected side.
   b. Have the patient lie down with head dependent (as in Dix-Hallpike maneuver).
   c. Rotate the head 90° with chin upwards and maintain dependent position.
   d. Ask patient to roll onto side while holding head in this position.
   e. Rotate the head so that it is facing obliquely downward, with nose 45° below horizontal.
   f. Raise patient to a sitting position while maintaining head rotation.
   g. Simultaneously rotate the head to central position and move it 45° forward (return to normal position).

Alternate: Antihistamines: **meclizine (Antivert)** 25 mg po q 6 hrs as needed.
Advanced: Valium 10 mg po q 6 hrs as needed.

Meniere's Disease:
Basic: Valium 10-15mg po q 4 hrs
Alternate: Valium 5-10 mg IV or IM every 10-15 minutes prn to max of 30 mg

Purulent Otitis Media
Basic: Amoxil 250 mg tid x 7 days
Alternate: Augmentin 250 mg tid or 500 mg bid x 7 days

Serous Otitis Media
Basic: Over-the-counter antihistamines and decongestants

Labyrinthitis
Basic: Reassurance and support

Patient Education
General: The symptoms of labyrinthitis will probably resolve in less than a week but may persist for up to a month. Serous otitis media is probably secondary to eustachian tube dysfunction from allergy, barotrauma

or viral infection.
Follow-Up: Return if dizziness persists beyond 7-10 days, or if symptoms worsen or if alteration of hearing is noted.

Evacuation/Consultation Criteria: Evacuate patients with persistent or recurrent symptoms of vertigo or dizziness, especially if there is an alteration of hearing, for neurological evaluation.

Symptoms: ENT Problems

To be published. Other references to ENT problems in this edition are in:
Symptom: Cough; Fever
Respiratory: Common Cold and Flu; Asthma
Dental Surgery: Herpetic Lesions; Aphthous Ulcer; Oral Candidiasis
Infectious Disease: Adenovirus; Infectious Mononucleosis; Streptococcal Infections
Dive Medicine: Barotrauma, Ears; Barotrauma, Other
Aerospace Medicine: Barosinusitis; Barotitis

Symptoms: Eye Problems: Acute Vision Loss without Trauma
MAJ Thomas Lovas, MC, USA & CAPT Frank Butler, MC, USN

Introduction: Many disorders may cause acute visual loss in a non-inflamed eye: retinal detachment, anterior ischemic optic neuropathy, optic neuritis, central retinal vein occlusion, anterior ischemic optic neuropathy, vitreous hemorrhage, significant high-altitude retinal hemorrhage, giant cell arteritis and central retinal artery occlusion. These disorders are difficult to diagnose and treat while deployed. In most cases all that can be done is to arrange for an expedited evacuation. Giant cell arteritis (GCA) and central retinal artery occlusion (CRAO) can be treated in the field. Vision loss in one eye due to giant cell arteritis is often rapidly followed by loss in the other eye if untreated. In addition, giant cell arteritis has a significant mortality. If vision loss is associated with trauma, see Eye Injury section.

Subjective: Symptoms
Sudden versus gradual loss of vision, eye pain, seeing bright spots, fever, headache, foreign-body sensation, increased sensitivity to light or photophobia (from irritation of cornea or iris), dry eye, jaw pain.
Focused History: Quantity: *Have you lost your central or peripheral vision or noticed a blind spot?* (A blind spot in the field of vision represents an area that is not receiving visual information due to disease.) **Quality:** *Has the sharpness of your vision decreased?* (may indicate an optic nerve disease such as optic neuritis) *Do you have pain in or behind your eye?* (Deep, dull ache in or behind the eye is most often due to uveitis, glaucoma, scleritis or other inflammatory processes affecting the anterior segment.) *Do you feel you have something in your eye?* (Foreign body sensation is due to irritation or trauma of the cornea or conjunctival epithelium.) **Duration:** *Did you have sudden or gradual loss of vision?* (gradual decreases in vision typically indicate a non-emergent process such as cataract or diabetic retinopathy) **Alleviating or Aggravating Factors:** *What makes your vision better?* (Improvement with rapid blinking suggests a tear abnormality such as dry eyes; improvement with squinting suggests a refractive problem.)

Pearl: Scotomas (blind spots) noticed by the patient are usually due to retinal hemorrhage, edema, or detachment, or due to optic nerve dysfunction (e.g., optic neuritis or optic neuropathies).

Objective: Signs
Partial or total loss of vision, fever, jaw tenderness, conjunctivitis, photophobia
Using Basic Tools: Inspect extraocular muscles. Have patient look in all directions and note any limitations that may indicate entrapment of a muscle, orbital fracture or palsy.
Flashlight: Look for possible residual metallic foreign body

Visual Fields: Defects seen in optic neuropathies, retinal detachment, GCA, others.
Color Vision: Red image less vivid in optic neuritis
Snellen Chart (if available): Loss of visual acuity may be seen in any of the conditions. Provides baseline to work from. Decreased vision indicates abnormality in anterior segment (cornea, crystalline lens or iris). If a Snellen chart is not available, reading the print in a book or other printed material will provide a rough measure of visual acuity.
Using Advanced Tools:
Ophthalmoscope: Look for intraocular foreign body. If intraocular foreign body found, evacuate immediately
Fundus exam: "Cherry red" spot for CRAO; "blood and thunder" for CRVO
Fluorescein Strip and UV light: Observe for stained tissue indicating corneal abrasion. UV Light intensifies staining from strips.

Assessment:
Differential Diagnosis
Central retinal artery occlusion (CRAO) - painless, sudden and profound loss of complete vision in one eye; loss can range from a subtle decrease (uncommon) to a complete loss of all vision; affected eye will sometimes demonstrate a "cherry red spot" (red spot in the center of a white, swollen retina); typical patient is sixty or older; profound loss of vision demands any and all measures be taken in attempt to restore any potential sight. Risk factors include smoking, diabetes, birth control pills and cardiac disease.
Giant cell arteritis (GCA; AKA arteritic anterior ischemic optic neuropathy or temporal arteritis)- gradual onset of decreased vision with associated temporal headaches, weight loss, jaw pain while chewing, fever, and/or joint pain; should be ruled out in any patient over 55 years of age with any combination of the above symptoms; treat quickly, since significant mortality and potential for loss of vision in other eye also.
Retinal detachment (RD) - painless, gradual loss of vision like a "shade," "veil" or "curtain" being pulled over the eyes.
Anterior ischemic optic neuropathy (AION) - sudden decrease in vision affecting the lower half of the visual field in patients up to age 55 years (above 55 years of age giant cell arteritis is more common).
Optic neuritis - central decrease in vision and a color vision deficit, peripheral neurologic signs.
Central retinal vein occlusion (CRVO) - sudden, less profound loss of vision than in CRAO; "blood and thunder" fundus picture (i.e., massive numbers of retinal hemorrhages)
Vitreous hemorrhage - "shower" of floaters accompanied by significant loss of vision; floaters represent blood cells casting shadows on the retina after a blood vessel ruptures; occurs in older diabetics, but can occur after significant trauma, Valsalva and high altitudes.

Plan:
Treatment
Primary:
1. Rule out CRAO with a trial of supplemental oxygen at the highest inspired fraction achievable as soon as possible after onset of vision loss. If supplemental oxygen is to be of any benefit a response is typically seen in a few minutes.
2. Start ocular massage immediately in an effort to dislodge any potential embolus (CRAO, CRVO). Continue for several minutes and abandon if no improvement in symptoms is noted.
3. **Acetazolamide** 500mg po initially, then 250 mg po q 6 hours thereafter to lower intraocular pressure (CRAO, CRVO, Vitreous Hemorrhage, others).
4. If GCA is suspected, give **prednisone** 80 mg a day (divided dose) and expedite evacuation.
Primitive: Hyperventilate to decrease the amount of retained carbon dioxide and increase available oxygen to tissues.

Patient Education
General: Patient has severe visual dysfunction and needs immediate care to have the best chance for vision recovery.
No Improvement/Deterioration: Return ASAP for evaluation.

Follow-up Actions
Return evaluation: Evacuate ASAP since the greatest potential to improve symptoms exists within the first 90 minutes after onset and significantly decreases after 24 hours.
Evacuation/Consultation Criteria: Immediate consultation if possible. Evacuate as above.

Symptom: Eye Problems: Acute Red Eye Without Trauma
MAJ Thomas Lovas, MC, USA & CAPT Frank Butler, MC, USN

Introduction: The differential diagnosis of non-traumatic acute red eye includes herpes simplex virus keratitis, corneal erosion, acute angle-closure glaucoma, scleritis, conjunctivitis, blepharitis, ultraviolet keratitis, episcleritis, conjunctival foreign body, dry eye and contact lens overwear syndrome. If red eye is associated with trauma, see Eye Injury Section.

Subjective: Symptoms
Fever, eye pain, loss of vision, redness, discharge, foreign-body sensation (especially in chemical injuries), increased sensitivity to light or photophobia (irritation of cornea or iris), dry eye, nausea and vomiting (if the intraocular pressure rises suddenly).
Focused History Questions: Quality - *Is the pain sharp or dull?* (Sharp pain usually indicates a corneal process; dull pain typically indicates iritis or scleritis) *Is your eye sensitive to light?* (Photophobia can indicate many anterior segment diseases form corneal abrasion to iritis). *Is there any discharge?* (Mucoid discharge may indicate viral infections, whereas purulent discharge may indicate bacterial infection)
Duration - *Was the onset of symptoms sudden or over hours or, days?* (Sudden onset is typically corneal; gradual onset indicates inflammatory processes such as iritis)
Pearl: *Does the patient notice a relief with one drop of topical anesthesia?* If so, this indicates corneal or conjunctival disease.

Objective: Signs
Using Basic Tools: Vital signs: Fever may indicate systemic infection.
Inspect extraocular muscles. Have patient look in all directions and note any limitations that may indicate entrapment of a muscle or palsy.
Pupil exam: Irregular pupil may indicate scarring from iritis or ruptured globe.
Flashlight: Look for injected conjunctival vessels: perilimbal (cornea-sclera junction) injection indicates iritis; diffuse injection indicates infection or corneal disease. Look for discharge: Mucoid discharge may indicate viral infections, whereas purulent discharge may indicate bacterial infection
Snellen Chart (if available): Decreased visual acuity to between 20/40 and 20/100. Decreased vision indicates abnormality in anterior segment (cornea, crystalline lens or iris). If a Snellen chart is not available, reading the print in a book or other printed material will provide a rough measure of visual acuity.
Using Advanced Tools: Fluorescein Strip and UV light: Observe for stained tissue indicating corneal abrasion. UV Light intensifies staining from strips.
Topical anesthesia: 1 drop **proparacaine** or **tetracaine** 0.5% Pain relief can indicate ocular surface disease (conjunctivitis, keratitis, etc.).

Assessment:
Differential Diagnosis: Both traumatic causes (see Eye Injuries Section) and non-traumatic causes, such as:
Herpes simplex virus keratitis - dendritic figure on fluorescein staining; no trauma; often a history of previous episodes
Corneal erosion - abrasion or ulcer noted on fluorescein exam

Acute angle-closure glaucoma - age over 40; decrease in visual acuity; history of previous episodes of eye pain
Scleritis or iritis - often associated with systemic inflammatory disorders
Conjunctivitis - acute onset; the presence of a discharge
Blepharitis - chronic recurrent eyelid inflammation; more common in older patients; usually bilateral, but one eye may be worse
Ultraviolet keratitis - bilateral eye pain; sunburned face; maximum intensity several hours or longer after exposure; fluorescein staining typically reveals numerous small dots of stain uptake called superficial punctate keratitis or SPK.
Episcleritis - benign and self-limited inflammation of the episclera (the lining of the eye between the conjunctiva and the sclera); identified by sectors of redness, no discharge and often a history of previous episodes; discomfort is typically mild or absent.
Conjunctival foreign body - identification of the foreign material
Dry eye - usually bilateral and may result in secondary tearing; history of previous episodes; occurs in dry environments.
Contact lens overwear syndrome - as in dry eye, except that the symptoms are magnified by the presence of contact lenses
Subconjunctival hemorrhage - bleeding often seen with coughing or retching; innocuous and self-limited

Plan:

Treatment
Herpes simplex keratitis: Expedited evacuation; do not use steroids; patch eye.
Corneal erosion (abrasion and ulcer): See Eye Injury Section.
Angle-closure glaucoma (ACG): Acetazolamide 250 mg qid po; **emergently evacuate** since markedly elevated intraocular pressures may result in permanent damage to the optic nerve in 24 hours or less.
Scleritis or iritis: Prednisolone 1%, 1 drop q1 hour continuously until evacuated. Patch eye; **scopolamine** 0.25%, 1 drop bid; expedite evacuation; if no improvement in 24-48 hours and not yet evacuated, start **prednisone** 80 mg qd until evacuated.
Conjunctivitis: Ciprofloxacin ophthalmic drops 1 drop qid for 5 days.
Blepharitis: Bacitracin ophthalmic ointment applied to the lid margins q hs x 3-4 weeks; apply qid for 1 week in more severe cases; warm compresses for 10 minutes bid-qid. Follow by gently wiping away of the inflammatory material on the eyelashes.
Ultraviolet Keratitis: Bacitracin ophthalmic ointment qid until signs and symptoms resolve; sunglasses; patch severely affected eyes for comfort; **scopolamine** 0.25%, 1 drop bid or systemic analgesia may be required for pain relief; monitor daily until epithelial staining resolves to ensure that they do not develop a corneal ulcer.
Episcleritis: Usually resolves without treatment over several weeks; use **prednisolone** 1% drops qid x 3 days if persistent and patch eye.
Foreign body: (See Eye Injury section.)
Dry eye: Artificial tears prn to relieve symptoms; systemic rehydration; sunglasses or goggles for protection.
Overwear syndrome: Re-wet contact lens and use sunglasses; if ineffective, remove the contact lenses and use glasses; if significant SPK are present on fluorescein staining, use **ciprofloxacin** or **ofloxacin** 1-2 drops qid until the SPK have resolved; do not replace contact lenses until the eye is symptom-free.
Subconjunctival hemorrhage: Will resolve without treatment over one to two weeks.

Patient Education
General: Discuss the level of injury with the patient but do not give prognosis in diseases that should be managed at a higher level of care.
Activity: As tolerated
Diet: As tolerated
Prevention and Hygiene: Keep eyes clean. Avoid spreading conjunctivitis to the other eye or other individuals. Contact lens wearers in the wilderness should always carry a pair of glasses that can be worn if contact lens problems arise.

Follow-up Actions
Return evaluation: Follow patients closely on daily basis for signs of improvement or worsening
Evacuation/Consultation Criteria: Evacuate as indicated above in treatment. Evacuate any patient not showing improvement within 24-48 hours. Consult with an ophthalmologist if available prior to using steroids in the eye.

Symptom: Eye Problems: Orbital or Periorbital Inflammation
MAJ Thomas Lovas, MC, USA & CAPT Frank Butler, MC, USN

Introduction: Orbital or periorbital inflammation can result from nasal cellulitis, orbital cellulitis, pseudotumor, insect envenomation or lacrimal gland inflammation (dacryocystitis).

Subjective: Symptoms
Periocular edema, erythema, pain, possibly sensing a foreign body in or near the orbit.
Focused History: *Has the sharpness of your vision decreased? Do you have pain in your eye or around your eye?* (typical symptoms) *Do you feel you have something in your eye?* (Foreign body sensation is due to irritation or trauma of the cornea or conjunctival epithelium.)
Risk Factors: *Has an insect bitten you? Have you had any recent infections in your teeth or sinuses? Have you had any recent trauma to this eye? Do you wear contact lenses?*

Objective: Signs

Using Basic Tools	Clinical Findings	Interpretations
Vital signs	Fever	May be indicative of orbital cellulitis
Printed material	Check visual acuity*	Corneal damage, discharge; dysfunction of some aspect of vision
Flashlight	Swollen eyelid(s); eye slightly protruding from orbit when compared to opposite side	Indicates orbital or preseptal cellulitis

Using Advanced Tools	Clinical Findings	Interpretations
Ophthalmoscope	Observe fundus for signs of retinal or optic nerve disease	May indicate advanced orbital disease
Fluorescein Strip	Staining?	May indicate corneal abrasion (initiation site for cellulitis)
UV Light	Enhances swelling	

Assess visual acuity with Snellen chart if available. Reading any printed material will provide a rough measure of visual acuity.
Pearl: Check eye movements (decreased eye movements indicate orbital process)

Assessment:
Differential Diagnosis
Preseptal cellulitis- associated with a history of periocular trauma or hordeolum (stye), no proptosis (protrusion of the eye), no restriction or pain with eye movement and no change in visual acuity.
Dacryocystitis- a specific type of preseptal cellulitis in which the source of the infection is an obstructed nasolacrimal duct. The erythema and inflammation are localized to the area overlying the lacrimal sac at the inferior nasal aspect of the lower lid.
Periocular insect envenomation- may have a papular or vesicular lesion at the site of envenomation.
Orbital cellulitis- associated with a history of sinusitis or upper respiratory tract infection, proptosis (protrusion of the eye), restricted extraocular muscle motility, decreased visual acuity and/or fever. It can progress to

meningitis.

Plan:
Treatment
1. Preseptal cellulitis: **levofloxacin** 500 mg po once a day, expedite evacuation if no improvement in 24-48 hours.
2. Dacryocystitis: as for preseptal cellulitis above. Use warm compresses on the inflamed eye.
3. Periocular insect envenomation: cool compresses and antihistamines; **levofloxacin** 500 mg po once a day if secondary infection is suspected based on increasing pain, redness, or swelling.
4. Orbital cellulitis: **life-threatening** disorder requiring emergent evacuation, **levofloxacin** 500 po mg bid and decongestants.

Patient Education
General: Discuss severity of the condition with the patient
Activity: Limited with IV antibiotics.
Diet: Regular as tolerated
Prevention: Good personal hygiene
Wound Care: Warm or cool compresses, depending on diagnosis.

Follow-up Actions
Return Evaluation: Return every 12 to 24 hours.
Evacuation/Consultation Criteria: Immediate evacuation for acute proptosis and/or decreased eye motility.

Symptom: Eye Problems: Eye Injury
MAJ Thomas Lovas, MC, USA & CAPT Frank Butler, MC, USN

Subjective: Symptoms
Eye pain, loss of vision, foreign-body sensation (especially in chemical injuries), increased sensitivity to light or photophobia (irritation of cornea or iris), nausea and vomiting (if the intraocular pressure rises suddenly).
Focused History: *Did something hit you in the eye?* (typical exposure) *What were you doing when the injury occurred?* (exposure) *Were you wearing glasses and/or safety glasses? If so, are there any broken fragments, etc?* (look for "missing pieces" which may have become intraocular foreign bodies) *Were you (or someone near you) hammering metal-on-metal?* (foreign bodies; i.e., initial pain as the high speed foreign body enters the eye, then resolution of symptoms over a brief period of time) *What type of chemicals were you using (in the case of chemical injury)?* (exposure) *Did you or someone else flush your eyes?* (can decrease extent of injury) *Are you sickle cell positive?* (can have significant impact on intraocular bleeding after trauma)

Objective: Signs
Using Basic Tools: Do not palpate potentially ruptured globe! obvious wound (including lid laceration), decreased visual acuity, possible ruptured globe, conjunctivitis, extra-ocular muscle derangement, non-circular (irregular) pupil, deviated (disconjugate) gaze, eye drainage, photophobia; tangential light exam may reveal blood in anterior chamber (hypema), corneal ulcer (white or gray spot).
Using Advanced Tools: Ophthalmoscope: retinal (red) reflex, blood in anterior (hyphema) or posterior chamber, visible foreign body; fluorescein strips (only if globe intact), ophthalmic anesthetic (1 drop) if available and UV light: look for corneal abrasion or ulcer.

Assessment:
Differential Diagnosis
Hyphema - blood seen in anterior chamber
Orbital fracture - detected by extra-ocular muscle derangement or new onset gaze derangement

Occult ruptured globe - suspect with history of blunt or impaling injury, dark uveal tissue exposed at junction of cornea and sclera, a distorted pupil, or a decrease in vision
Traumatic iritis - pain and photophobia
Subconjunctival hemorrhage - bright red area of blood overlying the sclera
Also consider corneal abrasion, corneal ulcer, foreign body, obvious ruptured globe,

Plan:

Treatment

1. Cover all injured eyes with a metal shield or other device to prevent further injury.
2. **Obvious ruptured globe: levofloxacin** 500 mg po bid and **evacuate immediately**.
3. **Occult ruptured globe:** An occult ruptured globe also entails the possibility of endophthalmitis. Treat in the same manner as an obvious open globe. If an open globe is suspected, protect the eye until definitive treatment is obtained.
4. **Corneal abrasion:**
   a. **Bacitracin** ophthalmic ointment qid.
   b. Small or less painful abrasions need not be patched.
   c. **Diclofenac** 0.1% drops qid or systemic analgesics for pain control.
   d. Sunglasses reduce irritation from light if the eye is not patched.
   e. Remove the patch daily to check for the development of a corneal ulcer and to repeat the fluorescein stain to monitor healing. The abrasion should heal within 1-3 days.
   f. If the trauma causing the abrasion is related to contact lens wear or insertion, there is a higher incidence of secondary infection with gram negative organisms and the eye should NOT be patched. Use topical **ciprofloxacin** or **ofloxacin** 1-2 drops qid until the abrasion is healed and watch the eye closely for development of a corneal ulcer.
5. **Corneal ulcer:** topical **ciprofloxacin** or **ofloxacin** as follows: 1 drop every 5 minutes for 3 doses; 1 drop every 15 minutes for 6 hours; then 1 drop every 30 minutes. **Scopolamine** 0.25% 1 drop bid, or systemic analgesics may be added for pain control if needed. A corneal ulcer is a vision-threatening disorder that may progress rapidly despite therapy, so evacuate should emergently if pain and inflammation continue to increase or expedite evacuation even if the ulcer is responding to therapy.
6. Subconjunctival hemorrhage requires no treatment, but carefully inspect the eye for associated injuries. If the subconjunctival hemorrhage is massive and causes outward bulging of the conjunctiva (called chemosis), then suspect an occult ruptured globe and manage as described above.
7. **Hyphema:** The primary concerns in this disorder are associated globe rupture, increased pressure in the eye and permanent damage to vision. Expedite evacuation. Restrict activity to walking only. Rest in a foot-dependent position to encourage blood to settle in the bottom of the anterior chamber. Do not let these individuals read. **Do not treat them with NSAIDs or aspirin.** If evacuation is delayed use **prednisolone** ophthalmic drops qid in the affected eye for 3 days.
8. **Traumatic iritis** typically resolves without treatment in several days. Severe cases may be treated with topical **prednisolone** 1% drops qid for three days if evacuation is not available and no lesion is noted on fluorescein exam.
9. **Lid Laceration:** Any laceration that is full-thickness, involves the lid edge, or is in the medial or lateral corners (epicanthal folds) should be repaired by an ophthalmologist. Suspect other trauma beneath the laceration. Repair partial thickness lacerations that are not in the areas mentioned above with simple 6-0 proline sutures, sterile technique, and limited 2% **lidocaine** in the lid. Do not expose the eye to surgical scrub or preparation solutions. Do not puncture underlying globe!
10. **Foreign Body:** Apply topical anesthesia (1 drop 0.5% **proparacaine** or **tetracaine**). Locate and remove foreign body using enhanced lighting and magnification if available. Evert upper eyelid with a cotton-tipped applicator to identify foreign bodies there and remove them with a cotton-tipped applicator moistened with **tetracaine**. Stain eye with fluorescein to check for a corneal abrasion. If symptoms persist, irrigate vigorously with artificial tears or sweep conjunctival corners with a moistened cotton-tipped applicator after applying topical anesthesia.
11. Although topical steroids should not be given except by ophthalmologists, **prednisolone** drops will probably not cause any significant adverse effects in an individual with a fluorescein-negative eye disorder if used no longer than three days.

Patient Education
General: Discuss the level of injury with the patient but do not give prognosis in diseases that should be managed at a higher level of care.
Activity: Tailor activity level to severity of injury.
Diet: Keep patient NPO for obvious and occult globe ruptures and lid lacerations that will be evacuated.
Prevention and Hygiene: All eye patients should maintain a high level of hygiene while recovering from their injury.
Wound Care: Keep eye patched, clean and dry. Change dressings daily.

Follow-up Actions
Return evaluation: Follow patients closely on daily basis for signs of improvement or worsening.
Evacuation/Consultation Criteria: Evacuate patients as indicated in Treatment above and any patient who does not show improvement within 24-48 hours. Consult with an ophthalmologist if available prior to using steroids in the eye.

Symptom: Fatigue
CPT Brooks Morelock, MC, USA

Introduction: Fatigue is a nonspecific complaint for which no precise diagnosis may be found. The patient must be detailed in describing their fatigue. Broadly assess body systems when taking a history because fatigue may be due to an underlying endocrine disorder, systemic illness, sleep disorder, drug side effect (recreational or prescribed), or psychiatric disorder. If the patient's history does not yield a specific etiology, a careful psychiatric history for depression should be performed. Expect to find more psychiatric illness (usually depression) than organic disease if the patient has a primary complaint of fatigue rather than fatigue as part of a clustering of symptoms.

Subjective: Symptoms
Focused History: Sleeping: *Have you had trouble with sleeping or had a change in your sleep pattern?* (Fatigue may be a presenting symptom for undiagnosed COPD.) *Early awakening/difficulty getting to sleep?* (associated with depression—see below.) *Snoring?* (may indicate Obstructive Sleep Apnea [OSA] if coupled with Chronic Fatigue. See Respiratory: Apnea) *Told they stop and start breathing when they sleep?* (hallmark sign of OSA) *Wake up with headache on most days?* (secondary symptom of OSA due to nocturnal hypoxemia) *Shift work?* (disordered circadian rhythm) *Sleep during the day?* (Daylight lowers secretion of melatonin, a hormone that regulates the sleep cycle.) **Medication:** *Do you take any medication?* (Starting or stopping many medications can cause fatigue. Starting an antihypertensive or stopping thyroid replacement are the most common.) **Drugs/Alcohol:** *Do you drink alcohol, use tobacco or use any other type of recreational drugs?* (Alcohol, drugs can alter normal sleep patterns.) **Weight:** *Have you lost weight?* (may indicate occult malignancy or systemic illness) **Duration:** *How long have you had fatigue?* (6 months to be Chronic Fatigue) **Night sweats:** *Do you have night sweats?* (can be symptom of tuberculosis, lymphoma, malaria, others) **Exposure:** *Have you had any known exposure to HIV, hepatitis, or mononucleosis?* (all can have fatigue as presenting complaint) *Have you been flying or diving recently?* (possible decompression sickness–see Dive Medicine) **Muscle/Joint Pain:** *Do you have muscle or joint pain?* (seen in connective tissue disorders such as rheumatoid arthritis and lupus) **Depression:** *Does the patient have any symptoms related to depression? Difficulty with getting to sleep or early awakening? Lost interest in activities that he/she used to enjoy? Does patient feel guilty about things that they could not control? Decreased energy? Does the patient have difficulty with concentration? Absent appetite? Does the patient feel like their thoughts are slower than usual, or seem to be fleeting? Has the patient had a decrease in libido? Has the patient contemplated suicide?* (All of these questions try to elicit symptoms of depression, although none are comprehensive. The more symptoms the patient has, the more likely he has depression. Any time the diagnosis of depression is entertained, screen the patient for suicidal or homicidal ideation.)

Objective: Signs (See Fatigue: Table 3-3)

Table 3-3: Fatigue: Objective Signs

Basic Tools	Clinical Findings	Interpretations
General	- Assess interaction and general appearance	- Depressed patient will be less likely to engage the clinician and may have decreased personal hygiene. Vital signs
Vital signs	- Check pulse ox - if available	- Hypoxemia may cause isolated fatigue rather than dyspnea. Hypoxemia may also be a marker of underlying sleep apnea
HEENT	- Jugular venous distention - Pallor of the mucous membranes - Pharynx erythematous +/- exudate - Thyroid goiter	- Associated with congestive heart failure (CHF). - May suggest anemia. - Mononucleosis can cause severe pharyngitis and persistent fatigue.
Cardiovascular	- Systolic ejection murmur - S3 / S4 Gallop	- Hypothyroidism is a classic etiology of fatigue. - Anemia will often have a soft ejection murmur. - These abnormal heart sounds will be heard in CHF.
Pulmonary	- Increased AP diameter of chest - Rhonchi / rales / pleural rub	- Possible underlying pulmonary disease (Emphysema / COPD) - Possible bronchitis / pneumonia
Abdomen	- Splenomegaly	- Possible mononucleosis
Neurologic	- Sluggish relaxation phase of DTRs - Abnormal muscle tone / muscle girth	- Common finding in hypothyroidism - Possible neuromuscular disorder
Lymphatic	- Lymphadenopathy.	- Possible lymphoma, mononucleosis or chronic infection
Rheumatic	- Inflamed joints	- Possible lupus, rheumatoid arthritis, or Lyme disease

Advanced Tools: Labs: CBC, rule out anemia hematologic malignancy, or chronic infection
With a negative history and a normal physical exam, labs are not likely to help with diagnosis. However, a routine set of basic labs should be drawn prior to initiation of treatment of depression or other psychiatric illness.

Plan:
Treatment
Stopped Medication
Resume medication or consider alternate if side effects are a problem.

Altered sleep patterns
Primary: **Ambien** (**zolpidem**) or **Sonata** (**zaleplon**) 5-10 mg po at bedtime
Alternate: Antihistamines (more side effects) such as **Benadryl** or **Atarax**

Melatonin helps restore abnormal circadian rhythms resulting from shift work or jet lag.
Primitive: Sunglasses that block UA/UVB; blackout curtains in sleeping quarters

Snoring and/or Obstructive Sleep Apnea
Primary: Over-the-counter decongestants
Alternate: Flonase 1-2 sprays each nostril q hs
Primitive: Use a sleep wedge to prevent sleeping on the back

Depression (See Symptom: Depression)
Primary: (See Mental Health chapter); serotonin specific re-uptake inhibitor (SSRI), such as **Prozac**
Alternate: Different SSRIs such as **Zoloft**. **Paxil** or tri-cyclic antidepressants should be avoided as these often have the side effect of fatigue!

Patient Education
General: Symptom reduction- help the patient to learn to cope with their symptoms and maintain their highest possible functioning level. This requires a good medic-patient alliance and a mutual understanding of the diagnosis and treatment goals.
Follow-Up: Follow patient on a routine schedule to assess for improvement of symptoms with treatment interventions. Change therapy if symptoms continue. **NOTE:** The symptom reduction in depression may take 4-6 weeks; therefore, until the patient has taken therapeutic doses of a drug for at least 2 months, therapy should not be discontinued.

Evacuation/Consultation Criteria: Acute evacuation is not usually necessary, unless the patient is unstable or non-mission capable. If labs remain unrevealing and the patient has persistent symptoms of fatigue for greater than 6 months, then a diagnosis of chronic fatigue may be entertained and the patient referred. Note that chronic fatigue is NOT the same as chronic fatigue syndrome, which has specific diagnostic criteria.

Symptom: Fever
COL Naomi Aronson, MC, USA

Introduction: Low-grade fevers are non-specific, commonly of viral etiology, generally do not have a large impact on morbidity and mortality in adults and will not be addressed here. Significant fever (oral temperature >101°F in an adult) is a common symptom of infectious disease, but it can be seen with other conditions such as malignancy, heat-related illness, drug reactions, rheumatologic conditions, or hyperthyroidism. Travel history is very important because infectious disease patterns are dynamic. The Armed Forces Medical Intelligence Center (AFMIC) and electronic networks such as PROMED and MEDIC should be consulted to identify diseases occurring specific areas, current outbreaks and emerging disease patterns.
  Several principles apply to the initial approach of a febrile patient:
- Assess first for infections that can be quickly life-threatening (the duration of fever, place and type of exposure, related clinical symptoms can be helpful in further directing you).
- Avoid delays in starting empiric broad-spectrum treatment for most possible infections in ill-appearing patients.
- Treat fever in the tropics or in a traveler from the tropics as malaria until proven otherwise.
- Remember to consider infections that are contagious and/or have public health implications
- Intravenous fluids (normal saline or Lactated Ringer's solution) are recommended for hypotensive febrile patients.

Subjective: Symptoms
Fever > 101°F
Incubation period of febrile diseases:
Acute (1-14 days): Malaria, pneumonia, arboviruses, hemorrhagic fevers, diarrhea illness, UTI, rickettsial infections, leptospirosis, measles
Sub acute (15-30 days): Hepatitis, HIV, CMV, typhoid fever, schistosomiasis, mononucleosis, rabies, rubella,

relapsing fever, rheumatic fever
Chronic (>30 days): Tuberculosis, leishmaniasis, visceral amebic liver abscess, trypanosomiasis, brucellosis, bartonellosis, deep fungal infections, filariasis, Q fever
Specific symptoms (combined with high-grade fever) that differentiate between febrile illnesses include hemorrhage, hypotension, rash, diarrhea, lymphadenopathy, altered mental status or other central nervous system symptoms.

Focused History: Fever: *How high is your temperature?* (expect to document a temperature >100°F) *Is the fever constant, or does it come and go, or only come at night?* (nocturnal fever pattern with some malaria, TB; recurrent fever with some malaria, relapsing fever, leptospirosis) **Duration**: *How long have you felt sick?* (differential diagnosis of causes of fever based on chronicity) **Other Sx:** *Have you noticed [select: rash, headache, diarrhea, swollen lymph nodes, localized pain, stiff neck, cough]?* (identify affected systems). *Do you know other persons with similar symptoms?* (suggests contagious illness or point source outbreak) *Are you pregnant?* (changes medications that can be used) *Are you taking any medications?* (Antimalarials may decrease your ability to diagnose malaria. Fever may be sign of an allergic reaction to medication.)
NOTE: Micromedex software includes Martindale's "The Extra Pharmacopoeia," which is a good reference for drugs worldwide.

Objective: Signs
Using Basic Tools:
Vital signs: Temperature > 101°F (also look for relapsing pattern or pulse-temperature dissociation as hints to specific types of *infection*), respiratory rate over 14/minute, heart rate over 100 beats/minute, systolic blood pressure <100 mm
Inspection: Evaluate for pallor, diaphoresis, rigors, mental status changes, rash (especially any petechiae or purpura), jaundice, unwillingness to move body parts (stiff neck, limb, back etc); evaluate for pharyngitis
Palpation: Feel for inflammatory changes (warmth, tenderness) in areas of symptoms; test range of motion in areas where patient voluntarily restricts movement to see if tenderness is elicited; check for enlarged lymph nodes and assess if fluctuant, tender or draining; evaluate abdomen for hepatosplenomegaly, peritoneal signs, tenderness.
Auscultation: Listen for rales in lungs (egophony or dullness suggest pneumonia; rales that clear after cough have traditionally been associated with tuberculosis); listen for heart murmurs, S3 that could suggest carditis (such as rheumatic fever), or endocarditis; evaluate for abnormal bowel sounds (hyperactive in gastroenteritis, hypoactive in ileus/intra-abdominal abscess)
Using Advanced Tools:
Lab: Thick and thin blood smears (malaria, filaria, babesia and other parasites); WBC with differential; urinalysis; Gram's stain of any pus or sputum; stool for fecal leukocytes and O&P if diarrhea; pregnancy test for females of childbearing age (less likely if breastfeeding)
CXR if available (PA and Lateral)

Assessment:
Clinical Questions - Complete attached algorithm and follow indicated protocol as appropriate for mission situation.
1. Does patient need outpatient treatment or evacuation/hospitalization for intravenous therapy, resuscitation, possible end organ support (ventilator, dialysis) if available?
2. Does patient need isolation to prevent contagion (such as meningitis, hemorrhagic fever viruses, plague, pulmonary tuberculosis)?
3. Should the patient receive empiric treatment for several conditions simultaneously (e.g., bacterial infection) or should he be treated more specifically? (e.g., for malaria).
4. Is the patient at increased risk: elderly, young child, immunocompromised, wounded, concurrent or chronic illness (e.g., sickle cell disease, HIV, diabetes, cancer, malnourished)? If so, he will require evacuation and/or more resources.
5. Does the patient have signs of sepsis syndrome present (fever or hypothermia, heart rate >90/minute,

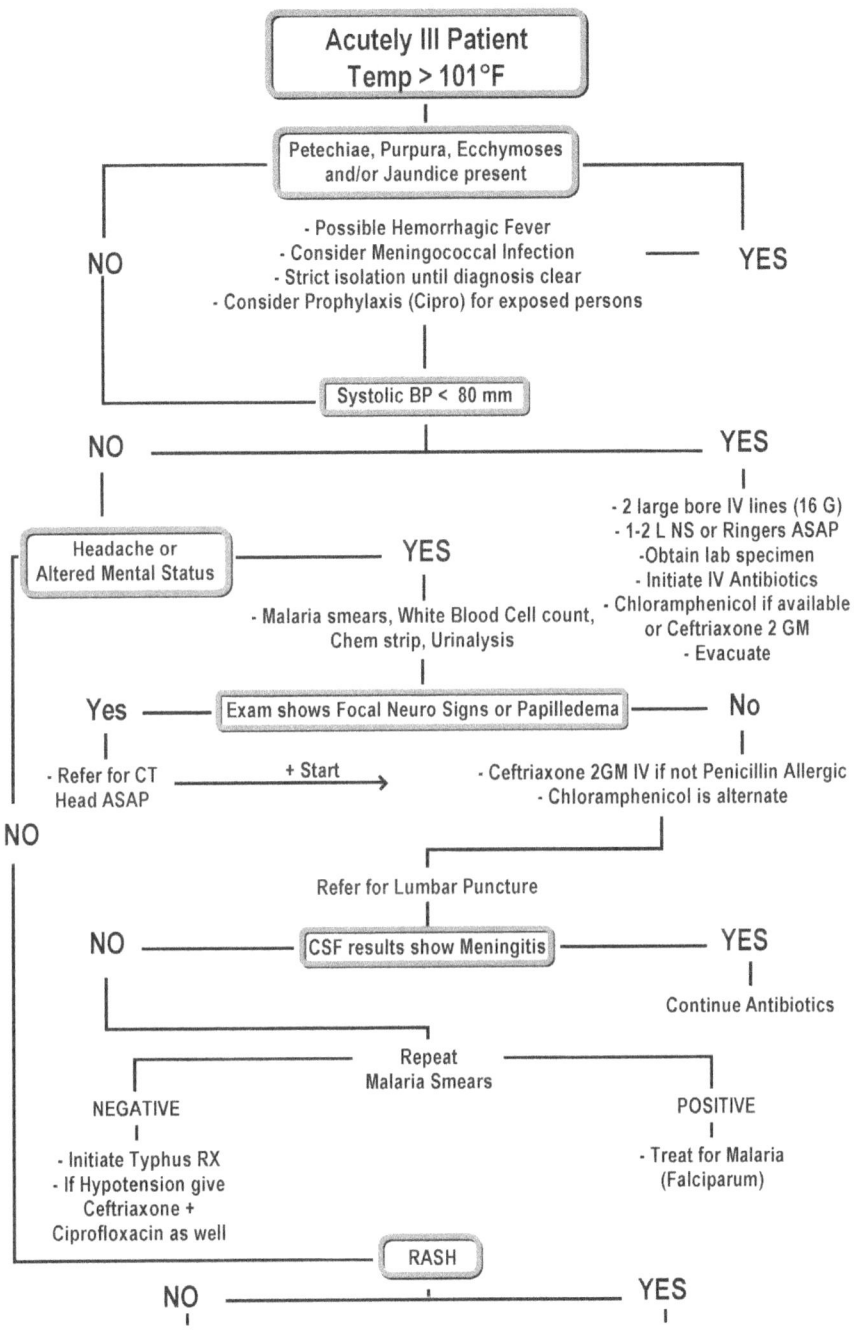

3-33

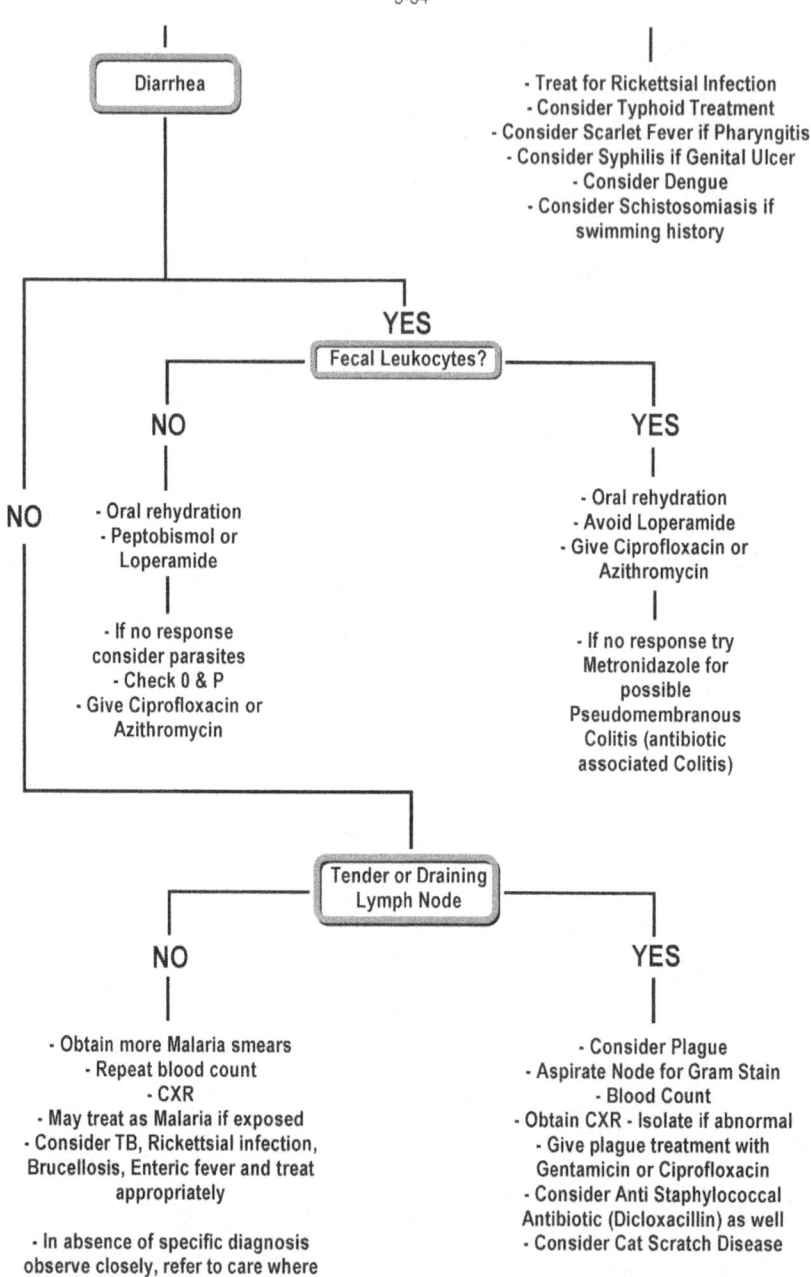

respirations>20/minute and at least one sign of end organ dysfunction; mental status changes, pulse oximetry <90%, urine output less than 30 ml/hr)? If so, he will probably require IV antibiotics, evacuation and/or more resources.

Plan:
Treatment
See individual sections of this book for appropriate treatment. Always check for medication allergies and pregnancy before selecting medication for treatment.

Patient Education
General: If fever does not respond to treatment in 3-5 days or symptoms progress seek further medical care
Activity: As tolerated unless patient requires isolation.
Diet: Regular; avoid alcohol.
Medications: Tylenol is preferred for symptom control but avoid if patient is jaundiced.

Symptom: GYN Problems: Female Pelvic Examination

MAJ Ann Friedmann, MC, USA

When: As indicated per differential diagnosis, which primarily is for abdominal and pelvic complaints. Most females in the U.S. have one pelvic examination per year from the age of 18 to assess for gynecologic malignancies or benign conditions such as uterine fibroids. The Pap smear, normally a part of this exam, provides cellular material to screen for cervical carcinoma. However, the Pap smear is not normally done in the field and will not be discussed here. Rectal evaluation may or may not be a part of each examination. Rectovaginal or rectal exam can aid in assessing the posterior aspect of the uterus, the uterosacral ligaments and posterior cul-de-sac. A stool sample can be obtained and tested for occult blood and lower rectal masses can be palpated. Stool occult blood sampling is recommended with pelvic examination after the age of 40-50.

For the field medic who will only be performing problem-based pelvic examinations, the exam may be tailored to the specific problem and need not always include bimanual, rectovaginal and rectal exam. For example, if the patient complains only of abnormal vaginal discharge without pelvic pain, a vaginal exam, KOH and wet prep may be all that is necessary

For females this examination is the most sensitive and fraught with concern. Discussion before and after the exam what will happen and what was found. It is necessary to discuss each part of the exam prior to its performance – prior to touching or moving any structure the examiner should say, for example, "I am going to touch the right side of the labia now" or "Now I will move the cervix." The exam should be performed expeditiously yet thoroughly. A female chaperone is absolutely necessary for male examiners in the United States. In field situations, a trusted female friend of the patient, her partner or spouse and a male assistant to hand you equipment could be an acceptable alternative. Keep the patient as covered as possible. If the examination targets the vulva ask the patient to undress only from the waist down. Cover the abdomen with a sheet or drape. As soon as the exam is done, allow the patient to dress.

What You Need:

For all exams: Good light source – preferably a mobile light, non-sterile gloves, water-soluble lubricant, vaginal speculum of the correct size (small for patients who are virginal, medium Graves are appropriate for most sexually active females and for parous women. Obesity may necessitate a large Graves due to redundant tissue in the vagina). A speculum may be improvised out of two spoons joined by a rubber band, or two bent spoons.

Figure 3-1

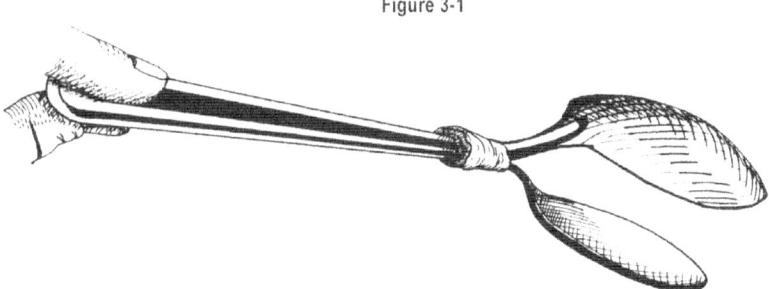

On certain exams: Culture medium for gonorrhea and *chlamydia* testing, large cotton-tipped swabs, pH paper, screening test for fecal occult blood, set-up for wet mount and KOH prep (see Lab Procedures).

What To Do:

The exam described below will be appropriate for triage screening and diagnosis. It is not necessary that a meticulous exam be performed in asymptomatic areas. Position the patient in low lithotomy. The head may be elevated 30-60° to allow the patient visual access to the exam. If a lithotomy table with stirrups is not available, the patient may flex her knees to her chest or an exam table can be improvised with a litter, litter stands, IV poles and small battle dressings.

Figure 3-2

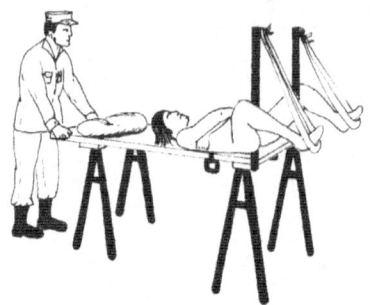

External Genitalia Examination:
Prior to touching the perineum inspect it visually. Observe for symmetry, bulges, rashes or lesions
If asymmetry or mass is noted these areas should be palpated to confirm mass, induration or other abnormality.

Vaginal Examination:
Select the appropriate sized speculum. Speculum may be lubricated with warm water prior to insertion. Remember to discuss the examination with the patient. Ask the patient to consciously relax the muscles at the opening of the vagina. Gentle downward pressure with the tip of the speculum or with the finger may help the patient relax, although many times this is not necessary. Place the speculum at the opening of the vagina and gently push it in and downwards until the vaginal apex is reached. Insert the speculum obliquely through the introitus and then rotated to the horizontal plane. Always control the speculum blades, holding them shut with one hand until the blades are opened as the vaginal apex is reached. Observe the vaginal sidewalls and the cervix. Inspect vaginal discharge for quantity, color, consistency and odor. Evaluate the cervix for erosion, lesion, infection, laceration, polyps, ulceration and tumors. To perform cervical cultures for gonorrhea and chlamydia, place each swab into the cervical os and allow to sit for 20-30 seconds. Replace in culture tube. Assess vaginal pH by obtaining a sample of vaginal discharge from the sidewall or from any pooled discharge in the posterior fornix using a small cotton swab. Touch the cotton swab to the pH paper and look for color change. Obtain discharge samples for KOH and wet mount with a similar technique. Blood or cervical mucus will be basic (high pH) and will give false readings, so avoid these while sampling.
Bulging of the bladder and/or rectum into the vagina may be seen in patients with pelvic relaxation. Women who have had children vaginally will have some relaxation of the rectovaginal and vesicovaginal septum. The cervix may descend into the vagina with Valsalva.

Bimanual Examination:
With a gloved hand, lubricated with water or gel, insert the index and middle fingers of the dominant hand along the posterior wall of the vagina (Figure 3-3). Place the other hand on the patient's abdomen in the midline above the symphysis pubis. The vaginal fingers should encounter the cervix, which should feel firm and circular. Push the cervix upward in the cervical canal. This will move the body of the uterus. If the uterus is anteverted or midline it will be palpable with the abdominal hand. If the uterus is retroverted it will move toward the rectum and the abdominal and vaginal hands will meet in the midline as they come together. If possible feel the body of the uterus for shape, symmetry and mass. A nulliparous uterus will be quite small (pear-sized). A parous uterus can be the size of a small grapefruit. Assess the uterus for mobility by moving

it side to side by grasping the cervix between the 2 examining fingers. The ovaries may be palpated by moving the vaginal hand to the right or left and reaching upward to the side of the uterus. Sweep the abdominal hand down to meet the vaginal hand along the uterus. Feel the tissue to the side of the uterus. The ovary will normally be pulled down in to the vaginal hand by sweeping along the side of the uterus. Premenopausal ovaries are 3-5 cm in length and 2-4 cm in width. The ovaries are normally tender to palpation. This tenderness is localized to the time of exam and is usually described as an aching sensation.

Rectovaginal Examination:
Always change gloves (so no blood is carried to the rectum) and obtain a large dollop of lubricant prior to rectovaginal examination. Ask the patient to relax the anal sphincter. The middle finger is inserted into the rectum and the index finger into the vagina. It is helpful to have the patient bear down as the rectal finger is inserted slowly. The rectovaginal septum can be felt between the index and middle finger. Palpate the posterior aspect of the uterus, uterosacral ligaments and posterior cul-de-sac along with the anorectal area. Note masses, nodularity and pain. A simple rectal examination is the only way to assess the pelvis in an infant or child. Test a sample of fecal material for occult blood.

What Not To Do:
Do not be insensitive, unprofessional or humorous with the patient during the exam.
Do not perform routine pelvic exams during pregnancy. Most women will not require such exams unless there is a strong suspicion of vaginal pathology. The exam may introduce harmful organisms into the vagina or otherwise endanger the fetus.
Do not perform the exam in stages. Do a complete exam the first time and allow the patient to get dressed.
Do not perform an occult blood test with gloves that have been used to examine the vagina. The test will likely be falsely positive.

Symptom: GYN Problems: Abnormal Uterine Bleeding
MAJ Ann Friedmann, MC, USA

Introduction: Abnormal uterine bleeding (AUB) is symptom that can have a variety of etiologies, ranging from simple to life-threatening illnesses. The work-up of abnormal uterine bleeding is systematic and will usually lead to diagnosis. AUB is characterized by changes in the interval and duration of menstrual flow. The normal menstrual interval is 28 days with a range of 21-35 days. The normal duration of the menstrual flow is 2-7 days with average blood loss < 80 cc. AUB can present as vaginal hemorrhage, which is the primary focus of this section.

Subjective: Symptoms
Significant change in menstrual pattern or amount of bleeding. Acute, excessive menstrual blood loss can lead to symptomatic anemia, if so, patient will complain of fatigue, orthostatic changes, and heart palpitations. Excessive blood loss may be described as "gushing"; large clots may be passed; pregnant patients may complain of associated early pregnancy symptoms such as nausea and vomiting, breast tenderness and missing last menstrual cycle.

Objective: Signs
Using Basic Tools: May have large amounts of blood issuing from the vagina; the cervix may be obscured with blood; uterus may be misshapen by a uterine fibroid, a common cause of heavy menses; patients with acute anemia will have heart rate and blood pressure changes based on the volume of blood loss.
Using Advanced Tools: Lab: Assess anemia via hematocrit; perform pregnancy test

Assessment:
Differential diagnosis: For many of these etiologies of AUB, physical findings will be minimal and limited

Figure 3-3

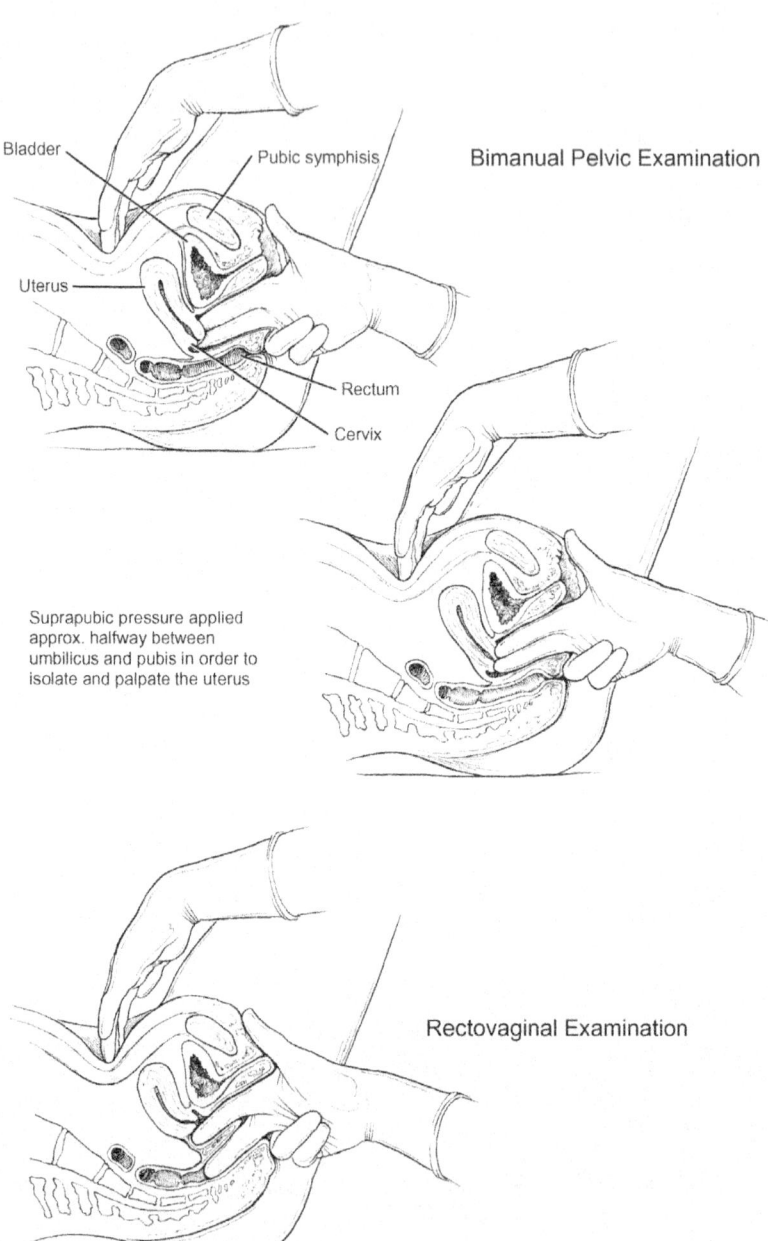

to pelvic examination.
Pregnancy - always rule out first
Anatomic abnormalities - uterine fibroids, uterine and cervical polyps and large ovarian cysts may secrete estrogen, which disrupts normal menstrual function. Uterine fibroids are a common cause of AUB in women from ages 30 through menopause. After menopause, most fibroids become asymptomatic. A postmenopausal woman with AUB and an enlarged irregular uterus has cancer until proven otherwise.
Infection - cervicitis and endometritis, pelvic inflammatory disease
Cancer and pre-cancerous lesions - post-coital bleeding is a presenting symptom of cervical cancer. Postmenopausal bleeding is a primary presenting symptom of uterine carcinoma and endometrial hyperplasia/ dysplasia. Ovarian cancer may present with bleeding due to estrogen secretion by the tumor.
Endocrine disorders - disorders of prolactin secretion, hyper and hypothyroidism, adrenal dysfunction. This category includes anovulation which results in irregular/heavy and occasional absent menses. Stressful conditions such as basic training may cause anovulation in female soldiers.
Hematologic - suspect coagulopathies in a young, newly menstruating female with abnormally heavy flow

Plan:
Treatment
Significant vaginal hemorrhage:
1. Stabilize patient (ABCs, etc.); monitor vital signs closely and transfuse if necessary
2. Maintain patient on bed rest
3. If evacuation delayed, give oral contraceptive pill qid (estrogen can stabilize the uterine lining) - anticipate nausea and treat with oral or IV antiemetic.
4. Give antibiotics (see below) liberally for: febrile patient (start immediately), tender uterus (suspect infection of the uterine lining), foul-smelling discharge.

Minor menstrual irregularities:
1. If HCG negative and HCT and physical examination are normal, treatment may be delayed until appropriate consultative services are available.
2. NSAIDs (**ibuprofen** 800 mg po tid or **Naprosyn** 500 mg po bid) may reduce blood flow.
3. Oral contraceptive pills are the most effective way to control menstrual irregularities (see Contraception section on CD-ROM).

Treatment of other AUB is dependent on appropriate diagnosis that will not be obtainable in the field.

Antibiotic regimens:
Primary: **Ampicillin/sulbactam** 3 gm IV q 4-6 hours
Alternate: **Cefotetan** 1-2 gm IV q 12 hours or **piperacillin** 3-4 gm IV q 4 hours, or **ticarcillin/clavulanate** 3.1 gm IV q 6 hours, or **gentamicin** 1.5 mg/kg load then 1.0 mg/kg IV q 8 hours and **clindamycin** 900 mg IV q 6 hours (if patient remains febrile after 48 hours, add **ampicillin** 2 gm IV q 6 hours).
Empiric: If IV therapy is not available, treat the patient as per oral PID protocol with IM **ceftriaxone** 250 mg and 1 gm of oral **azithromycin** or 100 mg doxycycline po bid.

Follow-up Actions
Evacuation/Consultation Criteria: Evacuate after initial stabilization for significant hemorrhage. Consult OB/GYN expert as needed for continued or recurrent symptoms.

Symptom: GYN Problems: Pelvic Pain, Acute
MAJ Ann Friedmann, MC, USA

Introduction: Internal gynecologic pathology is a common cause of pelvic and abdominal pain. Acute pain may be secondary to an ectopic pregnancy, a ruptured ovarian cyst, torsion of the ovary, pelvic inflammatory disease or a non-GYN cause such as appendicitis. All of these situations require close observation for possible surgical intervention or transfer to an acute care facility.

Subjective: Symptoms
Abrupt onset of abdominal/pelvic pain. Character of pain and other symptoms vary depending on the cause.
Ruptured ovarian cyst: acute lower pelvic pain, often mid-cycle (day 12-16); occasional postcoital onset of pain due to disruption of cyst during intercourse; should not have associated fever; may present acutely and improve over 8-12 hours; minimal blood loss.
Torsion of the ovary: intermittent severe pelvic pain initially localized in right or left lower quadrant then becoming more diffuse as the ovary necroses. The ovary will often torse and de-torse prior to a final torsion so the patient will describe this same pain lasting for shorter intervals prior to the event that brings the patient in for care. Not initially associated with fever; pain often radiates down inner thigh on affected side; hematologically stable. Fever and elevated white blood cell count are seen if the ovary has become necrotic (prolonged torsion - usually greater than 8 hours although no good data is available on exact time).
Ectopic pregnancy: late, short or missed menses; suddenly worsening abdominal pain which may radiate to a shoulder due to irritation of the diaphragm; circulatory collapse due to internal bleeding; history of vaginal bleeding, infertility, prior PID or pelvic surgery to include infertility procedures; increased risk if prior ectopic pregnancy.
Pelvic Infection: See PID section.

Objective: Signs
Using Basic Tools:
Ruptured ovarian cyst: tender adnexa, normal ovaries after rupture. Fullness in posterior cul-de-sac may suggest blood in pelvis. Localized guarding but no rebound until rupture, no abnormal vaginal discharge, normal uterine exam, afebrile, normal vitals.
Torsion of the ovary: Very tender mass in right or left lower quadrant; rebound and guarding; afebrile early with normal WBC, but both elevated later; nausea and vomiting; rare anorexia; NO leg numbness or weakness to accompany pain (consider disc herniation if present); hemodynamically stable.
Ectopic pregnancy: If ruptured: acute abdomen with peritonitis; nausea and vomiting; hemodynamically unstable with tachycardia, hypotension and anxiety; slightly enlarged, tender uterus with severe cervical motion tenderness. If not ruptured: unilateral, palpable, tender mass without peritonitis; early ectopic will not be palpable and may have intermittent, severe, cramping pain; mild spotting through a closed cervical os (open os with significant vaginal bleeding is a miscarriage).
Pelvic Infection: Lower abdominal tenderness, bilateral adnexal tenderness (see PID section).
NOTE: Perform rectal examination with Hemocult for blood. Change gloves before rectal/ Hemocult exam if the patient is having vaginal bleeding. Positive Hemocult does not occur with the above diagnoses without co-existing GI disease.
Using Advanced Tools: Lab: WBC count (elevated in later torsion and in PID), urine HCG (ectopic), CBC (anemia due to hemorrhage), stool Hemocult (guaiac), cervical cultures for gonorrhea, type blood (pending transfusion if needed).

Assessment:
Differential Diagnosis (see Symptom: Abdominal Pain and GI chapter)
Appendicitis- frequently confused with gynecological acute pathology. Appendicitis begins in the epigastrium, migrates to the periumbilical region and then settles in the right lower quadrant (RLQ) after 6 to 8 hours, with rebound tenderness and RLQ tenderness to palpation (most common finding). Anorexia, nausea and vomiting are common. Prodromal symptoms include indigestion and irregularity of the bowels. The WBC count is often NOT elevated until the patient has had symptoms for over 24 hours.
Diverticulitis- pain due to infected diverticulum is usually left lower quadrant. Past history includes diarrhea and bloody stools, low-grade fever, elevated WBC counts and age over 40.
Severe Constipation- acute cramping pain; anorexia and nausea or vomiting; common in the second and third trimesters; treatment with fluids and fiber will be sufficient for many pregnant women. Acute constipation can indicate underlying disease.

Inflammatory Bowel Disease or Irritable Bowel Syndrome- abdominal pain rarely sudden in onset; will have a past history of intermittent symptoms including diarrhea and bloody stools.

Plan:
Treatment
Primary:
1. Stabilize patient (airway, breathing, circulation, etc).
2. Start 2 large bore IVs and initiate fluid resuscitation for unstable patients (see Shock: Fluid Resuscitation).
3. If PID, then treat per PID section (antibiotics and bedrest).
4. Initiate transfer/evacuation if patient has: possible ovarian torsion, ectopic pregnancy or is hemodynamically unstable.
5. If the patient's diagnosis is consistent with ruptured ovarian cyst and she is hemodynamically stable, she can be placed on bedrest. Repeat vital signs and physical examination q 4 hours. She should improve and be ambulatory in 6-12 hours, NSAIDs may be given, as well as mild narcotics (should only be necessary for the 1st 12-24 hours, if at all. If a significant narcotic need exists beyond 12 hours the patient should be evacuated).

Primitive: If you are unable to evacuate the patient immediately, institute bedrest and fluid resuscitation with lactated Ringer's or normal saline. Blood transfusion may be life saving and should be given if patient has failed resuscitation with crystalloid or is otherwise showing signs of inadequate tissue perfusion. Type and crossmatched vs. type specific vs. O negative depending on urgency and availability

Patient Education
General: Reassure patient and discuss treatment plan even in emergency situations.
Activity: Bedrest
Diet: NPO initially
Medications: Give adequate narcotics to patients whose care is delayed or who must be transferred. Do not over-sedate.
No Improvement/Deterioration: Return for evacuation.

Follow-up Actions
Return evaluation: Start a patient with recurrent ovarian cysts (see Mittelschmerz - chronic pelvic pain section) on oral contraceptive pills to suppress ovulation. A good pill to start would be **Ortho-Novum 1/35** if available.
Evacuation/Consultation Criteria: Acute pelvic pain can only be observed if the facility has the capability to intervene surgically. If not, the patient must be transferred. These patients often need abdominal ultrasound or CT, which is not available in primitive areas.

Symptom: GYN Problems: Pelvic Pain, Chronic
MAJ Ann Friedmann, MC, USA

Introduction: Defined as pelvic pain of greater than 6 months duration, chronic pelvic pain is usually multifactorial in etiology. Pain is often not proportional to amount of disease present. A multidisciplinary approach is most effective in treating this disorder. For the field medic, chronic pelvic pain cannot be cured, but simple, helpful treatments can be initiated. The four most commonly seen diagnoses are discussed below. For a more complete differential diagnosis see the CD-ROM. Also see the Abdominal Pain and Acute Pelvic Pain sections for more information. In non-western countries where routine medical care is not available, female patients with chronic pelvic pain may have more serious disease such as uterine, cervical or ovarian cancer. These women are more likely to have weight loss, fatigue, loss of appetite, night sweats, frequent abdominal pain not related to menses, heavy menstrual bleeding, irregular vaginal bleeding and bleeding after intercourse. Masses will often be noted on the pelvic exam.

Assessment: Differential Diagnosis of Chronic Pelvic Pain

Endometriosis	Dysmenorrhea	Mittelschmerz	IBS
Age of onset			
Early 30s but may be seen as early as late teens.	With first menstrual cycles, usually age 12-14.	May occur at any time after the start of menses.	May occur at any time in life cycle.
Subjective: **Symptoms**			
New or gradual onset of menstrual pain. May start 7-10 days prior to menstrual cycle, deep dyspareunia - worse during menses, sacral backache with menses. Pain often radiates to inner thighs. Menses may be heavier.	Pain (cramps), usually start and end with menstruation. Crampy pain, usually radiating to sacrum, vagina and inner thigh area. Ranges from mild to severe.	Mid-cycle pain days 12-16 of menstrual cycle (count the first day of bleeding as day #1). Gradual onset of crampy lower pelvic pain which peaks in 24-36 hours. Often unilateral. Pain is occasionally sudden in onset. Some light vaginal bleeding may occur.	Symptoms wax and wane. Pain-cycles lasting weeks to months. Colicky pain associated with a feeling of rectal fullness. Pain improves with bowel movement and flatus. Alternating constipation and stressors. "Bloating" frequent complaint.
Objective: **Signs**			
Abd exam may elicit mild tenderness. Palpation of uterus and ovaries often very painful. Occasionally may feel modules behind the uterus, will not have fever or elevated WBC count.	Physical exam will be WNL unless patient is menstruating - Patient will have significant discomfort with exam during menses but no abnormalities of uterus or overies shoudl be found. Afebrile, nml WBC.	May have diffuse mild abd tenderness. Enlarged, tender ovary may be present or ovaries may be WNL Afebrile, WBC count WNL.	Diffuse tenderness on abd exam, greater in LLQ. Often note excessive discomfort with rectal exam. Exam otherwise nonspecific.
Plan: **Treatment**			
NSAIDs for pain. Start with first pain, even if mild initially. **Motrin** 800 mg tid or **Naprosyn** 500 mg bid. 24 hour dosing of NSAID important. For acute relief - **Toradol** 30-60 mg IM. Occasional use of mild narcotic OK (**Tylenol** #3). Oral Contraceptive or **Depo-Provera** will bring relief to many but may not be effective for 3 months. Start ASAP.	As per Endometriosis.	For acute pain: **Toradol** 30-60 mg IM narcotic as needed. For chronic treatment NSAIDs and OCPs as per endometriosis.	**Metamucil** or other bulk-forming agent = cornerstone of therapy - take on a regular basis. Increase fluid intake. Identify, avoid food triggers. Relaxation techniques may be effective. Increase aerobic exercise. Warm baths.

Primary Dysmenorrhea - a major cause of chronic pelvic pain and the easiest to diagnose. Dysmenorrhea (painful menstruation) is classified as primary when there is no underlying organic cause other than prostaglandin release from the uterus itself during the time of menstruation.

Symptoms: Once a woman's cycles become ovulatory, anywhere from 6 months to 2 years after the start of her periods, she can experience dysmenorrhea and most do. Age of onset is therefore 6-24 months after the start of menstruation. The pain ranges from very mild to quite severe. It is described as cramping in nature and is felt in the sacral area, low pelvis and inner thigh area. It usually starts and ends with menstruation. The patient feels well throughout the remainder of her cycle. Some will have a day or two of premenstrual pain. Many young women will "grow out of" their primary dysmenorrhea. Women may have associated nausea, vomiting and diarrhea (due to excessive prostaglandin release from the uterus). Some may be severely fatigued, pale and ill appearing. Occasionally vasovagal loss of consciousness may occur -usually with the early years of menstruation only. Fever is not present; anorexia is rare other than with the first day of a severe menstrual cycle.

Diagnosis: Based on history. Physical examination including pelvic should be normal. If the patient is examined during her menstrual cycle, her bimanual examination may be notable for a tender uterus that is of normal shape and size.

Treatment: Involves prostaglandin release suppression with NSAIDs and/or hormonal suppression of ovulation with birth control pills. Any available NSAID is appropriate; it is important to prescribe at the maximal level and to have the patient take them regularly either once the pain starts or 1-3 days prior to the onset of her periods. Birth control pills are *very effective* at decreasing menstrual pain and should be prescribed to all who fail NSAIDs. Regular exercise also decreases menstrual pain and premenstrual tension. Patients who do not respond to the above treatments should be referred to a gynecologist. They may have endometriosis.

Endometriosis - very common cause (60-70%) of chronic pelvic pain in premenopausal women. Caused by the presence of functional ectopic endometrial glands, which may be located in the ovaries, uterus, uterosacral ligaments or any area within the pelvis. Essentially small bits of the uterine lining are growing in areas where they should not be - the body reacts to these implants causing tissue damage. The usual age of onset is in between age 30 and 40.

Symptoms: Dysmenorrhea (pain with menstruation) will occur in most women with endometriosis. This is usually a change for them with worsening from the normal minor menstrual discomfort; it will often start at least a week prior to the onset of menstruation and may last a few days after blood flow stops. This pain often radiates to the rectum and inner thighs. Pain with intercourse (dyspareunia) is common and sometimes the only complaint: it becomes worse during menses. Patients are often reluctant to engage in sexual activity because of this pain. It is not uncommon for women with endometriosis to have daily pelvic pain - these women will often have more severe disease. At least 30-40% of women with infertility problems have endometriosis. There also appears to be a certain genetic component with 5-10% having a family history positive for the disease.

Focused History for Endometriosis: *Does the pain worsen with menstruation? Does the pain occur at times other than during menses? Have you had problems getting pregnant (infertility)? Do you have pain with intercourse?* (Painful intercourse [dyspareunia] caused by endometriosis is described as "deep inside" or high in the vagina. Pain with initial penetration that occurs at the entrance to the vagina is of other origin.

Pelvic Examination: Palpating the uterus and ovaries often reproduces the pain. The pain may be reproduced with deep abdominal palpation but this is not a reliable finding. As endometriosis can cause scarring in the pelvis, one can find that the uterus and ovaries are immobile due to adhesions. Endometriosis can also cause ovarian cysts. Often the examination is unremarkable other than in the fact that you can reproduce the patient's pain. This is because very small areas of endometriosis can cause great pain.

Treatment: As endometriosis is a common cause of pelvic pain it is best keep this disease high in the differential diagnosis. Many women are not diagnosed for years and so treatment is delayed. In a patient with any history consistent with the above review it is best to start NSAIDs, birth control pills and refer to Gynecology as surgery (laparoscopy) gives the definitive diagnosis and often alleviates symptoms. **Ibuprofen** 800 mg po tid or **Naprosyn** 500 mg po bid should be helpful. Birth control pills suppress ovulation, which will decrease the activity of the endometriosis implants.

Mittelschmerz - ovulatory pain. Some women have midcycle pain due to either distension of the ovarian capsule or spillage of the ovarian contents at the time of ovulation. This pain usually coincides with the 12th-16th day of the menstrual cycle (count the first day of bleeding as day #1). Women will sometimes have a small amount of vaginal bleeding during this time.
Symptoms: Gradual or rapid onset of pelvic pain that will usually peak in 24 hours and then remit. Occasionally the pain will be acute in onset and more painful then usual. This may be a ruptured ovarian cyst. The most significant piece of history is the timing of the pain- Mittelschmerz will usually be on the 12th-16th day. In women with irregular and/or infrequent periods the diagnosis will be more difficult.
Pelvic Examination: Often the exam is only significant for generalized lower pelvic discomfort that is mild to moderate in nature. The ovary will sometimes be enlarged (a woman ovulates from only one side each month so the pain is often lateralizing and changes sides month-to-month).
Diagnosis: A menstrual diary and pain scale are very helpful. The patient can mark the first day of her cycle and then each day that she has pain. If the pain occurs only during midcycle it is Mittleschmerz. Pain that occurs frequently throughout the month will fall into another category.
Treatment: NSAIDs such as **ibuprofen** 800 mg po tid will help to alleviate discomfort. Primary treatment is ovulatory suppression with birth control pills.

Irritable Bowel Syndrome (IBS) - may be the source of 50% of cases of chronic pelvic pain or may occur in conjunction with diseases such as endometriosis. IBS is a disease of abnormal bowel motility triggered by situational stress and certain substances (lactose). Studies show that patients with IBS have increased colonic contractions particularly in response to meals. IBS is often worse the week prior to and during menses and may cause dyspareunia. The discomfort that occurs is often left lower quadrant and lower abdominal - causing many women to interpret their symptoms as related to the uterus and/or ovaries.
Symptoms: Colicky abdominal pain with a sensation of rectal fullness and bloating. Pain is often relieved with bowel movement and exacerbated by meals. The symptoms wax and wane in a cyclic fashion, sometimes lasting for months. The cycles often parallel physical or emotional stress. Abdominal pain is usually accompanied by diarrhea and/or constipation but occasionally may be the only complaint.
Physical Examination: In patients with IBS the uterus and ovaries should be WNL unless a coexisting gynecological problem exists. It is best not to examine these patients in the week prior to and during menses as they may have increased sensitivity to examination. A patient in the midst of an IBS attack may have slight abdominal distension due to gas and mild discomfort to palpation.
Treatment: Involves behavior and dietary modifications. Patients often respond to increased dietary fiber (**psyllium powder/Metamucil**). Fluid intake is often inadequate and should be increased, caffeine should be minimized. Patients should identify and avoid food triggers. Common food triggers include fried and other excessively fatty foods, milk products, rice and beans (in patients not used to a primarily vegetarian diet). Increased aerobic activity may be helpful. Warm baths or heating pads to the abdomen are often helpful during acute exacerbations.
See Abdominal Pain section for more in depth discussion of the above.

Plan:

Diagnostic Tests
1. Urine culture, cervical cultures to rule out gonorrhea and chlamydia.
2. A pain diary is extremely helpful and diagnostic in many cases. The patient should chart the days of menstruation. She should note pain on a scale of 1-10 and any other accompanying symptoms, including physical and psychological symptoms. If she did something that made the pain better or worse it should also be noted. Episodes of intercourse are important to mark down. Women with significant pelvic pain will often limit their intercourse.
3. If available, radiographs of the pelvis and lumbo-sacral spine can identify other potential explanations of chronic pelvic pain.

Treatment
Primary: See Differential Diagnosis Chart
Primitive: Warm compresses, rest and warm baths can be helpful for many types of chronic pain. Relaxation and meditation have helped many women deal with and decrease pain.

Patient Education
General: Most chronic pain can be successfully treated in a systematic fashion. There is no quick fix or "magic bullet" for chronic pelvic pain. Comply with therapeutic suggestions.
Activity: As tolerated. Maintain a normal lifestyle.
Diet: As tolerated. For patients with constipation, recommend high-fiber with adequate fluids.
Medications: A three-month trial of OCPs is necessary to initiate pain suppression. They will not work immediately.
No Improvement/Deterioration: Referral to gynecologist.

Follow-up Actions
Return evaluation: 3 months if placed on OCPs
Evacuation/Consultation Criteria: Evacuate as needed for acute pain, unclear diagnosis, unresolving pain

Symptom: GYN Problems: Vaginitis
MAJ Ann Friedmann, MC, USA

Gynecology may seem to be a complicated challenge for the field medic based on the simple lack of essential equipment such as a gynecological exam table with stirrups and a vaginal speculum. Given these limitations, proceed per Table 3-4 to diagnose and treat vaginitis. There are more complete details in the additional sections on candida vaginitis (GYN), bacterial vaginosis (GYN) and trichomonas (STDs). Always remember that simple vaginitis does not cause pelvic pain or systemic signs of illness such as fever, nausea and vomiting, or pelvic pain. If these are present, the diagnosis is most likely pelvic inflammatory disease. Monitor vaginal pH by testing secretions with urine dipsticks.

Symptom: GYN Problems: Bacterial Vaginosis
MAJ Ann Friedmann, MC, USA

Introduction: Bacterial vaginosis is caused by a vaginal overgrowth of several indigenous bacterial species. Absolute risk factors have not been identified. Treatment of male sexual partners has not been shown to prevent recurrence.

Subjective: Symptoms
Symptoms are localized to the vagina rather than throughout the pelvis: a gray-yellowish, thin vaginal discharge with a foul-fishy odor made worse after intercourse; vulvar burning and irritation; pain during and after intercourse due to vaginal irritation.

Objective: Signs
Using Basic Tools: Pelvic exam: Thin, homogenous, gray or greenish-yellow discharge adherent to side walls of the vagina; pooled fluid in the posterior vaginal cul-de-sac; normal vaginal epithelium; amine (fishy) odor to discharge; erythema of external genitalia; normal uterus and ovaries.
Using Advanced Tools: Lab: Examine discharge, prepare wet mount/KOH slides (see Lab Procedures), test pH with urine dipstick. Consider STD and pregnancy evaluation.

Assessment:
Diagnosis based on the discharge having three of the following four characteristics: pH greater than 4.7;

Table 3-4: GYN Problems: Vaginitis

INFECTION	SYMPTOMS	DIAGNOSIS	TREATMENT	NOTES
Candida Vaginitis	- Vaginal itching - Perineal itching - Swelling and redness of the vulva with irritation +/- dysuria - Clumpy, white to yellow-white discharge - ODORLESS - No pelvic pain - Dyspareunia may be present because of vaginal irritation	- Vaginal pH <4.5 - Discharge visualized either on external genitalia or in the vagina is consistent with diagnosis - Patient will not have tender uterus and will not complain of pelvic pain - KOH prep will show hyphae	- Intravaginal antifungal, such as: **clotrimazole** & **miconazole** - If oral **fluconazole** is available, give 150 mg po x 1	- A patient with complaints consistent with Candida may be treated empirically without examination. Follow-up if discharge worsens despite treatment. - treat with **Flagyl** if patient fails anti-fungal and has a foul-smelling discharge, also consider STD testing.
Bacterial Vaginosis	- Fishy smelling vaginal discharge causes vaginal and perineal irritation due to abnormal pH - Patient will not have pelvic pain or fever - Dyspareunia may be present because of vaginal irritation - External dysuria may be present	- Vaginal pH >4.5 - Odor should be apparent at time of exam - Thin yellow-white discharge present on vulva and vagina - Vagina may be tender on exam, but uterus and ovaries will be WNL - Wet prep reveals clue cells	- **Flagyl** 500mg po bid x 7 days (1st choice) - **Clindamycin** 300mg po bid x 7 days - **Ampicillin** 500mg po qid x 7 days (last choice, cure rate only 60%)	- If unable to perform speculum exam and microscopic exam of discharge, the diagnosis can be suggested based on the type of discharge, the patient's symptoms and the abdominal exam - A bimanual exam is necessary if patient complains of pelvic pain.
Trichomonas	- Large amount of yellow-green discharge which may or may not smell fishy, causes significant irritation of the vulva and vagina - Patients often have dyspareunia and external dysuria (the urine pH irritates) - No pelvic pain or fever	- Vaginal pH >4.5 - Vulva may appear red and swollen - discharge present on vulva with or without odor - Cervix may appear red, vagina will be tender - Uterus and ovaries will be WNL - Wet prep reveals motile, flagellated organisms	- **Flagyl** 2 grams po x 1 - **Flagyl** 500mg po bid x 7 days - **Flagyl** is the only recommended therapy - if the patient cannot take it, give vaginal **clotrimazole** (48% cure rate)	- As Trich is sexually transmitted, partners must be treated and intercourse should be stopped until treatment is complete - Patients should be tested for STDs if possible

homogenous, thin appearance; fishy odor with the addition of 10% KOH; presence of clue cells (vaginal epithelial cells with their cell borders obscured by bacteria). In bacterial vaginosis, 2-50% of epithelial cells will be clue cells.
Differential Diagnosis: Candida vaginitis or trichomonas (see Vaginitis chart), GC/chlamydia (see STD section), PID (may have fever, chills, abdominal/pelvic pain as well as vaginitis complaints).

Plan:
Treatment
Primary: **Metronidazole** 500 mg po bid x 7 days
Alternative: **Clindamycin** 300 mg po bid x 7 days or **ampicillin** 500 mg po qid x 7 days (cure rate only 60%)

NOTE: If patient is breastfeeding or in 1st trimester of pregnancy, give **clindamycin**. Alternatively, use vaginal **clindamycin** gel or **metronidazole** gel in the first trimester of pregnancy.

Patient Education
General: Take medications as prescribed, abstain from intercourse during treatment period.
Activity: Regular
Diet: As tolerated
Medications: No alcohol consumption (including mouthwash or topical alcohol-containing products) during treatment with **Metronidazole** due to Antabuse-like effect (extreme fatigue, vomiting, anxiety, etc.).
Prevention and Hygiene: None
No Improvement/Deterioration: Return immediately

Follow-up Actions
Return evaluation: If symptoms do not resolve, the most likely cause of persistent disease is noncompliance with medical therapy. If patient has been compliant, may re-treat with **metronidazole** 500 mg po bid x 14 days. Consider that patient may have trichomonas and be reinfected from a sexual partner. Treat the partner as well with **metronidazole** 500 mg po bid x 7 days. The couple must abstain from intercourse during the treatment period. If the patient has any suggestion of STD/PID treat immediately.
Consultation Criteria: Worsening/possible PID.

Symptom: GYN Problems: Candida Vaginitis/Vulvitis
MAJ Ann Friedmann, MC, USA

Introduction: Candida vaginitis and vulvitis are inflammatory conditions caused by Candida yeast. At least 25% of women with vaginitis will be diagnosed with candida infections. Other than the localized symptoms there are no long-term or immediate sequelae of vaginal/vulvar candidiasis although a small percentage of females will have frequent recurrence requiring prolonged treatment. Risk factors include pregnancy, diabetes, immunosuppression (includes HIV) and antibiotic use.

Subjective: Symptoms
Vulvar and vaginal itching are the most common complaints; thick, curdy white discharge - increased from baseline; external irritation and occasionally dysuria and pain with intercourse; no systemic symptoms (i.e., fever or abdominal pain) unless there is another illness; no foul-smelling vaginal discharge.
Focused History: *Have you had this before?* (may be "exactly like" her last infection). *Do you have a new partner?* (Increased intercourse can change vaginal pH and predispose to candida vaginitis and bacterial vaginosis. Consider STD screening- see PID section.)

Objective: Signs
See pelvic exam procedure and KOH prep lab procedure sections.

Using Basic Tools: Thick, white, curdy discharge adherent to side walls of the vagina (may look like cottage cheese or be thick and white/yellow); later signs include erythema and edema of the vulva/vagina (perineal rash with red, shiny appearance); fissures of the vulva; self-inflicted scratches of the vulva; swelling and redness of the labia; abdominal exam will be benign.
Using Advanced Tools: Lab: KOH (potassium hydroxide) wet-mount examination yields yeast hyphae in 50-80% of patients (See Lab Procedures: Wet Mount and KOH Prep). Vaginal pH: will be < 4.7 in patient with candida (test with urine dipstick).

Assessment:
Differential Diagnosis: Bacterial vaginosis, trichomonas, gonorrhea or chlamydial infection, atrophic vaginitis (see Vaginal Discharge Table and STD chapter).

Plan:
Treatment
Patients with vaginal or vulvar itching only may be treated without physical examination. A thorough disease-specific history must be taken to evaluate for complicating factors such as pelvic pain, lesions, fever and risk factors for sexually transmitted disease. If any of these are present, evaluate accordingly; if not, prescribe intravaginal therapy.
Primary: Various intravaginal azole agents used for 3-7 days (**miconazole**, **clotrimazole** are most common)
Alternative: Oral **fluconazole** 150 mg X 1
NOTE: Intravaginal antifungals may be used throughout pregnancy. Avoid oral fluconazole. Both intravaginal and oral therapy are safe during breastfeeding.

Patient Education
General: Complete all medication as prescribed since incomplete treatment is a reason for recurrence.
Activity: Normal
Diet: Regular - some theorize that a low-sugar diet may be preventive in certain individuals.
Medications: Burning and erythema (sensitivity to meds) may accompany treatment; discontinue and treat with oral fluconazole.
Prevention and Hygiene: Wipe urethra/vagina from front to back. Wear cotton underpants and loose clothing. Wash underwear in hot water.
No Improvement/Deterioration: Return immediately for reevaluation

Follow-up Actions
Return evaluation: Treat with standard intravaginal medication for 2 weeks. Reevaluate for possible coexisting infection or misdiagnosis (bacterial vaginosis, trichomonas, GC/Chlamydia). Consider testing for diabetes and HIV.
Consultation Criteria: Symptoms which do not respond to therapy; 4 or more recurrences per year; complicating symptoms (pelvic inflammatory disease - patients with candida only do not have pelvic pain).

Symptom: GYN Problems: Pelvic Inflammatory Disease
MAJ Ann Friedmann, MC, USA

Introduction: Pelvic Inflammatory Disease (PID) results from organisms spreading directly from the cervix to the endometrium (uterine lining) and then to the Fallopian tubes. Fallopian tube damage from untreated PID is a major cause of infertility. Risk factors for PID: Sexual exposure to gonorrhea or chlamydia, uterine instrumentation (IUD insertion, D & C, abortion, endometrial biopsy), bacterial vaginosis, prior history of PID or STD. Diagnosis should be based on clinical exam and diagnostic tests. Many women perceived to be at increased risk will not have PID, and many who do not have the typical risk profile will have PID. Many women with PID may exhibit subtle, vague or mild symptoms. A high index of suspicion is necessary. PID is most common in sexually active, menstruating, non-pregnant women.

Subjective: Symptoms
Lower abdominal pain with or without signs of peritoneal irritation; severe and continuous pain in both lower quadrants, increased by movement and intercourse (dyspareunia); abnormal vaginal bleeding in 15-35%; fever in less than 50%; onset of symptoms likely within 7 days of onset of menses.

Objective: Signs
Using Basic Tools: Lower abdominal tenderness, bilateral adnexal tenderness, cervical motion tenderness; also fever >101°F, mucopurulent cervicitis.
Using Advanced Tools: Lab: Cervical culture positive for gonorrhea (chocolate bar), WBC count > 10,500; Gram stain of cervical discharge with gram-negative intracellular diplococci and >10 WBC/hpf.

Assessment:
Minimum criteria for clinical diagnosis of PID: Lower abdominal tenderness, bilateral adnexal tenderness, cervical motion tenderness
Differential Diagnosis: Appendicitis, pregnancy, constipation, UTI, ovarian torsion, endometriosis, ruptured ovarian cyst

Plan:
Treatment
Outpatient:
1. **Ceftriaxone** 250 mg IM plus **doxycycline** 100 mg po q 12 hours X 14 days or **cefoxitin** 2 gm IM plus **probenecid** 1 gm po plus **doxycycline** 100 mg po q 12 hours X 14 days.
2. **Tetracycline** 500 mg po qid X 10-14 days may be substituted for doxycycline.
3. Pregnant patients or those with GI intolerance to doxy/tetracycline may use **erythromycin** 500 mg po qid X 10 days.
4. **Azithromycin** 1 gm x 1 day may be substituted for doxy/tetracycline or erythromycin.
5. **Strict rest is mandatory** during the first 72 hours of therapy. Family members should facilitate patient compliance.
6. Treat partners after patient provides identity.
7. Due to the significant consequences of PID, it is better to treat the patient at risk that has only uterine tenderness or lower pelvic discomfort. This is particularly true in the field setting where all diagnostic testing will not be available.

Inpatient (unstable, pregnant or unresponsive to outpatient therapy in 72 hours, unable to tolerate oupatient medication, if 72 hour follow-up cannot be arranged or guaranteed):
1. Refer immediately. While awaiting transfer:
2. Start antibiotic treatments: Either **cefoxitin** 2 grams IV q 6 hours plus **doxycycline** 100 mg po/IV q 12 hours or **cefotetan** 2 grams IV q 12 hours plus **doxycycline** 100 mg po/IV q 12 hours.
3. Alternatively, the IM **cephalosporin** regimen (above) with **azithromycin** or **doxycycline** can be used. Consider repeating the IM **ceftriaxone** in 24 hours if the patient cannot be transferred, is worsening and IV therapy is not available.
4. Start IV hydration with one or two large bore IVs.
5. Maintain the patient on bedrest and NPO in case surgery will be necessary.
6. If you have any doubt regarding compliance, always treat the patient with **ceftriaxone** or **cefoxitin** IM and the oral **azithromycin** 1 gm dose which the patient can take under observation.
7. Treat partners after patient provides identity.

Patient Education

General: Comply with follow-up care. Discuss STD prevention and condom use.
Activity: Rest
Diet: As tolerated
Medications: As above. If patient develops GI intolerance, pre-medicate with any oral antiemetic before antibiotic dose.
Prevention: Patient must not have intercourse with untreated partner even if he is asymptomatic. Avoid high-risk sexual behaviors. Use barrier methods to prevent STD transmission.
No Improvement/Deterioration: Return for transfer/hospitalization/evaluation for alternative diagnosis.

Follow-up Actions
Return evaluation: 48-72 hours
Evacuation/Consultation Criteria: Patients whose diagnosis is unclear or whose pain is worsening despite treatment will require immediate evacuation for extensive workup, including laparoscopy, pelvic ultrasound, CT scan, cervical cultures for gonorrhea and chlamydia, CBC, and testing for HIV, Hepatitis B and for syphilis.

Symptom: GYN Problems: Bartholin's Gland Cyst/Abscess
MAJ Ann Friedmann, MC, USA

Introduction: The mucus-secreting Bartholin's glands drain by way of a 2 cm duct into the vaginal vestibule immediately outside the hymenal ring at about the 4:00 and 8:00 position. Blockage of the duct causes secretion accumulation and cyst formation in the gland itself. Blockage may be a slow, chronic process leading to an asymptomatic vulvar mass as the gland gradually accumulates fluid, or it may be an acute inflammatory process leading to a painful, infected mass. An acute abscess may occur as the result of infection (chlamydia, gonorrhea, perineal aerobes and anaerobes), vaginal surgery, episiotomy or trauma as in childbirth.

Subjective: Symptoms
Pain with walking, tight clothing or intercourse; mass presents right or left outside of the vagina.

Objective: Signs
Using Basic Tools: Tender, cystic mass in the area of the Bartholin's duct and gland; warm, red overlying skin; purulent drainage from duct. A chronic duct obstruction will often be asymptomatic and an incidental finding on pelvic exam (no treatment required).
Using Advanced Tools: Lab: Gram stain may disclose gonococci or clue cells. Culture gonorrhea from exudate.

Assessment:
Differential Diagnosis: Adenopathy, gumma, bubo, trauma, tumor, scar.

Plan:
Treatment
Primary
1. Antibiotics (per pelvic inflammatory disease protocol): **ceftriaxone** 250 mg IM and **doxycycline** 100 mg po bid x 10-14 days.
2. Incision & drainage (see procedure description below).
3. Pain control with **ibuprofen** 800 mg tid (or other NSAID). Occasionally patient will need a mild narcotic for the first 24-48 hours (**Tylenol** #3 or **Percocet**).

Alternative: Any of the other antibiotic regimens as listed in the PID section.
Primitive: Antibiotics, warm compresses or sitz baths, pain medications and transfer for further care. A Bartholin's cyst may resolve with only this treatment but it is best to drain an abscess.

Patient Education
General: Wear loose clothing.
Activity: Avoid sexual activity, douching or tampons until wound healed. Limit physical activity for 48-72 hours.
Medications: Avoid doxycycline in pregnancy, breastfeeding and children. Avoid sun exposure with doxycycline.
No Improvement/Deterioration: Return immediately for purulent or foul-smelling drainage, increased pain over baseline discomfort, redness and persistent heat in the area.

Follow-up Actions
Wound Care: Sitz baths or warm compresses tid. Keep area clean and dry.
Return evaluation: 48 hours and two weeks post-op.
Evacuation/Consultation Criteria: Evacuate immediately for worsening pain or signs of expanding infection. Evacuate if no improvement in 48 hours or if abscess recurs.

NOTES: If lab evaluation positive for gonorrhea or chlamydia, treat as in Pelvic Inflammatory Disease Section. Women over age 40 with new-onset Bartholin's cyst have a slightly increased risk of Bartholin's gland cancer. Biopsy is necessary.
Asymptomatic cysts in women under 40 may be observed. A gynecologic exam should be performed yearly.

Incision and Drainage of Bartholin's Gland Abscess

When: When the Bartholin's cyst becomes an abscess.

What You Need: Sterile prep and drape, local anesthetic agent such as 1% **lidocaine** (with or without epinephrine), 5 cc syringe, 18 gauge needle to draw up the lidocaine, 22-26-gauge needle for injection, scalpel with 15 or 11 blade, Kelly clamp or other instrument to insert into abscess and break up any loculations or adhesions, Foley catheter (if available), suture with needle (if available).

What To Do:
1. Discuss and describe procedure with the patient. Keep her informed as you move along.
2. Give dose of antibiotics as above.
3. Gather materials and set up surgical site
4. Fill 5 cc syringe with **lidocaine** using 18-gauge needle; replace 18 gauge with smaller gauge needle for injection.
5. Protect yourself – abscess may spray when opened and decompressed (mask, gloves, gown if you have one).
6. Place patient in low lithotomy position (patient lies on her back with her buttocks at the end of the table and her feet supported in stirrups). If table with stirrups not available then patient may lie on table or bed with feet drawn up to buttocks and ankles together in midline. It may help to place a pillow underneath the buttocks.
7. Apply sterile prep and drape
8. Inject anesthesia (3-5 cc of **lidocaine** should be sufficient) in triangular pattern around abscess. Wait 5 minutes.
9. Test for numbness by pinching skin lightly with Adson forceps or other sharp object. Inject 2-3 cc more anesthetic if necessary.
10. Identify incision site – inside the vaginal mucosa outside the hymenal ring near the usual duct opening
11. Make a vertical incision in the vaginal mucosa approximately 1 cm long. It must enter the abscess cavity. If pus is not draining out you are not there yet. Gradually deepen the incision until reaching the abscess.
12. Insert Kelly clamp into the abscess to a depth of 2-3 cm if possible, and break up any loculations or adhesions. Allow the abscess to drain.

Figure 3-4

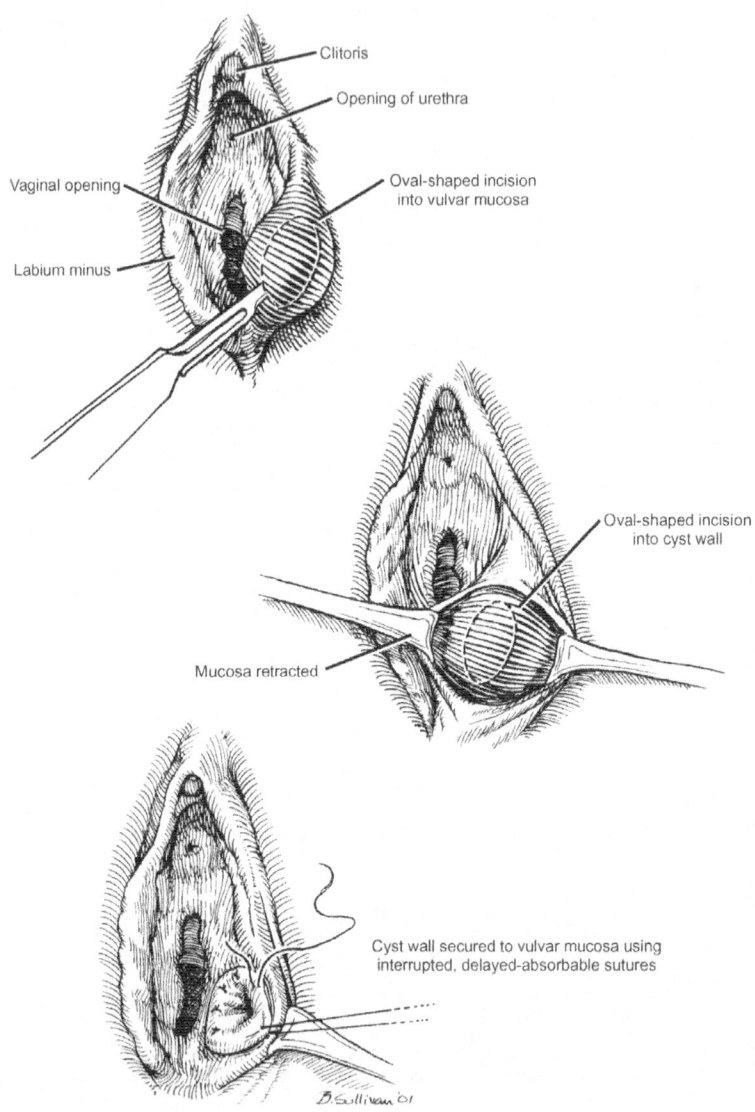

Marsupialization of Bartholin's (Greater Vestibular) Gland

13. Evacuate patient to a gynecologist for definitive care. If evacuation not possible, proceed with the following steps.
14. Cut a 15cm length of Iodoform gauze and loosely pack the cavity with it. This will keep the incision open and allowing continuous drainage over the next few days. Have the patient return in 24 hours for reassessment.
15. If the abscess recurs, marsupialize the gland. Open the cyst again as in before, then suture the everted edges of the gland to the vaginal mucosa. This allows for continuous drainage and a permanent open path (Figure 3-4).

What Not To Do:
Do not open abscess through normal skin. Enter via the vaginal mucosa.
Do not make too large or deep an incision.
Do not attempt marsupialization of the gland unless evacuation is not available to a trained gynecologist.

Symptom: Headache
CDR David Llewellyn, MC, USN

Introduction: There are virtually hundreds of conditions that can cause headaches. Any disease process affecting anything from the neck up can have headache as a symptom. Pain can be referred from the eyes, ears, nose, throat, teeth, sinuses, neck, tongue. Therefore a good examination of HEENT, neck, and the nervous system is crucial. Generalized illness, such as viral syndromes, can also cause secondary headaches. If the headache has an identifiable cause such as cerebral hemorrhage or meningitis, it is termed a secondary headache. Most headaches, such as migraine, tension-type and cluster headaches, are not due to any clearly identifiable cause and are termed primary headaches-- they are real but have no identifiable cause.

Subjective: Symptoms
Head pain, which can vary in severity, and be accompanied by virtually any symptoms; fever, rash, neck stiffness; loss of consciousness or altered mental status.
Migraine: Pounding/throbbing pain, usually but not always unilateral, moderate to severe in intensity, often with nausea/vomiting, often with light or noise sensitivity; routine activities make it worse, patient wants to lie down in a quiet, dark room; builds up over minutes to hours and lasts hours to days; some patients have an "aura", such as flashing lights; women affected more than men.
Tension-type: Global, squeezing headache; less severe than migraine; can last hours to weeks; no nausea or aversion to light and sound.
Cluster: Less common, but affect young men predominantly; severe, short-lived unilateral headaches, usually around the eye, lasting at most a few hours; can occur many times in a day and even wake the patient at night; may want to pace the halls (compare to migraine).
Focused History: *Have you had a similar headache before?* (This makes a benign headache more likely). *Have you taken headache medications? What have you tried in the past?* (use same treatment if effective) *Do you have other symptoms such as nausea, vomiting, fear of light or sound?* (typical migraine) *Have you been hit, or been in an accident? Do you have a fever, stiff neck or rash?* (affirmative answer suggests meningitis, stroke, other CNS damage) *Do you have any neurologic symptoms such as: trouble thinking/talking, loss of consciousness, visual blurring, double vision, vertigo, numbness, weakness etc?* (Positive answers suggest an intracranial problem)

Objective: Signs
Using Basic Tools: Possible tachycardia and hypertension; fever in infection or some CNS trauma; neck stiffness in meningitis; neck tenderness in tension-type headache; abnormal neurological examination (including mental status examination [MMSE is adequate]; see Appendix) suggests significant CNS or PNS; papilledema suggests intracranial swelling; neurological exam may change over time, indicating a worsening condition.
Using Advanced Tools: Lab: WBC for infection.

Assessment:
Differential Diagnosis: Most headaches are benign and are probably migraine or tension-type headaches Cluster headaches are more unusual. Secondary headaches can be referred pain from HEENT, etc. Finally there are the serious, life-threatening conditions such as hemorrhage, infection, or increased intracranial pressure from an acute hydrocephalus or a tumor. Most brain tumors do not initially cause headaches.

Plan:
Treatment
1. Meningitis should be treated as soon as suspected. (See Neurology: Meningitis)
2. First or worst headache needs emergent evacuation (CT, MRI and/or spinal tap may be needed).
3. Treat source of secondary headache, such as sinusitis, if recognized.
4. Treat primary headache symptomatically. Some patients need all three methods.
    a. Behavioral/nonpharmacologic: Ensure patients sleep regularly, get aerobic exercise, manage stress constructively and eat a healthy diet. Avoid caffeine or analgesic withdrawal.
    b. Prophylactic medications (oral): **Inderal** 40 – 160 mg/d, **Pamelor** 25 – 75 mg q hs, **Neurontin** 300 mg tid, or **Depakote** 500 mg bid. Start at a low dose and increase every 2 weeks to effect. Ask women about pregnancy before prescribing.
    c. Abortive or acute therapy:
        (1) Pain relief: 2 – 3 adult aspirin tablets, or 1000 mg **Tylenol**, or 800 mg **Motrin** with food, or 500 mg **Naprosyn** with food works for most headaches. Caffeine in coffee or cola sometimes helps. **Midrin** (2 initially, then 1 q 1 h to max of 5 in 12 h) is a combination medication with acetaminophen. **Fiorinal** and **Fioricet** (1-2 po q 6 h prn) each have caffeine and a mild barbiturate as well as aspirin or Tylenol respectively. IM **Toradol** 60 mg is an option for nausea. The "Triptans", such as **sumatriptan** (**Imitrex**) 50 – 100mg po, 20 mg in a nasal spray, or 6 mg in a sq auto-injector or **rizatriptan** (**Maxalt**) 5 –10mg po are the most effective migraine medications, but nothing works for everyone. Narcotics are rarely needed. Beware that overuse of analgesic medications can produce rebound headaches.
        (2) Nausea or vomiting: **Reglan** 10 mg po q 8 h prn or **Compazine** 5-10 mg IM q 3-4 hrs, max 40 mg/day.

Patient Education
General: Migraine is not curable and will recur. Medications and behavioral interventions can decrease frequency, severity or duration. Most prophylactic medications take weeks to months to work, therefore patience is necessary. Smoking can worsen migraine. For women, menses often worsen migraines. Most migraine medications are FDA categories C (toxic in animals) and D (risk of human toxicity) for Pregnancy, and Ergots (**Cafergot** and others) are X (contraindicated in Pregnancy.)
Activity: Regular aerobic exercise is helpful. Regular, adequate sleep is advised
Diet: Avoid red wine, cheeses, NutraSweet, preservatives in bologna and salami, and chocolate.
Medications: NSAIDs (FDA C) can cause GI problems. The Triptans (FDA C) can cause a chest tightening sensation and should not be used in those with a risk of vascular disease. Ergots (FDA X) should also not be used in those with vascular disease or risk factors. Ergot products and Triptans should not be given within 24 hrs of each other. Prophylactic medications like **Pamelor** (FDA D) or **Neurontin** (FDA C) can cause sedation, usually manageable if titrated up slowly. **Inderal** (FDA C) can cause fatigue, bradycardia, and is contraindicated in asthma. **Depakote** (FDA D) can cause weight gain (a good reason to prescribe exercise!)
No Improvement/Deterioration: Always reconsider your diagnosis if the patient does not do well. Unfortunately, some patients have no response to treatment or find the side effects intolerable. Refer those who do not respond for imaging and neurology evaluation.

Follow-up Actions
Return evaluation: Reevaluate to assess effectiveness of treatment. Some patients may be depressed, and pain may be a manifestation of their depression. Treating the depression is the best treatment of the headache. Be careful not to cause a bigger problem by getting the patient hooked on narcotics.
Evaluation/Consultation Criteria: Evacuate if first or worst headache. Evacuate in other situations if service member unable to perform mission due to pain. Refer if patient fails adequate trials of standard prophylactic medications.

Symptom: Jaundice
COL (Ret) Peter McNally, MC, USA

Introduction: Jaundice is the condition characterized by yellow discoloration of the skin, sclera of the eyes and mucous membranes. It results from systemic accumulation of bilirubin due to significant dysfunction of the liver &/or biliary tract. Under normal conditions, the heme protein from old red blood cells (RBCs) is broken down in the reticuloendothelial system (macrophages and other phagocytic cells) into unconjugated bilirubin, which is transported by the blood to the liver. There the liver cells (hepatocytes) conjugate the bilirubin and excrete it into the bile, where it is eventually eliminated from the body in the stool. Bilirubin is an end-product of heme degradation.

Normal Bilirubin Metabolism:

Reticuloendothelial System	Liver	Hepatocyte/canniculi
RBC→heme→unconjugated bilirubin	→conjugation of bilirubin	→excretion into bile→passage into stool
(makes stool brown)		(Normally there is little or no renal excretion of bilirubin)
	Renal bilirubin excretion occurs when the liver fails	
	(makes dark urine)	

Abnormal Bilirubin Metabolism:
1. Hemolysis: RBC destruction and unconjugated bilirubin production exceeds the ability of the liver to conjugate and excrete the bilirubin.
   Result: Serum: - unconjugated hyperbilirubinemia
2. Hepatitis: Normal RBC destruction with normal levels of unconjugated bilirubin cannot be sufficiently conjugated and excreted by the acutely or chronically damaged liver.
   Result: Serum: - unconjugated and conjugated hyperbilirubinemia
   Urine: - conjugated bilirubin (dark urine)
   Stool: - bilirubin (clay-colored stool)
3. Biliary Obstruction: Normal RBC destruction leading to normal amounts of unconjugated bilirubin, which is conjugated by the liver, but obstruction of the biliary tree prevents bile excretion into the stool.
   Result: Serum: - conjugated hyperbilirubinemia
   Urine: - conjugated bilirubin (dark urine)
   Stool: - bilirubin (clay colored stool)

Subjective: Symptoms
Itching, confusion, abdominal pain, fever, weight loss, fatigue

Objective: Signs
Using Basic Tools: Inspection: Frank jaundice of skin and scleral icterus (icterus of one eye only may indicate other eye is glass); fatigue, confusion and suppressed sensorium in fulminant hepatitis; spider angiomata over the blush area of the upper thorax and gynecomastia in chronic liver disease.
Auscultation: Usually normal
Palpation: Enlarged and tender liver due to inflammation; spleen may also be enlarged. Ascites is uncommon

in acute hepatitis, but may occur if fulminant hepatic failure occurs. A small, hard and nodular liver is typical of chronic hepatitis, cirrhosis and submassive necrosis.
Percussion: Shifting abdominal dullness and fluid wave is suggestive of ascites.

Comparison of Common Causes of Jaundice

Symptom/Sign	Acute Hepatitis	Chronic Hepatitis	Obstruction –stones	Obstruction – cancer
Viral prodrome	Yes	No	No	No
Pain	Tender liver	No	Yes	No
Fever	Low grade	No	Yes	No
Hepatomegaly	Yes, often tender	Often shrunken, may be enlarged and nodular	Normal	Enlargement suggests metastasis
Weight loss	Mild, anorexia	Variable	No	Yes, >10%
Stigmata of cirrhosis: Spider angiomata, gynecomastia, ascites.	Usually not	Invariable, yes	No	Usually not

Using Advanced Tools: Lab: CBC for anemia, infection; urinalysis for urobilinogen

Assessment:
Differential Diagnosis
Hemolysis or hemoglobinopathy (sickle cell, G6PD, thalassemia) (see Hematology Chapter)
Hepatic Causes (see ID: Hepatitis):
    Acute or subacute hepatitis - usually no signs of cirrhosis
    a. Hepatitis A - influenza-like illness (fever, myalgia, nausea, vomiting, etc); tender hepatomegaly; splenomegaly in 10%.
    b. Hepatitis B - 2 week pre-icteric phase (fever, rash, arthritis); icteric phase (1 month); then resolution.
    c. Liver cell toxicity - drugs (isoniazid, methyldopa, acetaminophen and chlorpromazine), Amanita mushroom poisoning, ETOH
    Chronic Hepatocellular Disease - signs of cirrhosis (small, hard liver; ascites; esophageal varices; etc.)
    Hepatitis B or C (never hepatitis A) - ethanol, autoimmune hepatitis, other rare diseases
Obstructive:
    a. Stones in the gall bladder (cholecystitis), or stone in the bile duct - abdominal pain and/or fever
    b. Primary Sclerosing Cholangitis (PSC) - 80% will have inflammatory bowel disease (diarrhea)
    c. Neoplasm - associated weight loss; usually painless.
    d. Inflammation - primary Biliary Cirrhosis usually seen in females 40-60 yrs, drugs (chlorpromazine, erythromycin)
    e. Infiltrative - tuberculosis, sarcoidosis, lymphoma; evidence of invasion of other tissues

Plan:
Treatment
1. Prevent spread of infectious hepatitis by using body fluid precautions, and managing all acute cases in an isolation area.
2. Encourage po fluid and nutrition intake.
3. Discontinue medications that can cause hepatitis or jaundice. Avoid all Tylenol and alcohol.
4. For suicide attempt with acetaminophen overdose, give **N-acetyl cysteine** (**NAC** or **Mucomyst**) or evacuate immediately.
5. If clinical picture is consistent with obstructive jaundice with cholangitis (fever, RUQ abdominal pain), start IV antibiotics: **ticarcillin** or **piperacillin** 4 gm IV q 6 h and **Flagyl** 1 gm q 12 h; or **imipenem** 0.5 gm IV q 6 h.
6. Signs of hepatic decompensation (bleeding, encephalopathy, ascites) or infection should prompt evacuation.

Patient Education
General: Follow body fluid precautions (see ID: Hepatitis). Obtain vaccination for hepatitis A and B.
Diet: High carbohydrate diet as tolerated. May need to limit protein to 1 mg/kg per day if hepatic encephalopathy is present.
Medications: Vitamin K 5-10 mg po qd for hypo-prothrominaemia. **Lactulose** for hepatic encephalopathy 15-30 ml bid/tid (titrate to 3 loose bowel movements per day). Avoid acetaminophen.
Prevention and Hygiene: Pre and post exposure vaccination for hepatitis A and B (see PM: Immunizations). Household contacts, pre-school children, barracks mates should be vaccinated for Hepatitis A. Good hand washing and sanitation are mandatory. Avoid promiscuous sexual contact.
Wound Care: Blood and secretions are potentially infectious.

Follow-up Actions
Return evaluation: Acute viral hepatitis usually requires 2-3 months of convalescence to recover. Monitor symptoms.
Evacuation/Consultation Criteria: Evacuate urgently if signs of encephalopathy, bleeding, easy bruisability, ascites, peripheral edema, or fever with RUQ abdominal pain. Evacuate more stable jaundiced patients when possible. Consult with GI specialist early and as needed.

Symptom: Joint Pain
CAPT Robert Johnson, MC, USN

Introduction: Arthritis (joint inflammation) is not only painful, but also causes fear of disability and deformity. The pain may be localized or diffuse. It may be related to trauma, overuse, degenerative processes or systemic inflammatory disease. The pain may come from the joint itself, supporting soft tissue structures, or may be referred from neurovascular structures. Over 120 different conditions have been called "arthritis" but most musculoskeletal or joint pain can be characterized as either mechanical or inflammatory. Mechanical processes can usually be treated conservatively with rest, ice, heat, other physical therapy modalities and rehabilitative exercise. Inflammatory conditions tend to be more chronic, limiting, and require referral for specialty management. History, physical exam and evolution of the process over time are generally sufficient to distinguish mechanical from inflammatory disease. Psychosocial stresses may aggravate musculoskeletal pain. See individual sections on various joint pains (e.g., Ankle Pain, etc.).

Subjective: Symptoms
Joint pain, fever, loss of appetite, fatigue, weight loss, rash, joint stiffness and swelling.
Focused History: *Where does it hurt?* (pain around a joint, with tenderness in soft tissue is likely muscle, tendon, ligament or bursa damage) *Does it hurt when you move the joint?* (joint centered pain, worse with motion suggests localizes joint process– true arthritis) *Did something hit the joint, or did you bend it, twist it or otherwise traumatize it?* (traumatic arthritis) *Is there redness, warmth or swelling around a joint or joints?* (seen in infection--cellulitis, septic bursa or septic joint; or inflammatory arthritis like rheumatoid arthritis) *How long does it take you to limber up in the morning to reach your best for the day?* (stiffness resolving in < 1 hr., likely mechanical, e.g., osteoarthritis; > 1 hr. suspect inflammatory disease) *Does the pain get better with exercise?* (pain at rest, with stiffness after prolonged immobility, gets better with exercise--tends to be inflammatory; pain worse with exercise or at the end of the day tends to be mechanical) *How many joints hurt?* (more joints, more likely systemic disease) *How long have you had pain?* (>6 weeks is chronic)

Pearls:
1. Symptoms of knee giving way or locking is usually mechanical (cartilage or ligament tear or loose body).
2. Complaints of numbness and tingling related to a sore joint are not typical arthritis complaints. Consider alternate diagnoses.
3. Red, hot, swollen joint is a septic joint until proven otherwise. Gout and other crystal arthritides can mimic infection.

4. Only a few conditions are so painful that the patient can put no weight on the extremity: fracture, septic joint, gout.

Objective: Signs
Joint tenderness, redness, warmth and/or edema; fever; weight loss; rash; and morning stiffness.
Using Basic Tools:
Skin Rash: A new rash associated with arthritis usually indicates systemic disease, either infectious or inflammatory. Lesions include: hives (hepatitis B, C, or other viruses); diffuse maculopapular rash (allergic reaction, drug reaction, serum sickness or virus); scaly, red, hypertrophic lesions (psoriasis; discoid lupus –on face scalp or elsewhere; Reiter's syndrome–on palms and soles; circinate balanitis—on penis); maculopapular rash on palms and soles (syphilis, Rocky Mountain Spotted Fever); papules that progress to vesico-pustules to larger hemorrhagic or bullous lesions (gonorrhea); large, red, tender subcutaneous nodules on shins that may coalesce into a more diffuse, red, swollen lower extremity, resembling cellulitis or ankle periarthritis (Erythema Nodosum).
Eye: Conjunctivitis and iritis (Reiter's syndrome, vertebral arthritis); dry eyes and dry mouth (Sjögren's syndrome, and other connective tissue diseases).
Oral Cavity: Painless ulcers (Reiter's syndrome, lupus, syphilis).
Lymphadenopathy: Diffuse lymphadenopathy (inflammatory diseases like lupus and sarcoidosis).
Chest: Pleuro-pericarditis (lupus-like connective tissue disease); heart block (Lyme disease, vertebral arthritis); heart murmur (endocarditis).
Abdomen: Hepatosplenomegaly (connective tissue diseases).
Extremities: Nodules on extensor surfaces (esp. elbows) (gout, rheumatoid arthritis (RA), and rheumatic fever); intense inflammation of the entire digit (sausage digit), including joints and tendons, (vertebral arthritis).
Joint exam:
1. Palpate joint:
    a. Temperature with the back of your hand and compare to adjacent muscle, other joints, same joint on the other extremity and to your joints. The joint should be the coolest part of the extremity. Redness (septic; rheumatoid arthritis) is unusual but warmth is common in the inflamed joint.
    b. Swelling, which may be joint centered, in joint effusion, periarticular from edema or cellulitis, or bony in nature as in osteoarthritic nodes of the hands.
    c. Crepitus (snap, crackle, pop; audible or palpable; joint grinding with motion) may be due to degenerative or inflammatory joint disease. Crepitus without pain or limitation is not usually significant.
    d. Tenderness may be joint centered or in periarticular structures (tendon, ligament, bursa, muscles).
2. Assess:
    a. Range of motion (ROM): is it full; with or without pain; can you move the joint further than they can or is it ankylosed (stuck in one position); compare their ROM with your ROM.
    b. Joint stability: laxity with valgus or varus stress on a joint (usually elbow or knee) or a drawer sign (usually knee or ankle) or repeated joint dislocations is generally a sign of ligament or tendon injury.
    c. Attempt to reproduce pain by joint motion or palpation to localize source to either the joint itself, the periarticular soft tissue, or to sources outside of the musculoskeletal system (skin, vessels, etc.).

Spine:
1. Inspect for symmetry, scoliosis, pelvic tilt.
2. Palpate for muscle spasm, vertebral tenderness, or SI joint tenderness (see maneuvers that stress the SI joint below). Pain increased with maneuvers that torque the sacroiliac joint imply sacroiliitis.
    a. FABER maneuver: with patient supine and one hand on an anterior iliac crest, cross the opposite leg into a figure 4 position (**F**lexion, **Ab**duction and **E**xternal **R**otation at the hip) then press down on the bent leg. This distracts the SI joint opposite the crossed leg. Sacroiliitis is suggested if this maneuver causes pain in the lower back over the SI joint opposite the bent leg. Pain in the back or hip on the same side as the crossed leg does

not count.
  b. Pelvic compression: with the patient supine on a firm surface press downward and inward on the anterior superior iliac crests. This stresses the SI joints. In a patient with sacroiliitis this should cause pain localized to the SI joint.
3. Assess ROM:
  a. Limited lumbar mobility as measured by the Schober maneuver: place a mark on the back in the midline at the level of the presacral dimples, approximately L5. Measure upward and make another mark 10 cm toward the head on the midline of the spine. Have the patient bend forward and touch his toes. The best measurement should be after the third try, with repeated bending providing maximum soft tissue stretch. The distance between the marks should be 5 cm greater in a bent position compared to straight position.
  b. A normal Schober maneuver with the measurement increasing from 10 – 15 cm goes against the diagnosis of ankylosing spondylitis. A measurement of 0 mm would be seen in a totally fused spine. A measurement of 1 - 3 cm suggests ankylosing spondylitis. Some limitation at 4 cm may be seen in either ankylosing spondylitis or a process with severe pain and muscle spasm.

Using Advanced Tools: Laboratory plays a limited role and X-ray changes may be delayed for months to years.
If there is significant joint effusion in a single accessible joint, it should be aspirated (see Procedure: Joint Aspiration). Joint aspiration is not essential in a polyarthritis. Septic joints are almost always monoarticular.
Lab:
1. Synovial fluid cell count with hemacytometer:
  a. WBC < 2000 with < 75% PMNs = noninflamatory fluid - Osteoarthritis, Trauma, Viral infection
  b. WBC > 2000, but <50,000 (5000 - 15,000 common) with 75 - 90% PMNs = inflamatory fluid - gout, pseudogout, viral, Lyme disease, rheumatoid arthritis, other arthritis
  c. WBC > 50,000 = septic joint until proven otherwise
  d. RBC: TNTC RBC = hemarthrosis - major trauma including fracture, internal derangement (ACL, meniscal tear), bleeding disorder
2. Gram Stain: Positive if infected with staph or strep; may be negative if infected by gonococcus 50% of the time
3. WBC: Elevated WBC with a left shift favors infection but does not rule out gout or inflammatory process. Low WBC with lymphocytosis suggests viral illness.
4. Urinalysis: Proteinuria, hematuria, and RBC casts are seen in glomerulonephritis, which is associated with many types of polyarthritis including lupus, and some infections like endocarditis, Hepatitis B&C, osteomyelitis, HIV
5. Monospot in the appropriate clinical setting can confirm the diagnosis of mononucleosis.
6. RPR in the appropriate clinical setting can confirm the diagnosis of syphilis.
   - remember false positive VDRL (+ RPR with NEG FTA-abs) can be seen in lupus-like connective tissue diseases

Assessment:
Differential Diagnosis: Pain in one or more joints is a common presenting symptom of decompression sickness (see Dive Medicine Chapter).
Acute Monoarthritis:
Tendonitis, bursitis and other soft tissue inflammation usually involves one joint region and can be distinguished from true arthritis by physical exam.
Mono- or oligoarthritis (very few joints) implies trauma, infection or crystal induced arthritis. History of trauma suggests fracture or hemarthrosis.
Joint-centered tenderness, redness, and swelling most common in gout or septic joint.
  a. A family history of gout or kidney stones and history of similar previous episodes suggests gout. Abrupt onset with maximum pain in 12-24 hrs is likely gout or other crystal induced process.
  b. Chills, sweats, and fever more common with septic joint but does not rule out gout. Gradual onset over several days suggests infection, with or without fever.

Chronic mono- or oligoarticular pain (>6 weeks) is unusual. Degenerative disease (osteoarthritis) presents without inflammation or effusion. Inflamed joints are likely septic with fungus or TB as the most likely organisms.

Polyarthritis: Involves both small and large joints in a symmetric fashion.
Acute polyarthritis is often self-limited and resolves within 6 weeks without clear-cut diagnosis (viral, post-viral or post-vaccination arthritis).
If the process last >6 weeks it is most commonly rheumatoid arthritis.

Oligoarthritis: (arthritis of very few joints)
The spondyloarthropathies, or vertebral arthritides, share common features of asymmetric oligoarthritis, inflammatory back pain, mucocutaneous lesions, and inflammation of attachment tendons and ligaments. The group includes ankylosing spondylitis, reactive arthritis, psoriatic arthritis, and the arthritis associated with inflammatory bowel disease. Reactive arthritis typically follows an infection of the genitourinary tract or bacterial dysentery.
Inflammatory low back pain awakens patient at night, causes profound morning stiffness lasting > 1 hr, localizes to sacroiliac joint with tenderness, improves with exercise, radiates into the buttocks, posterio thigh and knee (not below the knee) and may alternate buttocks.
Reiter's syndrome begins with an infection, often a nonspecific urethritis, a post gonococcal urethritis or a bacterial bowel infection with bloody diarrhea. In the weeks following the infection the patient develops conjunctivitis and/or shallow, painless oral ulcers. These symptoms may be mild and overlooked or not reported. Shortly after the patient develops an asymmetric oligoarticular inflammatory arthritis, usually involving large joints of the lower extremity, e.g. knee or ankle. There is also a tendency for the Achilles tendon and plantar fascia to be involved. Small joints of the hands and feet may be involved with inflammation of every joint in a digit as well as the tendons and ligaments of that digit (a dactylitis,) giving a sausage digit appearance.
Ankylosing Spondylitis presents as insidious onset of inflammatory low back pain in young men (onset < age 40 yr.) and is often mistaken for mechanical lumbar sprain or strain for years before the diagnosis of inflammatory arthritis disease is entertained. Also involves uveitis of the eyes.

Mechanical low back pain: Common syndrome of pain, which may radiate into the buttocks or legs and is often associated with sensory derangement (numbness, tingling, etc.), history of trauma, morning stiffness which improves, and pain with lifting.

Arthralgias/myalgias without physical findings:
Acute arthralgia/myalgia is seen in viral illness, overuse. Chronic polyarthralgias/polymyalgias may be seen with fibromyalgia, overuse syndromes, hypothyroidism, diabetes and more rarely, metabolic bone disease such as hyperparathyroidism. Myositis, or inflammatory muscle disease, presents with insidious onset proximal muscle weakness and elevated muscle enzymes. Rhabdomyolysis, the massive breakdown of muscle cells due to a variety of causes (trauma or compression, ischemia, infection, drug toxicity, metabolic disease, heat injury), may be associated with muscle swelling and pain. Myoglobin release into the blood stream is toxic to the kidneys. Patients may present with brown urine and renal failure. This illness, presenting with acute onset muscle pain is a medical emergency.

Plan:
Treatment
General (mechanical, as well as some inflammatory causes of arthritis):
1. Rest, compress and elevate inflamed joints.
2. Apply ice and/or heat (alternating for 20 min each). Heat may feel better on a sore joint, but do not use within 48 hours of an acute injury.
3. Immobilize and protect joint with splint, brace or cast and crutches or cane, as appropriate.
4. Perform (active or passive) gentle range of motion exercises bid/tid early in treatment to retain mobility.
5. Later, perform strengthening and isometric exercises for supporting muscles followed by isotonic and weight bearing exercises as soon as tolerated.

6. Provide analgesia:
   a. **Acetaminophen:** up to 1gram po qid prn
   b. Severe joint pain may require narcotics (**codeine** or **oxycodone**) until patient can be transported for further evaluation.
   c. Nonsteroidal Anti-inflammatory drugs (NSAIDs) will also decrease inflammation at full strength and regular dosing.
      1) **Ibuprofen** 800 mg po tid
      2) **Naproxen** 500 mg po bid (often provides better compliance with less frequent dosing)
      3) **Piroxicam** 20 mg po qd (better compliance but slower onset)
      4) **Indomethacin** 50 mg po tid (preferred for intense inflammation as in gout or spondyloarthropathy)
      5) **Tolectin DS** 400 mg op tid (good alternative if indomethacin not tolerated due to CNS side effects)
7. Consider muscle relaxants, which may reduce muscle spasm and back pain associated with sacroiliitis, and may improve sleep disturbances and early morning pain and stiffness. **Flexeril** (cyclobenzaprine) 10 mg po tid, or q hs for sleep effects.
8. Aspirate knees for comfort and to assist ambulation.

Septic joint:
1. Immobilize joint and allow no weight bearing or exercise until infection treated.
2. Antibiotics: 2 – 3 week course (GC responds rapidly)
   Gram positive cocci: **Nafcillin** 9gm IV daily, dosed q 4 hrs
   Gram negative organisms: **Ceftriaxone** 1-2 gm IV daily
   Gram stain negative: **Ceftriaxone** 1-2 gm IV daily prophylactically

Gout:
1. Current preferred treatment is **indomethacin** 50 mg po tid, or other NSAID as above.
2. Alternatively, use **colchicine** 1 mg po q 2 hours prn until pain is relieved, or until diarrhea or vomiting presents. No more than 7 mg should be given in 48 hours for a given attack. On average, about 5 mg is required.
3. Obtain consultation before initiating chronic therapy with **probenecid** or **allopurinol** to prevent attacks (suppress serum uric acid levels).

Other Inflammatory Arthritis (non-infectious):
1. **Prednisone** 40 mg given once in the morning tapered (decrease 5 mg/day) over a week to 15 mg daily. The Medrol dose-pack does this over a 5-day period.
2. If evacuation is delayed, prednisone can be further tapered by 2.5 – 5 mg per week as tolerated.
3. Rheumatoid and lupus patients often respond well to 5-15 mg daily.
4. Once the patient has been on prednisone for more than a week it should be tapered slowly and not stopped suddenly.
5. Do not give steroids in trauma or for ankylosing spondylitis.
6. Steroids may benefit some patients with peripheral joint inflammation by improving their ability to function and ambulate until they can be transported for definitive care.
7. Alternate taper: **Prednisone** 40 mg X 5 days, then 30 mg X 5 days, then 20 mg for 5 days etc.
8. Patients with potentially life threatening illnesses like systemic lupus erythematosus, allergic reactions with anaphylaxis, and drug hypersensitivity reactions (serum sickness like responses) need high dose steroids (**Prednisone** 1 mg/kg po in split dose, example: 20 mg tid – qid) and immediate transport.

Patient Education
General: Discuss appropriate illness with patient.
1. Natural history of gout: attacks of 7 – 10 days average duration with multiple recurrences.
2. Acute polyarthritis: May be the reaction to an infection or an exposure; frequently resolves within weeks with unknown cause. If it is symmetric, involves small joints of the hands and wrists and persists >6 weeks, it may be rheumatoid arthritis.

3. Ankylosing spondylitis may lead to significant limitation of spinal motion. The initial inflammatory phase may cause severe pain and stiffness over many years. While symptoms wax and wane, the disease seldom goes into spontaneous remission with return to unrestricted function.
4. Reiter's disease flares often lasts 6 – 24 months. 25% of the time it is a chronic disease never completely remitting. Avoiding chlamydial STDs may reduce the risk of recurrences.

Activity: Rest the inflamed joints but include range of motion, strengthening, and aerobic exercise as tolerated. The extent to which a patient can continue their job depends on the physical requirements of the job, the affected joint/s and the severity of the inflammation in the joints involved.

Diet: Chronic gout can be controlled through a diet low in purines (animal proteins) and alcohol. Follow a Heart Smart diet.

Medications: NSAIDs: gastritis, ulcer, GI bleed if not taken with food; **indomethacin** may cause CNS side effects (HA, drowsiness, confusion)

Flexeril: daytime drowsiness if used tid; side effects tend to lessen after 2 weeks on the drug.

Prednisone: The goal of steroid treatment is to use the lowest dose that is necessary for as short a time possible. Steroid side effects are multiple and potentially severe: weight gain, increased appetite, peripheral edema from salt and water retention, hypertension, hyperglycemia, osteoporosis, avascular necrosis of bone, nervousness, emotional lability, psychosis, yeast infections.

Prevention and Hygiene: Practice safe sex and use a condom use to decrease the risk of chlamydial STD and/or Reiter's flare.

No Improvement/Deterioration: Return for prompt reevaluation.

Follow-up actions
Follow any acutely inflamed joint until resolution or referral.

Evacuation/Consultation criteria: Promptly evacuate patients with septic joints, pain undiagnosed after 6 weeks or not adequately controlled with conservative therapy, steroid treatment required, acute tendon/muscle rupture, or severe internal derangement.

Joint Pain: Joint Dislocations
COL Roland J. Weisser, MC, USA and LTC Winston Warme, MC, USA

Introduction: Joint dislocations occur when joints are stressed beyond the normal range of motion. Although dislocations occasionally occur spontaneously (e.g., patella), they are usually associated with some degree of trauma. A dislocation is a complete joint disruption such that the articular surfaces are no longer in contact. Dislocations may be associated with marked swelling/edema and may cause injury to adjacent blood vessels and nerves. For this reason, most dislocations should be reduced as soon as possible. This minimizes the morbidity to the patient, but caution is required because there may be associated fractures. Gentle examination of the distal limb for crepitus or abnormal motion due to fracture is prudent prior to attempting a reduction maneuver. In this instance, you can cause more harm and you probably need X-rays and an orthopedic consultation. The following table presents an overview of common dislocations and their management.

Body Part	History and Usual Mechanism of Injury	Objective Findings	Treatment (Details below)
Shoulder: Anterior	Common; may be recurrent; frequently associated with athletics, hyperabduction/ hyperextension most common or from direct impact on posterior shoulder	Pain and splinting of the extremity. Arm slightly abducted; unable to move rm across chest; inability a to rotate arm; anterior "fullness" from anterior, medial, inferior displacement of humeral head to subcoracoid position	Reduce ASAP (technique below), providing no crepitus is noted with gentle IR/ER of the arm

Shoulder: Posterior	Rare; direct blow to front of shoulder, translational injury from falling on outstretched arm, ballistic movements with internal rotation during convulsion/seizure activity or electric shock	Pain and splinting of the extremity. Arm held at side; near total loss of ability to rotate externally; posterior "fullness" when compared with contralateral side	As Above
Elbow: Posterior	Fall or forceful impact on outstretched hand with arm in full extension; generally posterior and lateral displacement of radius and ulna	Foreshortening of the forearm with posterior deformity at the elbow; marked pain; rapid swelling. Ulnar N. injury possible from valgus stretch.	Longitudinal traction followed by anterior translation of the forearm relative to the humerus (technique below); posterior splint
Elbow: Anterior	Direct impact on the posterior forearm with the elbow in flexion. Anterior displacement of radius and ulna	Fullness/ deformity of antecubital fossa; inability to flex/extend forearm; may have associated brachial artery injury and /or ulnar, median and radial neuropraxias	Longitudinal traction followed by anterior translation of the humerus relative to the forearm. Apply posterior splint
Radial head subluxation (RHS) in children	AKA "Nursemaid's elbow"; commonly caused by a sudden jerk or yank on a child's wrist or hand during discipline.	Pain, refusal to use arm; able to flex and extend elbow but unable to fully supinate; may have minimal swelling and no visible deformity Be alert sign/Sx of abuse!	Reduce (technique below); place in sling, advise parents that the problem may recur until age five
Hand: PIPJ or DIPJ	Direct trauma in athletics	Obvious, usually dorsal deformity	Longitudinal traction to reduce. Splint in alumifoam or buddy tape. X-ray for associated fractures when able.
Hip: usually posterior	Massive impact to knee while hip is flexed and adducted; common dashboard injury to front seat passengers during auto accidents	Hip flexed, adducted, internally rotated, and shortened; may have associated fracture of femur; sciatic nerve injury is common as nerve lies posterior to joint	Reduce (as below) if unable to MEDEVAC STAT.
Patella: Spontaneous	May occur spontaneously following predictable specific leg movements in people with loose connective tissues and/or abnormal anatomy;	Knee flexed; patella palpable lateral to femoral condyle	Reduce; reduction may occur spontaneously; immobilize with long leg splint or knee immobilizer. Refer to ortho if possible.
Patella: Traumatic	Associated with trauma: "cutting" laterally while sprinting, etc.	Moderate swelling, tenderness especially medial to the patella; patella palpable lateral to the femur.	Reduce gently as above. Refer for X-rays, possible arthroscopy and surgical repair of soft tissue injury or fractures
Knee (rare)	Direct, massive blow to	Dislocation cannot occur	Immediate distal neuro/

	upper leg or forced hyperextension of knee. Exam of a knee that is reduced post injury but really swollen and "floppy" probably represents a dislocation that has spontaneously reduced and should be treated as such!	without ligamentous and capsular disruptions; inability to straighten leg; injury to peroneal nerve and popliteal artery are common	vascular assessment and gentle reduction with longitudinal traction. Stabilize with long leg splint and frequent vascular (q15min) re-checks MEDEVAC ASAP! LIMB THREATENING INJURY!
Subtalar Joint: Medial	Most common type. AKA "basketball foot" due to frequency of injury when one player lands on another's foot and inverts it. Dislocation can be associated with osseous or osteochondral fractures	Swelling, tenderness, and medial displacement of the foot in plantar flexion	Flex the knee and have an assistant hold the thigh. Apply longitudinal traction and gently evert, abduct and dorsiflex foot. Splint once reduced; may require open reduction
Subtalar Joint: Lateral	Rare	Similar to above but with lateral displacement of the abducted and plantar flexed foot	Flex the knee and have an assistant hold the thigh. Apply longitudinal traction and gently adduct and dorsiflex foot. Splint once reduced; may require open reduction

TECHNIQUES

Anterior and Posterior Shoulder Dislocations:
1. Assessment: Examine the affected joint and determine the sensation of over the deltoid muscle to rule out injury to the axillary nerve. Also confirm the sensation, circulation, and motor function of the forearm and wrist. Repeat these evaluations post-reduction.
2. Differential Diagnosis: Fractures as above, combined fracture/dislocation; muscular contusion; brachial plexopathy; acromioclavicular separation
3. Diagnostic Tests: X-ray if available to confirm dislocation and rule out fractures of the humerus, clavicle, and scapula. (True AP scapula; scapular lateral; and an axillary view.)
4. Procedures:
   a. Many dislocations can be reduced without anesthesia, especially if the reduction is performed immediately after the dislocation occurs. However, if anesthesia is required, develop a sterile injection site, and with a 1-1/2 in 20 Ga. needle, inject 20 ml of 1% **lidocaine** inferior and lateral to the acromion process in the depression left by the displaced humeral head. Allow the local 10 min. to set up and the patient to begin to relax.
   b. Reduction maneuvers: Reassure the patient that you will not make sudden, unexpected moves, and that you will stop momentarily if pain occurs. Halt all efforts at reduction if there is marked pain, or crepitus noted that would suggest a concomitant fracture.
      1) Self-Reduction: Instruct the patient to sit on the ground with his back to a tree, wall, etc. and flex the knee to the body on the same side as the injured shoulder. Have the patient grasp the knee with both hands (fingers interlaced) and then slowly extend the leg. The slow extension of the leg may provide all the traction required to reinsert the humeral head back into the glenoid. This technique works especially well if instituted early with acute injuries, or on recurrent dislocators and requires minimal assistance. However, for the first time dislocator, assistance may be required as detailed below.
      2) Stimson: Place patient on table in prone position with affected arm hanging off the table. Hang about 10-15 lbs from the wrist (holding weights diminishes capacity to relax shoulder) to provide

traction and slowly (in about 10-15 minutes) return the shoulder to its normal configuration.
   3) "Dirty Sock" or "Water Ski" Technique: Instruct the patient to lie flat on the ground. Sit down beside the patient on the affected side with your hip touching the patient's hip. Place your sock covered foot in the patient's axilla, grasp the patient's wrist with both hands, and slowly lean back as if rowing or water skiing. Maintain slow, steady traction along an axis directly parallel to the patient's leg. (Do not pull laterally. Remember that the anterior shoulder dislocation is anterior and inferior). The firm, steady traction with your foot in the axilla providing counter traction will gradually overcome the shoulder muscle spasm and the arm will often "clunk" into the socket after several minutes.
   4) Successful reductions will be recognized by marked pain relief, an audible "clunk", restoration of normal anatomic appearance of the shoulder, and return of more normal range of motion. Verify reduction by having the patient touch the opposite shoulder with his hand.
5. Post Treatment: Post reduction x-rays
6. Patient education:
   General: Avoid the motion that contributed to the injury. Consider use of sling +/- swath to restrict abduction and external rotation. Refer to ortho if possible. Prevention: The recurrence rates for anterior shoulder dislocations vary markedly with age: < 20 yo = > 80 % recurrence rate; > 40 yo = < 25 % rate; higher if very active. If young (<35) and active in sports, etc. consider orthopedic consultation ASAP for acute arthroscopic stabilization to alter the natural history. Instruct in self-reduction technique described above.
   Activity: Restore motion and strength/endurance with physical therapy
   Diet: No limitations
   Medications: Control pain and inflammation with NSAIDs on regular basis **(ibuprofen** 800 mg tid x 7 days or **diclofenac sodium** 150 mg AM and hs x 7 days)
7. Follow-up actions
   Consultation criteria: Failure to reduce the shoulder is a requirement for reduction under anesthesia and possible surgical stabilization.

Elbow Dislocation:
1. Assessment: Examine the affected and confirm the distal sensory, motor, and circulatory status. Repeat these evaluations post reduction.
2. Differential Diagnosis: Elbow fractures; combined fracture/dislocation; muscular contusion; ulnar nerve injury.
3. Diagnostic Tests: AP and Lateral X-rays if available to confirm dislocation and rule out fracture.
4. Procedure:
   a. Reduction of a dislocated elbow is straightforward. A local injection of 10 cc of 1% **lidocaine** w/o epi in the posterior prominence from the lateral side can be helpful if the reduction has been delayed. Premedication with an opiate or benzodiazepine may be desirable.
   b. Apply longitudinal traction with the patient in the prone position.
   c. Position the patient so that an assistant can grasp the upper arm or torso and apply countertraction. Grasp the forearm at the wrist and apply anterior traction along the axis of the forearm with the elbow slightly flexed and the forearm at the original degree of pronation or supination and attempt to move the forearm anteriorly, relative to the humerus, to its proper position. In the case of an anterior dislocation, move the humerus anterior relative to the forearm.
   d. If the reduction is not complete after three vigorous attempts, or if there is evidence of nerve or vascular injury, splint the arm for comfort and evacuate ASAP.
   e. Successful reductions may be recognized by marked pain relief, an audible "clunk", restoration of normal anatomic appearance and some mobility in the joint
5. Post Treatment:
   a. Insure integrity of sensory, motor, and circulatory structures. Damage to the median, ulnar, and radial nerves has been reported, and entrapment of the median nerve following the reduction of a dislocated elbow has also been reported. Most injuries occur to the ulnar nerve, which sustains a valgus stretch during dislocation.
   b. Apply a posterior splint with the elbow in 90° of flexion and the forearm in the neutral position.

c. Obtain post reduction x-rays when possible.
6. Patient education:
   a. General: Avoid the motion that contributed to the injury.
   b. Medications: Control pain and inflammation with NSAIDs on regular basis (**ibuprofen** 800 mg tid x 7 days or **diclofenac sodium** 150 mg am and hs x 7 days)
7. Follow-up actions:
   a. **Consultation criteria**: Failure of all the above maneuvers suggests the requirement for reduction under anesthesia or surgical stabilization. Acute indications for surgery include open or irreducible dislocations, those associated with vascular injuries, and fractures, or entrapment of bony/ligamentous fragments in the joint space.

Radial head subluxation (RHS):
1. Assessment: Usually there is no history of significant trauma. Be mindful to pick up clues of non accidental trauma in all injured children! Parents may admit that the toddler has received a sudden tug or pull on the arm, usually by an adult. The injury appeared to cause instant pain, and the child holds the arm motionless. Although comfortable at rest, the arm is splinted limply at the side with mild flexion in the elbow and pronation of the forearm. Examination reveals no deformity, discoloration, crepitation, or swelling present. There is little palpable tenderness present. However, the child is likely to begin crying and splinting the arm with any forced movement, especially supination. Further exam of the arm should be entirely normal.
2. Differential Diagnosis: Elbow fractures; combined fracture/dislocation; muscular contusion.
3. Diagnostic tests: X-ray the elbow
4. Procedure:
   a. Advise the parents that you believe the child's elbow is slightly "out of joint" and that this is a common and not serious problem but may hurt the child for a few moments until it snaps back into place.
   b. To elicit parental assistance, ask the mother or father to hold the child comfortably in their lap.
   c. Place your thumb directly over the head of the radius on the "tender spot" and press down gently while you smoothly supinate the forearm and extend the elbow. Then, fully flex the elbow as you continue to press against the radial head in the supinated forearm.
   d. You will recognize a "click" or ""thunk" under the pressing thumb which will indicate that the radial head has slipped back into normal position. The child is also likely to begin using the limb normally.
   e. Advise the parents that you think the problem is resolved
   f. Return when the child is calm and reevaluate the elbow.
5. Post Treatment: Avoid repeating the mechanism that caused it. There is no reason to immobilize of the arm.
6. Patient education:
   a. General: Advise the parents that the dislocation is a common problem and admonish them to handle the child gently. If problem is recurrent, consider the potential of abuse.
   b. Medications: Acetaminophen, dose according to age, if child is "fussy"
   c. Prevention: Do not lift child by outstretched arms
   d. Recurrence or no improvement: return ASAP for recurrence
7. Follow-up actions:
   Consultation criteria: Consider potential of abuse if multiple recurrences. Examine for other evidence to support suspicion.

Hip Dislocations (Anterior and Posterior):
1. Assessment: Posterior dislocations of the hip result from direct blows to the front of the knee or upper tibia, typically in an unrestrained passenger in a motor vehicle. If the leg is adducted at the time of impact, the pure dislocation is more common. If the leg is abducted, then posterior wall acetabular fractures are more likely. Typical appearance of patient with a posterior hip dislocation is with the hip flexed, adducted, internally rotated, and resistant to movement in marked pain. If a fracture is present, this posture is less likely. Anterior dislocations result from forced abduction and external rotation. They are quire rare (only 5% of all hip fractures) and tend to present showing the abduction and external rotation.

2. Differential Diagnosis: Combined fracture/dislocation; muscular contusion; concomitant sciatic nerve injury; pelvic fractures.
3. Diagnostic Tests: AP pelvis film and cross table lateral X-rays needed.
4. Procedure:
    a. Follow ATLS protocols. If able to MEDEVAC, DO SO STAT. If not, and injury is uncomplicated by ipsilateral fractures distal or pelvic fractures, proceed as below.
    b. Evaluate for crepitus with gentle IR/ER/log rolling of limb. Sedation with **morphine** and **Valium** if available. Place patient on the ground/floor and have an assistant stabilize the pelvis by pushing down on the iliac wings. Flex the hip and knee 90° each. Position yourself between the patient's leg and grasp the affected leg in both arms and pull the hip and leg on a 90° axis to the floor. This will take lots of force and may take 5-10 minutes of work, so use your legs and gently IR/ER the leg as you distract. Reduction should be manifested by a satisfying clunk. If you are unable to reduce the joint, it may be more complicated than you expect, so splint the patient in situ with pillows and make an EVAC happen.
    c. Reduction may require general anesthesia. Rapid reduction in less than eight hours is necessary to minimize the risk for neurologic dysfunction and avascular necrosis of the femoral head.
5. Post Treatment: Even if you get the hip to reduce, keep the patient non-weight bearing until a CT exam is acquired to rule out incarcerated bony fragments.

Patellar Dislocation:
1. Assessment: History may reveal that there was direct blow to the medial aspect of the patella, or the ailment began suddenly, following a "cutting movement" away from the fixed foot, which causes contraction of the quadriceps and external rotation of the tibia on the femur. Patient will usually present in considerable pain with the knee slightly flexed, and the patella obviously located adjacent to the lateral femoral condyle. Some patients may report that this is a periodically repeated occurrence.
2. Differential Diagnosis: Fractures; soft tissue contusion
3. Diagnostic Tests: X-ray with patellar views if available, to rule out concomitant fractures (28-50%)
4. Procedures:
    a. If traumatic, perform ATLS protocols.
    b. Provide immediate reduction
        1) Address patient's concerns that you will not execute any sudden painful movements.
        2) Anesthesia/analgesia is generally not required, but you can aseptically put 15-20 cc of 1% **lidocaine** in the knee make them more comfortable. Inject right under the laterally displaced patella after a good prep.
        3) Gently grasp the patella while stabilizing it with mild lateral traction and maintaining its position to prevent sudden movement.
        4) Support the limb and flex the patient's hip to relax the quads. Request the patient to slowly extend the knee.
        5) When knee is fully extended, release traction on the patella, which will usually slip comfortably back into its normal anatomic location.
5. Post Treatment:
    a. Provide patient with a knee immobilizer or a long leg splint and crutches to maintain straight leg position. Weight bearing as tolerated is authorized
    b. Provide NSAIDs
6. Patient education:
    a. General: Avoid the motion that contributed to the injury.
    b. Medications: Control pain and inflammation with NSAIDs on regular basis (**ibuprofen** 800 mg tid x 7 days or **diclofenac sodium** 150 mg am and hs x 7 days)
    c. Recurrence or no improvement: return NLT 24 hours, or ASAP after recurrence
7. Follow-up actions
    a. Return evaluation: ASAP if recurrence
    b. Consultation criteria: Schedule orthopedic evaluation within 72 hours if available and particularly if related to trauma.

NOTE: Although it is possible to reduce a patellar dislocation by simply "pushing" the patella back into the original position, resist the temptation— it may cause cartilage damage and it is unnecessarily painful for patients.

Symptom: Joint Pain: Shoulder Pain

CDR Scott Flinn, MC, USN

Introduction: The shoulder is an inherently unstable, complex and intricate joint that can be injured through trauma or overuse. A good history and physical will guide most proper initial management. Shoulder pain is usually due to one of the four main groups of shoulder structures; the muscles, the bones and acromioclavicular (AC) joint, and the glenohumeral joint.

Shoulder pain may also be referred from a neck injury, such as nerve impingement, heart problems, such as a heart attack or pericarditis, and lung problems such as a pneumothorax. Diagnostic differentiation is often difficult. Some of these entities and clinical syndromes represent primary disorders of the cervicobrachial region; others are local manifestations of systemic disease.

Risk Factors: Patients with "loose joints" may have more intrinsic motion in the glenohumeral joint, known as multidirectional instability (MDI). MDI predisposes people to certain injuries such as shoulder dislocations, subluxations, and impingement syndrome. A young, otherwise healthy active person who traumatically suffers an anterior dislocation has roughly an 80% chance of recurrence.

Subjective: Symptoms

Acute shoulder pain (immediate onset) is usually due to a recent traumatic cause.
Chronic or recurrent intermittent pain is usually due to overuse syndromes or the sequelae of recurrent trauma.
Constitutional
If there is night pain, think of rotator cuff injury, or, rarely, tumors.
Fever and chills may be present in the rare case of septic shoulder joints. If there is diaphoresis, shortness of breath, and pain in the left shoulder, consider an acute myocardial infarction, neoplasm, or lung injury.
Numbness or tingling suggests referred nerve impingement pain.
Location
AC joint injury presents with pain at/over the A/C joint. A rotator cuff impingement or tear produces deep pain in the anterior shoulder. Fractures produce pain that may be hard to localize but, is aggravated with movement. Shoulder dislocations usually present with obvious deformity and severe pain. Subluxations and labrum (cartilaginous ring around glenoid fossa) tears produce a less intense but deep, aching pain and a painful click.
Focused History: Was there an acute injury? (trauma suggests dislocation, subluxation, muscle tear or fracture). *Is the pain worse with overhead motion and at night?* (This suggests impingement syndrome). *Is there numbness, tingling, or weakness?* (This is consistent with a nerve impingement). *Can the arm be moved?* (If not, the possibility of fracture or dislocation is very high). *Is there a painful, reproducible click?* (Clicks that are associated with pain suggest a labrum tear).

Objective: Signs
Using Basic Tools
Inspection: Look for obvious deformity, suggesting a fracture or dislocation. An anterior dislocation will have a prominent acromial process as the humeral head has slipped inferiorly and anteriorly out of the glenohumeral joint. Erythema and edema may be the result of a contusion or fracture. A prominent clavicle may be the result of an AC separation or clavicle fracture. Have the patient attempt to touch their opposite shoulder with the affected arm. If they are unable to do so, think anterior shoulder dislocation or fracture.
Auscultation: A painful click may represent a labrum tear or a loose body in the joint.
Palpation: Palpate for edema from contusions or fractures. Feel along the clavicle from midline to the AC joint for obvious deformity or crepitus from a clavicle fracture. Feel the AC joint for point tenderness or deformity. If acute, tenderness may represent an AC joint separation, if chronic it could be due to AC joint capsulitis. Check sensation on the lateral deltoid (axillary nerve root, C5 nerve root). Check shoulder strength for internal and external rotation, abduction, and in the "empty can" position (thumb down at about six or seven o'clock) to check the rotator cuff muscles. This sign is positive in rotator cuff tendonitis (impingement syndrome).

Assessment:
Differential Diagnosis
Acute pain is usually due to trauma. Anterior shoulder dislocations occur when the arm is forcefully abducted and externally rotated. Fractures occur from forceful, often direct blows.
AC joint separations occur from landing on the shoulder, as when a football or rugby player is tackled and driven into the ground on their shoulder.
Acute biceps tears present with acute pain and deformity after a sudden lift or catching activity.
Rotator cuff tears occur in the young, healthy population following trauma. Older individuals can get rotator cuff tears from chronic overuse and impingement if not adequately treated.
Overuse injuries include rotator cuff impingement (tendonitis), AC joint capsulitis, degenerative joint disease (DJD, osteoarthritis), and subacromial bursitis.

Plan:
Diagnostic Tests
1. AP and Lateral x-rays are used to rule out fractures. In high-speed trauma this should include cervical spine x-rays. There is debate about the need for x-rays prior to reducing an anterior shoulder dislocation. In the field the shoulder should be empirically reduced. (see Joint Dislocation)
2. In cases of suspected septic arthritis, the shoulder joint may have to be aspirated to examine the fluid with Gram stain (see Procedure: Joint Aspiration).

Procedures
Injection of a local anesthetic into the subacromial region or AC joint (injecting the AC joint can be very difficult) may confirm impingement syndrome or AC joint capsulitis respectively.
Anesthetic and corticosteroid injection may provide longer-term relief, but should not replace PRICEMM (Protection, Relative Rest, Ice, Compression, Elevate, Medication, Modalities).

Treatment
Overuse injuries without significant damage such as rotator cuff tendonitis (impingement syndrome), AC joint capsulitis, sprains and strains can be treated with PRICEMM; protect from harm, appropriate activity modification, icing for twenty minutes three times a day if available, and administration of non-steroidal anti-inflammatory medicines if not allergic.
Long-term administration of anti-inflammatories may cause serious bleeding ulcers, liver and kidney damage. Although somewhat controversial, most anterior shoulder dislocations can be reduced prior to obtaining x-rays. There are many techniques to do this (see Joint Dislocations). Two of the easiest that require no equipment and no additional help other than the reducer are the "water ski" technique and the external rotation technique. Both are easy to master and have low risk of complicating matters. Other techniques are available and may be used depending on the provider's training. Following reduction, the arm is usually put in a sling for a minimum of two weeks and then gradual rehabilitation is performed over the next 6-8 weeks.

For almost all fractures, initial treatment should consist of placing the injured extremity in a sling and swathe and administration of pain medicines. Grossly deformed fractures or those causing neurovascular compromise may need reduction by in line traction. Clavicle fractures and AC joint separations should likewise be placed in a sling and swathe. Pinning or somehow affixing the arm sleeve to the shirt just above the navel can accomplish this if no sling or other material is available.
Open fractures should be cleaned of gross debris and covered with a sterile dressing if possible. Do not reduce open fractures; splint them until definitive surgical care is available. Empirically administer broad-spectrum antibiotics such as **ceftriaxone**.

Patient Education
General: The severity of the injury will dictate the length of time necessary for full recovery
Activity: The activity level should be modified to prevent further injuring, often pain can be the guide ("doc, it hurts to do this" - "so don't do that")
Diet: Must eat to take non-steroidal anti-inflammatories.

Medications: Pain medicines cause sedation; non-steroidal anti-inflammatories may cause fatal stomach or duodenal ulcer bleeds, kidney, or liver damage. All medications can produce allergic reactions.
Prevention and Hygiene: Avoid offending activities for overuse injuries; perform an appropriate rehabilitation program (strengthening).
Wound Care: Grossly contaminated wounds should have the material removed and a sterile dressing placed over them until definitive care is available.

Follow-up Actions
Return evaluation: For overuse injuries, follow-up in 2-3 weeks if no resolution is appropriate. For acute injuries, most fractures heal in 4-6 weeks. Anterior shoulder dislocations should be in a sling for two weeks, then gradual range of motion and strengthening exercises instituted.
Consultation Criteria: Fractures, rotator cuff tears, and suspected septic joints should be referred ASAP for surgical evaluation and treatment in an operating room.

NOTES:
Rehabilitation Guide for overuse*:
Shoulder Rehabilitation Exercises for Strengthening   Use light weights and someone to help "spot" as a safety.

Dumbbell Shoulder Flies   **stop immediately if pain is felt with any exercise**
1. Hold the dumbbells down at the waist with thumbs pointing down
2. Raise the dumbbells up at 10 o'clock and two o'clock to shoulder height.
3. Perform three sets of 10 every other day.
4. As strength improves, increase the weight, and keep repetition number the same.

Dumbbell Shoulder lateral Flies
1. Hold a dumbbell in the hands at waist level, thumb pointing forward.
2. Slowly raise the dumbbell to shoulder height.
3. Perform three sets of 10 every other day.
4. As strength improves, increase the weight, not the number of repetitions.

Bench press
1. Lie on the back with weights in a rack.
2. Lift the barbell off the rack with hands shoulder width.
3. Slowly lower the weight to the mid chest and then push it up again slowly during exhalation.
4. Perform three sets of 10, every other day.
5. As strength improves, increase the weight.
*After the initial injury has healed, a strengthening program is essential to prevent additional injury.

Symptom: Joint Pain: Hip Pain
CDR Scott Flinn, MC, USN

Introduction: Hip and pelvic injuries may occur as a result of overuse or trauma. Knowing the mechanism of injury and evaluating the degree of functional impairment provides the basis for appropriate treatment. Sudden onset of pain with an inability to bear weight is an obviously worrisome presentation. Acute pain may result from trauma, change in degenerative disease or infection. Chronic pain may be due to osteoarthritis, bursitis, referred pain or aseptic necrosis of the femoral head. The need for more advanced diagnostic tests requiring removal from the operational environment is based on the history and exam. Risk Factors: Recent increases in activity/training, biomechanical or anatomic variations, and females with the "female athlete triad" (amenorrhea, eating disorder, and osteoporosis) are predisposing factors that contribute to overuse injuries including stress fractures.

Subjective: Symptoms
Constitutional: Fever (joint infection); non-ambulatory.
Local: Traumatic nerve damage causes loss of sensation, cold leg; ecchymosis.
Focused History: *Can you walk?* (determines if serious problem requiring crutches and evacuation) *How quickly did the pain come on?* (sudden onset of severe pain and inability to walk suggest either fracture or completed stress fracture) *Did you fall or get hit on your hip or back?* (suggests mechanism of injury) *Do you have back pain, or numbness or tingling in your leg?* (suggests pinched nerve from herniated disc or nerve damage from other etiology).

Objective: Signs
Using Basic Tools: Inspection: deformity indicates obvious injury.
Palpation: tenderness over the greater trochanter, anterior superior or inferior iliac spines (hip flexors), or deep in the joint area (tendonitis, bursitis, fracture); pain with gentle logrolling (suspect hip fractures); absent or diminished distal pulses (fracture or dislocation); diminished muscle strength relative to normal side (suggests muscle strain, tendonitis, or nerve injury).
Range of Motion (ROM): limitations in active or passive full ROM suggest possible serious injury. Always try active range of motion first, then passive range of motion should be done gently and stopped if patient is experiencing pain.
Using Advanced Tools: X-rays: imperative for a definitive diagnosis if fracture suspected; Lab: Gram stain joint fluid if infection is suspected

Assessment:
Differential Diagnosis
Traumatic fractures - deformity, pain, inability to ambulate
Femoral neck stress fracture - pain at the extremes of internal and external hip rotation
Greater trochanteric bursitis and others - chronic pain over particular bursa
Hip joint infection (extremely rare) - unable to conduct any active or passive ROM without severe pain
Osteoarthritis - history of chronic overuse (e.g., excessive weight bearing) or trauma, with degeneration of the joint on x-ray.
Strains and tendinitis, such as hip flexor strain - point tenderness and diminished strength of muscle
Pyriformis syndrome - chronic posterior hip and thigh pain, numbness or tingling; pain with passive internal rotation; no back pain
Herniated disc with nerve impingement - back pain; numbness and tingling in distribution of a nerve, including sciatic nerve (sciatica) and others
Referred Pain - lower leg ailments (injuries, malalignments, etc.) may place abnormal stresses on the hip
Aseptic Necrosis - chronic pain and/or limp typically in pediatric patients; also seen in sickle cell disease.
Other injuries and degenerative changes of the hip, which can be diagnosed only with x-rays or specialty referral.

Plan:
Treatment
1. If the patient is unable to ambulate, use either crutches or a litter for transport.
2. Give pain medication, including **morphine**, as required (see Procedure: Pain Assesment and Control).
3. Treatment for specific conditions:
   A. Trochanteric bursitis: After sterile preparation, inject a mixture of 1cc **Lidocaine**, 1cc **Marcaine**, and 1cc **Kenalog** using a 25 gauge 1½ needle using a lateral approach. Insert the needle directly over the palpable greater trochanteric bursae on the lateral proximal thigh, push in until the greater trochanter is reached, and then slightly withdraw off the bone. Inject the mixture after aspirating slightly to ensure the tip of the needle is not in a vessel. The mixture should flow in very easily. If not, it is in muscle or tendon tissue and should NOT be forced. If resistance is met, reposition the needle by going back down to bone, backing off only slightly, and repeating the attempt to inject. Injection may produce excellent pain relief for a prolonged period of time.
   B. Strains, tendonitis, arthritis and other bursitis: Use stretching, rest, compression (strains, tendonitis)

ice if available, and NSAIDs (see Joint Pain section). Always check for allergies prior to giving medications. Injecting bursae as above may provide long lasting pain relief.
C. Fractures: Apply traction splint, give IV fluids (see Shock: Fluid Resuscitation) and evacuate urgently.
D. Herniated disc: Use rest (48 hours max), ice, NSAIDs (see Joint Pain section). Later use heat, stretching and ROM exercises.
F. Aseptic Necrosis: Apply splint, allow no weight bearing, give NSAIDs (see Joint Pain section) and ROM exercises, evacuate.
G. Septic arthritis: Do not aspirate hip joint. Start antibiotics as below and evacuate urgently:
H. If suspected (most common in young adults) or demonstrated gonorrhea (Gram stain), treat as for PID (see STDs chapter).
I. Otherwise give **nafcillin** 1-2 gm q 4h IV or IM, or **oxacillin** 1-2 gm q 4h IV, or **cloxacillin** 0.25-0.5 gm q 6h ac (use higher doses if gram stain positive for staphylococcus); or alternate: **cephazolin** 0.25 gm q 8h - 2.0 gm q 6h IV or IM, or **ciprofloxacin** 750 mg po + **Rifampin** 300 mg po bid. For Gram negative organisms: **ceftriaxone** 1 gm qd IV or **cefotaxime** 1 gm q8h IV, or **ceftizoxime** 1 gm q8h IV

Patient Education
General: If a fracture is suspected, inform patient of probable need for surgical treatment.
Activity: If a femoral neck stress fracture is suspected, allow no further weight bearing to prevent complete fracture that may necessitate surgical correction. Recommend crutches and total non-weightbearing on that hip until x-rays and/or bone scan or MRI can be obtained.
Diet: Additional calcium if suspect stress fracture
Medications: All medications may produce allergic reactions. Long-term use of anti-inflammatories may produce GI bleeds or kidney problems.
Prevention and Hygiene: Proper training may prevent development of stress fractures.

Follow-up Actions
Return evaluation: If pain persists, reconsider diagnosis and consult specialist or evacuate patient.
Evacuation/Consultation Criteria: Evacuate cases of joint infection, fracture or suspected fracture, and aseptic necrosis. Also, evacuate any unstable patient or any team member unable to complete the mission without burdening the team. Consult orthopedics for any patient to be evacuated, and for others as needed.

Symptom: Joint Pain: Knee Pain
CDR Scott Flinn, MC, USN

Introduction: Knee pain is a very common complaint with a broad differential diagnosis. Narrow the diagnostic possibilities based on the mechanism of injury (acute/trauma vs. chronic/overuse) and signs and symptoms. Acute knee pain, within minutes to hours, is usually due to trauma or infection. Chronic knee pain occurs without a specific initiating event, but may be preceded by a long history of minor complaints. Anterior knee pain is usually due to an overuse condition such as patellofemoral syndrome (PFS) or patellar tendinitis. Posterior knee pain may be due to a Baker's cyst. Pain from injuries to the collateral ligaments or menisci are referred to the side of injury. In addition to local causes, knee pain may result from referred pain such as femoral shaft stress fracture. Risk Factors: Malformations or variations in anatomic structures may predispose to overuse injuries.

Subjective: Symptoms
Constitutional: Limp or inability to walk; fever (suggests an acute single joint septic arthritis caused by gonorrhea until proven otherwise).
Local: Swelling, grinding or popping noises, buckling or giving way, ecchymosis.
Focused History: *Can you show or tell me your position when the injury occurred?* (direct trauma suggests structural injury; overuse syndromes suggest tendinitis or PFS) *Did it swell immediately?* (suggests cruciate ligament tear or fracture) *Did it swell within 12-24 hours?* (suggests meniscal tear, osteochondral defect, or capsule tear with patellar dislocation) *Did you feel or hear a "pop"?* (suggests anterior cruciate ligament (ACL) tear) *Does the knee give way or buckle due to pain or is it unstable?* (instability suggests ACL tear or

more rarely meniscal tear) *Does the knee lock or catch?* (suggests meniscal tear, loose body, or ACL tear with stump "catching" in joint) *Did you hit your knee on something or have you been crawling on your knees?* (suggests prepatellar bursitis) *Can you walk and/or bear weight?* (inability suggests fracture or knee dislocation)

Objective: Signs

Using Basic Tools: Inspection: Look for swelling, erythema, bruising, ambulation and full range of active motion.
Palpation: Compare knees throughout exam for temperature differences (increased warmth suggests infection or prepatellar bursitis). Note any swelling in the joint (trauma [blood] or inflammation [gout, arthritis, infection or reactive effusion from PFS]). Perform passive range of motion with hand on knee feeling for abnormal limitations in motion, clicking or popping (cartilage, ligament or meniscal injury). Perform Lachman (drawer) test for ACL and posterior sag test for posterior cruciate ligament (PCL) to indicate tears of the respective ligaments. Assess meniscal integrity (tears cause joint line tenderness, effusion, and positive McMurray sign [pain with passive extension of knee while externally rotated]). Palpate the collateral ligaments. If tender, bend the knee medially and laterally at 0 and 30° of flexion (increased opening suggests tear). Palpate patella (apprehension suggests patellar dislocation or subluxation). Evaluate mobility of patella (PFS) by pushing it medially and laterally while quadriceps is flexed. Feel the patellar tendon for intactness and tenderness (rupture or tendinitis). Assess distal pulses.
Using Advanced Tools: Lab: CBC and Gram stain of aspirated joint fluid to rule out infection; Urinalysis: urate crystals in sediment under polarized light microscopy in gout; X-ray: For fractures, dislocations, alterations of joint space.

Assessment:
Differential Diagnosis
Acute traumatic knee pain can be due to damage to ligaments (ACL, PCL, MCL, LCL), bone (fracture or osteochondral defect), muscle (quadriceps tear), cartilage (meniscal tear) or capsule (patellar subluxation or dislocation).
Infection and other inflammatory conditions (gout) often present acutely.
Chronic knee pain may be due to an old untreated injury, PFS, iliotibial band syndrome (ITBS), arthritis, tendonitis, bursitis or stress fractures.

Plan:
Treatment
1. If patient unable to walk, use crutches, cane and/or splint (may be field expedient).
2. Reduce/inhibit swelling in the injured joint: Rest, ice, compression (wrap, brace or splint), elevation (RICE).
3. Use NSAIDs as necessary for pain and inflammation. For severe pain, see Procedure: Pain Assessment and Control.
4. Gonococcal arthritis is the most likely cause of infection in an otherwise intact knee without a history of trauma. If septic arthritis is suspected, treat with **ceftriaxone** 1 gm IV or IM q 24h until systemic symptoms are clearly resolving. Then, continue therapy for least 7 days with **cefixime** 400mg bid or **ciprofloxacin** 500mg bid. Because of high probability of simultaneous infection with chlamydia trachomatis as an STD, treat with **azithromycin** 1 gm po in a single dose, or **doxycycline** 100 mg po bid x 7 days
5. Aspirate pus/fluid and consider injecting anesthetic to enable member to walk out in combat conditions (see Procedure: Joint Aspiration). Injecting steroids is contraindicated--steroids may allow infection to rapidly worsen.

Patient Education
General: Unless truly catastrophic, most knee injuries will resolve through conservative treatment, rehabilitation and/or laparoscopic surgery.
Activity: Depending on severity of injury, gradually advance range of motion, then add strength program with weight bearing as tolerated.
Medications: Any medicine may cause an allergic reaction. NSAIDs may cause bleeding ulcers, kidney and liver damage.

Prevention and Hygiene: Overuse injuries and recurrences can be prevented with proper stretching, rest and conditioning. Review sexual history and provide appropriate treatment for sexual contacts of patients with gonococcal arthritis.
No Improvement/Deterioration: Return to clinic if symptoms persist for 3-4 weeks.

Follow-up Actions
Return evaluation: Repeat exam. Refer for recurrent, persistent or occult injuries.
Evacuation/Consultation Criteria: Evacuate those unable to complete the mission or keep up with the team. Aspiration of the joint may be indicated in instances of delayed evacuation. Consult Orthopedics for cases of severe knee pain or recurrent, persistent or occult injuries. Knee dislocations (tears of ACL, PCL and one or both of the collateral ligaments) need referral for evaluation of the popliteal artery.

Symptom: Joint Pain: Ankle Pain
CDR Scott Flinn, MC, USN

Introduction: Ankle pain may be due to nerve injury in back, gout, or trauma. Risk factors are previous injury, parachute landings, or walking in rough terrain.

Subjective: Symptoms
Acute pain (immediate onset to a few hours) is usually due to trauma, or more rarely infection or severe inflammation. Subacute pain occurs with inflammation. Chronic pain is usually due to old recurrent trauma causing degenerative joint disease. It may also be due to other arthritides. Constitutional: Acute constitutional symptoms of infection could include fever, malaise, chills, and nausea. Other joints may be involved if there is a chronic arthritic component. Location: Ankle trauma will cause sprains, fractures and rarely, dislocations. Other joints that may be involved include the midfoot (Lisranc's joint), metatarsals (especially the fifth), the tarsal navicular, and fibular head.
Focused History: *How did you hurt your ankle?* (twisting injury - think sprain versus fracture; no trauma suspect infection or inflammation) *Did you hear a "pop" or feel a tearing sensation? Can you walk?* (tells extent of injury) *Have you had this before?* (for example, gouty arthritis in a patient known to have gout)

Objective: Signs
Using Basic Tools: Inspection: Examine the ankle and document swelling, deformity, erythema, ecchymosis, and range of motion.
Palpation: Palpate for a sensation of warmth (suggesting infection or inflammation) and edema (suggesting trauma). Palpate the posterior aspect of the medial and lateral malleolus, palpate any area of tenderness, but especially, the base of the fifth metatarsal
Using Advanced Tools: X-rays (when if available), Lab: Gram stain of aspirated joint fluid if infection or gout is suspected.
Ottawa Ankle Rules - Always obtain x-rays to rule out fracture when any of the following are present:
1. Pain/tenderness on posterior aspect or tip of medial malleolus
2. Pain/tenderness on posterior aspect or tip of lateral malleolus
3. Unable to walk immediately after injury and when evaluated

Assessment:
Differential Diagnosis
The history and physical will almost always lead to an accurate diagnosis. Joint infections are very rare without pre-existing trauma and the patient will not want to move their ankle at all.
Sprain, fracture/dislocation, infection, inflammatory joint disease such as gout or pseudogout, degenerative joint disease, other arthritis.

Plan:
Treatment
1. If patient is ambulatory, encourage ambulation. Keeping boot on may help reduce swelling and provide

support. Splint or cast if necessary.
2. If unable to walk or bear weight, and even with suspected fracture, ambulating with improvised crutches or cane is preferred over a litter patient from an operational perspective.
3. If infection is suspected, and MedEvac is unavailable, treat with antibiotic regimen (see Knee Pain section)
4. Provide pain relief as needed- NSAIDs for sprains, narcotics for fracture.

Patient Education
General: Describe level of injury - suspected fracture vs. sprain
Activity: As tolerated. Rest and elevation if possible.
Diet: No special requirements
Medications: NSAIDs may cause bleeding ulcers, narcotics may cause respiratory depression, all drugs may cause allergic reactions
Prevention and Hygiene: If compound fracture, try to keep as clean as possible to prevent infection.
No Improvement/Deterioration: If infected joint or compound fracture, concern for systemic infection/sepsis. Monitor blood pressure, mental status
Wound Care: Keep wounds clean, dry

Follow-up Actions
Return evaluation: If available, start rehab program for ankle sprain ASAP, return to full duty when able to run a figure eight at full speed pain-free. The theory is to use the ankle as rapidly as possible while protecting from reinjury. Under normal conditions, this would take 3-6 weeks depending on the severity of the sprain and number of previous sprains. Providing ankle support such as with an Aircast or slide-on brace with laces may help speed return to activity. Obviously, if under operational constraints, ankle sprains are not life threatening and the injured person may use the ankle to the best of their ability as tolerated.
Evacuation/Consultation Criteria: Suspected fractures should have x-rays as soon as practical. Surgical incision and drainage ASAP treatment of choice for infected joint. If compound fracture, refer for definitive care ASAP. Infected joints and compound fractures wreak much havoc quickly, making the earliest available MedEvac most appropriate for definitive care.

Symptom: Male Genital Problems: Genital Inflammation
CAPT Leo Kusuda, MC, USN

Introduction: Genital ulcers and urethral discharge are covered in the STD chapter. Most inflammation of the penis is related to the presence of foreskin and may be an early sign of diabetes mellitus. Skin infections in the genital area are similar to cellulitis in other parts of the body, in that they present with pain and redness and are usually caused by staphylococcal or streptococcal organisms. Skin inflammation/infection in this region can lead to urethral stricture or perirectal abscess. This later infection can involve multiple organisms, including gram-negative rods, that can lead to life threatening necrotizing fasciitis (Fournier's gangrene), particularly in the severely injured or diabetic patient. With severe inflammation of the penis, patients may have difficulty voiding or may experience symptoms of septicemia: fever, fatigue and shock.
Phimosis/paraphimosis: Inflammation of the foreskin in the uncircumcised male. Phimosis patients will be unable to retract their foreskin, and in severe cases, the glans and urethral meatus cannot be seen. In paraphimosis, the foreskin is trapped behind the glans with a doughnut-shaped swelling of the foreskin between a tight constricting band in the penile skin and the glans. Both conditions are a result of scar tissue forming on the foreskin at the most distal aspect of the foreskin when the foreskin is extended.
Balanitis: Inflammation of the glans penis and foreskin occurs primarily in the uncircumcised male but is rare in the circumcised male. The glans will look wet, red and may have multiple small red bumps and a whitish material on the surface consistent with yeast. This condition should raise suspicion for diabetes mellitus. If the foreskin cannot be retracted easily, leave it extended. There are a number of non-infectious causes of a wet, red patch of skin on the glans penis. Differentiation often requires a biopsy. When in doubt treat for infection,

and refer for biopsy if the condition does not resolve. Common organisms: yeast and *Gardnerella*.
Thrombosed Penile Vein and Sclerosing Lymphangitis: The shaft of the penis just under the skin and on the surface of the erectile bodies contains numerous large veins that can develop clots. They will appear as dark, hard, raised bumps that follow the course of the vein. Lymph channels can also become hard cords, but will be more clear or lack color. These conditions probably result from overly vigorous intercourse.
Fournier's Gangrene: This condition is life threatening and is most likely to occur in the severely injured patient with poor circulation or diabetes. It presents as a rapidly spreading skin inflammation with development of necrotic/purplish tissue.
Candidal Infection: Most cases will present as balanitis. Occasionally, the patient will present with a red scrotum with satellite red lesions. The rash will be itchy, painful and tender.
Cellulitis: Patients will have diffusely red and painful scrotal or penile skin. The skin may be weeping, thickened and have pustules. When the skin lacks pustules, it is important to determine if the skin changes are a reaction to underlying inflammation such as epididymo-orchitis or torsion of the testis. In the latter cases, the testis and epididymis are markedly tender. The patient can usually differentiate testis pain from skin pain. If the inflammation shows dark areas suggestive of necrosis, the patient is developing Fournier's gangrene.
Contact Dermatitis: Contact with chemicals and even some ointment may cause a profound inflammation of the scrotal skin. The skin may have the appearance and tenderness of cellulitis. History of exposure is extremely important.

Objective: Signs
Using Basic Tools: Swelling, tenderness and redness; purulent discharge in the case of severe phimosis; swollen lymph nodes.
Using Advanced Tools: Lab: Urinalysis for presence of glucose, leukocytes, blood or nitrite; KOH prep of the weepy material on the skin may show the presence of yeast elements such as budding yeast or strands called hyphae.

Assessment:
Differential Diagnosis
Sexually transmitted disease lesions are usually much more focal than these inflammatory conditions.

Plan:
Treatment
Phimosis: Keep the penis clean with soap and water several times per day. Broad-spectrum antibiotics such as **ciprofloxin** 500 mg po bid or **Keflex** 500 mg po qid x 1 week may be used if purulent discharge is noted from the meatus. If the phimosis is symptomatic and severe, perform a dorsal slit (procedure below) and refer for circumcision later.
Paraphimosis: Try to reduce the foreskin by pushing the glans in and pulling the edematous skin forward. You may be able to decrease some of the swelling with direct compression prior to reducing the foreskin. If the constricting band around the penis is very tight, it may be necessary to put some local anesthetic and incise the band with a dorsal slit (procedure below). Refer for circumcision.
Balanitis: Wash the penis several times a day and apply antifungal cream such as **Nystatin, Mycolog** or **Lotrimin** bid. If a wet prep shows no yeast elements, can give **Flagyl** 500 mg po bid x 1 week.
Thrombosis of Penile Vein and Sclerosing Lymphangitis: Refrain from any sexual activity. Use NSAIDs such as **ibuprofen.**
Fournier's Gangrene: Perform emergent aggressive surgical debridement. Broad-spectrum IV antibiotics such as **ampicillin** 2 gm q 8 h, **gentamicin** 5 mg/kg qd and **Flagyl** 500 mg q 6 h are usually warranted.
Candidal Fungal Infection: Keep skin dry and antifungal medications such as **Nystatin, Mycolog** or **Lotrimin** bid for 1 week or single dose **fluconazole** 150 mg po. See ID: Candidiasis section for more information.
Cellulitis: Treat with **dicloxacillin** 500 mg po q 6 h or **Keflex** 500 mg po q 6 h for mild cases. For severe cases, treat with **oxacillin** 2 gm IV q 4 h or **vancomycin** 1 gm IV q 12 h.
Contact Dermatitis: For mild, treat with topical 1% **hydrocortisone** and oral **diphenhydramine (Benadryl)**. In severe cases, add **prednisone** 50 mg po qd and wean by 10 mg /day over 5 days. If there is a question of infection, treat for cellulitis also.

Alternative: Yeast infections can be treated with alkaline washes to the glans penis, if no oral agent is available. Dissolving a few sodium bicarbonate tablets in a small container of water can make an alkaline solution, which can be directly applied to the skin.

Patient Education
General: If the patient has balanitis, phimosis or paraphimosis and has not been circumcised, he should have it done electively.
Activity: Until the inflammation has subsided, activity that would cause the genital region to be wet should be avoided. Refrain from any sexual activity until healed.
Prevention and Hygiene: Circumcision and general cleanliness will largely prevent much of the inflammatory problems. **IT IS VERY IMPORTANT TO KEEP THE GENITAL REGION DRY.**
No Improvement/Deterioration: If the genital swelling and erythema spreads rapidly, return for immediate reevaluation.

Follow-up Actions
Wound Care: Keep area clean and dry.
Return evaluation: Re-evaluate in 1 week: Phimosis/paraphimosis, balanitis, genital fungal infections. Re-evaluate as needed: Thrombosis of penile vein (takes several weeks to months to resolve). Rapidly spreading inflammation may be Fournier's gangrene, requiring immediate evacuation.
Evacuation/Consultation Criteria: Evacuate as above for Fournier's gangrene, phimosis/paraphimosis. When lesions on the skin persist, refer for biopsy. Consult urology or dermatology as needed.

Dorsal Slit Procedure
Essential: If the patient has severe phimosis where the foreskin has scarred down to a small hole and the patient is having significant pain and discharge from the penis, the foreskin needs to be incised (dorsal slit). Similarly, if paraphimosis is severe, excessive circumferential swelling may compromise blood flow in the penis, which can be relieved with a dorsal slit.
1. Attempt non-surgical reduction with anti-inflammatory medications, ice water and lubricants. Evacuate patient to a trained provider for this procedure if possible. If there are signs of systemic infection (fever, nausea, fatigue, etc.), and prompt evacuation is not available, perform a dorsal slit.
2. Assemble equipment: 1% **lidocaine** (w/o Epi), needle and syringe, clamp, forceps, scalpel or surgical scissors, needle driver, 4-0 suture, prep solution, alcohol.
3. Prep the penis as with any surgical procedure (sterile scrub, **Betadine**, drape), and attempt to clean between the head and the foreskin, especially on the dorsal side.
4. Use 1% **lidocaine** and a small needle (25-26 gauge) infiltrate the skin about mid-shaft and extend the wheal at least halfway around the shaft of the penis.
5. Confirm the top of the penis (dorsal side) is numb with forceps or needle.
6. Use a straight clamp to crush the skin from the phimotic area back to the glans (head). Make sure the jaw stays between the glans and the foreskin. Do NOT pass the jaw into the meatus. The glans will still have sensation and the patient should be able to tell you if the meatus is being cannulated.
7. Leave the clamp on for 5 minutes to compromise blood flow in the area to be incised.
8. Remove the clamp and use a scissors to cut the crushed skin where the clamp had been. Do not incise the glans.
9. This should expose the glans. Control bleeding with figure 8 stitches using 4-0 non-absorbable sutures.
10. Clean the penis with sterile prep solution between the head and foreskin, then wipe prep solution away with alcohol. Allow to air dry and apply a sterile dressing leaving the meatus clear.
11. Monitor the patient, as this maneuver is only temporary and the slit can contract. He should have elective circumcision later.

Symptom: Male Genital Problems: Testis/Scrotal Mass
CAPT Leo Kusuda, MC, USN

Introduction: Testis masses are alarming to the patient since they may represent cancer. Testis cancer is quite curable in its early stages. However, the cancer can grow rapidly so early detection and referral is necessary to avoid treatment delays.

Subjective: Symptoms
Testicular pain, enlarged scrotum or testis. Increased risk of benign tumors with history of trauma and vasectomy.

Objective: Signs
Using Basic Tools: Tender testis; palpable mass in testis, spermatic cord or epididymis; mass may appear smooth and spherical, be located on the surface or deep in the testis, enlarge with standing, transilluminate with a bright flashlight.
Using Advanced Tools: Lab: Urinalysis: Nitrite and leukoesterase positive urine suggest infection.

Assessment:
Differential Diagnosis
Solid, non-transilluminating mass that is >4 millimeter size, located below the testicular surface and inseparable from the testis must be considered to be cancer until proven otherwise.
Transilluminating smooth spherical masses are benign and are hydrocele (around the testis), spermatocele or loculated hydrocele (above the testis). Rarely it can be a cyst adenoma.
Small 1-4 mm size nodules on the surface of the testis are benign.
"Wormy" mass above the left testis that gets smaller when patient shifts from the standing to the supine position are varicoceles. Right-sided varicoceles need elective referrals.
Other masses in the scrotum, either on the cord, in the scrotal skin or in the midline area near the penis are almost always benign.
Painful area behind the testis is usually an indication of epididymitis (see Epididymitis section).
If the urinalysis shows leukoesterase or heme positivity, assume that there may be an infection (see UTI section).
If the pain in the scrotum is severe, refer to sections on epididymitis and torsion.

Plan:
Treatment
Primary: Give 2-week course of NSAIDs (e.g., **Naprosyn** 375 mg po bid or **ibuprofen** 800 mg po tid with food). If epididymitis is suspected or cannot be eliminated, add **doxycycline** 100 mg po bid x 14 days. Elevate scrotum with comfortable athletic supporter and decrease activity.

Patient Education
General: Watch benign masses, such as a hydrocele. Perform monthly testicular self-exams.
Activity: Avoid activities that increase testicular pain until the pain is gone. Avoid lifting more than 10-15 lbs., prolonged standing (greater than 30 min.) or walking greater than ¼ mile. Use an athletic supporter.
Medications: Avoid sun exposure if on doxycycline.

Follow-up Actions
Return evaluation: Check patient in 2-4 weeks for change in mass.
Evacuation/Consultation Criteria: Urgently evacuate patients with suspected cancer. Any mass that prevents examination of the entire testis, that is increasing in size, or appears to be inseparable from the testis should be referred for further evaluation.

Symptom: Male Genital Problems: Prostatitis
CAPT Leo Kusuda, MC, USN

Introduction: Prostatitis is used liberally to describe voiding problems and pain associated with the prostate.

Prostatitis is commonly due to an infection, so an empiric trial of antibiotics is useful. Urinalysis is suggestive but not conclusive.

Subjective: Symptoms
Difficulty urinating: Obstructive symptoms include slow start, low flow and dribbling; irritative symptoms include frequency (> q 2 hours) and/or urgency; pain in the head of the penis or under the scrotum; low back pain; fever.

Objective: Signs
Using Basic Tools: Tender prostate with/without tender pelvic floor or coccyx (palpate 360° on rectal exam); distended bladder
Using Advanced Tools: Lab: Urinalysis: heme and leukoesterase positive urine (infection).

Assessment:
Differential Diagnosis
Irritative voiding symptoms with or without fever - urinary tract infection until proven otherwise, distal ureteral stone, urethral stricture, bladder neck dysfunction, bladder or prostate cancer, foreign body in bladder, overflow incontinence.
Obstructive voiding - enlarged prostate, urethral stricture, and neurologic disease of the spine or peripheral nerves.
Painful prostate - urinary tract infection, bladder neck dysfunction/prostatodynia/pelvic floor dysfunction, musculoskeletal pain, coccydynia, seminal vesiculitis

Plan:
Treatment: Infection
Primary:
1. If the patient is lethargic and febrile, begin high dose IV **ampicillin** 1-2 gm IV q 6-8h and **gentamicin** 5 mg/ kg IV qd. If penicillin allergic, use IV fluoroquinolones (**Levaquin** 250 mg IV q day or **Cipro** 400 mg IV bid) or **vancomycin** 1 gm IV bid and **gentamicin** 5 mg per kg IV qd. Hydrate aggressively. When the patient is afebrile, switch to oral fluoroquinolone (see UTI section) for a total of 30 days.
2. If the patient is alert and manifesting either low-grade fever or no fever, treat with fluoroquinolones (see UTI section).
3. Treat any male suspected of having an infection for 30 days regardless of the location of symptoms (kidney, prostate or scrotum). Infected urine can easily reflux into prostatic ducts, therefore assume the prostate is infected.
4. If symptoms persist and urinalysis continues to be abnormal without improvement after 3-5 days, suspect bacterial resistance and change antibiotics.
5. If a bladder is palpated, attempt to pass a Foley catheter (Procedure: Bladder Catherization). Inability to pass catheter suggests a stricture. If patient's symptoms worsen, consider suprapubic aspiration (see Procedure: Suprapublic Bladder Aspiration).
6. If the entire pelvic floor is tender, pain is not from prostate alone. Treat for musculoskeletal pain with NSAIDs.

Alternative Antibiotics:
Septra DS po tid until fever resolves, followed by **Septra DS** po bid x 30 days, **Augmentin** 500 mg po bid x 30 days, or **Keflex** 500 mg po qid x 30 days. **Doxycycline** or **Vibramycin** are not as effective since they are bacterio-static, and should only be used (100 mg po bid) if there is no other alternative. **Nitrofurantoin** has minimal tissue penetration and should not be used in prostatitis.
Primitive: If patients are unable to void, have them sit in a tub of warm water and ask them to void in the tub. The warm water can relax the perineum, decrease pain and encourage voiding.

Treatment: No Infection

Primary:
1. Treat initially with a course of antibiotics, with or without alpha-blockers.
   Levaquin 500 mg po qd has broad coverage for both *C. trachomatis* and urinary pathogens. Treat for 30 days.
   Alpha blockers include the following:
   Hytrin (terazosin) 1-5 mg po q hs (start at low dose and titrate up over several weeks)
   Cardura (doxazosin) 1-4 mg po q hs (start at low dose and titrate up over several weeks)
   Flomax (tamsulosin) 1 po q d (No titration necessary)
   Minipress 1-5 mg PO q hs
2. Patients who complain of frequency should be given bladder antispasmodics with caution since they can cause urinary retention. **Hyoscyamine** 1 po bid prn, **Ditropan** 5 mg po TID prn, **Elavil** 10-25 mg po qd, or **Flavoxate** 1 po bid can all be used.

Primitive: Sitz baths or sitting in a tub of warm water can relax the bladder and decrease pain.

Empiric: If the patient has had multiple sexual contacts and initial symptoms of urethritis (discharge) after finishing **Levaquin**, a trial of **erythromycin** 500 mg po qid for 30 days can be given. If symptoms persist, try **metronidazole** (**Flagyl**) 250 mg po tid for 30 days. The empiric usage of Levaquin, erythromycin and metronidazole covers most sexually transmitted diseases that would affect the lower urinary tract. Without urethral discharge or abnormalities on urinalysis or physical exam (no genital ulcers, lymphadenopathy, etc) STDs are very unlikely.

Patient Education
Activity: Bedrest for septic patients with close monitoring of vital signs.
Medications: Alpha-blockers may make sinus conditions worse. Likewise, sinus medications may make voiding more difficult.
Prevention and Hygiene: Use condoms. Most cases are not associated with any underlying abnormality, however elective evaluation by a urologist may prevent recurrences.
No Improvement/Deterioration: If fever persists beyond 48-72 hours, return promptly. Some prostatitis is slow to resolve.

Follow-up Actions
Return evaluation: Fever can be expected to last 1-3 days.
Evacuation/Consultation Criteria: Aggressively rehydrate unstable patients in shock and evacuate immediately. If urinary frequency and urgency persist after initial 30 days, continue treatment for another 30 days. Stable patients should eventually follow-up with a urologist electively to assess the need for further evaluation.

Symptom: Male Genital Problems: Testis Torsion
CAPT Leo Kusuda, MC, USN

Introduction: Rapid identification and treatment of torsion is necessary to preserve testis function. Such patients require tacking or fixation of the opposite testis to the scrotal skin since it is also at risk. Loss of both testes not only results in sterility but loss of testosterone, which requires lifelong supplementation for normal body function. Salvage of the affected testis can be achieved if reduced in 4 hours. Some salvage of testis function can occur with reduction of the torsion 2 days out, but this is unusual.

Subjective: Symptoms
Acute (< 2 hr): Severe scrotal pain, onset can be at night while asleep, may have prior history of scrotal pain lasting less than 1 day, may have nausea/vomiting, testis is extremely painful, spermatic cord may be tender
Sub-acute (2-48 hr): Scrotal pain increases over several hours. After 24 hours, some of the pain may start to

subside **Chronic (>48 hr):** History of acute onset of pain. Testis pain is improved but not gone.

Objective: Signs
Using Basic Tools: Extreme, diffuse tenderness of the entire testis; edema; vomiting.
Using Advanced Tools: Lab: Urinalysis: Strongly heme or leukoesterase positive sample suggests kidney stone or infection.

Assessment:
Differential Diagnosis
Orchitis - fatigue, muscle aches, sore throat or other flu-like symptoms with gradual onset and normal spermatic cord. Mumps orchitis is extremely rare with modern vaccinations.
Severe epididymo-orchitis - voiding symptoms with leukoesterase and nitrite positive urine.
Testis tumor - considered when there is a mass with mild to moderate pain
Torsion of the appendix testis/epididymis - point tenderness on the superior portion of the testis/epididymis with the remaining testis being non-tender.
Ruptured testis - history of trauma
Spermatic cord torsion - tender testis with a non-palpable vas deferens.
Kidney stone - especially if lodged just below the kidney will present with scrotal pain, but a non-tender, normal scrotal exam.
Incarcerated hernia - most of the discomfort will be above the testis. A hydrocele may be present, making it difficult to examine the testis. If omentum is in the hernia, there may not be any bowel symptoms.

Plan:
Treatment
Primary: Manual detorsion, with or without injection of the spermatic cord with local anesthesia. Torsion may be 180-720° (2 full twists).
1. Attempt to detorse the testis first by rotating the testis outward (like opening a book). If the pain worsens or does not improve, rotate the other direction. The torsion may be 2 full twists. If the testis hangs lower but pain persists, continue untwisting the cord.
2. If **lidocaine** is available, inject into the cord using a long needle or spinal needle in the spermatic cord. This can be accomplished by straddling the cord on the affected side between two fingers just as the cord crosses over the pubic bone lateral and superior to the penis. Make multiple passes through the cord and down to the pubic bone injecting a total of 10 cc of 1% **lidocaine** local anesthetic. This should numb the testis. This may relieve the pain, causing the cremasteric muscles to relax and may result in spontaneous de-torsion. If the pain is gone, check the testis for descent to the normal position and if the testis has become less tense. Also try to palpate the vas deferens posteriorly. This tube is about the consistency and size of uncooked spaghetti and is located behind and is easily separable from the bulk of the spermatic cord. If the tube is in its normal location, spermatic cord torsion is unlikely. You can use the vas deferens as a guide to untwist the cord. The vas deferens should lie posteriorly to the cord. Palpate the vas high in the scrotum and try to follow it down.
3. If unable to detorse the testis, treat with narcotics and empirically with antibiotics (**Cipro** 500 mg po bid or **Keflex** 500 mg po qid or **Septra** DS 1 po bid) until pain has resolved. Most pain from a dead testis will improve after 48 hours. Increasing pain would suggest the presence of infection or testis tumor or rupture of testis (suspect if there is a history of blunt trauma).

Patient Education
General: Wear a scrotal support and await definitive surgery to prevent recurrence.
Activity: Light activity with supporter until problem surgically corrected.
Diet: NPO if surgery is imminent
Prevention and Hygiene: Preventive surgery is required to avoid similar torsion on the opposite side. Examine testis regularly.
No Improvement/Deterioration: Prompt evacuation and surgical correction.

Follow-up Actions
Wound Care: Light activity for 3-4 weeks after surgery.
Return evaluation: If the testis was salvaged, the risk for shrinkage or atrophy increases with the length of time the testis was torsed.
Evacuation/Consultation Criteria: All cases of suspected torsion should be referred. The patient is at risk for torsion on the opposite side as well. He should be evacuated as soon as possible to prevent this calamity, particularly if one testis was not salvaged. The testes both need to be surgically explored and fixed to the scrotal wall to prevent rotation. Loss of both testes results in significant hormonal changes and infertility and should be avoided.

Symptom: Male Genital Problems: Epididymitis
CAPT Leo Kusuda, MC, USN

Introduction: The epididymis, usually located behind each testis, is the site of final maturation and storage of sperm. It can become painful from either mechanical or infectious irritation. Treatment involves both medication and scrotal support which may require strict bedrest in severe cases. Prior vasectomy is a risk factor (epididymal pain can develop 8-10 years post-vasectomy)

Subjective: Symptoms
Pain in the scrotum behind the testis with tenderness of the epididymis, and without pain in the testis.

Objective: Signs
Using Basic Tools: Marked swelling of the hemi-scrotum, urethral discharge, frequent and urgent urination, fever. Use penlight or otoscope to transilluminate the scrotum to differentiate swelling due to a mass vs. fluid (bright, diffuse glow; seen with spermatocele or hydrocele).
Using Advanced Tools: Lab: Urinalysis: nitrite and leukoesterase positive urine (infection). Do urine culture if available and dipstick is positive. Gram stain urethral discharge to screen for gonorrhea and chlamydia.

Assessment:
Differential Diagnosis (see the appropriate topics in this book)
Pain and tenderness in other areas of the scrotum, such as the cord or groin, of equal or greater severity would suggest other causes of the pain such as hernia, varicocele, musculoskeletal pain or entrapped nerve.
An abnormal testis on physical exam may suggest tumor, appendix testis, viral orchitis, testis trauma or testis torsion. Recurrent symptoms lasting less than 1 day are much more suggestive of intermittent testicular torsion.

Plan:
Treatment
Primary:
1. Scrotal support/elevation, bedrest.
2. **NSAIDs** such as **ibuprofen** 800 mg po tid with food.
3. If the urine is nitrite and leukoesterase positive, treat with antibiotics.
   a. If a fever is present and there is no urethral discharge, give:
   Levaquin 500 mg po bid x 10 days or **Septra** DS 1 po bid x 30 days.
   In severe cases **ampicillin** 1 gm IV q6h plus **gentamicin** (loading dose of 1.5 mg/kg followed by 1 mg/kg IV q8h) or **Rocephin** 5 mg/kg IV qd (**ceftriaxone** 500 mg to 1gm IV bid) should be given until fever resolves, then convert to oral antibiotics (above).
   b. If there is a urethral discharge and no fever, give **ceftriaxone** 250 mg IM (or **Cipro** 500 mg po or **Floxin** 400 mg po) single dose to treat gonorrhea, and follow with 7-10 days of **doxycycline** 100 mg po bid (or **Floxin** 400 mg po qd or **Levaquin** 500 mg po qd).
   c. If both fever and urethral discharge are present, treat for disseminated gonorrhea: **ceftriaxone** 1 gm

IV qd (or **spectinomycin** 2 gm IV q 12 hours) until symptoms improve, followed by **Cipro** 500 mg po bid (or **Levaquin** 500 mg po qd) for 7 more days.
  d. If neither fever nor discharge are present, empirically treat for chlamydia: **doxycycline** 100 mg po bid for 7-10 days (or **Floxin** 400 mg po qd for 7-10 days or **Levaquin** 500 mg po qd for 7-10 days).

Patient Education
General: Use condoms.
Activity: Light duty until swelling has resolved.
Diet: Regular
Medications: Be alert for allergic reactions to some antibiotics. Avoid sun exposure if on doxycycline.
Prevention and Hygiene: The epididymis is susceptible to repeated inflammation, so recommend a comfortable athletic supporter for all high-impact activities. If symptoms occur on the other side, be suspicious that the first episode was torsion.

Follow-up Actions
Return evaluation: Follow up in 1 month for reevaluation of the scrotum. Refer patients with progression of symptoms for exploration and possible orchiectomy if significant abscesses are found.
Evacuation/Consultation Criteria: Evacuation is not usually necessary. If there is a persistent mass in the scrotum, refer patient for ultrasound. Consult urologist as needed.

Symptom: Memory Loss
LCDR Christopher Jankosky, MC, USN

Introduction: Memory dysfunction is the most common cognitive problem brought to the clinician and is common in deployed active duty forces. 34% of Persian Gulf veterans indicated in a survey that they had symptoms of memory problems. Determining the etiology of the symptoms and whether they represent a life threatening condition can be difficult. Memory loss may occur in isolation, or may be associated with difficulties in attention, concentration, naming, or language. Sorting out these symptoms is very complex. Head injury is the most common cause of amnesia.

Subjective: Symptoms
Vary widely, including headache, neck pain, weakness, fatigue, delusions, hallucinations, changes in sleep pattern, and physical/emotional stresses.
Focused History: *What is your name?* (rare for a patient to forget his own name, and if this occurs it is unlikely to represent an organic etiology) *Have you had this problem before? Have you had any physical or psychological illnesses in the past?* (can suggest diagnosis, or related illnesses; also tests memory) *Have you had any recent head injuries, infections, fevers, medication change, alcohol use, or toxic exposure?* (can explain etiology of symptoms)

Objective: Signs
Using Basic Tools: Perform vital signs assessment, HEENT exam, and a complete neurological exam, including a Mini-Mental State Examination (MMSE) (see Appendix). Assess memory function in detail. May have obvious head injury, odor of alcohol, fever. Difficulty with immediate recall is generally seen in patients with confusion or delirium. Test this aspect of memory with the simple test of digit span. A normal individual should have a digit span of five or more. Slowly recite to the patient a string of 5 digits, and then immediately ask them to repeat them to you. Document the longest string that he can correctly repeat. Additional tests can include counting backwards from 20, or reciting months in reverse order. Normal patients should be able to perform these tests correctly. A patient with amnesia may have difficulty with the learning ability function of memory. Ask the patient to retain 3 unrelated words for 5-10 minutes, during which you perform other testing. Although an anxious patient may need prompting to recall one of the three words, most patients should be

able to recall all three. Testing of retrieval ability is more difficult because the examiner may not know the patient's fund of knowledge. Testing can include standard information questions such as national leaders and dramatic news events. Knowledge specific to his individual military unit can also be asked. If the patient is able to successfully pass these memory tests, then his primary problem is unlikely to be with memory function. Further assessment of other cognitive functions will need to be performed as indicated by the clinical situation and his response to the MMSE.
Using Advanced Tools: Lab: WBC for infection; pulse oximetry for hypoxia. Papilledema should cause immediate concern for intracranial hemorrhage, swelling, or mass.

Assessment:
Differential Diagnosis
Traumatic brain injury - memory problems following head trauma is common, so patient may not remember the injury. Intracranial bleeding may lead to severe symptoms within minutes, but may also progress slowly over hours to days, requiring close follow-up.
Infection - herpes simplex encephalitis is the most common infection causing predominantly memory problems. There is usually associated behavioral deviations, disorientation, seizures, or weakness.
Stroke (ischemic from blocked vessel, or hemorrhagic from bleed) - uncommon in young individuals without other neurologic signs or symptoms. Ask about stroke risk factors: hypertension, smoking, diabetes, positive family history, hyperlipidemia, oral contraceptives, binge alcohol drinking, atrial fibrillation, and coronary heart disease.
Seizure disorder - rarely manifested as isolated memory and cognitive problems. (See Neurology: Seizure Disorders)
Hypoxia - can result in permanent damage to the memory systems of the brain. History may reveal a recent hypoxic event.
Inflammatory - consider multiple sclerosis or CNS sarcoidosis, although these rarely cause isolated memory problems.
Transient global amnesia - uncommon in young adults; no associated symptoms, and the condition resolves within 24 hours
Migraine - memory problems are transient (hours), and usually associated with a headache; prior history of migraines.
Metabolic (such as thiamine deficiency), toxic, degenerative, or neoplastic causes may also be considered
Psychiatric - causes can include sleep deprivation, stress, anxiety, and depression. Psychiatric etiologies usually present as a retrograde amnesia. Other psychiatric symptoms are usually present. This category should be strongly considered if the patient does not know his own name.
Decompression Sickness history of recent diving and/or flying; may have other associated neurological findings, including subtle (soft) signs

Plan:
Treatment
1. Secure all weapons.
2. Provide supportive treatment: rest, fluids, reassurance, and observation will improve or stabilize many conditions.
3. Closely supervise military personnel who return to duty. If given significant independent responsbilities, they can be a danger to themselves or others if memory problems remain.
4. Herpes simplex encephalitis has a high mortality and demands early and aggressive treatment with IV antiviral medication, such as **acyclovir** if available.
5. Intracranial bleeding, swelling, or mass requires neurosurgical intervention.

Patient Education
General: Significant memory problems may be due to an organic or psychiatric etiology.
No Improvement/Deterioration: Return for evaluation daily initially, as symptoms may worsen rapidly in some illnesses.

Follow-up Actions
Return evaluation: If traumatic head injury occurred and symptoms worsen over days, suspect a slow intracranial bleed. Medically evacuate ASAP. Repeat funduscopic exams to look for papilledema (an indication of increased intracranial pressure) on follow up exams.
Evacuation/Consultation Criteria: Evacuate if patient unable to function, if unstable or if deteriorates. Consult psychiatry or neurology if needed.

NOTE: Definitive diagnosis may require lumbar puncture, neuroimaging, EEG or use of specialty consultation.

Symptom: OB Problems: Pregnancy
MAJ Marvin Williams, MC, USA

Subjective: Symptoms
In women with regular menstrual cycles, a history of one or more missed cycles (periods) is suggestive of pregnancy. Associated symptoms include fatigue, nausea/vomiting, breast tenderness, frequent urination (caused by the enlarged uterus compressing the bladder), and "quickening" (first movements of a fetus felt in utero at 16-20 weeks). Calculate the EDC (estimated date of confinement) by adding 7 days to the first day of the last normal menses and subtracting 3 months.

Objective: Signs
Spider angiomata (branched capillaries on the skin, shaped like a spider) and blotchy or patchy palmar erythema (more than 50-60% of patients), regress after delivery.
Striae gravidarum (stretch marks) develop in 50% of pregnant woman.
Hyperpigmentation of nipples, areola, umbilicus, axillae, perineum and midline of lower abdomen (linea nigra).
Breast enlargement due to increased hormone levels, which later causes release of colostrum (thin, yellowish fluid seeping from the nipple) and lactation.
Other common signs and symptoms include lightheadedness, backache, dyspnea, urinary symptoms (frequency, urgency, and incontinence), hemorrhoids, heartburn, ankle swelling, varicose veins, abdominal cramping and constipation.
Using Basic Tools: Fetal heart tones can be heard via auscultation (bell side of stethoscope) at or beyond 18-20 weeks of gestation.
Using Advanced Tools: Pregnancy tests can be used in the field. Most are nearly as accurate (97-99%) as a laboratory test on serum.

Assessment:
Palpation of fetal parts and the appreciation of fetal movement and heart tones are diagnostic.

Plan:
Treatment: See appropriate sections in this chapter.

Symptom: OB Problems: Vaginal Delivery
MAJ Marvin Williams, MC, USA

When:
Labor is defined as progressive dilation of the uterine cervix in association with repetitive uterine contractions resulting in complete dilation (10 cm) and effacement (thinning) of the cervical os. Normal labor is a continuous process.
First stage: Onset of cervical changes and uterine contractions through full dilation and effacement of cervix
Second Stage: Full cervical dilation and delivery of the infant
Third Stage: Interval between the delivery of the infant and delivery of the placenta

Fourth Stage: Recovery of the uterus after delivery of the placenta

What You Need: 1% **Lidocaine** without epinephrine (approx. 20-30cc), sterile gloves (several pairs), gauze-bandages and prep solution, 2-0, 3-0, and 4-0 absorbable suture (Vicryl or Chromic) for the repair, scissors or a scalpel to make the episiotomy incision, sterile (or clean) towels, suction device (bulb syringe), suture forceps, and needle holders

What to Do:
During the First Stage of Labor:
1. Assess mother's gestational age by asking the date of the first day of her last normal menstrual period, subtracting 3 months and adding 7 days (40 weeks +/- 2 weeks). Palpate and measure the height of the uterine fundus (top) from the pubic bone. Up to about 36 weeks, this distance in centimeters approximates gestational age (i.e., 32 cm from pubic bone to top of fundus equals approx. 32 weeks gestational age). If not >36 weeks by dates or measurement, see Preterm Labor section.
2. Listen for fetal heart tones. Normal rate is approximately 160 beats/minute. Periodically reassess during labor. The rate normally decreases during contractions but should recover.
3. Encourage the mother to walk as gravity and motion will encourage cervical dilation.
4. Read this section and others related to birth (Episiotomy, Breech Delivery, etc.).
5. Periodically assess progression of labor by timing contractions. Eventual delivery is likely when contractions are <3 minute apart.
6. Check the birth canal with a sterile gloved hand once before birth to ensure the cervix is fully dilated and effaced.

During the Second Stage of Labor: (see figures 3-6 & 3-7)
1. As the cervix progresses to complete dilation, place the patient in the dorsal lithotomy position (patient is on her back with her thighs flexed on the abdomen). Women from some cultures may prefer to squat.
2. With each contraction the patient should be urged to push and the care provider should perform perineal massage.
3. As the fetal head crowns (i.e., distends the vaginal opening), consider an episiotomy if massage has failed to stretch the tissue adequately. An episiotomy facilitates delivery of a large infant, or one with shoulder dystocia. It is better to cut an episiotomy than to have the baby tear the perineal tissue into the rectum. See Episiotomy Procedure guide.
4. As crowning continues, it is very important to support the fetal head via a modified Ritgen maneuver. This is accomplished by placing one hand over the fetal head while the other exerts pressure through the perineum onto the fetal chin. Use a sterile towel to avoid contamination of this hand by the anus.
5. After the head is delivered, suction the nose and mouth with the bulb syringe.
6. Check the neck for the presence of a umbilical cord around it, which should be reduced if possible. If the cord is too tight, it should be doubly clamped and cut.
7. Place your hands on the chin and head, applying gentle downward pressure, delivering the anterior shoulder. Avoid injury to the brachial plexus; avoid excessive pressure on the neck.
8. Deliver the posterior shoulder by upward traction on the fetal head.
9. Delivery of the body should occur easily (see figure 3-8).
10. Cradle the fetus in your arms, suction once again and the umbilical cord is clamped and cut. If no clamps are available, suture may be used.
11. To avoid significant heat loss, dry the newborn completely and wrap in towels or blankets.
12. If the mother (or another person) can hold the baby safely, put the baby to the mother's breast. This will help the uterus contract.

During the Third Stage of Labor:
1. Soon after delivery of the infant, the placenta will follow. Placental delivery is iminent when the uterus rises in the abdomen, the umbilical cord lengthens, and a "gush" of blood is noted. AVOID excessive traction on the umbilical cord to avoid uterine inversion (pulling the uterus "inside-out") which will cause profound blood loss and shock. Instead, wait up to 30 min. for spontaneous delivery of the placenta.

Figure 3-5

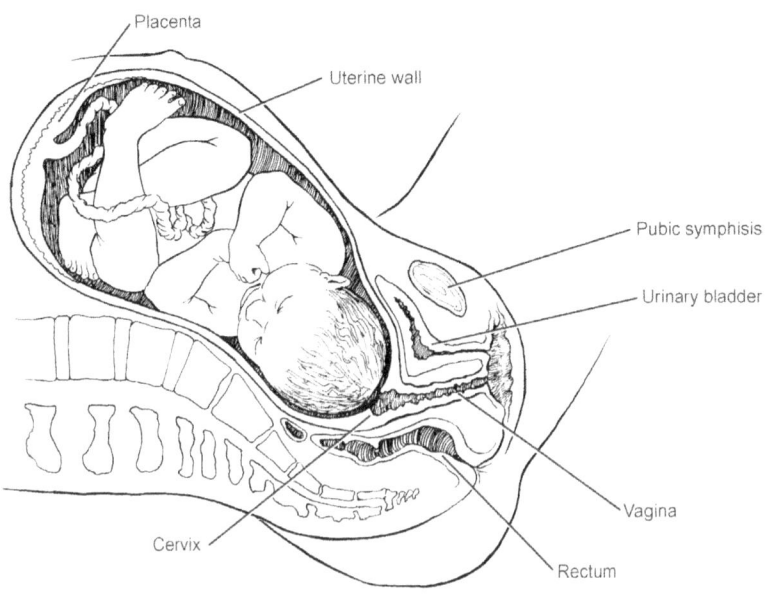

Normal delivery

Figure 3-6

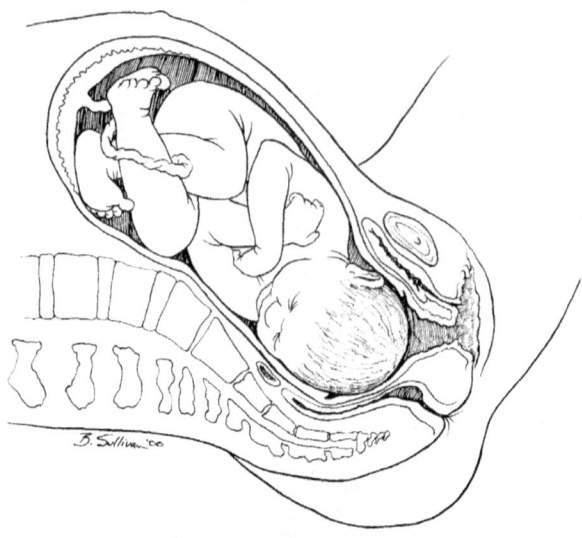

Normal delivery

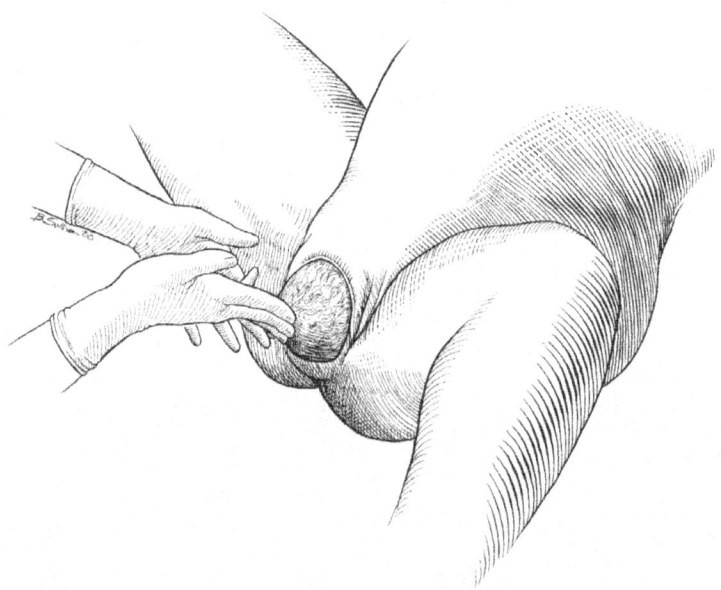

Normal delivery, Crowning of the Head

Figure 3-7

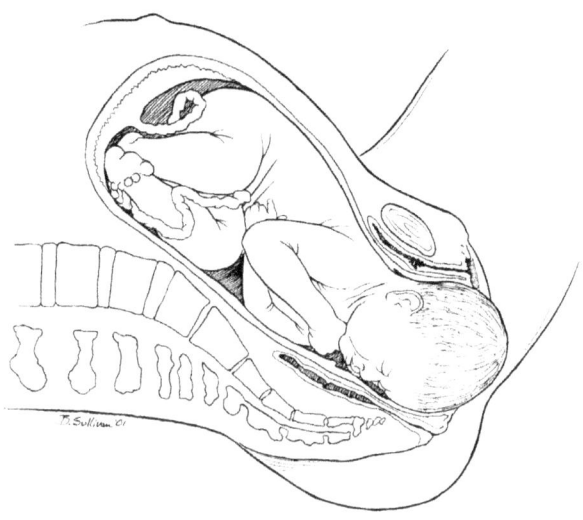

Normal delivery, Crowning of the Head

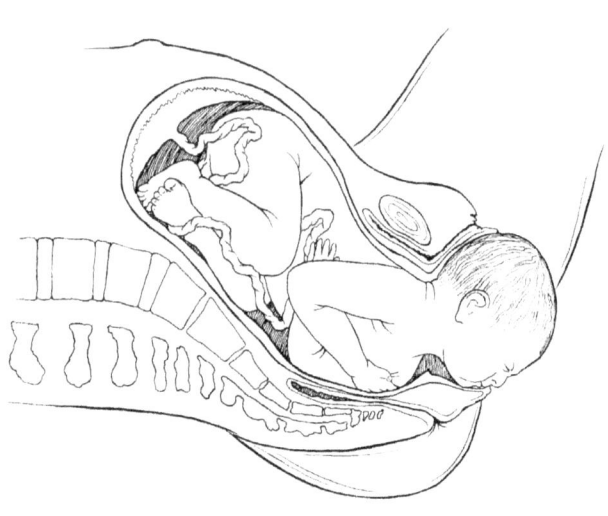

Normal delivery

Figure 3-8

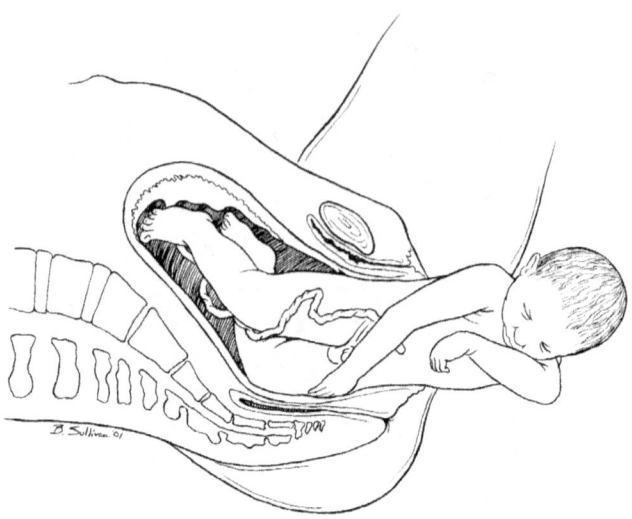

Normal delivery

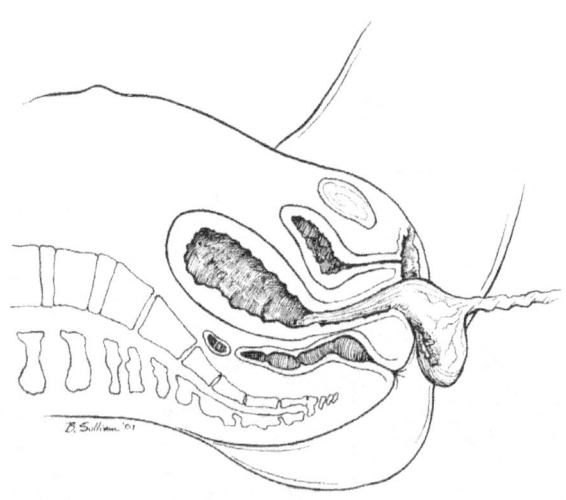

Normal delivery of Placenta

(see fig. 8)
2. If spontaneous placental separation does not occur, remove the placenta manually. Pass a gloved hand into the uterine cavity and gently apply traction to the umbilical cord, using the side of the hand to develop a cleavage plane between the placenta and the uterine cavity.
3. Inspect the placenta to ensure it is complete. Inspect the cord for the presence of the expected two umbilical arteries and one umbilical vein.
4. After the delivery of the placenta, palpate the uterus to ensure that it has reduced in size and become firmly contracted.
5. Inspect the birth canal in a systematic fashion. Evaluate lacerations of the vagina and/or perineum and extensions of the episiotomy and repair if necessary (refer to Episiotomy section).

During the Fourth Stage of Labor:
1. The likelihood of serious postpartum complications is greatest in the first hour or so after delivery. Palpate the uterus to ensure that it is firm. Repeat uterine palpation through the abdominal wall frequently during the immediate postpartum period to ascertain uterine tone.
2. Monitor pulse, blood pressure and the amount of vaginal bleeding every 15 minutes for the first hour, then every 30 min. for 3 hrs after delivery to identify excessive blood loss.
3. Manage pain with NSAIDS or low dose narcotics (i.e., **Motrin** 800 mg 1 tablet tid with food or **Tylenol #3** 1-2 tablets q 4-6 hrs).
4. Apply ice to the perineum for 20-30 minutes every 4-6 hrs to decreasing swelling after the delivery.

What Not to Do:
Do not contaminate the birth canal prior to birth.
Do not allow the episiotomy to tear into the rectum.
Do not forget to reduce umbilical cord.
Do not forget to suction the baby's nose and mouth.
Do not forget to clamp and cut the umbilical cord.
Do not forget to clean and dry the baby.
Do not forget to deliver the placenta.
Do not forget to repair an episiotomy and any tears.
Do not forget to monitor the mother's vital signs.

Symptom: OB Problems: Preterm Labor (PTL)
MAJ Marvin Williams, MC, USA

Introduction: Preterm birth as a result of preterm labor is the most common cause of infant morbidity and mortality. Complications include respiratory distress syndrome (RDS), intraventicular hemorrhage (IVH), necrotizing enterocolitis, sepsis, and seizures. Long term morbidity associated with PTL and delivery includes chronic lung problems (bronchopulmonary dysplasia) and developmental abnormalities. Preterm labor is defined as regular uterine contractions occurring with a frequency of 10 minutes or less between 20 and 36 weeks gestation, with each contraction lasting at least 30 seconds. When contractions are accompanied by cervical effacement (thinning), dilation (opening), and/or descent of the fetus into the pelvis, it becomes increasingly difficult to stop labor. The cause of preterm labor is unknown but many factors have been associated with it and some include: dehydration, rupture of membranes, infections, uterine enlargement (twins), uterine distortion (fibroids), and placental abnormalities (previa and abruption), smoking and substance abuse. Approximately 10-15% of women will rupture the amniotic membrane around the fetus >1 hour prior to the onset of labor. This is called premature rupture of membranes, or PROM. After PROM, labor must begin promptly or infection will develop as bacteria ascend through the birth canal. Delivery should be completed within 24 hours of PROM to avoid infection in fetus and mother. One of the most important causes of early onset neonatal infection is group B streptococcus (GBS).

Subjective: Symptoms
Menstrual-like cramps, low back pain, abdominal or pelvic pressure, painless uterine contractions, and

increase or change in vaginal discharge (mucus, watery, light bloody discharge).

Objective: Signs
Using Basic Tools: Palpable uterine contractions; palpable cervical dilation and effacement
Using Advanced Tools: Lab: Urinalysis and a saline wet preparation to evaluate for bacterial vaginosis. Fern Test (amniotic fluid dries on a slide to resemble the leaves of a fern plant) to assess for PROM.

Assessment: Assess for PTL risk factors, and treat those found.
Differential Diagnosis
Low back pain/spasm - palpate for back spasm; evaluate for associated neurological symptoms (leg tingling, radiation, etc.)
Infection - evaluate for urinary tract infection (urinalysis, fever), GI infection (diarrhea, fever) and vaginosis (saline wet prep).
Ureter/kidney stone - evaluate flank pain, fever; perform urinalysis
True labor - verify dates of last menstrual period.
PROM - history of vaginal gush or leak, positive Fern Test
Constipation - history of infrequent bowel movements; retained fecal material
Diarrhea - infection (above), food poisoning, and others.

Plan:
Treatment
1. Arrest labor:
Primary: Magnesium sulfate, 4 gm loading dose over at least 5 minutes, followed by 2 gm/hour in a steady IV drip. Watch for magnesium toxicity with diminished reflexes and respiratory depression. Counteract with calcium gluconate 1 gm slow IV push over 2-3 minutes if magnesium toxicity is encountered.
Alternates: Terbutaline 0.25 mg SQ, q 1-4 hours x 24 hours, total dose not to exceed 5 mg in 24 hours. May also be given po in 2.5 - 7.5 mg doses q 1.5 - 4 hours. Target maternal pulse rate is > 100 and < 120 BPM
Indomethacin (Indocin) 50 mg po (or 100 mg PR), followed by 25 mg po q 4-6 hours for up to 48 hours. Watch for gastric bleeding, heartburn, nausea and asthma.
NOTE: Contraindications to Tocolytic Therapy:
Maternal: Significant hypertension (see Preeclampsia section), antepartum hemorrhage, cardiac disease.
Fetal: Gestational age > 37 wks, fetal death, chorioamnionitis (intrauterine infection)
2. Prevent infection (PROM; PTL > 12 hours; history of UTI, vaginal infections in last 2 weeks):
Primary: Penicillin 5 million units IV initially, then 2.5 million units q 4 hrs until delivered
Alternates: Ampicillin 2 g IV q 4-6h or **Rocephin** may also be used, 1 gram IV q 12 hrs.
For those patients who are penicillin allergic, **clindamycin** 600 mg q 6 h or 900 mg q 8 h or **erythromycin** 1-2 g q 6 h or vancomycin 500 mg q 6 h or 1000 mg q 12 h.
3. Help fetus mature: After postponing delivery, many fetuses less than 34 weeks gestation will benefit from administering steroids to the mother. The effect of the steroids on the fetus is to accelerate fetal lung maturity, lessening the risk of respiratory distress syndrome at birth.
Dexamethasone 6 mg IM q 12 hours x 4 doses.
4. Evacuate the mother: Keep her rolled over on her left or right side, with a pillow between her knees, with an IV securely in place. If IV access is lost during a bumpy truck or helicopter ride, it will be nearly impossible to restart it without stopping or landing. Consider tocolytic therapy in all mothers being transported unless contraindications exist or greater than 37 weeks gestation.

Patient Education
Activity: Decreased activity and bedrest may be required to avoid further PTL.
Prevention and Hygiene: Avoid emotional stress.
No Improvement/Deterioration: Report recurrent symptoms immediately. Early intervention is more effective in stopping PTL.

Follow-up Actions
Return evaluation: Follow weekly to assess for recurrent symptoms and risk factors for PTL.
Evacuation/Consultation Criteria: Evacuate after initial PTL symptoms if possible. Avoid emergent evacuation if possible. Consult expert if available.

Symptom: OB Problems: Relief of Shoulder Dystocia
MAJ Marvin Williams, MC, USA

What: Shoulder dystocia is a labor complication caused by difficulty delivering the fetal shoulders. Although this is more common among women with gestational diabetes and those with very large fetuses, it can occur with babies of any size. Unfortunately, it cannot be predicted or prevented. Improperly relieving the dystocia can result in unilateral or bilateral clavicular fractures.

When: After delivery of the head, the fetus seems to try to withdraw back into the birth canal (the "Turtle Sign"). Further expulsion of the infant is prevented by impaction of the fetal shoulders within the maternal pelvis. Digital exam reveals that the anterior shoulder is stuck behind the pubic symphysis. In more severe cases, the posterior shoulder may be stuck at the level of the sacral promontory.

What To Do:
1. Check for nuchal cord (the umbilicus wrapped around the baby's neck) and relieve it.
2. Suction the infant's mouth to clear the airway of amniotic fluid and other debris.
3. Do not apply excessive downward traction on the head to get the baby out. This action can injure the nerves in the neck and shoulder (brachial plexus palsy) and must be avoided. While most of these nerve injuries heal spontaneously and completely, some do not.
4. Otherwise cut a generous episiotomy following proper technique (see Episiotomy Procedure in this chapter) unless a spontaneous perineal laceration has occurred, or if the perineum is very stretchy and offers no obstruction.
5. Initially apply gentle downward traction on the chest and back initially to try to free the shoulder. If this has no effect, do not exert increasing pressure. Try some alternative maneuvers to free the shoulder.
6. Place the mother in the MacRobert's position, and apply gentle downward traction on the baby again. Maneuver involves flexing the mother's thighs tightly against her abdomen. This can be done by the woman herself or by assistants. By performing this maneuver, the axis of the birth canal is straightened, allowing a little more room for the shoulders to slip through.
7. If the MacRobert's maneuver fails have an assistant apply downward, suprapubic (above the bony pubic arch) pressure to drive the fetal shoulder downward, to clear the pubic bone. Again apply coordinated, gentle downward traction on the baby.
8. If pressure straight down is ineffective, have the assistant apply it in a more lateral direction. This tends to nudge the shoulder into a more oblique orientation, which usually provides more room for the shoulder. Again apply coordinated, gentle downward traction on the baby.
9. Often, the baby's posterior arm has entered the hollow of the sacrum. Reach in posteriorly, identify the posterior shoulder, follow the humerus down to the elbow and identify the forearm. Grasping the fetal wrist, draw the arm gently across the chest and then sweep the arm up and out of the birth canal, freeing additional space and allowing the anterior shoulder to clear the pubic bone. Once again apply coordinated, gentle downward traction on the baby.
10. An electric light bulb cannot be removed by simply pulling it out- it must be unscrewed. This concept can be applied to shoulder dystocia problems. Rotate the posterior shoulder, allowing it to come up outside of the subpubic arch. At the same time, bring the stuck anterior shoulder into the hollow of the sacrum. Continue rotating the baby a full 360 degrees to rotate (unscrew) both shoulders out of the birth canal. Two variations on the unscrewing maneuver include:
    - Rotating/shoving the shoulder towards the fetal chest ("shoving scapulas saves shoulders"), which compresses the shoulder-to-shoulder diameter, and
    - Rotating the anterior shoulder first rather than the posterior shoulder. The anterior shoulder may be

easier to reach and simply moving it to an oblique position rather than the straight up and down position may relieve the obstruction.
11. Applying fundal pressure in coordination with the other maneuvers may, at times, be helpful. Applied alone, it may aggravate the problem by further impacting the shoulder against the pubic symphysis.
12. If these measures fail, return to number 4 above and consider cutting or extending the episiotomy, then progress through these maneuvers again. Do not attempt a C-Section. Since the baby's head is delivered, there should be no need to do so.

Symptom: OB Problems: Breech Delivery
MAJ Marvin Williams, MC, USA

Breech presentation is an abnormality in which the buttocks or legs of the fetus, rather than the head, appear first in the birth canal. There are several breech variations, including buttocks first, one leg first or both legs first. "Frank breech" means the buttocks are presenting and the legs are up along the fetal chest—the safest position for breech delivery. In any breech birth there are increased risks of umbilical cord prolapse and delivery of the feet through an incompletely dilated cervix, leading to arm or head entrapment. These risks are greatest when a foot is presenting ("footling breech").

When: Because of the risks of breech delivery, many breech babies are born by cesarean section (see Cesarean section template) in developed countries. In operational settings, cesarean section may not be available or may be more dangerous than performing a vaginal breech delivery. It is up to the care team to decide which option will be the safest mode of delivery for both mother and infant.

What To Do: (see figures 3-9 through 3-11)
1. The simplest breech delivery is a spontaneous breech. The mother pushes the baby out with normal bearing down efforts and the baby is simply supported until it is completely free of the birth canal. These babies essentially deliver themselves. This works best with smaller babies, mothers who have delivered in the past or frank breech presentation.
2. If a breech baby gets stuck halfway out or if you need to speed the delivery, perform an "assisted breech" delivery. It is very helpful to have a second person assist you.
3. A generous episiotomy will give you more room to work, but may be unnecessary if the vulva is very stretchy and compliant.
4. Grasp the baby so that your thumbs are over the baby's hips. Rotate the torso so the baby is face down in the birth canal. A towel can be wrapped around the lower body to provide a more stable grip.
5. Have your assistant apply suprapubic pressure to keep the fetal head flexed, expedite delivery and reduce the risk of spinal injury.
6. Exert gentle outward traction on the baby while rotating the baby clockwise and then counterclockwise a few degrees to free up the arms.
7. If the arms are trapped in the birth canal you may need to reach up along the side of the baby and sweep them one at a time, across the chest and out of the vagina.
8. It is important to keep your hands low on the baby's hips. Grasping the baby above the hips could easily cause soft tissue injury to the abdominal organs including the kidneys.
9. During the delivery, always keep the baby at or below the horizontal plane or axis of the birth canal. If you bring the baby's body above the horizontal axis, you risk injuring the baby's spine. Only when the baby's nose and mouth are visible at the introitus is it wise to bring the body up.
10. At this stage, the baby is still unable to breathe and the umbilical cord is likely occluded. Without rushing, move steadily toward a prompt delivery. Place a finger in the baby's mouth to control the delivery of the head. Try not to let the head "pop" out of the birth canal. A slower, controlled delivery is less traumatic.

What Not To Do:
Do not assist too early. Only intervene if a breech baby gets stuck part way out of the pelvis.
Do not place hands too high on the abdomen. Keep hands low on the baby's hips.
Do not raise baby above the horizontal plane until the nose and mouth are delivered.

Figure 3-9

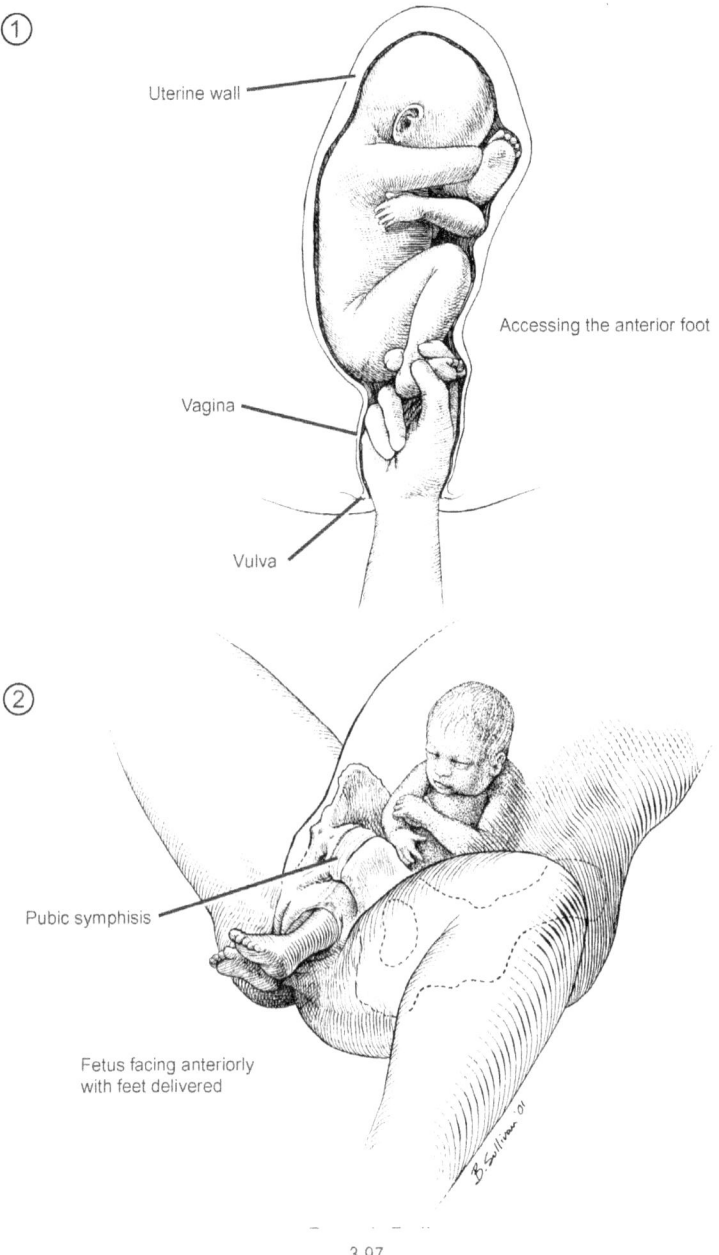

Figure 3-10

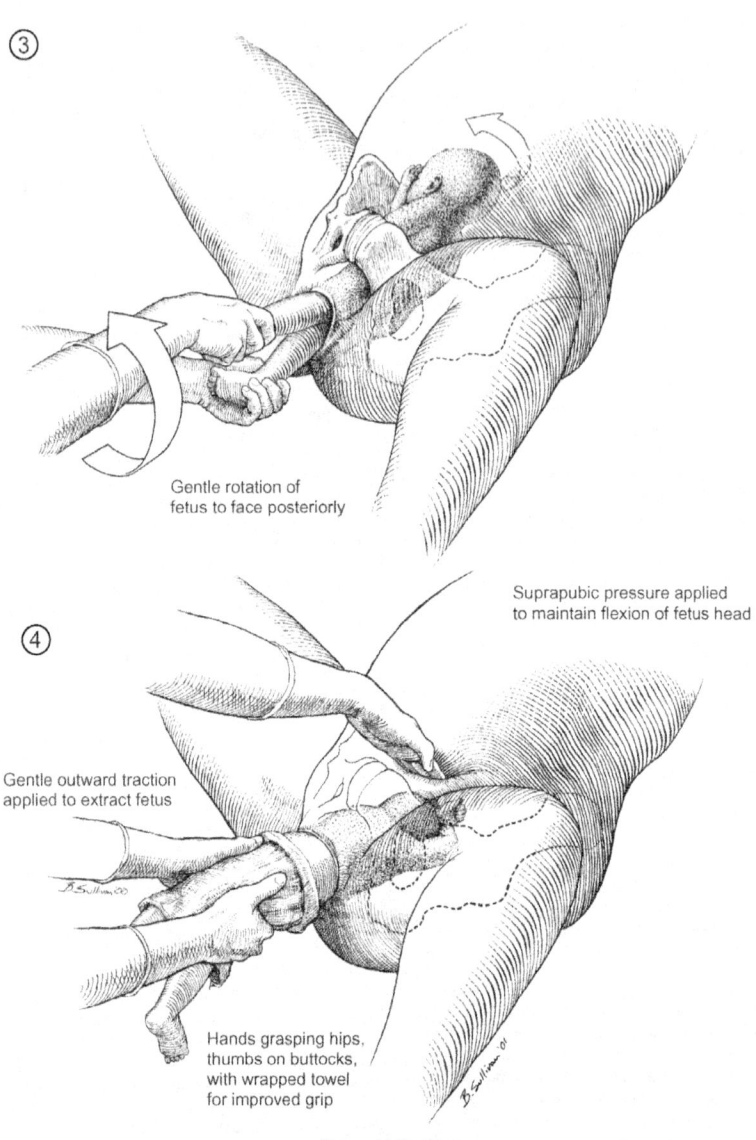

Breech Delivery

Figure 3-11

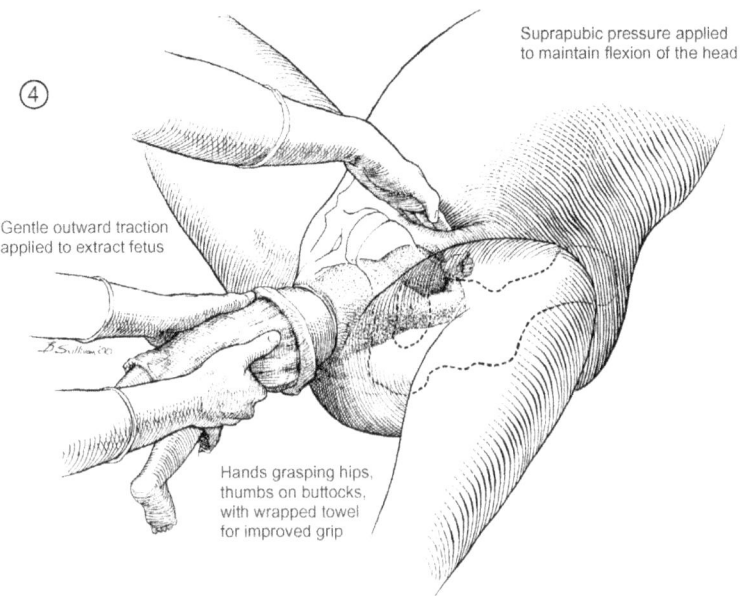

Breech Delivery

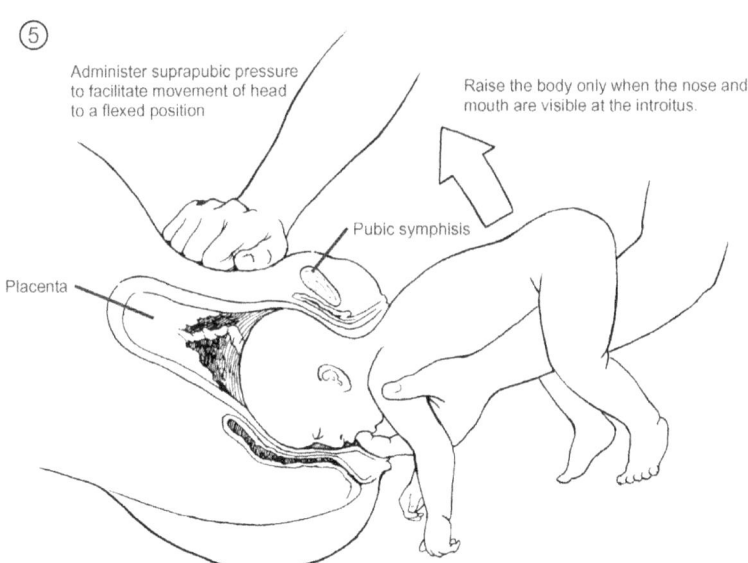

Wigand Maneuver for Breech Delivery of the Head

Symptom: OB Problems: Cesarean Section
MAJ Marvin Williams, MC, USA

What: The delivery of a fetus by abdominal surgery (laparotomy) requiring an incision through the uterine wall (hysterotomy).

When: Perform this procedure only when it is absolutely necessary, and is the only life saving measure for mother or infant! The decision to perform a C-Section must be based on the health and stability of the mother and fetus. Recognize that performing the procedure in the field as an untrained provider is extremely dangerous and will likely result in significant morbidity or mortality for both mother and infant. If a C-Section is anticipated, prepare equipment and read this material, since the procedure must often be performed emergently when vital signs become unstable.

The following are relative indications for cesarean delivery under field conditions:

1. **Fetal Complications:**
    a. Non-vertex (not head first) such as a transverse lie or breech presentation. Attempt vaginal delivery first. (See Breech Birth section).
    b. Multiple gestation: triplets or greater, twins in which the first twin is not head first (vertex). Attempt vaginal delivery of at least one fetus. (See Breech Birth section).
    c. Large Fetus: Attempt vaginal delivery first.
    d. Fetal distress: Fetal heart rate (<90 bpm) for more than 1 minute.
2. **Placental Complications:**
    a. Placenta Previa (placenta lies over cervical os)
    b. Placental Abruption (premature placental separation).
    c. Although these conditions cannot be diagnosed in the field, any large vaginal hemorrhage during labor, or hemorrhage accompanied by fetal distress should be reason to suspect them and consider C-Section.
3. **Uterine Complications:**
    a. Previous cesarean section with midline incision (classical).
    b. Other surgery (e.g., for fibroids) where the uterine cavity was entered.
    There is an increased risk for uterine rupture in this situation. C-Sections with low transverse incisions are much less risky. Attempt vaginal delivery first.
4. **Vaginal Complications:**
    a. Obstructive conditions (e.g., genital warts, cervical cancer). If noted, these conditions may hinder dilation of the cervical os and descent of the fetus through the birth canal. Attempt vaginal delivery first.
    b. Vaginal Infections (e.g., Group B Strep, genital herpes). Delivering the infant through the birth canal will greatly increase the baby's risk of contracting these infections, which will likely be deadly.
5. **Maternal Complications:** Any medical condition that worsens during labor in which a delayed delivery would harm the mother (e.g., eclampsia).
6. **Abnormal labor (failure to progress):** Labor that does not progress (continued cervical changes and fetal descent) over several hours endangers the fetus by increasing the risk of cord compression and neurological damage.

What You Need:
Surgical assistants (two if possible); prep solution; sterile gloves; 0 (Vicryl and Chromic), 2-0 non-absorbable, and 3-0 or 4-0 Vicryl absorbable suture; scalpel to make the incision; IV access; anesthesia: (See Anesthesia: Total Intravenous Anesthesia [TIVA] procedure), or use 1-2% **Lidocaine** with **Epinephrine** ONLY if TIVA is not available. Do not exceed 300cc or 5 mg/kg (usually requires 30-50cc be aware of lidocaine toxicity). One gram of **Ancef**, **Rocephin** or any equivalent IV antibiotic; administer **Oxytocin** 20-40 units, 30-45 minutes prior to surgery if possible; sterile (at least clean) bandages; sterile surgical instruments: needle driver, forceps, retractors, clamps, scissors and a Foley catheter if available.

What to Do:
Surgical Procedure
1. Place the patient in supine position with a roll under her left side (leftward tilt for uterine displacement).
2. Prep her with some sort of cleaning solution (**Betadine** or equivalent), and catheterize her with a Foley catheter, if available.
3. Provide anesthesia. If TIVA sedation is not available, use **Lidocaine** to infiltrate just below the skin and into the subcutaneous tissue following a vertical pattern from 2-3 cm above the pubic bone to 1-2 cm below the umbilicus.
4. Take the scalpel blade and make a vertical incision beginning 1-2 cm below the umbilicus to approximately 2-3 cm above the pubic bone. The incision should initially cut through the skin and some of the subcutaneous tissue (fat). Carefully cut through the remaining fat with shallow strokes. Be careful not to cut directly through the uterus!
5. Once you reach the rectus fascia (shiny white tissue), make a shallow midline vertical incision through it, being careful not to injure the rectus muscles. Have your assistant (if you have one) elevate the fascia with a retractor while you cut. Separate the abdominal muscles in the midline.
6. The next layer you encounter is the peritoneum, a clear, thin layer of tissue. If you look closely, you may be able to see bowel (intestines) through it. Pick up the peritoneum with forceps or a clamp; with scissors, make a small incision into the peritoneum. Extend the incision vertically (both superiorly and inferiorly) exposing the abdominal contents. Make sure you can visualize everything before cutting to avoid potential injury to bowel or bladder.
7. Visualize the uterus and notice the shiny peritoneal surface located on the lower aspect of it (uterovesical peritoneum). Grasp and elevate this area in the midline with forceps or clamp. Make a small incision laterally. With your fingers, bluntly dissect the peritoneum off of the uterus, creating a bladder flap, which decreases the chance of injury to the bladder. After the bladder flap has been developed, use a retractor to retract the bladder anteriorly and inferiorly to facilitate exposure of the intended incision site.
8. Make an incision at the inferior margin of the lower segment of the uterus and insert your first 2 fingers toward the fundus (top of the uterus). QUICKLY extend the incision toward the fundus by cutting between the spread fingers (vertical incision) with scissors (or a scalpel). Have an assistant frequently suction or wipe the area with each cut to help you visualize the incision. Be careful not to cut the infant! Perform the remainder of steps prior to giving oxytocin quickly to avoid massive blood loss.
9. Remove all retractors and insert a hand into the uterine cavity to elevate and flex the fetal head through the incision. Should the head be deeply wedged into the pelvis, an assistant can apply upward pressure through the vagina to dislodge the head.
10. Once the head is present through the incision, suction the infant's nose and mouth.
11. When suctioning is complete, deliver the baby by applying moderate fundal pressure on the uterus from the abdomen.
12. Doubly clamp the cord and hand the infant off to your assistant (see section on Vaginal Delivery for care of the newborn).
13. Following delivery of the newborn, 20-40 units of **oxytocin** can be mixed with a 1 L bag of Lactated Ringer's and run via IV (do not exceed more than 500-600 cc/hr with the first bag. Decrease the rate to 100-125 cc/hr for the second bag).
14. Remove the placenta manually by applying gentle traction on the cord until the placenta is expelled from the uterus. Exteriorize (lift it out of the abdominal incision) the uterus and place it on the abdomen. Cover the top of the uterus (fundus) with a sterile, moist sponge (bandage). Use a sterile, dry sponge to wipe the uterine cavity clean of all clots and placental debris. Do not expose the uterus or other intraperitoneal structures to any non-sterile objects, if at all possible.
15. Inspect the uterine incision and control any bleeding points temporarily with clamps (Ring/Sponge forceps or Allis clamps). Then take a number 0 suture and begin just inferior to the lower margin of the incision, tie your suture and with subsequent stitches, run them toward the fundus in a continuous locking manner (see figure). Place stitches 1 cm from the edge of the incision, 1 cm apart and attempt to keep them out of the uterus. Use a second inverting layer only if hemostasis is not obtained with the first layer. The bladder peritoneum need not be closed.
16. Inspect for any bleeding from the incision, and control it with interrupted figure-of-eight stitches. The

uterus can then be returned to the abdomen. Irrigate the pelvis and lower abdomen with at least 1 L of sterile fluid. Make sure that all sponges (bandages) and needle counts are correct or have been accounted for prior to closing the abdomen.
17. There is no need to reapproximate the peritoneum. Using 0 Vicryl (not chromic due to its inability to maintain tensile strength) begin closing the fascia. The initial suture should be placed inferior to the lower margin of the vertical incision and in a running fashion, close the fascia. Irrigate and inspect the subcutaneous tissue for bleeding. If there is significant fat tissue, reapproximate the subcutaneous tissue with several interrupted stitches with 3-0 or 4-0 Vicryl. Close the skin using staples (if applicable) or 2-0 non-absorbable sutures in an interrupted fashion. Apply a sterile dressing and leave it in place for approximately 24 hrs.
18. If the bladder is inadvertently lacerated, use a two-layer technique to close it (running layers like the fascia) as well as leaving the Foley in for 7-10 days. Put the patient on a prophylactic antibiotic (**Macrobid** or **Zithromax**) while the Foley catheter is in place.
19. Counsel the patient that she will not be a future candidate for a trial of labor. SHE WILL ALWAYS HAVE TO HAVE A CESAREAN SECTION FOR EACH SUBSEQUENT PREGNANCY, or risk uterine rupture and death for both mother and fetus.

Post-Operative Orders:
1. Diet: NPO except sips of water. May begin clear liquids 24 hrs after surgery if have bowel sounds, then advance diet as tolerated.
2. List Allergies to Medications
3. Initial vital signs (BP, P, RR) every 15 minutes for the first hr., then VS every 2 hrs X 2 then every 4 hrs X 72 hrs.
4. Bedrest for 8 hrs, then out of bed with assistance.
5. Turn, cough and deep breath every 2 hrs while awake.
6. Ice pack to the incision every 4-6 hrs for 30-45 minutes
7. Strict monitoring and recording of intake and output (fluids).
8. Leave the Foley catheter in place for 24 hrs to monitor urine output. A patient should make at least 30 cc/hr.
9. **Demerol** 25-50 mg IM every 3-4 hrs for pain
10. **Phenergan** 25 mg IM every 8 hrs for nausea and vomiting
11. Once patient tolerating clear liquid diet well, remove IV, change IM pain management to oral, and remove Foley catheter.
12. **Motrin** 800 mg po q 8 hrs with food
13. **Hydromorphine** or equivalent 1-2 tabs po every 3-4 hrs for pain.
14. CBC 6-8 hrs post op and again at 24 hrs post-op to check for bleeding (low HCT), infection (increased WBC, bands).
15. Remove stitches at 7 days

What Not To Do:
Do not get too excited. It will impede decision-making.
Do not cut into the intestines, bladder or baby.
Do not forget to always retract the bladder once you have developed your flap so as not to injure it.
Do not expose the uterus or other intraperitoneal structures to any non-sterile objects if at all possible.
Do not operate too slowly. Once the uterus is entered the baby must be delivered and the uterus closed quickly to achieve control of bleeding before significant hemorrhage endangers the mother's survival.

Symptom: OB Problems: Episiotomy and Repair
MAJ Marvin Williams, MC, USA

What: Incision of the perineum to enlarge the vaginal opening, and subsequent repair.

When: At the time of delivery, perform an episiotomy when 3-4 cm of fetal scalp is visible at the vaginal opening. The most common practice is to repair the episiotomy after the delivery of the placenta.

What You Need: 1% **Lidocaine** without epinephrine (approx. 20-30cc), sterile gloves, gauze bandages, prep solution, 2-0, 3-0, and 4-0 absorbable suture (Vicryl or Chromic) for the repair, scissors or a scalpel to make the episiotomy incision, suture forceps and needle holders.

What to Do:

Episiotomy: Indicated to assist in the delivery of a large infant or one with shoulder dystocia after perineal massage has failed to stretch the tissue adequately. It is better to cut an episiotomy than to have the baby tear the perineal tissue into the rectum.

1. If no anesthesia has been given, administer 5-10 cc of **Lidocaine** along the midline of the perineum and posterior vagina. Remember not to administer more than 50cc of **Lidocaine** total, and to aspirate prior to injecting. If the perineum is very thin and well stretched, anesthesia may not be necessary.
2. Place the first two fingers into the posterior vagina between the fetal head and the vaginal wall, with one finger on either side of midline.
3. Cut between the fingers, through the vaginal wall and the perineum, but do not to incise the anal sphincter or its capsule. (Figure 3-12)
4. If pressure from the fetal head does not control bleeding, press a 4x4 bandage against the incised tissue to stop hemorrhage.
5. An alternate site for the episiotomy is the mediolateral position (approximately 4:30 on a clock face).

Classifications of perineal episiotomies and lacerations

First Degree: Extends only through the vaginal and perineal skin
Second Degree: Extends deeply into the soft tissues of the perineum down to, but not including, the external anal sphincter
Third Degree: Extends through the perineum and anal sphincter
Fourth Degree: Extends through the perineum, anal sphincter, and rectal mucosa to expose the lumen of the rectum

First Degree Episiotomy or Laceration Repair

1. Test the area to be sutured for residual sensation. If necessary, administer **Lidocaine** into the tissue as with typical wound repair.
2. If the edges of lacerated tissue are less than 1 cm apart and not bleeding, repair is not necessary.
3. Place the first stitch (2-0 or 3-0) 1 cm deep (proximal) to the end of the vaginal portion of the episiotomy (or laceration) and tie it.
4. Continue with locking, continuous sutures 1 cm from each wound edge, 1 cm apart and 0.5 cm deep through to the introitus or hymenal ring (or distal end of the laceration). Ensure the edges of the hymenal ring lay approximated. Do not cut the suture in episiotomy repair.
5. With another suture, sew a running subcuticular suture from the anal end of the skin wound back up toward the vagina to close the perineum.
6. Take the end of the suture back under the hymenal ring and tie it in the vagina.

Second Degree Episiotomy Repair: (see figure 3-13 and 3-14)

1. Repair vaginal wound as in First Degree Repair down to the hymenal ring. Do not cut the suture.
2. With another suture, sew 3 - 4 deep, interrupted stitches in the subcutaneous fascia, muscle and fat of the perineum.
3. Take the suture from the vagina under the hymenal ring and approximate the edges of the perineal fascia with continuous non-locking suture sewing toward the anus.
4. Sew a running subcuticular suture back up toward the vagina to close the perineum.
5. Take the end of the suture back under the hymenal ring and tie it in the vagina.
6. Do not suture any tear near the ureter without inserting a Foley catheter to avoid inadvertent urethral closure.

Third and Fourth Degree Repair: (see figure 3-15)
1. Reapproximate the rectal mucosa with interrupted, fine 4-0 sutures (usually two layers), taking care not to puncture the mucosa and to leave the ends of the suture in the tissue, not the rectal lumen.
2. Repair the torn ends of the doughnut-shaped anal sphincter with four well-spaced interrupted sutures that traverse through the capsule of the muscle.
3. Then repair the wound as in a second-degree laceration or an episiotomy (above).

Management after Episiotomy
1. Apply ice to the affected area tid to control swelling.
2. Relieve pain with **Tylenol #3** 1-2 tabs po every 3-4 hours.
3. Pain may be an indication of a large hematoma or perineal cellulitis. Examine these areas carefully if pain is severe or persistent.
4. Give stool softeners for approximately 7-14 days and direct increased intake of fiber and water.
5. Do not give enemas.

What Not To Do:
Do not fail to restore anatomical features.
Do not use too many sutures.
Do not fail to achieve hemostasis.

Symptom: OB Problems: Preeclampsia/Eclampsia
MAJ Marvin Williams, MC, USA

Introduction: Preeclampsia is maternal hypertension accompanied by proteinuria or edema, seen from the 20th week of gestation through delivery. If these symptoms are complicated by seizures or coma, the mother has **eclampsia**. Hypertensive disorders are the most common medical complications of pregnancy, effecting approximately 5-14 percent of pregnancies, and are more common in first-time mothers. The etiology of preeclampsia is unknown and it can be defined as mild or severe. Approximately 1% of patients with preeclampsia develop eclampsia.

Subjective: Symptoms
Visual disturbances (usually irregular luminous patches in the visual fields after physical or mental labor), headaches, nausea, vomiting, epigastric pain and generalized edema, seizures or coma.

Objective: Signs
Mild Preeclampsia (1 of the following):
1. Blood pressure changes (measure on two occasions at least 6 hours apart):
a. Systolic blood pressure (SBP) of 140 mm Hg or greater **OR**
b. Diastolic BP (DBP) of 90 mm Hg or greater (< 110) **OR**
c. Mean arterial BP (MAP) (calculated as 1/3 the difference between SBP and DBP, plus the DBP) of 105 mm Hg and/or an increase of 20 mm Hg over baseline
2. Proteinuria 2+ or > on a urine dipstick
3. Pathologic edema: generalized or involving the hands or face (**Note:** Moderate edema is a feature of approx. 70-80% of normal pregnancies). Weight gain greater than 4 pounds/week in the third trimester may be one of the first signs of preeclampsia.

Severe Preeclampsia (1 of the following):
1. Blood pressure changes (measure on two occasions at least 6 hours apart):
a. SBP of 160 mm Hg or greater **OR**
b. DBP of 110 mm Hg or greater
2. Proteinuria 3+ **OR** 4+ on dipstick
3. Severe edema, including pulmonary edema

Figure 3-12

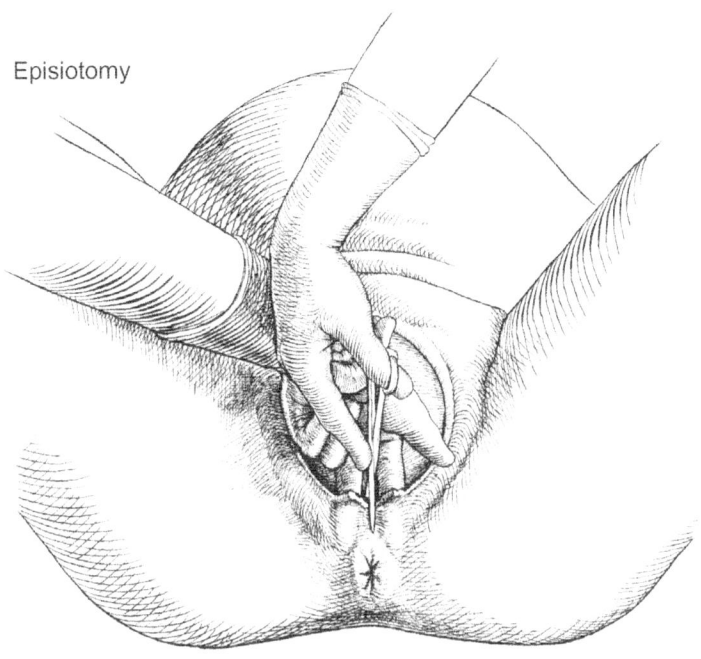

Episiotomy

4. Evidence of end organ compromise (cerebral or visual disturbances)
5. Persistent abdominal pain with nausea and vomiting.
   May also see:
   a. Oliguria (< 400 ml in 24 hr).
   b. Decreased platelet count (thrombocytopenia <100,000)
   c. Hyperreflexia

Eclampsia
1. Convulsions (seizures) during pregnancy with history of preeclampsia, or without other explanation.
2. Hypertension.
   May also see weight gain, edema, proteinuria, visual disturbances, and right upper quadrant/epigastric pain.

Figure 3-13

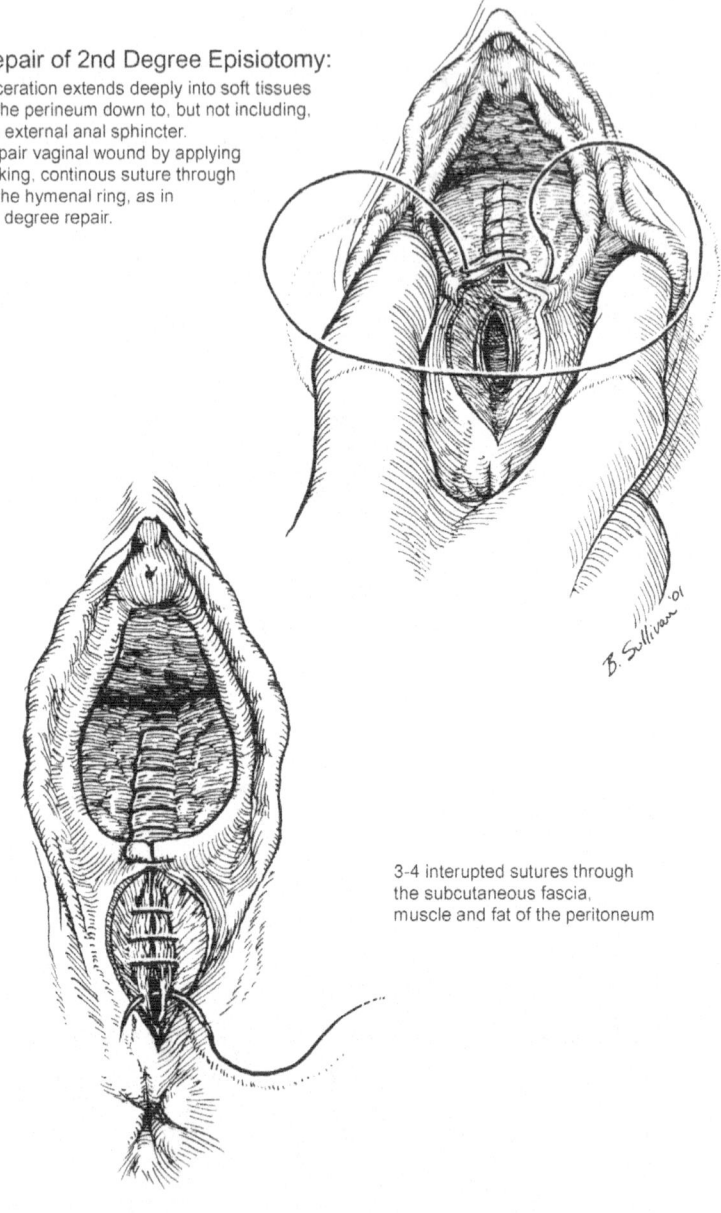

Repair of 2nd Degree Episiotomy:
Laceration extends deeply into soft tissues of the perineum down to, but not including, the external anal sphincter.
Repair vaginal wound by applying locking, continous suture through to the hymenal ring, as in 1st degree repair.

3-4 interupted sutures through the subcutaneous fascia, muscle and fat of the peritoneum

Figure 3-14

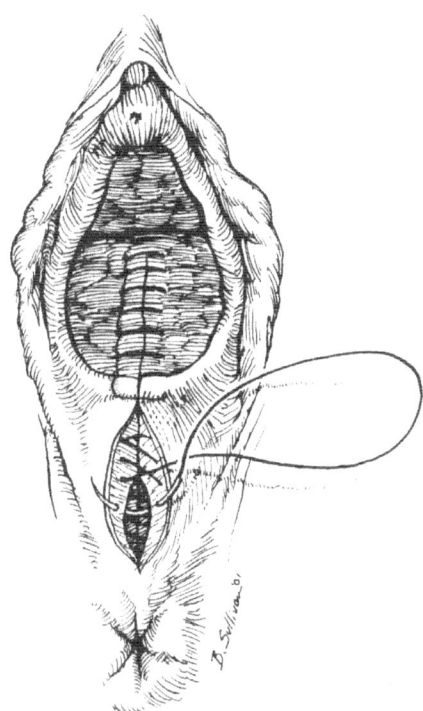

Approximate the edges of the perineal fascia with continuous, non-locking suture sewing towards the anus

Repair of 2nd Degree Episiotomy, cont'd

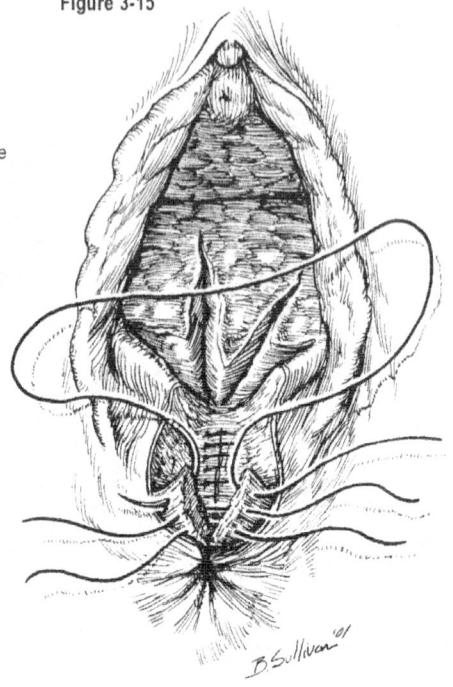

Reapproximate rectal mucosa with interupted, fine 4-0 sutures (usually two layers), taking care not puncture the mucosa and to leave the ends of the suture in the tissue, not the rectal lumen.
Repair the torn ends of the donut-shaped anal sphincter with four well-spaced interrupted sutures that traverse through the capsule of the muscle.

Repair of 3rd and 4th Degree Episiotomy

Using Advanced Tools: Lab: Urinalysis for urine protein (dipstick); platelet count.

Assessment:
Differential Diagnosis
Appendicitis, diabetes, gallbladder disease, gastroenteritis, glomerulonephritis, hyperemesis gravidarum (excessive vomiting in pregnancy), kidney stones, peptic ulcer, pyelonephrits, lupus, viral hepatitis. See appropriate sections of this book.

Plan:
Stabilize and evacuate. Definitive therapy in the form of delivery is the only cure for preeclampsia. The difficulty in therapy is deciding when to deliver the infant. The decision to deliver will depend on the severity of the disease, the status of the mother and the fetus, and the gestational age at the time of the evaluation. Take the severity of the condition and the fetal gestational age into consideration, and either deliver the pregnancy or place the patient on bed rest. Perform close surveillance until the pregnancy reaches term or the preeclampsia worsens, dictating the need to deliver. There is no advantage to cesarean delivery over vaginal delivery for preeclampsia. Therefore, delivery route should be based on obstetric indications (worsening condition).

Treatment
Mild Preeclampsia:
1. Observe for worsening signs of the disease.

2. If patient's condition does not progress, discharge and follow on a twice weekly basis.
3. In a field or remote setting, manage aggressively with **dexamethasone** 6 mg IM q 12 hours for a total of 4 doses to prevent fetal respiratory distress syndrome and maternal thrombocytopenia, and to improve perinatal outcome in severe preeclampsia.
4. Give **magnesium sulfate** (to prevent seizures) by a controlled continuous infusion with a loading dose of 4-6 gm in 100 ml over 15-20 min. followed by a continuous infusion of 2-3 gm/hr. Toxic levels cause muscle weakness, respiratory paralysis, and cardiac arrest. Administer **calcium gluconate** 1 gm slow IV push over 2-3 minutes to counteract magnesium toxicity.

Severe Preeclampsia:
1. With SBP > 180 mm Hg or the DBP > 110 mm Hg, the possibility of intracerebral damage increases warranting antihypertensive medication. Give Hydralazine (**Apresoline**) 5-10 mg IV every 20 minutes as indicated, or **labetalol** 20 mg IV q 10 min with a max dose of 300 mg, to reduce BP. Monitoring BP q 5 minutes for at least 30 min. after giving the drug. Please note that some of the side effects with labetalol are maternal tachycardia, headache and flushing.
2. Give magnesium sulfate as above.
3. Evacuate.

Eclampsia:
1. Give magnesium sulfate as above.
2. Provide oxygen and airway support as needed.
3. Evacuate.
4. If evacuation not feasible deliver the fetus **after** seizure activity has abated.
5. If magnesium sulfate is not available, consider cesarean section as only option to save both mother and fetus (see Cesarean Section procedure).

Patient Education
Activity: Remain at bedrest on left side to minimize symptoms.
Prevention: Increase water intake, but maintain normal salt intake.

Follow-up Actions
Evacuation/Consultant Criteria: Evacuate early to avoid complications of eclampsia. Consult experts for management in remote settings (to include C-Section if necessary).

Symptom: Palpitations

CAPT Kurt Strosahl, MC, USN

Introduction: Palpitations are the sensation of a change in the heartbeat. A patient may complain of a sensation of skipped, missed, strong, fast, or accelerating heartbeats. The majority of palpitations are unlikely to cause harm but cause concern for the patient. Ventricular Tachycardia and Fibrillation (VT/VF) are followed immediately by collapse and imminent death without appropriate resuscitation. Premature Ventricular Contractions (PVCs) and Premature Atrial Contractions (PACs) feel like skipped or single hard beats and are found in all normal healthy persons. While PVCs may be harbingers of a bad outcome in patients with known heart disease, they do not require a search for heart disease in the otherwise healthy person. Short runs of Paroxysmal Supraventricular (or Atrial) Tachycardia (PSVT or PAT) are frequent, particularly when straining against a load or drinking cold beverages after heavy physical exertion. Unless there is loss of consciousness, they can be readily aborted by the vagal maneuver (bearing down as if you are having a bowel movement). Frequent episodes of palpitations may require further evaluation when the mission is completed, but should not remove the service member from the field.

Subjective: Symptoms

Focused History: Quantity: *How often does this occur?* (PVCs and PACs may occur daily, hourly, or rarely. They cause anxiety, which heightens the member's awareness of them and increases the frequency, since they are driven by adrenalin.) *How long do they last?* (Most runs of PSVT are self-limited and the member finds that lying down or bearing down stops them. If the rapid heart beating lasts for hours, they usually are gone after a nap.) **Quality:** *Do you feel dizzy or lightheaded with them?* (If the heart rate goes over 180, the member usually must sit or lay down. If the heart rate is <180, the member usually can continue with their activities, affected only by their concern over what is happening.) **Alleviating factors**: *What makes them go away or lessen?* (Sitting down, lying down, or taking a nap will often stop the symptoms. Bearing down will abruptly terminate PSVT. Splashing cold water on the face [the diving reflex] may also abruptly stop rapid beating.) **Aggravating factor:** *What seems to bring these on?* (Caffeine, alcohol, nicotine, over-the-counter cold remedies and lack of sleep are the most common aggravating factors. Some members utilize anabolic steroids, weight reduction supplements or other health food supplements that contain ephedrine or similar drugs. Asthma medications are largely derivatives of adrenaline and can cause palpitations. Non-sedating antihistamines and hypermotility agents such as **Propulsid** cause PVCs and VT. Discontinuation of these factors will usually make the symptoms go away. Fear and anxiety worsen the symptoms but not necessarily the rhythm disturbance. Most people incorrectly associate palpitations with a heart attack, which is VERY rarely the case.) **Associated symptoms**: Nausea, thirst, frequent urination, and complaints of pain in the chest are non-specific and of little value in determining the cause.

Objective: Signs

Using Basic Tools: Take pulse for 60 seconds to determine the rate, and assess for skipped beats (PVC or PAC) or irregular beating (atrial fibrillation).
Altered vital signs: Hypotension with dizziness indicates the rhythm is compromising cardiac output.
Orthostatic hypotension (BP drop of 20mm and HR increase of 10 comparing standing to supine) suggests volume depletion as cause of arrhythmia
Respirations >30 may result from anxiety and hyperventilation, which can cause arrhythmias
Lungs: Pneumonia or pneumothorax is a cause of tachycardia
Heart: Irregular beating suggests atrial fibrillation. Skips indicate PVC/PAC, which cannot be differentiated in the field. Click(s) between heart sounds result from mitral valve prolapse (associated with benign PVCs)
Neck: An enlarged, tender thyroid, associated with tremor of the hands strongly suggests hyperthyroidism as the cause of arrhythmia.
Using Advanced Tools: Lab: CBC for anemia (cause of tachycardia); EKG to identify the rhythm (25% of reported symptomatic palpitations correlate with normal sinus rhythm during 24 hrs continuous ambulatory EKG monitoring).

Assessment:
Differential Diagnosis
Tachycardia - normal if demand for cardiac output is high as in anemia, fever, exercise, anxiety, dehydration, hypoxia. Over 150 at rest is abnormal unless hypovolemic.
Bradycardia - normal athletic adults have HR as low as 40 with skips due to second degree AV block.
PVCs - wide (>3 little boxes) QRS complexes that occur out of synchronization with the other beats.
PACs - narrow (<3 little boxes) QRS complexes that occur out of synchronization with the other beats.
Heart Block - dropped or missing QRS where one would have been expected
PSVT - narrow QRS occurring at a rate >150
Atrial Fibrillation - QRS complexes occurring irregularly irregular without a P wave in front of them.

Plan:
Treatment
1. REASSURANCE. Make the patient calm.
2. Prevention: Remove causes (caffeine, sleep deprivation, OTC medications)
3. Valsalva maneuver: Bear down like having a bowel movement; this will stop or slow all supraventricular tachycardias, at least momentarily.
4. Carotid massage: After auscultation for bruits, massage the right carotid for 15 seconds
5. Diving reflex: Place the member's face in a basin of cold or ice water
6. Medications: **Diazepam** 5 mg po. **Propranolol** 20-40 mg po qid (hyperthyroidism may require up to 480 mg/day). **Verapamil** 40-80 mg po qid if asthma precludes use of **propranolol**. Do not use in wide complex tachycardias.
7. Treat atrial fibrillation (AF) with **propranolol** or **verapamil** in increasing doses until the heart rate is under 100 during a slow jog in place. AF is not a heart attack, can be controlled in the field, and does not require evacuation UNLESS it is caused by a heart attack or pulmonary embolism.
8. Treat acute coronary cause as in the Symptom: Chest Pain section.

Patient Education
General: Palpitations are ALMOST always not life threatening. The patient is not having a heart attack.
Activity: If no syncope or near syncope symptoms, then no restriction is indicated. When the symptoms of rapid heart beating come on, the member should sit down and bear down.
Diet: Avoid caffeine
Medications: **Propranolol** may cause tiredness and **verapamil** may cause constipation.
Prevention: Avoid those medications and foods that cause symptoms. Get plenty of rest.
No improvement/Deterioration: If symptoms persist, worsen, or are associated with syncope or pre-syncope return to the clinic.

Follow-up Actions
Return evaluation: Increase dose of propranolol up to 360 mg total per day. If symptoms continue, remove the member from the field.
Evacuation/Consultation Criteria: Evacuate if unable to perform duties, if symptoms are not controlled or underlying heart disease is suspected. The presence of stable PVCs, PACs, PSVT, or AF should not be the sole cause to remove the member from the field while performing critical mission executions. Consult cardiology or internal medicine as needed.

NOTE: See service specific Aviation and Diving Medicine guidelines for operators performing special duty.

Symptom: Rash and Itching

Rash with a Fever
MAJ Daniel Schissel, MC, USA

The sudden appearance of a cutaneous eruption and a fever is frightening and an often intimidating clinical challenge. Rarely does one have to rely solely on their morphologic assessment as when confronted by an acutely ill patient with a fever and a rash. If a diagnosis is not established quickly in certain patients (e.g., septicemia), life-saving treatments may be delayed unnecessarily. Furthermore, rapid diagnosis and isolation of patients with contagious disease prevents spread to other persons and preserves the fighting force.

Contagious cutaneous diseases presenting with fever and a rash include viral infections (e.g., adenovirus, echovirus, herpes simplex, measles, rubella, varicella/chicken pox) and bacterial infections (e.g., meningococcal, staphylococcal, streptococcal, and secondary syphilis.) As mentioned in the Dermatology Introduction (see Skin chapter), the morphologic diagnosis of cutaneous eruptions is a discipline based on detailed observation, with precise identification of the primary and secondary skin lesions being paramount.

The differential diagnosis of an acutely ill febrile patient with a rash may be broken down into three main categories according the primary lesion(s) observed. Refer to the specific illness in the Infectious Disease or Dermatology chapters for more diagnostic and treatment information.

Disease	Macules or Papules	Vesicles, Bullae, or Pustules	Purpuric Macules, Papules, or Vesicles
BACTERIAL:			
Cat-Scratch	X	X	
Gonococcemia			X
Meningococcemia	X	X	X
Pseudomonas aeruginosa		X	X
Staphlococcal			
Erysipelas	X	X	X
Scalded Skin Syndrome	X	X	
Toxic Shock Syndrome	X	X	
Streptococcal (Scarlet fever)	X		X
Tularemia	X	X	
RICKETTSIAL:			
Boutonneuse Fever	X		
Richettsialpox	X	X	
Rocky Mountain Spotted fever	X	X	X
Trench Fever	X		
Typhoid Fever	X		
Typhus, Louse-borne/epidemic	X		X
Typhus, Murine/endemic	X		
Typhus, Scrub	X	X	
VIRAL:			
Adenoviral infections	X		
AIDS / HIV	X		X
Enterovirus infections			
Echo, Coxsackie	X	X	X

Disease	Macules or Papules	Vesicles, Bullae, or Pustules	Purpuric Macules, Papules, or Vesicles
Erythema Infectiosum	X		
Erythema Multiforma (HSV)	X	X	
Herpes Simplex / Disseminated		X	
Herpes Zoster / Generalized		X	
Measles (rubeola)	X		
Rubella (German measles)	X		
Varicella (Chicken Pox)	X	X	
OTHER:			
Drug Hypersensitivity	X	X	X
Kawasaki Disease (mucocutaneous lymph node syndrome)	X		
Lyme Disease	X		
Psoriasis			
Secondary Syphilis	X		
Serum Sickness	X		
Steven Johnson Syndrome	X	X	
Systemic Lupus Erythematosus	X	X	X
Toxic Epidermal Necrosis	X	X	

Pruritus (Itching)

Lt Col Gerald Peters, USAF, MC

Introduction: Rash and pruritus sometimes present together and may be related (see Rash Symptom Section for additional guidance). The intensity of itching can vary from mild to severe, and can interfere with sleep at night and the ability to focus during the day. Suicide due to intractable pruritus has been reported. In some cases, the cause may be obvious, such as exposure to an irritant or allergen known to the patient. Other cases, especially the more chronic and generalized ones may be more difficult to understand and/or treat. A thorough history and physical is often needed to unravel such cases. Begin by determining whether the pruritus is due to a local skin condition or a systemic problem.

Subjective: Symptoms (in addition to itching)
Variable: Pain, insomnia, rash (virtually any type of skin lesion), fever, erythema
Focused History: Have you been in contact with plants, chemicals, shoes or other sources of leather or rubber? (suspect contact dermatitis). Do you have any personal history or family history of allergy, asthma, hay fever or childhood eczema? (suggests atopic allergic diagnosis) Have you changed medications or dosages lately? Have you had significant, recent exposure to sunlight, biting insects or sick people? (may suggest diagnosis) Have you had a viral illness in the last 2-6 weeks? (Pityriasis rosea is characterized by herald patch and pruritus.) Have you noticed any hair loss, dry skin or pigment changes? (consider thyroid disease). Have you had any jaundice, or clay-colored stools? (consider liver disease) Where have you noticed the rash? (Specific sites are more likely in some conditions, e.g., finger web spaces, axillae, nipples, umbilicus and genitals for scabies; sun-exposed areas for sunburn, some drug reactions.)

Objective: Signs
Using Basic Tools: Agitation, excoriation, thickened and/or hyperpigmented skin, fatigue, possible rash (varied presentations, but if there are no lesions where the patient cannot reach [e.g., mid-back], the condition

is likely systemic), erythema, fever, lymphadenopathy, jaundice, hives; others also possible.
Using Advanced Tools: Labs: CBC with differential for infection (eosinophils in parasitic infestations), anemia; Urinalysis for urobilinogen, protein, casts, sugar; Stool specimens for occult blood and/or O&P; KOH to identify tinea infections; Gram stain and/or culture for cellulitis; CXR: to rule out complications.

Assessment:
Differential Diagnosis
Pruritus With Rash
Xerosis (dry skin is itchy skin!) - dry, flaky, macular lesions over large area.
Atopic dermatitis - affects the popliteal and antecubital fossae; dry, hyperkeratotic, confluent papules.
Psoriasis - scalp, external auditory canals, genitals and superior aspect of the intergluteal cleft; plaques of dry, hyperkeratotic skin
Drug eruptions - erythematous, generalized, fine rash; history of drug/medication use
Urticaria - edema, erythema, wheal reaction to allergen/irritant; may be generalized
Seborrheic dermatitis - oily, flaking skin; often in the hair
Sunburn - sun exposed areas; erythema and edema, with pain
Contact dermatitis (allergic or irritant) - edema, erythema, pain on areas exposed at work (hands, face, etc.)
Pityriasis rosea - history of viral illness; oval, macules and papules on trunk and extremities
Insect bites - single or multiple erythematous, edematous macules
Scabies - linear burrows with excoriations in finger web spaces, axillae, nipples, umbilicus and genitals
Pediculosis (head, body or pubic lice) - excoriations in appropriate areas; visible lice or nits
Fungal infections - erythematous, hyperkeratotic plaques in moist body areas; KOH positive scrapings
Folliculitis - cellulitis around hair shaft
Lichen simplex chronicus - single or confluent papules with mosaic pattern
Pruritus ani or vulva - often due to unknown causes, or from infections (peri-anal strep), infestations (pinworms), trauma, contactants, cancer, lichen planus or lichen sclerosis, or psychological factors.
Contagious diseases - fever and rash (usually macular/papular); include viral infections (e.g., adenovirus, measles, rubella, varicella/chicken pox); bacterial infections (e.g., meningococcal, staphylococcal, streptococcal, and secondary syphilis).
See separate sections on scabies, lice, psoriasis, contact dermatitis, fungal/tinea infections and many of the infectious cutaneous diseases.

Pruritus Without Rash
Thyroid disease - thinning of the lateral eyebrows, loss of skin pigment.
Uremia/kidney failure - proteinuria, casts on urinalysis (UA)
Obstructive biliary disease - urobilinogen on UA, jaundice, abdominal pain
Diabetes mellitus - sugar on UA, polyphagia, polyuria, polydipsia.
Microcytic anemia
Drug side effects - history of medication use.
Neurologic disorders - such as paroxysmal pruritus in multiple sclerosis
Several cancers - lymphadenopathy in many types, low blood cell lines
Pregnancy
See separate sections on jaundice, hypothyroidism, diabetes, anemia.

Plan:
Treatment
1. Treat infectious agents (tinea, parasitic, viral and bacterial) as discussed in their respective sections.
2. Treat other rashes (psoriasis, etc.) and systemic diseases (diabetes, etc.) as discussed in their respective sections.
3. Treat dry skin with a gentle regimen of less bathing (every other day instead of daily) with warm, not hot water, for short duration (<5 minutes), and either gentle soap like Dove for Sensitive Skin or a soap substitute (Cetaphil lotion). Cleanse only the areas needing it, like the face, axillae and anogenital region. Pat dry gently with a towel, taking care not to rub. Immediately apply an emollient like Moisturel or Eucerin or Vaseline or Crisco. Aveeno Colloidal Oatmeal bath can help, but be sure to rinse off the residual matter.

4. Relieve itching.
    a. Sarna lotion (camphor and menthol) and cool compresses can relieve itch for short periods of an hour or so.
    b. Avoid extensive applications of topical steroids when the etiology of pruritus is unclear.
    c. Antihistamines like Atarax 25-50 mg po q hs or antidepressants like Doxepin (75 mg po q hs) can be helpful at bedtime, but tend to cause drowsiness, so use with caution.
    d. Ultraviolet Sunlight (UVB or PUVA) can help. Avoid midday sun with burning infrared rays.
    e. Anxiety and stress, as well as depression can elicit or worsen pruritus. Psychiatric consultation may be helpful, but work hard to get the patient's confidence before even suggesting this.

Patient Education
General: Do not use water to moisten skin. It causes further drying. Take medication as indicated. Do not scratch.
Prevention: Keep skin moist during winter by avoiding hot water, excessive washing, harsh soap.

Follow-up Action
Evacuation/Consultant Criteria: Evacuation is not normally necessary. Consult dermatology as needed.
NOTE: Skin biopsy by dermatologist is sometimes necessary for accurate diagnosis.

Symptom: Shortness of Breath (Dyspnea)
COL Warren Whitlock, MC, USA

Introduction: Dyspnea or "shortness-of-breath" is an uncomfortable sensation of difficulty in breathing. It is a symptom, for which the sign may be a rapid respiratory rate. There are several types (see below). **Stridor** is a physical finding (usually loud enough to be heard at some distance) associated with upper airway obstruction and is a reason for medical concern.

Subjective: Symptoms
Focused History: Exposure History: Affirmative answers to any of the following place patient at risk for dyspnea. *Have you had trauma to head, neck or chest? Have you been flying or diving recently?* (decompression sickness) *Did the symptoms start while you were eating or holding something in your mouth? Do you smoke?* (risk for COPD, tension pneumothorax, MI) *Are you taking any medicines or drugs?*
Past Medical History: Affirmative answers to any of the following place patient at risk for dyspnea. *Have you ever had heart problems, diabetes, vascular disease, asthma, COPD? If so, did you fail to take your medications? Have you ever had any mental health problems, including suicide attempts?* (anxiety, drug abuse) *Have you had a recent respiratory illness?*
Specific Symptoms: *Have you had chest pain?* (MI, trauma, occasionally pulmonary embolus) *or radiating pain?* (MI, aneurysm) *or fever?* (pneumonia, sometimes pulmonary embolus) *or cough?* (pneumonia [productive]; pulmonary embolus, CHF, COPD, asthma [may be productive]) *or difficulty breathing at night?* (MI, CHF) *or trouble swallowing?* (oropharyngeal obstruction due to foreign body, tumor, glottic edema, epiglottitis, retropharyngeal mass) *or weight loss?* (tumor, COPD, drug abusers) *or loss of control of muscles or feeling in part of your body?* (stroke, brain tumor, drugs) *or muscle spasms?* (possible psychogenic etiology)
Onset: *Did the symptoms start suddenly?* (foreign object, pneumothorax, MI, stroke, epiglottitis, pulmonary embolus, drug abuse) *or progressively worsen over hours to days?* (cardiac tamponade, pericardial effusion, COPD, asthma, diabetes, pleural effusion due to inflammatory process, aneurysm)

Objective: Signs
Using Basic Tools:
Vital Signs: Pulse: Tachycardia except if drug or Central Nervous System (CVA, tumor) related, and in MI.
Respiration: Tachypnea EXCEPT POSSIBLY in Central Nervous System or drug-related dyspnea.
Blood Pressure: Hypotension: cardiac cause, pulmonary embolus. Hypertension: other causes, but especially chronic vascular disease-related dyspnea or in psychogenic dyspnea. Either: CVA, brain tumor, drugs. Decreased pulse pressure seen with cardiac tamponade

Temperature: Elevated with infectious processes (pneumonia, epiglottitis); low grade fever may seen with pulmonary embolus, MI and in those patients "working" to breathe (asthma, COPD, psychogenic)
Neurological Exam:
Mental status: Variable depending on degree of hypoxia (drugs, stroke, head trauma, diabetic coma).
Other findings: Common with Central Nervous System and drug-induced causes; also loss of peripheral sensation in diabetics.
Inspection: General: Usually anxious and sitting upright, unless there is mental status alteration (e.g., due to drugs, head trauma, stroke, diabetic coma). Psychogenic dyspneic patients (as well as those which are drug-related) may appear tachypneic, tachycardic and diaphoretic (sweating). There may be evidence of trauma with open wounds, distorted anatomy, bruising, swelling.
Head/Neck: Foreign body evident in mouth/throat; tracheal deviation (tension pneumothorax, cervical tumor or hematoma); large epiglottis, tumor, mass visible in oropharyngeal area; jugular venous distention (tamponade, MI, CHF, tension pneumothorax); perioral cyanosis (indicates severity of respiratory compromise); pupils sluggish and dilated or pinpoint (drug induced, head trauma, stroke)
Chest: Barrel chest (asthma, COPD); unequal expansion (pneumothorax, splinting due to rib fracture)
Abdomen: Visible pulsating mass (aneurysm)
Extremities: Pedal or pretibial edema (CHF); skin ulcerations (chronic diabetes)
Palpation:
Neck: Cervical lymphadenopathy (retropharyngeal abscess)
Chest: Tender chest wall (rib fracture and risk for pneumo/hemothorax; cardiac tamponade)
Abdomen: Palpable pulsating mass (aneurysm)
Pulses: Asymmetric pulses between arms or between arms and feet (aortic dissection)
Percussion of Chest: Tympanic (pneumothorax); resonant (COPD); dull over dependent side of chest (pleural effusion, empyema, hemothorax); diffuse chest dullness (pneumonia).
Auscultation of Chest: Diminished breath sounds: pneumonia, COPD, asthma, atelectasis, pneumo/hemothorax.
Stridor: Upper airway obstruction (epiglottitis; tumor; retropharyngeal mass; foreign body)
Rales: Pneumonia
Wheeze: Obstruction due to mass or inhaled foreign body; COPD; asthma
Auscultation of Heart:
Murmur, gallop (abnormal sounds): MI, CHF, valvular disease
Faint heart sounds: Cardiac tamponade.
Using Advanced Tools: Pulse oximetry (> 92% is normal); CXR: pulmonary edema (increasd vascular markings) in CHF, infiltrate (increased local density) in pneumonia, effusion (fluid in pleural space) with pulmonary embolus; neck x-ray (if available) for epiglottal swelling or foreign body in airway; EKG (see Procedure: EKG) for MI, tamponade

Assessment:
Differential Diagnosis
Pulmonary - airflow restriction or obstruction, whether at the upper airway level (e.g., edema of the glottis, tumor, inhaled foreign body, trauma, or retropharyngeal abscess or hematoma) or within the pulmonary structure (e.g., asthma, COPD, pneumonia, atelectasis, pneumothorax, pleural effusion, hemothorax) causes an increase in the work required to breathe, producing dyspnea.
Cardiac - problems that slow or impede the delivery of oxygenated blood cause dyspnea. Etiologies include valvular malfunction, infarction, tamponade, pulmonary embolism, and heart failure. Chemical: Metabolic changes (diabetes, drugs) can cause a change in pH, resulting in an increased rate and depth of respiration to blow off CO_2, and perceived dyspnea.
Central Nervous System - diseases or processes that depress the brain's respiratory control center can result in dyspnea. Some causes include narcotics, stroke or head trauma.
Psychogenic - anxiety is a well-known cause of tachypnea. Patients may interpret tachypnea as dyspnea, which further increases their level of anxiety. They may develop muscle (myoclonic) spasms.

Treatment
Pulmonary:
Asthma, COPD, Pneumonia: See respective sections in Respiratory chapter.
Atelectasis: Oxygen (O_2); systematic chest percussion may loosen mucus plug
Pneumothorax/Tension pneumothorax/hemothorax: O_2; needle thoracotomy; address underlying condition; place chest tube (see Procedure: Chest tube and Needle Thoracotomy)
Pleural effusion: O_2; needle thoracentesis to drain pleural effusion; address underlying illness
Airway obstruction: Provide O_2 in all these cases; be prepared to perform emergency surgical airway (cricothyroidotomy) if patient becomes cyanotic or excessively dyspneic or tachypneic
    Epiglottitis/Edema of glottis – Antibiotics (**cefuroxime** 1 gm IV q 8 h, or **cefotaxime** 2 gm IV q 4-8 h); do not attempt intubation
    Tumor – May require intubation, nasal airway or emergency airway (tumor location may influence choice of airway)
    Retropharyngeal abscess/hematoma – Intubate or try nasal airway if possible; do NOT drain abscess or hematoma; antibiotics (**cefuroxime** 1 gm IV q 8 h, or **cefotaxime** 2 gm IV q 4-8 h)
    Foreign object – Remove if possible (Heimlich Maneuver or sweep/grasp object); do NOT intubate with object in airway
    Neck Trauma – Intubation or nasal airway if possible after check for foreign bodies in airway
Cardiac: Provide O_2 in all these cases
MI, CHF, pulmonary embolism: see appropriate sections
Tamponade: see Cardiac: Pericarditis.
Valvular malfunctions - follow Fluid Resuscitation section guidelines as needed to maintain stable vital signs; follow CHF section treatment guidelines to mobilize fluid in pulmonary and peripheral edema
Chemical: Diabetes (see Endocrine: Diabetes)
Central Nervous System: Include: Narcotics, stroke or head trauma.
Drug induced: May require intubation; O_2, **dextrose** 50 IV, fluid resuscitation, **Narcan** or other agent-specific antidote (if known, if available); supportive care
Stroke/Cerebral hemorrhage/trauma: O_2; intubate, Foley catheter, **dextrose** 50 IV, fluid resuscitation and protective restraints if unconscious; bedrest if conscious
Psychogenic: Have patient breathe into a bag (will reverse myoclonic spasms in 10-15 minutes). Rule out underlying medical condition. Treat anxiety with an anxiolytic (**Valium**, 10 mg IV/IM), or alcohol.

Patient Education
General: Remain calm. Anxiety will worsen the symptoms.

Follow-up Actions
Evacuation/Consultant Criteria: Most of these patients will require evacuation for definitive treatment and advanced procedures. Evacuate when stable and capable of travel.

Symptom: Syncope (Fainting)
CAPT Kurt Strosahl, MC, USN & CPT Brooks Morelock, MC, USA

Introduction: Syncope (fainting) is the sudden, unexplained loss of consciousness with loss of motor tone. Most causes are cardiac or neurologic in nature and include: hypoperfusion of the brain caused by blood pooling in the lower extremities (neurocardiac or vasovagal); decreased intravascular volume (blood loss, adrenal insufficiency); seizure; autonomic dysfunction in Shy-Drager syndrome or recurrent heat exhaustion; tachycardia (>180) or bradycardia (<40); hypoglycemia or psychological disorders. Malignant arrhythmia (e.g., ventricular fibrillation) is one of the most worrisome causes of syncope and can be life threatening.

Subjective: Symptoms
Sudden, unexplained loss of consciousness possibly preceded by light-headedness, nausea, sweating, sudden fatigue, hunger or "seeing stars."

Objective: Signs
Using Basic Tools: Cool, clammy skin; depressed consciousness; weak pulse and hypotension. Tonic-relaxation movement of the extremities (forceful muscular contraction followed by passive relaxation) represents hypoperfusion seizures, not tonic-clonic epileptic seizures. Neurologic exam: assess patient according to Glasgow Coma scale (GCS) (see Appendix) if not fully conscious.
Using Advanced Tools: Lab: Urinalysis for glucose, hematocrit for anemia, EKG for arrhythmia.

Assessment:
Differential Diagnosis
Acute myocardial infarction - ST elevation on EKG
Rhythm disturbance - EKG shows tachycardia >180, bradycardia <40, ventricular tachycardia
Seizure Disorder - Evidence of tongue biting, urinary and bowel incontinence
Hypoglycemia - ChemStrip urinalysis positive for ketones in starvation, diabetic ketoacidosis (DKA), others; in DKA, urine should be positive for glucose also (glucose in blood, but cannot be absorbed by cells)
Hypovolemia - Tilt test positive with BP dropping >20mm and HR rising >10 bpm
Psychiatric - consciousness returns abruptly if nose and mouth are held closed
Many medications can cause a loss of consciousness, especially if taken in combination with alcohol.
Alcohol and recreational drugs are also a leading cause of loss of consciousness in young persons without a history of previous syncope.
Heat injury - history of exposure; other patients from unit

Plan:
Treatment
Primary:
1. Protect patient from injury and place with feet elevated.
2. Protect airway: a good rule of thumb: on GCS, less than 8, intubate.
3. Start IV and deliver D5NS bolus infusion of 500 cc. Continue 200 cc/hr until systolic BP >90.
4. Determine EKG rhythm. If HR >150 and QRS is greater than three little boxes and the patient is hypotensive, defibrillate as for ventricular tachycardia. If the HR is >200 and the QRS is less than three little boxes, bolus with 500 ml NSS and give **propranolol** 0.1 mg/kg IV. If the rhythm is sinus and less than 50 bpm, give 1mg **Atropine** IV and may repeat in 5 minutes or until the HR rises above 70. If **propranolol** IV is used, start on **propranolol** 40 mg qid. If **Atropine** IV is used, place transdermal **scopolamine** patch.
5. If hypoglycemia is suggested by the ChemStrip or the history, give a bolus of D50 IV.
6. If no response to above, give **Narcan** 1 ampoule IV.
7. If patient is seizing, see Neurology: Seizure Disorders.

NOTE: Ventricular tachycardia that responds to **lidocaine** should be treated as Unstable Coronary Syndrome.
Primitive: Elevate the feet and cover with warm blanket.
Empiric: Start IV with NSS as a bolus of 1000ml and give 1 ampoule of **Narcan** if member remains unconscious.

Patient Education
General: A simple faint may be vasovagal and is not serious. If recurrent (more than one per month) or results in bodily injury, then see consultant for preventive medications. When an aura occurs, lie down or sit down and place your head between your knees.
Activity: Patient should not stand duty alone for 48 hours. They should not jump; drive or dive after the second event until they have been further evaluated at a higher echelon of care.
Diet: No salt restrictions, drink plenty of fluids, restrict refined sugars, avoid alcohol.
Medications: Propranolol may cause tiredness; **scopolamine** may cause dry eyes and dry mouth (urine retention in males).
Prevention and Hygiene: Maintain hydration. Advise unit to prevent other heat or drug related injuries.
No Improvement/Deterioration: Return for reevaluation promptly, particularly if problem recurs.

Follow-up Actions
Return evaluation: Repeat evaluation, including EKG, ChemStrip, blood pressure, hematocrit.
Consultation Criteria: Any life-threatening rhythm disturbance or seizures should be referred to higher level of care immediately. More than two syncopal episodes, or dysrhythmias producing syncope should be evaluated further once out of the field.

PART 4: ORGAN SYSTEMS
Chapter 1: Cardiac/Circulatory
Cardiac: Acute Myocardial Infarction
Lt Col Robert Allen, USAF, MC

Introduction: Acute myocardial infarction (AMI) is the leading cause of death in the US and in most of the western world. AMI is usually thought of as a disease of older people, but almost half of the cases of AMI in the US occur in people under the age of 65. AMI can and does occur in people in their 20s and 30s. Early aggressive management of AMI can significantly improve mortality and morbidity. Approximately 25% of patients with AMI die within one hour of symptom onset, usually due to a malignant cardiac arrhythmia. Another 30-40% of patients with AMI die immediately or within several days. While some of the diagnostic and therapeutic procedures noted in the chapter are currently not possible in the SOF environment, advances in miniaturization of EKG machines, blood chemistry machines and pharmacology allow remarkably advanced diagnosis and treatment even in austere environments.

Cardiac Risk Factors: Prior AMI or history of coronary artery disease, smoking/tobacco use, hypertension, AMI before age 55 in parents or siblings, diabetes, high total cholesterol, low HDL cholesterol, obesity and history of peripheral vascular disease.

Subjective: Symptoms (See Symptom: Chest Pain)
Chest pain associated with AMI is generally dull, diffuse, and often described as a pressure sensation. Chest pain that lasts for a few seconds is not usually cardiac-related. Angina usually lasts less than 5 minutes, while AMI chest pain lasts more than 5 minutes. Classical angina pain is brought on by exertion and is relieved by rest. Chest pain from an AMI can start during exertion or at rest, and can radiate to the left arm, jaw or epigastric area. AMI pain is frequently accompanied by diaphoresis, shortness of breath, nausea and feelings of dread. The presence of three or more cardiac risk factors increases the likelihood of AMI.

Objective: Signs
Using Basic Tools: Tachycardia or bradycardia, hypertension or hypotension; diaphoresis in association with chest pain; inspiratory rales and S-3 gallop (left-sided cardiac failure); hepatojugular reflux, jugular venous distension and peripheral edema (right-sided cardiac failure).

Using Advanced Tools: Electrocardiogram (EKG): ST segment elevation is the hallmark of myocardial infarction, while ST depression and T-wave inversion are signs of myocardial ischemia. Use patterns of ST segment elevation to identify the location of the infarction. Elevation in leads I, AVL, and V-1 through V-3 indicate an anterior MI. Elevation in leads II, III and AVF usually indicate an inferior MI. ST elevation in V-2 through V-6 indicates an anterolateral MI. In up to 40% of AMIs the initial EKG does not show acute ST segment changes. **A 'normal' EKG does not rule out AMI!** The EKG may not reflect ST changes early in the AMI. Repeat the EKG as the chest pain changes over the first few hours to ensure ST changes are captured. (Figure 4-1: EKG ST Segment Changes)

Assessment:
Differential Diagnosis: Presumptively diagnose AMI if there is ST segment elevation of 1 mV(mm) or greater in contiguous leads. Symptoms consistent with AMI accompanied by a new-onset left bundle branch block (LBBB) are also considered presumptive evidence of AMI.

ST segments can be falsely elevated in several conditions, including myocarditis, left ventricular hypertrophy, ventricular aneurysms, early repolarization, hypothyroidism, and hyperkalemia.

Other causes of chest pain are reviewed in the Symptom: Chest Pain section, including thoracic GI causes (esophagitis, hiatal hernia), abdominal GI causes (pancreatitis, cholecystitis), musculoskeletal causes (costochondritis), vascular causes (pulmonary embolus) and others.

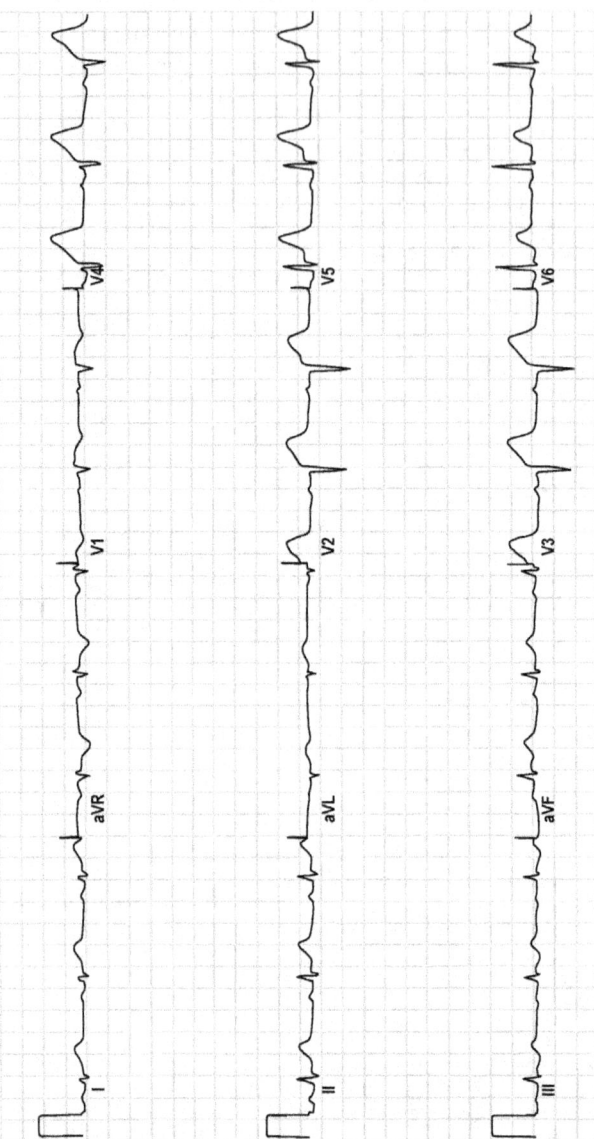

Figure 4-1
EKG showing ST segment elevation in leads V-2 through V-5.
Elevation in V-2 through V-5 indicates anterior MI.

Plan:
Treatment
1. Put the patient at bedrest to reduce myocardial oxygen demand.
2. Provide supplemental oxygen to bring the oxygen saturation above 95%.
3. Attach cardiac monitor if available and treat any malignant arrhythmias (See Cardiac Resuscitation procedure).
4. **Nitroglycerin** 0.4 mg sublingual q 5 minutes X 3 doses, or until pain relief
   -Check BP between doses, do not give if systolic BP below 90
   -If BP drops below 90, give 500 cc normal saline boluses until BP returns to >100 systolic
5. IV of normal saline at 100 cc/hr
6. **Aspirin** 325 mg po q 24 hours if not contraindicated.
7. For severe pain, **Morphine** 2 mg IV q 5 minutes until pain relief or sedation. Monitor respiratory status closely.

Patient Education
General: Healthy lifestyle may help prevent AMI.
Activity: Maintain bedrest to minimize cardiac oxygen demand.
Diet: Sips of water only for first 6-12 hours, then small, light meals, preferably clear liquids only.
Prevention: Make lifestyle changes to minimize cardiac risk factors.

Follow-up Actions
Evacuation/Consultant Criteria: Evacuate patients suspected of AMI immediately after stabilization treatment. Keep patient under close observation at all times. Establish phone or radio contact with medical control as early as possible and continued until the patient reaches definitive care.

NOTE: Several enzyme markers are routinely used for diagnosis of AMI but they are not available in the field. Thrombolytic therapy is considered only in cases of confirmed AMI.

NOTE: The Killip classification is a useful tool for assessing prognosis in association with AMI:
Killip Class I: No clinical signs of heart failure. Prognosis good, < 5% mortality.
Killip Class II: Rales bilaterally in up to 50% of lung fields, isolated S-3. Prognosis good
Killip Class III: Rales in all lung fields, acute mitral regurgitation. Prognosis fair to poor.
Killip Class IV: Cardiogenic shock. Decreased LOC, systolic BP 90 or less, decreased urine output, pulmonary edema. cold, clammy skin. Prognosis poor, mortality near 80%.

Cardiac: Congestive Heart Failure
(cardiac vs. non-cardiac pulmonary edema)
CAPT Kurt Strosahl, MC, USN

Introduction: Congestive Heart Failure (CHF), characterized by fluid in the alveoli of the lungs, is defined as the inability of the heart to pump enough blood to meet the demands of the tissues. CHF may be seen following any of these conditions: pulmonary embolism, sepsis, anemia, thyrotoxicosis in pregnancy, arrhythmias, myocarditis, endocarditis, hypertension and myocardial infarction. Other associated causes include fluid overload due to acute renal failure, shock lung due to toxic fumes / smoke / heat inhalation of a fire or blast that causes destruction of the alveolar surfactant, and mountain sickness (high altitude pulmonary edema). It is easier to treat CHF after recognizing and treating any of these precipitating causes.

Subjective: Symptoms
Shortness of breath (SOB); dyspnea on exertion (DOE), dyspnea when lying down (orthopnea), or when awakening from sleep (paroxysmal nocturnal dyspnea, or PND); swelling of the ankles and legs (pedal edema); fatigue and nausea.

Objective: Signs
Using Basic Tools:
Vitals: Tachycardia >120, tachypnea >30, low BP or 10mm variation in systolic pressure between inspiration and expiration (pulsus paradoxicus), fever >101 (valvular infection).
Head/Neck: Jugular venous distension (JVD) with meniscus >5cm above the sternal notch when patient sitting at 30°; lip cyanosis.
Lungs: Fine, crepitant rales (bubbling of alveolar fluid) spreading from the bases to all lung fields; wheezing; productive cough with pink, frothy fluid; pleural effusion.
Heart: sounds of horse galloping (all four hoofs striking the ground S4S1-S2S3); murmur throughout systole (mitral regurgitation); murmur throughout diastole (aortic regurgitation).
Extremities: Cyanosis, pitting edema of the ankles and legs.
Using Advanced Tools: EKG: look for ST elevation or new bundle branch block;
Lab: Elevated WBC suggests infection; casts on urinalysis suggest renal failure.

Assessment:
Differential Diagnosis
Acute myocardial infarction - >1mm ST elevation in 2 or more EKG leads
Myocarditis - ST elevation in all EKG leads
Acute infective endocarditis - loud murmur of mitral regurgitation or aortic insufficiency with a fever and WBC >14K
Hypertensive emergency - diastolic BP >110
Acute blast injury and/or smoke inhalation - sputum contains carbon particles
Multilobar pneumonia - fever, WBC>15, thick yellow/green sputum with WBCs
Mountain Sickness - history of recent arrival at >5000 feet

Plan:
Treatment
Primary:
1. Oxygen at 3-5 L/min to raise oxygen saturation above 90%.
2. **Furosemide (Lasix)** starting at 20 mg IV and doubling every 30 minutes until diuresis ensues, up to 200 mg total. There is no additional benefit above 200 mg if diuresis has not ensued.
3. **Nitroglycerin** 0.4 mg sublingual, repeating q 5 minutes for a total of 3 doses.
4. Sit patient up with legs dangling to ease dyspnea.
5. Monitor urine output and weight (2.2 pounds, or 1 kilogram, equals a liter of fluid loss).

Alternative: Intubation with positive pressure breathing (bag-valve-tube forced inhalation) if oxygen saturation remains <85%.
Primitive: Phlebotomy of 500 cc and rotating tourniquets to decrease venous return.
Empiric: Antibiotics if infective endocarditis is suspected from fever, heart murmur, and red spots on the fingers. **Cephalothin** 2 gm IV qid and **gentamicin** 1.5mg/kg IV q 8 hrs

Patient Education
General: Salt restriction is necessary and weight should not increase over 2 pounds a day.
Activity: Rest, sitting upright until diuresis ensues.
Diet: Clear liquids until oxygen sat >95% and breathing easier, then allow low sodium foods.
No Improvement/Deterioration: Evaluate immediately for evacuation.

Follow-up Actions
Evacuation/Consultation Criteria: Evacuate immediately if unable to achieve diuresis or oxygen sat >90%, for persistent cyanosis or if HR remains over 150 (dysrhythmia). Sedation, intubation with 100% oxygen by bag-valve-tube ventilation and EKG monitoring (if available) is required for transport. A low altitude flight is preferable, if possible. Even a healthy 28 yr. old can not remain in CHF over 2-4 hours without risk of developing life-threatening heart rhythms.

Cardiac: Hypertensive Emergency
CAPT Kurt Strosahl, MC, USN

Introduction: Most hypertension does not require intervention in the field but should have further evaluation and treatment upon completion of the mission. A hypertensive emergency is defined as acute hypertension with damage resulting to other organs. The diastolic BP is usually over 110 mg Hg. Marked hypertension may accompany head trauma.

Subjective: Symptoms
Headache; blurred vision; neurologic deficits; decreased urination; shortness of breath while walking or sitting that worsens when lying down; fatigue; nausea; lack of energy; confusion and chest pain.

Objective: Signs
Using Basic Tools: BP >220/120; bounding pulses; swelling of the legs; cyanosis; tachypnea, tachycardia or bradycardia; blurring of the optic disc or red splotches of hemorrhage on the retina on fundoscopic examination; lungs may be clear or have rales of CHF or pulmonary edema; forceful heartbeat on chest wall; loud bruit just above the umbilicus in renal artery stenosis; decreased urinary output from renal dysfunction (as either a cause or an effect); neurologic defects. Organ damage of note includes effects on heart, brain, eyes and kidneys.
Using Advanced Tools: Lab: Protein and casts on urinalysis (renal failure); HCT (anemia suggests hemorrhage)

Assessment:
Differential Diagnosis
Acute renal failure - casts seen in the urine sediment
Stroke - focal neurologic deficit or depressed consciousness
Aortic dissection - >10 mm difference in systolic BP between the arms
Closed head trauma with elevated intracranial pressure - history of trauma, wound on the head, pupils different in size.

Plan:
Treatment
Treat acute hypertension if evidence of organ damage, or if BP reaches 220/120. Below 200/110, there is no acute benefit to treatment in the absence of end-organ damage.
Primary: Sedation: **diazepam** 5 mg; Diuresis: **furosemide** 20 mg IV (or po) and double every 30 minutes until diuresis occurs.
Alternative: Use antihypertensives to lower the diastolic pressure to 100-110, not to under 90, in the first 24 hours: **lisinopril** 10 mg po or **enalapril** 5 mg po qid; **clonidine** 0.1 mg po q 1hour up to 6 doses.
Primitive: Sitting or elevating the head is essential if head trauma is suspected. Phlebotomy of 500 cc (withdraw one unit by gravity into an empty IV fluid bag) if pulmonary edema is present and diuresis not possible.

Patient Education
General: Hypertension is a common, chronic disease that CAN be controlled with diet, exercise and medications to minimize the risk of stroke, kidney failure or heart failure. There is no need to acutely treat hypertension unless organ damage is evident.
Activity: Rest will lower the BP. Strenuous physical exertion will raise the pressure and should be minimized until BP stable.
Diet: Avoid salt and salty foods.
Medications: Antihypertensive medications are often sedating and can cause orthostatic hypotension.
Prevention and Hygiene: Exercise, weight control, and salt avoidance.
No Improvement/Deterioration: Return for reevaluation.

Follow-up Actions
Wound Care: Clean and dress any wounds appropriately.
Return evaluation: Daily BP checks after initial lowering until resolution of end organ findings. If evidence of end organ damage is present or worsening, additional medications not available in the field will be needed to treat the member.
Evacuation/Consultation Criteria: Progression of neurologic deficits, CHF or anuria should prompt urgent consultation and evacuation if possible.

Cardiac: Pericarditis
CAPT Kurt Strosahl, MC, USN

Introduction: Acute pericarditis is inflammation of the pericardial sac surrounding the heart that results in chest pain. The majority of cases are idiopathic or post-viral. Other causes are acute myocardial infarction, uremia, bacterial infection, tuberculosis, collagen-vascular disease, neoplasm (lung, breast, melanoma, lymphoma, leukemia) or trauma.

Subjective: Symptoms
Precordial chest pain with a pleuritic component (worse when breathing in and out), pain is worse lying down and better sitting up or leaning forward, fever, shortness of breath on exertion or rest, fatigue and malaise.

Objective: Signs
Using Basic Tools: Pericardial friction rub (squeaky leather sound) loudest leaning forward on held expiration; diaphoresis, pallor, neck vein distension greater than 5cm above the sternal notch; possible pleural rub (sound of Velcro). A 10 mm or greater difference in the systolic BP between inspiration and expiration (pulsus paradoxicus) suggests tamponade.
Using Advanced Tools: EKG: Diagnostic ST elevation in most leads and PR depression in II, III, aVf. Lab: WBC >15 suggests infectious cause; urinalysis showing protein and casts suggests uremic cause.

Assessment:
Differential Diagnosis
Pleurisy - pleural rub without pericardial rub or EKG abnormality
Aortic dissection - different pulse pressures between the arms
Pulmonary embolism - typically unilateral calf tenderness and swelling consistent with phlebitis
Pneumothorax - absence of breath sounds on one side is typical
Acute Myocardial Infarction - EKG shows typical ST elevation in 2 or more contiguous leads, not all leads
Pericardial tamponade - falling BP with rising neck veins and signs of hypovolemic shock

Plan:
Treatment
Primary: Rest. **ASA** 650 mg po q 4-6 hours or **ibuprofen** 800 mg tid or **indomethacin** 50mg po tid. Perform pericardiocentesis for tamponade (see Procedure: Pericardiocentesis).
Alternative: Prednisone 60 mg po qd
Primitive: Morphine: Titrate dosage beginning at 2mg IV and repeating q 5 minutes, until pain relief without over-sedation
Empiric: If bacterial infection is suspected, the most common cause is staph. Give **nafcillin** 2 gm q 4 hours plus **gentamicin** 1mg/kg IV q 8 hours.

Patient Education
General: Inflammation of the pericardial sac is often idiopathic or viral and self-limiting over 5-7 days. Pericarditis is not life threatening unless fluid starts to accumulate in the sac. Exertion, even though it hurts, will not worsen the condition.

Activity: Rest and limit heavy exertion when possible.
Diet: Low sodium
Medications: Can cause GI upset. Tinnitus (ringing in the ears) on **ASA** suggests the maximum dose has been exceeded
No Improvement/Deterioration: Return for reevaluation if pain persists >7 days on **ASA**. Pericardiocentesis may need to be repeated for recurrent tamponade.

Follow-up Actions
Return evaluation: Consider tamponade if increasing SOB and fatigue. Perform pericardiocentesis if falling BP and rising neck veins associated with a pericardial rub.
Evacuation/Consultation Criteria: Recurrent tamponade, after pericardiocentesis or continued symptoms despite time and treatment.

Cardiac: Cardiac Resuscitation
Lt Col Robert Allen, USAF, MC

Resuscitation of sudden cardiac death
Sudden cardiac death is a common presentation of coronary artery disease, and may be its first sign. Approximately 75% of sudden cardiac death is due to cardiovascular disease. Sudden syncope may or may not be preceded by chest pain, fluttering sensation in chest, diaphoresis or dizziness. Frequently, arrest is due to malignant cardiac dysrhythmias, most commonly ventricular fibrillation (V-Fib) and pulseless ventricular tachycardia (V-Tach). Early defibrillation of V-Fib or pulseless V-Tach is closely correlated with neurologically intact survival. The most important goal in the treatment of sudden cardiac death is to provide diagnosis and electrical defibrillation of V-Fib/pulseless V-Tach as soon as possible after onset.

When: A patient is unresponsive and in cardiac arrest.

What You Need: Monitor/Defibrillator, or Automatic External Defibrillator (AED), oxygen, airway adjuncts as needed and ACLS drugs as needed.

What To Do:
1. Do a rapid scene survey/tactical assessment to determine any threats in the immediate area.
2. Establish that the patient is unresponsive.
3. Send for help: send for monitor/defibrillator or AED.
4. Open the patient's airway, check for breathing.
5. If the patient is not breathing, give rescue breaths that cause the chest to rise.
6. Check carotid pulses.
7. If no pulse, begin CPR, continue until monitor/defibrillator or AED is available.
8. Intubate and give oxygen if possible.
9. If AED arrives first, attach leads to patient and turn on AED as per instructions.
9. Stop CPR while AED analyses rhythm.
10. Deliver shocks as advised, or if no shock advised, check pulses and continue CPR if pulses are absent.
11. When monitor/defibrillator arrives, attach leads to patient, and hold CPR while checking rhythm on the monitor.
12. Determine cardiac rhythm and initiate the appropriate resuscitation treatment algorithm.
13. Transport victim to highest-level of medical care available as soon as possible.
13. Establish IV access en route if not already done.

Remember the goal in cardiac resuscitation: preventing ischemic brain injury while restoring the normal circulatory action of the heart. When evaluating a possible cardiac patient have your resuscitation medications and equipment set up and ready to go. If the patient goes into arrest, the appropriate action can be taken with a minimum of confusion.

What Not To Do:
Do not initiate CPR on an obviously dead patient.
Do not initiate CPR under direct enemy fire. A patient in cardiopulmonary arrest during a firefight is dead.
Do not touch the patient while AED is analyzing the cardiac rhythm.
Do not use the 'Analyze' function on an AED while in a moving ground vehicle or during moderate/heavy turbulence if in an aircraft.
Do not leave your patient alone. Patients with acute myocardial infarctions can go into a malignant arrhythmia (frequently V-Tach or V-Fib) with no warning.
Do not assume that a 'normal' EKG rules out heart disease as a cause of chest pain.
-Do not assume that all 'heartburn' pain is due to indigestion. In particular, do not assume that eliminating chest pain by administering a 'GI Cocktail' rules out cardiac origin chest pain.
-Do not withhold aspirin if there is a chance of cardiac chest pain. Most people can tolerate a single dose of aspirin without difficulty. Exceptions are cases of true aspirin allergy, asthma with aspirin sensitivity, and active ulcer, GI bleeding or hemorrhagic stroke.

Resuscitation Algorithms

Cardiac resuscitation algorithms have been developed by the American Heart Association, but could not be reprinted here for copyright reasons.

Chapter 2: Blood
COL Richard Tenglin, MC, USA

Blood is made up of solid (cellular) and liquid (plasma) components. Cellular elements originate from bone marrow, and may be broken down as follows: white blood cells (WBCs) that fight infection, red blood cells (RBCs) that transport oxygen and platelets that stop bleeding. Symptoms result due to low numbers of cells or deficient cell function (which paradoxically may occur with increased numbers of abnormal cells, such as with leukemia), or when cell numbers build up to such a point that they obstruct blood flow. Low cell numbers are caused by decreased production, or increased loss (bleeding), consumption or destruction. Plasma contains the soluble coagulation factors, immunoglobulins, electrolytes, protein and water. Evaluation of blood disorders often requires performing a spun hematocrit and a Wright (Cameco Quick Stain) stained peripheral smear (see Lab Procedures Section).

Anemia

Introduction: Anemia refers to an abnormally low amount of the oxygen-carrying protein hemoglobin (may also have low number or volume of red cells) in peripheral blood. "Hematocrit" is the percent volume of whole blood occupied by the red cells, and is determined by spinning a sample of blood and measuring the volume of the "packed" red cells divided by the total volume of the sample. It is a rough measure of the amount of oxygen-carrying protein (hemoglobin) in the sample, but is subject to many problems with technique that can lead to numbers that do not reflect the true hemoglobin content of blood. Modern Coulter Counters used in clinical labs actually measure the amount of hemoglobin, and calculate, but do not actually measure, the hematocrit. The SOF medic will actually measure the hematocrit, but must be aware that some diseases, or bad technique (not spinning the sample sufficiently) will give results that do not reflect the actual hemoglobin content of the patient's blood. Anemia may be acute (traumatic blood loss) or chronic (due to chronic disease), and results from either increased loss/destruction of red cells or failure of the bone marrow to produce sufficient quantities of hemoglobin/red cells to make up for normal red cell loss. Anemia can be determined by a spun hematocrit. The procedure for obtaining a hematocrit and "normal" values are specific to the machine used. Normal hemoglobin levels differ among ethnic populations and between men and women, with males and whites averaging higher values. Worldwide, the most common cause of acquired anemia is iron deficiency due to chronic blood loss from hookworm and menstruation. Other important causes are lack of important nutrients (protein, Vitamin B12, Folic Acid) and suppression of the bone marrow from chronic infection or inflammation. The causes of anemia are extensive and beyond the ability of the medic to accurately diagnose in the field environment.

Subjective: Symptoms
Acute: Lightheadedness, pallor, shock, syncope, altered mental status.
Chronic: Lethargy, fatigue and decreased energy, rapid heartbeat and shortness of breath/dyspnea with exertion.

Objective: Signs
Using Basic Tools: Acute: shock, hypotension, weak pulse, syncope, altered mental status. **Chronic** (compensated for intravascular volume loss): pale skin, mucous membranes (eyelids, under the tongue), nail beds and palm creases (compare color of normal palmar crease to patient's).
Using Advanced Tools: Lab: spun hematocrit or hemoglobin; stool for occult blood; peripheral smear (see Laboratory Procedures Chapter)

Assessment:
Differential Diagnosis: (see Color Plates: Identification of Cellular Blood Components):
Elevated total white blood cell (WBC) count with great preponderance of polymorphonuclear leukocytes (PMNs), along with fever and chills, suggests a bacterial infection.
Significantly low lymphocytes suggest viral infection, including HIV and others.
Increased eosinophils suggest either infection (parasites, especially visceral larval migrans, or chlamydia), hypersensitivity or allergic reactions.
Increased basophils are so unusual as to suggest a problem with the stain or staining procedure.
Macrocytosis suggests folic acid or B12 deficiency*.
Microcytosis suggests iron deficiency or thalassemia*.
Aniso, Poikilocytosis and target cells suggest thalassemia*.
Banana or "sickle" shaped cells suggests one of the sickle cell conditions*.

*****NOTE:** Confirmatory testing for these anemias is beyond the scope of the SOF medic. Unusual hemoglobins or hemoglobin levels may be common within certain ethnic groups. Attempting to correct these presumed anemias is inappropriate.

Plan:
Treatment
1. Iron supplementation is appropriate only for menstruating females and patients whose stool is positive for occult blood, pending further evaluation.
2. Iron supplementation is not otherwise appropriate without laboratory determination of iron deficiency.
3. Correct obvious nutritional deficiencies and treat infections or inflammation.
4. Blood replacement in the face of rapid loss is addressed in Procedure: Field Transfusion*.
5. Treat acute sickle crisis with:
    a. Fluids - orally if possible and IV if needed--3-4 liters/day in adults;
    b. Liberal use of medications for pain (Selection of medications is determined by the severity of the pain. Non-steroidal anti-inflammatory drugs like **ibuprofen**, **acetaminophen** with **codeine**, or intravenous **morphine** are appropriate for mild, moderate and severe pain, respectively, and should be continued until pain levels decrease.)
    c. Treatment of predisposing conditions, such as infections.
    d. Steroids and antibiotics are not routinely indicated.

*****NOTE:** No non-US sources of blood can be trusted for accuracy of blood types, freedom from infection, or deterioration from improper storage. US Embassies can often provide information on where the safest blood products may be obtained within a specific country or region, but use of these products always entails an significant risk of both immediate and delayed life threatening consequences.

Follow-up Actions
Evacuation/Consultation Criteria: Evacuate patients with acute anemia and acute sickle crisis after initial stabilization. There is usually no need to evacuate patients with other anemias. Consult hematologist as needed.

Chapter 3: Respiratory

Respiratory: Common Cold and Flu
COL Warren Whitlock, MC, USA

Introduction: Over 200 kinds of viruses and bacteria infect the mucous membranes, leading to such symptoms as nasal congestion, sore throat, and coughing (see also ID sections on Adenovirus, Infectious Mononucleosis) Several typical pediatric diseases, such as diptheria, that can infect unprotected adults). This section focuses on the relatively mild, viral, acute respiratory tract infections (contrast with Acute Respiratory Distress Syndrome Section later in this chapter), typically called colds. The frequency of these infections generally decreases with age. Acute local infections generally occur at the site of viral infection - the nose and throat. The "flu" is a viral infection of the nose, throat, bronchial tubes and lungs caused by influenza viruses A or B, typically presenting with sudden fever, chills, headache and fatigue. The fever often runs higher than 101°F, and usually subsides within three days. Muscle aches of the back, arms and legs are especially prominent symptoms. Patients generally continue functioning with a cold virus, but with flu their activity is more seriously curtailed. Elderly patients, infants, expectant mothers (3rd trimester) the immunosuppressed and patients with chronic heart and lung diseases will have more frequent life-threatening complications, such as pneumonia. Approximately 10,000 people die annually in the US due to influenza.

Subjective: Symptoms
General: Malaise, fever, nasal congestion, clear secretions, sneezing, scratchy or sore throat, cough, hoarseness, and headache
Focused History: Quality: *Is your cough productive?* (Viral coughs are generally non-productive or produce only clear mucus.) *How high has your fever been?* (characteristically normal or low-grade temp; patients commonly "just feel hot") *How bad is your headache?* (Mild headache that worsens upon standing is typical; severe headache signals other potential illness.) *Where is the headache?* (If located over sinuses and worsens when head is lowered, may have sinusitis with or without cold.) *Is the cough worse at night?* (typical for post-nasal drip from cold, sinusitis, allergic or irritant rhinitis) **Duration:** *How long have you had the symptoms?* (Colds usually do not last longer than a few days, and not over 2 weeks.) *How often do you get colds?* (frequent colds may suggest allergies, increased susceptibility to infection due to immunocompromise, anatomical defect, etc.) *Have you had the influenza vaccination?* (Flu symptoms should be less severe and shorter in duration if vaccinated for the infecting strain.) **Alleviating or Aggravating Factors:** *What makes the symptoms better, or worse?* (Medications including decongestants and **acetaminophen** may improve symptoms.)

Objective: Signs
Using Basic Tools: Inspection: Secretions in nose; red, irritated throat which may have exudate; normal to low-grade fever
Palpation: Non-tender sinuses (tender sinuses with sinusitis)
Auscultation: Clear chest (wheezes occur in patients with known asthma or in 10% of cases of viral influenza)
Using Advanced Tools: Otoscope: Eardrums appear normal with no fluid behind them; Lab: Monospot negative; WBC with differential is normal or may demonstrate an increase in atypical lymphocytes

Assessment:
Differential Diagnosis: See appropriate sections in this book for more information on many of these conditions.

Allergic rhinitis - seasonal history of itching in nose, roof of mouth, throat, and eyes; increased tearing, sneezing and watery discharge; sore throat due to postnasal drip (granular appearance to posterior pharynx); nasal membranes swollen and red.
Irritant rhinitis - non-seasonal history of exposure to irritant, rhinorrhea without ocular symptoms or sore throat.
Influenza - more systemic symptoms than cold, including muscle aches, severe headache, chills, and higher fever to 103°F.
Atypical pneumonia (See Pneumonia) - fatigue, mild respiratory symptoms.
Mononucleosis - positive Monospot; longer illness; higher fever, chills, malaise, sore throat, and swollen cervical lymph nodes.
Sinusitis - tender sinuses; red, swollen nasal membranes; green or yellow discharge from nose and throat; can follow a cold.
Rubeola - characteristic rash; Koplik's spots on mucous membranes; seen typically in children but can attack non-immune adults.
Mumps - seen typically in children but can attack non-immune adults; swollen salivary glands.
Pertussis - characteristic whooping cough; seen typically in children but can attack non-immune adults.
Diphtheria - seen typically in children but can attack non-immune adults; gray membrane maybe seen on pharynx.
Adenovirus - fever, pharyngitis, and/or conjunctivitis; usually in epidemics in non-immune recruits or displaced persons
Strep Pharyngitis - palatal petechiae, red beefy uvula, and scarlatiniform rashes are typical for Group A streptococcal pharyngitis.
Gonococcal Pharyngitis - often asymptomatic, but may have sore red throat, painful swallowing and history of exposure

Plan:
Treatment
1. See appropriate sections for treatment of pneumonia, infectious mononucleosis, sinusitis and other illnesses that present similarly to cold/flu.
2. Treat cold symptomatically: Rest; fluids (higher "insensible" losses due to fever); **acetaminophen**; warm, salt water gargles (sore throat); warm steam from a tea kettle or shower, and saline nasal drops (nasal congestion); no tobacco and alcohol products; hot chicken soup.
3. Antibiotics: Only indicated in patients that may have or are at high risk for a secondary bacterial infection (see pneumonia).
4. Moderate to severe nasal congestion: Nasal **ipratropium bromide** (nasal **Atrovent** 0.06% in adults, 0.03% in children) tid or q hs 2 sprays/nostril has been shown to reduce local symptoms and shorten duration of rhinorrhea symptoms by one day. **Oxymetazoline** (e.g., **Afrin**) nasal spray 0.05% solution (use for no more 3 days, or just at night, to avoid rebound congestion) - adults and children ages 6 and up: 1-3 sprays/nostril q 12 hrs. **Pseudoephedrine** tabs or liquid –Children 6-12 years: 30 mg po q 6 hours, Children 2-6 years: 15 mg q 6 hours, Adults: 60 mg po q 6 hours.
5. Allergic rhinitis: **Diphenhydramine** - Adults and children over 12: 25-50 mg po tid/qid; children under 12: 5 mg/kg/day po in divided doses qid.
6. Irritant rhinitis: Use saline nose drops to lavage nasal mucosa, followed by **pseudoephedrine** 30-60mg po q4-6h to decrease mucus membrane swelling.
7. Cough: See Symptom: Cough.
   Nonproductive Cough: **dextromethorphan** – adults and children over 12: 10-20 mg po q 4h, or 30 mg q6h; children 6-12 years: 5-10 mg po q 4h or 15 mg q 6-8 hrs; children 2-6 years: 2.5-5 mg po q 4h, or 7.5 mg every 6-8 hours
   Productive Cough: Do not suppress a productive cough unless it interferes with obtaining adequate rest/sleep or jeopardizes your mission. **Codeine** q hs can be used for severe cough and will cause drowsiness (use no more than 3 nights). Expectorants like **guaifenesin** are often used but have not been proven effective.
8. Influenza: Give **Relenza (zanamivir)** for patients ages 7 years and older within two days of the onset of symptoms. The drug is less effective in patients whose symptoms are not severe and do not include

fever. Administer bid for 5 days via plastic inhaler (Diskhaler). If patients develop wheezing, discontinue the drug and be prepared to treat symptoms (see Respiratory: Asthma). **Amantadine** or **rimantadine** shorten duration of symptoms by 50% and are recommended for patients at high risk for complications from infection. 100 mg bid po for 3-5 days if started within the first 48 hours. Give symptomatic treatment for cough and nasal symptoms as well.

Patient Education
General: Infections can spread via airborne droplets (cough, sneeze) and contact (contaminated hands, lips and objects). The usual course of a cold is 6-10 days, and about half that length for uncomplicated influenza.
Activity: Rest is important to speed recovery. Strenuous activity can delay recovery.
Medication: Aspirin is not recommended, especially in children, due to the risk of Reye's syndrome – a life-threatening form of kidney failure.
Prevention: Hand washing can reduce transmission. Vaccination will prevent specific strains of influenza, but not all influenza. The vaccine cannot cause influenza, but some side effects (myalgia, headache) may mimic mild influenza or cold symptoms.
No Improvement/Deterioration: Return if symptoms worsen or do not resolved in two weeks, or if fever over 103°F develops.

Follow-up Actions
Return Evaluation: Evaluate for alternative diagnoses and complications, including secondary bacterial infection, if still symptomatic after 72 hours of treatment (particularly if in high-risk group).
Evacuation/Consultation criteria: Evacuation not usually necessary, except for moderate to severe influenza. Consult primary care physicians as needed.

Respiratory: Pneumonia
COL Warren Whitlock, MC, USA

Introduction: Pneumonia, an infection of the lungs, is a leading cause of death worldwide. There are three important principles for successful treatment:
1. Accurately assess severity and initiate appropriate treatment-- outpatient or inpatient administration of IV antibiotics. Hospitalize any patient with more than 2 of the following characteristics: age >65, immunosuppression due to chronic disease, significantly altered vital signs (temp >102°F, tachypnea, hypotension) or mental status changes, lab or x-ray findings as below. Mortality for these patients can be 10-25% versus < 1% for other patients.
2. Avoid delays in treatment, which can negatively affect the patient outcome.
3. Treat appropriate organisms: typical (bacterial) and atypical *(mycoplasma, chlamydia, legionella, viruses)* organisms

Subjective: Symptoms
Fever (over 101°F), rigors (shaking chills), malaise, shortness of breath, cough (productive and non-productive), occasional myalgias, chest pain– generally pleuritic (rarely upper abdominal).
Focused History: Quality: *Do you cough up anything? What color is it?* (Colored sputum is a good indicator of bacterial infection.) *Does coughing or deep breathing make your chest hurt?* (Chest pain of pneumonia characteristically worsens with cough or deep breathing.) *Where does it hurt?* (If located over rales, unilateral rhonchi, or a pleural friction rub, indicates probable "lobar" or whole lobe pneumonia.) *Do you have fever or shaking chills?* (Atypical pneumonia can present with low-grade fever but typical pneumonia classically presents with high, spiking fevers that follow rigors or shaking chills.) *Do you have any trouble breathing?* (Shortness of breath or difficulty breathing is typical.) *When is the cough worse?* (Post-nasal drip cough is worse at night.) **Duration:** *When did the symptoms start?* (Typical pneumonia develops suddenly and patient presents within hours. Atypical [viral, mycoplasma, etc.] usually begins with a prodrome of low-grade fever and malaise for several days.) **Alleviating or Aggravating Factors:** *Is the pain better in certain positions?* (Chest pain that

changes with body position, as in going from supine to upright or vice versa, is indicative of a pleural effusion or empyema. Pain severe enough to cause the patient to lie on the affected side for hours is suggestive of an empyema – see Empyema section.)

Objective: Signs
Using Basic Tools: Vital Signs: Fever over 101°F, Respiratory rate over 14, Resting Pulse over 90 beats/min
Inspection: Cyanosis (bluish skin color may be normal in some dark pigmented people, especially around the lips and nails); splinted respirations; lying with pleuritic side down.
Palpation: Warm over dull-sounding area - empyema
Auscultation: Rales indicates an infiltrate; rhonchi indicate airway secretions; dullness may indicate lobar consolidation, collapsed lung, or a pleural effusion.
Using Advanced Tools: Pulse oximetry: < 90% Lab: WBC < 4000 (atypical pneumonias or immunocompromised patients) or > 30,000; Hematocrit <30; Creatinine over 1.2 mg/dl or BUN over 20 mg/dl (if available); Sputum: Gram stain (> 15 WBC/HPF indicates infection);
CXR (if available): findings other than single-lobe involvement (i.e. multiple lobes, cavitation, and rapid progression from a prior film, or the presence of a pleural effusion).

Assessment:
Differential Diagnosis
Lobar (typical) pneumonia, viral pneumonia (see Common Cold and Flu, Adenovirus), mycoplasma pneumonia, tuberculosis (see TB Section), other atypical pneumonias.
Atelectasis can resemble pneumonia but is caused by a mechanical airway obstruction, chest wall abnormality or a loss of normal lung space. The treatment for atelectasis focuses on opening up the alveoli with aerosol bronchodilators, cough induction, and antibiotics if infection is present. Untreated atelectasis is a significant risk factor for developing pneumonia.

Plan:
Treatment: Always check for medication allergies!
1. **Outpatients on po antibiotic**: Generally preferred: macrolides, fluoroquinolones, or **doxycycline**
   Azithromycin: 500 mg po, then 250 mg po qd x 4 days (safe and effective – children and pregnant women)
   Erythromycin: 250-500 mg/day po qid for 10 days (30% have gastrointestinal side effects)
   Levofloxacin: 500 mg po qd x 7 – 10 days
   Doxycycline: 100 mg po bid x 10 days (not in children under 12 years or pregnant women)
2. **Hospitalized Patients on Intravenous Antibiotics:** Start treatment as soon as possible. Change to single agent oral therapy 24 hours after the patient clinically improves.
   Ceftriaxone[A] 1 gm IV qd, plus **azithromycin**[B] 500 mg IV qd (separate IV infusions, children: 15 mg/kg/day)
   Alternatives: **levofloxacin**[C] 500 mg IV qd – single agent, **Cefuroxime**[A] 750 – 1.5 g (children 50-100 mg/kg/day) IV plus **erythromycin**[B] 500 mg IV (1 g for *Legionella*)
   [A] – Beta lactam: second or third generation **cephalosporin**;
   [B] – Macrolide: only **azithromycin** and **erythromycin** are IV;
   [C] – Fluoroquinolone: must be broad spectrum: **levofloxacin, omnifloxacin, grepafloxacin**
   Supplemental oxygen
   Postural drainage and frappage (systematic, forceful percussion of back) to help remove secretions.

Patient Education
General: Stop smoking.
Medication: Complete a full course of the appropriate antibiotic therapy.
Activity: Bedrest while on IV antibiotics. Limited activity for one to two weeks (temporary profile, T-2).
Prevention: Vaccinate all personnel for influenza and adenovirus, and vaccinate those without spleens for pneumococcus and haemophilus.
No Improvement/Deterioration: Return if symptoms do not resolve or improve in 48 hours.

Follow-up Actions
Return Evaluation: Reevaluate or refer patients with delayed recovery. If poor response or late deterioration (5-10% of patients), suspect inadequate antibiotic dosing (e.g., changing from IV prematurely) or a complication such as empyema or pleural effusion. Identify the organism if possible. If available, do follow-up chest x-ray in 6-8 weeks to evaluate resolution of infiltrate.
Evacuation/Consultation Criteria: Evacuate if unstable: respiratory rate >30/min, falling BP, increasing tachycardia, fatigue or drowsiness, cyanosis or decreasing saturation of O_2 despite oxygen therapy. Consult pulmonologist, internist or infectious disease specialist as needed and for pleural effusion (see Pleural Effusion section), empyema and hemoptysis (possible occult malignancy).

Common Causes of Pneumonia (in developed countries)
Streptococcus pneumoniae 40-70%
Mycoplasma pneumoniae 10-20%
Legionella species 10-15%
Haemophilus influenzae 10-15%
Influenza virus 5-10%
Chlamydia pneumoniae, Moraxella catarrhalis, Staphylococcus aureus, Gram negatives, other 5-10%

Respiratory: Pleural Effusion
COL Warren Whitlock, MC, USA

Introduction: A pleural effusion is fluid in the space between parietal pleura on the chest wall and visceral pleura around the lungs. If the fluid accumulation is large (>1/3 of the hemithorax or over 1-2 liters), it can interfere with the mechanical ability to breathe. The two major types are transudative effusions, which are passive fluid accumulations, and exudative effusions due to irritation and inflammation. Transudative effusions are usually bilateral, slightly greater on the right side and are usually caused by heart failure, low albumen in circulation and rapid loss of albumen in the urine (nephrotic syndrome). Exudative effusions are caused by inflammatory involvement (including infection) of the overlying visceral pleura, which often results in acute pleurisy and the leakage of serous fluid into the pleural space. The type of cells in the fluid may indicate the cause of the effusion. A large number of neutrophils containing bacteria indicate an early empyema (see Empyema) while large cells of abnormal shape may indicate cancer. Atypical lymphocytes can occur with viral infections such as influenza or Coxsackie virus.

Subjective: Symptoms
Stabbing chest pain with breathing or cough (pleuritic pain), or chest pressure or tightness that changes with position (mimicking angina).
Focused History: *Have you had a recent respiratory illness?* (typical for an effusion). *Do you have chest pain during every deep breath?* (Chest pain worsens with inspiration.) *Have you had fever?* (indicates a complicated or infected effusion or empyema) *What makes the symptoms better?* (Pain may be minimized by shallow breathing, minimal talking or exercise, or by holding/lying on the affected side.)

Objective: Signs
Using Basic Tools:
Inspection: May lean towards or lie on affected side, and use arms to support and minimize chest movement (splinted respirations).
Palpation: Warm skin over affected area. Abnormally large or small liver with possible ascites. Peripheral edema in heart failure.
Percussion: Dullness means a pleural effusion or empyema.
Auscultation: Chest: Clear, although occasionally a pleural friction rub may be heard. If rales and rhonchi are heard, a pneumonic process such as pneumonia is likely. Heart: Extra sounds (murmurs, rubs and gallops) may indicate signs of cardiac failure.
Using Advanced Tools: Labs: Elevated WBC or increased neutrophils on differential supports infection;

sputum for Gram stain and culture; Effusion fluid for gram stain and WBC differential; CXR: effusion shadow overlying lungs, and possible enlarged heart.

Assessment:
Differential Diagnosis
Transudative effusion - congestive heart failure, liver failure (any cause), nephrotic syndrome (any cause). Exudative effusion - infection– bacterial (empyema), fungal, tuberculosis; cancer (lung or metastatic); collagen vascular disease/rheumatoid arthritis, lupus; vascular – pulmonary embolus; unknown – granulomatous

Plan:
Treatment
1. Treat the primary disorder if possible.
   a. Give antibiotics for pneumonia (see Pneumonia Section).
   b. If a transudative effusion is suspected, give a trial of **Lasix** 20-60 mg po qd-bid
2. Perform thoracentesis (see following section) to improve breathing, if **Lasix** ineffective or in the face of unimproving pneumonia.
   a. Withdraw 30 cc of fluid for diagnostic laboratory evaluation or
   b. Draw off the effusion if fluid accumulation compromises respiratory status. Try not to remove more than 1000-1500 cc of fluid in the first 24 hours (can repeat procedures). Removing too much fluid can cause rapid fluid shifts in the lung tissue, which worsens hypoxemia (newly expanded lung is poorly perfused) and causes hypotension.
   c. A chest tube thoracostomy MUST be performed if the fluid is infected (see Procedures chapter).
3. Administer 30-40% oxygen since these patients are commonly hypoxic.

Patient Education
Activity: Bedrest with indwelling chest tube initially.
Diet: High protein diet unless liver failure is present, then diet must be modified to avoid hepatic encephalopathy

Follow-up Actions
Return Evaluation: Refer patients that do not improve for specialty care and additional special studies.
Evacuation/Consultation Criteria: Evacuate unstable patients, or those who require on-going thoracenteses. Consult internist or pulmonologist.

Respiratory: Thoracentesis
COL Warren Whitlock, MC, USA

What: Thoracentesis the removal of pleural fluid percutaneously by needle aspiration to determine the cause of fluid accumulation or to relieve the symptoms associated with the fluid accumulation.

When: Perform a diagnostic thoracentesis when the presence of fluid in the pleural space is confirmed by physical examination (and preferably by CXR), and the likelihood of bacterial infection in the fluid is high (worsening condition despite broad spectrum antibiotics, lying with affected side down, progressive fever and lethargy).
Risks: Thoracentesis is a relatively safe procedure; however, some relative contraindications include history of coagulopathy (increase risk of bleeding), pleural effusion of insufficient volume (little fluid layering on lateral decubitus chest film), and underlying severe respiratory disease. Complications of thoracentesis include pneumothorax, bleeding, infection, puncture of abdominal organs, and pulmonary edema of the reinflated lung. The most common major complication is pneumothorax. Thoracentesis can cause a pneumothorax in two ways: by introducing air through the back of the syringe or needle hub into the pleural space (it does not progress to complete pneumothorax and does not require treatment), or by an accidental puncture of the lung. If the patient is symptomatic, keep him under observation and follow the patient's progress with a serial CXR.

Normally the puncture in the lung seals and air is absorbed spontaneously. More severe leaks are caused by coughing or needle movement, which causes a larger tear in the lining of the lung. These injuries may require a chest tube to reinflate the lung.

What You Need: Essential: 1½ inch needle 18 – 21 gauge (21 may be too small if pus is in the pleural space), 10- 30 cc syringe to aspirate fluid, topical antiseptic (iodine-based cleanser followed by alcohol wipe). Recommended: 1-2% **lidocaine** for SC anesthesia in a 10 cc syringe with 23 gauge needle, sterile drape, sterile gloves, clamp, sterile laboratory tubes.

What To Do:
1. Determine the point of entry. The posterior approach is most common because the interspaces between ribs are wider in the back. The ideal location is the **7th or 8th** interspace posteriorly, midway between the posterior axillary line and midline. This site avoids possible accidental puncture of the liver, spleen or diaphragm. Tap with a finger and listen with or without a stethoscope to identify where the percussion becomes dull (height of pleural fluid accumulation). Mark this location by pressing the tip of an ink pen (point retracted) into the skin below where dullness begins and inferior to any underlying rib (avoid the neurovascular bundle immediately below the inferior rib margin). Gently apply pressure for 30 seconds to leave a small red circle that will last during the procedure. Loculated or small effusions may not always be accessible with this approach and should be evacuated if possible for advanced care.
2. Have the patient straddle a chair backwards; resting their arms on the back of the chair.
3. Disinfect the skin around the insertion site and drape the area.
4. Anesthetize the tissues. Begin by anesthetizing the skin at the mark. Aspirate to ensure no blood return before injecting lidocaine, then advance slightly and repeat. Aim the needle towards the upper margin of the rib and anesthetize the top of the rib, then the parietal pleura. Advance the needle gently and carefully while keeping suction, then stop and inject **lidocaine**, and advance again. The anesthesia needle is generally a 23 – 25 gauge, and you can use it to withdraw several cc's of fluid if you enter the pleural space, confirming your landmarks for introduction of the larger needle and syringe.
5. Insert the thoracentesis needle with syringe. Aim for the top of the rib below your mark and inch your way past, continuing at a 30° angle downwards toward the pleural. A slight "give" will indicate that you have pierced the parietal pleura. Aspirate 50 cc's of fluid or more (see Pleural Effusion). The clamp may be used to stabilize the needle at the skin to prevent accidental additional penetration of the needle down to the lung.
6. Withdraw the needle and syringe.
7. Write a procedure note. Be sure to describe the site and approach used, the appearance of the fluid and how much fluid was removed.
8. Complications appear with in the first 24 hours. Have the patient remain in bed for at least 2 hours after the procedure, avoid coughing or lifting objects for 24 hours, and inform you immediately if they cough up blood, experience shortness of breath, dizziness, a tight feeling in the chest, or any other problems.
9. Send sample of fluid for the most important tests first, which are gram stain and differential count of inflammatory cells in a field setting.
10. Repeat as needed.

What Not To Do:
Try not to move the plunger end of the syringe laterally during the procedure. This swings the needle around inside the patient, tearing the pleura and causing a large pneumothorax.
Do not take off the syringe and leave the needle hub in the patient. This can also result in pneumothorax, allowing air to enter the pleural space. If it is necessary to change syringes while leaving the needle in, have the patient "hum" to produce positive pleural pressure.

Respiratory: Empyema
COL Warren Whitlock, MC, USA

Introduction: Empyema, a pleural effusion of pus caused by progression of infection into the pleural cavity, is usually a life-threatening complication of pneumonia. It can also arise from inoculation of the pleural cavity after penetrating chest trauma, esophageal trauma, thoracentesis or chest tube placement.

Subjective: Symptoms
Gray skin, sweating, chills, malaise, fever, chest pain, cough, emaciation, and poor appetite.
Focused History: *Have you had pneumonia or a respiratory illness?* (Empyema is a complication of untreated or ineffectively treated pneumonia.) *How high is your fever?* (Anaerobic empyemas may have low-grade fever, but high fever is more common.) *What makes the symptoms better or worse?* (Shallow breathing and holding or lying on the affected side minimizes chest pain.)

Objective: Signs
Using Basic Tools: Vital signs: Fever > 101.5°F, tachycardia, hypotension
Inspection: Generally toxic appearance. Patient may be somnolent or gravely ill. Mental status changes are common in impending sepsis. Lying on the affected side for several days may predispose to empyema.
Palpation: Warmth over local area of chest suggests an empyema or other effusion. Enlarged lymph nodes may indicate infection or cancer.
Percussion: Dullness means a pleural effusion or empyema.
Auscultation: Rales and rhonchi may be heard from surrounding areas of pneumonia. Pleural friction rub may be heard. Dull respirations with shifting margins of dullness (change in position) indicate fluid effusion; unchanging pattern of dullness is consistent with empyema
Using Advanced Tools: Lab: Pus in effusion fluid (thoracentesis) is virtually diagnostic of empyema (damage to thoracic duct causing chylothorax is exception); elevated WBC with left shift on differential (bacterial infection); sputum for Gram stain and culture. CXR: upright, flat and decubitus views (loculated pleural effusion suggests empyema). Consider PPD skin test for TB in resistant cases.

Assessment:
Differential diagnosis
Other causes of pleural effusion.

Plan:
Treatment
Primary:
1. Antibiotics: **cefuroxime** 750 mg to 1.5 gm IV or IM (if necessary) q8 hours, or **ceftriaxone** 1-2 gm IV once/day plus anaerobic coverage such as **clindamycin** 300-600 mg IV q8 hours.
2. Chest tube drainage (see Procedure Section) is the most important treatment. Give IV antibiotics before inserting chest tube.
3. Supportive: Oxygen, hydration, chest physiotherapy (systematic, forceful percussion of back) (see Nursing Chapter on CD-ROM).

Primitive: Clear liquids for 24 – 48 hours, humidified environment, ANY antibiotic, prevent lying on one side.

Patient Education
Activity: Restrict activity
Diet: High calorie with adequate protein to replace body stores

Follow-up Actions
Return Evaluation: Refer patients that do not improve for specialty care and additional special studies.
Evacuation/Consultation Criteria: Evacuate unstable patients. Consult internist or pulmonologist.

NOTE: Pleurectomy (pleural stripping) in cases of "trapped lung" is a definitive procedure only done by surgeons when a large scar forms and impairs normal lung function.

Respiratory: Allergic Pneumonitis
COL Warren Whitlock, MC, USA

Introduction: Hypersensitivity pneumonitis begins with allergic lung inflammation from repeated inhalation of dust particles consisting of animal proteins, plant proteins or reactive inorganic compounds. It is different from inflammatory disease of the airways (asthma) in that it diffusely affects the lungs and symptomatically presents as a sub-acute, progressive or recurrent pneumonia. In cases where the allergen is inhaled repeatedly, recurrent pneumonia can be sudden and life-threatening. *Thermophilic actinomycetes* is a mold that causes several types of hypersensitivity pneumonitis: farmer's lung or silo filler's lung (exposure to moldy silos), air conditioner lung (exposure to moldy air filter) and bagassosis or cotton worker's lung (inhalation of fibers or moldy cotton). Other types include bird breeder's lung (inhalation of avian protein, blood or dander), isocyanate lung (exposure to toluene diisocyanate [TDI] or methylene diisocyanate (MDI) used in polyurethane, plastics and some spray paints) and washing powder lung (*Bacillus subtilis* enzymes). Chronic exposure to the allergen can result in permanent restrictive lung disease.

Subjective: Symptoms
Acute illness (within 6 hours of exposure): cough, dyspnea, malaise, and body aches (mimics an acute infectious pneumonia).
Chronic illness: progressive condition without acute exacerbation, cough, dyspnea and exercise limitation, anorexia, weight loss, and fatigue.
Focused History: Quality: *Do you get sick after a specific activity or exposure? Do symptoms go away when on vacation or visiting relatives in a distant city or state?* (Diagnose based on history of symptoms only occurring at work or in a certain environment. Sometimes a patient will need to keep a diary to log all their activities and exposures.) **Duration**: *How long do your symptoms last?* (Generally, pneumonitis start to improve over 72 hours, unless additional exposure.)

Objective: Signs
Using Basic Tools: Acute illness: Vital Signs: Fever up to $104^\circ F$, tachypnea, tachycardia. Inspection: acute sputum production (clear, white or colored), cyanosis. Auscultation: Fine, mid- to end-inspiratory crackles in chest. Signs may improve without treatment if removed from the offending antigen.
Chronic illness: Vital Signs: Afebrile or low-grade fever. Inspection: Progressive cyanosis and clubbing of fingers. Auscultation: Fine, mid- to end-inspiratory crackles in chest; right heart failure with extremity swelling. Signs will not acutely improve when removed from the offending antigen due to lung scarring from chronic exposure. Diagnosis can only be made by laboratory testing at this point in the disease.
Using Advanced Tools: CXR: Pneumonia appears as a localized or diffuse infiltration or granulomas in the peripheral areas of the lung. Pulmonary function studies (if available) may show restriction and reduction in diffusing capacity of the lung

Assessment:
Definitive diagnosis can only be made by laboratory testing for allergies (hypersensitivity panel).
Differential Diagnosis: Acute illness resembles typical and many atypical pneumonias including influenza and mycoplasma
Chronic illness: Non-resolving or recurrent pneumonia, tuberculosis, sarcoidosis, fungal infections, and *Pneumocystis carinii* pneumonia with HIV.

Plan:
Treatment:
Primary:
1. Initial therapy: **Avoid exposure**.
2. Corticosteroids: **Prednisone**, 2 mg/kg/day or 60 mg/m2/day po, or other comparable corticosteroid. Initial course of 1-2 weeks with progressive withdrawal of medication. If exposure cannot be discontinued, alternate day therapy may help, but may not prevent progression.
3. If symptoms have progressed to pneumonia, give antibiotics **(Macrolide, Vibramycin)** and bronchodilator **(albuterol)** as discussed in Pneumonia and Asthma Sections respectively.

Primitive: Avoid allergen. Breath humidified air from steam kettle or shower.

Patient Education
General: Avoid the allergen. There is risk of irreversible lung damage with continued exposure. Note that chronic exposure may lead to a loss of acute symptoms previously experienced on exposure, i.e., patient may lose awareness of exposure-symptom relationship.
Activity: Restrict if symptoms worsen after exposure to antigen
Prevention: Use appropriate masks and filters when exposed to allergen. Keep ventilation systems clean and well maintained.
No Improvement/Deterioration: Return for worsening symptoms or those that do not resolve after 3-4 days of treatment.

Follow-up Actions
Return Evaluation: Symptoms that do not improve should be referred for specialty care and additional special studies.
Evacuation/Consultation criteria: Evacuate patients who are not able to complete the mission, or whose symptoms do not resolve. Consult internist or pulmonologist as needed.

Respiratory: Asthma
COL Warren Whitlock, MC, USA

Introduction: Asthma is usually an acute allergic response triggered by inhaled particles (dander, pollen, mold, and dust), fumes, drugs or food. It is characterized by continuous or paroxysmal breathing, wheezing, coughing or gasping caused by narrowed airways in the lungs. This narrowing is due to spasm of bronchial smooth muscle, edema and inflammation of the bronchial mucosa, and production of mucus. Asthma can occur at any age but develops most commonly in children, with 7-19% of children experiencing asthma at some time. 50% of cases are children under age 10. Asthma attacks may have a slow onset or they may occur suddenly, causing death in minutes. Some cases may become continuous (called status asthmaticus), and can be fatal. Intermittent symptoms are usually brought on by exercise, cold air or respiratory tract infections. Nocturnal asthma attacks occur in up to 50% of all asthmatics and may be the only symptoms presented by the patient. Smoke, other inhaled pollutants, respiratory tract infections (especially viral), aspirin use, tartrates, exercise, sinusitis, gastroesophageal reflux, and stress are aggravating factors.

Subjective: Symptoms
May be paroxysmal or constant: Coughing, labored breathing, wheezing, gasping, feeling of constriction in the chest.
Focused History: Quantity: *How many nights are you awakened by wheezing?* (few times per week indicates moderate disease; almost nightly indicates severe disease). *How many days of work or school have you missed in the last month because of asthma?* (provides gauge to improvement and control)
Duration: *Have you been admitted to the hospital for asthma?* (Patients with histories of hospitalization tend to progress very rapidly and should be treated aggressively and early.) **Alleviating Factors**: *What typically triggers an asthma attack?* (Any recognized trigger should be diligently avoided.)

Objective: Signs
Using Basic Tools: Vital signs: respiratory rate >18; potentially unstable BP— pulsus paradoxus (systolic BP varies by more than 10 mmHg during breathing ; can be seen in other intrathoracic diseases, such as tumor or pneumothorax also)
Inspection: Nasal cavity: Nasal polyps (seen in cystic fibrosis and asthmatics with aspirin sensitivity). Accessory respiratory muscle use in severe cases. Auscultation: Respiratory system: wheezing, prolonged expiration are common. Decreased breath sounds may mean worsening!
Using Advanced Tools: Labs: Eosinophils on Gram stain of nasal secretions or blood; Chest x-ray: rule out other diseases; Pulmonary function tests or peak flow meter (if available) documents airflow obstruction and serial improvements predicts better response. The response on peak flow or pulmonary function tests after administration of a bronchodilator can be helpful from a diagnostic, as well as therapeutic, view point. Remember the pearl: "All that wheezes is not necessarily asthma!"

Assessment:
Differential Diagnosis
Foreign body aspiration (always consider in children), viral respiratory infections (croup, bronchiolitis), epiglottitis, chronic obstructive pulmonary disease, rarely vocal cord dyskinesis (paradoxical vocal cord motion as a post traumatic stress disorder or panic attack).

Plan:
Treatment
Primary: MDI = Metered Dose Inhaler
1. Emergency Treatment : Measure initial peak flow *if possible*, provide a baseline for repeated measures (doubling of the initial peak flow value, measured hourly is a reliable indicator of improvement).
Initial therapy:
   a. Inhaled beta-agonist (**albuterol**) to reverse obstruction (1 ml **albuterol** neb) or MDI (2-4 puffs, .36 mcg, every 1-2 hours until clear)
   Children under 2 - nebulizer or MDI with valved spacer and mask; children 2-4 years - MDI and valved spacer; children over 5 years - MDI or powder inhaler
   b. Short course of oral corticosteroids, 2 mg/kg po q am for 5-7 days
2. Non-emergent treatment based on NIH asthma severity categories.
   a. Mild intermittent asthma: Brief wheezing once or twice a week
Give **albuterol** (short-acting beta-agonist) in Metered Dose Inhaler (MDI), the preferred method, or nebulizers, or oral syrup.(Inhaler: 2-4 puffs .18 – 36 ug q 6 hr prn) (Nebulizer: Dilute ½ ml (1 ml of 5% **albuterol** contains 5 mg of **albuterol**) to 3 ml total volume, with sterile normal saline for nebulizing). (Syrup: Adults and children over 14 years, 2 mg or 4 mg (1 teaspoonful = 2 mg) tid to qid; children 6 to 14 years: 2 mg po tid to qid; children 2 to 6 years of age, start at 0.1 mg/kg of body weight tid and do not exceed 2 mg tid. Alternative for stable patients: Long acting beta-agonists (e.g., **salmeterol [Serevent]** 2 puffs bid).
   b. Mild persistent asthma: Symptoms >2 times a week, but < 1 time a day; affects activity. Add long-term control medication - choose from:
   Inhaled Steroid: **Beclomethasone dipropionate** or equivalent: 2 inhalations (84 micrograms) given tid to qid or alternatively, 4 inhalations (168 micrograms) can be given bid. Or **zafirlukast:** adults and children 12 years of age and older: 20 mg po bid; children 7-11years of age: 10 mg po bid or **montelukast**: adults 15 years of age and older: 10 mg po q evening; children 6-14 years of age: one 5 mg chewable tablet q evening
   c. Moderate persistent asthma: Weekly symptoms interfering with sleep or exercise; occasional ER visits; peak flow is 60-80% of predicted.
   Increase inhaled steroids (**beclomethasone dipropionate** or equivalent) to 12 to 16 inhalations a day (504 to 672 micrograms) and adjust the dosage downward according to the response of the patient. Add additional long-term control medications (consider **theophylline** but blood levels are required to prevent toxicity). Consider adding inhaled 2-4 puffs qid **ipratropium bromide** (anticholinergic drug)
   d. Severe persistent asthma: Daily symptoms affecting activity; occasional ER visits; peak flow < 60 %

predicted
Re-examine environment for source of allergens and triggers. Consult specialist in allergy, pulmonary, or internal medicine.
3. Treat as an outpatient if no severe history, and patient is able to talk and achieve > 70% of peak flow after initial therapy. Otherwise, evacuate for intensive bronchodilator therapy.
Alternative: Ipratropium nebulizer or MDI instead of **albuterol**.
Primitive: Eliminate irritants and allergens if known, caffeine in coffee has been shown to have some bronchodilation effects (40-180 mg/cup brewed)

Patient Education
General: Understand disease medications, inhalers, nebulizers and peak flow meters. Monitor symptoms, peak flow rates. Have a pre-arranged action plan for exacerbations or emergencies. Give a written action plan and school plan to caretakers of asthmatic children.
Prevention: Investigate and control triggering factors (pollutants, exercise, house-dust mite, molds, animal dander) if symptoms are severe. Get annual influenza immunization. Avoid **aspirin** and **aspirin** containing medications. Avoid sulfites and tartrazine (food additives).
No Improvement/Deterioration: Return immediately if symptoms worsen.

Follow-up Actions
Return Evaluation: Evaluate for on-going control of symptoms, and alter medications as outlined above.
Evacuation/Consultation criteria: Evacuate severe asthmatics and those with a history of emergent attacks, once they are stable. Evacuate moderate asthmatics that are not able to complete the mission, since they may worsen and require intensive therapy during the mission. Consult primary care physician, internist or pulmonologist as needed.

Respiratory: Chronic Obstructive Pulmonary Disease
COL Warren Whitlock, MC, USA

Introduction: Chronic obstructive pulmonary disease (COPD) encompasses several disease processes including emphysema, chronic bronchitis and a mixture of these two (including long-standing, poorly controlled asthma). COPD reduces the lungs' capability to ventilate by obstructing the airway through different mechanisms. Emphysema occurs because of airway collapse on exhalation, causing air-trapping. Only about 25% of cigarette smokers develop emphysema, but those that show early disease will continue to lose function for as long as they smoke and for some time after they quit. In chronic bronchitis and in some long-standing asthma, airways are narrowed by reactive smooth muscle constriction, mucus and secretions. The clinical criteria used to diagnose chronic bronchitis is a productive cough for 3 months during 2 consecutive years, and spirometry to confirm expiratory airflow obstruction (reduced FEV1/FVC ratio). Patients with chronic bronchitis (many smokers) usually have a mixed obstructive airway disease including emphysema and recurrent respiratory tract infections.

Subjective: Symptoms
Recurrent or persistent shortness of breath, wheezing, dry or productive cough and smoking history.
Focused History: Quantity: *How long have you smoked tobacco, or when did you quit?* (Most will have a long smoking history.) **Quality:** *Do you cough up any mucus?* (emphysema—dry, non-productive cough; chronic bronchitis-- almost always productive) *What color is your sputum?* (White secretions generally suggest no infection; green or yellow indicates a bacterial infection.) *How long have you been coughing up mucus or sputum?* (Productive cough daily for 3 months in 2 years is evidence of chronic bronchitis.) **Duration:** *Have you had the influenza and pneumonia vaccinations?* (important preventive regimens for all COPD patients). **Alleviating Factors:** *What seems to improve your symptoms?* (Bronchodilators will help partially; avoiding smoking is the best alleviator.)

Objective: Signs
Using Basic Tools: Vital signs: RR > 18
Inspection: Respiratory system: Labored breathing; use of accessory muscles; "barrel" shaped chest or increased in diameter due to constant struggle for deep breath (indicates obstructed airways); costal margins may be pulled paradoxically inward during inspiration.
Auscultation: Rhonchi (secretions in the airway); breath sounds may be diminished
Percussion: Excursion of diaphragm with inspiration/expiration is reduced 2-4 cm.
Using Advanced Tools: Labs: CBC for polycythemia (HCT over 52% suggests chronic hypoxia or nocturnal hypoxia); sputum for gram stain and culture not generally helpful unless a resistant pathogen is suspected. Chest x-ray: Maybe normal in mild to moderate COPD, but diaphragms usually flattened in moderate to severe disease (evaluate for pneumonia or lung cancer also). Pulse oximetry: < 90% saturation. Pulmonary function tests (if available): evaluate for airway obstruction.

Assessment:
Differential Diagnosis
Moderate to severe persistent asthma - reversible with appropriate treatment, COPD is NOT completely reversible
Gastroesophageal reflux disease (GERD) - those with recurrent aspiration present with symptoms resembling chronic bronchitis
Bronchiectasis - a form of chronic scarring of the airways causing frequent bouts of bacterial bronchitis.
Bronchogenic carcinoma - symptoms may improve with treatment of bronchitis since both may be present

Plan:
Treatment
Primary:
1. Give bronchodilators as first line therapy: **metaproterenol (Alupent)**, **albuterol (Proventil, Ventolin)**, 1-2 puffs from the metered dose inhaler q 4-6 hrs, which may be increased to q 3 hrs in more severe cases. (Use of spacer device (AeroChamber, InspirEase) may be beneficial.) The toxicity of **theophylline** precludes its use in the field. **Ipratropium (Atrovent)** 2-4 puffs qid and prn can be used as an alternate bronchodilator.
2. Long-acting bronchodilators, such as **salmeterol (Serevent)** 2 puffs bid, and corticosteroids, such as **prednisone** 1.0 mg/Kg/day, should be used in patients with reversible obstruction, as measured by peak flow meter or pulmonary function tests.

Primitive: **Caffeine** has some bronchodilation effects and can be effective in some patients.
Belladonna plant (deadly nightshade) was administered in the past by smoking the dried plant for the anticholenergic effects of the atropine found in the plant. (Not recommended—atropine can have severe nervous system side effects)
Empiric: Oxygen (low flow 1-2 liters/min) if pulse oximetry shows < 90% saturation. Antibiotics (see Pneumonia Section).

Patient Education
General: Avoid inhaled pollutants. Stop smoking tobacco. This treats emphysema better than medications.
Prevention: Immunize with pneumococcal vaccine and influenza vaccine.

Follow-up Actions
Return Evaluation: Symptoms that do not improve should be referred for specialty care and additional special studies.
Evacuation/Consultation criteria: Evacuate unstable patients. Consult primary care physician, internist or pulmonologist as needed.

Respiratory: Pulmonary Embolus
COL Warren Whitlock, MC, USA

Introduction: A pulmonary embolism (PE) is an obstruction of the lung's arterial circulation. It usually occurs when a thrombus (blood clot) in the deep venous system of the legs dislodges and travels to the lung, causing a loss of oxygenation of the blood flowing to that area of the lung (hypoxemia). PE presents as three different syndromes: embolism without infarction (most common and causing acute unexplained dyspnea until clot is lysed), pulmonary infarction (complete obstruction of a distal branch of the pulmonary arterial circulation) or acute cor pulmonale (a massive clot obstructing a majority of both the pulmonary arteries and right ventricle of the heart, causing right heart failure).

Subjective: Symptoms
Three different clinical presentations are possible, depending on which PE syndrome is present.
1. Embolism: Acute unexplained shortness of breath without other significant symptoms.
2. Infarction: Chest pain associated with labored breathing, anxiety, occasional low-grade fever and cough (possibly with bloody sputum) for which no other cause (chest trauma, pneumonia, angina, etc.) can be determined.
3. Massive PE: Patients that have risk factors for venous thromboembolism (sedentary, post-surgical, obese, elderly or infected patients or those with a blood disorder) and have sudden, unexplained loss of consciousness.

Focused History: *Did the shortness of breath start suddenly?* (PE is an acute condition. Symptoms may progress over several days, but it starts suddenly.) *Do you feel anxious?* (A sense of foreboding or anxiety without clear reason is common.) *Have you had any recent lower extremity or pelvic injury?* (PE is usually associated with stasis or an injury to a great vein.) *Did you strain with a bowel movement before the symptoms began?* (Straining can dislodge lower abdominal clot.) *Have you had blood clots before?* (may be prone to form clots—hypercoaguable) *Have you recently become active after a time of bedrest?* (PE typical after 2-3 days bedrest)

Objective: Signs
Using Basic Tools: Vital Signs: Low-grade fever if any (< 101°F), respiratory rate >18, resting pulse over 90 BPM
Inspection: Cyanosis, hypotension and distended neck veins may indicate massive PE; anxiety, dyspnea and splinted breathing (due to pleuritic chest pain) is typical; peripheral edema suggests source of emboli or right heart failure
Auscultation: Area of rales or absent breath sounds may indicate location of an infarct or large PE. S3 gallop may indicate PE or CHF. Distant heart sounds with tamponade. Rub suggests pericarditis or pleural effusion.
Using Advanced Tools: EKG: To rule out other diagnoses including pericarditis or MI; CXR: low infiltrates suggest infarction, massive PEs cause "pruning" of the lung blood vessels

Assessment:
Differential Diagnosis
Embolism without infarction - anxiety attack or hysteria (dyspnea not typical, although tachypnea may be).
Infarction - septic shock (hypotension and altered mental status typical), MI (typical history and EKG changes), tamponade (distant heart sounds, elevated neck veins, hypotension)
Massive PE - pneumonia (toxic appearance with auscultory and CXR findings—see Pneumonia section), atelectasis (increased density [whitening] in wedge of lung on CXR—see discussion in Pneumonia Section); congestive heart failure (peripheral edema, orthopnea and other findings—see Cardiac: CHF); pericarditis (typical chest pain with EKG findings and possible pericardial rub)

Plan:
Treatment
Primary:
1. Prevent PE in high-risk post-op patients by frequent and early ambulation; use of 5000 U **heparin** IV or SC (low-molecular weight heparin) every 8-12 hours; external pneumatic compression, and gradient elastic stockings.
2. Evacuate patients with PE and those with risk factors and suspicion of PE.
3. If the risk of bleeding is low, use **heparin** as above while en route or awaiting evacuation with severe PE.
4. Administer oxygen 2L via nasal cannula. Use mask and increased rate in massive PE.
5. Administer IV fluids (see Shock: Fluid Resuscitation) to help maintain cardiac function in massive PE.

Patient Education
General: Avoid thrombus in high-risk patients.
Activity: Early ambulation to avoid PE. No activity during PE until anticoagulated at referral facility.
Medications: Avoid anticoagulants such as aspirin or NSAIDs while on heparin to minimize bleeding risks.
Prevention: Avoid lower extremity activity or intra-abdominal strain that could dislodge additional thrombus if already have PE. Maintain blood flow during prolonged travel or bedrest by walking and stretching to avoid PE.

Follow-up Actions
Evacuation/Consultation Criteria: Emergently evacuate patients with PE, and those with risk factors and suspicion of PE. Consult pulmonologist or internist.

Respiratory: Acute Respiratory Distress Syndrome
COL Warren Whitlock, MC, USA

Introduction: Acute Respiratory Distress Syndrome (ARDS - was previously Adult Respiratory Distress Syndrome) can result from either direct trauma to the lung (infection, pulmonary contusion, aspiration, inhalation of toxic substances – including nitrogen mustard) or secondary to systemic pathology, such as shock or septicemia, which causes serous fluid from the blood stream to "leak" into the lungs. ARDS is characterized by a sequence of events, beginning with diffuse pulmonary edema, that are now thought to be part of an inflammatory cascade that causes multiple organ failures. The pulmonary edema resembles diffuse pneumonia or congestive heart failure on CXR but is not directly due to heart failure. ARDS is sometimes referred to as non-cardiac pulmonary edema, wet lung syndrome, or Da Nang lung (seen in Vietnam War). Renal failure, diffuse intravascular coagulation (DIC) or liver failure may precede or follow lung failure. ARDS cases rapidly deteriorate and die (100% mortality) without ICU and ventilatory support. Mortality is 50% even with such care. There are 4 recognized phases:
- Phase 1: Acute injury: Normal physical exam, normal chest x-ray, tachypnea, tachycardia and respiratory alkalosis (low CO_2 partial pressure and high blood ph)
- Phase 2: Latent phase (6-48 hours after injury): Hyperventilation, hypocapnia (low CO_2), increase in work of breathing, hypoxemia
- Phase 3: Acute respiratory failure (48-96 hours after injury): decreased lung compliance, diffuse infiltrates on CXR due to pulmonary edema
- Phase 4: Multi-organ failure: Severe hypoxemia unresponsive to therapy, metabolic and respiratory acidosis (high CO_2 partial pressure and low blood ph)

Subjective: Symptoms
See phases above. Difficulty breathing and severe shortness of breath, rapid progression (2-24 hours) of respiratory failure, increased agitation.
Focused History: (Patient may be too ill to provide comprehensive information.) *Duration: How long ago did you start feeling sick?* (ARDS usually begins 24-48 hours after the initial insult i.e., septic shock.) **Alleviating or Aggravating Factors**: *Did symptoms get better then worse?* (In some cases symptoms seem to be improving, then worsen— like appendicitis, which wanes after rupture then progresses to septic shock.)

Objective: Signs
Using Basic Tools: Vital signs: Fever > 101.5°F, RR > 18, HR > 100, BP < 110/60
Inspection: Toxic and unstable, cyanotic (blue/gray/purple skin discoloration), use of accessory muscles in impending respiratory failure.
Respiratory System: Normal in Phase I and rales in Phase II.
Neuro exam : Agitation, followed later by lethargy and obtundation.
Using Advanced Tools: Pulse Oximeter: Oxygen saturation < 90% and not responsive to oxygen therapy; CXR: Normal at 2- 24 hours, then diffuse, fluffy infiltrates & pulmonary edema at 48- 96 hours.

Assessment:
Differential Diagnosis
Other causes of Pulmonary Edema - drowning (typical history); drug overdose (history of exposure and mental status changes); non-septic shock (see Shock chapter and eliminate causes of shock): congestive heart failure (peripheral edema, orthopnea and other findings—see Cardiac: CHF).
Diffuse infectious pneumonia (see Respiratory: Pneumonia).

Plan:
Treatment
Primary:
1. **Evacuate for ICU and ventilator support.** If evacuation is not possible, skip steps below, make patient comfortable and treat expectantly.
2. Pending evacuation and en route, administer oxygen– start with low flow 2 L/min and increase as needed.
3. Treat the underlying specific etiology, if recognized.
4. Administer fluid and blood products sparingly to minimize severity of pulmonary edema.
5. Administer broad-spectrum antibiotics (see Respiratory: Pneumonia).

Patient Education
Activity: Limit activity.
Prevention: Suspect this complication with severe injuries and evacuate early.

Follow-up Actions
Evacuation/Consultation Criteria: Urgently evacuate patients with severe injuries, particularly those who are elderly, very young, have underlying chronic diseases or are immunocompromised. Consult pulmonologist or internist.

Respiratory: Apnea
COL Warren Whitlock, MC, USA

Introduction: Apnea is the cessation of breathing for ≥ 10 seconds at a rate of >20 times an hour, and it occurs most commonly during sleep. Apneic episodes may occur up to 5 times per minute in normal adults, usually during rapid eye-movement (REM) sleep. Apnea is also known as "obstructive sleep apnea" or as "Pickwickian Syndrome" in the obese. The obstruction is usually due to enlarged pharyngeal tissues in the obese, inflamed tonsils, low-hanging soft palate or uvula, or craniofacial abnormalities that narrow or close the airway. Relaxation of pharyngeal and palatal muscles from alcohol, sedatives, muscle relaxants or other CNS depressants can contribute to snoring and airway closure at night. Apnea is associated with hypoxia and frequent nocturnal arousals (60-100 per hour), contributing to excessive daytime sleepiness. Apnea generally does not cause shortness of breath unlike other conditions associated with a narrow upper airway, such as epiglottis. Hypertension is the most common medical condition accompanying sleep apnea. Underlying lung disease (COPD) worsens apnea. Risk Factors: Obesity, nasal obstruction (due to polyps, deviated septum, old trauma), hypothyroidism, upper airway narrowing, sedative drugs and alcohol.

Subjective: Symptoms
Early: Excessive daytime sleepiness, disrupted sleep, recent weight gain, repetitive awakenings with transient sensation of shortness of breath or for unclear reasons, tired and unrefreshed upon AM awakening, poor concentration, memory problems, decreased libido; spouse or others may witness apneas, loud snoring, irritability or short temper.
Late: Morning headache, depression
Focused History: Quantity: *How much weight have you gained in the last year?* (> 20 pounds is a significant risk factor). **Quality:** *Do you fall asleep during the daytime?* (Excessive daytime sleepiness is the hallmark of this syndrome). *Do you snore? How long have you snored?* (loose or excessive pharyngeal tissue can cause chronic snoring and apnea) **Aggravating Factors:** *What seems to worsen your symptoms?* (Alcohol or sedatives, and additional weight gain will worsen sleep apnea).

Objective: Signs
Using Basic Tools:
Vital signs: Hypertension, tachycardia or bradycardia seen in chronic apnea.
Inspection: Narrowed airway, large tonsils, low-hanging soft palate or uvula may predispose to airway blockage at night. Neck: Inspect for "bull neck" indicating possible intrathoracic disease, tumor, pneumothorax.
Auscultation: Turbulent airflow during sleep may produce rhonchi that may be localized to neck or nasopharynx. Extra heart sounds may suggest heart failure.
Using Advanced Tools: Lab: High hematocrit (polycythemia) on CBC may be a consequence of hypoxemia.

Assessment:
Differential Diagnosis
Excessive daytime somnolence - narcolepsy (usually associated with sudden loss of muscle tone during emotional moments, and/or hallucinations on awakening), inadequate sleep (review history), depression/anxiety disorder (see Symptom: Depression and/or Symptom: Anxiety).
Nocturnal awakenings - asthma, COPD, CHF. See respective sections in this book. If occurs suddenly, see panic attacks in Symptom: Anxiety).
Hypothyroidism causes sleep disturbances and sluggishness (see Endocrine: Thyroid Disorders).
Gastroesophageal reflux may also cause awakenings with transient sensation of shortness of breath or for unclear reasons (see GI: Acute Gastritis).

Plan:
Treatment
Primary:
1. Treat the underlying specific etiology, which may mean weight loss in the obese.
2. Treat with nasal Continuous Positive Airway Pressure (CPAP) to prevent apneas until body weight is lost. The extra oxygen can have the dramatic effect of "waking" someone up that has been partially asleep for years during the daytime.
Alternative: CNS stimulants provide some short-term effect. **Protriptyline** 10-30 mg or **fluoxetine** 20-60 mg po can occasionally be helpful for mild to moderate sleep apnea.
Primitive: Apnea is related to sleep position (on the back), so have patient sleep on his side and elevate the head of the bed.

Patient Education
General: Treat obesity with behavior modification. Regaining lost weight will generally cause a return of symptoms.
Activity: Encourage exercise after ensuring cardiorespiratory system is healthy enough to tolerate the stress.
Prevention: Avoid sedatives and alcohol, which act as central nervous system depressants and worsen sleep apnea.

Follow-up Actions
Return Evaluation: Long-term compliance is not high for CPAP unless the patient has severe sleep apnea syndrome. Consider alternate therapies listed above in these patients.

Evacuation/Consultation criteria: Evacuation is not typically necessary. Consult primary care physician, internist or pulmonologist as needed. Enlarged tonsils and anatomic abnormalities usually require surgical correction by an ENT surgeon.
NOTE: A diagnostic sleep study is recommended to make definitive diagnosis.

Chapter 4: Endocrine

Endocrine: Adrenal Insufficiency
Col Stephen Brietzke, USAF, MC

Introduction: Primary adrenal insufficiency, characterized by deficient production of cortisol and aldosterone, can occur acutely due to hemorrhage or an infarction involving the adrenal circulation. Proximate causes of such an event include gram-negative bacterial sepsis and blunt or penetrating abdominal trauma. Acute adrenal insufficiency is a medical emergency, heralded by severe orthostatic hypotension, shock, hyponatremia and often hyperkalemia. Sub-acute or chronic primary adrenal insufficiency is usually caused by autoimmune disease (Addison's disease) or metastatic cancer in developed countries, but in the developing world, replacement of normal adrenal tissue by tuberculous infection is more prevalent. Secondary adrenal insufficiency, characterized by deficient production of cortisol but normal production of aldosterone, is due to some form of hypothalamic or pituitary gland disease. Chronic or sub-acute causes include tumors of or near the pituitary gland. Acute causes include transection or infarction of the gland due to closed or penetrating head trauma.

Subjective: Symptoms
Acute: Severe orthostatic hypotension or shock; severe, poorly localized abdominal pain; nausea; vomiting; weakness; mood change; confusion or psychosis. **Sub-acute and chronic:** Fatigue, malaise, weight loss, poor appetite, nausea, postural faintness or lightheadedness, loss of libido, depression, anxiety, confusion or acute psychosis.
Focused History: *Do you feel faint or pass out when you stand up?* (low blood pressure) *Do you have nausea and vomiting?* (poor general condition) *Have you lost interest in sex? Are you having trouble thinking clearly? Do you feel sad or depressed?* (prominent symptoms) *Do you have any abdominal pain? Where?* (usually, diffuse and poorly localized) *How long have you felt ill?* (if less than four weeks, may reflect an urgent emergency) *Do you feel better if you eat salty foods? Do you crave salty food and drink?* (Since sodium loss through the urine is responsible for volume depletion, salt craving is an adaptive response.)

Objective: Signs
Using Basic Tools: Acute Presentation (< 2-4 weeks onset): Orthostatic hypotension*, tachycardia, flank ecchymoses, fever > 100.4°F, confusion/disorientation, ileus/abdominal tenderness (mimicking "acute abdomen")
Sub-acute/Chronic Presentation (>4 weeks): Orthostatic hypotension* (mild), hyperpigmentation (especially palmar creases and scars), loss of muscle mass/loose skin folds, vitiligo
*BP < 100/60 or supine BP > 20 mm Hg higher than standing BP
Using Advanced Tools: Lab: Acute: Elevated WBC, platelets on CBC, hypoglycemia on urine dipstick. **Chronic:** elevated platelets on CBC, hypoglycemia.

Assessment:
Definitive diagnosis will be beyond the capabilities of field laboratories (low sodium, high potassium, others).
Differential Diagnosis
Acute - other hypotensive states, including blood loss hypovolemia, volume depletion from gastroenteritis-related vomiting and diarrhea, pancreatitis, and diabetic ketoacidosis.
Sub-acute/chronic - chronic infections, such as TB or malaria, metastatic cancer, diabetes mellitus,

hyperthyroidism, depression or psychotic states such as bipolar disorder or schizophrenia.

Plan:
Treatment
Primary: Acute: Rapidly infuse normal saline solution (2 liters rapidly, then 250-500 cc hour, adjust rate of infusion based on pulse, blood pressure, and overall state of well-being); administer **dexamethasone** 4 mg intravenously as a single dose. Replacement therapy with **hydrocortisone**, 20 mg each morning and 10 mg each evening should be administered after resolution of acute symptoms until a definitive medical evaluation by a physician.
Alternative: Prednisone 5-7.5 mg once daily in the morning may be substituted as replacement therapy in lieu of hydrocortisone.
Primitive: If no glucocorticoid medication is available, attempt hemodynamic stabilization by aggressive intravenous hydration using normal saline solution at 250-500 cc per hour or more.
Empiric: In any case of shock or severe hypotension without obvious blood loss, render empiric treatment to cover the possibility of adrenal insufficiency. Accompany wide-open intravenous infusion of isotonic saline solution with the administration of **dexamethasone** 4 mg as an IV bolus every 24 hours (or, alternatively, **hydrocortisone** 100 mg may be given IV, every 8 hours).

Patient Education
General: Taking medication daily is essential to preserving health. In the event of any illness, double the daily dose of steroid medication for the duration.
Activity: No restrictions.
Diet: No restrictions. If steroid medication is unavailable, a high-salt diet can help minimize symptoms, preserve blood pressure and functional status.
Medications: Chronic steroid use can result in weight gain and other side effects.
Prevention and Hygiene: In developing countries, test the patient and their close contacts for TB, the most common cause of this syndrome in developing countries.
No Improvement/Deterioration: Seek medical care promptly for any acute illness resulting in vomiting or if an illness persists for more than a day on double-dose steroid therapy.

Follow-up Actions
Return evaluation: Expect rapid improvement in symptoms after initiating steroid therapy. After starting maintenance therapy, reassess symptoms and vital signs, including weight and blood pressure within one week. If improved, re-evaluate every 1-3 months.
Evacuation/Consultation Criteria: Referral to a medical center for appropriate confirmatory testing, and treatment.

Endocrine: Diabetes Mellitus
Col Stephen Brietzke, USAF, MC

Introduction: Diabetes mellitus (DM), the most common disease of the endocrine system, is characterized by abnormally high blood glucose levels. Diabetes mellitus results from either absolute deficiency of insulin **(type 1 diabetes)**, or from subnormal target cell response to insulin (insulin resistance) combined with failure to compensate for this insulin resistance by producing higher concentrations of insulin (relative insulin deficiency) **(type 2 diabetes)**. Type 2 diabetes accounts for up to 80-90% of all cases of diabetes. In the Americas and Western Europe, most cases of type 2 diabetes are associated with obesity, a sedentary lifestyle and/or a genetic predisposition. Type 1 diabetes is most often caused by autoimmune destruction of the insulin-producing beta cells within the pancreatic islets. Even though there are genetic factors which confer susceptibility, it is unusual to identify multiple first-degree relatives with type 1 disease. Patients with type 1 diabetes may exhibit other autoimmune diseases, such as hypothyroidism, rheumatoid arthritis or hyperthyroidism due to Graves' disease. DM due to another endocrine disease is called secondary diabetes mellitus. Gestational DM encountered during pregnancy is reversible if not prior to or post-gestation. These women are often overweight, and have increased risk of type 2 diabetes in middle age or later.

Subjective: Symptoms
Classically: Excessive thirst (polydipsia), excessive urination, especially at night (polyuria/nocturia), weight loss despite increased appetite and food intake (polyphagia), blurred vision. Patients with new onset diabetes frequently have no symptoms or may present with complications of diabetes, such as foot ulcers or gangrene.
Focused History: Quantity: *How many times do you wake up to urinate each night?* (>1 is suspicious). *How much weight have you lost?* (weight loss despite increased food intake is suspicious) *Is your appetite increased? Are you unusually thirsty?* (polyphagia/polydipsia) *Do you notice any blurring of your vision? Have you had sores on your feet or other wounds that are slow to heal?* (typical symptoms) *Do your feet feel as if they are asleep or as if they are not part of your body?* (Loss of sensation suggests neuropathy, a possible complication of diabetes.) *Is it difficult or impossible for you to have an erection (males)?* (50% male DM have impotence) **Duration:** *When did your symptoms begin?* (more symptomatic, more recent the onset of symptoms) **Other:** *Does anyone in your family have diabetes?* (There is a strong family predisposition particularly with type 2 diabetes.)

Objective: Signs
Physical examination is usually unreliable for diagnosis of DM.
Using Basic Tools: Vital Signs - BP drop > 20 mm Hg systolic comparing standing vs. supine position (orthostatic hypotension), tachycardia
Inspection: Central obesity ("beer belly" or "apple on a stick" configuration), dry mucous membranes (reflecting volume loss/dehydration); ulcers on the soles of the feet; vitiligo (de-pigmented regions of skin can be associated with type 1 diabetes and other autoimmune endocrine diseases)
Palpation: "Tenting" of the skin (suggests volume depletion); reduced or absent light touch sensation in distal legs and feet (reflects peripheral neuropathy, associated with diabetes)
Percussion: Absent ankle jerk or knee reflexes (may reflect peripheral neuropathy, associated with diabetes)
Using Advanced Tools: Lab: Test for glucose and ketones on urine dipstick (DM diagnosis).

Assessment:
Differential Diagnosis:
Weight loss/increased appetite - malabsorption states, protein/calorie malnutrition, hyperthyroidism
Polyuria - diabetes insipidus, urinary tract infection, prostatic hypertrophy
Polydipsia - diabetes insipidus

Plan:
Treatment
Primary: Give patients with severe hyperglycemia (> 250 mg/dl) and/or large ketonuria **NPH** insulin at an empiric starting dose of 0.25 unit/kg body weight SC twice daily. Give an oral hypoglycemic agent for fasting blood sugar in the 200-250 mg/dl range: **glyburide** 5 mg/day or **glipizide** 5 mg/day are effective and widely available. Evacuate profoundly symptomatic patients; look for orthostatic hypotension, nausea and vomiting, "large" ketones on urine dipstick as major indications to evacuate. Give high volume fluid therapy and intravenous insulin (10 units initially, followed by 5 units per hour) while en route. Give newly diagnosed diabetics appropriate dietary and exercise regimens (see Patient Education below).
Primitive: Severely symptomatic patients who are volume depleted can be treated with aggressive isotonic saline infusion intravenously (1-2 liters over one hour, followed by 150-250 cc per hour), pending transport to a definitive care facility.
Empiric: Empiric drug or insulin therapy is potentially dangerous in the absence of blood glucose testing and should be avoided.
Patients with positive urine ketones and a rapid respiratory rate probably have profound metabolic acidosis (ketoacidosis).

Patient Education
General: DM is almost always a permanent condition, but careful self-management offers long-term benefit in minimizing occurrence and severity of microvascular complications.
Activity: Stable patients benefit from 40 to 60 minutes of moderate aerobic exercise daily. Walking 2-3 miles daily is usually well tolerated in the absence of foot lesions.
Diet: Limit intake of simple sugars in favor of complex carbohydrates. Obese patients should receive a total calorie prescription with a goal of long-term weight reduction; 1500-1800 calories a day are usually sufficient.
Prevention and Hygiene: Patients should return for a urinary dipstick test bid until urinary glucose and ketones are negative or trace.
No Improvement/Deterioration: Patients should be referred for definitive medical care.

Follow-up Actions
Return evaluation: Daily to twice weekly initially, as dictated by severity of hyperglycemia and ketonuria.
Evacuation Consultation Criteria: Immediately evacuate all severely hyperglycemic, severely ketonuric or pregnant patients. All other patients with diabetes mellitus should be evacuated at the earliest availability.

Endocrine: Hypoglycemia
Col Stephen Brietzke, USAF, MC

Introduction: Hypoglycemia is an abnormally low blood glucose level. Normal body function depends glucose, the primary energy source for most cells. Metabolism of glucose is mediated by glucagon and epinephrine (which stimulate the liver to change stored glucagon into glucose for use as an energy source) and by cortisol and growth hormones. Insulin does the opposite, promoting removal of excess blood glucose for storage of in the liver as glycogen. Hypoglycemia is caused by an imbalance between insulin and glucagon, epinephrine, cortisol and growth hormone.

Subjective: Symptoms
Abrupt decline in mental status function, level of consciousness, amnesia, bizarre behavior, hemiparesis, poor coordination, double or blurred vision. Anxiety, generalized sweating, and tremor may occur prior to other neurologic symptoms.
Focused History: Quality: *Do you feel shaky or nervous?* (suggests adrenergic nervous system response to hypoglycemia) *Do you have a craving for sugar or foods?* (appetite stimulated by falling blood glucose level) *Are you having difficulty thinking clearly?* (CNS dysfunction results from significant hypoglycemia) **Duration:** *When did your symptoms begin? Has it ever happened before?* (True hypoglycemia begins abruptly, and can recur in similar circumstances over time.) **Alleviating or Aggravating Factors:** *When was your last meal? Were you more active today than usual?* (Decreased food intake or unaccustomed exercise are common precipitants of hypoglycemic attacks.) *Do you have diabetes? Are you taking insulin or pills to control your diabetes?* (Most patients with hypoglycemia are receiving drug treatment for diabetes mellitus.) *How long after eating did your symptoms improve?* (Hypoglycemia reverses rapidly with ingestion of carbohydrate.)

Objective: Signs
Using Basic Tools: Notoriously non-specific symptoms that resolve WITHIN MINUTES after giving IV or oral glucose. Vital Signs: Tachycardia, hypertension, tachypnea
Inspection: Diaphoresis, dilated pupils, confusional or psychotic state, drowsy or comatose, ataxic gait, coma, generalized seizure
Auscultation: Aortic or pulmonic flow murmurs
Palpation: Left- or right-sided facial/upper extremity/lower extremity weakness or paralysis; decreased visual acuity or visual fields
Percussion: Brisk deep tendon reflexes
Using Advanced Tools: Lab: Glucose and ketones on urine dipstick.

Assessment: Always consider hypoglycemia as an easily treatable form of mental status impairment. ALWAYS check for it in patients presenting with coma, seizure, confusion or focal neurologic signs.
Differential Diagnosis - drug overdose with sedative or narcotic agents, alcohol intoxication, idiopathic seizure disorder, closed head trauma, CNS infection (meningitis, encephalitis), and a variety of metabolic insults (uremia, metabolic alkalosis, respiratory acidosis or severe hyponatremia) can all produce severe confusion or coma and cannot be visually distinguished from hypoglycemia. At times hypoglycemia can provoke focal neurologic signs such as hemiparesis which reverse with treatment of hypoglycemia.

Plan:
Treatment
Primary: One ampule of 50% **dextrose** (D50%) should be injected IV push, rapidly.
Alternative: Mild symptoms and cooperative: 8 oz of sweetened fruit juice, non-diet colas or sports drink (i.e., Gatorade).
Empiric: 1 amp of D50% rapid IV push for any form of mental status impairment when blood glucose testing is unavailable.
Emergent: Glucagon 1 mg may be reconstituted (comes as two vials, which must be combined) and injected intramuscularly if IV access is difficult or impossible.

Patient Education
General: Wear a medical alert bracelet or necklace if prone to hypoglycemia.
Activity: Except as noted, normal unrestricted activity is permitted.
Diet: Normal diet unless frequent hypoglycemia, then add mid-morning, mid-afternoon, and bedtime snacks.
Medications: Patients who had coma, seizure or focal neurologic signs need glucagon 1 mg for IM self-injection.
Prevention and Hygiene: Do not miss meals or exercise strenuously after 4 or more hours of fasting.
No Improvement/Deterioration: Give patients with severe symptomatic presentations (coma, seizures, focal neurologic signs) IV 5% dextrose (in normal or half-normal saline) after recovery. Observe 12-24 hours. Recurrent hypoglycemia following treatment mandates additional 50% dextrose and evacuation. Consider pituitary or adrenal insufficiency, renal or hepatic failure.

Follow-up Actions
Return evaluation: Routine diabetes-oriented care should suffice for most patients.
Evaluation/Consultation Criteria: Severe (coma, seizure, focal neurologic presentation) or frequent (> 1 severe episode per month) hypoglycemia should be evacuated.

Endocrine: Thyroid Disorders
Col Stephen Brietzke, USAF, MC

Introduction: Goiter is an enlargement of the thyroid gland, which can be appreciated visually or by palpation. In part, goiter is an adaptive process, reflecting increased size and number of thyroid follicles in an attempt to overcome deficient production of thyroid hormones by individual cells. Other causes of thyroid enlargement include chronic inflammation and scarring. Worldwide, a common cause of simple goiter is iodine deficiency. This condition is not expected in island or coastal regions where seafood or kelp (iodine-rich foods) is consumed regularly, but may occur inland in large continents. Areas of the world where iodine deficiency is known to be a significant problem include mountainous regions, parts of sub-Saharan Africa and central China.
Hyperthyroidism (overactive thyroid) is most commonly due to Graves' disease, an autoimmune disease caused by an antibody directed against the thyroid stimulating hormone (TSH) receptor (the "on/off" switch) on the hormone producing cell. Another cause, multinodular goiter, results when thyroid cells lose the normal "on/off" switch control of TSH and produce thyroid hormone independently. Finally, hyperthyroidism can result

from leakage of thyroid hormones from a gland damaged in trauma or from viral infection in a self-limited process called thyroiditis. **Hypothyroidism** (underactive thyroid) most commonly results from autoantibodies directed against thyroid enzymes, resulting in decreased production of thyroid hormones. Another cause is inherited defects in thyroid cell function.

Subjective: Symptoms

Simple goiter: Mass in the anterior neck; dysphagia; dysphonia (hoarseness); stridor, cough or wheezing as a result of compression of the esophagus, recurrent laryngeal nerves or trachea by the goiter.

Hyperthyroidism: Excessive sweating, intolerance of hot temperature, decreased stamina and endurance, nervousness, irritability, tremor, weight loss, increased size and frequency of bowel movements (hyperdefecation or diarrhea), palpitations, insomnia, and eye irritation/discomfort. See notes below for comments on pregnancy and thyroid storm.

Hypothyroidism: Fatigue, depressed mood, excessive sleepiness/reduced alertness, intolerance for cold temperature, weight gain, hair loss, constipation, and muscle cramps.

Focused History: Quantity: Goiter: *Are you aware of a lump in your neck?* (Many patients with goiter are unaware.) **Hyperthyroidism:** *Do you have more bowel movements per day than you did before?* (increased frequency suggests hyperdefecation) *Does your heart race at night or does a rapid heart beat wake you from sleep?* (Persistent rapid heart rate is typical.) *Have you become thinner or lost weight?* (typical) **Hypothyroidism:** *How many hours are you sleeping each day? How long have you required this much sleep? Do you fall asleep or feel tired during the day?* (increased sleep requirement is typical) *Are you having fewer bowel movements each week than you did before?* (Constipation is typical.)

Quality: Mass Effect of Goiter: *Do you have any difficulty when you breathe in?* (stridor) *Do you cough or feel as if you are choking when you lie down?* (compressing trachea) *Do you have difficulty swallowing solid food like meat?* (esophageal compression) *Do you have persistent hoarseness or change in your voice?* (compression of vocal cord nerves). **Hyperthyroidism:** *Do you feel "hotter" than other members of your family and do you feel sick when it is hot?* (heat intolerance) *Do you feel nervous or irritable most of the time? Do your hands tremble most of the time?* (typical symptoms) **Hypothyroidism:** *Do you notice any change in your hair? Is it falling out by handfuls? When did you become aware of hair loss? Do you seem "colder" than other members of your family?* (typical symptoms) *Do you have trouble concentrating or being alert while you are working?* (decreased state of alertness and mental slowing). *Do you have cramps in your legs when you walk or run?* (neuromuscular dysfunction).

Duration: Goiter: *How long have you been aware of a mass in your neck?* (benign goiter may be present for many years; sudden appearance suggests possibility of cancer) *Do other members of your family have a lump like this?* (benign goiter may manifest heritable thyroid dysfunction). *When did you first begin to feel different from normal?* (sudden onset suggests thyroiditis or a dietary exposure, such as iodine)

Alleviating and Aggravating Factors: *Do you feel better and function better when it is warm (or cold)? Do you feel worse and function worse when it is cold (or hot)?* (hypothyroidism: poor tolerance for cold, hyperthyroidism: poor tolerance for heat)

Objective: Signs

Simple Goiter

Using Basic Tools: Vital signs should be normal.

Inspection: Look for an obvious lump or mass in the anterior neck, below the thyroid cartilage ("Adam's apple").

Palpation: Enlarged, smooth or nodular, fleshy or firm to hard mass in the anterior neck, between the thyroid and cricoid cartilages.

Hyperthyroidism

Using Basic Tools: Vital signs: Tachycardia > 100/min; BP > 140 systolic with diastolic < 80.

Inspection: Fine tremor of hands at rest, wide-eyed stare, proptosis (projecting globe of the eye, creating a "bug-eyed" appearance).

Palpation: Diffuse or nodular goiter (thyroid enlargement), diaphoresis (warm, moist palms and generalized increased sweating).

Auscultation: Bruit over thyroid gland or supraclavicular space, systolic ejection murmur (reflect increased

cardiac output).
Percussion: Brisk deep tendon reflexes (reflects increased neuromuscular irritability).

Hypothyroidism
Using Basic Tools: Vital signs: Temperature < 97° F; bradycardia < 60/min
Inspection: Pale, thin skin; "droopy" eyelids; loss of lateral eyebrow hair (cutaneous changes are prominent in advanced hypothyroidism); coarse voice; slow response to questions; depressed affect
Palpation: Diffuse, firm goiter (thyroid enlargement) in anterior neck; cool, dry skin
Percussion: Delayed deep tendon reflex return phase, especially at ankles (return is like a "ratchet")

Assessment:
Patients with goiters require serum TSH and thyroid hormone measurement, which are not available in the field, for accurate differentiation between hyper- and hypothyroidism.

Differential Diagnosis:
Goiter - other anterior neck masses are usually either tumors or benign cysts. Goiter will usually be distinguished by moving up and down with swallowing, whereas other neck masses remain fixed.
Hyperthyroidism - anxiety states, starvation/malnutrition, pheochromocytoma, stimulant drug abuse (cocaine, amphetamines), and congestive heart failure may produce similar symptoms. The finding of a goiter and the absence of prior psychiatric history and drug abuse history strongly favors hyperthyroidism.
Hypothyroidism - depression, chronic fatigue syndrome, bereavement, hypothermia, sedative drug abuse (barbiturates, benzodiazepines, etc) may cause similar symptoms. Most patients with hypothyroidism are not severely ill, and can begin empiric treatment on a cautious basis. Very sick individuals require urgent evacuation, including patients in coma and those with profound hypothermia (temperature < 96° F).

Plan:
Treatment
Goiter
Primary: Definitive treatment is dependent on the results of blood tests not available in the field.
Primitive/Empiric: Lodine supplementation (low dose if in goiter-endemic area)

Hyperthyroidism
Primary: Propranolol 10-40 mg po qid to render pulse < 80/min

Hypothyroidism
Primary: Definitive treatment is dependent on the results of blood tests not available in the field.
Primitive: Provide warmest possible environmental temperature and encourage maximum physical exertion.

Patient Education
General: Definitive blood testing is an important aspect of care and follow-up is essential to well-being.
Activity: Normal activities are permitted without restriction for patients with simple goiter. Patients with suspected or confirmed hyperthyroidism should not perform strenuous exercise or employment-related activity until the condition is improved. Acceptable activities include light duty in a cool (preferably indoor) environment, and clerical duties.
Diet: Avoid dietary goitrogens (see note below) in patients with simple goiter or hypothyroidism.

Follow-up Actions
Return evaluation: Most patients with minor symptoms can be followed at two to four-week intervals. For patients with hyperthyroidism, emphasis should be on heart rate control (goal is < 80 beats per minute) and tremor control.
Evacuation/Consultation Criteria: Severe symptoms, including high fever, confusion or delirium, congestive heart failure, hypothermia, coma, severe bradycardia, or local compressive symptoms in the neck (especially choking or stridor) require urgent evacuation to a referral center attended by an endocrinologist.

NOTES: Dietary goitrogens: Cassava meal, cabbage, rutabaga, and turnips impair the action of thyroid

peroxidase, and may exacerbate simple goiter and induce hypothyroidism.
Pregnancy: Normal pregnancy results in mild symptoms that can be confused with hyperthyroidism. Severe symptoms suggest the presence of hyperthyroidism in pregnancy and expedient consultation is required if at all possible. Slight enlargement of the thyroid gland is expected in normal pregnancy.
Thyroid Storm: Most cases of hyperthyroidism are not emergencies, but the medic should be alert to the possibility of the syndrome of thyroid storm. Thyroid storm is a symptom constellation including fever, delirium (or confusion, also known as encephalopathy), very marked tachycardia, and a generally "toxic" appearance. Individuals with these features require emergent evacuation to a skilled medical facility. Interim treatment in the field should include beta-blocker medication (**propranolol**), antipyretic therapy with **acetaminophen** (theoretically, aspirin may worsen the condition by releasing thyroid hormone from binding sites on plasma proteins) and/or ice packs, and sedation using benzodiazepines or **haloperidol**.

Chapter 5: Neurologic

Neurologic: Seizure Disorders and Epilepsy
CDR Robert Wall, MC, USN

Introduction: A seizure is an uncommon event that can be caused by many different ailments and processes. Not all convulsions become an epileptic condition, and most are brief and self-limited. Once a "diagnosis" of epilepsy is documented, it will follow the patient for the rest of their life and greatly impact their employability, insurability, driving status and many other areas.

Subjective: Symptoms
Abrupt onset of abnormal muscle activity, or prodrome of confusion, déjà vu, peculiar behavior, automatisms, or other psychic phenomena preceding onset.

Objective: Signs
Using Basic Tools: Sudden onset of loss of consciousness, followed by abnormal motor activity such as tonic rigidity, clonic rhythmic movements of the limbs, urinary incontinence, frothing at the mouth, and biting the tongue and mouth; may last seconds to minutes, and is usually followed by a period of weakness, somnolence and confusion (post-ictal state); will spontaneously stop without any intervention after a few minutes.
Using Advanced Tools: Lab: WBC for infection; urinalysis for glucose level; EKG for arrhythmia etiology for syncope.

Assessment:
Differential Diagnosis - the differential diagnosis of a convulsive event is extensive: idiopathic epilepsy, alcohol or drug associated seizures, post concussive syndrome, convulsive syncope, heat stroke, infectious (meningitis), brain mass lesions, nerve gas exposure and metabolic abnormalities. See index and appropriate sections of this book for discussions of most of these conditions.

Plan:
Treatment: Many of these medications and procedures may not be available in the field.
Primary: Symptomatic treatment initially.
1. Remove the patient from an area where he could injure himself or others. Keep sharp and breakable objects away from the patient. Pad objects if possible to avoid injury.
2. Do not put anything in the patient's mouth. Never put your fingers in the patient's mouth.
3. MEDICATIONS ARE RARELY INDICATED FOR A FIRST TIME SEIZURE.
4. After the seizure, evacuate the patient to an appropriate treatment facility for a neurological examination and further evaluation.

The exam will usually be normal, other than confusion and somnolence in the immediate post-ictal period, which may last for hours. After focal motor seizures, there may be a period of Todd's Paralysis, which is focal weakness of the affected limb.

Alternate: For recurrent seizures:
1. If the seizure lasts more than 10 minutes, immediate medical intervention is indicated.
2. Begin an IV access line.
3. EEG monitoring if available.
4. Give 5 to 10 mg IV **Valium**.
5. If this does not stop the seizure, consider **Dilantin** 1000 mg IV as an infusion over 30 minutes, not to exceed 50 mg/min. When giving Dilantin as an infusion, do not mix it with D5W because it will precipitate. Clear the IV tubing with normal saline first.
6. If this does not stop the seizure, consider **Phenobarbital** 10 mg/kg IV over 10 minutes. May be repeated one time. Must have a secure airway and closely monitor breathing.
7. If this does not stop the seizure, general anesthesia or barbiturate coma may be required. Advanced care will be required.
8. Transport to the nearest MTF for further evaluation and disposition. Use **Dilantin** 300 mg po or IV qd for MEDEVAC transport.

Always monitor the airway as these drugs may cause respiratory suppression. If IV unavailable, **Phenobarbital** may be given IM. Do not give **Dilantin** or **Valium** IM.

NOTES: If seizure lasts more than ten minutes, there is the possibility of Status Epilepticus. These seizures must be stopped ASAP. This is a life-threatening event and may produce significant brain injury if the patient survives. Emergency medical assistance and intervention must be rapidly sought

Patient Education
General: NO DRIVING OR OTHER DANGEROUS ACTIVITIES UNTIL MEDICALLY CLEARED. DMV reporting per state requirements.
Activity: Normal as tolerated. Avoid sports/activities such as scuba diving, skiing, horseback riding, or activities where there could be injury to self or others should a seizure occur. Weight lifting with a spotter only.
Diet: Avoid alcohol.
Prevention and Hygiene: Low stress, good diet, exercise, and good sleep hygiene (8 hours per night, regularly).

Follow-up Actions
Return evaluation: In 2 to 4 weeks as necessary.
Evacuation/Consultation Criteria: Urgent evacuation is not normally required. Patients should ultimately be referred for a non-emergent, ROUTINE Neurological Consultation.
NOTE: Though epilepsy afflicts up to 1% of the population, non-epileptic convulsive events are considerably more common.

Neurologic: Meningitis
CDR Robert Wall, MC, USN & COL Naomi Aronson, MC, USA

Introduction: Meningitis is an acute, life-threatening infection of the lining of the brain and spinal cord. It can be caused by a virus, bacteria, fungus, parasite, or more complex organism. Travel to exotic places, especially those with questionable sewage and pest control, increases the risk of acquiring and disseminating this disease. Bacterial meningitis is rapidly progressive and should be considered an emergency. Aseptic meningitis (normally caused by viruses) has a slower course. Meningitis is a treatable and potentially curable disease if diagnosed and treated early. However, delays in diagnosis and treatment can lead to permanent neurological disability and possibly death.

Subjective: Symptoms
Fever, stiff neck, headache, photophobia, malaise; later: delirium, coma, seizures, nausea, vomiting, dizziness
Focused History: *How fast did your symptoms progress?* (period of hours for bacterial meningitis, but longer

for viral) *Have you been exposed to others who have been ill or had meningitis?* (Ear infections, sinus disease, pneumonia, UTIs, bronchitis, sepsis, infected wounds may harbor meningitis organisms). *Have you taken any antibiotics recently?* (presentation may be masked) *Have you had meningitis in the past?* (increased susceptibility) *Do you have any immune deficiency?* (increases susceptibility) *Does it hurt to bend your neck or touch your chin to your chest?* (typically positive in infection) *Have you noticed any new rash?* (petechiae in early meningococcal meningitis)

Objective: Signs
Using Basic Tools: Fever to 104°F; cervical meningismus (stiff neck, painful to move); prostration; toxic appearance; positive Kernig's sign (inability to completely extend the knees straight – stretches spinal cord); positive Brudzinski sign (forward flexion of the head produces flexion at hip and knee – stretches spinal cord); rash (may indicate activation of the clotting cascade—hemorrhagic fevers; or petechial with meningococcus). Chronic infection: deafness.

NOTE: Do complete neurological examination, including Glasgow Coma Scale (see Appendix: Neuro Exam), looking for alterations of mental status or ambulation, and focal neurological deficits. Perform a thorough examination for possible sources of infection, such as middle ear, sinus, lungs, urinary tract, and wounds.

Using Advance Tools: Lab: WBC count or blood smear (may show leukocytosis in bacterial meningitis), urinalysis; RPR, blood cultures; chest X-ray

Assessment:
Differential Diagnosis: Other than by using the presence of systemic signs of infection and meningeal signs, these diagnoses will be very difficult to distinguish in the field. If in doubt, treat for bacterial meningitis. Meningitis comes in many forms that are infectious, the most common including bacterial and viral. Fungal forms also may occur in immuno-compromised patients. Other more rare forms include tuberculous, parasitic, spirochetal.

Rickettsial infection - usually no leukocytosis; no meningeal signs; "tache noire" lesion (ulcer covered with black, adherent crust).

Leptospirosis - look for conjunctival discharge; history of exposure to water which might be contaminated with animal urine

Cerebral malaria - no meningeal signs; positive blood smear; thrombocytopenia

Malignancy - variable symptoms based on lesion location.

Severe viral or bacterial sepsis with headache, high fever, but without meningeal seeding.

Brain abscess - focal neurologic findings; low-grade temperatures; no neck stiffness or tenderness

Subdural/Epidural hematoma - history of trauma with rapid or progressive development of symptoms.

Subarachnoid hemorrhage - often low-grade temperature; acute onset preceded by severe head aches; focal neurologic findings.

Stroke - variable symptoms based on lesion location.

Plan:
Treatment: Availability of certain procedures and medications may be limited in the field.
1. Begin antibiotics as soon as possible if bacterial meningitis is suspected (time to antibiotic administration is correlated with outcome). **Empiric Choices: Penicillin**: 24 million units/day in 6 divided doses, **ampicillin**: 12 gm/day in 6 divided doses, **vancomycin**: 2 gm/day in 2 divided doses, **ceftriaxone**: 6 gm/day in 3 divided doses.
2. Evacuate immediately.
3. Airway support and oxygen. Intubate as needed.
4. Fluid hydration with IV NS or LR.
5. Control fever with **Tylenol**.
6. If viral meningitis is suspected (slower onset, less severe symptoms), give **acyclovir** 12.5 mg/kg/day IV divided tid x 10 days.
7. Consider steroids (**Decadron** 0.4 mg/kg q 12 hour for 4 doses), with first dose prior to starting antibiotics, if bacterial meningitis is likely present (based on an acute presentation).

Patient Education
Activity: Bedrest
Diet: As tolerated.
Prevention and Hygiene: Meningitis may be contagious and good personal hygiene is mandatory. Respiratory isolation for first 24 hours of therapy (possible droplet spread of organism). Consider meningococcal vaccine pre-deployment. Give intimate/household contacts of meningococcal meningitis patients prophylaxis with **ciprofloxacin** 500 mg x one dose in adults. Alternate regimens: **ceftriaxone** 250 mg IM, **rifampin** 600 mg q 12 hours x 4 doses or **azithromycin** 500mg one time.
No Improvement/Deterioration: Return to the medic for persistent fever or mental status changes.

Follow-up Actions
Return evaluation: Return to medic 3 to 5 days after discharge for reevaluation, including repeat neuro exam.
Evacuation/Consultation Criteria: Evacuate immediately after starting antibiotics, if meningitis is suspected. Refer to Infectious Disease or Internal Medicine for definitive treatment. Consult Neurology in difficult cases. A lumbar puncture with evaluation of spinal fluid is the definitive test to diagnose meningitis, which is not available in a field environment.
NOTE: Penicillin allergic patients may also have allergic reactions to ceftriaxone.

Neurologic: Bell's Palsy (Idiopathic Facial Nerve Palsy)
CAPT Elwood Hopkins, MC, USN

Introduction: Bell's palsy is a common peripheral mononeuropathy involving the seventh cranial nerve. It usually follows a benign course, has no obvious underlying cause and is a condition from which nearly all patients recover fully.

Subjective: Symptoms
Abrupt onset of ear pain followed by weakness in muscles of facial expression, slurred speech and drooling when drinking; diminished or altered taste, increased sensitivity to sound on the involved side; evolves over 1-2 days; bilateral involvement and numbness are rare.

Objective: Signs
Using Basic Tools: Unilateral weakness (paresis) of the entire face, slurred speech and drooling

Assessment: Abrupt onset of unilateral facial muscle weakness in a young adult without other explanation is likely to be Bell's palsy.
Differential Diagnosis: Bilateral involvement can occur, but is rare and suggests more serious disease such as sarcoidosis, Lyme disease or Guillain-Barre syndrome.
Idiopathic neuropathy
Herpes zoster - simultaneous characteristic vesicular eruptions in the ear canal or on the face.
Lyme disease - a history of tick bite and characteristic rash (erythema migrans).
Sarcoidosis - paralysis of additional parts of the nervous system.
Guillain-Barre - absence of reflexes are typical of Guillain-Barre.
Myasthenia gravis - weakness of additional muscles (especially the eye muscles, causing double vision).
Peripheral Nerve System (PNS) Lesion - weakness of the forehead corrugator muscles and absence of involvement in other parts of the nervous system suggest a more serious disorder such as a peripheral nerve problem from a central lesion in the cerebrum or brainstem.

Plan:
Treatment:
1. Protect the eye from exposure keratitis (dryness, erythema, poor vision) and foreign bodies by wearing

eyeglasses when outdoors and taping or patching the eye during sleep.
2. Instill artificial tears several times throughout the day and viscous artificial tears (if available) at bedtime will help keep the eye surface lubricated and free of debris.
3. **Prednisone** 60mg/day po with taper over 10 – 14 days for severe cases. Most young adults will make a full recovery with no treatments.
4. For herpes zoster, give **Acyclovir** 800 mg po five times a day x 5 days.

Patient Education: Expect full recovery in several weeks. Protect the eye until able to close it fully.
No Improvement/Deterioration: Weakness may worsen during the first few days but then stabilize.

Return Evaluation: Evaluate patient once a week or until recovery is imminent.
Evacuation/Consultation Criteria: Refer to ophthalmology if signs of exposure keratitis develop. Refer to neurology for gradual worsening (over several days to weeks), failure to improve by three months and/or involvement of other parts of the nervous system.

Chapter 6: Skin

Introduction to Dermatology

MAJ Daniel Schissel, MC, USA

Classical Elements of the Clinical Approach to Dermatologic Disease Diagnosis and Disposition

Subjective:
Gather information just as in the approach to other organ systems, including **skin symptoms** like pain, pruritis and paresthesia, and **constitutional symptoms** like fever.

Objective: Diagnose skin eruptions visually based on primary and secondary type, shape, arrangement, and distribution of skin lesions. Always include a thorough evaluation of all the mucous membranes, hair and nails.
I. Type of Skin Lesion (see Color Plates Picture 20)
    a. Primary Lesions
        i.   **Macule:** A circumscribed area of change in normal skin color that is flat and less than 1 cm in diameter. Example: freckles.
        ii. **Patch:** A circumscribed area of change in normal skin color that is flat and > 1 cm in diameter. Examples: café-au-lait spots, port-wine stains.
        iii. **Papule:** A solid lesion, usually dome-shaped, <1 cm in diameter and elevated above the skin. Examples: verrucae, molluscum contagiosum.
        iv. **Nodule:** A solid lesion, usually dome-shaped, > 1 cm in diameter and elevated above the skin. Examples: neurofibromas, xanthomas, and various benign and malignant growths.
        v.   **Plaque:** An elevation above the skin surface occupying a relatively large surface area in comparison with its height. Frequently formed by a confluence of papules. Examples: lichen simplex chronicus and psoriasis.
        vi. **Vesicle:** A circumscribed, thin walled, elevated lesion < 1 cm in diameter and containing fluid. Examples: herpes, dyshydrotic eczema, varicella, and contact dermatitis
        vii. **Bullae:** A circumscribed, thin walled, elevated lesion > 1 cm in diameter and containing fluid. Examples: burns, frostbite, pemphigus.
        viii. **Comedone:** Retained secretions of horny material within the pilosebaceous follicle. Examples: open (blackheads) and closed (whiteheads), the precursors of the papules, pustules, cysts and nodules of acne.
        ix. **Pustule**: A circumscribed elevation containing pus. Examples: sterile lesions as in pustular psoriasis or bacterial as in acne and impetigo.
        x.   **Cyst:** A circumscribed, thick walled, slightly elevated lesion extending into the deep dermis and

subcutaneous fat. Examples: epidermal inclusion and pilar cysts.
- xi. **Wheal/Hive:** A distinctive white to pink or pale, red, edematous, solid elevation formed by local, superficial, transient edema. They characteristically disappear yet may reappear within a period of hours. Examples: dermographism, insect bites, and urticaria.
- xii. **Telangiectasia:** Blanchable (fades with fingertip pressure), small, superficial dilated capillaries. Examples: rosacea, lupus erythematosus and basal cell skin cancer.
- xiii. **Purpura:** Non-blanchable, purple area of the skin that may be flat/nonpalpable or raised/palpable. Examples: hemorrhagic lesions of some fevers.

b. **Secondary lesions** represent evolution (natural) of the primary lesions or patient manipulation of primary lesions. Although helpful in differentiating lesions, they do not offer the same diagnostic descriptive power as the primary lesion.
- i. **Atrophy:** Thinning and wrinkling of the skin resembling cigarette paper
- ii. **Crusts** (Scab): Dried serum, blood, or pus.
- iii. **Erosion:** Loss of part or all of the epidermis that will heal without scarring.
- iv. **Ulcer:** Loss of epidermis and at least part of dermis that results in scarring.
- v. **Excoriation:** Linear or hollowed-out crusted area caused by scratching, rubbing, or picking.
- vi. **Lichenification:** Thickening of the skin with accentuation of the skin lines.
- vii. **Scales:** Accumulation of retained or hyperproliferative layers of the stratum corneum
- viii. **Scar:** Permanent fibrotic changes seen with healing after destruction of the dermis.

II. **Shape of Lesion**
  a. **Annular:** Ring shaped, round or circular
  b. **Nummular:** Coin shaped
  c. **Oval:** Oblong
  d. **Polycyclic:** Rings within rings
  e. **Polygonal:** Geometric shaped
  f. **Serpiginous:** Snake-like

III. **Arrangement of the lesions in relationship to each other**
  a. **Grouped Arrangement**
  i. **Annular:** Circular, round or like a ring
  ii. **Arciform:** Shaped in curves
  iii. **Herpetiform:** Like or in the shape of a herpetic lesion
  iv. **Linear:** Geometrically forming a straight line
  v. **Reticulated:** Net-like
  vi. **Serpiginous:** Shape or spread of lesion in the fashion of a snake
  b. **Disseminated Arrangement:** Diffuse involvement without clearly defined margins or scattered discre telesion.

IV. **Distribution of Lesions**
  a. Isolated single lesions
  b. Localized to specific body region
  c. Universal over the entire body surface
  d. Patterned
  i. Sun-exposed areas
  ii. Symmetrical
  iii. Follicular based
  iv. Flexure or extensor surfaces

Assessment:
Synthesize, integrate, and form a hypothesis by combining the history and the primary and secondary characteristics of the lesion(s) together with their shape, arrangement and distribution.

Plan:
Find treatments that allow the service member to complete their mission, treat the process, and prevent recurrence. Explain the disease process and treatment thoroughly in words the patient can understand.

Bacterial Skin Infections
Skin: Disseminated Gonococcal Infection
MAJ Daniel Schissel, MC, USA

Introduction: Gonococcemia is a systemic infection with *Neisseria gonorrhoeae* following the bloodborne spread (dissemination) of the gram-negative diplococci from infected sites (1% of untreated cases).

Subjective: Symptoms
Prodrome of fever and chills, anorexia, malaise during the 7 to 30 day incubation period. May have migratory polyarthritis.

Objective: Signs
Using Basic Tools: Lesions: 1.0 -5.0 mm erythematous macules (2 to 20 lesions) that evolve to slightly tender, deep-seated, hemorrhagic pustules within 24-48 hrs; center may become necrotic; located on arms and hands more often than legs or feet, in regions near the joint spaces. (see Color Plates Picture 4). Tenosynovitis of the extensors / flexors of the hands and feet is common. Septic arthritis occurs with asymmetrical, erythematous, hot, tender knee, elbow, ankle or metacarpophalangeal joints. Other organ systems may also be infected: hepatitis, carditis, meningitis, and others. Evaluate for other STDs.
Using Advanced Tools: Lab: Gram stain of mucosal surfaces may yield gram-negative diplococci; culture mucosal sites (80-90% yield).

Assessment:
Differential Diagnosis: Meningococcemia (CNS effects); other bacteremias; psoriatic arthritis (plaque lesions, multiple joints); systemic lupus erythematosus (multiple joints, multiple organ effects) See appropriate sections in this book.

Plan:
Treatment:
Primary: **Ceftriaxone** 1 gm IM or IV q 24 hrs
Alternative: **Ceftizoxime** or **cefotaxime** 1 gm IV q 8 hrs, or **spectinomycin** 2 gm IM q 12 hrs
After initial symptoms resolve, the uncomplicated patient should be treated for 1 week with oral **cefuroxime axetil** 500 mg bid, or **amoxicillin** 500mg with **clavulanic acid** tid, or **ciprofloxacin** 500 mg bid if patient is not pregnant.

Patient Education
Prevention: Re-educate patient on safe sexual habits

Follow-up Actions
Evacuation/Consultant Criteria: Evacuate patients if possible. Consult as needed.

Skin: Meningococcemia
MAJ Daniel Schissel, MC, USA

Introduction: *Neisseria meningitides* is a gram-negative coccus found in the nasopharynx of approximately 5 – 15% of the general population. It invades the blood stream, causing acute meningococcal septicemia and meningitis. Transmission is through person-to-person inhalation of droplets of infectious nasopharyngeal secretions. The highest incidence is observed midwinter in children ages 6 months to 1 year, while the lowest is in adults over 20 years during the midsummer. Infants, asplenics, immunodeficient or complement (blood proteins important in immune response) deficient individuals are considered at increased risk.

Subjective: Symptoms
Prodrome of spiking fever, chills, myalgia, arthralgia; rash, photophobia, headache

Objective: Signs
Using Basic Tools: Abnormal vital signs: high fever, tachypnea, tachycardia, mild hypotension; rash: small, palpable, petechial lesions with irregular borders and pale gray, vesicular centers most commonly observed on the trunk and extremities (but may be seen anywhere, including the palms, soles and mucous membranes); posterior neck rigidity and tenderness with stretching; photophobia; altered consciousness; severely ill patients may display ecchymosis and coalescence of the purpuric lesions into bizarre shaped gray-to-black necrotic areas (see Color Plates Picture 16) associated with disseminated intervascular coagulation.
Using Advanced Tools: Lab: Gram stain scrapings from lesions to identify characteristic organism. Culture blood to identify organism.

Assessment:
Differential Diagnosis: Rocky Mountain Spotted Fever, other rickettsial diseases, staphylococcal toxic shock syndrome, enteroviral infections and acute bacteremia.

Plan:
Treatment: Initiate treatment immediately if meningococcemia is suspected and evacuate ASAP.
Primary: Cefotaxime 2.0 gm IV q 4-6 hrs + **vancomycin** 1.0 gm q 6-12 hrs
Alternate: Ceftriaxone 2 gm IV q 12 hrs + **vancomycin** 1.0 gm q 6-12 hrs

Patient Education
General: Recovery rate is >90% if adequately treated, and 50% or lower if not treated.
Prevention and Hygiene: Exercise protective measures for patient and provider by using a surgical mask (or other respiratory protection) on both the patient and support staff exposed. Close contacts of patient and others exposed should receive prophylaxis:
Ciprofloxacin (adults): 500mg po x 1
Ceftriaxone (adults): 250 mg IM x 1
(child <15): 125 mg IM x 1
Spiramycin (child): 10 mg/ kg po q 6h x 5d

Follow-up Actions
Evacuation/Consultation Criteria: Evacuate cases after instituting immediate therapy.

Skin: Erysipelas
MAJ Daniel Schissel, MC, USA

Introduction: Erysipelas is most commonly an acute, dermal and subcutaneously spreading cellulitis caused by Group A beta-hemolytic *Streptococcus pyogenes* or *Staph. aureus*. It is characterized by an erythematous, warm, raised, tender area of the skin. Inoculation is through a break in the skin barrier (puncture, laceration, abrasion, surgical site) an underlying dermatosis (pitted keratolysis, tinea, or stasis dermatitis/ulcer), or through the middle ear or nasal mucosa in children. Risk factors include prior surgery resulting in lymph-edema, diabetes mellitus, hematologic malignancies and other immunocompromised states.

Subjective: Symptoms
Prodrome of malaise, anorexia, fever and chills is occasionally observed. More common is the rapid development of high fever and chills.

Objective: Signs
Using Basic Tools: The primary lesion is a bright erythematous, edematous, raised, warm, tender plaque with sharp, palpable leading margins (see Color Plates Picture 5). The distribution of lesion varies from the face to the lower extremities. Usually seen with an associated regional lymphadenopathy.
Using Advanced Tools: Lab: WBC count for infection

Assessment:
Diagnose based on clinical findings.
Differential Diagnosis: Early allergic or irritant contact dermatitis; fixed drug eruption; deep venous thrombosis; thrombophlebitis; rapidly progressive necrotizing fasciitis (a well-demarcated dusky purpuric lesion that is caused by thrombosis of the vessels, which is usually palpable). See related topics in this book.

Plan:
Treatment:
Primary: Dicloxacillin 500 mg po q 6 hrs for early mild cases
Or **nafcillin** or **oxacillin** 2.0 grams IV q 4 hrs for more severe cases
Alternative: Erythromycin 500mg po q 6 hrs

Patient Education
Prevention: Keep wounds clean, dry and protected.

Follow-up Actions
Evacuation/Consultation Criteria: Evacuate urgently. Consult dermatology or infectious disease if possible.

Skin: Staphylococcal Scalded Skin Syndrome
MAJ Daniel Schissel, MC, USA

Introduction: Staphylococcal scalded skin syndrome is a fairly distinctive pediatric dermatosis caused by an epidermolytic (epidermis-destroying) toxin. The reason for the association with children appears to be related to the fact that most adults and children over the age of 10 can localize, metabolize, and excrete the toxin more efficiently. They may develop bolus impetigo instead, but will limit the hematogenous dissemination of the toxin. This condition is most common in children 5 years of age and younger.

Subjective: Symptoms
Prodrome: Fever, malaise, extreme irritability, and anorexia; irritable child with low-grade fever.

Objective: Signs
Using Basic Tools: Generalized macular erythema, with fine, stippled, "sandpaper" appearance, rapidly progressing to a tender scarlatiniform phase over 24-48 hrs; spreads from the intertriginous and perioral facial areas to the rest of the body. The exfoliative phase is heralded by a characteristic perioral crusting that often cracks in a radial fashion. Within 48-72 hours, the upper epidermis may become wrinkled or slough off with light stroking of the skin (Nikolsky's sign). Shortly thereafter, flaccid bullae and desquamation of the upper layers of the epidermis are noted (see Color Plates Picture 12). Unless subsequent infective processes are present, the entire skin will re-epithelialize with scarring with in 2 weeks.
Using Advanced Tools: Lab: *Staph aureus* cultured only from colonized site of infection; umbilical stump,

external ear canal, conjunctiva, or nasal mucosa.

Assessment:
Diagnosis based on clinical criteria and verified by culture from a primary infection site.
Differential Diagnosis - Erythema multiforme, drug-induced toxic epidermal necrosis, and pemphigus vulgaris. See related topics.

Plan:
Treatment:
Primary: Supportive: Reliable home care, including cool baths or compresses and oral fluid replacement.
Antibiotics: Dicloxacillin 30–50 mg/kg/day po in divided doses. In newborns and infants where extensive sloughing has occurred: IV **oxacillin** 200 mg/kg/day q 4 hrs. Apply **mupirocin** ointment or **silver sulfadiazine** (**Silvadene**) for more irritated and inflamed areas.

Follow-up Actions
Reevaluation: Change antibiotics or evacuate if not improving. Ensure normal healing is taking place at follow-up exams.
Evacuation/Consultant Criteria: Evacuate if unstable or not responding. Consult dermatology as needed.

Skin: Impetigo Contagiosa
MAJ Daniel Schissel, MC, USA

Introduction: Impetigo is an acute, contagious, superficial infection caused by *Staphylococcus aureus* (bolus/ulcerative), or group A beta-hemolytic streptococci (vesiculopustular), or both. Although seen in all age groups, impetigo is most common in infants and children, occurring most frequently on the exposed parts of the body, especially the face, hands, neck and extremities. Predisposing factors include crowded living conditions, neglected minor wounds, and poor hygiene.

Subjective: Symptoms
Itching, weeping lesions

Objective: Signs
Using Basic Tools: Lesions: 1– 2 mm erythematous macules, which quickly develop into vesicles or bullae surrounded by a narrow halo of erythema. The vesicles rupture easily and release a thin, yellow, cloudy fluid which subsequently dries to a characteristic "honey crust." Scattered, discrete lesions located most frequently on the exposed parts of the body, especially the face, hands, neck and extremities. Groups of lesions may have satellite autoinoculated lesions at the periphery (see Color Plates Picture 8). There may be associated regional lymphadenopathy.
Using Advanced Tools: Lab: Gram stain of early vesicular lesions reveals gram-positive intracellular cocci in clusters or chains.

Assessment:
Differential Diagnosis
Varicella, herpes simplex, bullous tinea, allergic contact dermatitis (see appropriate topics in this book).

Plan:
Treatment
Primary: Dicloxacillin 250-500 mg po qid x 10 days
Alternative: Keflex (250 to 500 mg bid to tid); **erythromycin** (250 mg po qid)
Empiric: A high bacterial load may stimulate a super antigen reaction and aggravate the disease process. To decrease the bacterial load, wash the area with **Hibiclens** soap (**chlorhexidine gluconate**) once daily until cutaneous lesions clear.

Patient Education
Wound Care: Keep lesions covered and moist with antibiotic ointment or petrolatum to aid in rapid healing

Follow-up Actions
Evacuation/Consultation Criteria: Evacuation not usually indicated. Consult dermatology or infectious disease as needed.

Skin: Cutaneous Tuberculosis
Lt Col Gerald Peters, USAF, MC

Introduction: Cutaneous tuberculosis (TB) commonly represents skin manifestations of underlying pulmonary TB. It can also mean primary skin inoculation with TB, which is very rare. (see Respiratory: TB). Mycobacterium tuberculosis infects almost 2 billion people in the world (about 1 out of 3 living humans). It is an acid-fast, aerobic, gram-positive bacterium with both human and bovine forms. The BCG vaccine (not used in the US) is derived from the latter. Only about 1% of all TB patients will have skin lesions. Most cutaneous TB is indicative of systemic TB infection and can be the first sign of TB, especially in patients with HIV infection.

Subjective: Symptoms
Most are asymptomatic, some have a painless nodule progressing to painful ulcer, itching is uncommon.

Objective: Signs
Using Basic Tools: Primary skin inoculation: Non-tender, well-circumscribed ulcer; non-tender lymphadenopathy (3-8 weeks later); these characteristics are the ulceroglandular complex
Cutaneous manifestations of pulmonary TB: Variable presentations: boggy, indurated, erythematous skin and purulent ulcerations overlying infected subcutaneous tissue and enlarged lymph nodes (scrofuloderma); brown-red papules, soft and apple-jelly like, most common on the face, ears, buttocks and breasts (lupus vulgaris); discrete, blue-red to brown, tiny papules, some capped with minute vesicles, which burst to leave a crust and may affect all body parts, but the trunk, thighs, buttocks and genitalia are predisposed (Miliary TB); and others.
Using Advanced Tools: Place a PPD. A positive test does not indicate active TB, only exposure.

Assessment:
Differential Diagnosis
Primary Skin Inoculation - primary syphilis, tularemia, cat scratch disease, sporotrichosis, and others.
Cutaneous Manifestations of Pulmonary TB: extensive list including tertiary syphilis, deep fungal infections, chronic granulomatous disease, leishmaniasis, sarcoidosis, squamous cell carcinoma, and many others

Plan:
Treatment
Primary: There are at least 13 different agents to choose from. Many organisms are multi-drug resistant. See Respiratory: TB section for details.
Primitive: None effective
Empiric: Basic health measures, including clean and nutritious food and water, immunizations, and sanitation to help fight the infection.

Patient Education
General: TB is not an emergency. If exposed, the risk of developing TB is 5% in the first year. Get evaluated thoroughly by experts with in 1-2 months. There is a risk of re-infection from trauma and exposure to the organism in meat handlers, veterinarians and staff and persons involved in autopsies and undertaker duties.

Medications: Some medications can be toxic, so periodic blood tests are necessary during treatment.
Prevention and Hygiene: If exposed to an infected individual (several days of living or working in close quarters) test with PPD. Those who test positive need CXR to evaluate for active infection and should start prophylaxis. See Pulmonary TB section for details.

Follow-up Actions
Evacuation/Consultation Criteria: Evacuation is not usually necessary. Within 1-2 months, refer to dermatologist or infectious disease specialist for complete evaluation and choice of multiple drug therapy.

Skin: Leprosy (Hansen's disease)
MAJ Joseph Wilde, MC, USAR

Introduction: An inflammatory disease caused by infection with *Mycobacterium leprae*. It may be localized to the skin only or may involve internal organs. Infection of cutaneous nerves is very common. Leprosy is endemic in India, sub-Saharan Africa, South and Central America, the Pacific Islands and the Philippines. India, Myanmar and Nepal account for 70% of all the cases in the world. It is found in the Southeastern US and Hawaii. Most patients in the United States have a history of exposure to armadillos, a natural host for *M. leprae*.

Subjective: Symptoms
Hypopigmented or reddish skin lesions with decreased or no sensation.

Objective: Signs
Using Basic Tools: Circular patches and plaques with variable color including erythema, hyperpigmentation, or hypopigmentation; tissue swelling with nodules or ulcerations; lesions are common on the face, ears, and extremities.

Assessment:
Differential Diagnosis
Tinea corporis, mycosis fungoides, cutaneous TB, other causes of erythema nodosum (GI infections, drug reactions, sarcoidosis, others).

Plan:
Treatment
Dapsone, rifampin and **clofazimine** in combination as per specific protocols. Treatment lasts 6-12 months or longer. The relapse rate is very low (0.1% per year on the average).
Prevention: Isolate known patients until therapy is instituted.

Follow-up Actions
Evacuation/Consultation Criteria: Evacuate to receive specialty care. Diagnosis may be confirmed by skin biopsy (acid-fast stain) or slit skin smears.

Viral Skin Infections
Skin: Ecthyma Contagiosum
MAJ Daniel Schissel, MC, USA

Introduction: Ecthyma contagiosum infections (Milker's nodules, human orf) are caused by a genus of parapoxviruses which normally infect animals and only cause human infections when one handles infected livestock or works in infested areas where the virus may be harbored on fences, feeding basins, and other surfaces in the livestock areas. Human-to-human infection does not occur.

Subjective: Symptoms
One usually reports a history of contact with ungulates (sheep, goats, yaks), or cattle. After a 5 to 7 day incubation period one notes pruritus in an abraded area. This lesion enlarges and becomes painful. One may note systemic complaints of fever and chills on occasion.

Objective: Signs
Using Basic Tools: In Milker's nodule primary lesion is a deep erythematous papule (or small group of papules that enlarges) gradually into a firm, smooth, hemispherical nodule varying in size up to 2 cm in diameter. As with other viral processes the lesions usually become umbilicated. In human orf infections (HO) there are six classic clinical stages of the usually solitary lesion that last approximately 1 week; erythematous papular stage, targetoid plaque with central crusting, tender nodular, acute weeping nodular stage, regenerative stage, and regressive stage that heals without scaring. An ascending regional lymphangitis may be observed. The most common location is the dorsum of the index finger on the dominant hand due to handling the livestock or items within the livestock area. Other exposed skin sites of the arm, face, and leg are also at risk of infection.

Assessment:
Diagnosis based on clinical morphologic criteria and history of exposure to infected bovine or ungulate livestock or livestock areas.

Differential Diagnosis
Erysipelas, erysipeloid, atypical mycobacterium, bacillary angiomatosis, cat-scratch disease, leishmaniasis, pyogenic granuloma.

Plan:
Treatment
Primary: Symptomatic treatment for pain and pruritus.
Alternative: Antibiotic for secondary wound infections.

Patient Education
General: There are no effective vaccines for Milker's nodules or human orf available for livestock. The highest risk of HO is in workers slaughtering sheep, infecting approximately 5% on workers. Spontaneous, non-scaring, healing occurs generally in 4 to 6 weeks.
Prevention and Hygiene: Keep wounds clean and covered to aid in healing and decrease secondary infection.
No Improvement/Deterioration: Referral for biopsy and histological confirmation of process.

Follow-up Actions
Reevaluation: Refer for biopsy and histological confirmation of diagnosis.
Evacuation/Consultation Criteria: Evacuation not necessary. Consult as needed.

Zoonotic Disease Considerations
Contagious Ecthyma (Orf)
Agent: Orf virus (Parapox)
Principal Animal Hosts: Sheep, goats
Clinical Disease in Animals: Lesions on skin of lips and oral mucosa of lambs; udders of nursing ewes
Probable Mode of Transmission: Occupational exposure
Known Distribution: Worldwide; common

Skin: Herpes Zoster (Shingles)
MAJ Daniel Schissel, MC, USA

Introduction: Herpes zoster is an acute, localized, recurrence of the varicella-zoster virus (VZV) most commonly seen in patients over the age of 50. Less than 10% of cases are under the age of 20.

Subjective: Symptoms
A prodrome of pain, tenderness, itching, burning, and/or tingling in a dermatomal distribution precedes the eruption by 2-7 days. Intense pain in the dermatome usually persists throughout the eruption and resolves slowly (post herpetic neuralgia). Constitutional symptoms of headache, fever and chills occur in approximately 5% of patients.

Objective: Signs
Using Basic Tools: Erythematous papules or plaques, followed by umbilicated vesicles and bullae that commonly evolve to pustules (progression over 48-72 hours) and crust over by 7-10 days; new lesions may continue to appear for up to 1 week; lesions typically cluster in the distribution of a dermatome. (see Color Plates Picture 9).
Using Advanced Tools: Lab: Tzanck smear of vesicle (undersurface of the vesicle or bullae has the highest yield) with multinucleated giant epidermal cells.

Assessment:
Differential Diagnosis
Pain can be intense and may resemble that of cardiac disease, an acute abdomen or vertebral disk herniation. The eruption of zoster can resemble allergic contact dermatitis, irritant contact dermatitis or a localized bacterial infection.

Plan:
Shorten course of illness and subsequent development of notalgia paresthetica (back pain numbness and tingling) - a painful prolonged sequelae
Treatment
Primary: High dose oral **acyclovir** 800 mg po 5 times a day x 7 days.
Prevention: Prompt treatment during prodrome can lessen severity and shorten course of illness.

Follow-up Actions
Evacuation/Consultation Criteria: Evacuate soldiers whose condition interferes with mission performance.
Refer patients with: Ophthalmic zoster: Look for vesicles on the eyelids and tip of the nose - occur in 30% of patients with involvement of the nasociliary branch. **Requires URGENT MEDEVAC.** Blindness may develop if not treated appropriately.
Ramsay Hunt syndrome: Zoster involvement of the facial nerve and auditory nerves resulting in same-side facial paralysis.

Skin: Molluscum Contagiosum
MAJ Daniel Schissel, MC, USA

Introduction: Molluscum contagiosum is endemic in school age children through casual contact and spread of the poxvirus. Lesions are discrete, umbilicated, pearly, red papules. Involvement of the diaper area, trunk, face, and axilla is common. In the adult population, molluscum contagiosum is usually transmitted sexually, and may resolve spontaneously after several months or may reappear. HIV infected patients may have hundreds of small (2-3 mm) papules or develop giant 1-2 cm lesions.

Subjective: Symptoms
Asymptomatic, slow growing papules; commonly irritated by trauma or scratching, which can cause local spreading and/or secondary infection.

Objective: Signs
Using Basic Tools: Sharply circumscribed single or multiple superficial, pearly, dome-shaped papules with a characteristic umbilication seen easily with a hand-held lens. They initially present as pinpoint papules, increase slowly to 2-3 mm in size (see Color Plates Picture 11), and are often found in the genital region.

Assessment:
Differential Diagnosis See these topics in Dermatology Chapter.
Flat warts (Verruca Plana), condylomata acuminata (venereal warts), squamous cell or basal cell skin cancers, sebaceous gland hyperplasia (seen in elderly), epidermal inclusion cyst (on CD-ROM).

Plan:
Treatment
Primary: Treatment depends on location on body. Facial lesions should be treated less aggressively to decrease the risk of scarring. **Retinoid** gel 0.01%-0.25% applied to affected area q hs will cause a minor irritation to the lesion and works well for the face, as does **potassium hydroxide** topically. Apply to lesion daily until cleared.
Alternative: Curettage, followed by **imiquimod** 5% cream applied to the lesion each day until cleared.

Patient Education
General: Patient will likely continue to develop new lesions (in or around the area of old ones) that are not clinically visible at the initial visit. Return for additional treatment as needed
Medication: Avoid sun exposure when using retinoid gel.
Prevention: Keep all skin surfaces well hydrated/emolliated to avoid spread of the lesions.

Follow-up Actions
Return Evaluation: Consider alternate treatment or diagnosis, or refer for biopsy to rule out cancer.
Evacuation/Consultation Criteria: Evacuation not usually necessary. If not responding to appropriate therapy, refer to dermatologist or primary care physician for biopsy to rule out cancer.

Skin: Warts (including Venereal Warts)
MAJ Daniel Schissel, MC, USA

Introduction: The common wart is a benign growth in the epidermis seen in 7-10 % of the population. It is caused by the human papilloma virus (HPV) and commonly presents as papules and plaques in school age children. Some HPV strains have been associated with malignant transformation.

Subjective: Symptoms
Usually slow growing; commonly irritated with minor trauma or excoriations

Objective: Signs
Verruca vulgaris: Appear predominately on the dorsal aspect of the hands and periungual region of the nail, but may occur anywhere. They vary from solitary isolated lesions to vast numbers in any given individual. The normal progression of the primary lesion is from a small, round, discrete, flesh colored papule to a larger yellowish tan to black lesion that measures from several millimeters to a couple centimeters. The surface of the lesion commonly takes on a rough finely papillomatous (verruciform) surface with many characteristic

"reddish-black seeds" (thrombosed capillary loops).
Verruca plana: Appear predominately as smooth, flesh-colored to slightly tan, elevated papules, 2 to 5 mm in diameter, with a round or polygonal base on the face, neck, arms, and legs. In the bearded area of men and on the legs and axilla of women, irritation from shaving tends to cause the warts to spread in linear arrays (Koebner effect)
Verruca plantaris: Appear initially as small, shiny, skin colored, sharply marginated papules that evolve to plaques with a rough hyperkeratotic surface. They commonly present over the weight bearing points of the foot and therefore are commonly tender to palpation. They may be distinguished from calluses by noting the loss of the normal dermography (skin lines) that calluses usually retain. (see Podiatry: Plantar Warts).
Venereal Wart / Condyloma Acuminatum: Appear initially as tiny, pinpoint, flesh-toned, papules that may grow rapidly to cauliflower-like masses in any region of the anogenital area and the oral mucosa.

Assessment:
Differential Diagnosis See appropriate sections in this book for most conditions listed.
Verruca vulgaris - seborrheic keratosis, lichen planus, simple callus, molluscum contagiosum, carcinoma with verrucous (wart-like) appearance
Verruca plana - simple callus, foreign body, lichen planus, carcinoma with verrucous (wart-like) appearance
Venereal warts - condyloma lata (syphilis), intraepithelial neoplasm (Bowen's disease), invasive squamous cell skin cancer, molluscum contagiosum, lichen planus, skin tag

Plan:
Treatment
Verruca vulgaris: Trim warts; apply liquid **nitrogen** to the lesion (for 5 seconds, slowly thaw, repeat x 1); or apply 40% **salicylic acid** to area once daily after trimming lesions; apply duct tape if unable to keep acid on lesion; may need to treat for weeks until dermography (skin lines) returns.
Verruca plana: Trim warts; apply liquid **nitrogen** to the lesion for 5 seconds, slowly thaw, repeat x 1; or apply 40% **salicylic acid**, **retinoic acid**, or **cimetidine** to the lesions as described above
Venereal Warts: Apply **podophyllin** (10-25%), **imiquimod** cream to the lesions as described above; or apply liquid nitrogen as above.

Follow-up Actions
Return Evaluation: Consider alternate treatment or diagnosis, or refer for biopsy to rule out cancer.
Evacuation/Consultation Criteria: Evacuation not indicated, unless venereal warts cannot be treated and are large enough to interfere with mission. Consult dermatologist as needed.

Superficial Fungal Infections
Skin: Dermatophyte (Tinea) Infections
MAJ Daniel Schissel, MC, USA

Ringworm - Tinea corporis
Athlete's foot - Tinea pedis
Finger/Toenails - Tinea unguium
Palms and Soles - Tinea manuum

Jock itch - Tinea cruris
Some Dandruff - Tinea capitis
Dandruff of Beard - Tinea barbae/faciale

Introduction: Dermatophyte infections are superficial, caused by fungi that invade only dead outer layers of the skin or its appendages (stratum corneum, nails, hair). *Microsporum, Trichophyton* and *Epidermophyton* are the genera most commonly involved. Some dermatophytes produce only mild or no inflammation. In such cases the organism may persist indefinitely, causing intermittent remissions and exacerbations of a gradually extending lesion with a scaling, slightly raised border. In other cases an acute infection may occur, typically causing a sudden, vesicular and bullous reaction of the feet or an inflamed, boggy lesion of the scalp (kerion) due to a strong immunologic reaction to the fungus. Transmission of dermatophyte infections may be grouped into 3 main categories: from animals, from the environment, and from one another. Infections are seen worldwide.

Subjective: Symptoms
Rash, itching and flaking; severe cases: Painful, pruritic blisters.

Objective: Signs
Using Basic Tools:
Tinea corporis: Scaling, sharply demarcated plaque (with or without vesicles or pustules) with central clearing and peripheral enlargement producing an annular configuration, commonly on forearm and neck. (see Color Plates Picture 2)
Tinea pedis: Various patterns: Slightly erythematous plague on the plantar surface of the foot; dry, superficial, white scale is observed in an arciform pattern; moist patch of erythema, small fissures and erosions usually localized to the third and fourth interdigital spaces and the lateral sole.
Tinea unguium (onychomycosis): Whitish/yellow/brown, thick, dry, subungual (under nail) accumulation of friable keratin debris; the great toe is commonly affected first. (see Color Plates Picture 1)
Tinea capitis: 3 main groups: Focal patch of alopecia with minimal scale and erythema; "Black dot" appearance of broken of hair shafts also seen with minimal scale and erythema; kerion- a boggy, purulent, inflamed group of pustules that is often tender to touch and heals with severe scarring.
Tinea cruris: Similar to Tinea corporis but restricted to the intertriginous area of the groin.
Tinea barbae: Similar to Tinea corporis but restricted to the facial area.
Tinea manuum: Similar to Tinea corporis but restricted to the hands and feet.
Using Advanced Tools: Lab: KOH preparation for characteristic fungal hyphae.

Assessment:
Differential Diagnosis
Pityriasis rosea, discoid eczema, and psoriasis. See appropriate sections and index in this book. Diagnosis requires a KOH preparation (see Color Plates Picture 3 and Lab Procedures).

Plan:
Treatment
Primary: **Lac-Hydrin** and **Clotrimazole** topically bid for 2-4 weeks for tinea pedis and onychomycosis.
Lac-Hydrin decreases the dryness and increases the barrier properties of the skin.
Griseofulvin 500 mg bid with fatty meals (aids absorption) for 2 weeks for tinea corporis to 2 to 3 months for tinea capitis. Not effective against candidiasis or tinea versicolor.
Clotrimazole topical preparations (cream and solution) are effective against most fungal infections when applied bid/tid to affected areas and washed off before reapplication.
Alternate: **Itraconazole** and **terbinafine** are not recommended in the field due to the inability to properly evaluate side effects. A more effective treatment of onychomycosis is prevention and topical **ciclopirox** nail lacquer (**Penlac**).
Note: **Nystatin** treats only candidal infections not dermatophytes

Patient Education
Prevention and Hygiene: Good foot maintenance: change socks frequently, dry out boots, use antifungal soaps, use shower shoes.

Follow-Up Actions:
Reevaluation: Repeat KOH evaluation and clinical assessment.
Evacuation/Consultation Criteria: Evacuation should not be necessary. Consult for difficult cases.

Zoonotic Disease Considerations
Dermatophytosis (Ringworm)

Principal Animal Hosts: Dogs, cats, cattle
Clinical Disease in Animals: Dogs - focal alopecia, scaly patches with broken hairs; cats - focal alopecia, scaling, crusting; cattle - scaling patches of alopecia with gray-white crust; calves - periocular lesions
Probable Mode of Transmission: Direct contact with fomites and infected animals

Skin: Pityriasis (Tinea) Versicolor
MAJ Daniel Schissel, MC, USA

Introduction: Pityriasis versicolor (tinea versicolor) is a chronic, asymptomatic fungal infection caused by *Pityrosporum orbicularis*, a normal resident of the skin. The fungus, found worldwide, is seen most commonly in young adults in temperate zones, and accounts for up to 5% of all reported fungal skin infections. The fine scales of this lesion are teeming with hyphae and spores that transmit the disease. Factors that predispose to infection include warm humid climate, genetic predisposition, high plasma cortisol levels (i.e., patients taking corticosteroids), serious underlying disease or immunocompromise, pregnancy.

Subjective: Symptoms
Asymptomatic to slightly pruritic depigmentation of skin.

Objective: Signs
Using Basic Tools: White, tan, brown, or pink coalescing macules and patches with a fine scale; found most commonly on the upper torso and neck (or face in children).
Using Advanced Tools: Lab: KOH prep of scales reveals numerous short, straight or ring-shaped hyphae.

Assessment:
Differential Diagnosis
Allergic contact or irritant dermatitis, various tinea, post inflammatory changes (hyper- or hypo-pigmentation), vitiligo (familial, autoimmune hypopigmentation, with only cosmetic effects). See appropriate sections and index in this book.

Plan:
Treatment
Primary: Ketoconazole 400 mg po with orange juice to aid in absorption, wait one hour, exercise, leave sweat on body till the morning, then wash with **selenium sulfide** suspension. Repeat in one week
Alternative: Fluconazole 200 mg po as above.
Primitive: Apply **Propylene glycol** 50% in water bid for 2 weeks.

Patient Education
General: Relapse may require weekly washing with **selenium sulfide** suspension. Repigmentation of the affected areas takes up to 90 days.
Diet: No alcohol use with oral medications
Prevention and Hygiene: Use antifungal soaps. Wash weekly with **selenium sulfide**.

Follow-up Actions
Evacuation/Consultation Criteria: Evacuation is not necessary. If lesions do not respond to appropriate therapy refer patient to dermatologist or to endocrinologist to rule out endocrine disorders.

Parasitic Infections
Skin: Loiasis (loa loa)
MAJ Joseph Wilde, MC, USAR

Introduction: A parasitic infection transferred to humans by the bite of blood-sucking tabanid flies (deer fly, horse fly, or mangrove fly), found in damp, forested areas of West and Central Africa.

Subjective: Symptoms
Migrating sensations in the skin described as itching, tingling, pricking or creeping; recurrent temporary edema of hands and upper extremities (Calabar swellings); painful irritation of the conjuctiva and eyelids.

Objective: Signs
Using Basic Tools: Recurrent swelling of the upper extremities; sighting of adult worms in the conjunctiva.
Using Advanced Tools: Lab: CBC for eosinophilia; peripheral thick blood smear for microfilaria (see Color Plate Picture 31)

Assessment:
Differential Diagnosis
Gnathostomiasis – May also cause recurrent hand swelling; due to the ingestion of certain types of raw fish
Toxocariasis (Ocular and Visceral Larva Migrans)—Hepatomegaly, chronic abdominal pain, pneumonitis
Onchocerciasis—Subcutaneous nodules on head and shoulders; pigment changes; atrophy; edema of skin

Plan:
Treatment
Primary: Ivermectin 200 mcg/kg po x 1
Alternative: Diethylcarbamazine 2mg/kg tid after meals x 14 days; not available in US

Patient Education
Prevention: Personal protective measures against insect bites (Permethrin, DEET, insect netting, etc.).

Follow-up Actions
Evacuation/Consultant Criteria: Evacuation not usually indicated. Consult dermatology or infectious disease as needed.

Skin: Myiasis
MAJ Joseph Wilde, MC, USAR

Introduction: A parasitic infestation of the skin caused by the larvae of flies. Infestation can occur in any open wound or exposed intact skin. The fly eggs are carried on the abdomen of mosquitoes, which then deposit them on to human skin. Eggs then develop into larvae that burrow under the skin.

Subjective: Symptoms
Painful boils on exposed skin

Objective: Signs
Using Basic Tools: A localized group of furuncular (boil-like) lesions typically seen on the face, arms, legs, or scalp, with a central punctum draining serosanguinous fluid. Close observation of the central punctum reveals the posterior portion of a fly larva.

Assessment: Diagnose by clinical presentation.

Differential Diagnosis
Bacterial furunculosis - no larvae in lesion
Deep fungal infection - KOH positive
Atypical mycobacterial infection - no larvae in lesion

Plan:
Treatment
Primary: Place occlusive material such as petrolatum, mineral oil, butter, or raw pork to cause the larvae to exit the skin (also diagnostic).
Alternative: Surgical excision of the larvae may be required in some cases.

Patient Education
Prevention and Hygiene: Proper use of insect repellents and mosquito netting will decrease transmission.

Follow-up Actions
Evacuation/Consultation Criteria: Evacuation not necessary. Consult dermatology or infectious disease as needed.

Skin: Onchocerciasis (River Blindness)
MAJ Joseph Wilde, MC, USA

Introduction: Onchocerciasis is a filarial parasitic disease of humans caused by *Onchocerca volvulus*. Black flies, the vectors of the disease, require fast-flowing streams or rivers for reproduction. The worm affects inhabitants of central Africa, Yemen, Central America, southern Mexico, and South America.

Subjective: Symptoms
Severe and diffuse pruritus which worsens with scratching; deep skin nodules; ocular symptoms (photophobia, excessive tears, pain, blurred vision)

Objective: Signs
Using Basic Tools: Deep subcutaneous nodules on the scalp (in Central and South America) or over bony protuberances (in Africa). Acute cases may present only with pruritus, which is often worse on the buttocks, abdomen, and lower extremities. Chronic skin changes, such as lichenification and scarring, are seen in long-standing cases.
Using Advanced Tools: Lab: CBC for eosinophilia.

Assessment: Diagnose with clinical findings and travel to endemic areas.
Differential Diagnosis
Scabies - common in groin area and fingers
Insect bites - pruritus eases with scratching
Miliaria rubra - small papules and vesicles at opening of sweat glands

Plan:
Treatment
Primary: Ivermectin 100-200 mcg/kg single dose
Primitive: Excise nodules and examine (histologic examination) for adult worms

Patient Education
Prevention and Hygiene: Avoid river areas, especially riverbanks, in endemic areas. Maximize personal protective measures (insect repellents and mosquito netting, etc.).

Follow-up Actions
Evacuation/Consultation Criteria: Evacuation not necessary. Consult dermatology or infectious disease as needed.

Skin: Swimming Dermatitis
(Sea Bather's Eruption [salt water], Swimmer's Itch [fresh and salt water])
MAJ Daniel Schissel, MC, USA

Introduction: Swimming dermatitis occurs globally and takes two forms. Sea bather's eruption is caused by contact with larvae of a marine jellyfish that release a toxin when trapped between clothing and the skin. It typically presents in areas covered by clothing or swim wear, unlike swimmer's itch, which usually occurs in exposed areas. Cercarial dermatitis is more commonly known as swimmer's itch in the fresh waters of the north central United States, and clam digger's itch along the coastal salt waters. It is caused by the penetration by immature forms (cercariae) of a schistosome that normally infest birds, into the skin of an unsuspecting swimmer or bather. The cercariae die later in the skin, self-limiting the infection.

Subjective: Symptoms
Prickly eruption (rash, itching) within a few minutes to hours where the larvae sting or the cercaria penetrate; repeated exposure (allergic response) cause larger, longer lasting, and more pruritic lesions in cercarial dermatitis.

Objective: Signs
Using Basic Tools: Fine, papular rash in covered areas (sea bather's eruption) or exposed areas (swimmer's itch); lasts hours (swimmer's itch) to days; scratches often become secondarily infected; repeated cercaria exposures cause larger, longer lasting papules, that may advance to pustules and vesicles over a 3 to 4 days.

Assessment:
Diagnosis based on clinical presentation and history of exposure in infested waters.
Differential Diagnosis
Swimmer's itch, allergic or irritant contact dermatitis

Plan:
Treatment
Primary: These are self-limited diseases. Treatment consists of symptomatic relief of pruritus with antihistamines and prevention of secondary infection in areas of excoriation. Topical steroids can alleviate more advanced allergic reactions to repeated exposures to cercaria.

Patient Education
General: Prevent cercarial dermatitis by avoiding prolonged immersion in infested waters and treating infested fresh water streams and lakes with a mixture of copper sulfate and carbonate, or sodium pentochlorphenate. Dry briskly after potential exposure to remove cercaria before they have sufficient time to penetrate. Apply 20% **copper sulfate** solution to the skin and allow to dry prior to potential exposure.

Follow-up Actions
Evacuation/Consultant Criteria: Do not evacuate patients. Generally no need to consult specialists.

Bug Bites and Stings
Skin: Bed Bugs
MAJ Daniel Schissel, MC, USA

Introduction: The bed bug (*Cimex lectularius*) is a 3-5 mm, wingless, 6 legged, reddish brown, flattened, oval bodied, blood-sucking insect. It hides in crevices, bedding, or furniture, and normally emerges to feed at night in the dark. It is capable of traveling long distances in search of its blood meal, often from one house to another. Under normal conditions, it feeds about once a week but has been known to survive 6 months to a year without feeding. It characteristically leaves 3 bite marks in succession on its victim: "breakfast, lunch and dinner."

Subjective: Symptoms
Bite is not felt while sleeping; small asymptomatic macule at bite site; if sensitized (having developed an allergic reaction to the salivary secretions) have intensely pruritic papules that may evolve into a nodule that persists for weeks.

Objective: Signs
Using Basic Tools: Lesion is variable from a small, erythematous macule in non-sensitized individuals to an intensely pruritic papule or wheal, often with a central hemorrhagic dimple, in sensitized individuals. Characteristically 2-3 lesions grouped in a linear fashion on the exposed areas of the face, neck, arms, or hands. Bullae are more common in younger patients. Secondary infection of the excoriated lesions often clouds the clinical presentation.

Assessment:
Diagnosis based on clinical morphologic criteria and history of exposure.
Differential Diagnosis - other arthropod assault, nummular dermatitis, irritant or allergic dermatitis (see appropriate sections).

Plan:
Treatment
Relieve pruritus with antihistamines (e.g., **Benadryl**) and prevent secondary infection (topical antibiotic lotion, e.g., **Neosporin**). Use topical steroid cream (1% **hydrocortisone**) on more advanced cases.

Patient Education
General: Eliminate the bug from the environment with insecticides (consult preventive medicine).

Follow-up Actions
Evacuation/Consultation Criteria: Evacuation is not necessary. Consult as above.

Skin: Centipede Bites
COL Roland J. Weisser, Jr., MC, USA

Introduction: Centipedes are elongated, cylindrical arthropods having a single pair of legs ending in claws on each body segment (millipedes have two leg pairs per segment). Centipedes have been reported up to 26 cm long, and are frequently more colorful (red, yellow, black, and blue) than millipedes and thus more likely to be sought as trophies. They are widely distributed, especially in warm, temperate and tropical regions. Centipedes live in moist environments, most commonly in forests amongst leaf litter or rotting timber, in caves, along the seashore (under damp seaweed and other detritus). Most have very poor or non-existent vision. Unlike millipedes, centipedes are carnivorous. The legs on the first body segment are modified into fangs that bite and channel venom into prey. In addition to the venom, some species exude defensive substances from glands along the body segments that may cause skin vesication like millipede exposure. A single fatality has been reported in a child bitten on the head by a large centipede. Most small species are innocuous.

Subjective: Symptoms
Severe local pain, swelling and redness; swollen, painful lymph nodes; headache; palpitations; nausea/vomiting; anxiety.

Objective: Signs
Local: Edema, erythema, tenderness and local necrosis around bite; lymphangitis/lymphadenopathy
General: Significant anxiety, possible systemic toxic reaction (unlikely)

Assesment:
Diagnose based on the bite history or identification of the centipede.
Differential Diagnosis - Anaphylaxis (see Shock); other bug bites and stings including hymenoptera and spiders (see Skin: Bug Bites and Stings); cellulitis (fever, hot wound, advancing erythema)

Plan:
Treatment: Supportive: Apply ice/cold (some prefer heat), **acetaminophen** or NSAIDs (see Symptom: Joint Pain for NSAIDs doses). There is no known antivenin.

Patient Education
Prevention: Never handle centipedes. Use caution when turning soil and when moving or climbing over rocks. Use work gloves in endemic areas.
No Improvement/Deterioration: Return for fever, or reddening or blackening of the skin.

Follow-Up Actions
Return Evaluation: Observe for potential secondary infection or tissue blackening (necrosis is uncommon).
Evacuation/Consultation criteria: Evacuation not necessary unless bite complicated by necrosis. Consult primary care or preventive medicine physician, or entomologist as needed.

Skin: Millipede Exposure
COL Roland J. Weisser, Jr., MC, USA

Introduction: Millipedes are elongated, worm-like arthropods having two pairs of legs on each body segment (centipedes have one pair of legs per segment). They range in size from almost microscopic to 30 cm in length, with 100-300 pairs of legs. They are generally brown/black/gray in color, slow moving, nocturnal herbivores that live in humid environments and can be found in soil, leaf litter, under stones or decaying wood. Millipedes sense primarily with their antennae, having only rudimentary eyesight. When threatened, they coil up into a ball to protect their more vulnerable underbelly. Millipedes do not have biting mouthparts or fangs, but they secrete an irritating, repellent liquid from pores along the sides of their bodies when they feel threatened. In large doses these secretions can be corrosive and cause blistering of skin. No deaths have been reported from millipede exposures, and it is unlikely that any such exposure, even to a small child, would prove fatal.

Subjective: Symptoms
Painful, irritated skin, eye irritation and pain (ocular exposures).

Objective: Signs
Brown staining of the skin at the site of contact, along with erythema, mild edema and vesicle formation; skin may later crack, slough and heal; conjunctivitis may progress to ulceration of the conjunctiva and cornea (ocular exposures).

Assesment: Diagnose based on the history of millipede handling or identification of the specimen.
Differential Diagnosis: Centipede bite and caterpillar (Lepidoptera) "sting" (see topics in this chapter and on CD-ROM).

Plan:
Treatment
1. Irrigate exposed eye promptly with copious amounts of water or saline to dilute toxin. If conjunctival ulcer is noted, see Symptom: Eye Problems: Eye Injuries.
2. Wash exposed skin thoroughly with soap and water to remove any remaining toxin.
3. Apply topical steroid cream (1% **hydrocortisone**) as needed to skin. Do not use steroids in the eye.
4. Supportive therapy with ice/cold; acetaminophen or NSAIDs (see Symptom: Joint Pain for NSAIDs doses) may be comforting.

Patient Education
Prevention: Avoid handling millipedes. Use caution when turning soil and when picking up or climbing over rocks.
No Improvement/Deterioration: Return promptly for continuing eye pain or deteriorating vision.

Follow-Up Actions
Return Evaluation: Examine involved eye(s) daily until healed. No further exams needed for skin lesions.
Evacuation/Consultation criteria: Evacuation not necessary unless conjunctival ulcer is large or does not heal in 24-48 hours. Consult ophthalmologist or primary care physician, or entomologist as needed.

Skin: Hymenoptera Stings
(bees, wasps, hornets, yellowjackets, and ants)
COL Roland J. Weisser, Jr., MC, USA

Introduction: The stinging insects of the order Hymenoptera include bees, wasps, hornets, yellow jackets and ants. Hymenoptera stings are a nuisance for most victims who usually recover without sequelae. One study reports 17-56% have a local reaction, 1-2% have a generalized reaction, and 5% seek medical care. However, because the Hymenoptera are so ubiquitous and live in such close proximity to humans, they are responsible for more human deaths each year than all other venomous animals combined. 50% of fatalities occur within the first hour, and 75% occur within four hours after the sting. The venom load from 30 wasp or 200 honeybee stings may be sufficient to cause death. Alternately, a single sting may provoke a generalized anaphylactic reaction (the proteinaceous venom is a potent activator of the immune system) and death in a sensitized individual, particularly if there was an earlier, milder generalized reaction. The shorter the time interval since the previous challenge, the more likely a severe subsequent reaction. Additionally, cross-reactions to the venoms of various members of the Hymenoptera family have been reported. For example, an individual who suffers an anaphylactic reaction to a wasp bite may also simultaneously develop anaphylaxis to ant bites. Hymenoptera are social insects that live in colonies or hives located in caves, hollow trees or in the ground. They are most often found among flowers and fruit where they feed, and are probably attracted by bright colors, perfumes and colognes.

Bees have a stinger with a specialized tip that not only penetrates the skin and delivers venom, but possesses a barb that anchors the stinger in the skin. The bee is able to sting only one time, because the barbed stinging apparatus remains in the victim when the bee flies away causing evisceration and death of the insect. The so-called "killer bees" are hybrids of species (e.g., European bees and African bees) that evolved under different environmental conditions. While the individual venom and sting are no different from other species, "killer bees" tend to be overtly aggressive, are prone to "gang up" on victims, and may chase intruders up to 150 meters.

Wasps, hornets, and yellow jackets are generally more aggressive than bees, and have a barbless stinger that allows them to inflict multiple stings. Yellow jackets are carnivores that congregate around dead meat or foodstuffs.

Fire ants typically bite and hold onto the victim with their mandibles and then swivel their abdomen in an arc around the fixed mouthparts, inflicting multiple stings. The venom of most ants is less potent than that found in flying Hymenoptera, causes much less tissue destruction and is much less likely to elicit a generalized allergic reaction (about 80 anaphylactic deaths reported). The venom of fire ants is almost 95% alkaloid and exerts a direct toxic effect on human and animal systems. A number of human deaths from the toxins contained in multiple fire ant bites have been reported, especially in old and debilitated persons. However, other individuals have been known to survive as many as 5,000 fire ant stings.

Subjective: Symptoms
Instantaneous stinging pain, warmth (vasodilation) and pruritus at site of sting(s), nausea and vomiting, visceral pain following ingestion of insect and stings to the GI tract.

Objective: Signs
Local: Rapidly spreading edema (as large as 10-15cm) and urticaria near sting site; compromised distal circulation from edema; stinging apparatus from bees and bleeding may be seen in the wound; distal sensory loss if stung over peripheral nerve; corneal ulceration from corneal sting.
Generalized: Rapid onset of symptoms, urticaria, confluent red rash, shortness of breath, wheezing airway (tongue, soft palate, etc.) edema, weakness, syncope, anxiety/confusion, chest pain, tachypnea, tachycardia, hypotension, delirium, shock and cardiorespiratory arrest

Assessment:
Diagnose based on history of exposure and/or captured specimens.
Differential Diagnosis - angina (see Cardiac: Acute MI), rheumatoid arthritis (see Symptom: Joint Pain), corneal abrasion/laceration (see Symptom: Eye Problems: Eye Injuries), snakebite (see Toxicology), cat scratch disease (see ID: Bartonellosis); cellulitis; honey exposure in susceptible individual

Plan:
Treatment:
Anaphylactic or Generalized Toxic (Anaphylactoid) Reaction: see (Shock: Anaphylactic)
Single Sting from Flying Hymenoptera (bees, wasps, hornets, and yellowjackets):
1. Remove stinger and venom sac intact as quickly as possible (stinging apparatus may actively injects venom into the wound for one minute), regardless of method.
2. Apply ice or cold water for anesthesia and to control swelling.
3. Apply local analgesic, antibacterial or steroid ointments as desired (see Symptom: Rash). Other remedies to include ammonia, sodium bicarbonate, and papain (meat tenderizer) have minimal proven effectiveness.
4. Elevate extremity to limit spread of edema.

Fire Ant Stings:
1. Do not unroof vesicles.
2. Apply topical antibiotic and/or anesthetic creme for secondary infections (see Symptom: Rash). Use prophylactic antibiotics for children with >30 fire ant stings.
3. NSAIDs may reduce the degree of inflammation; antihistamines may diminish itching but promote somnolence, topical anesthetics, e.g., **Dermoplast**

Massive Multiple Stings:
1. Be prepared to treat for anaphylaxis or anaphylactoid reaction from venom load (see Shock: Anaphylactic).
2. Do NOT use massive doses of parenteral steroids or antibiotics in individuals not having anaphylactic or toxic reaction.
3. Give oral NSAIDs to reduce inflammation (see Symptom: Joint Pain) and antihistamines to reduce pruritus (see Symptom: Pruritus/Itching).
4. Use topical hydrocortisone cream and topical anesthetics as needed (e.g., **Dermoplast Spray = benzocaine 20%/menthol 0.5%).**

Patient Education
General: If attacked or stung by flying bees or wasps, do not flail arms, etc. Crushing one insect may incite others to attack even more vigorously. Although the insects may defend an area up to 150 meters from their nests, they can only fly about four miles per hour. Therefore, healthy individuals can easily outrun the swarm and escape.
Medication: Any individual with a generalized reaction should be referred for allergy testing and desensitization, and should thereafter wear a medic-alert tag and carry an emergency medical kit containing at least an antihistamine and aqueous epinephrine in a pre-filled syringe for immediate self treatment. Continue immunotherapy indefinitely as long as risk of exposure is substantial.
Prevention: Avoid nests, hives, bee trees, locations around flowers, fruit, etc. Avoid bright colored clothing, perfume/cologne, etc. Wear shoes (ground nests are common). Do not use noisy equipment (mowers, etc.) in vicinity of "killer-bee" colonies.

Follow-Up Actions
Return Evaluation: Be alert for rebound anaphylaxis as medication levels diminish. Observe stings for secondary infection. Be alert for serum sickness up to 14 days post-sting.
Evacuation/Consultation Criteria: Immediately evacuate cases with anaphylactic or generalized reactions. Other cases do not require evacuation, even with multiple stings. Consult primary care physician and allergist as needed.

Skin: Mites
MAJ Daniel Schissel, MC, USA

Introduction: Mites are tiny parasites that burrow under or attach themselves to the skin where they inflict small bites that cause much larger rashes. The best known, the mite *Sarcoptes scabiei* or scabies, is covered in another section of this of this chapter. Other common mites are dust mites, which cause respiratory allergic symptoms, and chiggers. Chiggers hide in tall grass or undergrowth waiting to attach to a passing victim. When they meet an obstacle in the clothing, like a belt or boot top, they inject an irritating secretion that causes the itching sensation, and then drop off or are scratched off. The pruritus peaks on the second day and gradually subsides in a week.

Subjective: Symptoms
Local pruritic, burning or stinging sensation accompanying erythematous lesions.

Objective: Signs
Using Basic Tools: Chiggers: Discrete, 1-2mm, erythematous papules (often with a hemorrhagic center) commonly seen along the belt line or boot top. The primary lesion of other mites follows a spectrum from erythematous papules, pustules, vesicles, to general urticaria. Secondary linear excoriations are common with all mite infestations. In children these eruptions are often widespread, with urticaria, and even bullae formation. The pruritus may persist for weeks and may progress to impetigo in children.

Assessment: Diagnosis based on clinical morphologic criteria and history of exposure.
Differential Diagnosis - irritant or allergic contact dermatitis, drug or viral reaction.

Plan:
Treatment
Primary: Relief of pruritus with oral antihistamines (i.e., **diphenhydramine**), cool baths or compresses, topical steroids or topical antipruritics (**calamine lotion, aloe or Chig-a-rid**).
Primitive: Clear nail polish.

Patient Education
General: Material employed for protection against mites function more as toxicants than true repellents. DEET (diethylmethyltoluamide) provides the best protection when applied to the clothing. Benzyl benzoate is an excellent chigger toxicant and remains effective after rinsing, washing, or submersion in water.

Follow-up Actions
Evacuation/Consultant Criteria: Evacuation is not normally necessary. Dermatology consultation may be helpful for further treatment options.

Skin: Spider Bites
(Black Widow [Latrodectus mactans], Brown Recluse [Loxosceles reclusa])
MAJ Daniel Schissel, MC, USA

Introduction: Two dangerous spiders bite in the United States, the black widow and brown recluse. The **Black Widow** (BW) spider may be found from southern Canada to Mexico and Cuba. The female is easily recognized by her coal black globular body and red-orange hourglass marking on the underbelly. It favors cool, dark, little-used places to set its web, including outdoor toilet seats. The **Brown Recluse** (BR) has a 1cm oval light tan to dark brown body, and a leg span over 2.5 cm. A classic dark brown violin-shaped dorsal marking extending from the 3 sets of eyes (rather than 4 seen in other spiders) to the abdomen differentiates it from other brown spiders. It is found across the United States, and, like the BW, only bites in self-defense. It is commonly found in storage closets, old shoes or boots, rock bluffs and barns. BR venom is hemolytic and necrotizing and contains a spreading factor. Other species of *Latrodectus* and *Loxosceles* are found in other areas of the world. Treat their bites similarly. Most other spider bites should be treated supportively as with the BW, or with wound care and shock precautions as with BR. Some spiders have neurotoxic venom, which should be treated with antivenin if available.

Subjective: Symptoms
BW: Pinching bite followed by local swelling and burning at the puncture site; abdominal cramping and pain begins within 10 min to an hour, peaking at about 3 hours.
BR: Painful, stinging bite with gradual development of severe pain in 2-8 hours; slow progression over the following weeks of a necrotic ulcerating process spreading from the bite; 25% of patients will also have a systemic reaction also, with nausea, vomiting, fever, chills, muscle aches and pains.

Objective: Signs
Using Basic Tools:
BW: Bite with two, red, punctate markings on an erythematous plaque, most commonly found on the buttocks or groin area.
BR: Tender, swollen bite that progresses to hemorrhagic vesicle or bullae, and later (5-7 days) to a slowly enlarging gangrenous eschar with a border of erythema and edema; lymphangitis; wound granulates, leaving a large fibrous scar; systemic symptoms (25% patients) include generalized erythema, purpuric macular eruption, thrombocytopenia, hemoglobinuria, hemolytic anemia, renal failure and shock.

Assessment: Diagnosis based on clinical morphologic criteria and history of exposure.
Differential Diagnosis
BW - acute abdomen, tick or other arthropod bite.
BR - necrotizing fasciitis for localized wounds; anaphylaxis for systemic reactions.

Plan:
Treatment
BW: Pain control, **calcium gluconate** 3 mEq IV and tetanus prophylaxis.
BR: Tetanus prophylaxis, ice and elevate bite site to decrease the localised reaction; pain control; good

wound care to minimize scarring; **Dapsone** 100 mg po to decrease the necrotic reaction; treat for shock as needed (see Shock chapter).

Patient Education
General: Most patients recover fully in 2-3 days from a BW bite, but it has been fatal in children.

Follow-up Actions
Evacuation/Consultation Criteria: Urgently evacuate unstable patients including those with systemic reaction to BR bite. Advanced treatment of bites, including antivenin, requires evacuation to a medical treatment facility.

Skin: Scabies
MAJ Daniel Schissel, MC, USA

Introduction: Scabies is a transmissible parasitic skin infection caused by the mite (*Sarcoptes scabiei*) and is characterized by superficial burrows, intense pruritus and secondary infections. The female mite tunnels into the epidermis layer and deposits her eggs along the burrow. Scabies is most commonly transmitted by skin-to-skin contact with an infected person and has a worldwide distribution.

Subjective: Symptoms
Continuous low-grade pruritus of the genital areas (to include nipple region in females) with increased itching at night.

Objective: Signs
Using Basic Tools: Papules, vesicles, and linear burrows intermingled with or obliterated by scratches, dried skin, and secondary infection. The burrow is the home of the female mite, the papules are the temporary invasion of the developing larvae, and the vesicular response is believed to be a sensitization to the invader. The primary locations of invasion include the web spaces of the fingers and toes, the axillae, the flexures of the arms and legs, and the genital regions (to include the nipple region of females). The head and face are commonly spared. The papules of the genital region may persist for weeks to months after the mite has been cleared.

Assessment: Diagnosis based on clinical exam and laboratory/provider isolating evidence from the patient of an infestation-"scabies prep".
Differential Diagnosis - irritant or allergic dermatitis, arthropod bite reaction, eczematous dermatitis (see appropriate sections).

Plan:
Treatment
Primary:
1. Apply **permethrin** 5% cream **(Elimite)** from the neck down and leave on the skin overnight.
2. Clip the nails and scrub under the distal nail to dislodge any excoriated mites.
3. Change and wash all undergarments and bedding in hot water prior to showering off the **permethrin** cream. Dry-clean (or seal in an airtight bag for 2 weeks) clothing items that cannot be washed.
4. Treat all family members and personal contacts at the same setting.

Alternative: 6-10% **sulfa** in **Vaseline**, 10 % **crotamiton (Eurax)**, 1.0% **gamma benzene hexachloride (lindane/Kwell)**; oral **Ivermectin** has recently been approved.
Secondary: Relieve pruritus with oral antihistamines, cool baths or compresses, and topical steroids. Topical antipruritics like **saran** lotion or **prameGel** are alternatives.

Patient Education
General: Do not clean the hair or body excessively, as this can lead to excessively dry skin and a

secondary focus of pruritus. Often the pruritus persists despite normal hygienic routines if the patient has a hypersensitivity to the mite or its products.

Follow-up Actions
Reevaluation: Repeat examination for those with continued nocturnal exacerbation of their pruritus. This is the feeding time of the scabies mite and will help differentiate between a hypersensitivity reaction and persistent infestation. Consult dermatology.
Evacuation/Consultation Criteria: No need to evacuate. Consult dermatology as needed.

Zoonotic Disease Considerations
Principal Animal Hosts: Cattle, dogs, and cats
Clinical Disease in Animals: Intense pruritus, lesions start on head, neck and shoulders and can spread to the rest of the body.
Probable Mode of Transmission: Contact with infected animals.

Skin: Pediculosis (Lice)
(crab lice, head lice, body lice)
MAJ Daniel Schissel, MC, USA

Introduction: A parasitic infestation of the skin—scalp, trunk, or pubic areas—that usually occurs in overcrowded dwellings. Head and pubic lice are found on the head and in the pubic area. Body lice are seldom found on the body (only getting on the skin to feed), but can be found in the seams of clothing.

Subjective: Symptoms
Pediculus humanus capitis (head louse): pruritus of the sides and back of the scalp. *Pediculus humanus corporis* (body louse): localized or generalized pruritus on the torso. *Pthirus pubis* (crab louse): asymptomatic or mild to moderate pruritus in the pubic area for months.

Objective: Signs
Using Basic Tools: Head Lice: <10 organisms usually identified with naked eye or hand lens. The nit (1 mm oval, gray, firm capsule) cemented to the hair is the egg remnant of a hatched louse. New, viable eggs have a creamy yellow color. The infestation can be dated from the location of the nit, since they are deposited at the base of the hair follicle and the hair grows 0.5 mm daily. Also: excoriation, secondary infection and adenopathy.
Body Lice: Excoriated, small erythematous papules localized to the torso area. Nits and lice are found in the in the seams of clothes.
Crab Lice: 1-2 mm brown to gray specks in the hair-bearing areas of the genital region. Nits appear as tiny white adhesions to the hair. Small erythematous papules at the sites of feeding, especially in the periumbilical area. Secondary excoriation and lichenification may be present. Maculae caeruleae are non-blanchable blue to gray macules, 5-10 mm in diameter, at the site of a bite that result from the breakdown of heme by the louse saliva. These lesions may be found from the groin to the eyelash. Adenopathy may be observed with secondary infections.

Assessment: Diagnose based on clinical findings and confirm with identification of lice or nits.
Differential Diagnosis - irritant or allergic dermatitis, arthropod reaction, seborrheic dermatitis, scabies, eczematous dermatitis, folliculitis. Differentiate by finding lice or nit.

Plan:
Treatment
Primary:
1. Wash bedding in hot water. Clothing items that cannot be washed should be sealed in an airtight bag for 2 weeks or dry-cleaned.
2. Apply **permethrin** 1% rinse to the scalp and wash off after 10 min.

3. Examine and treat all family members and personal contacts at the same time. Remove nits with a very fine-toothed nit comb. "**Step-2**" or Clear Lice Egg remover Cleansing Concentrate applied to the hair will help aid in nit removal (no pediculocide is 100% ovicidal).

Alternative: **Lindane** shampoo 1.0% (may not be available in US), and **pyrethrins** shampoo 0.3% used in a similar fashion as above.

Secondary: Relieve pruritus with oral antihistamines, cool baths or compresses, and topical steroids. Topical antipruritics like **saran** lotion or **prameGel** are alternatives.

Patient Education
General: Do not clean the hair or body excessively, as this can lead to excessively dry skin and a secondary focus of pruritus.

Follow-up Actions
Evacuation/Consultant Criteria: No need to evacuate. Consult dermatology as needed.

Spirochetal Diseases
Skin: Yaws

MAJ Joseph Wilde, MC, USAR

Introduction: *Treponema pallidum* subspecies *pertenue* causes this chronic relapsing infectious disease. It is found primarily in warm rural areas of the tropics. Yaws is transmitted by broken skin (i.e., cut, abraded, or inflamed) coming in direct contact with active skin lesions.

Subjective: Symptoms
Single, exophytic skin lesion that tends to ulcerate and crust; may be followed by period of healing, then reappearance or multiple raspberry-like lesions. Finally, untreated patients may have bone involvement, resulting in joint pain, difficulty walking or fractures.

Objective: Signs
Using Basic Tools: Yaws may have three clinical phases. The primary stage shows a single erythematous, infiltrated plaque, which eventually heals with scarring. The secondary stage emerges rapidly, with multiple papules that ulcerate and form yellowish crust. The tertiary stage develops after several years and shows deep ulcerated nodules with underlying involvement of bone.

Assessment: Diagnose based on clinical findings in an endemic region and confirm with darkfield microscopic exam of the exudates from skin lesions.
Differential Diagnosis - syphilis, paracoccidiomycosis, leishmaniasis. See appropriate sections.

Plan:
Treatment
Primary: **Penicillin** 2.4mU single IM dose
Alternative: Oral antibiotics such as **tetracycline** 500 mg bid or **erythromycin** 500 mg qid x 14 days.

Patient Education
General: Avoid contact with infected persons having active lesions.

Follow-up Actions
Evacuation/Consultation Criteria: No need to evacuate. Consult dermatology as needed.

Skin: Pinta
MAJ Joseph Wilde, MC, USAR

Introduction: Pinta is a chronic infectious disease affecting of the skin caused by *Treponema pallidum* subspecies *carateum*. It is found only in low-altitude tropical areas of Central and South America. Transmission occurs by direct skin or mucous membrane contact with infected individuals. It is usually acquired during childhood.

Subjective: Symptoms
Nonspecific, diffuse, red scaling papules which may coalesce and become hypopigmented over several years.

Objective: Signs
Using Basic Tools: Acute: Multiple erythematous macules that may be slightly raised on exposed skin.
Chronic: After a lapse of months or years, mottled hypopigmentation skin appears.

Assessment: Definitively diagnose by darkfield microscopy.
Differential Diagnosis - vitiligo (autoimmune hypopigmentation), tinea corporis (KOH positive)

Plan:
Treatment
Primary: Penicillin 600,000 U IM
Alternative: Tetracycline or **erythromycin** 250 mg qid x 14 days.

Patient Education
General: Avoid contact with infected persons.

Follow-up Actions
Evacuation/Consultation Criteria: No need to evacuate. Consult dermatology as needed.

Skin Disorders
Skin: Psoriasis
MAJ Daniel Schissel, MC, USA

Introduction: Psoriasis is a multifactorial genetic disorder of the skin that affects approximately 2% of the population in western countries, with onset before age 20 in 1/3 of cases. There are many clinical manifestations, but the most common (vulgaris) is typically expressed as chronic scaling papules and plaques in a characteristic extensor surface distribution.

Subjective: Symptoms
Chronic history (months to years) of itching, especially in the anogenital crease and scalp; acute exacerbations occur in guttate psoriasis and generalized pustular psoriasis; fever, chills, arthritis, and weakness will accompany acute onset of generalized pustular psoriasis. Subtle cases may be suspected in patients with only a slight gluteal crease "pinkening" and nail findings.

Objective: Signs
Using Basic Tools: Skin: sharply demarcated, "salmon pink" erythematous, round to oval papules and plaques with marked "silvery-white" scale. The arrangement ranges from a few scattered discrete lesions to diffuse involvement without identifiable borders. Lesions on the extensor surfaces are usually quite symmetrical. When one excoriates this scale, there is pinpoint bleeding (Auspitz sign). Fingernail: pitting, subungal hyperkeratosis (thickening of the nail material), onycholysis (loosening of the nail plate from the nail bed), and "oil spot" (yellowish-brown) spots under the nail.

Using Advanced Tools: Lab: KOH of scale from lesion to insure the process is not fungal.

Assessment: Diagnose based on the characteristic clinical presentation.
Differential Diagnosis
Seborrheic dermatitis - dry scalp
Lichen simplex chronicus - small, very itchy, papule(s)
Candidiasis - KOH positive
Drug reaction - history of drug exposure

Plan:
Treatment: Only with topically administered agents.
Primary:
1. NEVER GIVE ORAL STEROIDS as they may cause a systemic pustular eruption and kill the patient.
2. Use topical fluorinated corticosteroids (**betamethasone, flucinolone, clobestasol**) in an ointment base. Apply after soaking off the scale in a salt-water bath bid x 2 weeks (then move to non-fluorinated steroid ointment). Apply the ointment to the skin when still wet then pat dry.
3. Never apply fluorinated steroid to the face or in occluded areas like the groin or axilla
Alternative: Triamcinalone ointment
Symptomatic: **Hydroxyzine** (**atarax**) 25-75 mg po q4 hrs for pruritus.
Empiric: Ultraviolet exposure (20 min exposure to noonday sun) will accelerate the resolution of the lesions.

Patient Education
General: Lesions can be exacerbated by stress and illness
Medications: NO oral steroids, beta-blockers, lithium, NSAIDs. All can exacerbate the lesions.
Prevention and Hygiene: Use antifungal soaps.

Follow-up Actions
Reevaluation: If lesions do not start to thin in 2-3 weeks referral is needed
Evacuation/Consultation Criteria: Referral is not usually indicated, unless unstable. Consult dermatology as needed.

Skin: Pseudofolliculitis Barbae
MAJ Daniel Schissel, MC, USA

Introduction: Pseudofolliculitis barbae (PFB) is a common disorder of the pilosebaceous unit of the beard. It is caused by multiple factors and is more common in those with very curly beard hair. Affected persons may have a genetic predisposition due to abnormal formation of the hair follicle. When hair is lifted and shaved it retracts into the pilosebaceous unit. Curly hair can then penetrate the side of the follicular unit and cause a mechanical irritation in the skin, or the curly hair may exit appropriately and then curl back into the surface of the face again, causing an irritation.

Subjective: Symptoms
Rapid development of papules and pustules in the beard area after shaving.

Objective: Signs
Using Basic Tools: Follicular-based papules and pustules below the jawline and on the anterior neck. Long-standing lesions may become nodular, cystic or granulomatous.

Assessment: Diagnose based on history and clinical findings.
Differential Diagnosis
Acne - lesions also in other areas
Irritant contact dermatitis - lesions also in other areas.

Plan:
Treatment
Primary: Manage mild cases with proper shaving techniques. Allow the hair to grow out onto the surface of the skin and then trim with a safety razor or clipper. Gently lift out remaining buried ingrown hair tips onto the surface and clip-- do not pluck or pull.
Alternative: A chemical depilatory.
Empiric: Minocycline 100 mg po bid will help decrease the irritation and secondary infection.

Patient Education
General: Shave gently with "bump fighter" razor, without pulling the skin taut or repeating over the same area; shave "with the grain" of the hair. Close shaving promotes oblique penetration of the sharpened hairs into the skin and should be avoided whenever possible.
Prevention and Hygiene: Apply moist heat after shaving, followed by a moisturizer (like razor bump fighter), and avoid strong astringents like alcohol that will only dry the face and cause more irritation.

Follow-up Actions
Evacuation/Consultation Criteria: No need to evacuate. Consult dermatology as needed.

Skin: Skin Cancer
(Basal & Squamous Cell Carcinoma, Malignant Melanoma)
Lt Col Gerald Peters, USAF, MC

Basal Cell Carcinoma (BCC)
Introduction: BCC is by far the most common cancer in the world, with over 700,000 cases each year in the U.S. alone. Early detection and treatment are paramount in order to avoid extensive tissue destruction, damage to adjacent structures, and complex surgery and reconstruction. The good news is that this cancer is virtually 100% curable if approached early and properly. Metastasis is very rare. Sun exposure and fair complexion, light-colored hair and eyes are the main risk factors for skin cancer. Unfortunately about 80% of all the ultraviolet radiation (sun) exposure comes before age 18 when most people think they are immortal and not affected by skin cancer. Patients who have had a BCC are at a 50% risk of developing at least one more within the subsequent 5 years.

Subjective: Symptoms
Very slow-growing, small, pearly or waxy papule, usually in a sun-exposed area. Sometimes the presenting complaint is that of a sore that will not heal. There may be a history of trauma preceding the lesion.

Objective: Signs
Using Basic Tools: Waxy or pearly papule (2-3mm diameter) that can grow to several cm over time; peripheral telangiectasias (small, dilated blood vessels); superficial erosion; some lesions are flat scars, usually without a history of trauma; sun-exposed areas are the most common sites: ears, periauricular skin, eyelids and periocular skin, nose, cheeks, temples, forehead, upper chest and back, and arms and forearms; can occur even in protected areas like the axilla, so skin exams should be thorough and complete.

Assessment:
Differential Diagnosis - benign lichenoid keratosis, intradermal nevus, neurofibroma, irritated seborrheic keratosis, amelanotic melanoma, tricholemmoma. Differentiating these conditions in the field is nearly impossible, since they require expert microscopic evaluation of a biopsy.

Plan:
Treatment: Defer treatment of team members until return from mission (tumors are very slow growing). Observation is preferable to any treatment. For local nationals, if evacuation or referral cannot be accomplished within the coming 6 months, perform full thickness excision with 5mm margins all around the tumor.

Freeze or store tumor in formaldehyde if possible for later study.
Prevention: Education about aggressive sun precautions is vital: avoid the sun during hours when the individual's shadow is shorter than their height; wear long sleeves, pants and broad-brimmed hats (brim > 4 inches wide); apply sunscreen of SPF 30 or greater (with special attention to the nose, ears and lips).

Follow-up Actions
Evacuation/Consultant Criteria: Evacuate non-urgently (Routine status) if on long deployment. Otherwise delay treatment until return from mission. Consult dermatology as needed.

Squamous Cell Carcinoma (SCC)
Introduction: SCC is the second most common skin cancer, affecting over 100,000 Americans each year. The risk of metastasis is very real in invasive SCCs, especially on the lip or ear. Rates of metastasis range from about 1-5 %. Non-invasive or in situ SCC can become invasive SCC with time. The variant of SCC in situ known as Bowen's disease has a 5% risk of invasion. Sun-exposed skin is at highest risk, and people with fair complexions are at higher risk than are those who are darker-skinned. Other risk factors: a history of radiation exposure or arsenic ingestion. In sun-protected areas, chronic ulcers or scars predispose to SCC, with a high metastatis rate (around 30%). There is a variant of SCC called keratoacanthoma (KA), which grows very quickly and may spontaneously regress. Mid-facial KAs can be especially aggressive and destructive. Patients who are immunosuppressed after an organ transplant operation are at especially high risk, and tend to develop more SCC than BCC, with a reversed incidence ratio.

Subjective: Symptoms
KAs grow rapidly, while SCCs usually are slow growing and much more common in sun-exposed areas; usually painless lesion; history of chronic sun exposure is common, as is an inability to tan, with many severe sunburns.

Objective: Signs
Using Basic Tools: Red, scaly (hyperkeratotic) papules are most characteristic for SCC. Size varies from a few mm to several cm. SCC in situ can occur on the glans penis or within the foreskin, and usually has a soft, red, velvety appearance, without hyperkeratosis (Bowen's disease), since it arises from mucosa. It is associated with a genital wart virus (HPV type 16) and can appear similar to genital warts, with dark flat-topped verrucous papules. Cervical and penile cancers are known to be caused by a genital wart virus. KA has a distinctive appearance, similar to that of a volcano, with a central core of keratin surrounded by domed, rolled borders, usually ranging in size from 1-3 cm. Metastatic SCC often presents with palpable lymphadenopathy.

Assessment:
Differential Diagnosis - BCC, ulcers, chronic ulcerative herpes, benign adnexal tumors, contact dermatitis, Bowen's disease. Differentiating some of these conditions in the field is nearly impossible, requiring expert microscopic evaluation of a biopsy.

Plan:
Treatment: Evacuate and refer to dermatology. If evacuation is not possible in the foreseeable future (1-2 months—relatively slow growing tumor), perform full thickness excisions with wide margins (5 mm) all around the tumor. Freeze or store tumor in formaldehyde if possible for later study.
Prevention: Education about aggressive sun precautions is vital: avoid the sun during hours when the individual's shadow is shorter than their height; wear long sleeves, pants and broad-brimmed hats (brim > 4 inches wide); apply sunscreen of SPF 30 or greater (with special attention to the nose, ears and lips).

Follow-up Actions
Evacuation/Consultation Criteria: Evacuate non-urgently (Routine status) if less than 2 months before return from deployment. Otherwise delay treatment until return from mission. Consult dermatology as needed.

Malignant Melanoma

Introduction: Incidence of malignant melanoma is rising faster than all other cancers in the United States. The lifetime risk for developing melanoma in U.S citizens is currently 1 in 75 and rising, up from about 1in 250 twenty years ago. About 70,000 people in the U.S. develop melanoma each year, and about 7,000 die. The 5-year survival rate is about 83% (double what it was 50 years ago), but since melanoma is becoming more common, the death rate is still rising (3 per 100,000/year). Melanoma is a killer, but is 100% curable with early detection and surgery. Severe sun damage is worrisome, and should lower your threshold for recognizing these lesions. Acral locations (hands and feet) are at risk for acral lentiginous melanoma. The nail beds and nail matrix (which is just proximal to the cuticle and between the skin and the bone of the distal phalanx) are also at risk, so have a high index of suspicion for pigmentation changes in these areas. Amelanotic melanomas are difficult to diagnose because they do not have much pigment and can look nothing like the above description. Amelanotic melanomas can be flesh-colored or hypopigmented papules or plaques and are often thought to be some other entity when they are biopsied.

Subjective: Symptoms
High-risk history: family history of melanoma, childhood history of sunburns, personal history of many atypical nevi. **Focused History:** *Have you had melanoma in your family/childhood sunburns/ unusual moles or freckles?* (typical risk factors)

Objective: Signs
Using Basic Tools: Use the mnemonic ABCD for lesion: "A"–asymmetry, "B"– border irregularity, "C"– color variegation or change, and "D"– diameter greater than 6mm. The earliest of these is probably color variegation, and the colors red, white or blue are most worrisome. Irregular areas which are very dark, or which become very light in color are also bad signs. Asymmetry and border irregularity come from uncontrolled growth of abnormal melanocytes at the edges of the lesion. Notched, grooved or scalloped borders are suspicious, and are usually apparent in a lesion of 6mm diameter or more. Early melanomas tend to be flat with nodular components indicating that the cancer is progressing, and a higher risk for metastasis. Lymphadenopathy may suggest metastasis.

Assessment:
Differential Diagnosis: Seborrheic keratosis, pigmented basal cell carcinoma, atypical nevus, solar lentigo. Differentiating these conditions in the field is nearly impossible, since it requires expert microscopic evaluation of a biopsy.

Plan:
Treatment: Evacuate immediately (aggressive tumor) for evaluation and biopsy, preferably by a dermatologist. If emergent evacuation is not possible, perform initial excisional biopsy with wide margins (5-10 mm) around the entire tumor. Freeze or store tumor in formaldehyde if possible. Evacuate patient with biopsy specimen as soon as possible.
Prevention: Perform self-exam of the entire skin at least monthly. Avoid ultraviolet exposure.

Follow-up Actions
Evacuation/Consultation Criteria: Evacuate immediately. Consult dermatology immediately.

Skin: Seborrheic Keratosis
Lt Col Gerald Peters, USAF, MC

Introduction: Seborrheic keratosis is a very common pigmented, benign tumor that typically presents on the torso after the age of 30.

Subjective: Symptoms
Strong hereditary predisposition; slightly more common in males; rarely pruritic or painful unless irritated or secondarily infected after trauma.

Objective: Signs
Using Basic Tools: Lesion: 1 mm to 3 cm, round to oval, slightly elevated, "stuck-on" appearing, papule or plaque with variable pigmentary change; surface of lesion commonly has "warty" (verrucoid) appearance as it matures and grows; face, trunk and extremities are common sites.

Assessment:
Diagnose based on clinical criteria
Differential Diagnosis: Early lesions: actinic keratosis, nevus. Later lesions: malignant melanoma, pigmented basal cell carcinoma. Differentiating these conditions in the field can be very difficult without expert microscopic evaluation of a biopsy.

Plan:
Treatment
Primary: None is required for this benign lesion

Patient Education
General: This is a local benign proliferation of keratinocytes.

Follow-up Actions
Evacuation/Consultation Criteria: No need to evacuate. Consult dermatology as needed.

Skin: Contact Dermatitis
Lt Col Gerald Peters, USAF, MC

Introduction: The two main categories are allergic contact dermatitis (ACD) and irritant contact dermatitis (ICD). The main difference is that an allergen will only cause problems such as dermatitis in those sensitized to it. Even a tiny amount of allergen can cause a reaction in an allergic person, whereas an irritant will irritate anyone, without previous sensitization, and the effects tend to be dose-related. ACD accounts for about 20% of all contact dermatitis, and ICD for about 80%. Both ACD and ICD can be acute or chronic. History and patterns of eruption can provide clues to the nature of the contact. Plants such as poison oak can leave linear streaks of itchy, red papules and vesicles, corresponding to the leafy contact made with the skin. Nickel is the most common ACD causative agent, often affecting the earlobes, from earrings, or the belly, from a belt buckle. Nickel is part of the alloy in many metals. Shoes and boots have leather with traces of tanning chemicals, as well as rubber and adhesives, all of which can cause ACD. Preservatives and fragrances in beauty and health care products are often a problem. **Neomycin** is a very common cause of ACD (sensitizing almost 10% of all people exposed to it). Beware of products containing **neomycin**, such as **Neosporin** ointment. Formaldehyde, present in dry-cleaned clothes and released by some preservatives, is also a common allergen. Many cases of ICD are puzzling because there is no problem upon exposure during warm humid months, but in the colder, drier months of fall, winter and early spring, the barrier function of the skin is compromised, increasing susceptibility to irritation. This is especially true of detergents, such as hand or laundry soaps. Some cases of detergent ICD arise from overly vigorous cleansing due to the mistaken notion that there is a fungal infection or some type of infestation. Some products, like cutting oil, can serve as both an irritant and an allergen. Some patients with ACD or ICD know exactly what the offending agent is. Others need education about some of the possibilities to allow them to figure out the problem later, particularly after keeping a journal to correlate symptoms and exposures.

Subjective: Symptoms
Various skin reactions including wheals, erythema, hives, edema, papules, vesicles and others, depending on the product and level of sensitization. Geometric or linear arrangements implicate a contactant substance.

Objective: Signs
Using Basic Tools: Asymmetric findings, like a skin reaction where a person carries the wallet or holds their

car's gear shift knob, are more likely ACD; not all sites exposed to an allergen will always erupt; look for patterns - if the skin is flared up around the sites of the elastic bands in underwear (beltline and proximal thighs), think of the "Bleached Rubber Syndrome", which results from contact of elastic with household bleach, creating a sensitizing chemical not present in new underwear unexposed to bleach. The patient must throw away all underwear that has been bleached, and buy new undergarments, never exposing them to bleach. If the agent is also a systemic allergen (e.g., latex), dyspnea, wheezing and other respiratory symptoms may be seen (this is very rare).

Assessment:
Differential Diagnosis - other dermatitides such as atopic, seborrheic, xerotic, and stasis dermatitis, as well as tinea, impetigo, erysipelas, cellulitis, or even Bowen's disease (carcinoma in situ). (See appropriate sections of this book).

Plan:
Treatment:
1. **AVOIDANCE IS KEY!** Protective clothing can help, but a change of occupation, hobby or substances used may be necessary.
2. Topical steroids are very useful, in higher potencies (**fluocinonide** or **triamcinolone ointment**), in order to calm the skin while identifying offending agent.
3. For exudative, weeping areas, a soothing astringent (drying) treatment such as **Domeboro** compresses bid/tid can help. Minimize wet-dry cycles and avoid over-cleansing the skin. In generalized cases, bathe just every other day, with lukewarm water for less than 5 minutes. Use only a mild cleanser like **Cetaphil** lotion or Dove Sensitive Skin soap. Avoid scrubbing the affected areas.
4. Bland emollients like white petrolatum (**Vaseline**) or Crisco vegetable shortening will help to moisturize and protect.
5. Oral antihistamines like **Atarax** (**hydroxyzine**) in doses of 25-50 mg tid/qid, or up to 100mg at bedtime can alleviate much of the itch. The antipruritic effects of **Atarax** last 24 hours and the drowsiness usually only lasts 8-10 hours, a decided advantage over **Benadryl**.
6. Oral steroids can be needed in the most severe cases, used in a tapering fashion for 3 weeks, starting at 60 mg each morning for a week, then 40 mg for a week, then 20 mg for the last week. Reserve oral steroids for the most widespread or bothersome cases. Three weeks of treatment is very important in order to outlast the hypersensitivity reaction in the skin.

Evacuation/Consultation Criteria: Evacuation is not necessary, except with systemic allergic symptoms. Consult as needed.

Chapter 7: Gastrointestinal (GI)

GI: Appendicitis
COL (Ret) Peter McNally, MC, USA

Introduction: Appendicitis is the most common abdominal surgical emergency. Between 5-10% of people develop this condition in life (lower percentage in developing world). Appendicitis can occur at ANY AGE, but is most common during from 20-40. Consider the diagnosis of appendicitis in anyone with an appendix that develops acute abdominal pain..

Subjective: Symptoms
Classic sequence: (1) generalized abdominal pain; (2) anorexia, nausea or vomiting; (3) localized pain over the appendix; (4) fever (Low-grade). Initially, the pain is usually colicky, vague and not severe. It reaches a peak at 4 hours only to gradually subside, and then reappear as a severe pain localized to the right lower quadrant (RLQ). The shift in pain from generalized to the RLQ (McBurney's March) is a diagnostic clue. About 95% of patients have anorexia, nausea or vomiting. Hunger or persistent eating is atypical in

appendicitis. The sensation of constipation or "gas stoppage" is common, but defecation does not bring relief of symptoms. With time the pain gradually increases, but may then subside for a period after the appendix perforates, and resume with greater intensity and generalization. In the field environment, peritonitis and death are likely at that point.

Objective: Signs
Using Basic Tools: Temperature: Fever (101-102°F) frequently develops over 24 hrs. Higher fever is atypical.
Inspection: Guarding, abdominal pain with cough
Palpation: Abdominal tenderness: more common in RLQ, may be localized over the appendix at McBurney's Point (2 inches from the anterior superior iliac spine along a line that intersects with the umbilicus); rebound tenderness; costovertebral angle tenderness (CVAT) in retrocecal appendicitis; positive psoas sign: pain extending the right hip while patient lies on his left side ; positive obturator sign: With the patient supine and the right hip and knee flexed, pain when right leg passively crosses over left (internal rotation).
Perform pelvic and rectal exams.
Using Advanced Tools: Lab: WBC with differential (>10,000/ml, in over 90% of appendicitis), pregnancy test, urinalysis.

Assessment:
Differential Diagnosis: Quite extensive (see Symptom: Abdominal Pain)
Industrialized nations:
Females - pelvic inflammatory disease, ovarian cysts, Mittelschmerz (pelvic bleeding from a ruptured ovarian follicle) and ectopic pregnancy (see Symptom: GYN Problems). History and abnormal pelvic exam can identify these conditions.
Gastroenteritis or mesenteric lymphadenitis - nausea, vomiting precede abdominal pain. Diarrhea, not a sensation of constipation, is common.
Ureteral colic, acute pyelonephritis - colicky pain, dysuria, abnormal urinalysis.
Constipation - LLQ pain; positive rectal exam
Food Poisoning - history, vomiting and/or diarrhea
Peritonitis - may have multiple etiologies; usually higher fever or rigors, different pain profile (see Symptom: Abdominal Pain section).
Bowel obstruction - vomiting, different pain profile
Developing countries:
Intussusceptions (a section of the bowel telescoping into another) are much more common and diverticulitis is much less common because of the high fiber diets. Colonic and even small bowel volvuli (twisting) are also common.
Typhoid fever - RLQ pain often with headache, fatigue, splenomegaly, normal WBC, and roseola-type rash.
Amebic colitis - often begins as RUQ pain but the hepatic abscess will progress and rupture. By gravity, it collects in the RLQ. Once the diagnosis is made antibiotics can quickly resolve the symptoms.
Ascaris infestation (Southern China, India and Central Africa) can lead to bowel obstruction or perforation, cholecystitis, and appendicitis. Similarly, filariasis (India) often mimics appendicitis but with higher (103-104°F) fevers, nausea, RLQ pain. It does not normally progress to peritoneal signs.
Sickle cell disease - usually abdominal pain accompanied by neurologic symptoms
Acute porphyria - 20-40 year old females, southern Africa, often precipitated by sulfa drugs, alcohol or barbiturates. Severe colicky pain with nausea, vomiting, and constipation, and neurologic symptoms. WBC often normal and abdominal exam more benign than the complaint of pain. The patients will have a low-grade fever and jaundice/dark urine.
Gonococcal (females) and pneumococcal peritonitis is becoming more common in developing countries.
Malaria - fever, chills, vomiting, history of travel to a malaria-endemic area
Tuberculous peritonitis/psoas abscess (Pott's disease)
Lead poisoning/colic - vague, persistent abdominal pain

Plan:
Treatment
1. IV fluids (see Shock: Fluid Resuscitation)
2. Antibiotics: **Cefotetan** 2 grams q 12 hours.
3. Evacuation: Elevate head and flex knees.
4. If evacuation is not possible or imminent, consider appendectomy (see following Appendectomy procedure guide). Perforation rate climbs steeply after 24 hours of pain. Evaluate abdominal pain expeditiously and explore promptly before perforation occurs.

Patient Education
General: Appendectomy should cure the patient of symptoms.
Diet: No dietary restrictions.

Follow-up Actions
Return evaluation: Reevaluation for symptoms of pain, fever, diarrhea. Return for evaluation if appendectomy does not lead to prompt restoration of baseline good health.
Evacuation/Consultation Criteria: Evacuate urgently for surgery. Needs routine postoperative surgical follow up, then primary care management.

GI: Emergency Field Appendectomy
LTC John Holcomb, MC, USA and SFC Dominique Greydanus, USA

What: The removal or drainage of a suppurative or perforated appendix through an emergency laparotomy

When: Only when the patient has failed 48-72 hours of appropriate antibiotic therapy, absolutely cannot be evacuated in time, is having high spiking fevers, has an elevated WBC count and peritonitis, and will die without the operation. Tell your commander this is a life or death maneuver, and the patient has only a small chance of living despite this operation.

Background: The ultimate goal should be to avoid operating in this environment. In a field setting without dedicated surgical support, acute appendicitis is optimally treated with IV antibiotics until evacuation is possible (you may avoid operating on up to 80% of such patients). If evacuation is not possible, the majority of acute appendicitis patients can still be treated with IV antibiotics (only 30% will recur later). The patient with perforated appendicitis presents more difficulty, however they can still be treated with IV antibiotics in a non-operative fashion, and only 50% of these patients will require an emergency operation. The decision to perform an appendectomy without the support of personnel proficient in intra-abdominal surgery is extremely dangerous. This is essentially a triage decision, maximizing your limited resources, personnel and surgical experience by treating the majority of patients with antibiotics alone. Once the decision has been made to operate, it is important to adequately prepare the personnel assisting you. Discuss all steps of the procedure extensively and review all reference material available. No one on the surgical team should have more than one job. Practice on an animal immediately before doing the appendectomy. After all this preparation, it is still likely that unintentional complications (and perhaps death) will result from this type of field surgery. These guidelines apply to both US and local national patients.

What You Need:
Personnel: A dedicated anesthesia technician is **required**, who is experienced or knowledgeable in performing intravenous anesthesia or general endotracheal anesthesia. The patient cannot move around and the abdominal wall must be relaxed during surgery. Two surgical assistants are **required**.
Supplies: Surgical preparation solution (for the abdomen), sterile gloves (>3 sets), silk ties or ligatures, 0 (zero)-Vicryl (for the fascia) on a taper needle, sterile bandages (to pack the wound), sterile gauze bandages (for incisional bleeding), a large volume (6-10 liters) of sterile saline or water (to irrigate the peritoneal cavity), suction device, NG tube, Foley catheter

Instruments: Scalpel, 2 needle drivers, 2 tissue forceps, 2 retractors, 6 clamps and scissors.
Additional things to do prior to surgery: Obtain a well-ventilated space with good lighting and a narrow tabletop that allows access to both sides of the patient. A headlight is extremely useful. Obtain good IV access and instill additional IV antibiotics, if not already given. Ensure that the NG tube and Foley catheter are in place prior to the first incision.

What To Do:
1. This procedure should not be performed without the radio consultation of a physician experienced in intra-abdominal surgery.
2. Stay calm.
3. Place the patient in the supine position on the table.
4. Have the anesthesia technician start the anesthetic.
5. Prep the patient's abdomen from pubic area to nipples and from side to side down to the level of the posterior axillary line.
6. Place NG tube and Foley if not already in place.
7. Place an incision between the umbilicus and the anterior superior iliac crest, transversely across the abdomen (Figure 4-2). Make a larger, instead of smaller, incision to see better (at least 4-6 inches long). The incision should cut through the skin, and then down through subcutaneous fat. Apply clamps to bleeders as required. Ensure you are clamping a vascular structure before you do so. Using the silk ligatures, tie off bleeders whose clamps obstruct the incision. Carefully deepen the incision down to the fascia, which is the shiny white tissue (gristle). Make an incision through the anterior fascial (Figure 4-3). Stay lateral to the rectus muscle seen beneath the fascia. At this point take a hemostat and spread in the tissue lateral to the rectus muscle along the line of your incision. Using the spreading motion, progress deeper through the lateral abdominal wall, and into the peritoneal cavity. Once into the peritoneal cavity, fluid should come out. Place both index fingers into the peritoneal cavity and spread in the direction of your incision to widen the peritoneal opening (Figure 4-4). Place the retractors at the medial and lateral portion of the incision, with the end of the retractor in the peritoneum. Have the assistants pull in opposite directions along the direction of the incision to enlarge access to the peritoneal cavity. Have your assistants keep the retractors in the wound.
8. The cecum is the part of the large bowel in the right lower quadrant, to which the appendix is attached. Following teniae (which are the longitudinal bands that are seen on the colon) on the colon down to the cecum, pull the cecum into the wound, and locate the appendix at the base of the cecum (Figure 4-5). This maneuver is not as easy as it sounds and may take some time.
9. If the appendix is easily seen, and is acutely inflamed (red, swollen) it must be removed. Dissect the mesoappendix (where the artery to the appendix lies) from the appendix, doubly ligate the mesoappendix, and cut in between the two ligatures (Figure 4-6). These ligatures must be tied down well to close the appendiceal artery running through the mesoappendix, and prevent significant bleeding. Now isolate the appendix by doubly ligating its base, adding a third ligature more distal to the proximal two and dividing between the ligatures (Figure 4-7). Remove and dispose of the appendix. Inspect the base of the appendix left on the cecum to make sure both ligatures are tight, as a loose ligature will fall off and cause a cecal fistula, resulting in worsening intra-abdominal sepsis and death.
10. If upon entering the peritoneal cavity, you discover an abscess and an abdominal cavity full of pus, it is strongly recommended that this abscess cavity be drained out the right flank by placing a stab wound incision (carefully avoid organs and important structures) lateral to the original incision, placing a drain into the abscess cavity and exiting it through the flank stab wound. Remember the iliac artery and the ureter also reside in the right lower quadrant; both are tubular structures and should be avoided. Irrigate the abscess and abdomen copiously with 4 to 6 liters of sterile fluid. There is no need to remove the perforated appendix.
11. If upon entering the peritoneal cavity you discover a localized abscess full of pus, drain it and irrigate as above.
12. Hemostasis is critically important. Closing the abdomen with ongoing bleeding will result in sepsis, hypotension and eventual death. Obtain hemostasis by tying the silk ligatures around any remaining clamped vessels. Search for other bleeding sites including those near the flank stab wound. Place a finger over the bleeding site, collect your thoughts and take a deep breath. This is a valuable maneuver

Figure 4-2

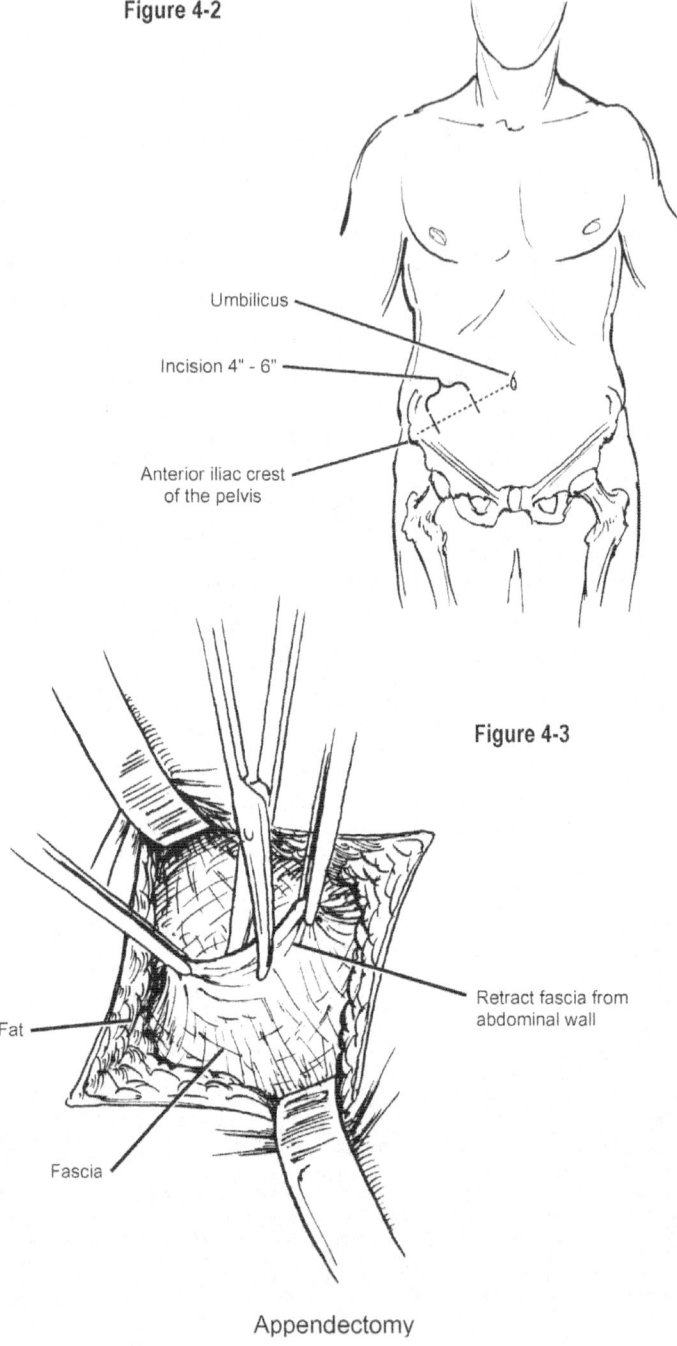

Figure 4-3

Appendectomy

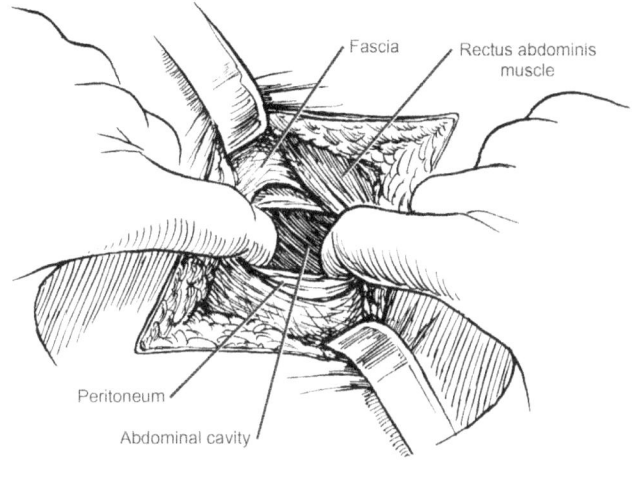

Figure 4-4

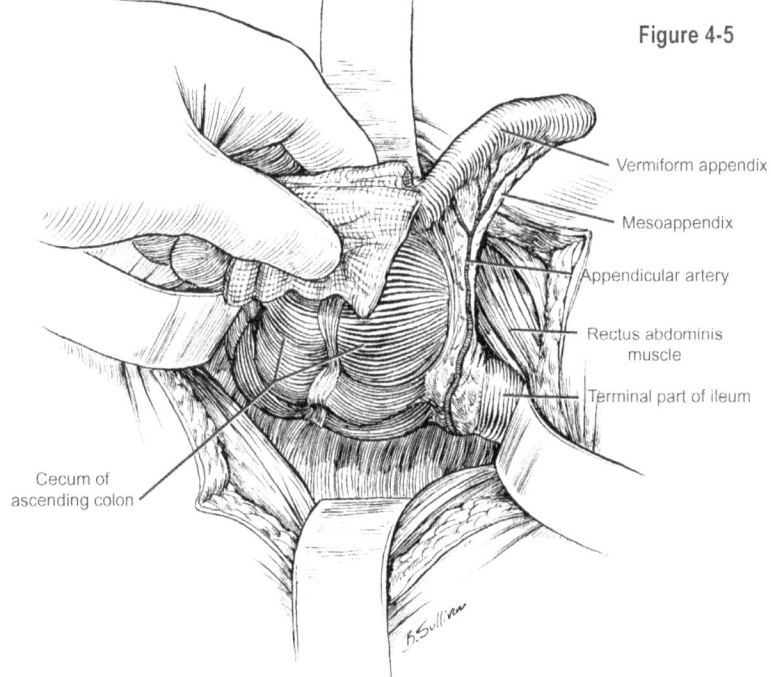

Figure 4-5

Appendectomy cont'd

Figure 4-6

Figure 4-7

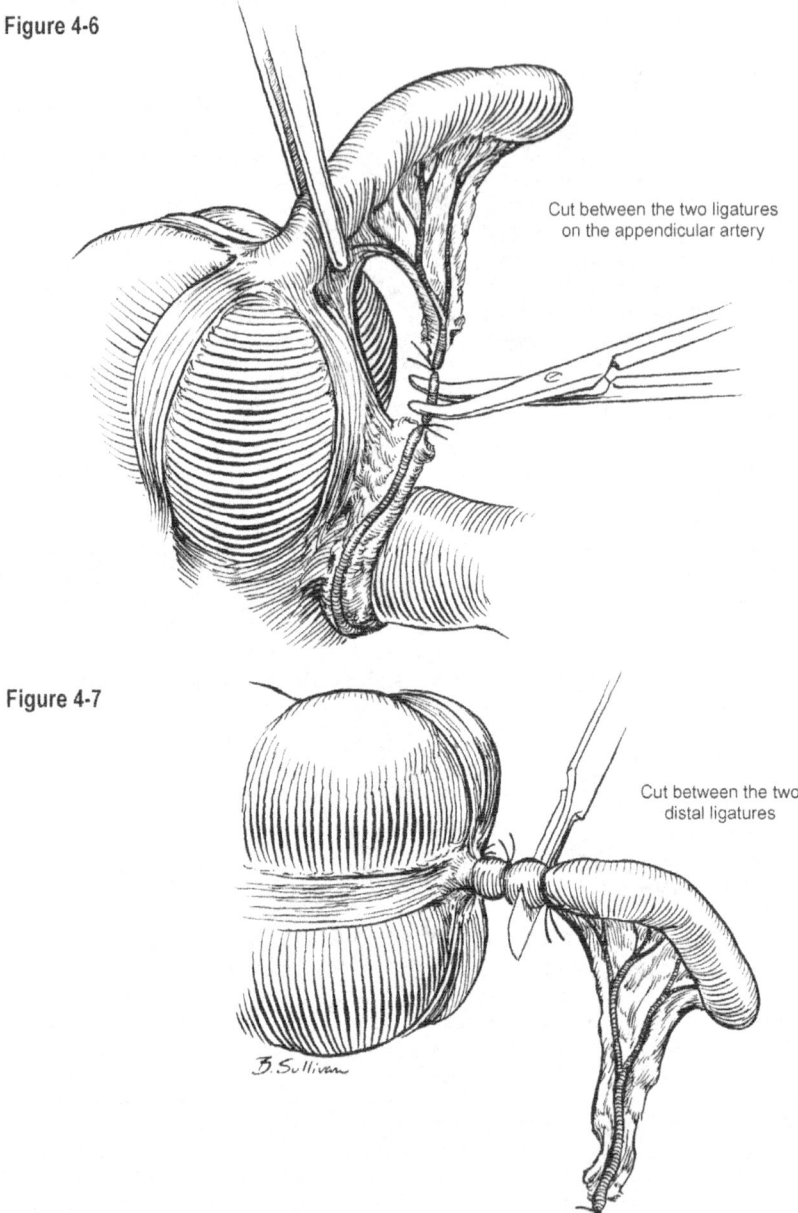

Cut between the two ligatures on the appendicular artery

Cut between the two distal ligatures

Appendectomy cont'd

prior to randomly clamping large, important structures (such as the iliac artery and the ureter), that may not be bleeding and will only cause further injury or damage if they are clamped.
13. If you enter the abdomen with the plan of doing an appendectomy and the appendix is normal, leave the appendix in place, and search for another obvious cause. However, there are not many easily correctable causes of intra-abdominal sepsis that would not have been effectively treated by 72 hours of antibiotics.
14. Close one layer of the abdominal wall, the fascia, to complete the operation. The difficulty here is closing without causing injury to the underlying bowel. This layer must be closed correctly, or the patient will eviscerate with coughing at a later time. Take your time, do not hurry, and watch every pass of the needle. Your assistants need to be as observant as you are. Take your O-Vicryl suture, start laterally, tie your knot, and work medially. Take bites of the fascia one centimeter back from the cut edge and advance only one centimeter at a time. It is imperative not to include loops of bowel in your sutures. It is not necessary to include muscle in your sutures —that weakens your repair. The only thing you need to put together is the fascia.
15. Once the fascia is closed, irrigate the wound again with a liter of sterile fluid and lightly pack the wound with saline soaked gauze. Place a sterile dry dressing over the top of the packs.
16. Do not close the skin. Perforated appendices create contaminated wounds that should not be closed, but should be allowed to heal by granulation. If evacuation becomes available in the next 48 hours, refrain from closing the skin to facilitate later abdominal exploration by a surgeon.
17. If the appendix was not perforated, the wound is clean without hemorrhage, and evacuation has not occurred by 5 days, close the skin and fat (as described in the C-Section Section) as a delayed primary closure. Use **lidocaine** as for other skin procedures, such as Procedure: Skin Mass Removal. Wounds handled by this approach are less likely to become infected than if closed immediately.
18. Have the anesthesia technician bring the patient out from under anesthesia.

Post-Operative Orders:
1. Keep the patient on bed rest for several days, then begin ambulation slowly and advance as tolerated.
2. Keep the NG tube in place until the patient has return of bowel function. The patient is very ill, so do not anticipate that they will be eating for several days after surgery. Provide stress ulcer (see GI: Acute Peptic Ulcer) and DVT (see Respiratory: Pulmonary Embolus) prophylaxis if available. Once bowel function returns (bowel sounds and passing gas) pull the NG tube and begin a clear liquid diet. Advance the diet as tolerated.
3. Monitor vital signs frequently, as often as every 2-4 hours. This patient should require almost constant attention the first 24 hours.
4. Monitor the I & O to make sure the IV fluids are at a high enough rate. Leave the Foley catheter in place as a way to monitor their output (should be 0.5-1.0 cc/kg/hr). Pull the Foley when patient is tolerating liquid diet. Ensure he is able to urinate after Foley has been pulled.
5. Provide pain control (see Procedure: Pain Assessment and Control).
6. Keep the patient from vomiting by using anti-emetic as needed (e.g., **Compazine** 5-10 mg IM q 3-4 hours, max 40 mg/d).
7. An elevated white blood cell count and spiking temperatures 3 days after operation may be concerning for evidence of continued sepsis. It is not unusual to have temperature spikes for a day or two after operation but these should decrease after 72 hours.
8. Drain output should decrease over 5 to 6 days, and the drain may be pulled at that time.
9. Continue antibiotics for 10-14 days after surgery.
10. Follow the principles in the Nursing: Wound Care section (on the CD-ROM) to keep the wound clean and allow it to granulate closed over several weeks. Use **Betadine** for the first 48 hours, and then switch to new dressings soaked in sterile saline until a pinkish layer of granulation tissue covers the fascia.

What Not To Do:
Take care not to cause unnecessary bleeding. There will be oozing from the surface of inflamed and cut tissue. Most of this will stop with pressure and time.
Do not cut into the intestines inadvertently. If you do, close it in a single layer with 000-silk or Vicryl suture.
Make sure your assistants retract and help you. You must see the layers of bowel as you close them.
Do not make your skin incision too small. If you have difficulty seeing, make the incision bigger.

Do not race through the operation. Take your time and operate safely. It is much better to be safe than to move too fast.
Do not operate without a dedicated anesthesia technician. The 'surgeon' will have enough on his mind.
Do not operate without 2 assistants for retraction. They will also be accessory eyes and brains during this procedure.

GI: Acute Cholecystitis
COL (Ret) Peter McNally, MC, USA

Introduction: Gallbladder stones are common in the United States, seen in 10-15% of adults. They are 2-3 times more common in women (4 Fs: fat, forty, female, and fertile). Cholecystitis (gall bladder inflammation) occurs more commonly in certain diseases such as malaria, sickle cell, and ascaris infestations. When gallstones become symptomatic, cholecystectomy is indicated.

Subjective: Symptoms
Biliary colic pain: usually located in the upper abdomen, frequently in the right upper quadrant (RUQ), may radiate to the right scapula; may be precipitated by a meal, but more commonly there is no inciting event; gradually increases over 15-60 min., plateaus for 1 or 2 hrs before slowly going away; if persists longer than 4 hrs it is unlikely to spontaneously resolve. 75% of patients have a history of previous attacks of biliary colic before acute cholecystitis. May include nausia and vomiting.

Objective: Signs
Using Basic Tools: Inspection: Patients with acute cholecystitis appear uncomfortable and ill. Febrile but < 102°F. Mild jaundice/icterus seen in 20%.
Auscultation: Bowel sounds should be present, unless gangrenous gallbladder or gallstone pancreatitis.
Palpation: Murphy's Sign - During palpation of the right subcostal region, pain and inspiratory arrest may occur when the patient takes a deep breath, bringing the examiner's hand in contact with the inflamed gallbladder. The obstructed and swollen gallbladder is palpable in 1/3 of acute cholecystitis.
Using Advanced Tools: Lab: WBC with differential will demonstrate infection.

Assessment:
Differential Diagnosis
Ascending cholangitis - fever, RUQ pain and jaundice (Charcot's Triad) - a surgical emergency!
Obstruction of the common bile duct (gall stone or tumor) - marked jaundice, dark urine, clay-colored stools
Pancreatitis - diffuse abdominal pain with ileus and vomiting
Peptic Ulcer Disease - vomiting, hematemesis or melena if bleeding
Cardiac Pain - angina, heart attack
Esophageal reflux, hiatal hernia - acid taste, pain relieved with antacid
Pleurisy/pneumonia - respiratory complaints, pain with deep inspiration
Liver Mass (abscess, tumor, cirrhosis) - jaundice, RUQ pain, no fever, no relationship to meals

Plan:
Treatment
1. IV fluids (see Shock Fluid Resuscitation)
2. Antibiotics **(ticarcillin** 4 gm IV q 6 hr plus **metronidazole** 500 mg IV q 6 hr, or **aztreonam** 2 gm IV q 8 hr plus **clindamycin** 450 mg IV q 8 hr).
3. Pain control (see Procedure: Pain Assessment and Control)
4. Antipyretics (**Tylenol**, etc.)
5. Antiemetic medications should be given as needed (e.g., **Compazine** 5-10 mg IM q 3-4 hours, max 40 mg/day)
6. Evacuate: High fever > 102°F, the presence of jaundice, persistent pain or vomiting—evacuate immediately; otherwise, worsening symptoms or failure to improve over 24 hours should prompt medical evacuation.

Remember — cholangitis (pus in the biliary tree) is a surgical emergency.

Patient Education
General: Half of acute cholecystitis will resolve within 7-10 days without emergent surgery. Left untreated, 10% will be complicated by localized perforation and 1% by free perforation.
Diet: Dramatic weight reduction programs are associated with development of gallstones in 25-50%. Avoid greasy or fatty foods.
Prevention: Avoid foods that precipitate symptoms.

Follow-up Actions
Return evaluation: Recurrent pain should be investigated promptly.
Evacuation/Consultation Criteria: See evacuation guidance above. Refer acute cholecystitis to a general surgeon.

GI: Acute Bacterial Food Poisoning
COL (Ret) Peter McNally, MC, USA

Introduction: Bacterial food poisoning is any illness caused by the consumption of food contaminating bacteria or bacterial toxins. The major recognized causes of bacterial food poisoning are limited to 12 bacteria: *Clostridium perfringens, Staphylococcus aureus* (see Color Plate Picture 24), *Vibrio cholera* (see Color Plate Picture 35) & *parahaemolyticus, Bacillus cereus, Salmonella, Clostridium botulinum, Shigella*, toxigenic *E. coli* (see Color Plate Picture 28), certain species of *Campylobacter, Yersinia, Listeria*, and *Aeromonas* (see Color Plate Picture 22). Most cases arise from ingesting contaminated food. The attack rates are high, with most persons ingesting the food becoming afflicted. Rapid onset of symptoms indicates the presence of pre-formed toxins liberated from contaminated food.

Subjective: Symptoms
Nausea, vomiting, crampy abdominal pain, fever, myalgias, headache, diarrhea (sometimes bloody).

Table 4-1 Food Poisoning

Organism	Source	Average Incubation (hours)	Clinical Features
Bacillus cereus	Fried rice, vanilla sauce, meatballs, barbecued chicken, boiled beef	2	Vomiting, crampy abdominal pain, diarrhea. Duration 1/2 - 1 day.
Clostridium perfringens	Beef, chicken, turkey	12	Diarrhea and crampy abdominal pain predominate symptoms. Duration 1 day.
Vibrio parahaemolyticus	Seafood	12	Nausea, vomiting, headache, fever, diarrhea and crampy abdominal pain. Duration 3 days.
Staphylococcus aureus	Ham, pork, canned beef	3	Vomiting, nausea, crampy abdominal pain, diarrhea.
Yersinia enterocolitica	Chocolate or raw milk, pork	72	Fever, crampy abdominal pain, diarrhea and vomiting. Keys: pharyngitis, arthritis, mesenteric adenitis and rash. Duration 7 days.
Listeria monocytogenes	Milk raw vegetables, cole slaw, dairy, poultry, beef	9-32 *2-6 wks	Diarrhea, fever, crampy abdominal pain, nausea, vomiting. Duration depends on severity. *Invasive disease

Campylobacter jejuni	Milk, beef, chicken, pet animals	48	Diarrhea, fever, nausea, vomiting, headache, myalgia. Key: may see bloody diarrhea. Duration 7 days.
Escherichia coli	Salad, beef	24	Diarrhea, crampy abdominal pain, nausea, headache, fever, myalgias. Keys: may see hemolytic uremic syndrome and bloody diarrhea. Duration 3 days.
Salmonella	Eggs, poultry, meat	24	Diarrhea, crampy abdominal pain, nausea, vomiting, fever, headache. Duration 3 days.
Shigella	Milk, potato, tuna, turkey salads	24	Crampy abdominal pain, fever, diarrhea, bloody diarrhea, headache, nausea, vomiting. Duration 3 days.turkey salads

Objective: Signs
Using Basic Tools: Inspection: Patients appear ill and dehydrated.
Auscultation: Abdominal bowel sounds are often hyperactive.
Palpation: Mild to moderate abdominal tenderness. Peritoneal signs (rebound, guarding, point tenderness) are atypical.
Using Advanced Tools: Lab: Elevated WBC with differential may indicate systemic infection; stool cultures, blood cultures.

Assessment:
Differential Diagnosis—see Symptom: Abdominal Pain
When the predominant symptoms are crampy abdominal pain and diarrhea, consider non-food-poisoning etiologies like acute *Giardia lamblia* or antibiotic diarrhea. When a recurrent past history of symptoms is present, consider inflammatory bowel disease.

Plan:
Treatment
1. Rest
2. Rehydration (oral or intravenous) correction of electrolyte disturbances (see Shock Fluid Resuscitation)
3. Anti-emetics as necessary (e.g., **Compazine** 5-10 mg IM q 3-4 hours, max 40 mg/day)
4. Antibiotics are unusually NOT necessary, in fact may promote increased carrier rates for *Salmonella*.

Patient Education
General: Be a cautious consumer. Food left at room temperature is a potential breeding source for bacteria. Proper hand washing is important in preparation of food, especially when handling raw meats, poultry and eggs. Wipe down counters before and after preparing food. Avoid wooden cutting boards because they cannot be easily disinfected.
Diet: If the food smells bad or has not been refrigerated, avoid it.
Prevention; For large outbreaks obtain stool cultures from patients. Obtain stool culture from food handlers when suspicious of *Yersinia enterocolitica*, *Salmonella* or *Shigella* to prevent further outbreaks.

Follow-up Actions
Return evaluation: Only if symptoms persist, worsen or relapse. For protracted illness or signs of sepsis, blood cultures may be necessary.
Evacuation/Consultation Criteria: Evacuation is not usually necessary. Consult infectious disease or internal medicine for systemic toxicity, bloody diarrhea, high fever.

GI: Acute Gastritis
COL (Ret) Peter McNally, MC, USA

Introduction: Inflammation of the stomach lining or acute gastritis, is common. The causes of gastritis are numerous, but it is most commonly caused by consumption of alcohol, aspirin, non-steroidal anti-inflammatory drugs (NSAIDs) or by acute infections.

Subjective: Symptoms
Dyspepsia (epigastric discomfort or burning), nausea, vomiting, postprandial fullness/bloating and occasional GI bleeding; history of excess alcohol consumption or ingestion of aspirin, NSAIDs, corrosives or poorly prepared or preserved food.

Objective: Signs
Using Basic Tools: Inspection: Nausea, vomiting and epigastric pain; appear pale and dehydrated; no fever. Palpation: Abdomen is usually soft but may have some mild to moderate tenderness in the epigastric region.
Using Advanced Tools: Lab: CBC for evidence of infection or anemia; and urinalysis (elevated bilirubin).

Assessment:
Differential Diagnosis
GI bleeding - hematemesis or melena should suggest alternative diagnosis such as bleeding peptic ulcer, Mallory-Weiss tear (esophageal tear caused by retching), etc.
Peritoneal signs (guarding, rebound, absent bowel sounds) - suggest intra-abdominal sepsis.
Icterus (jaundiced sclera) - suggest Gilbert syndrome (a benign disorder of bilirubin excretion) or hepatobiliary disease.
Significant weight loss - suggest possible malignancy.

Plan:
Treatment
Primary: Discontinue gastric irritants such as alcohol and/or medications (aspirin/NSAIDs). Rehydrate with oral or IV fluids. Give short course (4 weeks) of H2-blocker (e.g., **Pepcid** 40 mg q hs) to promote healing of gastritis.
Complications: If acute gastritis is associated with bleeding (hematemesis or melena), manage in the same manner as a bleeding ulcer. See section on Peptic Ulcer Disease.
1. IV fluid resuscitation, two large bore IVs ($\geq$ 18 gauge)
2. IV H-2 blocker (e.g., **Pepcid** 20 mg q 12 h)
3. Anti-emetic of choice (e.g., **Compazine** 5-10 mg IM q 3-4 hours, max 40mg/day)
4. Orthostatic vital signs, persistent systolic BP < 90 mmHg or tachycardia above 100 beats per minute after IV fluids should prompt medical evacuation.

Patient Education
Diet: Take clear liquids initially, then progress to bland diet and then back to regular diet.
Medications: Antiemetics may be needed. Use short course of over-the-counter antacids or acid-blocking medications.
Prevention and Hygiene: Consume only properly prepared and preserved food, clean water.
No Improvement/Deterioration: Return for additional evaluation if fail to improve within 24 hours or signs of abdominal pain, fever, icterus.

Follow-up Actions
Return evaluation: Repeat exams if worsening symptoms of pain, fever, signs of GI bleeding.
Evacuation/Consultation Criteria: Evacuation is not usually necessary unless GI bleeding develops..

Consult for worsening symptoms of pain or fever, or for signs of GI bleeding.

GI: Acute Pancreatitis
COL (Ret) Peter McNally, MC, USA

Introduction: Acute pancreatitis is an inflammatory process of the pancreas usually associated with severe pain in the upper abdomen. Gallstones and alcohol cause about 80-90% of acute pancreatitis. Most acute cases will spontaneously resolve, but severe, chronic pancreatitis has a 50% mortality rate.

Subjective: Symptoms
Pain: located primarily in the epigastrium; may be localized to the right upper quadrant and radiate to the back; pain reaches a maximum intensity rapidly over 10-20 minutes; described as unbearable, with little relief offered by position. Patients frequently assume a fetal position. Nausea and vomiting are common.

Objective: Signs
Using Basic Tools: Inspection: Appears acutely ill; mental status may be depressed, especially if associated with acute alcohol ingestion; fever, tachycardia and hypotension; ecchymosis along the flanks (Grey Turner's sign) or around the umbilicus (Cullen's sign) grave prognosis; jaundiced sclera (icterus); distended abdomen.
Auscultation: Decreased breath sounds from effusions or rales (ARDS)—grave prognosis; abdominal pain may cause splinting and shallow respirations; absent bowel sounds.
Percussion: The abdomen is tympanic and diffusely tender.
Using Advanced Tools: Lab: CBC with differential for evidence of anemia and infection; save the red top tube for future analysis for milky layer seen in hypertriglyceridemia. Pulse oximetry to evaluate ARDS. Three-way radiograph of the abdomen for GI conditions (if available).

Assessment:
Use Ranson's Criteria* as a scoring system to predict severity of pancreatitis. See below.
Differential Diagnosis - Peptic ulcer disease, cholecystitis, ischemic bowel, aortic aneurysm, bowel obstruction or perforation.

Plan:
Treatment - treat aggressively
1. NPO until pain resolved.
2. Pain control (avoid morphine - it may cause sphincter of oddi spasm, worsening pancreatitis).
3. Aggressive IV fluid resuscitation (D5-Lactated Ringer's solution)
4. NG decompression if vomiting or distended
5. Antibiotic (**cefotaxime** 2 gm q 8 h IV) if patient appears septic (fever > 102° F, rigors, or jaundice)
6. Evacuation to hospital.

Patient Education
General: If gallstones caused the pancreatitis, then definitive treatment (surgical removal) and cure may prevent future attacks. If the cause of pancreatitis is alcohol, abstinence is the key to prevent chronic relapsing pancreatitis.
Diet: NPO during acute pancreatitis, then as tolerated after resolution.
Medications Shown to Cause Acute Pancreatitis: 6-mercaptopurine, azathioprine, sulfonamides (**sulfasalazine** and **Bactrim**), oral **5-aminosalicylic acid**, antibiotics (**metronidazole**, **tetracycline**, and **nitrofurantoin**), valproate, corticosteroids, **furosemide**, estrogens, **Aldomet, pentamidine, didanosine**.
Prevention: Cessation of all alcohol. Stop medicines proven to cause pancreatitis indefinitely, see list above. Normalize high triglyceride or calcium levels to prevent additional attacks.

Follow-up Actions
Evacuation/Consultation Criteria: These patients need urgent evacuation. Any first attack of pancreatitis needs gastroenterology consultation. Mild isolated cases should undergo primary search for etiology. A

gastroenterologist should see all second attacks of pancreatitis.

NOTES: *Ranson's Criteria for Severity (1 point each)
At Admission (5 criteria)
 Age > 55 years
 WBC > 16,000/mm^3
 [If available: Glucose > 200 mg/dl, Lactate dehydrogenase > 350 IU/L, and Aspartate transaminase > 250 U/L]

During Initial 48 hours (6 criteria)
 HCT decrease of 10 mg/dl
 PaO2 < 60 mm Hg

 [If available: Blood urea nitrogen increase of > 5 mg/dl, Calcium < 8 mg/dl, Base Deficit > 4 mEq/L, Fluid sequestration > 6 L]

Number of Ranson's Criteria	Predicted MORTALITY Rate
< 2	5%
3-5	10%
> 6	60%

GI: Acute Peritonitis
COL (Ret) Peter McNally, MC, USA

Introduction: Acute peritonitis is a potentially catastrophic illness caused by infectious organisms attacking the peritoneum. It is usually characterized by rapid onset of symptoms and rapid medical deterioration. The five most common causes of acute peritonitis are appendicitis, cholecystitis, diverticulitis, pancreatitis, and bowel perforation. Each has a characteristic pattern of symptoms to suggest the etiology. When abscess or perforation complicates any of these causes, generalized peritonitis ensues. Generalized peritonitis requiring surgical intervention is caused by perforated peptic ulcer (40%), appendicitis (20%), gangrene of bowel/gallbladder (15%), post-op complications (10%) or other causes (15%). Exact details of the onset of the pain, and associated symptoms (e.g., change in bowel or menstrual habits) are helpful in drawing attention to the affected organ. Mortality is high in many groups, especially in the elderly and patients suffering organ failure before development of peritonitis. Peritonitis secondary to appendicitis or perforated duodenal ulcer is associated with >90% survival, whereas peritonitis from other causes, including postoperative peritonitis, has only approximately 50% survival.

Subjective: Symptoms
Pain and fever. See handbook sections on Appendicitis, Cholecystitis and Pancreatitis.
Appendicitis: Generalized abdominal pain that becomes localized to the right lower quadrant (and eventually McBurney's Point); anorexia; sensation of "gas blockage" and need for bowel movement, but no improvement after enema or defecation. **Cholecystitis:** 90% of patients will be symptomatic, with epigastric or right upper quadrant pain that peaks over 30 minutes, then plateaus for 1-2 hours before gradual decreasing; some relate pain to fatty meals, or radiation to the right scapula. **Diverticulitis:** More common in the elderly; pain may occur after straining to have a bowel movement, and is initially localized to the left lower quadrant (95%); associated with fever (60-100%) and elevated WBC count (70-80%). **Pancreatitis:** Chronic, excessive alcohol abuse and gallstones cause most pancreatitis, with acute onset of rapidly progressive, incapacitating, diffuse abdominal pain, radiating to the back; patients are typically in the fetal position for comfort. **Bowel Perforation:** Immediate onset of severe abdominal pain; several causes, including perforation of gastric or duodenal ulcer, appendix, diverticula, or other hollow viscus (due to foreign body ingestion, abscess, etc.).

Objective: Signs
Using Basic Tools: Vital Signs: Fever 100-101°F, tachycardia

Inspection: Patient in fetal position, because any movement worsens pain; visible peristalsis suggests bowel obstruction.
Auscultation: Absence of bowel sounds in all four quadrants suggests peritonitis. Always auscultate before doing percussion or palpation.
Percussion: Absence of dullness over the liver suggests free air and perforation.
Palpation: Begin with very gentle palpation away from the area of maximal symptoms; board-like abdomen is unmistakable and indicates obvious peritonitis; shake the pelvis to assess rebound tenderness; ileopsoas and obturator signs (see Appendicitis section) are suggestive for retroperitoneal inflammation.
Serial examinations: Diminishing bowel sounds with increasing tenderness and the development of rebound indicates peritonitis.
Using Advanced Tools: Lab: CBC with differential, urinalysis, blood cultures for infection. Abdominal X-ray (if available): free air, dilated loops of bowel, air-fluid levels, calcified gallstones (1/3) or pancreas.

Assessment:
Differential Diagnosis - see list above in Subjective.

Plan:
Treatment
1. Intravenous antibiotics must cover both aerobic and anaerobic bacteria:
   Single Agents: **Cefoxitin** 2 gm IV q 8 hr, **cefotetam** 2 gm IV q 12 hr, or **cefmetazole** 2 gm IV q 8-12 hr
   Combination Agents: **Aztreonam** 2 gm q 8 hr plus **metronidazole** 500 mg IV q 8 hr
2. IV fluids to compensate for respiratory and third space losses (see Fluid Resuscitation section). Use pressor agents at lowest dose needed to maintain adequate perfusion pressure, such as **Dopamine** 5 mcg/kg/min.
3. Pain control (see Procedure: Pain Assessment and Control) and antiemetic (e.g., **Phenergan** 25 mg IV, IM, or po) of choice.
4. Nasogastric tube decompression for significant abdominal distention or vomiting, and keep NPO.
5. Evacuate for definitive surgical treatment.

Patient Education
Activity: Bedrest.
Diet: Metabolic needs during acute peritonitis are great, equivalent to a 50% total body surface area burn. Caloric requirement is often in the 3000-4000 calorie range and must be given parenterally by IV (not available in field conditions).

Follow-up Actions
Return evaluation: Postoperative follow up is contingent upon operative findings, treatment and hospital course.
Evacuation/Consultation Criteria: Evacuate ASAP. Consult general surgery.

GI: Peptic Ulcer Disease
COL (Ret) Peter McNally, MC, USA

Introduction: Peptic ulcers are defects in either the gastric or duodenal mucosa. Almost all ulcers are caused by infection with *Helicobacter pylori*, consumption of **aspirin** or NSAIDs (**Motrin, Advil, Aleve, Clinoril, Feldene**, etc.) or severe physiologic stress (extensive trauma, burns or CNS injury). Some ulcers are related to ingestion of fish parasites. Most ulcers cause mid-epigastric pain, often associated with nausea or vomiting. Complications of ulcers include bleeding, perforation and obstruction. Generally, pain will herald the presence of an ulcer before complications occur. Ulcer pain is decreased by ingestion of alkali and patients often give a history of self-medication with bicarbonate of soda, antacids or over-the-counter acid blocking medicines.

Subjective: Symptoms
Gnawing epigastric pain between the umbilicus and the xiphoid, increased by food and relieved by alkali

(gastric ulcer); awakening from sleep with pain, that radiated to the mid back (duodenal ulcer); anorexia, nausea and vomiting.

Objective: Signs
Using Basic Tools: Tender epigastric area; vomiting bright red blood (hematemesis) or coffee grounds suggests active or recent bleeding from the upper GI tract; melena ("tarry" black, oily and odiferous stool that suggests upper GI tract bleeding); weight loss.
Vital Signs: Pulse > 100 bpm, systolic BP < 90: probable hypovolemia. Orthostatic change in VS (systolic BP drop of 20 mm Hg or pulse rise 20 bpm): significant hypovolemia.
Appearance: Pallor of anemia, diaphoresis: suggests significant blood loss.
Gastric Contents (check if melena or hematemesis): NG aspirate – bile, no blood or coffee grounds suggests no active bleeding. Coffee grounds: recent bleeding, bright red blood: active bleeding.
Abdomen: Absent bowel sounds, rigid exam, peritoneal signs: perforated or penetrating ulcer
Rectal Exam: Melena: recent UGI bleeding
Using Advanced Tools: Lab: Hematocrit.

Assessment:
Differential Diagnosis - dyspepsia, gallstones, pancreatitis, angina and malignancy.

Plan:
Treatment
1. Treat the uncomplicated ulcer:
    a. Stop **aspirin** or NSAIDs.
    b. Suppress acid secretion with oral therapy, or IV therapy until stable, then switch to oral therapy.
    IV therapy: **Cimetidine** 300 mg q 6 hr, **famotidine** 20 mg q 12 hr, or **ranitidine** 50 mg q 6-8 hr;
    Oral therapy: **Cimetidine** 400 mg bid, **famotidine** 20 mg bid, or **ranitidine** 150 mg bid for 8-12 weeks;
    Alternative antacids: **Omeprazole** 20 mg qd or **lansoprazole** 30 mg qd for 8-12 weeks
2. Manage bleeding ulcer
    a. Place 2 large bore IVs (>18 gauge) and give Lactated Ringers or Normal Saline to resuscitate and normalize blood pressure.
    b. Suppress acid with IV therapy.
    c. Evacuate and be prepared to perform blood transfusion (see Procedures).
3. Eradicate *Helicobacter pylori*: "triple therapy" includes many choices, but most treat for 10-14 day po course with: **omeprazole** 20 mg bid or **lansoprazole** 30 mg bid, plus **clarithromycin** 500 mg bid or **amoxicillin** 500 mg tid, plus **metronidazole** 500 mg bid

Patient Education
General: 90-95% of duodenal ulcers and ~80% of gastric ulcers are caused by infection with *Helicobacter pylori*. Most of the remaining ulcers are caused by ingestion of **aspirin** or NSAIDs. Emotional stress or food does not cause duodenal and gastric ulcers.
Diet: Consume a healthy diet and avoid foods that aggravate symptoms.
Medications: The ulcer should be treated with medicine to decrease stomach acid production. When *Helicobacter pylori* infection is suspected (recurrent ulcer disease), it should be treated with "triple therapy."
Prevention and Hygiene: Avoid **aspirin** and NSAIDs. Use **acetaminophen** instead for head, muscle or joint aches.
No Improvement/Deterioration: Return if symptoms worsen or persist after 2 weeks of treatment. Also, return immediately if vomiting blood or coffee grounds, or passing blood or tarry stools from the rectum (hematemesis or melena).

Follow-up Actions
Return evaluation: Evaluate worsening or persistent symptoms after 2 weeks of treatment with upper endoscopy to excluded complicated ulcer disease or malignancy. Refer those with hematemesis or melena. Relapse of symptoms after successful treatment suggests failure to eradicate or reinfection with *Helicobacter*

pylori or concurrent aspirin/NSAIDs use.

Evacuation/Consultation Criteria: Evacuate unstable and bleeding patients (melena, hematemesis). Consult gastroenterologist or internist for uncomplicated ulcer disease, and a general surgeon for patients with melena or hematemesis.

NOTE: Upper GI endoscopy or x-ray series may be required to confirm the diagnosis in garrison.

GI: Acute Organic Intestinal Obstruction
COL (Ret) Peter McNally, MC, USA

Introduction: Acute organic obstructions, which are partial or complete blockages in the bowels, are divided into small and large intestinal causes. Both will present with acute onset of severe abdominal pain, distention and nausea and vomiting. Prompt evaluation, decompression and surgical correction of the obstruction before bowel infarction or perforation occurs are the keys to management. The percentages listed below are for industrialized nations. Intussusception (20-30% of all obstructions in Africa and India where ascariasis is endemic) and volvulus are much more common than cancers and diverticulitis in the developing world. Etiology for small intestinal obstruction: adhesions (56%), hernias (25%), neoplasm (10%), other (9%). Etiology for large intestinal obstruction: neoplasms (60%), volvulus (20%), diverticular stricture (10%), other (10%).

Subjective: Symptoms
Acute onset of severe, crampy abdominal pain with associated vomiting (usually feculent due to increased bacteria in the gut) and abdominal distention; pain: in paroxysmal waves every 4-5 minutes for proximal obstructions (less frequent for distal obstructions), and continuously for strangulated bowel; rectal bleeding is consistent with mucosal ulceration from intestinal ischemia, inflammatory bowel disease or malignancy.

Objective: Signs
Using Basic Tools: Inspection: Febrile, toxic, dehydrated from vomiting, distended abdomen with visible peristaltic waves in small bowel obstruction.
Auscultation: Frequent, high-pitched bowel sounds occur in waves early, but the bowel may be silent later due to peritonitis or bowel infarction. Borborygmi (loud bowel rumblings audible without stethoscope) correspond to paroxysms of pain.
Percussion: Obstructed and dilated, gas-filled loops of bowel are often tympanic.
Palpation: A mass suggests the cause of obstruction. Check for hernias (inguinal, femoral, or umbilical), surgical scars (adhesions).
Using Advanced Tools: Lab: CBC with differential and urinalysis for infection. Abdominal X-ray (if available): free air, dilated loops of bowel, air-fluid levels demonstrating obstruction.

Assessment:
Differential Diagnosis - causes of peritonitis (see section on Peritonitis), including appendicitis, cholecystitis, peptic ulcer disease, and diverticulitis; various types of food poisoning and gastroenteritis; large neoplasms; labor (pregnancy)

Plan:
Treatment
1. Place NG tube to decompress and keep NPO.
2. IV fluids to restore fluid and electrolyte losses caused by vomiting (see Shock Fluid Resuscitation).
3. Give antiemetic (e.g., **Phenergan** 25 mg IV, IM, or po) of choice, but no pain meds until sure of diagnosis and awaiting evacuation (see Procedure: Pain Assessment and Control). Narcotics paralyze the bowel and can mask worsening symptoms that may precede perforation.
4. Prepare for medical evacuation if symptoms persist for > 12 hours or if fever or peritoneal signs develop.
5. IV antibiotics should be administered if peritoneal signs arise (must cover both aerobic and anaerobic bacteria)
Single Agents: **Cefoxitin** 2 gm IV q 8 hr, **cefotetam** 2 gm IV q 12 hr, or **cefmetazole** 2 gm IV q 8-12 hr

Combination Agents: **Aztreonam** 2 gm IV q 8 hr plus **metronidazole** 500 mg IV q 8 hr

Patient Education
General: Pain lasting more than 4 hrs should be evaluated by a health care professional.
Diet: Low roughage if a history of recurrent partial small bowel obstructions.
No Improvement/Deterioration: Persistent pain or pain associated with vomiting, dehydration, bleeding, or fever should be evaluated promptly.

Follow-up Actions
Return evaluation: Routine post-operative follow-up.
Evacuation/Consultation Criteria: Evacuate ASAP. Consult general surgery.

Chapter 8: Genitourinary (GU)

GU: Urinary Tract Problems
CAPT Leo Kusuda, MC, USN

Introduction: This section will provide tips for the assessment and disposition of major symptoms associated with the urinary tract, excluding trauma.

Examination Tips
1) Assessing a flank pain:
   Lightly tap or push with fingers on right or left lower chest wall. If there is significant kidney irritation, this will elicit increased pain.
2) Abdominal exam:
   a) Percuss the region superior to the pubic bone. A dull tap suggests a distended bladder holding a large volume of urine. In the female, the pubis is much lower and a smaller volume of urine can be appreciated on percussion or bimanual exam.
   b) Look for peritoneal signs: increased pain with light tapping on the abdomen, pain with shaking of the abdomen and hips, pain when suddenly releasing pressure on the abdomen (rebound).
3) Scrotal exam:
   a) If possible, always exam the patient in both the standing and supine positions.
   b) Testis position and varicoceles can only be appreciated in the standing position.
   c) Examine the testis of a patient complaining of pain in the scrotum while he is lying down. Increased pain in the testis may cause the patient to faint.
   d) If a bright light such as for an otoscope is available, transilluminate all scrotal masses to determine cystic (bright, diffuse glow) or solid nature of the mass.
4) Rectal Exam
   a) Prostates are generally the size of a walnut and no more than 2 finger breadths wide.
   b) With the patient standing and bending over the top of the prostate should be easy to reach. In young men, the presence of a large soft mass on rectal exam usually is the bladder. Some prostates in young men are difficult to palpate.

Urinalysis
1) Dilute urine with a specific gravity of 1.005 or less, or concentrated urine (dehydration, first morning void, etc.) of 1.015 or higher suggests normal renal function.
2) When there is visible blood in the urine, the protein from the blood can raise the urine dipstick protein value to 2+.
3) Nitrite positive urine can be from skin bacteria if the person (male or female) voided a small amount without doing a clean catch. (Avoid this problem by starting to void, then sliding the cup into the stream).
4) Infections can be nitrite negative.

5) Trace heme on a urine dipstick can be normal.
6) Trace leukocyte urine can look significantly positive when viewed under the microscope.
7) Cloudy urine in specimens with an alkaline pH (6 or higher) can be amorphous phosphate and be normal in young individuals.
8) The presence of crystals in the urine does not automatically mean that the person has kidney stones.

Normal voiding
1) Normal first urge to urinate occurs with about 5 ounces in the bladder.
2) Normal bladder capacity in an adult is 10-15 ounces.
3) Normal time between voiding averages greater than every 2 hours.
4) Average total 24-hour urine volume for adults is about 1 quart. Ideal would be 2 quarts/day. This translates to 40-80 ml of urine per hour.

Blood in the urine (hematuria)
1) Trauma and visible blood in the urine suggests possibility of major injury. Stabilize and transfer for evaluation. If patient is able to void, severe injury to bladder and urethra is much less likely.
2) Hematuria with irritative voiding symptoms should be treated initially as an infection. Exposure to bodies of fresh water in Africa or the Middle East may lead to schistosomiasis as a cause of blood in the urine.
3) Hematuria with flank pain and:
   a) No fever, no drug exposure and no trauma suggests a kidney stone.
   b) Fever, but no trauma should be treated for a possible kidney infection.
   c) High proteinuria, but no fever or drug or chemical exposure suggests nephritis.
4) Hematuria with painful scrotum should be treated initially as an infection.
5) Gross hematuria (visible blood) without any other symptoms can be a sign of cancer at any age.

Blood in the semen (hematospermia)
If there are no difficulties voiding, the physical exam (including rectal exam) is normal and the urinalysis several days after the event is negative, then this is a benign condition and no further workup is indicated.

Cannot control urine (leaking, incontinence) See Incontinence Section.

Cannot urinate (anuria)
Catheterization (see Procedure: Bladder Catheterization) is the best method of determining if there is an obstruction versus poor urine production as an explanation for anuria. In a patient with a very large bladder by palpation (dome of bladder extends more than half the distance between the umbilicus and the top of the pubic bone), rapid drainage of the bladder can result in the patient fainting. For a more detailed discussion, see the information on catheterization in the Prostatitis and Incontinence sections.

Discharge from the penis. Refer to section on STDs

Lumps in the genital region or swollen scrotum. Refer to section on Testis Mass and to STDs.

Pain in the side (flank). Refer to section on Urolithiasis

Pain in the scrotum
1) Tenderness located primarily in the testis: consider torsion, epididymitis.
2) Point tenderness on upper pole of testis: consider torsed appendix testis or cyst.
3) Tenderness primarily in cord above testis: consider varicocele.
4) Mass in testis: consider tumor.
5) Mass above testis or around testis that glows when a strong light is placed against it: hydrocele or spermatocele.
6) Large mass with history of direct blow to testis: fractured testis vs. hematoma.

Pain with urination
In most cases, it is safer to initially assume a urinary tract infection (UTI) and treat with antibiotics (see UTI section). Treat vaginitis if found (see Symptoms: GYN Problems).

Persistent erections (Priapism)
1) A tender, painful erection with no history of trauma is low flow priapism. This is an emergent condition best treated by a urologist. Although this condition may resolve spontaneously, cold water immersion and manual compression of the penis may be successful. A persistent erection greater than 4 hours may result in increasing tissue injury that may result in the loss of erectile function after the penis is decompressed.
2) A painless partial or full erection especially with history of pelvic trauma can be observed. Similar treatment can be used.

Skin lesions in the genital region
Ulcers (see Sexually Transmitted Diseases):
1) Ulcers that form immediately after intercourse are from trauma.
2) If always associated with the ingestion of one particular medication, the ulcer represents a fixed drug reaction.
3) Painful - chancroid, herpes
4) Painless - syphilis (hard or firm induration, chancre), granuloma inguinale, or LGV

Blisters and nodules
1) If there is any question of the diagnosis, assume it may be sexually transmitted (herpes) and avoid further sexual contact.
2) Persistent lesions should be evaluated electively to r/o cancer.
3) Most causes are benign and/or self-limited.

Generalized edema
1) Generalized swelling of the penile shaft skin with itching is usually either a contact allergic reaction or idiopathic. If an offending agent can be identified (or suspected), treat with antihistamines and avoid the chemical irritant.
2) Suspect a skin infection if there is significant erythema and pain, which may also involve the scrotum. In a sick individual with fever, this can represent a life threatening condition called Fournier's gangrene (see Symptom: Male Genital Inflammation).

Cannot Move Foreskin (Phimosis/Paraphimosis)
Inability to retract the foreskin (phimosis) or to pull it forward to its normal position (paraphimosis) can be problematic in the field. Often the foreskin is edematous from irritation or infection. Monitor this condition for excessive circumferential swelling which could compromise blood flow in the penis. Anti-inflammatory medications, ice water and lubricants may be helpful. If there are signs of systemic infection (fever, nausea, fatigue, etc.), and prompt evacuation is not available, a dorsal slit should be performed in the field. Most patients require circumcision later.
Dorsal Slit: Prepare the penis as with any surgical procedure (sterile scrub, Betadine, drape), and attempt to clean between the head and the foreskin especially on the dorsal side. Anesthetize the dorsum of the foreskin with **lidocaine** (NO EPINEPHRINE!) using the smallest gauge needle (25-26) available. Use forceps or needle to ensure dorsal foreskin is numb. Clamp the dorsal foreskin tightly beginning at the tip and working back to where the foreskin meets the shaft. Leave the clamp in place for several minutes, as this will compromise blood flow in the area to be incised. Remove the clamp, and using sterile scissors or scalpel, carefully incise the dorsum of the foreskin through its entire thickness, through the line of devascularized tissue formed by the clamp. Do not incise the head of the penis. Fold the two sides of the incised foreskin back and away from the penis. Clean the penis with sterile prep solution between the head and foreskin, then again with alcohol.

Allow to air dry and apply a sterile dressing.

GU: Urinary Incontinence
CAPT Leo Kusuda, MC, USN

Introduction: Incontinence, the inability to voluntarily control the flow of urine, is only a social nuisance in most cases. If the incontinence is not due to infection, and a physical exam including gross motor and sensory (numbness or muscle weakness) exam is normal, serious complications are unlikely. Incontinence is fairly common in women. Daytime incontinence in men is highly abnormal and suggests significant underlying disease.

Subjective: Symptoms
Uncontrollable loss of urine.
Focused History: *Do you leak urine when you cough, lift heavy objects or jump up and down? Do you go to the bathroom often to prevent urine leaking out?* (affirmative answers suggest stress incontinence) *When you have to go to the bathroom is the urge strong? Do you have a hard time holding your urine when you get the urge? When you leak, is it a lot?* (affirmative answers suggest urge incontinence) *Do you have a hard time emptying your bladder even though you feel like you have to? After you go to the bathroom do you feel like you still have to go again? After you go to the bathroom do you leak?* (affirmative answers suggest urinary retention, which may be accompanied by overflow incontinence). *Do you feel constantly wet? Do you feel that the wetness or dripping may be coming from your vagina?* (affirmative answers suggest fistula or hole between the bladder or ureter and the vagina). Patients may have mixed diagnoses, such as mixed urge and stress incontinence.

Objective: Signs
Using Basic Tools: Wet clothing; trauma or irritation to the vagina; neurologic deficits: difficulty walking, numbness in the perineum or increased deep tendon reflexes.
Using Advanced Tools: Lab: Urinalysis: moderately to strongly positive leukoesterase should be considered an infection. Moderate to strongly positive heme should be considered an infection initially, but may be cancer, urinary tract stone or other condition. Urinary catheterization (see Procedures) for suspected retention.

Assessment:
Differential Diagnosis - stress incontinence, urge incontinence, mixed incontinence, and retention as described above.
Trauma, with or without fistula - continuous leakage in the setting of trauma suggests laceration of the vagina and bladder either from a foreign body or bone fragment. Trauma frequently causes fistula formation.
Compression of the spinal cord from disk disease, spinal tumors and brain disease (e.g., stroke).
Multiple sclerosis or other neural tissue disease.
Renal obstruction with overflow incontinence.

Plan:
Treatment
Primary:
Treat any urinary tract infection (see Urinary Tract Infection section).
Treat specific type of incontinence:
Stress Incontinence: Empty bladder frequently. Wear diaper or tampon. Practice Kegel exercises (tighten the muscles around the vagina 40-160 times per day).
Urge Incontinence: Mild: **Hyoscyamine** 0.375 mg po bid, **Urised** 1 po qid or **Flavoxate** 1 po bid
Moderate/Severe: **Ditropan** 5 mg 3-4x/day
Mixed Stress and Urge Incontinence: Imipramine 10-25 mg po q hs
Retention with Overflow Incontinence: Patients with significant symptoms, especially those suspected to have overflow incontinence, should have a catheter passed into the bladder per urethra to determine if there

is significant residual urine. If there is greater than 200-300 cc, leave the catheter in place and monitor urine output. If urine output is greater than 200 cc/hour, suspect renal obstruction. Start an IV placed with NS running at a maintenance level with boluses for resting pulse rate greater than 100/min.

Decrease prostate resistance with alpha-blockers: **Hytrin** (**terazosin**) 1-5 mg po q hs (start at low dose and titrate up over several weeks), **Cardura** (**doxazosin**) 1-4 mg po q hs, **Flomax** (**tamsulosin**) 1 po qd or **Minipress** 1-5 mg po q hs
After removing urethral catheter, perform intermittent (self-) catheterization (see Procedures: Bladder Catheterization) q 4-6 hours to keep bladder volume under 300 cc.
For those patients in whom it is too difficult to pass a Foley catheter, use a straight or suprapubic catheter. (see Procedure: Suprapubic Catheterization)
Alternative: For frequency and urgency, **diazepam** (**Valium**) 5-10 mg po q 6h can be very helpful.
Primitive: None
Empiric: Antibiotics for chronic suppression of infection, such as **nitrofurantoin** 50 mg po bid, **Septra** 1 po q hs. **Cipro** 250 mg po q hs or **Keflex** 250 mg po q hs.

Patient Education
General: Avoid dehydration. Women tend to avoid fluids to minimize going to the bathroom and leaking, leading to significant dehydration.
Medications: Cold medications and antihistamines for sinus problems will counteract alpha-blockers and vice versa. Side effects of Ditropan include dry mouth, dry eyes and constipation.
No Improvement/Deterioration: If other neurologic symptoms such as visual disturbance, muscle weakness or sensory loss become apparent, refer patient to hospital for further evaluation

Follow-up Actions
Return evaluation: Evaluate for effectiveness of therapy and the necessity for referral to a urologist for surgery or other treatment.
Evacuation/Consultation Criteria: Evacuate all unstable patients and those with neurologic findings (i.e. cauda equina) as soon as possible. Additionally, refer at some time all patients with overflow incontinence, those with stress incontinence that is interfering with work, those with urge incontinence who deteriorate or fail to improve, or any patient with a continuing requirement for medication.

GU: Urolithiasis (Kidney Stones)
CAPT Leo Kusuda, MC, USN

Introduction: Ureteral stone pain is generally acknowledged as one of the worst pains a person can suffer. The majority of stones can be managed with hydration and pain control. Fever, vomiting and severe pain not controlled by oral medication requires intravenous treatment. Evacuate these patients with persistent symptoms beyond 24 hours.

Subjective: Symptoms
Intense, intermittent flank or inguinal pain radiating into the scrotum and not related to activity; nausea and vomiting; urinary frequency and burning (if stone at ureter/bladder junction); fever.

Objective: Signs
Fever, severe costovertebral angle (CVA) tenderness which waxes and wanes, vomiting.
Using Basic Tools:
1. Examine the patient between the lower chest and scrotum/pelvis.
2. Check for a tender liver by pushing under the anterior right ribs while the patient takes a large breath.
3. Check for a hernia.
4. Examine the scrotum for epididymitis or torsion.
5. Examine above the prostate on the rectal exam for any fullness on the side of symptoms.

6. Do a bimanual pelvic exam to check for adnexal tenderness.
7. Check for costovertebral angle (CVA) tenderness. Lightly thump the right and left lower ribs in the back. Increased tenderness suggests kidney pain.
8. Check for peritoneal signs. If pain increases with light tapping on the abdomen, shaking the abdomen, striking the heel of the foot, or there is significant irritation of the abdominal contents, then bowel inflammation/perforation (appendicitis, etc.) is suggested.
9. Light thumping over the right lower anterior chest wall would suggest gallbladder irritation. This, combined with increased pain on eating, especially in young, overweight women is suggestive of gallbladder disease.

Using Advanced Tools: Lab: Urinalysis may reveal casts, blood. Abdominal X-ray to assess for presence of stones.

Assessment:
Severe side pain not related to position, which waxes and wanes, without evidence of an abnormal genital exam or peritoneal signs strongly suggests ureteral stone.
Differential Diagnosis - any disease process between the lower chest and upper thigh can be considered.
Lower lobe pneumonia or pulmonary process - abnormal breath sounds
Abdominal causes (see Symptom: Abdominal Pain) - liver disease; cholecystitis/cholelithiasis (gallbladder) diverticulitis including Meckel's; appendicitis; mesenteric adenitis; abdominal aortic aneurysm.
Renal - waxing and waning pain excludes pyelonephritis, cysts, tumor or ischemic injury.
Musculoskeletal pain - this includes aches due to viral illness.
Inguinal hernia - distinguish by exam.
Urologic - Epididymitis - tender epididymis; testicular torsion - tender testis; congenital ureteropelvic junction obstruction
Gynecologic - abnormal pelvic exam (see GYN Problems section): ectopic pregnancy, pelvic inflammatory disease, torsion of ovary, ovarian cyst, tubo-ovarian abscess.

Plan
Treatment
Primary:
1. Pain control (in order of preference): **Ketorolac (Toradol)** 30 mg IM q 6 h is highly effective in relieving stone pain. Narcotics such as **morphine sulfate** 5-10 mg IM, **Demerol** 50-100 mg IM q3-4 h prn (can combine with **ketorolac**)
   Tylox 1-2 po q 4 h prn or **Demerol** 50-100 mg po q 4 h prn
   If above are not available, NSAIDs such as **ibuprofen** 800 mg po tid or **indomethacin** 25-50 mg po tid can help.
2. Hydration
3. Antibiotics when fever is present in a suspected urinary stone patient: Either **levofloxacin** or IV **ampicillin** plus IV **gentamicin** are acceptable (see pyelonephritis in UTI section).
4. Anti-emetics as needed

Primitive: None

Patient Education
General: Maintain good hydration.
Diet: Increased water and citrus juice intake may prevent further stone formation.
Prevention and Hygiene: See diet

Follow-up Actions
Return evaluation: All suspected stone patients need to eventually have an abdominal film taken to assess the presence of stones.
Evacuation/Consultation Criteria: Evacuate ASAP patients with fever, persistent severe pain and persistent vomiting. Refer patients with suspected stones for urologic consultation. Persistent symptoms require evaluation with an IVP or CAT scan after evacuation.

GU: Urinary Tract Infection
CAPT Leo Kusuda, MC, USN

Introduction: The causative organisms of cystitis, acute prostatitis and pyelonephritis are the same. The treatment of acute bacterial urinary tract infection (UTI) depends on the location of the infection and the presence of complicating factors. For the vast majority of infections, the fluoroquinolone antibiotics are highly effective. Bladder infections (cystitis) are treated for 3 days, kidney (pyelonephritis) for 14 days and prostate (prostatitis) for 30 days. In the male, it is practical to assume that any leukoesterase positive or culture positive bacterial urinary tract infection involves the prostate so treat for 30 days. Prostatitis and epididymitis are covered in separate sections. Urethral discharge suggests urethritis, usually sexually transmitted. Therefore, primary treatment is different, although the fluoroquinolones are a good alternative. Urethral discharge is covered in the STD chapter.

Subjective: Symptoms
Burning, frequency, urgency, fever, flank pain.

Objective: Signs
Using Basic Tools: Fever, flank tenderness, fatigue, nausea, vomiting, suprapubic tenderness.
Using Advanced Tools: Urinalysis: Pyuria (leukoesterase +) and nitrite positive indicates infection (some gram-positive organisms may be nitrite negative). Nitrite + and leukoesterase negative specimen is contaminated with skin; Gram stain: identify and quantify WBCs, gram-positive or gram-negative rods and epithelial contamination.

Assessment:
Differential Diagnosis:
Cystitis - burning or frequency, and leukoesterase-positive urine in a female.
Pyelonephritis - fever or flank pain, and leukoesterase-positive urine in a female.
Microhematuria without pyuria - pyelonephritis is less likely; patient may have other reasons for microhematuria (tumor, stone, etc.).
Peri-ureteral inflammation - inflammation around the ureter (e.g., appendicitis, PID) can result in an abnormal urinalysis.
Contamination - positive nitrite, negative leukoesterase and negative heme is likely skin contamination of the urine specimen.
Prostatitis - UTI symptoms in a male with leukoesterase-positive urine.

Plan:
Treatment
Cystitis
Septra DS 1 po bid, **nitrofurantoin** 100 mg po qid or **Macrobid 100** mg po bid, **Cipro 250** mg po bid or **Levaquin 250** mg po qd, **Keflex 250** mg po qid, or **Augmentin** 875/125 po bid or 500/125 po tid x 3 days. **Nitrofurantoin** is the safest drug in women since it is acceptable to give throughout a pregnancy.
If symptoms do not improve in 2 days, treat patient for 2 weeks. If Augmentin is used, candidal yeast infections frequently develop. Be prepared to treat with **fluconazole** 150 mg po single dose or **terconazole** vaginal suppositories qd x3 days.
Alternative: 1/3 of woman can clear their cystitis by increased hydration. **Doxycycline** 100 mg po bid
Cystitis with complicating factors
If patient has history of infections every 1-2 months, place on suppression (see below) until seen by urology. Women who are postmenopausal, especially greater than 60 years old, frequently take longer to eradicate cystitis. In such patients, treat for 7-10 days.
Pyelonephritis
Moderately ill:

Fluoroquinolones (**Levaquin** 500 mg po qd or **Cipro** 500 mg po bid or **Floxin** 400 mg po bid) x 2 weeks.
Alternative: **Augmentin** 875/125 mg po q12h or 500/125 mg po tid x 2 weeks or **Keflex** 500 mg po qid x 2 weeks.
Moderately ill but unable to tolerate po medications:
Ampicillin 1-2 gm IV q 6-8h and **gentamicin**.
Gentamicin, q d dosing at 5 mg/kg is preferred or 1.5 mg/kg loading and 1.0 mg/kg IV/IM q 8 h.
Alternative: cefotaxime 1.0 gm q12h IV up to 2.0 gm q4h IV or ceftriaxone 2.0 gm qd IV.
Once patient is clinically improved, treat with quinolones (as above) x 2 weeks.
Severely ill: Treat 2-3 weeks with same IV regimen as above. Do not progress to oral dosing.
Give IV fluids if there is dehydration and nausea.
Of the quinolones, **Levaquin** has a broader spectrum of coverage for UTI.
If quinolones are not available, use **Septra** DS tid until afebrile, then bid for 2 weeks.
Nitrofurantoin is not useful for deep tissue infections such as pyelonephritis.
Patients who are penicillin allergic should be treated with **vancomycin** 15 mg/kg q12h IV when **ampicillin** is indicated.
Pyelonephritis with complicating factors (recurrent UTIs or post-menopausal)
After initial treatment, begin suppression regimen until seen by urology: **Macrodantin** 50 mg po bid, **Septra** DS ½ po qhs, **Cipro** 250 mg po q hs or **Keflex** 250 mg po q hs.

Empiric:
Failure of symptoms and urinalysis to improve suggests resistance to the antibiotic being used. Antibiotics should be changed if there is no improvement after 3-4 days. Patients with a fever can be expected to take several days to become afebrile.
Recurrence of urinary tract infection within weeks of completing the initial course of antibiotics suggests an inadequate duration of treatment or reinfection. A longer course of antibiotics, possibly with the addition of 2-3 months of suppression is indicated.
Urine culture data is extremely valuable in both cases.

Patient Education
General: Hydrate well to ensure urination every 2 hours. Cranberry juice is a good fluid choice. Complete all antibiotics.

Follow-up Actions
Evacuation/Consultation Criteria: Evacuate unstable patients ASAP. Refer patients with pyelonephritis for urologic evaluation electively. Patients with cystitis that does not resolve within 3 days of initiating treatment should be referred for evaluation.

PART 5: SPECIALTY AREAS
Chapter 9
Podiatry: Heel Spur Syndrome
(heel spur, heel bursitis, plantar fasciitis)
CDR Raymond Fritz, MSC, USNR

Introduction: Heel spur syndrome is one of the most common foot problems seen in the special operations community. The term "heel spur syndrome" refers to any to heel pain with or without a spur that typically develops from excessive repetitive strain on the plantar fascia. The plantar fascia is loaded when weight is applied (standing), causing pain along the plantar fascia, particularly where the fascia connects to the heel tubercle. This condition is often a tolerable nuisance but it may be painful enough to make ambulation difficult. Chronic conditions may last for years if not properly treated. More than 90% of the cases in military personnel are due to faulty foot mechanics and increased activity demands.

Subjective: Symptoms
Insidious onset of heel pain, most severe in the morning or when standing up; may acutely follow an injury; pain can be bilateral.

Objective: Signs
Using Basis Tools: Point tenderness over medial tubercle of the calcaneus at the level of the plantar fascial attachment, which may radiate distally causing pain and swelling in the arch; more common in pronated foot type but heel pain can present in a high-arch foot type; distant symptoms due to compensatory gait changes; tight Achilles tendon.
Using Advanced Tools: X-rays: Spur presents 60% of the time; fracture, bone cyst or arthritic changes may be noted to explain symptoms.

Assessment:
Differential Diagnosis
Bursitis - palpate tenderness (inflamed bursa) directly below the calcaneal tubercle.
Nerve entrapment - point tenderness over nerve; pain radiating into heel; positive Tinel's sign*.
Tarsal tunnel syndrome - compression of the posterior tibial nerve; positive Tinel's sign*.
Referred pain from low back - L-4 L-5 extends to the heel as part of the area of distribution for this nerve root level; EMG/nerve conduction studies are helpful for diagnosis of nerve related heel pain.
Stress fracture - diagnose on x-ray; not common in calcaneous
Foreign body - usually an entrance portal visible
Arthritis (Reiter's, psoriatic, ankylosing spondylitis, rheumatoid) - See Symptom: Joint Pain section.

*Tinel's sign is pain radiating distally along the course of a nerve.

Plan:
Treatment
Primary:
1. Conservative: Ice (not heat) massage, Achilles stretching, heel pad (foreign body, bursitis, arthritides).
    a. Ice massage: Use ice directly on heel and arch but limit to 8-10 minutes 4-6 x day; use Dixie cup technique or frozen plastic water bottle or gel pack if available.
    b. Dixie cup technique when freezer available: Fill cup with water and freeze. Keep several ice cups on hand. Tear cup down to expose ice and use as an applicator to heel area.
    c. Achilles tendon stretching: Any limitation in ankle dorsiflexion increases force on plantar fascia.
2. Rest strap: Tape the foot to support the arch
3. Remove any splinter, glass or metal when the operational tempo permits.

4. Anti-inflammatories: **Motrin** 800 mg po tid with food; arthritides may need steroid injection. **Cortisone** injection for acute pain: Injection mixture: 1/2 cc long acting steroid i.e., **Celestone, dexamethasone acetate**, and 1cc **Marcaine** 0.5% plain. (See video on CD-ROM)
5. Consider a **Marcaine** block to the posterior tibial nerve if previous training and experience.
6. Rest is mandatory to allow healing.

Alternative: Arch supports, injection (2cc of **Marcaine** 0.5% mixed with 1/2cc of **dexamethasone acetate** or other long acting steroid could prove helpful for short mission if pain significant).
Primitive: Place soft, supportive material under boot insole arch area. (Ex. eye patch, 4x4 gauze cut to fit)

Patient Education
General: Get better arch support. Avoid walking barefoot if possible. For dive ops, use boot with fin if operational mission involves movement overland once exiting water.
Medications: Gastritis side effects with NSAIDs.
Prevention: Good shoe support and arch support. Prescription orthotics may be best measure when obvious faulty foot mechanics present. Good flexibility program.

Follow-up Actions
Return evaluation: Follow-up 1week or check more regularly if teammate. Try 2nd injection and stronger oral anti-inflammatory if not resolved. Recommend against narcotics if operational.
Evacuation/Consultation Criteria: Evacuation not normally necessary. If conservative measures fail to give any significant relief, consult podiatry or orthopedics. Custom orthotics will be the best consideration for the chronic recurrent case. Consult physical therapy for treatment modalities. Athletic trainer also great resource. Rheumatology consult if inflammatory etiology suspected (i.e., Reiter's syndrome)

NOTES: Remember to rule out referred pain.
Think mechanical – do not just treat symptoms.
Flexibility program a key factor in treatment and prevention.

Podiatry: Ingrown Toenail
CDR Raymond Fritz, MSC, USNR

Introduction: An ingrown nail occurs when the nail border or corner presses on the surrounding soft tissue. This condition is painful and often results in an infection once the skin is broken, with the offending nail corner acting like a foreign body introducing pathogens. An ingrown nail may result from improper trimming of nails, injury, tight shoes, genetic predisposition and fungal nail infections.

Subjective: Symptoms
Toe pain, especially in shoes; history of recurrent ingrown nails and infections, and previous procedures to remove the nail.

Objective: Signs
Using Basic Tools: Most commonly involves great toe; soft tissue penetration and secondary infection, with purulence, tenderness, erythema and edema; excessive granulation tissue in more chronic cases; malodorous wound when gram-negative bacteria involved.
Using Advanced Tools: C&S in a severe infection before beginning empiric coverage. X-rays are rarely considered but one should be aware that osteomyelitis secondary to a chronic ingrown nail infection is a possibility if the condition has been neglected or chronic. X-rays will also reveal a subungual exostosis (bony growth under the toenail) when present.

Assessment:
Diagnose this problem clinically in the field
Differential Diagnosis (may be secondary diagnosis)
Subungual exostosis - spur on the distal phalanx which pushes upward causing the nail to incurvate.

Fungal nail infection, subungual hematoma, foreign body reaction (granuloma)

Plan:
Treatment
Primary: Partial nail avulsion
1. Perform digital block using **Xylocaine** 1% or **Marcaine** 0.5% plain (no epinephrine for digits) - see Anesthesia: Local/Regional.
2. Use elevator to free nail from bed along border. Also free nail from overlying soft tissue.
3. Use an English nail anvil or nail clipper to remove the offending nail border. Scissors will also work.
4. Use curette to remove infected necrotic tissue or excessive granulation tissue (proud flesh) from the nail groove.
5. Dress with **Betadine** gauze and Kling. Coban or Elastoplast helps hold dressing in place.
6. Elevate foot and apply warm soaks or compresses tid.
7. Antibiotics for 7 days: **Dicloxacillin** 500mg po qid or **Keflex** 500mg po qid for broader coverage. **Erythromycin** 500mg po qid for penicillin allergic.
8. Pain control: **Motrin** 800mg po tid prn pain. Narcotics are not usually necessary.

Alternative: Remove nail corner with clipper, antibiotics.
Primitive: Lift side of nail corner and remove with small scissors.

Patient Education
General: Instructions on soaking: add few ounces of **Betadine** solution to water; remove loose necrotic tissue or scab covering with washcloth while soaking to promote drainage when infected and speed the healing process.
Prevention and Hygiene: ALWAYS cut nails straight across.
No Improvement/Deterioration: If recurrent problem, return for definitive procedure.

Follow-up Actions
Return evaluation: At 3-5 days, check for any remaining nail spicules (small, needle-shaped pieces); check cultures; consider X-ray.
Evacuation/Consultation Criteria: Evacuation not usually necessary. Partial nail avulsion should be considered in recurrent cases once the infection is resolved. This will destroy the nail matrix and prevent re-growth. The definitive procedure is not recommended in an operational setting. If problem recurs or fails to respond, consult podiatrist or dermatologist.

Podiatry: Plantar Warts
CDR Raymond Fritz, MSC, USNR

Introduction: Warts are caused by human papillomavirus viruses and can be found anywhere on the skin when the virus is introduced through a crack in the skin of a susceptible individual. When located on the sole of the foot, these warts are called plantar warts. A plantar wart can be found as a single lesion or grouped together (referred to as a mosaic wart). Most common areas include the ball of the foot and heel, where increased pressure and irritation is common. Discrete plantar corns are sometimes mistaken for warts. A wart has tiny dots in the center which are small vascular elements. These dots are often black (dried blood) due to irritation, when located on the plantar aspect of the foot. Warts are often ignored until they become painful.

Subjective: Symptoms
Pain, especially if wart is on prominent plantar area; may have tried over-the-counter preparations, other family or team members may have warts as well.

Objective: Signs
Using Basic Tools: Lesions tender to palpation and squeezing especially if located on weight-bearing area; callus may form over the wart, increasing pain.

Assessment:
Differential Diagnosis: Corn, callus, pyogenic granuloma, other lesions. See Podiatry: Corns and Callus.
A wart may bleed (pinpoint) with debridement but callus will not. Pyogenic granuloma bleeds easily.

Plan:
Treatment
Primary:
1. Debride overlying callus with #15 or 10 blade to allow medicine to reach wart. **See video on CD-ROM**
2. Apply aperture pad to keep topical preparation isolated over the wart. 1/8" felt padding with sticky back works well. Pre-cut felt pads are available, but if material is in sheets, cut and size to fit. Moleskin okay to cover but it will not relieve the load a tender area. The padding and aperture prevents adjacent skin irritation.
3. Apply 60% **salicylic acid** paste (or **monochloroacetic acid**) to wart. Tape to cover and hold in place for 3 days.
4. Repeat treatment in one week.
5. Surgical curettage should be reserved for unresponsive cases and is not recommended in the field. Curettage reduces the chance of plantar scarring since the procedure does not involve penetration below the dermis when done correctly.
6. A surgical excision of a wart using two semi-elliptical incisions is a consideration for a wart in a non-weight bearing area. Surgical excision should never be performed on weight bearing areas because of the risk of scarring and subsequent pain with ambulation.

Alternative: **Liquid nitrogen** (LN_2), **trichloroacetic acid**, many over-the-counter preparations.
Primitive: Pad around wart to increase comfort in the field. Hold other treatment if short-term mission.

Patient Education
General: The cause of the wart is a virus. Topical re-treatment may be required. Discontinue treatment for a few days if the area becomes too sore and painful. Also discontinue if the area becomes infected.
Medications: Use over-the-counter anti-inflammatories if pain significant.
Prevention and Hygiene: Use deck shoes or sandals in shower/pool areas to prevent spread among troops.

Follow-up Actions
Return evaluation: Follow up weekly until resolved
Evacuation/Consultation Criteria: Evacuation not normally necessary. Consult podiatry or dermatology for resistant cases.

Podiatry: Bunion (Hallux Abductor Valgus)
CDR Raymond Fritz, MSC, USNR

Introduction: A bunion is an enlargement at the 1st metatarsal head of the great toe, which deviates laterally. Often there is no bump, but rather an angulation of the first metatarsal (hallux abductor valgus) that makes the head of this bone more prominent. Genetic factors, foot mechanics and poorly fitting or excessively worn shoes are commonly blamed for the development of both deformities. Pain is a result of cartilage erosion, bursitis and neuritis in the effected joints.

Subjective: Symptoms
Pain near first metatarsal head, history of a progressive deformity over time.

Objective: Signs
Using Basic Tools: Bump, erythema and tenderness medially (tibial aspect) over the first metatarsal head; joint stiffness in more chronic cases, especially with excessive pronation (flat feet).
Using Advanced Tools: X-rays are helpful in evaluating angular relationships and joint integrity when

available, but are not required.

Assessment:
Diagnose by clinical presentation/appearance
Differential Diagnosis - rigid toe due to traumatic osteoarthritis (hallux rigidus or limitus). Toe joint displacement/swelling (metatarsalgia, sesamoiditis). Local toe irritation (shoe irritation in absence of deformity).

Plan:
Treatment: Primary:
1. Change to a wider shoe or soft sneaker if operationally permissible.
2. Use bunion pads. Over-the-counter bunion pads come in all shapes and sizes. A doughnut hole cut in felt or several layers of moleskin will work as a substitute for a bunion pad.
3. NSAIDs for pain relief. Ice massage if acute presentation.
4. Arch supports and orthotics in severely pronated feet.

Alternate: Inject 0.25 cc **dexamethasone acetate** (or other long acting steroid) and 0.5 cc 0.5% **Marcaine** SC just medial to the metatarsal head as a one-time temporary pain relief measure during an operation. Multiple injections could weaken joint structures, causing progression of the deformity. Shoe or boot pressure can irritate the cutaneous nerve running medially along the first metatarsal head, causing severe neuritis pain and making ambulation difficult.

Primitive: Cut a hole in the boot over the bump if pain is severe.

Patient Education
General: Although these are structural deformities, changing shoe style and size may provide the most relief in an operational setting when surgery is not an option.
Activity: Limit running for 1-2 days. If pain is severe, wear open sandals for 2 days if possible.
Medications: Motrin 800mg po tid with meals
Prevention and Hygiene: Avoid tight shoes. Wear larger and wider boots if necessary.

Follow-up Actions
Return Evaluation: 1-2 weeks
Evacuation/Consultation Criteria: Evacuation is not usually necessary. If no change with conservative measures, refer to podiatrist or orthopedic surgeon.

NOTES: Wider boot, shoe and running shoes are most important for pain relief. Postpone surgical correction until absolutely necessary. If deformity and symptoms are severe and conservative measures fail, elective surgery is an option.

Podiatry: Corns and Calluses
CDR Raymond Fritz, MSC, USNR

Introduction: A callus is a thickening of the outer layer of skin, in response to pressure or friction, that serves as a protective mechanism to prevent skin breakdown. The hyperkeratotic change for corns and calluses is similar except a corn involves a discrete pressure spot, typically over a bone. Foot and toe deformities are subject to higher pressures and shoe irritation. A boot may rub a hammertoe at the knuckle and result in a painful corn. Corns may also develop between toes where two bones press together. Typical callus patterns are seen in certain foot types.

Subjective: Symptoms
Pain history of a corn or callus in the same areas.

Objective: Signs

Thickened, dry skin over prominent bones (corn); larger patches of thickened, dry skin over friction areas from walking (calluses); tenderness, blisters, breakdown and infection after continued irritation.

Assessment:
Differential Diagnosis
Wart, foreign body

Plan:
Treatment
Primary: Trim areas with #15 blade or beaver blade. Trim with #10 blade for larger callus areas. Place felt with a doughnut-shaped hole cut in the middle (or pre-cut felt available over the counter) around area to relieve pressure and friction. Medicated pads are not recommended.
Alternative: Pad around prominent areas without trimming to protect and prevent irritation.
Primitive: Sand callus or corn gently with abrasive stone

Patient Education
General: Daily foot inspections in the field if possible.
Prevention and Hygiene: Trim corns and calluses. Safest technique is to file areas with callus stone. Inspect shoes for frayed seams or torn liner. Check shoe fit. Mitigate source of pressure and deformities.

Follow-up Actions
Evacuation/Consultation Criteria: Evacuation not normally necessary. Refer to podiatrist for orthotics or surgery to correct deformities, or for other advanced foot care.

Podiatry: Stress Fractures of the Foot
CDR Raymond Fritz, MSC, USNR

Introduction: A stress fracture may affect any bone. The most common stress fracture in the foot, known in the military as a march fracture, is the second metatarsal. Stress fractures are often seen in intense training programs around week four, when bone absorption exceeds bone-building activity. Improper preparation as well as errors in training (warm-up, stretching, program progression) are causative factors.

Subjective: Symptoms
Pain in a specific area that persists during and after exercise; history of increased activity in a new program; or a specific event, such as a long run, which significantly exceeds previous training.

Objective: Signs
Using Basic Tools: Point tenderness with palpation; (i.e., tibial stress fracture most common at junction of middle and lower thirds or middle and upper thirds of the bone); significant edema in the dorsum of the foot over metatarsal fracture; compensatory antalgic gait.
Using Advanced Tools: X-rays (if available) Initially normal but repeat study at 3-4 weeks after onset will often show slight callus formation.

Assessment:
Differential Diagnosis - metatarsal stress fracture: metatarsalgia, Freiberg's neuroma, capsulitis

Plan:
Treatment
Primary:
1. Conservative: Rest until point tenderness subsides; ice and NSAIDs.
2. Alternate exercise: Swimming or biking in place of running to maintain cardiovascular fitness. Gradually resume a running program once pain free.
3. Identify biomechanical and structural predisposing factors (i.e., tibial varum, cavus foot, flatfoot, long 2nd

or short 1st metatarsal) and treat with appropriate custom foot orthotics.
4. Short term immobilization if necessary, especially with non-compliant individuals.
Alternative: Arch supports, padding to decrease weight on specific area. A metatarsal pad or doughnut cutout will decrease weight on the metatarsal when correctly placed.
Primitive: For metatarsal stress fractures, duct tape two tongue blades transversely across boot just behind the metatarsal heads, or use other substitute material to fill arch area to get some weight off the involved metatarsal head.

Patient Education
General: Do alternate activities to maintain fitness. Return to running activities progressively after time off. Start with a walking program for one week once pain free. If still symptom-free, start running short distances the second week. Slowly increase distance and speed.
Diet: One **Tums** a day, and balanced diet with adequate calcium.
Medications: Be alert for gastritis with NSAIDs. Stay well hydrated.
Prevention: Perform proper warm up, stretching, warm down activities. Wear good-fitting, high-performance athletic shoes. Change running shoes every 3 to 6 months depending on mileage. The midsole will wear out long before the outer sole.

Follow-up Actions
Return evaluation: At 2 week intervals until released to full duty. Cannot resume full duty unless pain free. Consider immobilization
Evacuation/Consultation Criteria: Evacuation not necessary unless mission requires heavy weight-bearing or long hikes. Podiatry or orthopedic consultation recommended for stress fractures. Custom orthotics, highly recommended, especially in recurrent case.
NOTES: If stress fracture incidence or any specific injury statistically increases in any one team or unit take a closer look at the training program and cadre. Encourage troops to present early rather than suffering with the "suck it up...No pain, no gain" attitude until disabled.

Podiatry: Friction Foot Blisters
CDR Raymond Fritz, MSC, USNR & MAJ Daniel Schissel, MC, USA

Introduction: Friction blisters are a common injury in the military. Training programs subject individuals to high intensity activities, including high-mileage running and land navigation. Footwear is often new and sometimes ill fitting. Swim fins may also cause blisters. Hyperhidrosis (excessive sweating) of the feet may increase friction over pressure areas in the shoe. A high arch or cavus foot may be more susceptible to shoe rub and blister formation on the top of the foot as well as over the metatarsal head area.

Subjective: Symptoms
Sore feet, blister, history of high-level training or running

Objective: Signs
Obvious blisters over involved areas.

Assessment: Diagnosis Is based on clinical presentation.
Differential Diagnosis - genetic blister disease, epidermolysis bullosa (inherited disease in which bullae form from slight trauma), insect bite, or burn.

Plan:
Treatment:
Primary:
1. Prevent additional and future blisters.
2. Aspirate blister with a sterile needle.

3. Cleanse with **Betadine** and cover with moleskin.
4. Leave the "roof" of the blister in place to act as a biological dressing. This will decrease tenderness until new skin forms and matures in a few days.
5. Do not inject blister with benzoin (no longer the preferred method for blister treatment).
6. If infected, the blistered skin covering should be removed using a scalpel or scissors. Cleanse the area and apply a thin layer of **Neosporin** or **Bacitracin** followed by a thin non-adherent dressing. Then apply moleskin over the dressing and adjacent skin to hold everything in place.
7. **Coban** and **Elastoplast** also work well for holding dressings in place on the foot.
8. Avoid bulky coverings if an operator is in the field (in boots) and must continue with the mission.

Alternative:
Option 1: If the blister is already open as a result of repetitive irritation, the underlying skin is usually clean and red. Remove remaining loose skin, cleanse and treat open area as above. Place felt "donut" around blister to decrease pressure and irritation.
Option 2: Apply tincture of benzoin topically to toughen the skin and hold moleskin in place. Drain fluid with fine gauge needle (27-30). Apply an antibiotic ointment, a layer of DuoDerm over top and a doughnut pad to prevent rubbing. Blister roof may reattach to underlying skin, allowing rapid healing and return to duty
Primitive: Pop the blister if large and painful. Cleanse if conditions permit. Place moleskin or duct tape over the area and continue with the operation.

Patient Education
General: Continue with activities if possible. If infection or deeper ulceration develops, rest feet and eliminate pressure to allow healing.
Prevention and Hygiene:
1. Make sure boots/shoe are the right size and width (fit the larger, longer foot). Try shoes on and stand to check fit. If orthotics or other shoe devices are used, remember to try shoes on with the orthotics in them before purchasing. The longest toe should be one thumb width from the end of the shoe Try a short test run and then check your feet. Avoid wearing new boots for the first time on a field exercise.
2. Always carry extra socks in the field. Wear synthetic moisture wicking socks (i.e., polypropylene) next to the skin and wool as a second layer because it retains insulative properties when wet. Fit boots with the two-layer sock system at the time of the boot purchase. Do not wear cotton socks. Cotton retains moisture and increases the coefficient of friction. Change socks often to keep them clean and dry.
3. For hyperhidrosis, apply products such as **alum** or **Drysol** (drying prevents skin softening) to the soles of the feet three times a week as needed.
4. Apply moleskin patches to areas that previously blistered.
5. Pad hammertoes and other prominent areas with Silipos or other padding devices.
An optional method: spray clean, dry feet with **Aerozoin** (40% tincture of benzoin, 60% alcohol), let dry, then apply thin layer of **hydropel**. Use a single sock with this method.

Follow-up Actions
Return evaluation: Only if needed
Evacuation/Consultation Criteria: Evacuation is not usually necessary. Refer to dermatology or podiatry (structural foot abnormality) for evaluation of underlying foot or skin problems.

Chapter 10: Dentistry
LTC David DuBois, DC, USA

I. General Information

1. ANATOMY
   a. A tooth has two major parts: the crown, normally visible in the mouth, and the root or roots embedded in the socket and partially covered by soft tissue (Figure 5-1).
   b. The crown has five surfaces: the occlusal (biting) surface, the lingual or tongue side surface, the facial (buccal) or cheek side surface, and the two surfaces that come in contact with adjacent teeth (mesial- the contacting surface nearest the midline and distal- the farthest from the midline) (Figure 5-2).

2. SUGGESTED MINIMAL DENTAL KIT FOR THE FIELD

Surgery:
1 each: Tooth extraction forceps #150 - universal maxillary forceps
1 each: Tooth extraction forceps #151 - universal mandibular forceps
1 each: Tooth extraction forceps #17- mandibular "cowhorn" forceps
1 each: Tooth extraction forceps #53R and #53L- maxillary "cowhorn" forceps
2 each: Periosteal elevator - Woodson #1 and Molt #9
2 each: Straight elevator - #301 (small) and #34 (large)
5 each: #15 Scalpel blades
1 each: Bard Parker blade handle
5 each: 4-0 Chromic gut sutures
5 each: 4-0 Silk sutures
2 each: Dental aspirating syringe
50 each: 27 gauge dental needle
25 each: .5% **bupivacaine (Marcaine)** with 1/200,000 epinephrine anesthetic 1.8 ml carpule
10 each: 3% **mepivacaine (Carbocaine)** without epinephrine anesthetic 1.8 ml carpule
15 each: 2% **lidocaine (Xylocaine)** with 1/100,000 epinephrine anesthetic 1.8 ml carpule
1 each: Topical **benzocaine** 20%
Operative/General Dentistry:
2 each: Explorer #23
2 each: Periodontal probe
2 each: Spoon excavator
2 each: Dental spatula
2 each: Plugger, plastic filling, dental-Woodson #2
1each: Intermediate restorative material (IRM)
1 each: Glass ionomer (Ketac-fil, or Fuji IX-GP)
1 each: Cavity varnish (Copalite)
1 each: Calcium hydroxide (Dycal)
20 each: Cotton rolls and cotton gauze
1 each: mixing pad; parchment paper, dental

II. Oral and Dental Problems

1. TOOTHACHES
Toothaches are usually associated with one of the following: caries (decay); fractures of tooth, crown, or root; or acute periapical (root end) abscess. All tooth surfaces may be affected by dental decay.

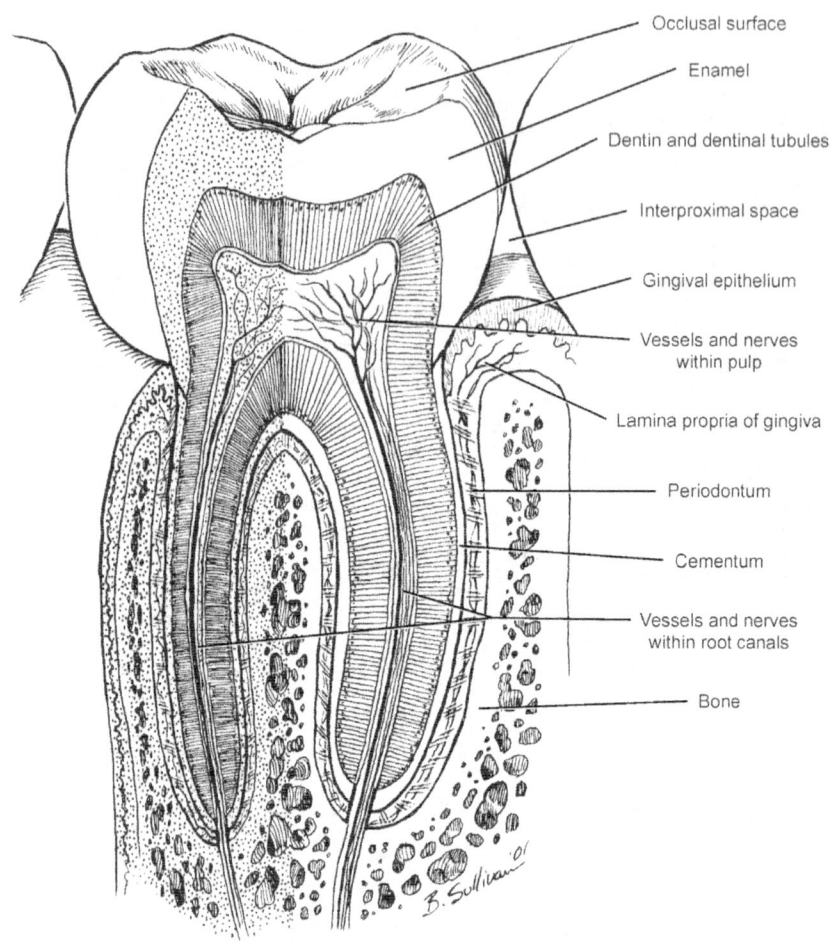

Figure 5-1
Structures of the Tooth

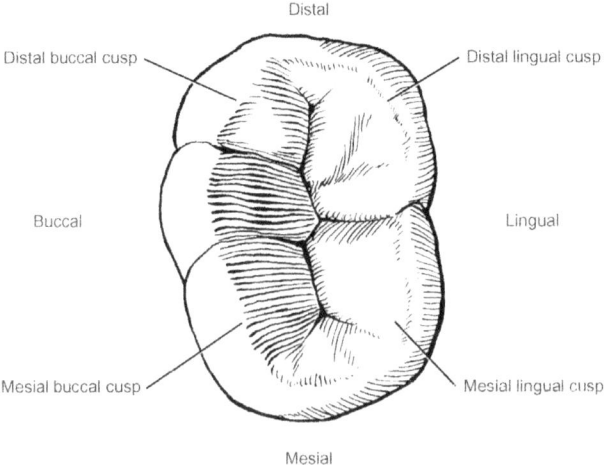

Figure 5-2

a. DENTAL CARIES ("cavities")

Subjective: Symptoms
Intermittent or continuous pain, usually intense. Heat, cold, sweet, acid, or salty substances may worsen the pain.

Objective: Signs
Finding the offending tooth may be difficult, but it will usually be grossly decayed, with the carious enamel and dentin area discolored. Tooth will be tender and sensitive to heat and cold. Tapping the tooth with an instrument will usually elicit pain. When conducting a thermal test, use a normal tooth as a basis for comparison. (see Thermal Test in Dental Procedures section). Check vitality: pain upon touching dentin indicates vitality. A vital tooth will give a painful response to cold.

Assessment:
Differential Diagnosis - caries in vital tooth versus dead tooth.

Plan:
Primary: Remove caries and place a temporary restoration. Local anesthetic may be necessary before applying a temporary restoration. (See Procedures for Dental Anesthesia and for Temporary Restorations.)
Alternate: *For teeth that are still vital:* **Eugenol** (IRM liquid) is an agent that will temporarily soothe hyperemic pulp tissue if treated indirectly (if not in direct contact with the pulp). If a mix of **zinc oxide** and **eugenol** is applied directly to vital pulp, it will kill the pulp. A dental officer must give definitive care in the near future.
Patient Education: Do not chew on the treated tooth.

b. TOOTH/CROWN FRACTURES - anterior (front) teeth are particularly susceptible to injuries that result in fracture of the crown.

Subjective: Symptoms
History of trauma or biting hard object; feels jagged tooth edge; finds tooth fragment; sensitivity to heat/cold.

Objective: Signs
Visibly broken or cracked tooth.

Assessment:
Look for other injury if related to trauma.

Plan:
Simple fractures of the crown involving little or no dentin, smooth the rough edges of the tooth with an emery board or small flat file.
Extensive fractures of the crown involving considerable dentin but not the pulp:
1. Wash the tooth with warm saline,
2. Isolate and dry the tooth with cotton gauze or rolls.
3. Then cover the exposed dentin using one of several methods:
   a. Cover the exposed dentin with **(IRM) zinc oxide-eugenol** paste (it is difficult to achieve retention on anterior fractures). An aluminum crown, trimmed and contoured to avoid lacerating the gingiva, can be filled with this paste and placed over the tooth.
   b. Incorporate cotton fibers into a mix of zinc oxide and eugenol (the fibers give additional strength) and place this over the involved tooth, using the adjacent teeth and the spaces between them for retention. Have the patient bite to be sure neither the bands or the "splint" interferes with bringing the teeth together.
   c. Glass ionomer cement can be used as a substitute for IRM. It has the advantage of readily bonding to teeth.

d. If a glass ionomer cement was not used, cover the calcium hydroxide or zinc-eugenol base and adjacent enamel with several coats of cavity varnish **(Copalite)**. Cavity varnish has low solubility in oral fluids.
4. This can provide protection for up to 6 weeks. Have the patient see a dentist as soon as possible.

Extensive fractures involving the dentin and exposed pulp:
1. Anesthetize the tooth. (See Procedures section.)
2. Wash gently with warm saline.
3. Isolate and dry the tooth with cotton gauze or rolls.
4. Cover the pulp and dentin with a mix of **(Dycal)** calcium hydroxide (DO NOT USE ZINC OXIDE AND EUGENOL-IT CAUSES NECROSIS OF THE PULP), and allow to harden. Apply several coats of cavity varnish to the calcium hydroxide base.
5. If a glass ionomer cement is available, it can be substituted for the (Dycal) calcium hydroxide. A condensable type of glass ionomer is preferable. Do not coat a glass ionomer cement with cavity varnish.
6. The efficiency of this treatment regimen depends on the size of the pulp exposure. If the exposure is larger than 1.5 mm. consider extraction. If all you have available is **zinc oxide** and **eugenol** you must also consider extraction.
7. Evacuate for consultant care.

c. ACUTE PERIAPICAL (ROOT END) ABSCESS

Subjective: Symptoms
Repeated episodes of pain that have gradually become more continuous and intense. The accumulating pus causes increased pressure and the tooth will feel "high" to the patient. It will seem to be the first tooth to strike when the teeth are brought together. Malaise and anorexia are sometimes noted.

Objective: Signs
Severe tooth pain on percussion* (very significant); swollen, tender gingival tissues around the tooth; fever; bright red elevation of the soft tissues in the area (parulis) due to untreated periapical abscess burrowing through alveolar bone.
*NOTE: *Always* begin percussion testing on a tooth that appears normal, then progress to the suspected tooth.

Assessment:
Pain on percussion of posterior maxillary teeth may indicate sinusitis.

Plan:
Drainage usually provides immediate relief from pain. (See Procedures for abscess drainage.) If abscess is severe, consider antibiotics only if fever, malaise and anorexia are present. (See Dental Antibiotics section.) Extracting the offending tooth should be a last resort. Evacuate for consultant care.

d. UNTREATED ACUTE PERIAPICAL ABSCESS
The typical course: Accumulation of pus and destruction of bone at the root end of the tooth, invasion of the marrow spaces and destruction of bone trabeculae, destruction of the cortex and displacement of the periosteum by suppurative material (subperiosteal abscess), rupture of the periosteum with resulting gingival swelling (gum boil or parulis) and finally spontaneous drainage by rupture of the parulis.

Subjective: Symptoms
Various presentations, depending on direction of spread of the abscess, which is usually toward the lateral aspect of the jaw, but may drain into the palate, mouth (rare), tongue or facial skin. Tongue infection can spread through the facial spaces of the neck, and grave, possibly fatal complications (e.g., Ludwig's angina) may result.

Objective: Signs
Erythema, swelling with fluctuance, breakthrough and purulent drainage; fever, malaise and anorexia are common.

Assessment:
Rapid progression of an oncologic process must be included in the differential diagnosis.

Plan:
1. Extract tooth after the acute symptoms subside. Positively identify the involved tooth.
2. Drain abscess surgically even with spontaneous drainage. Otherwise the infection will recur, especially during periods of lowered resistance.
3. Administer antibiotics for several days (See Dental Antibiotics section.)
4. Evacuate for consultant care.

e. LUXATED (DISLOCATED) TOOTH - a tooth moved from normal position.

Subjective: Symptoms
History of trauma or biting into hard object.

Objective: Signs
Visibly malpositioned tooth.

Assessment:
Look for other injury if related to trauma.

Plan:
Administer local anesthetic (See Procedure section.). Manually reposition tooth into normal occlusal scheme. Stabilize tooth with gentle pressure during splinting procedure. Splint to adjacent teeth with wire, heavy monofilament fishing line or an IRM/cotton fiber splint or glass ionomer splint (see Tooth/Crown Fractures).

f. AVULSED TOOTH - tooth completely removed from socket.

Subjective: Symptoms
History of trauma or severe dental caries; may present with tooth in hand.

Objective: Signs
Visible space or empty socket.

Assessment:
Look for other injury if related to trauma.

Plan:
1. If tooth has been saved, transport avulsed tooth in any clean liquid medium (saline, milk, and saliva). Do not let tooth dry out. Gently rinse tooth with 0.9% normal saline. Do not scrape off any debris or attempt to scale the tooth.
2. Administer local anesthetic. (See Procedure section.)
3. Replace tooth into its socket.
4. If blood clot prevents tooth placement, rinse socket with saline to remove blood clot.
5. Splint the tooth to adjacent teeth with dental wires, heavy monofilament, IRM/cotton fiber splint or a glass ionomer splint (see Tooth/Crown Fractures).
6. Provide pain relief. Use **acetaminophen** or **ibuprofen** for mild pain. Use **acetaminophen with codeine** for more severe pain. CAUTION: Do not use aspirin products if excessive bleeding is noted. (See Procedure section.).

7. Administer antibiotic regimen. (See Dental Antibiotics later in this chapter.)
8. Evacuate for consultant care.

NOTES: A partially avulsed tooth that is repositioned is usually permanently retained. A completely avulsed tooth may be permanently retained if replaced in the socket with minimal handling in less than one hour. When the replacement time exceeds one hour, the long-term retention rate drops and root resorption usually occurs.

2. PERIODONTAL ABSCESS - this acute suppurative process occurs in the periodontal tissues alongside the root of a tooth and involves the alveolar bone, periodontal ligament and gingival tissues. It most often is due to irritation from a foreign body, subgingival calculus (tartar, hard calcium deposits on the teeth) or local trauma, and subsequent bacterial invasion of the periodontal tissues.

Subjective: Symptoms
Deep, throbbing, well-localized pain of the soft tissues surrounding the tooth; tooth feels elevated in its socket.

Objective: Signs
Redness, tenderness and swelling of the surrounding gingiva; sensitivity to percussion; mobile tooth; cervical lymphadenopathy; fever; purulent exudate.

Assessment:
Differential Diagnosis - chronic apical abscess, necrotic pulp

Plan:
1. Carefully probe and drain the gingival crevice and locate any foreign body.
2. Spread the tissues gently and irrigate with warm water to remove remaining pus or debris from the abscess area.
3. Remove any foreign bodies.
4. Instruct the patient to use a hot saline mouth rinse hourly.
5. Administer antibiotic regimen if systemic conditions are present (elevated temperature, general malaise).

3. ACUTE NECROTIZING ULCERATIVE GINGIVITIS (Vincent's infection, trench mouth).
Necrotic gingival lesions result from ordinarily harmless surface parasites exposed to an altered environment. Virulent fusospirochetal organisms have been implicated, but the precise cause has not been proven. General health, diet, fatigue, stress, and lack of oral hygiene are the most important precipitating factors. This disease is not considered to be transmissible. Untreated lesions are destructive with progressive involvement of the gingival tissues and underlying structures.

Subjective: Symptoms
Constant gnawing pain, marked gingival sensitivity and hemorrhage, fetid odor, foul metallic taste, general malaise and anorexia.

Objective: Signs
Necrosis, ulcers with pseudomembrane cover, cervical lymphadenitis, fever. Advanced cases involve gingival tissues and underlying structures.

Assessment:
Differential Diagnosis - blood dyscrasias or vitamin deficiencies (scurvy), HIV-related periodontitis.

Plan:
Establish good oral hygiene in acute cases by following these steps:
1. First day: Wearing surgical or exam gloves if possible, swab the teeth and gingiva thoroughly with a 1:1 aqueous solution of 3% hydrogen peroxide on a cotton-tipped applicator twice. Instruct the patient to rinse

his mouth at hourly intervals with this same 1:1 solution. Issue the patient one pint of hydrogen peroxide. Caution him not to use this treatment for more than 2 days (due to possibility of precipitating a fungal infection). Place the patient on an adequate soft diet and advise a copious fluid intake. Adequate rest, food, and fluid are critical. Analgesics can be administered. Have patient return in 24 hours.
2. Second to third day: Patient will be much more comfortable. Using a soft toothbrush soaked first in hot water, clean the patient's teeth without touching the gingiva. Maintain the hourly hydrogen peroxide mouthwash regimen and have patient brush with a soft toothbrush soaked in hot water every hour. Have patient return in 24 hours.
3. Third to fourth day: Patient is essentially free of pain. Clean patient's teeth as before. Floss between all teeth. Discontinue hydrogen peroxide mouthwash regimen. May initiate **chlorhexidine (Peridex)** rinses twice daily for the next 4 to 5 days. Have patient brush 3-4 times a day and floss once a day.
4. After treatment the acute form subsides and the chronic phase ensues. Although clinical symptoms are minimal, tissue destruction continues until further corrective measures are completed. Definitive care consists of cleaning and scaling of the teeth, instruction in oral hygiene and, in some cases, re-contouring the tissues involved in the infection.

NOTES: Unless the patient develops systemic involvement, antibiotic therapy (including lozenges) should not be instituted. As in other oral disorders, the use of silver nitrate or other caustics is definitely contraindicated. Any case of gingivitis that does not respond well within 24 to 48 hours should be referred for evaluation for underlying blood dyscrasias or vitamin deficiencies.

4. HERPETIC LESIONS (COLD SORES, FEVER BLISTERS)
Predisposing factors include emotional stress, the common cold and other upper respiratory infections, gastrointestinal disorders, nutritional deficiencies, food allergies, and traumatic injuries to the oral mucosa. In females, menstruation and pregnancy often seem to trigger this process. The herpetic lesion is highly contagious.

Subjective: Symptoms
Intense pain, itching, burning; in children: greater pain, larger affected area, anorexia, dehydration.

Objective: Signs
Small, localized ulcerations (few blisters in mouth) with a bright red, flat or slightly raised border; later, ulcer covered by white plaque; generalized infections produce large area of fiery red, swollen, and extremely painful mucosa; children have more extensive and serious oral involvement resulting in anorexia and dehydration. (see Color Plates Picture 10)

Assessment:
Differential Diagnosis - herpes zoster (shingles), oral syphilis, burns, erythema multiforme (Stevens-Johnson Syndrome)

Plan:
1. Prevent spread of virus to eyes, fingers or other people.
2. Treat precipitating factors.
3. Antibiotics may prevent secondary infection in severe cases.
4. Force fluids to prevent dehydration.
5. Do not use topical steroids.
6. 5% **acyclovir** ointment can be applied to infected areas every three hours for the first few days. Topical anesthetic (20% **benzocaine**) can also be applied to reduce discomfort.

5. APHTHOUS ULCERS (CANKER SORES)
Aphthous ulcers are common manifestations of various systemic diseases such as Bechet's syndrome, HIV, autoimmune disorders, and Crohn's disease. They are not contagious or caused by an infectious agent, and will heal in 1-2 weeks without sequelae. Stress, acidic foods, and chemical sensitivities may trigger an attack.

Subjective: Symptoms
Burning, itching, or stinging.

Objective: Signs
Macule progressing to small ulcer surrounded by reddish halo (bull's eye) on loose tissues of the mouth (i.e., cheek, inner lip, tongue, soft palate, and floor of the mouth).

Assessment:
Differential Diagnosis - herpetic lesions. Traumatic ulcer. May be seen in more serious systemic and local diseases including Sutton's disease, Beçhet's disease, Reiter's syndrome, leukopenias, Crohn's disease and ulcerative colitis (see index).

Plan:
Apply topical steroids (**Kenalog in Orabase** gel, **Lidex** gel, **Decadron** rinses) to reduce pain and duration of lesions. Address underlying disorders and/or avoid triggers. Maintain healthy diet, ample fluids, and adequate rest.

6. PERICORONITIS
This is the acute inflammation of tissue flaps over partially erupted teeth, caused by trauma from opposing teeth, food and debris, and bacteria. Commonly seen with erupting wisdom teeth.

Subjective: Symptoms
Marked pain radiating to the ear, throat and the floor of the mouth; fever; general malaise; muscle spasm in jaw.

Objective: Signs
Red, swollen, tender, suppurative gums localized over tooth; fever; cervical lymphadenopathy; trismus of the masticator muscles.

Assessment:
Differential Diagnosis - periapical abscess, trauma from opposing tooth.

Plan:
1. Wrap the tip of a blunt instrument with a wisp of cotton. Dip the cotton in 3% peroxide and carefully clean the debris from beneath the tissue flap; pus may be released.
2. Flush the area using warm saline solution.
3. Instruct the patient to use a hot saline mouth rinse hourly.
4. Prescribe an adequate soft diet.
5. Repeat this treatment at daily intervals until the inflammation subsides.
6. Stress that oral hygiene must be maintained.
7. Extract the offending tooth if necessary. Extract the opposing molar if the inflammation does not subside.
8. Initiate antibiotic therapy if there is involvement of the cervical nodes, fever, and/or trismus of the masticator muscle.

CANDIDIASIS (THRUSH) See Infectious Disease chapter.

7. LOCALIZED OSTEITIS (DRY SOCKET)
Caused by breakdown and/or loss of blood clot at the site of an extracted tooth. Increases in frequency with patient age (>25 years old) and tobacco use.

Subjective: Symptoms
Constant moderate to severe pain, may involve entire side of mandible; occurs 3 to 5 days after extraction

of lower molar; bad taste in mouth

Objective: Signs
No fever, purulent exudates or other signs of systemic infection; visible open wound without clot

Assessment:
Differential Diagnosis - abscess, trauma, osteomyelitis (systemic signs)

Plan:
Antibiotics are rarely indicated.
1. Local anesthesia
2. Irrigation of extraction site with warm saline in syringe
3. Cut approximately 1 ½- 2 inch strip of iodoform gauze and apply 1 to 2 drops of (IRM liquid) eugenol. Place this eugenol/gauze gently into affected socket
4. Change medicated gauze daily until symptoms are gone (usually 3 to 5 days)
5. May also administer a NSAID for pain

Barodontalgia (Toothache induced with change in pressure.) See Aerospace Medicine section.

Injuries of The Jaw See Trauma chapter on CD-ROM: Maxillofacial Trauma.

8. DISLOCATION OF THE TEMPOROMANDIBULAR JOINT(S)
COL Glenn Reside, DC, USA

Dislocation of the Temporomandibular Joint (TMJ) occurs frequently, usually caused by mandibular hypermobility but can also be caused by trauma. The mandibular condyle translocates anteriorly in front of the articular eminence and becomes locked in that position. Muscle spasm may then prevent the patient from closing the jaw into normal occlusion. Dislocation may be unilateral or bilateral and may occur spontaneously after opening the mouth widely while yawning, eating, or during a dental procedure. Dislocation of the TMJ is usually painful, is not self-reducing, and usually requires professional management. Subluxation is a displacement of the condyle that is self-reducing and requires no medical management. Dislocations should be reduced as soon as possible.

Subjective: Symptoms
Pain in the preauricular area, accompanied by drooling and difficulty talking

Objective: Signs
Tender, bony prominence in preauricular area(s); malaligned dental bite; mouth locked in an open position

Assessment:
Differential Diagnosis - fractured mandible, tumor

Plan:
Treatment
Primary: Manually reduce the dislocated TMJ as soon as possible by moving the dislocated mandibular condyle inferiorly and posteriorly from in front of the articular eminence and then superiorly into the glenoid fossa.
1. Get behind the patient. If the patient is sitting in a chair, their head should be at the level of your waist. If the patient is lying down on their back on the ground, you should sit on the ground at the patient's head and cradle their head in your lap.
2. Wrap your thumbs in gauze or a small towel (to avoid being bitten) and put them inside the patient's mouth behind the last molar on each side of the lower jaw. All four fingers of each hand should be placed along the inferior border of the mandible with the forefingers near the chin and the little fingers near the angles of the mandible. Stabilize the patient's head against the back of the chair or your torso if the patient is sitting or in your lap if the patient is supine.
3. Apply firm, continuously increasing pressure downward (inferiorly) and backwards (posteriorly) on the

retromolar area with the thumbs and upward pressure on the chin with your forefingers. As the condyle passes the crest of the TMJ eminence it will slide easily into proper position in the glenoid (mandibular) fossa.
4. After the dislocation is reduced it is imperative to maintain pressure on the chin to hold the teeth together because the patient frequently will reflexively open their mouth and dislocate again. The muscles of mastication are frequently in spasm while the mandible is dislocated and will forcefully close the mouth when the condyle is replaced into the glenoid fossa. Protect your thumbs from being bitten.

Alternate: If the muscles of mastication are in such spasm that you cannot manually manipulate the condyle, sedate the patient with a muscle relaxant such as **diazepam (Valium)** or **midazolam (Versed)** prior to reducing the dislocation. In extremely rare cases the patient may require general anesthesia in conjunction with the reduction. If the patient tends to reflexively open the mouth and cause repeat dislocations, apply a dressing over the top of the head and under the jaw to inhibit the motion of the mandible.

Primitive: Warm, moist heat to the sides of the face to relax the temporalis and masseter muscles and allow the patient to reduce the dislocation by himself.

Patient Education
General: Activity: Avoid trauma to the mandible.
Diet: Avoid opening mouth too wide while chewing. Cut food into small pieces. Chew a soft diet until the preauricular tenderness resolves.
Medications: Over-the-counter analgesics, such as **acetaminophen** or **ibuprofen**, can be used.
Prevention and Hygiene: Avoid opening mouth wide while yawning, talking, eating, etc.
No Improvement/Deterioration: Refer to an oral and maxillofacial surgeon.

Follow-up Actions
Evacuation/Consultation Criteria: Evacuation is not usually necessary. If you are unable to achieve a proper reduction or if the patient presents on multiple occasions requiring emergent reduction, refer the patient to an oral and maxillofacial surgeon for evaluation and treatment.

III. DENTAL ANTIBIOTICS

A. Administer oral doses for 7-10 days duration. Administer IV antibiotics as needed and switch to oral drugs based on these criteria:
   1. Intercisal opening (distance between upper and lower front teeth when mouth is open) greater than 20mm
   2. Normal temperature
   3. Patient can eat a normal/acceptable diet
   4. No fluctuant area at affected site
B. Oral dosages (in order of preference):
   1. **PEN VK**
      a. Adults - normal loading dose is 1 gm then followed by 500 mg q 6 hours
      b. Children (< 25kg) 500 mg followed by 250 mg q 6 hours
   2. **Erythromycin**
      a. Adults - 1gm normal loading dose then followed by 500 mg q 6 hours
      b. Children - (<25kg) 500 mg followed by 250 mg q 6 hours
   3. **Clindamycin (Cleocin)**
      a. Adults - 600 mg loading dose followed by 300 mg q 6 hours
      b. Children - (<25kg) 300 mg followed by 150 mg q 6 hours
C. IV Dosages (in order of preference):
   1. **Aqueous PEN G**
      a. Adults - 2 million units q 4-6 hours

b. Children (<50kg) 100,000-300,000 units/kg/day q 6 hours, not to exceed adult dosage
2. **Cefadyl**
   a. Adults - 1gm q 6 hours
   b. Children (<50kg) 40-80mg/kg/day q 6 hours, not to exceed adult dosage
3. **Clindamycin** (Cleocin)
   a. Adults - 300-600 mg q 6 hours
   b. Children (<50kg) 150-300mg q 6 hours

IV. DENTAL PROCEDURES

1. THERMAL TEST FOR CARIES. If hot and cold are used, test a normal tooth as a basis for comparison. The application of cold to normal teeth elicits pain in most instances, but the response ceases soon after the stimulus is removed. A diseased tooth, compared to a normal tooth, varies in its reaction to the temperature test. For example, a reaction to cold could persist after application stops but the tooth responds very little to heat, or the reaction to heat persists after application but the tooth appears to respond very little or not at all to cold. Perform procedure as follows:
   a. Isolate the teeth to be tested from the saliva with gauze packs.
   b. Cold test: spray a cotton-tipped applicator with ethyl chloride (if available) then place the cold surface on the tooth crown (ice may also be used). Note the response and its duration. A vital tooth will give a painful response to cold that abates after the cold stimulus is removed.
   c. Heat test: heat an instrument (e.g., a mouth mirror handle) and touch against the tooth. Note the response and duration. Similar pattern to cold test.
   d. Test an unsuspected tooth for comparison.

2. DENTAL ANESTHESIA
   a. Anesthetic solutions:
      1. 2% **lidocaine** with 1/100,000 **epinephrine (Xylocaine)**
         (1) 1.8 ml carpule with red/ black lines
         (2) Duration of 2-4 hours
         (3) Max dose: 8 carpules for healthy 70 kg adult
         (4) Max dose: 2-3 carpules for child under age 12 (<50 kg)
      2. .5% **bupivacaine** with 1/200,000 **epinephrine (Marcaine)**
         (1) 1.8 ml carpule with blue/black lines
         (2) Duration of 6-8 hours
         (3) Max dose: 10 carpules for healthy 70 kg adult
         (4) Max dose: 2 to 3 carpules for child under age 12 (<50 kg)
      3. 3% **mepivacaine without epinephrine (Polocaine** or **Carbocaine)**
         (1) 1.8 ml carpule with tan/black lines
         (2) Duration of ½ to 1 hour
         (3) Max dose: 6 carpules for healthy 70 kg adult
         (4) Max dose: 1 to 2 carpules for child under age 12 (<50 kg)

NOTE: Anesthetic agents with vasoconstrictors (**epinephrine** or **neocobefrin**) should never be used in areas with compromised blood supply: fingers, toes, ears, nose and penis.
   b. Anesthesia for maxillary extraction.
      1. Infiltration of anesthetic will provide adequate analgesia of the maxillary teeth. (Analgesia – block pain impulses; Anesthesia – block all nerve impulses.)
      2. Technique. Both facial and palatal injections must be given.
         (a) Facial injection (see Figure 5-3).
            (1) Insert the needle into the mucobuccal fold directly above the tooth. Shaking the cheek and lip during needle insertion can help alleviate injection discomfort.
            (2) Advance the needle upward about three-eighths of an inch until the needle gently contacts bone (this should approximate the root end).
            (3) Aspirate to insure that the needle has not entered a blood vessel.
            (4) Slowly deposit three-fourths of the carpule's contents.

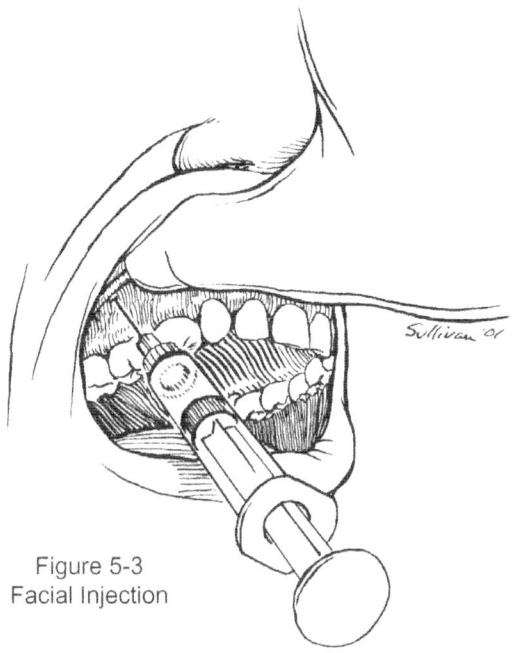

Figure 5-3
Facial Injection

    (b) Palatal injection (see Figure 5-4).
        (1) Apply firm pressure to palatal injection site with a finger or mirror handle for 3 to 4 seconds. This helps to minimize needle insertion discomfort.
        (2) Insert the needle one-half of an inch above the gingival (gum) margin of the tooth.
        (3) Deposit a small amount (< ¼ carpule) of anesthetic solution – **DO NOT BALLOON THE TISSUE.**

NOTE: The palatal injection is very painful.

  c. Anesthesia for mandibular extraction.
    1. Block the inferior alveolar nerve as it exits the mandibular foramen on the medial aspect of the ramus of the mandible. This foramen is located midway between the anterior and posterior borders of the ramus and approximately one-half inch above the biting surface of the lower molars. Estimate the width of the ramus at this level by placing the thumb on the anterior surface of the ramus (intraorally) and the index finger on the posterior surface extraorally. The inferior alveolar and lingual nerves are anesthetized by a single injection. (see Figure 5-5)
        (1) Place the index finger on the biting surface of the lower molars so that the ball of the finger will contact the anterior border of the ramus. The fingernail will then be parallel to the midline.
        (2) Place the barrel of the syringe on the lower bicuspids on the side opposite of the side to be anesthetized.
        (3) Insert the needle into the tissue of the side to be anesthetized in the apex of the V-shaped, soft tissue depression about one-half of an inch ahead of the tip of the finger on a line horizontally bisecting the fingernail.
        (4) Advance the needle to contact the medial surface of the ramus. A 1-inch soft tissue penetration will usually suffice to position the needle point in the area of the mandibular foramen.
        (5) Slowly deposit approximately two-thirds of the cartridge contents.
        (6) Swing the barrel of the syringe to the side of the mouth being injected (leaving the needle in the position described in (4) above) and inject the rest of the cartridge contents while withdrawing

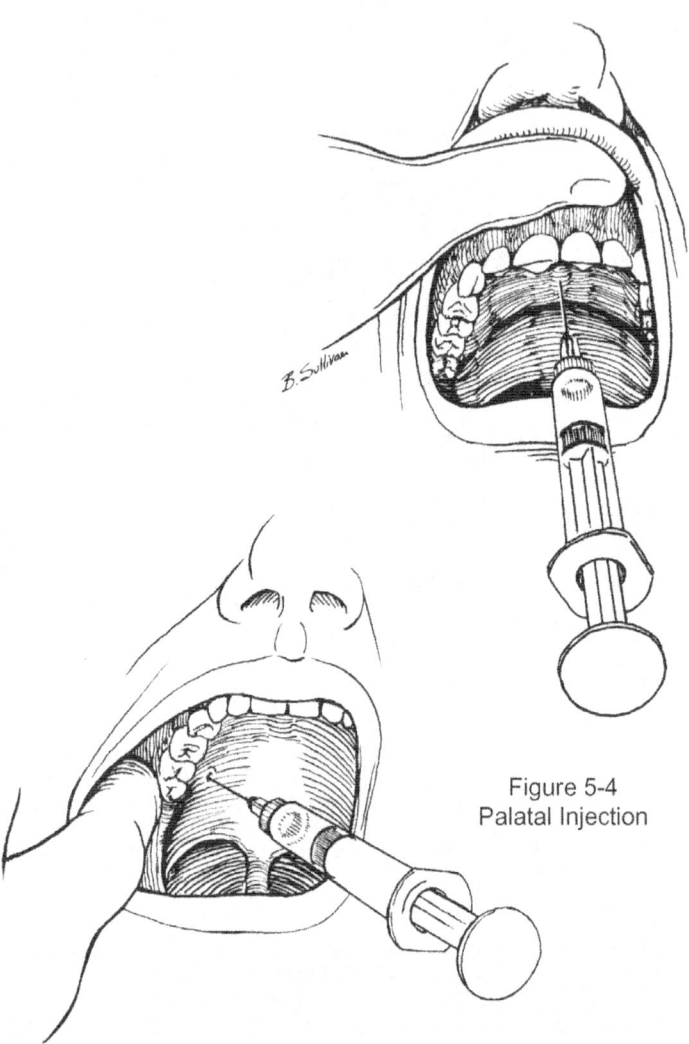

Figure 5-4
Palatal Injection

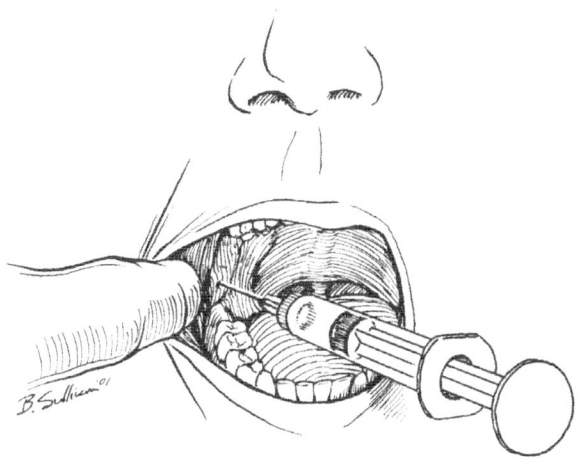

Figure 5-5
Blockade of the Inferior Alveolar and Lingual Nerve

    the needle. This should anesthetize the lingual nerve.
2. Finish anesthesia of the area with a long buccal injection (see Figure 5-6). Insert the needle in the mucobuccal fold at a point just anterior to the first molar. Gently pass the needle, held parallel to the body of the mandible, with the bevel down, to a point as far back as the third molar, depositing the solution slowly while advancing the needle through the tissue.
3. After a 5-minute interval, evaluate the results of the injections by checking the following symptoms:
   (a) Inferior alveolar nerve (supplies lower teeth, alveolar bone up to the midline).
      (1) Swelling and numbness extending to the midline of the lower lip on the injected side.
      (2) Numbness of the facial gingival tissue extending to the midline on the injected side.
   (b) Lingual nerve.
      (1) Swelling and numbness extending to midline of the tongue.
      (2) Numbness of the lingual gingival tissue extending to the midline.
   (c) DO NOT ATTEMPT EXTRACTION UNTIL FINDING THE SIGNS DESCRIBED ABOVE.

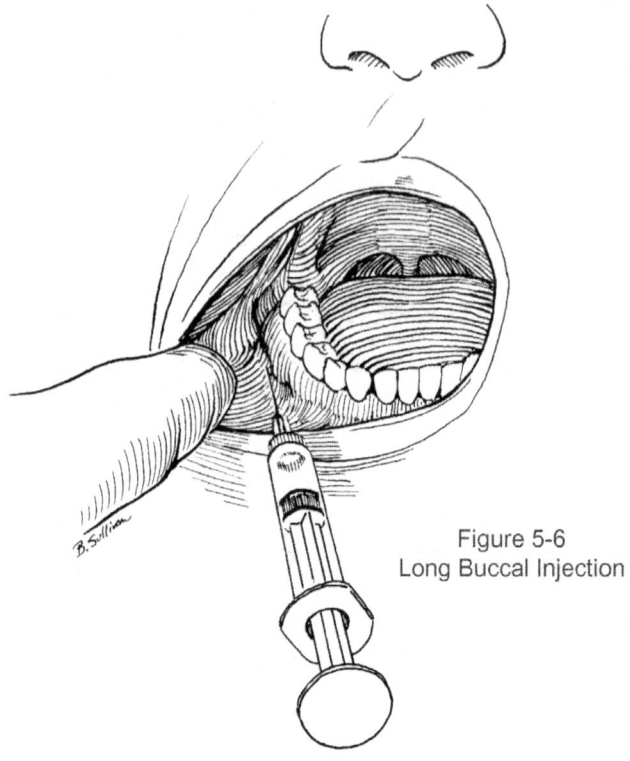

Figure 5-6
Long Buccal Injection

3. TEMPORARY RESTORATIONS
a. Remove as much of the soft decayed material as possible with a spoon-shaped instrument. If the patient is properly anesthetized he should feel no pain.
b. Irrigate the cavity with warm water until loose debris has is flushed out.
c. Isolate the tooth with gauze packs and gently dry the cavity with cotton pellets.
d. Mix the intermediate restorative material (IRM) zinc oxide powder with two or three drops of (IRM liquid) **eugenol** on a clean dry surface (parchment pad) until a thick puttylike mix is obtained. Adding a drop of water to the mixture will quicken setting.
e. Fill the cavity with the IRM putty, tamping it gently (use the Woodson Plastic Instrument #2 or #3 or a moistened cotton tip applicator).
f. Have the patient bite several times to compress the putty, and to avoid malocclusal problems with opposite teeth when dry.
g. Remove surplus filling material by lightly rubbing the tooth with a moist cotton pellet.
h. The pain should disappear in a few minutes and the putty will harden within 5-10 minutes. Caution the patient not to chew on the treated tooth.
i. If IRM is not available, a cotton pellet impregnated with **eugenol** may be left in the cavity. Glass ionomer cement is an excellent substitute for the IRM. A condensable glass ionomer is preferable, but any type will work. Glass ionomer can be placed directly against exposed pulpal tissue.
j. Instruct the patient that the procedure is temporary and a dentist must give definitive care.

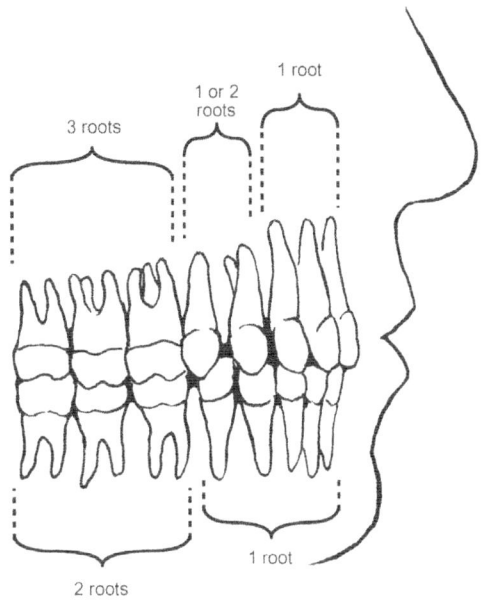

Figure 5-7
Associated Roots

4. TOOTH EXTRACTION

a. This section describes only one extraction technique. Although many types of extraction forceps are manufactured, the removal of any erupted tooth can usually be done with one of two instruments: The Maxillary Universal Forceps (150) or the Mandibular Universal Forceps (151).

b. Technique.
  1. Break the attachment of the gingival tissue to the tooth by forcing a blunt instrument (Periosteal Elevator, Woodson Plastic Instrument, etc.) into the crevice between the tooth and the gingiva, all the way around the tooth. The tooth-tissue attachment should be broken to the level of the alveolar bone.
  2. Use the free hand to guide the beaks of the forceps under the gingival margin on the facial and lingual aspects of the tooth and to support the alveolar process. Apply pressure toward the root of the tooth to force the tips of the forceps as far down on the root as possible.
  3. To loosen teeth with more than one root (molars and upper first bicuspids): slowly rock the tooth with progressively increasing traction in a facial-lingual direction. To loosen single-rooted teeth combine this rocking motion with rotation. (see Figure 5-7)
  4. When the tooth is loose, deliver it by exerting gentle traction. Note the direction in which the tooth moves most easily and follow this path for delivery.
  5. Inspect the extracted tooth to determine if the roots have been fractured.
  6. After the extraction, compress the sides of the empty socket (this repositions the bone that has

been sprung by the extraction fold) and place a folded dampened sponge or 2 x 2 over the wound. Instruct the patient to maintain light biting pressure on this compress for 60 minutes. Repeat if necessary to control hemorrhage. Caution the patient NOT TO RINSE the mouth for at least 12 hours since this may disturb the clot.

5. DRAINING A TOOTH ABSCESS
Two methods may be used to accomplish adequate drainage:

   a. For a "pointed" abscess, incise the fluctuant area using a stab procedure. Blunt dissection with a hemostat may help to establish drainage at the site. Suture a small drain or slice of surgical tubing in the wound to maintain drainage and leave for 2-3 days. Local anesthetic should be used but pain control may not be easy to obtain.

   b. If decay is evident, establish drainage through the tooth. Stabilize the tooth firmly with the fingers; remove the soft decay with a spoon-shaped instrument until an opening into the pulp chamber is made. Finger pressure on the gingiva near the root of the tooth should force pus out through the chamber opening.

6. PRESERVING/TRANSPORTING AN AVULSED TOOTH
If tooth has been saved, transport the avulsed tooth in any clean, liquid medium (saline, milk, and saliva). Do not let tooth dry out. Gently rinse tooth with 0.9% normal saline. Do not scrape off any debris or attempt to scale the tooth. A completely avulsed tooth may be permanently retained if replaced in the socket with minimal handling in less than one hour.

Chapter 11: Sexually Transmitted Diseases (STD)
STD: Urethral Discharges
COL Naomi Aronson, MC, USA

See Gynecology and Male Genital Inflammation sections for more detail on STDs.

Introduction: Gonorrhea and chlamydia both present with urethral discharges. Patients are often co-infected with both. *Neisseria gonorrhoeae* causes gonorrhea and appears on Gram's stain as a clump of gram-negative intracellular diplococci. Asymptomatic infection can occur especially in the cervix, rectum and oropharynx. Disseminated gonorrhea presents with infectious arthritis, tenosynovitis, and a characteristic gunmetal blue skin lesion surrounded by a red halo, usually on the extremities (arthritis-dermatitis syndrome). The incubation period is 2-14 days after exposure. Most nonspecific or non-gonococcal urethritis (NGU) and cervicitis is caused by *Chlamydia trachomatis*, but Ureaplasma, genital Mycoplasma, Trichomonas and Herpes simplex are also implicated. These are some of the most common sexually transmitted diseases in the U.S. and a leading cause of female infertility. Infant eye and lung infections are consequent to maternal genital infection with Chlamydia. Incubation period is 7-12 days after exposure. A thick mucus discharge with pain on urination and genital ulcer should suggest Herpes simplex. Incubation period is 2-12 days after exposure.

Subjective: Symptoms
Male: **Acute (< 3 days):** Dysuria with discharge/without discharge- usually NGU **Chronic (>10 days)** urethral stricture
Female: **Acute (< 3 days):** Dysuria or frequent urination, vaginal discharge, pain with intercourse, lower pelvic pain **Chronic (>10 days):** PID, infertility
Either: **Sub acute (3-10 days):** Painful joints, gun metal blue skin lesions (GC), tenosynovitis
Focused History: *Have you had unprotected sex with a new partner in the past 6 weeks?* (incubation period 2 weeks for GC; longer for NGU) *Have you had discharge stains in your underwear?* (suggests gonorrhea)
For males - *Are you having any difficulty retracting your foreskin?* (suggests phimosis with GC)

Objective: Signs
Using Basic Tools:
Inspection: **Acute (< 3 days):** Purulent yellow to green discharge (both GC and chlamydia), mucoid /scant discharge (more consistent with NGU), if oropharyngeal exposure, see red tonsils with exudate (GC), **Sub acute (3-10 days):** Infection of periurethral glands, epididymitis, if disseminated GC: red, tender swollen joints and tendon insertions, fever to 102°F.
Palpation: Pelvic exam may show cervical movement tenderness
Using Advanced Tools: Lab: Gram stain: Urethral discharge shows gram-negative intracellular diplococci (low diagnostic sensitivity in females), or many polymorphonuclear neutrophils but no organisms (NGU), pregnancy test (drug selection).

Assessment:
Differential Diagnosis - other causes of dysuria include:
Trichomonas - malodorous discharge in females
Vaginal candidiasis - pruritus, white curd-like discharge
Herpes simplex - presence of mucoid discharge with skin ulcer
Urinary tract infection - usually frequency, urgency; sometimes hematuria
The arthritis-dermatitis syndrome of disseminated gonorrhea may resemble:
Meningococcemia - GC rash on extremities has distinctive bluish lesions surrounded by erythema
Lyme disease - E. migrans rash has larger, red lesions

Plan:
Treatment: Treatment for both GC and chlamydia is recommended in an operational setting
Primary: Ceftriaxone 125 mg IM single dose **AND** azithromycin 1 gram po single dose
Alternatives: Choose one from column **A AND** one from column B:

Column A	Column B
cefixime 400 mg po single dose	**doxycycline** 100 mg po bid x 7 days
ciprofloxacin 500 mg po	**azithromycin** 1 gm po single dose
ceftriaxone 125 mg IM single dose	**erythromycin** 500 mg qid x 7 days

Or use **azithromycin** 2 gm in single dose (poor gastrointestinal tolerance) alone.
Pregnant patient: Use **ceftriaxone** or **cefixime** in a single dose AND erythromycin 500 mg po qid for 14 days. Do not use **doxycycline** in pregnant or nursing females.

Patient Education
General: Evaluate and treat recent sexual contacts
No Improvement/Deterioration: Always treat patient as if co-infected with chlamydia
Medications: Avoid taking doxycycline with antacids, milk, iron pills or multivitamins. Avoid sun exposure. Do not use doxycycline in children, or nursing or pregnant mothers.
Prevention and Hygiene: Use barrier protection (latex condoms) or abstinence for duration of treatment.
No Improvement/Deterioration: If relapse occurs, treat NGU/cervicitis for 21 days with **doxycycline** 100 mg po bid. Also evaluate and treat partner. For recurrent urethritis after treatment of patient and partner, give **metronidazole** 2 gm po in single dose and **erythromycin** 500 mg po qid for 7 days (discuss Antabuse effect of **metronidazole** and do not use during pregnancy).

Follow-up Actions
Return evaluation: Consider evaluating for HIV antibody in 4-6 weeks
Evacuation/Consultation Criteria: Evacuation is not usually required. Consult urology, gynecology, infectious disease or preventive medicine experts as needed.

STD: Genital Ulcers
COL Naomi Aronson, MC, USA

Introduction: Genital ulcers have a wide variety of sexually transmitted causes. **Chancroid**, caused by *Hemophilus ducreyi*, has been associated with an increased risk for HIV transmission and may co-exist with other causes of genital ulcer (such as herpes infection). The incubation period is 3-7 days. **Granuloma inguinale** (caused by gram-negative *Calymmatobacterium granulomatis*) causes beefy red granulomas that progress slowly but can cover the genitalia and heal slowly with scarring. Patients can spread lesions to other areas through autoinoculation. Incubation period is 1-12 weeks. **Lymphogranuloma venereum** (LGV), caused by variants of *Chlamydia trachomatis*, starts as a painless vesicle or nodule that then ulcerates and heals. Days to weeks later, regional lymph nodes become inflamed and tender. Suppuration, scarring, systemic infection, chronic elephantiasis and rectal strictures have been seen in untreated infection. Caused by *Treponema pallidum*, **syphilis** is known as the "great imitator". Left untreated, up to 1/3 of patients will develop tertiary syphilis (CNS infection, aortic aneurysm, gummas of the skin, bone and visceral organs). Syphilis is curable in all stages but treatment may yield a Jarisch-Herxheimer reaction with fever, rigors and intensification of the lesions 2-24 hours after initiating treatment. Incubation period of primary stage syphilis is 10 days to 3 months. **Herpes** is caused by Herpes simplex (HSV) type 1 and 2, both of which can cause genital ulcers. HSV 1 is usually milder and often associated with oral sores, "fever blister" (see Color Plates Picture 10), keratoconjunctivitis and encephalitis. HSV 2 causes a more severe initial episode (fever, exudative pharyngitis, toxic appearance, multiple genital lesions), and then recurs as a localized cluster of vesicles that ulcerate painfully. HSV2 is associated with aseptic meningitis and radiculitis. Vaginal delivery in pregnant women with active genital infection poses a high risk to her newborn (disseminated HSV and death). The incubation period for HSV is 2-12 days after exposure.

Geographic Associations: Herpes simplex virus and syphilis are worldwide. Chancroid is especially seen in Africa and Asia and is the most frequent cause of genital ulcer in the tropics. LGV is most prevalent in tropical and sub-tropical countries. Granuloma inguinale is most often associated with exposure in India, Australia, South Pacific, Brazil and South Africa.

Subjective: Symptoms
Painless ulcer: Syphilis, LGV, granuloma inguinale (Donovanosis)
Painful ulcer: Herpes simplex, chancroid, phagedenic ulcer (an otherwise painless ulcer secondarily infected with bacteria)
Constitutional: Acute (1-3 days): Fever (in LGV and 1st episode HSV)
Specific: Acute (1-3 days): Generally starts as papule that ulcerates, HSV: Paresthesias can precede outbreak, syphilis: rash including palms and soles **Chronic (>1 month):** LGV may be very chronic, HSV often recurs
Focused History: *Have you had this symptom before?* (HSV is often recurrent in same area) *Is the sore painful?* (see differential diagnosis) *Did the sore start out as a painless ulcer then become more painful?* (possible phagedenic ulcer) *How long has the genital sore been present?* (HSV usually resolves in ten days, syphilis in 3-6 weeks; others can be less likely to resolve without specific therapy.)

Objective: Signs
Inspection
HSV: Acute (1-3 days): Vesicles in clusters, 1st time may have fever **Sub acute (3-10 days):** Shallow painful ulcer (often multiple) **Chronic (>1 month):** In AIDS patients- huge non-healing ulcers
Chancroid: Acute (1-3 days): Papules **Sub acute (3-10 days):** Painful, shaggy edged, deep ulcers; suppurative regional adenopathy **Chronic (>1 month):** Phimosis
Granuloma Inguinale: Acute (1-3 days): Bright red, painless, satiny-surface, raised plaque, often in folds between scrotum or labia and thighs **Sub acute (3-10 days):** The border of the ulcer shows rolled edge
LGV: Acute (1-3 days): Painless vesicle or nodule that ulcerates and heals; may have fever **Sub acute (3-10 days):** Groove sign (lymph node adheres above and below inguinal ligament) **Chronic (>1 month):** Chronic elephantiasis of genitals, rectal strictures, rectal fistula

Syphilis: Acute (1-3 days): Painless ulcer with rolled border (chancre) heals without Rx in 3-6 weeks
Chronic (>1 month): Secondary syphilis: (see generalized 2-4 weeks after chancre, rash including palms and soles; can also see flat genital warts, loss of hair, patches on the mucous membranes
Tertiary or late (5-20 years) syphilis: gummas (tumor like masses in skin/bone/viscera) neurologic abnormalities- "neurosyphilis": posterior column findings such as slapped foot gait, loss of deep tendon reflexes, loss of position sense or hot/cold sense); also develop abnormal pupillary reflexes, dementia
Auscultation: **Syphilis:** In late syphilis may hear aortic regurgitation murmur
Palpation: Most patients with genital ulcers develop tender regional adenopathy
Using Advanced Tools: Lab: Gram's stain pus from bubo (enlarged fluctuant node): large numbers of small gram negative coccobacilli (see Color Plates Picture 15) in a "school of fish" pattern (chancroid); Giemsa stain edge of tissue scraping from edge of ulcer: intracytoplasmic bacilli in the macrophages (Donovan bodies) (Granuloma Inguinale); Giemsa stain of tissue scraping at base of the ulcer (Tzanck smear): observe for multinucleated giant cells (herpes); RPR (syphilis): may be negative during early stage, and should be repeated in 4-6 weeks.

Assessment: Diagnosing the cause of genital ulcer disease is mainly based on the clinical history and inspection.
Differential Diagnosis - Genital ulcers can have non-infectious etiology: fixed drug eruption (take medication history), Behcet's syndrome (recurrent symptoms with oral ulcers, conjunctivitis and uveitis), traumatic injury, malignancy. Secondary syphilis (rash) can be confused with infectious exanthems, drug reaction, Erythema multiforme. Helpful clues for syphilis are sexual history, prior healed chancre, rash on palms and soles, and absence of any skin lesions that look like targets.

Plan:
Treatment: Chancroid
Primary: Antibiotics: **Ceftriaxone** 250 mg IM single dose, needle aspirate fluctuant nodes to avoid rupture
Alternatives: Azithromycin 1 gm po single dose, **erythromycin** 500 mg po qid x 7 days or **ciprofloxacin** 500 mg po bid for 3 d
Prevention: Treat all sexual partners dating from 2 weeks prior to onset of symptoms
Return evaluation: Screen for HIV in 4-6 weeks

Treatment: LGV
Primary: Antibiotics: **Azithromycin** 500 mg po qd x 7 days or 1 gm po each week for 4 doses; drain fluctuant buboes by needle aspiration, avoid incision and drainage which can cause sinus tracts.
Alternatives: Doxycycline 100 mg po x 21-30 days, **ceftriaxone** 1 gm IM qd x 14 days, **ciprofloxacin** 750mg po qd x 21 days or **erythromycin** 1000 mg po bid x 21 days.
No Improvement/Deterioration: Drug resistant strains have been seen. Expect to see a treatment response by seven days but prolonged therapy is needed to avoid relapse.
Patient Education: Limit activity if possible during early week of antibiotics to decrease risk of strictures.

Treatment: Granuloma inguinale
Primary: Azithromycin 500 mg po qd x 7 days or 1 gm po each week for 4 doses
Alternatives: Doxycycline 100 mg po x 21-30 days, **ceftriaxone** 1 gm IM qd x 14 days, **ciprofloxacin** 750 mg po qd x 21 days or **erythromycin** 1 gm po bid x 21 days
No Improvement/Deterioration: Drug resistant strains have been seen. Expect to see a treatment response by seven days but prolonged therapy is needed to avoid relapse.

Treatment: Herpes simplex
Primary: Acyclovir 400 mg q 8 hours x 10-14 days if initial episode, for 5 days if recurrence
Alternative: Valacyclovir 1000 mg q 12 hours x 10 days (use 500 mg po qd for 5 days for recurrence), **Famciclovir** 250 mg po q 8 hours x 5-10 days (use 125 mg bid for 3-5 days for recurrence)

Patient Education: This virus can be sexually transmitted even in the absence of active lesions. Latex condoms are recommended.
Prevention and Hygiene: Health care workers should wear gloves to handle lesions to reduce risk of local inoculation to the hand (herpetic whitlow).
No Improvement/Deterioration: Resistant herpes has been described in HIV patients who have been maintained chronically on antiviral agents. Consider screen for HIV.

Treatment: Syphilis
Primary: Primary and secondary syphilis: 2.4 MU **benzathine penicillin** IM in a single dose
Neurosyphilis: **penicillin G** 2-4 MU q 4 hours IV x 10 -14 days
Latent: **benzathine penicillin** 2.4 MU q week for 3 doses
Alternative: Primary and secondary syphilis: **doxycycline** 100 mg po bid x 14 days
Neurosyphilis: **ceftriaxone** 2gm IV or IM qd for 14 days
Latent: **doxycycline** 100mg bid x 28 days

Follow-up Actions
Return evaluation: Refer for VDRL lab test at 6 and 12 months and expect titer to fall. If latent syphilis, another VDRL at 24 months is recommended. HIV test at 4-6 weeks.
NOTES: Congenital syphilis causes preterm delivery, snuffles (obstructed nasal respiration), rash, stillbirth. It can be asymptomatic initially but manifestations can include Hutchinson's teeth, saddlenose, saber shins, deafness. Suspect this if the umbilical cord is swollen and demonstrates a red/white/blue pattern like a barber pole.
Evacuation/Consultant Criteria: Evacuation is not usually required for any of these conditions in the acute phase. Consult urology, gynecology, infectious disease or preventive medicine experts as needed, particularly in chronic cases.

Molluscum Contagiosum
Information about this infection which may appear as a STD can be found in the Skin chapter under Viral Skin Infections.

Condyloma Acuminatum
Information about this infection which may appear as a STD can be found in the Skin chapter under Viral Skin Infections, Warts.

STD: Vaginal Trichomonas
MAJ Ann Friedmann, MC, USA

Introduction: *Trichomonas vaginalis*, a sexually transmitted fungus, is a common cause of abnormal vaginal discharge. The asymptomatic carrier rate in women is 10%. Sexual partners should be treated.

Subjective: Symptoms
Yellow-green discharge (may be frothy and malodorous but not usually fishy); vulvovaginal irritation and burning; dysuria.

Objective: Signs
See Symptoms: GYN Problems: Pelvic Exam and Lab Procedures: Wet Mount.
Using Basic Tools: Characteristic discharge not always present; vulva may be edematous and inflamed; redness of the cervix ("strawberry cervix"); tender vagina; no abdominal pain.
Using Advanced Tools: Lab: Vaginal pH: should be > 4.7 (test with urine dipstick); motile, flagellated trichomonads on wet prep.

Assessment:
Differential Diagnosis - bacterial vaginosis, gonorrhea or chlamydial infection, atrophic vaginitis, candida vaginitis (see Symptom GYN Problems: Vaginal Discharge Table and STD Chapter).

Plan:
Treatment
Primary: **Metronidazole** 2 gm po X 1 or **metronidazole** 500 mg po bid x 7 days (95% cure rate)
Note: Pregnancy: Oral therapy after the first trimester. Intravaginal **Metrogel** is recommended in the first trimester. If this is not available, consider vaginal **clotrimazole** or other antifungal (50% effective) if patient is very symptomatic, followed by oral **metronidazole** after the first trimester. In a mildly symptomatic patient in the first trimester of pregnancy, delay therapy until the 2nd trimester (after 12 weeks). During this time patient should abstain from intercourse. Douching is contraindicated in pregnancy.

Patient Education
General: Taking medication as directed is essential for cure. Partner must be treated with same regimen. Partner often asymptomatically carries trichomonas in the urethra.
Activity: Refrain from intercourse until treatment complete.
Diet: As tolerated
Medications: Refrain from alcohol and use of alcohol-containing products during treatment because of Antabuse-like effect (vomiting, anxiety, myalgia, etc.) with metronidazole
Prevention and Hygiene: Recommend condom use. Discuss STD risks and prevention.
No Improvement/Deterioration: Return for reevaluation.

Follow-up Actions
Return evaluation: Highly recommend GC/Chlamydia cultures - co-infection is common. Consider HIV testing.
Consultation Criteria: No improvement after first course of medication

NOTES: Always err on the conservative side - If there is any consideration of PID treat accordingly, including the partner.

Chapter 12: Zoonotic Diseases
MAJ Joseph Williamson, VC, USA and CPT William Bosworth, VC, USA

Diseases/Acquired From	Camels Llamas	Cattle	Sheep Goats	Horses	Pigs	Dogs	Birds
Diseases/Agents							
Anthrax		x	x	x	x	x	
Ascaris suum					x		
Blastomycosis						x	
Botulism					x		
Brucella canis						x	
Brucella suis					x		
Brucellosis	x	x	x	x			
Diseases/Acquired From	**Camels Llamas**	**Cattle**	**Sheep Goats**	**Horses**	**Pigs**	**Dogs**	**Birds**
Diseases/Agents							
Campylobacteriosis	x	x	x			x	x
Cheyletiellosis (Dermatitis)						x	
Chlamydia psittaci							x

Disease	1	2	3	4	5	6	7
Cryptosporidiosis	x	x	x	x	x		
Cutaneous larva migrans					x		
Demodex folliculorum					x		
Dermatophytosis					x		
Dipylidium caninum (dog tapeworm)						x	
Echinococcosis						x	
Equine morbillivirus			x				
Erysipelothrix rhusiopathiae							x
Escherichia coli (species and O157:H7)	x						
European tick-borne encephalitis	x	x					
Francisella tularensis		x				x	
Giardiasis	x	x					
Glanders			x				
Histoplasma capsulatum							x
Influenza				x			
Leptospirosis	x	x	x	x	x		
Lyme disease					x		
Mycobacterium bovis	x						
Neisseria spp					x		
Newcastle disease virus							x
Orf		x					
Pasteurella multocida						x	x
Pasteurella spp				x			
Plague	x				x		
Q-fever	x	x	x				
Rabies	x	x	x	x	x		
Rocky Mountain Spotted Fever					x		
Salmonellosis	x	x		x	x	x	x
Sarcosporidiosis					x		
Scabies					x	x	
Staphylococcus intermedius					x		
Streptococcus spp		x			x		
Strongyloides stercoralis					x		
Taenia saginata	x						
Taenia solium					x		
Trichinella spiralis					x		
Trichinosis						x	
Visceral larva migrans						x	
Yersinia enterocolitica	x	x			x	x	
Yersinia pseudotuberculosis					x		x
Yersiniosis				x			

Chapter 13: Infectious Diseases (ID)

Parasitic Infections
ID: Amebiasis
LTC Wortmann, MC, USA

Introduction: Third leading parasitic cause of death in developing nations. Caused by *Entamoeba histolytica*. Transmitted through contaminated food or water. Incubation: 2-4 weeks.

Subjective: Symptoms
Gradual onset of bloody diarrhea with associated abdominal pain and tenderness.
Focused History Questions: *Do you have bloody diarrhea and abdominal pain? (Abdominal pain and bloody stools suggests amebiasis or bacterial colitis.) Do you have a fever? (1/3 of patients)*

Objective: Signs
Using Basic Tools: Vital Signs: Temperature over 101°F
Inspection: Bloody, loose, mucus-containing stools
Palpation: The liver may be enlarged and tender.
Using Advanced Tools: Lab: Heme positive (Guaiac test) stool. O & P: Multiple examinations (minimum of 3) of stool to demonstrate *E. histolytica* trophozoites.

Assessment:
Differential Diagnosis
Diarrhea - giardiasis, viral gastroenteritis, bacterial gastroenteritis, cryptosporidiosis, isosporiasis, *E. coli* 0157:H7, shigellosis and inflammatory bowel disease.

Plan:
Treatment: **Metronidazole** 750 mg tid x 10 days followed by **paromomycin** 30 mg/kg/d in 3 divided doses x 10 days. Maintain oral fluids.

Patient Education
General: Maintain adequate oral intake of fluids to avoid volume depletion.
Activity: As tolerated.
Diet: As tolerated.
Medications: Metronidazole should not be used in the first trimester of pregnancy. Avoid alcohol to avoid an Antabuse-like effect (anxiety, vomiting, headache, etc.).
Prevention and Hygiene: Filters with pores < 1 micron or boiling for 1 min. can make water safe. Good hand washing and safe food preparation.

Follow-up Actions
Return evaluation: If diarrhea continues, consider other etiologies. Stool should be examined after successful treatment, as continued *E. histolytica* mandates re-treatment.
Consultation Criteria: Failure to improve after initiation of antibiotics.

ID: Ascariasis
LTC Glenn Wortmann, MC, USA

Introduction: The most common intestinal worm, *Ascaris* infects more than 1 billion people worldwide, who ingest eggs in contaminated food and drink. The eggs hatch in the small intestine, penetrate the intestinal wall and travel by venous circulation to the lungs. The larvae then pass into the trachea and are swallowed. They migrate through the GI tract becoming mature worms in the small intestine. Incubation period is 4-8 weeks. *Ascaris* is also known as roundworm, and is large enough to easily see without magnification.

Subjective: Symptoms
Abdominal pain (obstruction of bowel or bile ducts [biliary colic] with worms); wheezing and coughing (pneumonitis [Loeffler's syndrome]); occasional liver enlargement; fever. Worms (some larger than earthworm) pass from the anus, nose and mouth and are often brought for diagnosis.
Focused History: *Have you noted the presence of a worm in your stool?* (occasionally migrate out of the intestine) *How large was the worm?* (often the size of an earthworm)

Objective: Signs
Using Basic Tools: Cough and occasionally hepatomegaly; abdominal tenderness; decreased bowel sounds
Using Advanced Tools: Lab: Stool O&P (identify eggs or the adult worm); CBC (nutritional anemia)

Assessment:
Differential Diagnosis
Worm in stool - the presence of a visible worm in the stool is usually diagnostic.
Cough/wheezes - asthma and pneumonia can cause similar findings.

Plan:
Treatment:
Primary: Albendazole 400 mg once
Alternative: Mebendazole 100 mg bid for one day.

Patient Education
General: Wash hands thoroughly.
Activity: As tolerated
Diet: As tolerated
Medications: Occasional gastrointestinal side-effects
Prevention and Hygiene: Hand washing
No Improvement/Deterioration: Refer for evaluation

Follow-up Actions
Return evaluation: As needed
Consultation Criteria: Failure to improve.

ID: Babesiosis
LTC Glenn Wortmann, MC, USA

Introduction: Babesiosis is caused by *Babesia* species protozoa and is transmitted by deer tick bites. Infection is most commonly reported in the Northeastern U.S., but has also occurred in Europe. It is typically a mild illness in healthy people but it can be fatal, particularly in immunocompromised patients (especially splenectomized patients). Incubation period is a few days to weeks.

Subjective: Symptoms
Fever following tick bite, malaise, fatigue, chills, headache and possibly, jaundice.
Focused History: *Do you recall being bitten by a tick?* (tick bites cause babesiosis) *Do you recall the size of the tick that bit you?* (transmitted by the small deer tick) *How long ago were you bitten by a tick?* (The incubation period is a few days to a few weeks.)

Objective: Signs
Using Basic Tools: Fever, sometimes jaundice.
Using Advanced Tools: Lab: Giemsa or Wright stained thin or thick blood smears may confirm the presence of *Babesia* inside red blood cells, and significant hemolytic anemia.

Assessment:
Differential Diagnosis - malaria, viral infections or other tick-borne infections (Rocky Mountain spotted fever, relapsing fever) can cause similar findings.

Plan:
Treatment: Most patients with mild disease recover without treatment. Treat severe disease with **clindamycin** 600 mg IV q 6 hours and **quinine** 650 mg po q 6 hours x 10 days.

Patient Education
General: Avoid tick bites
Activity: As tolerated
Diet: As tolerated
Medications: Occasional gastrointestinal side effects.
Prevention and Hygiene: Avoid tick bites
No Improvement/Deterioration: Return for evaluation

Follow-up Actions
Return evaluation: As needed
Consultation Criteria: Failure to improve.

Zoonotic Disease Considerations
Agent: *Babesia microti, B. bovis, B. divergens*
Principal Animal Hosts: Cattle, wild rodents
Clinical Disease in Animals: Fever (106°F or higher), poor appetite, increased respiratory rate, muscle tremors, anemia, jaundice, weight loss
Probable Mode of Transmission: Bite of infected Ixodes tick
Known Distribution: Worldwide, rare; Europe (*B. divergens*)

ID: Clonorchiasis (Chinese liver fluke)
LTC Glenn Wortmann, MC, USA

Introduction: The Chinese or oriental liver fluke, *Clonorchis sinensis*, is acquired by eating raw or undercooked freshwater fish. Clonorchiasis is endemic to the Far East.

Subjective: Symptoms
Most infections are asymptomatic, but heavy worm burdens may cause right upper quadrant pain (worms block bile and pancreatic ducts), liver enlargement, loss of appetite and fever.
Focused History: *Do you have pain in your right upper abdomen?* (right upper abdominal pain is typical) *How long have you been experiencing abdominal pain?* (can be chronic)

Objective: Signs
Using Basic Tools: Right upper quadrant tenderness, liver enlargement and jaundice (rarely).
Using Advanced Tools: Lab: Identification of *Clonorchis* eggs in the stool on O&P evaluation.

Assessment: Travel to an endemic area suggests diagnosis of clonorchiasis
Differential Diagnosis - cholangitis, cholecystitis and fascioliasis

Plan:
Treatment:
Primary: Praziquantel 75mg/kg/day tid x 1 day
Alternate: Albendazole 10 mg/kg/day x 7 days

Patient Education
General: Avoid improperly prepared seafood.
Activity: As tolerated
Diet: As tolerated
Medications: Occasional gastrointestinal side effects
Prevention and Hygiene: Avoid improperly cooked fish.
No Improvement/Deterioration: Referral for evaluation

Follow-up Actions
Return evaluation: As needed
Consultation Criteria: Failure to improve

Zoonotic Disease Considerations
Principal Animal Hosts: Dogs, cats, swine, rats
Clinical Disease in Animals: Possible hepatic signs
Probable Mode of Transmission: Ingestion of raw or partially cooked infected freshwater fish
Known Distribution: Asia

ID: Cyclosporiasis
LTC Glenn Wortmann, MC, USA

Introduction: Cyclosporiasis is transmitted by fecal contamination of water or food. *Cyclospora* infections occur worldwide, and are an increasingly recognized cause of parasitic diarrhea. Transmission seems to be waterborne and more common during summer months. The incubation period averages one week.

Subjective: Symptoms
Watery (>6 stools per day) diarrhea, fatigue, abdominal cramps and fever (in 25%). Diarrhea can be prolonged (up to 45 days) but is generally self-limited.
Focused History : *Are you experiencing diarrhea? If so, is it watery and non-bloody?* (typical diarrhea) *How long have you been experiencing diarrhea?* (Diarrhea lasting >1 week is typical.)

Objective: Signs
Using Basic Tools: Watery diarrhea without blood, fatigue
Using Advanced Tools: Lab: *Cyclospora* oocysts in O&P examination of the stool.

Assessment:
Differential Diagnosis: There are many possible causes of diarrhea. Although the presence of watery

diarrhea suggests cyclosporiasis, it can also be seen with *Cryptosporidia*, Microsporidia or *Isospora*.

Plan:
Treatment: Most infections are self-limited, but **trimethoprim-sulfamethoxazole** (160 mg trimethoprim-800 mg sulfamethoxazole) given twice daily x 7 days is suggested in chronic infections.

Patient Education
General: Oral fluids to avoid volume depletion
Activity: As tolerated
Diet: As tolerated
Medications: Trimethoprim-sulfamethoxazole can occasionally cause a rash.
Prevention and Hygiene: Avoid contaminated food and water.
No Improvement/Deterioration: Return for reevaluation.

Follow-up Actions
Return evaluation: Consider alternative causes of diarrhea.
Consultation criteria: Failure to improve with conservative measures.

ID: Enterobiasis (Pinworm)
LTC Glenn Wortmann, MC, USA

Introduction: Enterobiasis is contracted by ingesting the eggs of the *Enterobius vermicularis* nematode. Pinworms inhabit the large intestines of humans. Enterobiasis occurs worldwide, particularly in temperate climates and is common among children.

Subjective: Symptoms
Perianal and perineal itching, as well as restless sleep
Focused History: *Have you noticed itching in the perineal or perianal area?* (Pinworms often cause itching.) *Is the itching worse at night?* (Nocturnal pruritus is common)

Objective: Signs
Using Basic Tools: Adult worms (about 1 cm long) may occasionally be seen in the perianal area).
Using Advanced Tools: Lab: O&P of stool should reveal pinworm eggs. Alternatively, apply Scotch tape to the perianal region first thing in the morning and then examine the tape microscopically for eggs.

Assessment: The presence of perineal/perianal itching, especially in a child, is very suggestive of pinworms.
Differential Diagnosis - candidal dermatitis can cause similar complaints.

Plan:
Treatment:
Primary: Pyrantel pamoate 11 mg/kg and repeat in 2 weeks
Alternative: Albendazole 400 mg once, repeat in 2 weeks; or **mebendazole** 100 mg once, repeat in 2 weeks

Patient Education
General: Treat all family members to avoid re-infection.
Activity: As tolerated
Diet: As tolerated

Medications: Occasional gastrointestinal side effects
Prevention and Hygiene: Wash bed linens and night clothes in hot water to destroy eggs.
No Improvement/Deterioration: Referral for evaluation

Follow-up Actions
Return evaluation: As needed
Consultation Criteria: Failure to improve

ID: Fasciolopsiasis
LTC Glenn Wortmann, MC, USA

Introduction: Fasciolopsiasis is caused by the giant intestinal fluke, *Fasciolopsis buski*, found in the Far East and Southeast Asia. Humans are infected by eating raw water plants (e.g., water chestnut/water bamboo) onto which the organism has attached. The parasite eggs can be found about 3 months after ingestion.

Subjective: Symptoms
Usually asymptomatic, although infection may cause diarrhea and abdominal cramping, with vomiting and anorexia.
Focused History: *Have you noticed diarrhea?* (most infections are asymptomatic, but diarrhea may occur) *Have you eaten any raw water plants, such as water chestnuts, in the previous few months?* (typical exposure) *How long ago did you eat a water plant?* (Symptoms usually occur several months after ingestion)

Objective: Signs
Using Basic Tools: Diarrhea. Massive infection can cause intestinal obstruction, edema of face/legs and ascites.
Using Advanced Tools: Lab: O&P of stool: Identification of the *F. buski* eggs or large flukes in the stool.

Assessment:
Differential Diagnosis: There are many potential causes of diarrhea. Clues to fasciolopsiasis are travel to an endemic region and diarrhea developing several months after the ingestion of raw water plants.

Plan:
Treatment: Praziquantel 25 mg/kg tid x 1 day

Patient Education
General: Avoid ingestion of raw water plants
Activity: As tolerated
Diet: As tolerated
Medications: Occasional nausea with praziquantel
Prevention and Hygiene: Avoid foods that may be contaminated
No Improvement/Deterioration: Return for reevaluation.

Follow-up Actions
Return evaluation: Investigate for other possible causes of diarrhea
Consultation Criteria: Failure to improve

ID: Fascioliasis
LTC Glenn Wortmann, MC, USA

Introduction: *Fasciola hepatica* infections are seen in sheep- and cattle-raising areas worldwide, with most cases occurring in South America, Europe, Africa, China and Australia. Humans become infected by eating aquatic plants (especially watercress) grown in water contaminated with feces from infected animals or humans (night soil). Most human infections are mild, although heavy infections can result in extensive liver damage. Even without treatment, many patients will have no symptoms. If flukes lodge in the extrahepatic biliary ducts, right upper quadrant (RUQ) abdominal pain may occur.

Subjective: Symptoms
In moderate-to-severe infections: Diarrhea, RUQ abdominal pain (biliary colic), weakness, malaise and night sweats, fever, jaundice.
Focused History: *Have you noted a fever? Have you noticed pain in your right upper abdomen?* (typical symptoms; right upper quadrant pain occurs as worms migrate to the liver) *How long ago were you in South America, Europe, Africa, China or Australia?* (Symptoms usually occur at least 12 weeks after ingestion of the eggs)

Objective: Signs
Using Basic Tools: Palpation: Hepatomegaly and splenomegaly
Using Advanced Tools: Lab: O & P of stool to identify *Fasciola* eggs.

Assessment:
Travel to an endemic area is critical for considering the diagnosis.
Differential Diagnosis: cholecystitis/cholangitis and clonorchiasis (see sections).

Plan:
Treatment: **Bithionol** 30-50 mg/kg per day in 3 divided doses on alternate days for a duration of 10-15 days (available from the Centers for Disease Control and Prevention). An alternate: **triclabendazole (Fasinex)** 10 mg/kg once.

Patient Education
General: Avoid improperly prepared food (particularly aquatic plants in endemic areas).
Medications: Bithionol can cause photosensitivity and gastrointestinal symptoms.

Follow-up Actions
Evacuation/Consultation Criteria: Evacuate unstable patients and those unable to complete the mission. Consult infectious disease specialist or internist.

Zoonotic Disease Considerations
Principal Animal Hosts: Cattle, sheep, and other large ruminants
Clinical Disease in Animals: Cattle – asymptomatic; sheep – distended painful abdomen, anemia, and sudden death
Probable Mode of Transmission: Ingestion of contaminated greens

ID: Filariasis (Elephantiasis)
LTC Glenn Wortmann, MC, USA

Introduction: Filariasis refers to infection by one of several nematodes found in the tropics and subtropics (see Onchocerciasis and Loiasis in Skin chapter). Lymphatic-dwelling nematodes are discussed here. They are transmitted by mosquitoes and cause three similar conditions: Bancroftian Filariasis (*Wuchereria bancrofti*), Malayan or Brugian Filariasis (*Brugia malayi*) and Timorean Filariasis (*B. timori*). Adult worms live in the lymphatics and release microfilaria, which can take up to a year to appear in the blood after infection, thus making diagnosis difficult.

Subjective: Symptoms
Early symptoms can include swollen lymph nodes (especially in the groin), headache and fever. Long-

standing cases may present with lymphedema (which causes swollen legs [elephantiasis], scrotum, breasts, genitalia, etc). Tropical pulmonary eosinophilia is a complication of filarial infection associated with intermittent nocturnal asthma, fever and interstitial lung disease.
Focused History: *Have you had repeated episodes of swollen lymph nodes?* (recurrent lymphadenitis is typical) *How long have/did you lived in this/that area?* (Most of the chronic complications of filariasis occur in long-term residents of endemic areas.)

Objective: Signs
Using Basic Tools: Intermittent fever and swollen lymph nodes. Chronic infection results in the swelling associated with elephantiasis.
Using Advanced Tools: Lab: Giemsa stained thick and thin blood smears may reveal microfilariae.

Assessment:
Diagnose by examining a blood smear for the presence of microfilariae. Most microfilariae are active at night, but many in SE Asia are best detected between 10 am and 2 PM. The presence of retrograde lymphadenitis is also helpful.
Differential Diagnosis - lymphadenitis may also be seen with acute bacterial or viral infections.

Plan:
Treatment:
Primary: Ivermectin 200-400 mcg/kg as a single dose may be used.
Alternate: Diethylcarbamazine citrate (DEC) 6 mg/kg daily x 2 weeks.
NOTES: Drugs only clear the microfilaria, not all adult worms. Relapses are common and may respond to repeated treatment.

Patient Education
General: Avoid insect bites
Activity: As tolerated
Diet: As tolerated
Medications: Occasional gastrointestinal side effects. **DEC** can have generalized side effects, requiring pain relief and steroids.
Prevention and Hygiene: Avoid insect bites
No Improvement/Deterioration: Return for evaluation

Follow-up Actions
Return evaluation: As needed
Consultation Criteria: Failure to improve

ID: Giardiasis
LTC Glenn Wortmann, MC, USA

Introduction: Giardiasis is a common water-borne cause of diarrhea throughout the world. It is transmitted by fecal contamination of food or water. Treating water with chlorine may not kill *Giardia* cysts, especially if the water is cold. Incubation period is 3-25 days, with an average of about one week.

Subjective: Symptoms
Diarrhea (often prolonged), abdominal cramps, bloating
Focused History: *Have you had watery diarrhea?* (common for giardiasis) *Do you taste rotten eggs when you burp?* (sulfuric belching is often noted) *How long have you been experiencing diarrhea?* (can extend for months)

Objective: Signs
Using Basic Tools: Watery diarrhea
Using Advanced Tools: Lab: Look for organism in stool O&P samples (repeat at least 3 times before considering negative).

Assessment:
Prolonged watery diarrhea and sulfuric belching suggest *giardia*.
Differential Diagnosis: Other causes of watery diarrhea: cholera, salmonella, etc.

Plan:
Treatment: Metronidazole 250 mg tid x 5-7 days, oral rehydration.
NOTE: Avoid **metronidazole** use in first trimester of pregnancy.

Patient Education
General: Maintain oral fluids to avoid volume depletion.
Activity: As tolerated
Diet: *Giardia* infection can result in transient lactose intolerance, so patients should avoid lactose-containing foods such as milk or cheese.
Medications: Avoid alcohol use while taking **metronidazole**.
Prevention and Hygiene: Boiling is probably the best water treatment. Most commercial water filters will not remove *giardia* from water. If using iodine or chlorine, treat water with iodine or chlorine for at least 20 minutes before use. Avoid fecal-oral contamination with good hygiene.
No Improvement/Deterioration: Return for reevaluation

Follow-up Actions
Return evaluation: Consider alternative causes of diarrhea.
Consultation Criteria: Condition not responding to treatment.

Zoonotic Disease Considerations
Agent: *Giardia lamblia*
Principal Animal Hosts: Dogs
Clinical Disease in Animals: May be asymptomatic; chronic diarrhea, steatorrhea, weight loss
Probable Mode of Transmission: Water
Known Distribution: Worldwide; common

ID: Hookworm and Cutaneous Larva Migrans
LTC Glenn Wortmann, MC, USA & MAJ Daniel Schissel, MC, USA

Introduction: Several species of hookworm can infect humans, most commonly *Necator americanus* or *Ancylostoma duodenale*. An estimated one fourth of the world's population is infected, and the geographic distribution is in the tropical and subtropical zones. Eggs in feces from infected people and animals are deposited on the ground and hatch into larvae. Larvae infect by direct penetrating skin, migrating to the lungs and up to the esophagus. They are then swallowed and travel to the small intestine. Symptoms become evident weeks to months after infection. Eosinophilia is common. Cutaneous larva migrans (CLM) is a unique serpentine lesion created by a canine or feline hookworm, *A. caninum, A. braziliense*, which migrates through the skin but is unable to penetrate the dermis. The lesions are very pruritic; thread-like; found most commonly on the feet, hands, and buttocks; and become progressively larger with time. The larvae usually die, but may require treatment.

Subjective: Symptoms
"Ground itch" (an itchy, red rash at the site of larval penetration through the skin), cough, abdominal pain,

diarrhea, and CLM lesions; chronic, long lasting infections can cause iron deficiency anemia and growth retardation; infection can also be asymptomatic.
Focused History: *Have you had a rash?* (ground itch and CLM are typical) *Are you more fatigued?* (due to anemia)

Objective: Signs
Using Basic Tools: Cough, diarrhea, ground itch at entry site, CLM rash (serpentine, slightly elevated, erythematous, palpable tract)
Using Advanced Tools: Lab: CBC: Anemia noted by hematocrit; Eosinophilia noted on WBC.
O&P of stool: Hookworm eggs may not be seen early in infection (see Color Plates Picture 18), so repeat test.

Assessment:
Differential Diagnosis:
Anemia - multiple potential causes, but the presence of iron-deficiency anemia in multiple patients in a community suggests hookworm infection.
CLM Rash - cutaneous Larva Currens rash of Strongyloides infection.
Other symptoms of human hookworm are non-specific and may be confused with many diagnoses.

Plan:
Treatment: Hookworm: Repeat Stool O&P and re-treat as needed.
Primary: Mebendazole 100 mg po bid x 3 days
Alternative: Albendazole 400 mg po once

Treatment: CLM
Primary: Ivermectin 150-200 mcg/kg once
Alternative: Albendazole 400 mg po once or topical **thiabendazole** 10% solution applied 1 cm around leading edge of the CLM lesion.

Patient Education
General: Avoid contaminated soil, especially sandy beaches with shady areas.
Medications: Watch for occasional gastrointestinal side effects (diarrhea and abdominal pain).
Prevention and Hygiene: Avoid contaminated soil.
No Improvement/Deterioration: Return for reevaluation.

Follow-up Actions
Evacuation/Consultation Criteria: Evacuation not normally necessary, unless patient fails to improve.

Zoonotic Considerations
Principal Animal Hosts: Dogs, cats, sheep, swine
Clinical Disease in Animals: *Ancylostoma* spp - anemia, unthriftiness, melena, emaciation, weakness of chronic disease

ID: Leishmaniasis
LTC, Glenn Wortmann, MC, USA

Introduction: Leishmaniasis is transmitted by a bite from an infected sandfly. Infection by Leishmiania species protozoa can result in cutaneous, mucocutaneous or visceral disease (kala azar). The incubation period can be long, up to 6 months after exposure. The visceral form, found in most tropical areas worldwide, is often fatal. Two cutaneous types: Old World disease in Asia, Africa, Middle East; and New World disease in Americas.

Subjective: Symptoms
Cutaneous: Non-healing skin lesion which is usually ulcerative. This skin sore often starts as a papule that enlarges and ulcerates. Sometimes trauma to the skin can initiate the infection at a site distant to the sandfly bite.
Visceral (kala azar): Fever often >104°F which can be intermittent, with chills, wasting, night sweats, nonproductive cough, abdominal complaints, fatigue and an enlarged abdomen.
Focused History: *Do you remember insect/sandfly bites?* (exposure) *Did you have trauma here before this sore developed?* (typical history) *How long have you noticed your skin sore(s)?* (tend to be chronic and non-healing)

Objective: Signs
Using Basic Tools:
Cutaneous: Chronic skin lesions, usually ulcerative (see Color Plates Picture 19), but can be infiltrative or papular; crust often forms over the surface and secondary bacterial infections can occur; usually on exposed portions of the body (frequently extremities and face); regional adenopathy.
Visceral: Fever, wasting, lymphadenopathy, skin changes, hepatosplenomegaly; late peripheral edema, renal failure and bleeding.

Assessment:
Differential Diagnosis
Cutaneous (chronic skin lesion) - sporotrichosis, syphilis, leprosy, neoplasm.
Visceral (fever and hepatosplenomegaly) - malaria, typhoid, typhus, acute Chagas'' disease and tuberculosis.

Plan:
Treatment: Primary:
Cutaneous: Usually self-limited. Old World cutaneous form may require **ketoconazole** and topical medications, or **pentavalent antimony** for very difficult cases.
Visceral: Pentavalent antimony (**sodium stibogluconate [Pentostam]**) from Walter Reed Army Medical Center.
Alternative: Liposomal amphotericin B can be used.

Patient Education
General: Avoid sandfly bites.
Activity: As tolerated
Diet: As tolerated
Medications: As tolerated
Prevention and Hygiene: Avoid sandfly bites.
Wound Care: Keep wound clean, dry and protected.
No Improvement/Deterioration: Return for referral to a higher level of care.

Follow-up Actions
Evacuation/Consultation Criteria: Cutaneous form requires non-urgent referral to a specialist in tropical medicine. Visceral leishmaniasis is potentially life threatening so patient should be transferred to infectious diseases/tropical medicine care urgently.
NOTES: Definitive diagnosis is made by identification of the cultured organism from biopsy of skin (cutaneous disease) or liver/spleen or bone marrow (visceral disease). Walter Reed Army Medical Center is the DoD site with comprehensive diagnostic capability for military beneficiaries.

Zoonotic Disease Considerations
Principal Animal Hosts: Dogs, wild canids

Clinical Disease in Animals: Skin lesions, weight loss, poor appetite, lymphadenopathy, ocular lesions, renal failure, epistaxis, lameness, anemia

ID: Malaria
LTC Glenn Wortmann, MC, USA

Introduction: Malaria is a tremendous problem in tropical, developing countries, causing 2 to 3 million deaths per year. There are 4 species of malaria protozoa which infect humans: *Plasmodium vivax* (incubation period of 12 days-10 months); *P. ovale* (similar to vivax); *P. malariae* (incubation period of 1 month); and P. falciparum (most deadly; incubation period 5-30 days). Mosquito bite, needlestick or a blood transfusion from an infected person transmits malaria.

Subjective: Symptoms
Headache, chills, sweats, and muscle aches are common; abdominal pain and diarrhea may occur.
Focused History: *Have you had a fever?* (Fever in a patient in or returning from a malarious area must be considered to be malaria until proven otherwise) *Do you have any other symptoms?* (Chills, low back pain and myalgias are often seen with malaria)

Objective: Signs
Using Basic Tools: Vital signs: Temperature over 100.4°F. Cyclic fevers (occurring every other day with *P. vivax* and *P. ovale* and every third day with *P. malariae*) may occur (although this is an unreliable finding).
Inspection: Sweats and rigors may be seen
Palpation: Enlarged liver and spleen may occur
Using Advanced Tools: Lab: Thick and thin blood smears stained with Giemsa. The thick smear is reported to be 30 times more sensitive than the thin smear, but the thin smear is required for species identification. Smears should be done 2-3 times a day for 48 hours to exclude the diagnosis of malaria. (see Color Plates Pictures 21, 30, 33, 36)

Assessment:
Differential Diagnosis:
Fever: See Symptom: Fever. Other causes of fever include leptospirosis, dengue, typhoid fever and bacterial meningitis.

Plan:
Treatment: Primary:
1. *P. falciparum:*
   a. In Haiti, the Dominican Republic and Central America west of the Panama Canal, uncomplicated *P. falciparum* can be treated with **chloroquine** 1 gm (600 mg base), then 500 mg (300 mg base) 6 hours later, then 500 mg (300 mg base) at 24 and 48 hours.
   b. In the remainder of world, *P. falciparum* has become resistant to chloroquine. In that case, treatment of uncomplicated malaria consists of **quinine sulfate** 650 mg po q 8 hrs x 3-7 days plus **doxycycline** 100 mg po bid x 7 days.
   c. For patients with severe malaria (parasitemia > 5 %, impaired consciousness, seizures, respiratory distress, substantial bleeding or shock), evacuation and therapy with intravenous **quinine** or **quinidine** is recommended.
2. **Species other than *P. falciparum*:**
   a. **Chloroquine** 1 gm (600 mg base), then 500 mg (300 mg base) 6 hours later, then 500 mg (300 mg base) at 24 and 48 hours.
   b. In Oceania, chloroquine resistant *P. vivax* has been reported, and therapy as per chloroquine resistant P. falciparum should be given.
   c. To prevent relapses with *P. vivax* and *P. ovale*, give **primaquine phosphate** 26.3 mg (15mg base) po

per day for 14 days.
3. **Unknown Malaria Species:** Treat as *P. falciparum*.
4. For children, the dose of chloroquine is 10 mg base/kg followed by 5 mg base/kg at 12, 24 and 36 hours. Mefloquine is dosed at 15 mg base/kg in a single dose. **Quinine** is dosed at 10 mg salt/kg every 8 hours for 7 days and **clindamycin** at 10 mg/kg twice daily for 3-7 days.
5. **Medication Contraindications:** Primaquine should not be given to pregnant women or newborn babies because of the risk of hemolysis. **Doxycycline** and tetracycline should not be given to pregnant women or children less than eight years old. **Chloroquine**, **quinine**, and **quinidine** are considered safe in all trimesters of pregnancy, and there is evidence that **mefloquine** is safe in the second and third trimesters.

Alternative: Alternative treatments include **quinine** plus **clindamycin** 900 mg po tid x 5 days, **mefloquine** 1250 mg as a single dose or **Malarone** 4 tablets po q d x 3 days.

Patient Education
General: Malaria is transmitted by mosquitoes that are active from dusk to dawn, so avoid outdoor activities during that time.
Activity: As tolerated
Diet: As tolerated
Medications: **Doxycycline** can cause photosensitivity, so avoid the sun or use sunscreen. **Primaquine** can cause hemolytic anemia in patients with glucose-6-phosphate dehydrogenase (G6PD) deficiency, and is contraindicated in severe deficiency.
Prevention and Hygiene: Prophylaxis:
1. For travel to areas without chloroquine-resistant *P. falciparum* (see above), give **chloroquine** 300 mg base weekly beginning 1-2 weeks before travel and continuing 4 weeks after returning.
2. For travel to areas with chloroquine-resistant *P. falciparum*, give **mefloquine** 250 mg weekly 1-2 weeks before travel and continuing 4 weeks after returning. **Doxycycline** 100 mg po q d beginning 1-2 days before travel and continuing 4 weeks after returning is an alternative regimen (watch for photosensitivity).
3. For patients with prolonged exposure to *P. vivax* and *P. ovale*, give **primaquine** phosphate 26.3 mg (15 mg base) po per day the last 2 weeks of the 4 week period of prophylaxis on return (may not be well tolerated by persons with G6PD deficiency).

No Improvement/Deterioration: Return for reevaluation promptly.

Follow-up Actions
Return evaluation: Repeat smears to assess effectiveness of treatment. Consider co-infections.
Evacuation/Consultation Criteria: The presence of severe malaria should prompt consultation with an expert in malaria. For complicated malaria (cerebral dysfunction, renal failure, very high parasitemia, ARDS) rapid evacuation to a higher echelon care facility is needed.

ID: Paragonimiasis
LTC Glenn Wortmann, MC, USA

Introduction: The only lung fluke that infects man, *Paragonimus* is found throughout the Far East (particularly China), in West Africa and in several parts of Central and South America. It is acquired from eating raw, salted, dried, pickled or incompletely cooked freshwater crabs, crayfish and shrimp.

Subjective: Symptoms
Most infections are asymptomatic. Heavier infections result in chronic productive cough, chest pain (pleuritic), hemoptysis and night sweats. Extrapulmonary disease can be found in subcutaneous tissues, liver, lymph nodes, others.
Focused History Questions: *Have you had a cough? If so, has your cough been bloody?* (typical symptoms) *How long have you had a cough?* (can have chronic bronchitis)

Objective: Signs
Using Basic Tools: Cough, hemoptysis, tender chest, rales and decreased breath sounds.
Using Advanced Tools: Lab: Eggs in O&P of sputum (particularly in colored flecks) or feces.
CXR: Increased markings and atelectasis on effected side (consolidation).

Assessment: Diagnose empirically by history of bloody cough in an endemic area.
Differential Diagnosis - chronic cough: pneumonia, tuberculosis, lung cancer and chronic bronchitis.

Plan:
Treatment: Praziquantel 25 mg/kg tid x 3 days.

Patient Education
General/ Diet/ Prevention and Hygiene: Avoid improperly cooked freshwater crabs, crayfish and shrimp.
Activity: As tolerated
Medications: Occasional gastrointestinal side effects.
No Improvement/Deterioration: Return for evaluation.

Follow-up Actions
Return evaluation: As needed
Consultation Criteria: Failure to improve.

Zoonotic Disease Considerations
Principal Animal Hosts: Dogs, cats, swine
Clinical Disease in Animals: Can migrate aberrantly and produce cysts in brain and spinal cord; may have neurological signs based on location of lesion.

ID: Schistosomiasis
LTC Glenn Wortmann, MC, USA

Introduction: A blood fluke (trematode) infection found in most tropical areas, particularly Asia, characterized by adult male and female worms living in the veins of a human host. Three major disease syndromes occur in schistosomiasis: dermatitis, Katayama fever and chronic infection. Infection occurs while swimming, wading, rafting, washing etc. in contaminated fresh water. Penetration of the skin by worm larvae causes dermatitis in the first 24 hours, but the clinical symptoms of acute schistosomiasis develop 2 weeks - 3 months after exposure. Chronic schistosomiasis can result in abdominal pain or liver failure (*Schistosoma mansoni, S. japonicum*) or hematuria or kidney problems (*S. hematobium*).

Subjective: Symptoms
Dermatitis: A pruritic rash known as swimmer's itch.
Katayama fever: Occurs 4-8 weeks after infection and presents with acute fever, chills, headache, sweating and cough.
Chronic infection: Can result in abdominal pain with diarrhea or hematuria.
Focused History: *Have you been exposed to fresh water (e.g., swimming in a lake or pond)?* (typical exposure in endemic areas) *Did you notice a rash after swimming?* (rash is caused by the organism invading the skin) *How long ago were you exposed to fresh water?* (Fever and lymphadenopathy from schistosomiasis occur 4-8 weeks after exposure.)

Objective: Signs
Using Basic Tools: Dermatitis: A papular rash; Katayama fever: fever and lymphadenopathy; chronic infection; enlargement of the liver and spleen may be noted, weight loss is common; hematuria may occur
Using Advanced Tools: Lab: Diagnose by finding schistosome eggs in O&P of feces or urine. (see Color

Plates Picture 34)

Assessment:
Differential Diagnosis:
Fever and lymphadenopathy - many possible etiologies to include secondary syphilis, mononucleosis and HIV.
Enlargement of liver - chronic liver diseases (Hepatitis B, hepatitis C and others)
Hematuria - kidney stones, urinary tract infections, bladder cancer and others

Plan:
Treatment: Praziquantel: 20 mg/kg bid po x 1 day for S. *mansoni* and S. *haematobium*, 20 mg/kg tid x 1 day for S. *japonica*. If Katayama fever is present, treatment may result in initial clinical deterioration. Use steroids for 5 days in conjunction with the praziquantel to prevent this. All infections should be treated to avoid the chronic complications of this parasitic illness.

Patient Education
General: Avoid contact with fresh water in endemic areas.
Activity: As tolerated
Diet: As tolerated
Medications: Praziquantel is usually very well tolerated
Prevention and Hygiene: Avoid contact with fresh water in endemic areas.
No Improvement/Deterioration: Refer to higher level of care.

Follow-up Actions
Return evaluation: Refer to higher level of care as needed.
Consultation Criteria: Failure to improve

Zoonotic Disease Considerations
Agent: *Schistosoma spp.*, Schistosome cercariae (swimmer's itch)
Principal Animal Hosts: Cattle, buffalo, swine, dogs, cats, sheep, goats (depending on species and location)
Clinical Disease in Animals: Ruminants - hemorrhagic enteritis, anemia, emaciation
Probable Mode of Transmission: Penetration of unbroken skin by cercariae from infected snails in water
Known Distribution: Worldwide, depending on species and location

ID: Strongyloidiasis (Cutaneous Larva Currens)
LTC Glenn Wortmann, MC, USA & MAJ Joseph Wilde, MC, USAR

Introduction: *Strongyloides stercoralis* is found worldwide in the warm, damp soil of the tropics and subtropics, especially in Southeast Asia. Larvae penetrate the skin and travel to the lungs, then are coughed up and swallowed where they pass into the small intestine and mature into adults. Incubation period is 2-4 weeks, but because of an autoinoculation cycle the parasite can be reactivated many years later when host is immunocompromised (steroids, chemotherapy, AIDS, advanced age). This may result in a hyperinfection syndrome that can result in overwhelming infection and death.

Subjective: Symptoms
Diarrhea, abdominal pain, nausea, cough, wheezing, SOB. Characteristic rash of cutaneous larvae currens (CLC)- migrating, thread-like, erythematous, pruritic, maculopapular, rapidly moving rash (several cm per hour) at the site of larval penetration, often occurs on the buttocks region from external autoinfection.
Focused History: *Have you noticed a rash around your anus or trunk?* (CLC) *Is the rash migrating?* (CLC migrates) *How long have you been having symptoms?* (can last for years)

Objective: Signs
Using Basic Tools: CLC rash, rales, wheezes, SOB and epigastric tenderness.
Using Advanced Tools: Lab: O&P of stool: identify *Strongyloides* larvae (see Color Plates Picture 14) in a fresh stool sample (multiple tests may be needed). CBC: Eosinophilia.

Assessment:
Differential Diagnosis
Hookworm - CLC appears similar to cutaneous larva migrans rash but moves faster and may be perianal.
Diarrhea - multiple causes

Plan:
Treatment: Repeat stool O&P and re-treat as needed.
Primary: Ivermectin 200 mcg/kg/d x 2 days
Alternative: Thiabendazole 25 mg/kg/d po bid (max 3gm/d) x 2 days

Patient Education
General: Avoid contaminated soil.
Medications: Watch for occasional gastrointestinal side effects (diarrhea and abdominal pain).
Prevention and Hygiene: Avoid exposure to contaminated soil.
No Improvement/Deterioration: Return for reevaluation.

Follow-up Actions
Evacuation/Consultation Criteria: Evacuation usually not necessary, unless patient fails to improve.

Zoonotic Disease Considerations
Principal Animal Hosts: Dogs, cats
Clinical Disease in Animals: Bloody-mucoid diarrhea, emaciation, reduced growth rate.

ID: Tapeworm Infections (Taeniasis)
LTC Glenn Wortmann, MC, USA

Introduction: Tapeworms infection occurs through eating infected fish, beef, pork or other contaminated food. Adult tapeworms (ranging from several millimeters to 25 meters long), live in the intestine. Encysted larvae enter the human host through raw or undercooked beef, pork or fish, or by contact with human feces (nightsoil) used as fertilizer.

Subjective: Symptoms
Most infections are asymptomatic. Heavy infections may result in abdominal pain, weight loss, nervousness, diarrhea, and a sensation of the contracting worms leaving the anus. The larval form of pork tapeworm can migrate to multiple areas of the body, to include the brain, causing seizures or death.
Focused History: *Have you noticed worms in your stool?* (may be tapeworm, pinworm or others) *How large are the worms?* (tapeworm segments seen in the stool are usually <1 cm)

Objective: Signs
Using Basic Tools: Segments of worm in stool; in cysticercosis, seizures may occur.
Using Advanced Tools: Lab: O&P of stool: Egg release is variable, so examine stool samples from several days.

Assessment: Diagnose by identifying eggs in the stool or identifying the proglottids (segments) in stool or on anal swab.

Differential Diagnosis:
Worms in stool - ascariasis (roundworm) can also be seen in the stool, although it is typically larger. Seizures - there are many possible causes of seizures, to include epilepsy and meningitis. The presence of seizures in a patient living in area endemic for pork tapeworm (Africa, South America, Eastern Europe, SE Asia) should prompt the consideration of cysticercosis.

Plan:
Treatment:
Primary: Niclosamide 2 gm as one dose. Dwarf tapeworm requires daily doses for one week.
Alternative: Praziquantel 5-10 mg/kg as a one time dose or **albendazole** are alternative regimens. For the dwarf tapeworm the dose of **praziquantel** is 25mg kg.

Patient Education
General: Avoid improperly prepared foods
Activity: As tolerated
Diet: As tolerated
Medications: Occasional gastrointestinal side-effects
Prevention and Hygiene: Avoid improperly cooked beef, pork or fish
No Improvement/Deterioration: Return for evaluation

Follow-up Actions
Return evaluation: As needed
Consultation Criteria: Failure to improve. For the diagnosis of cysticercosis, referral to a tertiary medical center for CT or MRI scan of the head may show multiple calcified lesions.

Zoonotic Disease Considerations
Fish tapeworm disease
Agents: *Diphyllobothrium latum, D. pacificum, D. dendriticum, D. ursi, D. dalliae, D. klebanovskii*
Principal Animal Hosts: Dogs, bears, foxes, minks, cats, dogs, pigs, walruses, seals, other fish-eating mammals
Clinical Disease in Animals: In fish, the larval stages encyst in the visceral organs and muscle while the adults are found in the intestinal tract, leading to possible mechanical obstruction in fish or mammals.
Probable Mode of Transmission: Ingestion of raw or undercooked fish
Known Distribution: Northern Hemisphere lakes region and sub-arctic, temperate or tropical zones where eating raw or undercooked fish is popular
Dwarf tapeworm disease
Agent: *Hymenolepis nana* (the only human tapeworm without an obligatory intermediate host)
Principal Animal Host: Humans, mice
Clinical Disease in Animals: Found in the small intestine of rats, mice and hamsters. Reduced growth or weight loss in rodents.
Probable Mode of Transmission: Ingestion of eggs in contaminated food or water; directly from fecal contaminated fingers
Known Distribution: USA, Latin America, Australia, Mediterranean countries, the Near East and India
Beef tapeworm disease
Agent: *Taenia saginata, Cysticercus bovis* (cyst form)
Principal Animal Hosts: Cattle, water buffalo
Clinical Disease in Animals: Fluid-filled vesicle in the skeletal and cardiac musculature
Probable Mode of Transmission: Ingestion of undercooked beef containing *Cysticercus bovis*
Known Distribution: Worldwide
Pork tapeworm disease
Agent: *Taenia solium, Cysticercus cellulosae* (cyst form)
Principal Animal Hosts: Swine, man

Clinical Disease in Animals: Fluid-filled vesicle in the skeletal and cardiac musculature
Probable Mode of Transmission: Ingestion of undercooked pork containing *Cysticercus cellulosae*
Known Distribution: Worldwide where swine are raised; rare in USA, Canada, UK, Scandinavia
Echinococcosis (Cystic hydrated disease)
Agent: *Echinococcus granulosus*
Principal Animal Hosts: Dogs, cattle, sheep, swine, rodents, deer
Clinical Disease in Animals: Typically asymptomatic
Probable Mode of Transmission: Ingestion of tapeworm eggs
Known Distribution: worldwide
Dipylidiasis (Dog tapeworm)
Agent: *Dipylidium caninum*
Principal Animal Hosts: Dogs, cats, fleas
Clinical Disease in Animals: Asymptomatic, flea infestation
Probable Mode of Transmission: Ingestion of fleas
Known Distribution: Worldwide

ID: Trichinellosis (Trichinosis)
LTC Glenn Wortmann, MC, USA

Introduction: Trichinosis develops when undercooked meat contaminated with *Trichinella spiralis* is ingested. Most infections result from eating undercooked pork, although bear or walrus meat can transmit the infection. Symptoms appear from a few to 15 days after ingestion.

Subjective: Symptoms
Most infections are asymptomatic. For heavier exposures, diarrhea, fever, periorbital edema, photophobia and muscle pain occurs.
Focused History: *Have you eaten undercooked pork in the last few weeks?* (Symptoms peak 2-3 weeks after ingestion) *Are your muscles sore and weak?* (myalgias and weakness are common) *How long have you been having symptoms?* (Trichinosis is usually a self-limited infection, lasting for a few weeks).

Objective: Signs
Using Basic Tools: Fever, splinter hemorrhages under the nails and conjunctivae, upper eyelid edema, muscle tenderness
Using Advanced Tools: Lab: Review of peripheral blood smears will show an increased number of eosinophils.

Assessment: Fever and myalgias after recent ingestion of pork is very suggestive of trichinosis
Differential Diagnosis: Fever and muscle tenderness - myositis, tetanus and schistosomiasis (Katayama fever)

Plan:
Treatment: Supportive therapy with bed rest and pain medication. In the rare event that a patient is known to have eaten infected meat within a week, **mebendazole** 200-400 mg po tid x 3 days, then 400-500 mg po tid x 10 days can be given. **Mebendazole** is only effective against intestinal worms. It does not kill muscle larvae, so it has no effect on established infections.

Patient Education
General: Avoid improperly prepared foods
Activity: As tolerated
Diet: As tolerated
Medications: Occasional gastrointestinal side effects

Prevention and Hygiene: Avoid improperly cooked pork.
No Improvement/Deterioration: Return for evaluation.

Follow-up Actions
Return evaluation: As needed
Consultation Criteria: Failure to improve. For definitive diagnosis, antibody testing (serology) for *Trichinella* is available at reference laboratories.

Zoonotic Disease Considerations
Principal Animal Hosts: Swine, rodents, bears
Clinical Disease in Animals: Asymptomatic
Probable Mode of Transmission: Ingestion of meat containing trichinella worms encysted in striated muscle.
Known Distribution: Worldwide, especially sub-arctic

ID: Trichuriasis (Whipworm)
LTC Glenn Wortmann, MC, USA

Introduction: Whipworm (*Trichuris trichiura*) is one of the most common human worm infections, with approximately 800 million cases occurring worldwide. It is spread by fecal-oral transmission or ingesting vegetables contaminated with whipworm eggs. Infection is generally asymptomatic, but patients with heavy worm burdens may present with anemia, bloody diarrhea, growth retardation or rectal prolapse. Children ages 5-15 are most commonly infected.

Subjective: Symptoms
Usually asymptomatic; may have abdominal pain, bloody diarrhea, malaise, and rectal prolapse.
Focused History: *Have you experienced bloody diarrhea? Have you had a decrease in your energy level?* (Whipworm can cause bloody diarrhea and iron deficiency anemia.)

Objective: Signs
Using Basic Tools: Bloody diarrhea, tender abdomen, rectal prolapse
Using Advanced Tools: Lab: CBC with low HCT; stool for O&P for characteristic lemon-shaped egg (see Color Plates Picture 29)

Assessment: Epidemic suggests diagnosis, and stool O & P will confirm it.
Differential Diagnosis: Bloody diarrhea/anemia - amebiasis, shigellosis and inflammatory bowel disease.

Plan:
Treatment:
Primary: Albendazole 400 mg po for one dose
Alternative: Mebendazole (Vermox) 100mg bid x 3 days

Patient Education
Prevention: Avoid uncooked vegetables in endemic areas.
Medications: For heavy infection, retreatment may be needed.

Zoonotic Disease Considerations
Principal Animal Hosts: Man, canids, and swine
Clinical Disease in Animals: Usually asymptomatic; can see melena, anemia, anorexia, unthriftiness in

heavy infestations
Probable Mode of Transmission: Ingestion of embryonated eggs

ID: African Trypanosomiasis (Sleeping Sickness)
LTC Glenn Wortmann, MC, USA

Introduction: Transmitted by tsetse fly bites, infection with *Trypanosoma brucei* may cause either West African (*T. brucei gambiense*) or East African (*T. brucei rhodesiense*) trypanosomiasis. Sleeping sickness is endemic in 36 African countries. In the more rapidly progressive *T.b. rhodesiense* infection the incubation period is 3 days to a few weeks, in *T.b. gambiense* the incubation period may last several months to years.

Subjective: Symptoms
A painful trypanosomal chancre may develop at the site of the tsetse fly bite.
West African: Fever develops weeks to months after the bite, followed by lymphadenopathy. Personality changes, intense headache and difficulty walking eventually occur. The final phase is marked by progressive neurologic impairment ending in coma and death.
East African: The onset of symptoms usually occurs more rapidly, with fever, malaise and headache occurring within a few days to weeks. Lymph node swelling is not as common. Without treatment, death usually occurs within weeks to months.

Focused History: *Do you remember a painful insect bite?* (The tsetse fly bite is usually painful.) *Have you noticed a rash?* (A rash is common with East African disease.) *When might have you been exposed to a tsetse fly?* (The incubation period for East African disease is usually a few days, while that for West African disease is weeks to months.)

Objective: Signs
Fever; tachycardia; painless, enlarged lymph nodes; painful chancre at bite site with surrounding edema.
Using Advanced Tools: Trypanosomes may be seen on examination of thick and thin peripheral blood smears. (see Color Plates Picture 32)

Assessment: Diagnose by identifying organism on blood smear.
Differential Diagnosis: Fever - many other diseases can cause similar symptoms, including tuberculosis and malaria. A history of travel to an area endemic for African Trypanosomiasis should prompt a diagnostic evaluation for that disease.
Altered mental status - meningitis, brain abscess

Plan:
Treatment: Requires evacuation to a medical center with infectious disease and tropical medicine support for definitive diagnosis and treatment. For patients without evidence of CNS infection (examine the cerebrospinal fluid), treatment with suramin is suggested. For CNS infection with East African disease, **melarsoprol** is recommended, while CNS infection with West African disease requires treatment with **eflornithine**.

Patient Education
General: Avoid tsetse fly bites by wearing protective clothing.
Activity: As tolerated
Diet: As tolerated.
Medications: Since medications have several severe side effects, they should only be given at a tertiary care center.

Prevention and Hygiene: Avoid insect bites.

Follow-up Actions
Consultation Criteria: All suspected cases should be referred for consultation.

Zoonotic Disease Considerations
Principal Animal Hosts: Dogs, ruminants, carnivores
Clinical Disease in Animals: Intermittent fever, anemia, weight loss; may be asymptomatic.

ID: American Trypanosomiasis (Chagas' Disease)
LTC Glenn Wortmann, MC, USA

Introduction: Transmitted by kissing bug bite, blood transfusion. *Trypanosoma cruzi* has been isolated from many wild and domestic animals from the southern US and through all of Latin America. Incubation period is 5-14 days after bite, 30-40 days after transfusion.

Subjective: Symptoms
Acute: Fever; malaise; red, swollen site of inoculation with an enlarged draining lymph node; unilateral, painless swelling of eyelid if inoculation was via the eye (Romana's sign); occasionally, CNS symptoms (e.g., seizures) or myocarditis. **Chronic:** Years later: heart failure; enlargement of the esophagus or colon
Focused History: Do you recall an unusual, red, swollen insect bite or swelling around one eye? (distinctive bite or Romana's sign suggestive of Chagas' disease) Did you ever live in Central or South America? (endemic to area)

Objective: Signs
Using Basic Tools: Variable fever; lymphadenopathy; Romana's sign; hard, red, painful nodule (chagoma) at the bite site.
Using Advanced Tools: Lab: Parasites in peripheral blood smears (thick and thin) can be found during febrile periods early in the course of infection. (see Color Plates Picture 32)

Assessment:
Differential Diagnosis
Skin lesion (chagoma) - leishmaniasis and bacterial skin infections.
Chronic Chagas' disease - other causes of heart failure (myocardial infarctions, hypertension), constipation and dysphagia.

Plan:
Treatment: Acute Chagas' disease: **Nifurtimox** or **benznidazole** from CDC in Atlanta, Georgia (for investigational use only) and from some hospitals in the endemic area.

Patient Education
General: Prevent insect bites (specifically kissing bugs, which are often found in thatched roofs and bite at night)
Activity: As tolerated
Diet: As tolerated
Medications: Per CDC guidelines.
Prevention and Hygiene: Avoid insect bites and infested areas; wear protective clothing.

Follow-up Actions
Evacuation/Consultation Criteria: Refer to tropical medicine/infectious disease specialist when possible.

Zoonotic Disease Considerations
Principal Animal Hosts: Dogs, cats, rodents
Clinical Disease in Animals: Intermittent fever, anemia, weight loss; may be asymptomatic
Probable Mode of Transmission: Contaminated bite wounds or contact with fecal matter of Reduviidae family of insects (kissing bugs).
Known Distribution: Western Hemisphere; Texas, Mexico, Central and South America

Mycobacterial Infections

ID: Tuberculosis
LTC Duane Hospenthal, MC, USA

Introduction: *Mycobacterium tuberculosis, M. Bovis* and others cause tuberculosis (TB), a chronic pulmonary infection that is seen worldwide. Infection is spread by airborne particles. Clinical disease (TB) develops in only about 10% of those infected. Others have latent tuberculosis infection (LTBI), since they do not have evidence of active disease. Disease in adults usually occurs secondary to reactivation of past infection. Disease occurs more frequently in children, and in adults with immunocompromised, including secondary to HIV infection, malignancy, chronic steroid therapy, uncontrolled diabetes, malnutrition, silicosis, or who are smokers. Extrapulmonary disease occurs in approximately 15% of infected persons and can affect virtually any organ system (see Skin: Cutaneous Tuberculosis) and can disseminate throughout the body. Purified protein derivative (PPD) skin testing can be used to document tuberculosis infection (active) or screen for exposure (LTBI). Many countries immunize infants and children with BCG (Bacillus Calmette-Guerin) vaccine, which may cause falsely positive reactions to PPD testing. These reactions wane with time, so a positive reaction to PPD testing in adults should not be dismissed as a reaction to BCG given as a child.

Subjective: Symptoms
Chronic productive cough (bloody), chest pain, fever, chills, night sweats, anorexia, weight loss, fatigue.
Focused History: *Do you cough up blood?* (other causes of chronic cough not usually associated with hemoptysis) *Do your night sweats drench your bedding or bedclothes?* (sweating in bed is normal; having "drenching" sweats is not) *How long have you been coughing?* (A cough lasting longer than 2-3 weeks is unlikely from bacterial or viral infection.)

Objective: Signs
Using Basic Tools: Vital Signs: Normal to low-grade fever, weight loss
Percussion: Unilateral, localized dullness over the upper lung fields
Auscultation: Decreased breath sounds or rales corresponding to percussed dullness (upper lung fields)
Using Advanced Tools CXR: consolidation or cavitary lesion in upper lung fields; purified protein derivative (PPD) skin testing to document tuberculosis infection.
NOTE: PPD may be negative in persons with active infections.

Assessment:
Differential Diagnosis
Chronic cough - chronic bronchitis or COPD not associated with progressive weight loss or night sweats
Lung cancer - usually seen in patient with smoking history
Fungal pneumonias (histoplasmosis, blastomycosis, paracoccidioidomycosis) - chronic pulmonary symptoms are rare and usually seen in smokers in endemic areas.

Plan:
Treatment

Primary: Base the selection of antimycobacterial drugs on knowledge of local resistance patterns. Usually, 3-4 drugs are initiated. The most common regimen is **isoniazid (INH)**, **rifampin (RIF)**, **pyrazinamide (PZA)**, and either **ethambutol (EMB)** or **streptomycin (SM)**. In cases of sensitive tuberculosis, these drugs are given for 8 weeks, followed by a 6-month course of **INH** and **RIF** only. Medicines may be given 2, 3 or 7 times each week. Use directly observed therapy (DOT) to assure compliance. Treat PPD positive contacts without active disease (LTBI) with **isoniazid**, 300 mg/day for 9-12 months. Children under 5 years of age who are contacts of an active pulmonary case should receive **isoniazid**, 10-20 mg/kg (300 mg maximum) even if initial PPD is negative. If the PPD remains negative on retesting after three months, **INH** may be stopped.

Table 5-1 Antimycobacterial Drugs

Drug	Daily dose, mg/kg (max dose)	Twice weekly dose, mg/kg (max)	Thrice weekly dose, mg/kg (max)	Adverse reactions	Monitoring
INH	5 (300 mg)	15 (900 mg)	15 (900 mg)	liver dysfunction, peripheral neuropathy	baseline liver enzymes
RIF	10 (600 mg)	10 (600 mg)	10 (600 mg)	drug interactions, liver dysfunction, bleeding problems	baseline complete blood count, liver enzymes
PZA	15-30 (2 gm)	50-70 (4 gm)	50-70 (3 gm)	liver dysfunction, hyperuricemia	baseline uric acid, liver enzymes
EMB	15-25	50	25-30	optic neuritis	baseline and monthly visual acuity and color vision testing
SM	15 (1 gm)	25-30 (1.5 gm)	25-30 (1.5 gm)	ototoxicity, renal toxicity	baseline and repeat hearing and renal function testing

Alternative: Alternate regimens are usually based on results of susceptibility testing.

Patient Education
General: Comply with the medication regimen to avoid developing active disease, and then spreading it to others.
Medications: See Table 5-1.
Prevention and Hygiene: Isolate patient for several weeks until not contagious. They should use a mask or cover their mouth with every cough. All contacts should be screened with PPD for active and latent infection. PPD positive contacts should get chest radiography to rule out active disease.

Follow-up Actions
Return evaluation: Monitor patient monthly for drug toxicity.
Evacuation/Consultation Criteria: Evacuation not necessary unless clinically unstable or patient develops significant medication side effect. Consult with pulmonologist, infectious disease specialist or primary care physician prior to treatment and as necessary.

NOTE: Multidrug resistant tuberculosis (MDR-TB) is defined as any *M. tuberculosis* that is resistant to both INH and RIF.

PPD skin testing is done as follows:
1. Placement of PPD: inject 5 IU (0.1 mL) intradermally into flexor surface of LEFT forearm (so you will remember where to look for reaction).
2. Interpretation of PPD reaction:
   a. Measure diameter (in mm) of INDURATION or swelling (not redness) of reaction, when viewed in cross-section at 48-72 hours.
   b. >5mm is positive in HIV patients, and in close or household contacts of a patient with an active TB infection.
   c. >10mm is positive for those personnel exposed to people at high-risk for having TB. This standard applies to medical personnel and team members.
   d. >15mm is positive for personnel with no risk factors for exposure. This standard applies to most Americans.

Zoonotic Disease Considerations
Agent: *Mycobacterium bovis*
Principal Animal Hosts: Cattle
Clinical Disease in Animals: Progressive emaciation, lethargy, weakness, anorexia, low-grade fever; chronic bronchopneumonia with moist cough, progressing to tachypnea and dyspnea.
Probable Mode of Transmission: Ingestion, inhalation (occupational exposure to farmers)
Known Distribution: Worldwide; rare in N. America, western Europe, Japan, Australia, New Zealand

ID: Nontuberculous Mycobacterial Infections (NTM)
LTC Duane Hospenthal, MC, USA

Introduction: Nontuberculous mycobacterial (NTM) or mycobacteria other than tuberculosis (MOTT) infectious diseases include lymphadenitis caused by *Mycobacterium avium* complex (MAC) and *M. scrofulaceum*; skin and soft tissue infection secondary to *M. fortuitum*, *M. abscessus*, *M. marinum*, and *M. ulcerans*; and pulmonary disease, most commonly secondary to MAC or *M. kansasii*. Lymphadenitis in children age 1-5 years is most commonly caused by *M. avium* complex. In adults, lymphadenitis is due to *M. tuberculosis* in 90% of cases. Pulmonary syndromes are usually chronic, often occurring in persons with other underlying pulmonary disease.

Subjective: Symptoms
Lymphadenitis (painless enlargement of the lymph nodes of the neck), usually unilaterally; skin and soft tissue infections - edema, erythema; pulmonary infection - chronic, productive cough with fever and weight loss; accompanied by malaise, night sweats and hemoptysis.
Focused History: *Do you have any swollen areas?* (lymphadenitis usually involves only one lymph node chain) *Have you been in contact with any kittens?* (Cat scratch disease/bartonellosis is spread by young cats; causes lymphadenitis)

Objective: Signs
Using Basic Tools:
Lymphadenitis: Enlarged, unilateral nodes of the neck (usually anterior cervical chain) that may spontaneously form sinus tracts and drain. No overlying erythema. Individual nodes are difficult to identify. Normal vitals.
"Swimming pool or fish tank granuloma" caused by *M. marinum* starts as a papule (usually on extremity) that slowly enlarges and ulcerates. These lesions are associated with water or fish exposure. Introduction of the organism is likely via an abrasion or puncture.
Pulmonary disease: Presents similar to tuberculosis (see ID: TB). Perform PPD testing to rule out tuberculosis.
Using Advanced Tools: CXR may reveal thin-walled cavities and more pleural thickening than tuberculosis.

Assessment:
Differential Diagnosis
Lymphadenitis - cat scratch disease (kitten exposure, pain), lymphoma, tuberculous lymphadenitis (also called scrofula; positive PPD or TB exposure)
Skin and soft tissue infection - nocardiosis (no water exposure), sporotrichosis (exposure to soil, organic gardening materials), leishmaniasis (endemic area, sandfly exposure)
Pulmonary disease - see ID: TB; see Respiratory: Pneumonia, COPD

Plan:
Treatment
Primary: Lymphadenitis (due to MAC or *M. scrofulaceum*) - excisional surgery without antimicrobial drugs.
Cutaneous lesions (due to *M. marinum*) - **doxycycline** 100 mg bid or **trimethoprim/sulfamethoxazole**, 160/800 mg bid for 3 months. Excisional therapy is also an option.
Other Syndromes: Therapy based on site of disease, organism and susceptibility testing results.
Alternative: A multitude of regimens exist for most of these infections. Consult an expert for guidance.

Patient Education
General: NTM infections are not contagious to others.
Activity: As tolerated.
Diet: No limitations.
Medications: Based on selected regimen.
Prevention and Hygiene: Avoid swimming with unhealed wounds.
No Improvement/Deterioration: Reevaluation and repeat culture and susceptibility testing.

Follow-up Actions
Wound Care: Local care (clean, dry, protect, topical antibiotics) to prevent secondary bacterial infection.
Return evaluation: Routine follow-up required for pulmonary infections.
Consultation Criteria: Management of chronic pulmonary infection usually requires specialty consultation. Although the acid-fast bacilli can be detected in lesional or sputum smears or biopsy material, culture is required to confirm diagnosis.

NOTES: Disseminated M. avium complex (DMAC) is another NTM disease that is virtually restricted to persons with late stage AIDS. This infection presents as a chronic febrile wasting syndrome with associated anemia and MAC bacteremia.

ID: Introduction to Fungal Infections (Mycoses)
LTC Duane Hospenthal, MC, USA

The fungal infections discussed in this subchapter, with the exception of the superficial presentations of candidiasis, are diseases that can rarely be diagnosed or treated in the field. Cryptococcosis is found worldwide but symptoms are most common in the immunosuppressed and are not acutely life-threatening. Blastomycosis, coccidioidomycosis, histoplasmosis, and paracoccidioidomycosis are endemic fungal infections that should be included in a differential diagnosis so individuals with potential infections may be removed or referred to higher echelons of care.

ID: Candidiasis (Thrush)
LTC Duane Hospenthal, MC, USA & MAJ Daniel Schissel, MC, USA

Introduction: *Candida albicans* is a yeast normally found in the mouth, intestines, and vagina. Overgrowth of this yeast can cause skin or mucosal diseases including oropharyngeal candidiasis (thrush), intertrigo (disease limited to moist skin folds), esophagitis, and vaginitis (see GYN section). In adults, disease commonly occurs in diabetics, the immunocompromised, and after antibiotic treatment for other disorders. Inhaled and oral corticosteroid preparations also increase risk. Disseminated, life-threatening infection can also occur in severely immunocompromised persons.

Subjective: Symptoms
Oral thrush: Usually asymptomatic; may cause mouth discomfort or difficulty swallowing. **Esophageal thrush:** Painful or difficult swallowing. **Intertrigo:** Local burning-like pain, often with pruritus. **Vaginal thrush:** Itching, dyspareunia (pain with intercourse) and change in the odor or consistency of vaginal discharge.
Focused History: *Do you have difficulty or pain with swallowing?* (suggests esophageal or oral lesions) *Do you have diabetes? Have you recently taken antibiotics or corticosteroids?* (hyperglycemia, antibiotic or steroid exposure may precede oral or vaginal disease)

Objective: Signs
Using Basic Tools: Inspection: **Oral/esophageal:** white plaques, which are scraped to reveal an erythematous base and are seen on any oral mucosal surface except the tongue. **Cutaneous** (intertrigo or vulvar): erythematous, shiny rash with small "satellite" lesions at its periphery. **Candidal vaginitis** is associated with a curd-like vaginal discharge (see Symptoms: GYN Problems: Candidal vaginitis).
Using Advanced Tools: Lab: Potassium hydroxide (KOH) wet mount of scrapings or discharge reveals typical yeast, usually with pseudohyphae and/or hyphae. (see Lab Procedures: KOH)

Assessment:
Differential Diagnosis
Oropharyngeal candidiasis - particulate debris secondary to poor oral hygiene (debris is usually easily removed)
Esophageal candidiasis - esophagitis due to herpes simplex, cytomegalovirus, aphthous ulcers, and toxins
Candidal vaginitis - trichomoniasis, bacterial vaginosis (can be differentiated with wet mount)

Plan:
Treatment
Primary: Oropharyngeal candidiasis - **nystatin** solution, 400,000-600,000 units qid po as a swish and swallow x 7 - 14 days. Esophageal candidiasis - **fluconazole**, 200 mg/day po or IV x 14 days. Intertrigo - **nystatin** powder or **clotrimazole** or **miconazole** cream twice daily until resolved. Candidal vaginitis - see GYN Problems section.
Alternative: Oropharyngeal candidiasis - **clotrimazole** troches (lozenges), 10 mg 5/day, oral **fluconazole**, 50-200 mg/day, **itraconazole**, 100-200 mg/day, or **ketoconazole**, 200 mg/day. Esophageal candidiasis - **itraconazole** 100-200 mg/day, or intravenous **amphotericin** B, 0.3-0.5 mg/kg/day, in refractory cases.
Primitive: Gentian violet applied topically.

Patient Education
General: This is a superficial infection that should resolve with standard therapy. It can occur in healthy people, but could indicate other disease such as diabetes or immunocompromise.
Medications: Topical antifungals have virtually no adverse effects associated with their use. The oral azoles, **fluconazole**, **itraconazole**, and **ketoconazole** are all well tolerated. These drugs may interact with other drugs processed through the liver, causes the levels of drugs such as oral diabetes, seizure, and anticlotting medications. **Ketoconazole** that is used long-term may affect steroid hormones, causing irregular menses in women and decreased libido or breast tissue enlargement in men. All may rarely cause severe liver damage. Malaise, nausea, vomiting, weight loss, and infusion site phlebitis (vein inflammation) may also occur. Decreased blood potassium and magnesium often complicate therapy. Intravenous use of **amphotericin B** is associated with infusion-related fever, headache, chills, myalgias, and rigors. Use of amphotericin B can also cause anemia and reversible kidney dysfunction.
Prevention and Hygiene: None necessary

No Improvement/Deterioration: Further evaluation is necessary if infection does not resolve within two weeks.

Follow-up Actions
Return evaluation: If lesions do not resolve consider alternate treatment.
Evacuation/Consultation Criteria: Evacuation is not required for most patients. However, those with recurrent thrush, disseminated infection or who require intravenous amphotericin B therapy should be referred to the appropriate higher echelon of care.

NOTE: Oral candidiasis (thrush) in a young adult should always raise the suspicion of immunocompromise, especially undiagnosed HIV infection.

ID: Blastomycosis
(North American Blastomycosis, Gilchrist Disease)
LTC Duane Hospenthal, MC, USA

Introduction: *Blastomyces dermatitidis* is a yeast-like fungus that causes a spectrum of disease including asymptomatic infection, acute and chronic pulmonary infection and disseminated infection of the skin, bone, GU tract, and rarely, the CNS. This infection is seen most often in central and southeast US in areas near rivers or streams. Approximately 1/2 of exposed persons will develop symptomatic disease. The incubation period is 30-45 days. Most individuals seeking care for this infection have progressive pulmonary disease or cutaneous lesions.

Subjective: Symptoms
Acute pulmonary infection produces fever, cough, and pleuritic chest pain. Chronic pulmonary disease presents with similar symptoms over a longer course. Skin lesions are typically painless or slightly tender.
Focused History: *Have you had any recent travel/exposure to rivers or streams in the central or southeast US?* (endemic area)

Objective: Signs
Pulmonary infection is associated with diffuse auscultatory findings and fever. Chronic pulmonary disease can also include hemoptysis, weight loss, and skin lesions. Skin lesions are most often located on the face, scalp, neck, and extremities. These begin as red papules or nodules that enlarge and then ulcerate or become verrucous. Associated adenopathy is uncommon.
Using Advanced Tools: Lab: Large (8-15 mm), thick-walled, broad-based, budding yeast cells may be visible on Gram stain of sputum or lesion.

Assessment:
Differential Diagnosis
Acute pulmonary infection - influenza, bacterial pneumonia
Chronic pulmonary infection - tuberculosis, lung cancer, other fungal pneumonias
Skin lesions - squamous cell carcinoma, mycosis fungoides

Plan:
Treatment
Primary: IV **amphotericin** B is required in life-threatening infections, all central nervous system infections, infections in immunocompromised patients and in pregnant patients. SOF medics should not administer this toxic agent in the field. **Itraconazole** can be used in all other infections at a dose of 200-400 mg/day po,

usually for 6-12 months.
Alternative: **Ketoconazole** 400-800 mg/ day or **fluconazole** 400-800 mg/ po day

Patient Education
General: Acute pulmonary infection may resolve untreated in 1-3 weeks. All other forms carry a high risk of death if not treated.
Activity: As tolerated.
Diet: No limitations.
Medications: See precautions listed for oral azoles (**itraconazole, ketoconazole** and **fluconazole**) and IV **amphotericin B** in the Candidiasis section.
Prevention and Hygiene: None necessary.

Follow-up Actions
Wound Care: Local care to prevent secondary bacterial infection.
Return evaluation: Observe patients over a 1-2 year period for resolution of infection.
Consultation Criteria: Refer all patients to a specialist for care.

Zoonotic Disease Considerations
Principal Animal Hosts: Dogs, cats, horses
Clinical Disease in Animals: Nonspecific, dependent on organ involvement; weight loss, coughing, anorexia, diarrhea, ocular disease, lameness, skin lesions, fever
Probable Mode of Transmission: Environmental or animal exposure
Known Distribution: Worldwide

ID: Coccidioidomycosis (Valley Fever, Desert Rheumatism)
LTC Duane Hospenthal, MC, USA

Introduction: *Coccidioides immitis* is a dimorphic fungus that causes disease ranging from self-limited pulmonary infection to chronic meningitis. Incubation period is 7-21 days. More than 60% of all infections are asymptomatic. Most symptomatic infections take the form of acute pulmonary disease. Untreated, acute infection resolves in 95% of patients. About 1% of those infected develop chronic pulmonary disease or disseminated infection to the meninges, skin, bone, or soft tissue. Geographic Associations: It occurs in the southwest deserts of the US and northern Mexico, and a few pockets of in Central and South America. It has frequently been reported in service members training at Fort Irwin, California. Incidence peaks during dry periods following rains, usually in summer and fall, and is often associated with wind and dust storms. Risk Factors: Filipinos, blacks, Hispanics, pregnant women, immunocompromised patients are at higher risk for dissemination and severe disease.

Subjective: Symptoms
Cough (usually dry), fever, pleuritic chest pain, malaise, headache, anorexia, myalgia and often rash; severe disease may present with a sepsis-like syndrome. Large joint pain may occur after asymptomatic infection, especially in white females (desert rheumatism). Meningitis presents with chronic headache, memory loss, lethargy, or confusion.
Focused History: *Have you traveled recently to the deserts of the southwest US or northern Mexico?* (endemic areas of disease)

Objective: Signs
Using Basic Tools: Vital signs: Fever and tachypnea
Inspection: Various rashes: Diffuse, faint erythematous rash lasting less than one week; or erythema multiforme (painless, diffuse rash consisting of rings and disks); or erythema nodosum (painful red nodules usually occurring on the shins).
Auscultation: Diffuse auscultatory findings (abnormal breath sounds).
Using Advanced Tools: Ophthalmoscope: Patients with meningitis may have papilledema on funduscopy.

Lab: Eosinophils may be seen on blood smear. Gram stain or KOH of sputum for spherules (10-80 μm round structures with 2-5 μm round endospores inside)

Assessment:
Differential Diagnosis
Acute pulmonary disease - influenza, "atypical" pneumonia, histoplasmosis, blastomycosis (see respective topics)
Meningitis - tuberculosis, syphilis, cryptococcosis (see respective topics); and CNS tumors (see Neurology: Seizure Disorders and Epilepsy)

Plan:
Treatment
Primary: Observation is the treatment of choice for acute pulmonary infection and for asymptomatic cavitary disease in patients not at increased risk for dissemination or chronic disease. **Amphotericin B** 0.5-1 mg/kg IV daily should be used in acute life-threatening infection. This can be followed with **fluconazole** 400-800 mg/day to complete 3-6 months of therapy. Meningeal infection is treated with **fluconazole**, 400-800 mg daily for life. All other forms of coccidioidomycosis are treated with long-term **fluconazole**.
Alternative: Itraconazole (400-600 mg/day) may be used in non-meningeal infections. Some authorities add intrathecal **amphotericin B** in the initial therapy of meningeal disease.

Patient Education
General: Acute pulmonary disease will likely resolve untreated in 6-8 weeks. Meningeal disease requires lifelong therapy.
Medications: See Candidiasis section for adverse effects of intravenous **amphotericin B** and azole antifungals.
Prevention and Hygiene: No human-to-human spread. Others should avoid inhaling dust where patient was exposed.

Follow-up Actions
Return evaluation: Patients should be evaluated frequently for progressive disease.
Evacuation/Consultation Criteria: Evacuate and refer all patients to a specialist for care.

ID: Histoplasmosis (Darling's Disease)
LTC Duane Hospenthal, MC, USA

Introduction: *Histoplasma capsulatum* is a dimorphic fungus that can cause disease ranging from asymptomatic pulmonary infection to life-threatening disseminated infection. Acute infection occurs 3-21 days after exposure. Most infection is asymptomatic or self-limiting pulmonary disease. Severity is dependent on patient's immunity and intensity of exposure. Chronic pulmonary disease, mediastinitis and disseminated disease are rare. Geographic Associations: Found worldwide, this infection is most common in the central US (Mississippi and Ohio River basins). Risk Factors: Outbreaks may occur with the removal of debris containing contaminated bird or bat droppings. Outbreaks in military personnel have been documented after clearing barracks and bunkers. Immunocompromised persons are at higher to develop disseminated disease.

Subjective: Symptoms
Acute (days): Malaise, fever, chills, anorexia, myalgias, cough, pleuritic chest pain. **Chronic (months):** cough
Focused History: *Have you traveled to the Midwest US recently?* (endemic in Ohio and Mississippi River valleys) *Have you been in caves or been near bird droppings lately?* (The fungus is found in debris and soil contaminated with bat or bird guano.)

Objective: Signs
Using Basic Tools: Inspection: Fever, weight loss; hypotension and shock in immunocompromised patient

(sepsis)
Auscultation: coarse breath sounds, pleural friction rub
Palpation: hepatomegaly and/or splenomegaly may be seen in disseminated infection
Using Advanced Tools: CXR: Hilar or mediastinal lymphadenopathy with or without patchy infiltrates. Lab: KOH identification on smear of sputum is usually quite difficult. Organism is a small, budding yeast (2-4 µm) often found inside macrophages.

Assessment:
Differential Diagnosis (see respective topics)
Acute pulmonary infection - influenza
Chronic pulmonary infection - tuberculosis, other fungal infections

Plan:
Treatment
Primary: Therapy is not needed in asymptomatic or acute pulmonary infection unless associated with hypoxemia or symptoms longer than one month. **Itraconazole** 200 mg daily for 6-12 weeks, can be given in those cases that do not spontaneously improve/resolve. For severe infection, including acute or chronic pulmonary disease, disseminated disease or meningitis, give **amphotericin B** 0.7-1 mg/kg IV daily. This therapy can be changed to **intraconazole** 200 mg once or twice daily, for 6-24 months when clinically stable or continued for 3-4 months (35 mg/kg total **amphotericin B**).
Alternative: **Ketoconazole** 200-800 mg/day can be used as an alternative to **itraconazole**.

Patient Education
General: Most acute pulmonary infections resolve spontaneously in 3-4 weeks.
Medications: See precautions listed for oral azoles (**itraconazole, ketoconazole**) and IV **amphotericin B** in the Candidiasis section.
Prevention and Hygiene: Encourage others to avoid areas where patient was exposed.

Follow-up Actions
Return evaluation: Follow-up is required in chronic infection and during long term anitfungal therapy.
Evacuation/Consultation Criteria: Evacuate all chronic and disseminated cases for referral to specialty care.

NOTES: Lung granulomas, and hilar and splenic calcifications are commonly seen on CXR of persons who have had acute pulmonary histoplasmosis in the past. Outside the endemic area however, lung granulomas and hilar calcifications more commonly represent inactive tuberculosis.

Zoonotic Disease Considerations
Principal Animal Hosts: Dogs
Clinical Disease in Animals: Nonspecific, dependent on organ involvement; emaciation, chronic cough, persistent diarrhea, fever, anemia, hepatomegaly, splenomegaly, lymphadenopathy; ulcerative lesions on skin; ocular disease
Probable Mode of Transmission: Environmental (aerosol) exposure primarily in river valleys
Known Distribution: Worldwide

ID: Paracoccidioidomycosis
(South American Blastomycosis)
LTC Duane Hospenthal, MC, USA

Introduction: *Paracoccidioides brasiliensis* is a dimorphic fungus that typically causes chronic, progressive, pulmonary disease in rural male workers. It may occur in individuals who live in or have visited the forests of Central or South America and southern Mexico, and present with mucocutaneous lesions of the face. Incubation may be prolonged up to 15 years.

Subjective: Symptoms
Chronic, productive cough, +/- bloody sputum; shortness of breath; weight loss; painful mouth or nose ulcers; hoarseness.
Focused History: *Have you ever lived in or visited rural South America?* (exposure) *How long have you been coughing?* (usually >3 weeks)

Objective: Signs
Vital signs: Normal
Inspection: Ulcerative lesions of the face, mouth, larynx, or pharynx
Auscultation: Rales or decreased breath sounds in the middle or lower lung fields

Assessment:
Differential Diagnosis
Pulmonary disease - tuberculosis (usually have night sweats, no oral lesions), COPD (usually have smoking history)
Mucocutaneous disease - leishmaniasis, leprosy, syphilis

Plan:
Treatment
Primary: Itraconazole 100-400 mg/day x 3-6 months.
Alternative: Sulfadiazine 4 gm/day for weeks to months, based on clinical response, then 2 gm/day for 3-5 years.
Other sulfa-based antibiotics can be used. **Amphotericin** B can be used in life-threatening and unresponsive infections. **Ketoconazole** 200-400 mg/day and **fluconazole** 600 mg/day have also been used.

Patient Education
General: Disease is chronic and progressive if not treated
Activity: As tolerated
Diet: No limitations
Medications: Hypersensitivity rashes and bone marrow depression can complicate use of sulfa-based drugs. See precautions listed for oral azoles (**itraconazole**, **ketoconazole**, **fluconazole**) and intravenous **amphotericin B** in the Candidiasis section
No Improvement/Deterioration: Relapse is common. Follow up if disease worsens or recurs

Follow-up Actions
Wound Care: Local care (clean, dry, protect, use topical antibiotics) to prevent secondary bacterial infection.
Return evaluation: Patients should be seen routinely for years.
Consultation Criteria: Required for diagnosis and as clinically indicated.

NOTE: Paracoccidioidomycosis has a less common juvenile form which causes acute, progressive, disseminated infection similar to acute disseminated histoplasmosis seen with AIDS and in younger individuals.

ID: Introduction to Viral Infections
LTC Niranjan Kanesa-thasan, MC, USA

Viruses are minute nucleic acid-containing particles with an outer protein coat that are invisible under a light microscope. Many hundreds of species of viruses live and replicate inside plants and animals. Fortunately, most human viral pathogens cause acute, self-limited illnesses for which symptomatic treatment is sufficient. A few of them however are sources of plagues including influenza, measles, smallpox, and human immunodeficiency virus (HIV). It is difficult to diagnose viral pathogens with certainty at the time of illness. Confirmation often requires a specialized viral culture, or recognition of the viral antigen or genome. There are few antiviral drugs and these are often reserved for use in immunocompromised individuals who are most at risk for severe or chronic disease.

ID: Adenoviruses
LTC Niranjan Kanesa-thasan, MC, USA

Introduction: Many adenoviruses cause febrile syndromes commonly associated with upper and lower airway diseases such as pharyngitis, bronchitis, and pneumonia (see these topics). Over 49 adenoviruses have been described as etiologic agents of acute respiratory distress syndrome (ARDS) but only a few (types 1-5, 7, 14, and 21) are major respiratory pathogens. These agents are extremely contagious, resulting in epidemic outbreaks worldwide in crowded quarters such as recruit training sites. Seasonal Variation: In temperate regions, adenoviruses appear more frequently in fall or winter months. In tropical areas, adenovirus infections occur during wet or cooler weather. Risk Factors: Age is a particular factor— infants and children are typically more susceptible than adults. Residing in close environments (training camps, institutions, shipboard) frequently gives rise to outbreaks of adenoviral ARDS because there is frequent exchange of respiratory secretions.

Subjective: Symptoms
Fever, headache, prostration, coryza (nasal mucous membrane inflammation and discharge), sore throat and cough after short (1-5 days) incubation period; usually occurs with constitutional symptoms of malaise, chills, anorexia; persists for 2-5 days then spontaneously resolves.
Focused History: *Do you have a productive cough?* (usually presents with dry, nonproductive cough) *When did you first feel sick?* (usually within past few days – if more prolonged consider bacterial causes of ARD) *Has any close contact been ill with similar illness in past few weeks?* (contagious illness that moves rapidly through closed communities)

Objective: Signs
Using Basic Tools: Vitals: fever up to 102°F
Inspection: Follicular, erythematous, or exudative pharynx/tonsils; swollen, tender cervical lymph nodes; conjunctivitis.
Auscultation: Possible rales
Using Advanced Tools: Lab: Monospot to rule out mononucleosis

Assessment:
Differential Diagnosis
Pneumonia - may be indistinguishable from viral ARD or ARDS (see Respiratory: Pneumonia).
Pharyngitis - ulcerative pharyngitis is associated with the enteroviruses; palatal petechiae, red beefy uvula, and scarlatiniform rashes are often associated with Group A streptococcal pharyngitis.
Influenza - typically results in more systemic disease, including sustained fever, malaise, and myalgia accompanying respiratory manifestations in adults
Infectious Mononucleosis - persistent illness for several weeks suggests IM (see ID chapter).

Plan:
Treatment
There is no specific therapy. Treat symptomatically with fluids and antipyretics until disease resolves.

Patient Education
General: Do not expose others to infected secretions. Cover mouth if cough or sneeze. Wash hands frequently.
Diet: Regular, but take extra fluids
Medications: Acetaminophen for discomfort or fever.
Prevention and Hygiene: Vaccination against types 4 and 7 in military populations previously reduced outbreaks of acute respiratory disease among recruits. However, susceptibility has returned after cessation of routine vaccination.

Follow-up Actions
Evacuation/Consultation Criteria: Evacuate any unstable patients. Consult primary care physician or pulmonologist as needed.

ID: Dengue Fever
LTC Niranjan Kanesa-thasan, MC, USA

Introduction: Dengue fever is a mosquito-borne viral infection especially prevalent in dense, urban centers in the tropics and subtropics. Most dengue infections are asymptomatic, but it may present as an acute, undifferentiated fever with headache, and myalgias. Classically, excruciating pains in the back, muscles, and joints ('breakbone fever') occur in adults. Most patients recovery fully, but some individuals with previous exposure to dengue will develop a more severe form called dengue hemorrhagic fever (DHF), with hypotension and bleeding, which if unchecked will progress to shock and death. Geographic Association: Wet tropical and subtropical areas in most of Latin America, Asia and the Pacific Islands. Seasonal Variation: Outbreaks typically follow rainy seasons in tropical regions, which produce increased densities of the mosquito vector. However, year-round transmission is found in endemic regions. Risk Factors: Travel to dengue-endemic area, with exposure to mosquito bites, is the principal risk factor.

Subjective: Symptoms
Sudden onset of fever, headache, and myalgias after a brief (1-2 days) prodrome of sore throat, nausea, and abdominal pain. Other symptoms: chills, malaise, prostration (similar to severe flu), retroorbital pain, photophobia. **DHF:** Rash, bleeding, mental status changes.
Focused History: *What symptom bothers you the most?* (severe headache, muscle pain, retroorbital pain, photophobia are typical) *When did you first feel sick?* (Typically, patient recalls exact time of onset of fever, headache, and prostration, usually within past several days.) *Have you traveled overseas within the past 2 weeks?* (look for travel to endemic areas [see above] to establish exposure)

Objective: Signs
Using Basic Tools: Vitals: Cyclical fevers to 104°F over days ('saddle-back fever'); **DHF:** Bradycardia, hypotension.
Inspection: flushing with conjunctival injection; prominent maculopapular, blanching rash over trunk and extremities, sparing palms and soles; no petechiae or purpura except with DHF (see Note below).
Palpation: Cervical lymphadenopathy, hepatomegaly; diffuse, abdominal tenderness without guarding
Using Advanced Tools: Lab: Neutropenia on WBC (<1.5 x 10^6/mm^3); blood smears x 3 to rule out malaria; serial hematocrit and platelet counts (hematocrit rising to >50%, decreasing platelet count to <100,000/mm^3 suggest DHF).
DHF: Perform a tourniquet test if DHF is suspected: inflate blood pressure cuff to a point midway between systolic and diastolic blood pressures, maintain for 5 minutes, release pressure and wait 2 minutes or more, then count the number of petechiae that appear in a quarter-sized area (2.5 cm diameter) on the skin distal to the cuff. More than 10 petechiae indicate vascular or platelet disorder and suggest DHF.

Assessment:
Differential Diagnosis

Malaria - rule out with serial blood smears.
Measles (rubeola) - coryza, respiratory symptoms, Koplik spots, discrete rash from face to trunk
Rubella - postauricular lymph nodes in children
Meningococcal fever - painful, palpable purpura and shock
Rickettsial or other bacterial fevers - vesicular or petechial rashes including the palms and soles.
Other viral hemorrhagic fevers - see ID: Yellow Fever and CBR: Viral Hemorrhagic Fever

Plan:
Treatment
There is no specific treatment. Treat symptoms including pain (e.g., Codeine, see Procedure: Pain Assessment and Control).

Patient Education
General: Use body fluid precautions with patient. Prevent mosquito access to patient.
Activity: Bed rest
Diet: Regular, maintain fluids
Prevention and Hygiene: Use personal protection against insect bites. Avoid exposure to mosquitoes at dusk, remove breeding sites (see Preventive Medicine chapter)
No Improvements/Deterioration: Consider hemorrhagic fevers including DHF or yellow fever.

Follow-up Actions
Evacuation/Consultation Criteria: Evacuate all DHF cases early and urgently, as well as all dengue patients who cannot complete the mission. Consult infectious disease experts for all DHF patients.

ID: Arboviral Encephalitis
LTC Niranjan Kanesa-thasan, MC, USA

Introduction: Arthropod-borne viruses (arboviruses), including Japanese encephalitis (JE), West Nile (WN), tick-borne encephalitis (TBE), St. Louis encephalitis (SLE), Kunjin, and Murray Valley encephalitis (MVE) flaviviruses have been associated with sporadic fatal meningoencephalitis in humans. Typically, many hundreds of asymptomatic infections occur for each clinical case of encephalitis. Less common causes of arboviral encephalitis include the alphaviruses: Western equine encephalitis (WEE), Eastern equine encephalitis (EEE), Venezuelan equine encephalitis (VEE); and the California group (CG) of bunyaviruses such as La Crosse virus (LAC). Japanese encephalitis is the most common and one of the most dangerous arboviral encephalitides (inflammation of the brain tissue), with over 50,000 cases reported annually. 25% of individuals with JE die from the disease and 50% are left with permanent neurologic or psychiatric sequelae. There are few clinical features to distinguish the types of encephalitis, so half the cases do not have a specific pathogen isolated. Birds (JE, WN, SLE), horses (WEE, EEE, VEE), and other animals play prominent roles as natural reservoirs for these pathogens, and humans are an accidental host. The alphaviruses and some flaviviruses (JE, SLE, MVE) are associated with epidemic disease in susceptible human populations. Case fatality rates from arboviral encephalitis range from <10% (WN, WEE, SLE, TBE) to 60% (EEE). Geographic Association: WN virus is widely dispersed through Asia, Africa, the Middle East, and recently the U.S. JE virus is distributed throughout Asia, including India and China. TBE is found in forested areas throughout Europe and Central Asia. SLE is widely distributed in the Americas. Kunjin and MVE are restricted to Australia and New Guinea. The alphaviruses and most CG viruses are principally found in the Americas. In highly endemic areas, adults are usually immune to these arboviruses through previous asymptomatic infection. Seasonal Variation: These diseases are associated with periods of vector (usually mosquito) abundance, typically warm and wet times of the year in the tropics. In the U.S., cases of encephalitis usually peak in the late summer/early fall. Risk Factors: Exposure to infectious viruses in vectors or animal hosts commonly occurs in rural or suburban areas (JE, SLE, CG, WEE, EEE), but SLE and WN viruses in particular may occur in urban outbreaks. Children especially those < 1year of age are at risk for severe disease with death or neurologic sequelae with WEE, EEE, JE, and LAC, while older adults > 55 years of age are at greater risk with SLE, WN, and VEE viruses.

Subjective: Symptoms
Sudden fever, headache, vomiting, and dizziness; rapid progression of mental status changes--disorientation, focal neurologic signs, seizures, stupor and coma; followed usually by recovery, or death (1-60% mortality) or severe sequelae.
Focused History: *Have you completed the full vaccination series for JE?* (significantly decreased risk of JE) *Was fever your first symptom?* (typically, see sudden rise of fever after a period of apparent recovery from acute febrile illness, or without any prodromal symptoms) *Have you traveled outside the country or been bitten by mosquitoes recently? If so, where?* (look for opportunity for infection in endemic area within past several weeks) *Have there been recent outbreaks of animal diseases in the area?* (look for epidemics of equine encephalitis [VEE, EEE, and WEE], pig abortions [JE] and bird deaths [WN, SLE])

Objective: Signs
Using Basic Tools: Neurological: Use Glasgow coma scale (GCS) to track progression of mental status changes, and gauge need for medical evacuation or consultation. (See Appendicies: GCS)
Vitals: Fever and respiratory insufficiency.
Inspection: Transient weakness, diminished sensorium, hyperactive deep tendon reflexes, sensory disturbances; limb paralysis (JE, TBE), paresis of the shoulder girdle or arms (TBE); tremors, abnormal movements, and cranial nerve abnormalities (gaze paralysis, speech disorders) (JE).
Palpation: Nuchal (neck) rigidity may be present
Using Advanced Tools: Definitive diagnosis may require lumbar puncture, CT, EEG and other advanced testing.

Assessment:
Differential Diagnosis
Herpes simplex encephalitis - focal, non-motor changes (personality, speech, temporal seizures)
Other Herpesviruses or HIV - see ID: Infectious Mononucleosis and HIV sections
Rabies, TB, meningitis- see appropriate sections
Subdural hematoma and other trauma - see Closed Head Injury on CD-ROM

Plan:
Treatment
There is no drug treatment for JE or other arboviruses. Closely monitor obtunded patients (seizures, aspiration, etc.) pending evacuation.

Patient Education
General: Arboviruses are not directly transmitted from person to person
Activity: Bedrest.
Diet: As tolerated.
Medications: Analgesics for fever or pain (see Procedure: Pain Assessment and Control).
Prevention and Hygiene: Vaccinate personnel against JE. Decrease exposure to mosquito vectors (see Preventive Medicine chapter).

Follow-up Actions
Return Evaluation: Decreasing Glasgow coma scale score, or onset of seizures or focal neurologic symptoms indicate disease progression and requirement for emergent evaluation. Onset of coma or respiratory failure necessitates intensive care for airway management and possible assisted ventilation.
Evacuation/Consultation criteria: Evacuate suspected cases of arboviral encephalitis early and urgently. Consult infectious disease specialists whenever this diagnosis is suspected.

Zoonotic Disease Considerations
Japanese encephalitis
Agent: Japanese encephalitis virus (flavivirus)
Principal Animal Hosts: Horses, swine, wild birds
Clinical Disease in Animals: Abortion in swine - teratogenic, hydrocephalus; stillbirth, mummification, embryonic death, and infertility (SMEDI)
Probable Mode of Transmission: Bites of mosquitoes (*Culex* spp)
Known Distribution: Asia, Pacific islands from Japan to the Philippines

ID: Viral Gastroenteritis (intestinal flu) See Symptom: Diarrhea.

ID: Hantavirus
COL Naomi Aronson, MC, USA

Introduction: Hantaviruses infect rodents worldwide and aerosolization of rodent excreta (especially urine) is responsible for transmitting infection. There are many hantaviruses, with the most important being the Hantaan and Seoul viruses (found in Korea, China, and far eastern Russia) which cause hemorrhagic fever with renal syndrome (HFRS). Dobrava virus (found in the Balkans) also causes HFRS. Puumala virus (found in Western Europe and Scandinavia) causes a milder form of HFRS. Sin Nombre virus, mainly found in the western U.S. and Canada, causes Hantavirus Pulmonary syndrome (HPS). The incubation period is generally 1 to 4 weeks.

Subjective: Symptoms
HFRS:
Constitutional: Acute (1-4 days): High fever, chills, myalgias, headache; **Sub-acute (5-14 days):** Low grade fever, apprehension; **Chronic (> 2 weeks):** Fatigue and lethargy
Specific: Acute (1-4 days): Abdominal pain, flushed face; **Sub acute (5-14 days):** Low urine output, back pain; **Chronic (> 2 weeks):** Diuresis, renal concentrating defect
HPS:
Constitutional: Same as above
Specific: Acute (1-4 days): Dizziness, abdominal pain, diarrhea; **Sub-acute (5-14 days):** Dyspnea, non-productive cough, shock
Focused History: *Have you recently seen evidence of mice/rats near or in where you live or sleep?* (typical exposure) *Have others in your family, village or unit had similar symptoms?* (outbreaks occur in others similarly exposed)

Objective: Signs
HFRS:
Using Basic Tools: Acute (1-4 days): Toxic appearance, fever to 104°F, conjunctival injection, flushed face/neck/ upper torso (blanches with pressure), dermatographism (drawing on skin leaves an exaggerated mark); **Sub-acute (5-14 days):** Temperature up to 101°F, truncal and axillary fold petechiae, lowered blood pressure, low urine output up to day 7, profound diuresis thereafter(up to liters/day)
HPS:
Using Basic Tools: Inspection: Acute (1-4 days): Increased respiratory rate, accessory muscle use for breathing, Fever to 104° F
Auscultation: Acute (1-4 days): Lungs often normal, tachycardia, mild hypotension; **Sub acute (5-14 days):** Diffuse "Velcro" rales
Using Advanced Tools:
HFRS: Urine is dilute (specific gravity 1.010) with proteinuria, hematuria, occasional red and white blood cell casts; may see elevated white blood count, thrombocytopenia, increased hematocrit (up to 55-65% in severe infection).

HPS: Pulse oximetry may demonstrate hypoxia even if CXR is normal in HPS. CXR may show bilateral whiteout, pleural effusion, Kerley B lines, increased vascular markings; may see elevated white blood count, thrombocytopenia, increased hematocrit (up to 55-65% in severe infection).

Assessment:
Differential Diagnosis
Leptospirosis - pulmonary hemorrhage presentation of leptospirosis (as seen in Hawaii) may present similarly to HPS. Travel history, conjunctival redness and skin contact with standing fresh water all suggest leptospirosis.
Hemorrhagic fever virus - more prominent bleeding (HFRS can have bleeding, but late in course) and rash seen with some types.
Typhus - responds to **doxycycline**, presents with a rash, lowered white blood cell count, and tache noire for some types. Also consider plague, tularemia.

Plan:
Treatment
Primary: Avoid excess fluids; consider blood transfusion and Trendelenburg position for shock; give O_2. For HFRS, **ribavirin** IV 2 gram loading dose, then 1 gram q 6 hours for 4 days, then 500mg q 8 hours for 6 days.

Patient Education
General: This infection has high mortality. Most deaths occur within the first 48 hrs.
Activity: Bedrest
Diet: As tolerated
Medications: In HFRS, once patient enters the polyuric phase (about day 8), replace urine losses carefully to avoid dehydration (use careful output measurements).
Prevention and Hygiene: Minimize human-rodent contact. Protect food source, keep rodents out of sleeping places, wet down deserted dwellings (preferably with detergent or disinfectant) to avoid aerosolization and clean out before living there. Use gloves to handle dead rodents and their nests.

Follow-up Actions
Consultation Criteria: Monitor oxygenation with a pulse oximeter. Be prepared to intubate for respiratory failure. Use fluids modestly in HPS to maintain cardiac output. Most deaths occur within the first 48 hours.
Evacuation: Patient needs to be transported to hospital where dialysis (HFRS) and ventilatory support (HPS) are available. Avoid air transport once patient enters the capillary leak syndrome presentation of this illness.

Zoonotic Disease Considerations
Principal Animal Hosts: Rodents
Clinical Disease in Animals: Asymptomatic
Probable Mode of Transmission: Aerosols from rodent excretions and secretions
Known Distribution: Worldwide

ID: Yellow Fever
LTC Niranjan Kanesa-thasan, MC, USA

Introduction: The yellow fever virus causes hemorrhagic fever (severe illness with bleeding), and treatment of these patients is challenging even in optimum environments. As with other viral hemorrhagic fevers (see CBR: Viral Hemorrhagic Fever), supportive fluid management with hemostatic monitoring is critical to recovery. The incubation period is typically 3-6 days, and 80-90% of cases recover completely. 10-20% develop jaundice and hemorrhagic disease, leading to death (50% mortality in these patients) or a chronic convalescent phase lasting for several weeks or months. Geographic Association: Yellow fever is uncommon in the US but occurs in jungle environments in Africa, and in jungle and possibly urban areas in Central and South America. Seasonal Variation: As with other arboviral illnesses, epidemics may follow the rainy season, particularly in areas contiguous to rain forests where jungle yellow fever is enzootic (monkeys). Endemic yellow fever occurs sporadically year-round. Risk Factors: Travel to yellow fever-endemic areas, especially if unvaccinated before exposure, is the major risk factor, especially among travelers. Occupational exposure among young adult males is responsible for much yellow fever in forest regions of tropical Latin America.

Subjective: Symptoms

Abrupt onset of fever, chills, headache, backache, vomiting for 2-3 days; some deteriorate over 3-10 days with coffee-ground hematemesis ('black vomit'), jaundice, and disorientation.
Focused History: *Have you ever been vaccinated against yellow fever?* (reduces chance of yellow fever infection to <1%) *When did you first feel sick?* (Typically patient recalls exact time of onset of fever, headache, and prostration, usually within past week) *Have you traveled overseas in the past 3 weeks? If so, where?* (look for travel to endemic areas—see above)

Objective: Signs

Using Basic Tools: Vitals: Fever to 102-104°F; hypotension and relative bradycardia; occasional arrhythmias Inspection: facial flushing with conjunctival injection, strawberry tongue (red edges with central prominent papillae); **Advanced disease:** Bleeding from orifices (epistaxis, hematemesis, and melena), jaundice. Palpation: Epigastric or RUQ abdominal tenderness
Using Advanced Tools Lab: Proteinuria on urinalysis (nephritis); CBC for neutropenia (typical), hematocrit and platelet count (see Dengue); type and cross match if bleeding; blood smear to r/o malaria

Assessment:
Sudden onset of high fever without prodromal symptoms or rash is characteristic, when accompanied by the classical triad of albuminuria, jaundice and hematemesis.
Differential Diagnosis
Rickettsial fevers - maculopapular rash that begins at the wrists and ankles and spread to the trunk.
Leptospirosis (see topic) - aseptic meningitis, encephalitis
Other hemorrhagic fevers (Lassa, Marburg or Ebola) - acquired in focal geographic areas (see Viral Hemorrhagic Fever Section) - pharyngitis and retrosternal chest pain.
Snake bite (viper) - bite and similar consumptive coagulopathy, but usually no jaundice or proteinuria.

Plan:

Treatment
1. There is no specific treatment for yellow fever.
2. Provide supportive care, including codeine for analgesia (see Procedure: Pain Assessment and Control) and anti-emetics (**dimenhydrinate** 50 mg IM prn, or **prochlorperazine** 25 mg PR bid, 5-10 mg IM q 3-4 or 5-10 po tid-qid).
3. Evacuate severe cases.
4. If unable to evacuate and bleeding or shock occurs, begin fluid resuscitation with IV isotonic fluid bolus, followed by intravenous colloids (see Shock: Fluid Resuscitation). Monitor vital signs, carefully insert nasogastric tube and Foley catheter, monitor fluid I & O, and serially repeat hematocrit and platelet count. Treat bleeding with **calcium gluconate** (1 g IV daily or bid) and transfusions as needed (see Procedures: Blood Transfusion).

Patient Education
General: Use body fluid precautions in hemorrhagic patients to avoid disease transmission.
Activity: Bedrest
Diet: NPO if hemorrhagic, until stable
Medications: Avoid **acetaminophen** (liver damage) and **aspirin** (bleeding)

Prevention and Hygiene: Immunize with licensed yellow fever vaccine. Booster doses are recommended every 10 years for travel to yellow fever-endemic regions. Avoid suspected foci of yellow fever or other hemorrhagic fevers. Practice personal protective measures against mosquitoes (see Preventive Medicine).

Follow-up Actions
Return Evaluation: Assess for onset of hemorrhagic signs, and evacuate if necessary.
Evacuation/Consultation Criteria: Urgently evacuate all suspected hemorrhagic cases (hematologic abnormalities, profound bleeding, or vascular instability). Consult infectious disease specialists for all cases of hemorrhagic yellow fever, and for any cases in team members.
NOTE: Serology may be performed to confirm diagnosis and for epidemiologic case definition.

ID: Hepatitis A
LTC Duane Hospenthal, MC, USA

Introduction: Hepatitis A virus (HAV) infection is spread by fecal contamination of food or water, or by person-to-person contact. Infection occurs worldwide with increased incidence in developing nations. Incubation period averages 28 days. Virus is excreted into the stool of infected individuals prior to the development of symptoms. Peak infectivity occurs two weeks prior to the development of jaundice. Most individuals recover spontaneously and completely. No "carrier state" exists. Children can have unrecognized infection and may shed virus for several months, making them a major source of infection to others. US service members should be protected from infection with pre-deployment immunization.

Subjective: Symptoms
Abrupt onset fever, nausea, anorexia and malaise, often following several days of nonspecific upper respiratory tract symptoms. Jaundice usually develops days later along with right upper quadrant abdominal pain, dark urine, light-colored stool and pruritus.
Focused History: *When did you notice you were turning yellow?* (Jaundice develops several days after other symptoms. In chronic liver disease, jaundice may develop more slowly, usually without fever or other acute symptoms.) *Is your urine darker than usual? Are your stools lighter than usual?* (typical symptoms) *Is anyone else ill?* (contamination from a common source) *Have you injected drugs or had unprotected sex with a new partner?* (Hepatitis B, C risk factors)

Objective: Signs
Using Basic Tools: Vital signs: Low grade fever
Inspection: Jaundice of skin, sclerae, and mucous membranes under tongue
Palpation: Smooth, tender, enlarged liver edge beyond costal margin
Using Advanced Tools: Lab: Urinalysis reveals positive urobilinogen

Assessment:
Differential Diagnosis
Hepatitis B, D, and C - usually will have parenteral or sexual exposure
Hepatitis E - may not be distinguishable
Mononucleosis (Epstein-Barr virus or cytomegalovirus) - usually associated with sore throat, more severe fever, malaise, and anorexia. May have positive Monospot
Leptospirosis - fresh water exposure, conjunctival suffusion, myalgias, more severe fever
Yellow fever - myalgia, more severe fever and malaise
Malaria - more severe, cyclic fever
Chemicals (including drugs and alcohol) - history of toxic ingestion, heavy alcohol use, no fever

Plan:
Treatment: Primary: Supportive care

Patient Education
General: Most persons with acute infection will recover within three weeks.
Activity: Bedrest
Diet: Refrain from use of all alcohol products.
Medications: Avoid medications that are cleared by the liver, including **acetaminophen**.
Prevention and Hygiene: Isolate infected persons up to one week after the onset of jaundice. Appropriate handwashing, food preparation, waste disposal (feces are highly infective), and water purification. Post-exposure treatment with **immune globulin**, 0.02 ml/kg, should be given to close contacts if less than two weeks from last exposure. Widespread immunization of susceptible people may be effective in stopping outbreaks.
No Improvement/Deterioration: Evacuate for evaluation of hepatic failure.

Follow-up Actions
Consultation Criteria: Refer cases of HAV that do not improve or that progress to encephalopathy.

ID: Hepatitis B (and Hepatitis D)
LTC Duane Hospenthal, MC, USA

Introduction: Hepatitis B virus (HBV) is spread via sexual intercourse, birthing and exposure to blood and blood products. HBV can cause both acute and chronic infection. 5-10% of acutely infected adults develop chronic infection. In contrast, perinatal infection leads to chronic infection in 70-90% of individuals. Chronically infected persons are the reservoir for this infection and are at risk to develop cirrhosis and hepatocellular carcinoma (HCC). The highest incidence of chronic disease is in Asia and Africa.

Subjective: Symptoms
Acute infection: See Hepatitis A. 10% of acute HBV infection is symptomatic. If symptoms do occur, they develop after an incubation period averaging 75 days. Chronic infection is only rarely associated with nonspecific symptoms such as malaise and fatigue.
Focused History: (for chronic hepatitis): *Have you ever been told you had hepatitis or jaundice?* (Chronic hepatitis may occur after acute hepatitis.) *How long have you had malaise and/or fatigue?* (These symptoms for greater than a month usually denote a chronic process.)

Objective: Signs
See Hepatitis A for signs of acute hepatitis.
Using Basic Tools (chronic disease):
Vital Signs: Normal
Inspection: Signs of chronic liver disease - telangiectasias (new blood vessel formation in the skin) over the upper chest, back and arms, reddened palms, gynecomastia, small testes

Assessment:
Differential Diagnosis
Acute hepatitis - same as listed in Hepatitis A
Chronic hepatitis - hepatitis C, autoimmune hepatitis, hemochromatosis, chronic alcoholic liver disease, and other primary liver disorders.

Plan:
Treatment
Primary: Supportive care for acute hepatitis.

Patient Education
General: Most adults recover from acute hepatitis within 4 weeks. Over 90% become immune and do not develop chronic infection.
Activity: Bedrest
Diet: Refrain from use of all alcohol products
Prevention and Hygiene: Sexual partners and children should be tested. Consider HBV immunization of close contacts if not immune. Patient should not donate blood, tissues or semen. Condom use decreases sexual transmission. Avoid sharing toothbrushes and razors.
No Improvement/Deterioration: Evacuate for evaluation of hepatic failure.

Follow-up Actions
Return evaluation/ Consultation Criteria: All patients suspected to have chronic hepatitis B should be referred to a specialist.

NOTES: Hepatitis D or delta hepatitis (HDV) is caused by a defective virus that only causes disease in the presence of HBV. HDV disease is similar to HBV, occurring in persons previously or concurrently infected with HBV. Found worldwide, endemic pockets occur in South America, Africa, the Middle East, and in the Pacific islands. HDV is often diagnosed when a person with known HBV infection is noted to have a flare up of disease or a second course of acute hepatitis.

ID: Hepatitis C
LTC Duane Hospenthal, MC, USA

Introduction: Hepatitis C virus (HCV) is a common cause of chronic hepatitis. This virus is transmitted by exposure to blood and blood products and less frequently, perinatally or by sexual intercourse. Acute infection (incubation period 6-7 weeks) is rarely diagnosed, but chronic disease develops in more than 60% of those infected. Persons with chronic disease are at risk to develop cirrhosis and hepatocellular carcinoma (HCC).

Subjective: Symptoms
Acute infection is asymptomatic or associated with nonspecific symptoms in most patients. Typical symptoms associated with hepatitis A virus (HAV) infection are only rarely present and then usually to milder degree. Chronic infection is only rarely associated with nonspecific symptoms such as malaise and fatigue.
Focused History Questions: *Have you ever been told you had hepatitis or jaundice?* (Chronic hepatitis may occur after acute hepatitis.) *How long have you had malaise and/or fatigue?* (These symptoms for greater than a month usually denote a chronic process.)

Objective: Signs
Jaundice may be seen in 20-30% of acutely infected individuals.
Using Basic Tools (chronic disease):
Vital Signs: Normal
Inspection: Stigmata of chronic liver disease - telangiectasias (new blood vessel formation in the skin) over the upper chest, back and arms, reddened palms, gynecomastia, small testes

Assessment:
Differential Diagnosis
Acute hepatitis - same as listed in Hepatitis A
Chronic hepatitis - hepatitis B, autoimmune hepatitis, hemochromatosis, chronic alcoholic liver disease, and other primary liver disorders.

ID: Viral Infections: Acute Hepatitis *

LTC Duane Hospenthal, MC, USA

Disease	Hepatitis A/ Hepatitis E	Hepatitis B/ Hepatitis D	Hepatitis C	Other Important Causes **	
				Other Infections (i.e., leptospirosis, yellow fever)	Toxins (i.e., Amanita mushrooms), Drugs (i.e., acetaminophen), or Alcohol
Route of acquisition (common)	Oral-fecal	Blood, perinatal, sexual	Blood	Disease-specific	Ingestion
Precautions to avoid spread	Enteric precautions ***	Universal precautions ****	Universal precautions	Most not spread person-to-person; disease-specific: including avoiding vectors	Not spread person-to-person
Prevention	Handwashing, proper food preparation and waste disposal; immunization and/or immune globulin prior to or during outbreak of Hepatitis A; isolation of infected patients	Avoid unprotected sexual contact, and exposure to blood; immunization and/or immune globulin for Hepatitis B	Avoid exposure to blood	Disease-specific: including immunization (yellow fever) and chemoprophylaxis (leptospirosis)	Avoidance of the specific toxin or drug
Treatment (acute)	Supportive	Supportive*****	Supportive *****	Disease-specific; usually supportive	Supportive

* See Infectious Diseases for detailed discussions of Hepatitis.   ** Must also consider gallbladder disease with acute pain (not commonly jaundiced), pancreatic cancer with painless jaundice, and many other diseases of the liver and hepatobiliary system.   *** Enteric precautions include use of gloves, gowns or other barriers when in contact with patient, patient waste, clothes, or linens.   **** Universal precautions include use of gloves, gown, mask and eyewear (or face shield) when drawing blood or coming into contact with blood or other body fluids.   ***** Chronic infection may be treated with **interferon** based regimens.

Table 5-2

Plan:
Treatment
Primary: Supportive care for acute infection

Patient Education
General: The natural history of chronic hepatitis C is currently not clear. All patients may not need specific therapy, but some will benefit from **Interferon** therapy.
Activity: As tolerated.
Diet: Refrain from all alcohol products.
Prevention and Hygiene: Sexual partners and children should be tested. Patient should not donate blood, tissues or semen. Sexual transmission likely occurs at a very low rate. Those infected should inform all sexual partners. Use of condoms in long-term monogamous couples is not absolutely required. Avoid sharing toothbrushes and razors.

Follow-up Actions
Return evaluation/ Consultation Criteria: All patients suspected to have chronic hepatitis C should be referred to a specialist.

NOTES: Other diseases associated with HCV infection include cryoglobulinemia, porphyria cutanea tarda and glomerulonephritis.

ID: Hepatitis E
LTC Duane Hospenthal, MC, USA

Introduction: Hepatitis E virus (HEV) causes infection that is spread by fecal contamination of water or food. Infection is endemic to India, Southeast and Central Asia, the Middle East, northern Africa and Mexico, with increased incidence in developing nations. Incubation period averages 4-5 weeks. Virus is excreted into the stool of infected individuals prior to the development of symptoms. Most individuals recover spontaneously and completely in 1-4 weeks. No "carrier state" exists.

Subjective: Symptoms
General: Flu-like illness with fever, nausea, anorexia and malaise. Jaundice usually develops a few days later, often accompanied by resolution of the flu-like symptoms. Pruritus may accompany jaundice.
Local: Right upper quadrant abdominal pain, dark urine, light-colored stool
Focused History: *When did you notice you were turning yellow?* (In acute hepatitis E, jaundice develops several days after the other symptoms. In chronic liver disease, jaundice may develop more slowly, usually without fever or other acute symptoms.) *Is anyone else ill?* (Hepatitis E can occur in outbreaks from common source contamination.) *Is your urine darker than usual? Are your stools lighter than usual?* (typical symptoms) *Have you injected drugs or had unprotected sex with a new partner?* (Hepatitis B is associated with parenteral exposures, hepatitis E is not.)

Objective: Signs
Jaundice and tender hepatomegaly.
Using Basic Tools:
Vital signs: Low grade fever
Inspection: Jaundice of skin, sclerae, and mucous membranes under tongue
Palpation: Smooth, tender, liver edge beyond costal margin; may also have splenomegaly

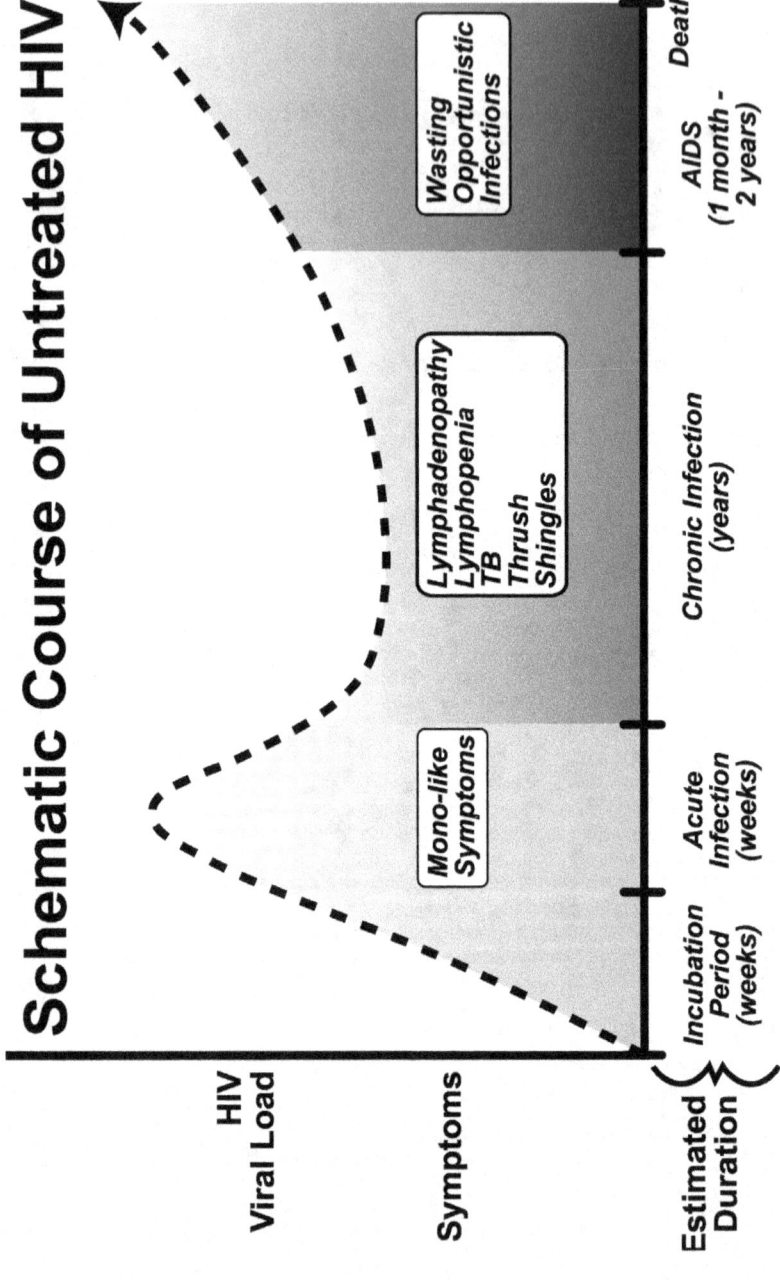

Table 5-3

Using Advanced Tools: Lab: Urinalysis reveals positive urobilinogen

Assessment:
Differential Diagnosis
Same as for Hepatitis A.

Plan:
Treatment
Primary: Supportive care

Patient Education
General: Most persons with acute infection will recover within four weeks.
Activity: Bedrest
Diet: Refrain from use of all alcohol products
Medications: Avoid medications that are cleared by the liver, including acetaminophen.
Prevention and Hygiene: Safe drinking water source is of the utmost importance. Handwashing, safe food preparation and waste disposal (feces are highly infective) are essential. Stool is infectious 1 week prior to symptoms and remains so as long as 2 weeks into the illness. Isolation of those infected from susceptible persons is not necessary because person-to-person transmission is low.
No Improvement/Deterioration: Return for reevaluation.

Follow-up Actions
Consultation Criteria: Refer cases of HEV that do not improve or progress to encephalopathy.
NOTES: The mortality rate of HEV is low (0.07% -0.6%) except for pregnant women in whom a high mortality is reported (15-25%). During outbreaks the attack rate is higher in pregnant women, highest in those in their second or third trimesters.

ID: Human Immunodeficiency Virus
COL Naomi Aronson, MC, USA

Introduction: Human Immunodeficiency Virus (HIV), a retrovirus, is transmitted through sexual contact, needlestick/sharps, perinatally, infected blood/body fluid contact with non-intact skin or mucous membranes, breastfeeding and blood transfusion. Infected individuals may be free of signs of disease for years after infection. Acquired Immune Deficiency Syndrome (AIDS) is the failure of the immune system seen during the late stage of HIV infection. Untreated infection is eventually fatal. The incubation period for HIV is several weeks (commonly) to months (less usual) after exposure.

Subjective: Symptoms
Acute (1-30 days): Fever to 102°F, malaise, myalgias, night sweats, sore throat, gastrointestinal symptoms, maculopapular rash, oral ulcers. **Sub-acute** (30 days-1year): Generalized adenopathy. **Chronic** (> 1 year): Fevers, sweats, fatigue, weight loss, oral thrush, shingles, symptoms of opportunistic infection, including wasting syndrome, recurrent *Salmonella* infections, HIV dementia, *Pneumocystis carinii* pneumonia, chronic fevers, Toxoplasmosis of brain, chronic diarrhea of >one month, Cryptococcal meningitis and more other unusual infections, oral and esophageal candidiasis, chronic Herpes simplex, severe seborrheic dermatitis, recurrent pneumonia, tuberculosis. These malignancies are also frequently seen in AIDS: invasive cervical cancer, Kaposi's sarcoma, brain lymphoma, non-Hodgkin's lymphoma
Focused History: *Are any of your sexual partners chronically ill? Have you tested positive for HIV? Have you had unexplained weight loss, whitish curds in mouth, fevers, sweats or chronic diarrhea?* (increased risk if answer YES to above) *Do you ALWAYS use protection* (such as condoms) *during sex?* (increased risk if answer NO) *When did you last test negative for HIV?* (document to assess time period for future testing).

Objective: Signs
Using Basic Tools: Inspection: **Acute** (1-30 days): Fever to 102°F, aphthous oral ulcers, maculopapular rash on neck and trunk. **Sub-acute** (30 days-1 year): Generalized adenopathy. **Chronic** (> 1 year): Fevers as high as 104°F, cachexia, oral thrush, Herpes zoster, signs of opportunistic infection (see Table 5-75)
Palpation: **Sub-acute** (30 days-1 year): Generalized adenopathy
Using Advanced Tools: Lab: Blood smear for lymphopenia and atypical lymphocytes.

Assessment:
Differential Diagnosis (in addition to AIDS and HIV-related illnesses)
Generalized TB - HIV patients are often co-infected with TB (see ID: TB), and may have chest X-ray findings.
Disseminated histoplasmosis - febrile wasting illness, unlikely to have oral thrush or shingles
Visceral leishmaniasis - chronic wasting, febrile illness with prominent hepatosplenomegaly (unusual for HIV alone), pancytopenia; history of travel to endemic area
Lymphoma - may present in a disseminated form like HIV symptoms; pronounced lymphadenopathy, splenomegaly
Acute viral illness - CMV, mononucleosis, hepatitis B, and adenovirus (see topics)

Plan:
Treatment
There is no cure. Evacuate newly diagnosed team member with HIV with an escort (suicide watch). Safeguard his weapon. Do not attempt to institute treatment, unless evacuation not available in the foreseeable future. Then test with PPD and treat with INH prophylaxis if >5mm (see ID: TB). Also treat active opportunistic infections (see appropriate topic). Refer local nationals to host nation medical resources.

Patient Education
General: Discourage pregnancy due to significant risk of sexual transmission to the partner, perinatal transmission to the fetus, and the likelihood that the mother will not survive to raise the child.
Diet: Maintain nourishment (AIDS patients are often severely malnourished). Replace vitamins. Note frequent lactose intolerance.
Prevention: Use latex condoms for sexual intercourse. Do not have unprotected sex, re-use needles, or be careless with body fluids. Do not breastfeed baby if formula is available, safe, cheap, and not dependent on unsafe water supply. Avoid contact with ill persons to avoid contracting diseases.

NOTES: Medics should use body fluid precautions: latex gloves and gown to handle fluids; add eye protection and mask for potential contact with blood and body fluids under pressure (needle stick, dental work, etc.). Wash skin well with soap and water if it becomes contaminated with body fluids.

Occupational needlestick/post exposure prophylaxis:
Zidovudine/lamivudine (Combivir) one pill bid and **efavirenz** 600mg po q hs
or **nevirapine** 200 mg po qd for 2 weeks
then 200 mg bid
or Protease inhibitor (**nelfinavir** 1250 bid,
indinavir 800 mg tid)
Retroviral must be started ASAP for maximal protection and continued for four weeks after needlestick/ exposure.

ID: Infectious Mononucleosis
LTC Niranjan Kanesa-thasan, MC, USA and COL Warren Whitlock, MC, USA

Introduction: Infectious mononucleosis (IM) is caused by one of the most common human viruses, the Epstein-Barr virus (EBV), a member of the herpes virus family. In the developed world, "mono" is particularly

common in young adults (ages 17-25 years), where it is transmitted by passage of infectious salivary secretions (kissing disease). In endemic regions, almost all children are infected by 3 years of age. EBV infects 80-90% of all persons by adulthood, but only 50% will develop clinical disease. Time from infection to appearance of symptoms is 4-6 weeks.
Risk Factors: Transmission is facilitated by crowded conditions allowing close contact, such as among military recruits. Because of prolonged excretion of infectious virus, transmission may be maintained in susceptible communities for months.

Subjective: Symptoms
Acute (2-7 days): Fever, chills, malaise, anorexia, severe sore throat **Sub-acute (1-2 weeks):** Fever, fatigue, malaise, severe sore throat, rash, swollen lymph nodes in neck **Chronic (2 Weeks to 3 months):** Fatigue, malaise

Focused History: *Do you have a cough?* (IM rarely has clinical pulmonary manifestations) *How long have you felt sick? What were your first symptoms?* (Usually, a patient presents with malaise lasting several days to a week, followed by fever, sore throat, and enlarged lymph nodes in the neck.) *Has anyone you live with or been sick with a similar illness in the past few months?* (indolent but contagious illness, requiring exchange of saliva)

Objective: Signs
Using Basic Tools
Vitals: Fever to 100.4 – 104°F, typically peaks in afternoon
Inspection: Nontoxic appearance, swollen neck, faint measles-like rash, pharyngitis with edema or exudative tonsillitis, palatal petechiae (red spots on back of throat)
Palpation: Splenic enlargement, +/- hepatomegaly; swollen cervical lymph nodes (discrete, firm to touch, tender but without fluctuance).
Auscultation: Stridor from upper airway obstruction (rare)
Percussion: Mild tenderness over liver
Using Advanced Tools: Lab: Monopot Test positive; differential reveals lymphocytosis with >10% atypical lymphocytes.

Assessment:
Differential Diagnosis
Pharyngitis
Other upper respiratory tract infections (see Respiratory: Common Cold and Flu). Mononucleosis persists longer.
Hepatitis A, B - prolonged malaise and fatigue, but jaundice is typically observed later
Hepatomegaly - cytomegalovirus and HIV are less likely to be associated with severe pharyngitis.
Toxoplasma and rubella - rare causes of IM-like syndromes, but may be a significant risk to the fetus if they occur during pregnancy.
Generalized lymphadenopathy - zoonoses such as brucellosis and leptospirosis are unlikely (see topics in ID).

Plan:
Treatment
1. Treatment is supportive, usually for relief of throat pain and fevers (see Procedure: Pain Assessment & Control).
2. For impending airway obstruction give **Prednisone** 1 mg/kg po qd x 3 days, then taper over a week. Alternatively, give **Prednisone** 50 mg po and taper off by 10 mg/d over 5 days.

Patient Education
General: No heavy lifting or contact sports or parachuting for 2 months after onset of illness to prevent splenic rupture.
Activity: Rest during acute illness

Diet: As tolerated. Increase fluids.
Medications: Acetaminophen (instead of aspirin) for pain and fever to avoid risk of Reye's Syndrome.
Prevention and Hygiene: Avoid close contact with others until well after secretions resolve.
No Improvements/Deterioration: Return immediately for sudden onset of severe abdominal pain, or fainting or lightheadedness after abdominal trauma.

Follow-up Actions
Return Evaluation: Examine liver and spleen carefully, for tenderness, enlargement or rupture.
Evacuation/Consultation criteria: Evacuate patient with enlarged spleen prophylactically, to avoid emergent evacuation (and poor prognosis) for a ruptured spleen. Otherwise, evacuation is not typically necessary, unless patient becomes unstable or is too fatigued to complete the mission. Consult primary care physician, infectious disease specialist or pulmonologist as needed, and consult a general surgeon urgently for a suspected splenic rupture.

ID: Poliovirus
LTC Niranjan Kanesa-thasan, MC, USA

Introduction: Poliovirus is an enterovirus spread through fecal or pharyngeal secretions and associated with outbreaks of paralytic poliomyelitis. The incubation period to illness is 7-14 days following respiratory or oral exposure to infectious poliovirus. Typically enterovirus infections are asymptomatic or minor febrile illnesses. Only about 1% of polio infections result in clinically apparent neurologic disease. Geographic Association: Polio is primarily a disease associated with poor sanitation and is found primarily in the developing countries of Asia and Africa. It has been eradicated from the Western hemisphere through immunization, except in rare cases of importation from endemic areas. Seasonal Variation: None in the tropics. In the past, clusters of infections occurred during fall months in temperate regions. Risk Factors: Outbreaks occur in unvaccinated populations, typically those living in poorer conditions or those objecting to immunization. While rare (<5% of symptomatic poliovirus infections), the frequency of paralytic polio increases with increasing age at time of infection. Secondary problems: Myelitis, peripheral neuropathy; Guillian-Barre syndrome (post-infectious polyneuritis)

Subjective: Symptoms
Malaise, headache, nausea, vomiting, and sore throat; uneventful recovery within several days (abortive poliomyelitis); 10-20% of symptomatic infections progress with severe muscle spasms, neck and back stiffness, and muscle tenderness lasting about 10 days with complete recovery (nonparalytic poliomyelitis); few develop paralytic poliomyelitis: asymmetric weakness or paralysis.
Focused History: *Have you completed the full polio vaccination series?* (drastically reduces chance of polio infection) *Did fever precede the limb weakness?* (Flaccid limb paralysis, especially if asymmetric, after acute febrile illness in a child or young adult is probable polio until proven otherwise.) *Have you traveled overseas or otherwise been exposed to poliovirus (including vaccine virus) recently?* (travel to endemic area [see above] or other contact within past several weeks establishes risk for infection).

Objective: Signs
Using Basic Tools: Vitals: Fever; rarely, respiratory embarrassment leading to paralysis
Inspection: Various findings: asymmetric flaccid paralysis from lower motor neuron damage (spinal poliomyelitis); paralysis of respiratory muscles or bulbar paralysis leading to respiratory embarrassment; cranial nerve palsies without sensory loss or dysphagia; deep tendon reflexes diminished or lost asymmetrically.
Palpation: Nuchal (neck) rigidity.

Assessment:
Differential Diagnosis
Mumps, tuberculous meningitis (TB) or brain abscess (see topics).
Two other enteroviruses are associated with asymmetric flaccid limb paralysis (in particular coxsackievirus A7 and enterovirus 71); in these rare cases, the clinical picture is identical to polio.

Acute Guillain-Barre syndrome (post-infectious polyneuritis) may be clinically similar in presentation, but usually has an afebrile, symmetric paralysis, often associated with sensory loss.

Plan:
Treatment
Supportive care is indicated, including analgesics for fever or pain (see Procedure: Pain Assessment and Control). No specific treatment exists for these viruses.

Patient Education
General: Do not expose others to infected body fluids (enteric precautions).
Activity: Rest on firm bed with footboard and sponge rubber pads or rolls. Physical therapy, with early mobilization after illness, is important during convalescence; brace and lightly splint affected limbs.
Diet: As tolerated.
Prevention and Hygiene: Immunize with inactivated poliovirus vaccine. Eradication of polio has been achieved in the Western hemisphere and a global campaign is underway using mass immunization to control outbreaks of disease.
No Improvements/Deterioration: Return for new, recurrent or worsening symptoms.

Follow-up Actions
Return Evaluation: Respiratory compromise or bulbar involvement requires intensive care (airway management, assisted ventilation).
Evacuation/Consultation criteria: Evacuate patients suspected of having of polio. Consult infectious disease or preventive medicine physician for any suspected case of polio.

NOTE: In endemic regions, 90-95% of cases occur before age 6 (median age for polio is 1 year).

ID: Rabies
LTC Niranjan Kanesa-thasan, MC, USA

Introduction Rabies is a fatal, acute, viral encephalomyelitis caused by the rabies virus (a Lyssavirus), which is endemic to areas of Asia, Africa, and South America. The disease is typified by fever with ascending paralysis and abnormalities of consciousness and behavior. Illness results from infection through bites, scratches or licks from an infected animal (dogs, bats or other mammals— rarely other humans). Recent fatalities in the US may have been exposed through contaminated aerosols in closed environments, e.g., fetid air in cave filled with bat guano. The incubation period is lengthy (usually 20-90 days) but varies. If recognized early, active immunization can abort the course of disease. Geographic Association: Worldwide, but has been successfully kept out of Australia, New Zealand, UK, Ireland, Scandinavia, Japan, Taiwan, Hawaii, other small islands. Risk Factors: An attack by rabid wild mammals, including bats, foxes, skunks, and other carnivores, is most suspect. A bite or other exposure from domestic pets, rabbits, or rodents generally conveys little risk. Worldwide, and particularly in developing countries, dogs are the major reservoir for transmission of rabies to humans.

Subjective: Symptoms
Tingling or pain at inoculation site (45% of cases); malaise, fatigue, headache or fever for 2-7 days; progression to apprehension, agitation, hyperactivity, bizarre behavior, hallucinations, nuchal (neck) rigidity, paralysis, coma and death (99% mortality) over 7-12 days.
Focused History: *Have there been behavioral changes or increased aggressiveness in the patient?* (supports the early symptoms of rabies, but not specific unless associated with ascending paralysis) *How long has the patient been ill?* (duration longer than 5 days with progression rules out intoxication and some viral encephalitides) *Is there a history of animal bite, exposure to bats, or travel to rabies-endemic area within past several months?* (if yes, increases clinical suspicion of rabies)

Objective: Signs
Using Basic Tools: Vitals: fever; rapid, shallow respirations; occasionally irregular RR and HR in later stages. Inspection: Agitated or frightened appearance is characteristic after CNS involvement with rabies. Classical spasms of pharynx or larynx during attempts to eat or drink ('hydrophobia') are seen in 50% of cases. Often ascending paralysis from bitten limb spreads to bulbar muscles and then causes coma.
Neurological: Use Glasgow coma scale to track progression of mental status changes, and help gauge need for medical evacuation or consultation. (see Appendices: GCS)
Auscultation: Late cardiac dysrhythmias coincident with myocardial involvement.
Using Advanced Tools: Confirmation of rabies diagnosis requires special clinical specimens (corneal scraping, skin biopsy, brain material) and specialized laboratory facilities for immunofluorescence or PCR unavailable in the field.

Assessment:
Differential Diagnosis: Pathognomonic (indicative) features - hydrophobia, inspiratory spasms
Polio - asymmetric ascending paralysis after minor febrile illness; encephalitic symptoms are rare
Viral encephalitides - respiratory symptoms not as prevalent as with rabies.
Intoxication (tetanus, botulism, drugs, etc.) - does not present with progression of CNS changes.

Plan:
Treatment after credible rabies exposure
1. Immediately scrub wounds or broken mucous membranes with soap or detergent and water.
2. Debride or irrigate wounds with water or sterile saline (preferred) using a 19 gauge blunt needle and a 35ml syringe to provide adequate pressure (7 psi) and volume. Flush individual punctures with approximately 200cc of irrigation solution. Treat with antiseptics. Do not close the wound.
3. Infiltrate around inoculation site with half of dose of **human rabies immune globulin** (HRIG 20 IU/kg) and give remaining half IM into gluteal region.
4. If patient is not immunized against rabies, give **human diploid cell rabies vaccine** (1 ml IM in deltoid x 5) beginning immediately. For individuals who have been previously fully vaccinated against rabies (including ID and IM protocols given to most SOF personnel), give 1 ml IM booster dose in deltoid immediately at presentation and again 3 days later. Pre-exposure vaccination does not guarantee protection against rabies, but it does buy time to get to definitive treatment if bitten, and it does decrease the number of post-exposure boosters required.
5. Give tetanus prophylaxis and antibiotic treatment (See Trauma: Human and Animal Bites).
6. Use narcotics or benzodiazepines judiciously for agitation (see Procedure: Pain Assessment and Control).
7. If possible, isolate suspected animal source and observe 10 days for signs of rabies.

Patient Education
General: Keep body fluids isolated from others (body fluid precautions).
Activity: Rest
Diet: As tolerated, but swallowing may be difficult with advanced disease.
Prevention and Hygiene: Pre-exposure prophylactic vaccination is strongly recommended for travelers in rabies enzootic for > 30 days, including most SOF personnel and is very effective.

Follow-up Actions
Wound Care: Usually no special care required after initial treatment.
Return Evaluation: Evaluate for progression of neurological signs.
Evacuation/Consultation criteria: Evacuate personnel suspected of exposure to rabies or a rabid animal. Consult infectious disease or preventive medicine specialists for any suspicion of rabies.

Zoonotic Disease Considerations
Principal Animal Hosts: Wild and domestic canids, raccoons, skunks, bats
Clinical Disease in Animals: Acute neurologic dysfunction, ataxia, progressive paralysis, absent reflexes; behavioral changes: anorexia, nervousness, irritability, hyperexcitability, uncharacteristic aggressiveness (wildlife lose fear of man; nocturnal animals seen during the daytime); furious form: pronounced aggressive-

ness ("mad dog" syndrome), progressing to ataxia, seizures and death; dumb or paralytic form: paralysis of throat or masseter muscles, profuse salivation, inability to swallow, progressing to coma and death.

Rickettsial Infections
ID: Rickettsial Infections
COL Naomi Aronson, MC, USA

Introduction: *Rickettsia rickettsia* is a tick-borne bacteria that causes **Rocky Mountain Spotted Fever** (RMSF) in the New World. Other tick-borne spotted fevers include African and Mediterranean tick fever (or **Boutonneuse fever**, *Rickettsia conorii*), Queensland tick typhus (*Rickettsia australis*). All these "spotted fevers" present similarly and are managed the same way. Fulminant RMSF is noted in blacks with G6PD deficiency. The incubation period is 3-12 days after a tick bite. *Rickettsia prowazekii* causes the severe illness **typhus**, and is transmitted worldwide between humans by the body louse. In the U.S, contact with flying squirrels has also been associated with transmission. A milder illness may recur years after the first attack, not associated with re-infection (Brill Zinsser disease). The incubation period is 7-14 days.
Orientia tsutsugamushi infection follows chigger bites in the Asiatic-Pacific area and causes **scrub typhus**. The dark punched-out skin ulcer at site of the bite is known as the "tache noire" (also in most spotted fevers), and is helpful to clinically suggest this infection. The incubation period is 7-21 days.
Murine typhus is caused by *Rickettsia typhi*, which is transmitted worldwide by flies that feed on rats. The illness resembles louse borne typhus but is much milder. Murine typhus peaks in the late summer/autumn and is prevalent in the urban environment. The incubation period is 6-18 days.
Rickettsialpox is caused by *Rickettsia akari* and is transmitted by rodent mite and chigger bites. The vesicular rash is similar to chickenpox, and is seen in urban areas of the northeastern US and some other areas in the world (Africa, Ukraine). The incubation period is 9-14 days.

Subjective: Symptoms
Spotted fevers: Acute (1-3 days): Fever >102°F, headache, myalgias, rash (typical but not in all cases), tache noire (except RMSF), mental status changes. **Sub-acute** (4-7 days): Same as acute. **Chronic** (> 1 week): continued myalgias, rash , and tache noire
Typhus Group: Acute (1-3 days): Fever (can last up to 14 days), chills, myalgias, headache **Sub-acute** (4-7 days): rash, ± cough **Chronic** (> 1 week): cough, rash
Focused History: *Have you had a recent tick bite?* (exposure) *Did you have a flu-like illness and fever a couple of days before the rash?* (c/w typhus group infection) *Have you been in the woods or traveled overseas in the last 2 weeks?* (exposure; see above for geographic clues). *Do you have a dark scab-like skin sore with surrounding redness?* (the tache noire of scrub typhus and other rickettsial infections)

Objective: Signs
Using Basic Tools: Inspection
All: Fever to 104°F, terminates by day 14 (except Brill Zinsser, which may recur), often ill-appearing.
Sub-acute/Chronic: Rash
Rocky Mountain spotted fever (other spotted fevers too): **Sub-acute** (4-7 days): maculopapular rash that starts on extremities includes palms/soles/face, petechiae and hemorrhagic lesions can occur, digital gangrene is rare **Chronic** (>1 week): may see late desquamation
Typhus (louse borne): **Acute** (1-3 days): Conjunctival injection **Sub-acute** (4-7 days): Small macules in axilla/trunk that spread over body and can become dark in color (hemorrhagic). Rare digital gangrene
Scrub typhus: Acute (1-3 days): Tache noire (see Color Plates Picture 6) **Sub-acute** (4-7 days): Spreading dull red maculopapular rash starts on trunk, spreads to extremities
Murine typhus: Sub-acute (4-7 days): Maculopapular rash that is sparse and discrete mainly trunk and extremities, can be palm/soles
Rickettsialpox: Acute (1-3 days): Tache noire **Sub-acute** (4-7 days): Maculopapular rash with vesicles that crust, spares palms and soles

Assessment: This is a clinical diagnosis.
Differential Diagnosis
Typhoid fever - usually rose spots are few and short-lived, elevated WBC
Malaria - no rash; parasites seen on blood smears
Leptospirosis - generally have more intense conjunctival redness, may see jaundice, rash is more petechial if present
Meningococcemia - petechial rash may progress to purpura, meningitis symptoms are common, rapid progression
Dengue - prominent myalgias and fever (breakbone fever)
Others - syphilis (usually no fever), drug reaction or side effect, rubeola or rubella (different rash; see topics).

Plan:
Rocky Mountain Spotted Fever (and **Boutonneuse fever**): Treat early based on clinical considerations.
Treatment: Doxycycline 100mg bid x 7 days (or at least 48 hours afebrile), or **chloramphenicol** 500 mg po qid x 7 days

Zoonotic Disease Considerations for RMSF
Principal Animal Hosts: Rabbits, field mice, dogs
Clinical Disease in Animals: Fever, anorexia, lymphadenopathy, polyarthritis, coughing, dyspnea, abdominal pain, edema of face or extremities
Probable Mode of Transmission: Bite of infected ticks, or crushing Dermacentor variabilis or andersoni ticks on skin
Known Distribution: Western Hemisphere

Zoonotic Disease Considerations for Boutonneuse Fever
Principal Animal Hosts: Dogs, rodents
Clinical Disease in Animals: Tick paralysis
Probable Mode of Transmission: Bite of infected tick
Known Distribution: Europe, Asia, Africa

Typhus
Treatment: Doxycycline 200 mg po load then 100 mg po q day for 10 days (or until 48 hours afebrile), or **Tetracycline** 25 mg/kg po in divided 6-8 hour doses until 48-72 hours afebrile, or **doxycycline** single dose 5mg/kg. Alternatives are **ciprofloxacin** and **azithromycin** (since resistance in Thailand is being seen to doxycycline and tetracycline). Use IV fluids cautiously.

Zoonotic Disease Considerations for Typhus
Principal Animal Hosts: Flying squirrels
Clinical Disease in Animals: Asymptomatic
Probable Mode of Transmission: Human louse; squirrel fleas or ticks suspected
Known Distribution: Worldwide; Eastern USA in squirrels

Scrub typhus
Treatment: Doxycycline 100 mg po bid for 7 days (resistance in Thailand to **tetracycline** and **chloramphenicol**)

Zoonotic Disease Considerations for Scrub Typhus
Principal Animal Hosts: Rodents
Clinical Disease in Animals: Asymptomatic
Probable Mode of Transmission: Bite of infected larval trombiculid mites
Known Distribution: "Typhus islands" in Asia, Australia, and East Indies

Murine typhus
Treatment: Doxycycline 100 mg/day until 2-3 days after defervescence

Zoonotic Disease Considerations for Murine Typhus
Principal Animal Hosts: Rats, cats, opossums
Clinical Disease in Animals: Asymptomatic
Probable Mode of Transmission: Infected rodent fleas; possibly cat fleas
Known Distribution: Worldwide

Rickettsialpox
Treatment: Tetracycline 15mg/kg/day in four divided doses x 3-5 days duration

Zoonotic Disease Considerations for Rickettsialpox
Principal Animal Hosts: Mice
Clinical Disease in Animals: Asymptomatic
Probable Mode of Transmission: Bite of infected rodent mites, *Liponyssoides* spp
Known Distribution: Eastern USA, Africa, and Russia; rare

Patient Education
Medications: Avoid taking doxycycline with iron, milk or milk products, multivitamins or antacids, and avoid the sun. Tetracyclines are often associated with mild gastritis and gastrointestinal upset. Take with food to help with this symptom. Avoid sulfa medications during this illness as they can make symptoms worse.
Prevention and Hygiene: Avoid ticks and control fleas, lice and rats (see Preventive Medicine chapter).

Follow-up Actions
Evacuation/Consultation criteria: Evacuate unstable patients and those unable to complete the mission, including those with RMSF. Consult infectious disease or primary care physician for RMSF cases and as needed.

ID: Q fever
COL Naomi Aronson, MC, USA

Introduction: Q fever is caused by *Coxiella burnetii*, a rickettsia that infects sheep, cattle, goats, and occasionally cats. Humans are infected by ingesting infected raw milk or by inhaling droplets or contaminated dust. Tick bite and blood transfusion are rare methods of infection. The incubation period is 9-28 days. Chronic infections cause hepatitis and sometimes jaundice.

Subjective: Symptoms
Constitutional Acute (1-7 days): Sweats, chills, fever (abrupt onset), severe retro-orbital headache. myalgias, malaise **Chronic (> 3 weeks):** Fever continues if endocarditis develops
Specific Sub-acute (8-20 days): Non-productive cough; 1% have neurologic symptoms-weakness, meningitis, sensory loss, paresthesias
Focused History: *In the past month have you had exposure to a pregnant animal that gave birth? Been working closely with ill animals? Have you ingested unpasteurized milk? Have you recently had a tick bite?* (all affirmative suggests exposure)

Objective: Signs
Using Basic Tools:
Inspection **Acute(< 1 week):** Fever to 104°F **Chronic (>3 weeks):** If fever persists 3 weeks, r/o endocarditis (rare): check for embolic events (splinter hemorrhages, Roth spots [advanced tools], Osler nodes), clubbed fingers
Auscultation **Acute(< 1 week):** Inspiratory crackles **Chronic (>3 weeks):** Diastolic murmur of aortic

regurgitation
Palpation **Sub-acute (8-20 days):** Hepatomegaly 5%, splenomegaly, rare neurologic findings: sensory loss, cranial nerve palsies, cerebellar signs
Using Advanced Tools: CXR may show pneumonitis (diffuse inflammation), 1/3 have pleural effusions.

Assessment:
Differential Diagnosis
Influenza - febrile respiratory infection of relatively short duration.
Salmonella - febrile illness with gastrointestinal symptoms which can persist for weeks, food/water borne.
Malaria - recurrent fever; blood smears show parasite.
Hepatitis - often less prominent fever; jaundice may be present; more prominent anorexia, malaise and fatigue
Atypical pneumonia - can mimic Q fever pneumonia; ask about relevant exposure history to lead to Q fever diagnosis
Other diagnoses to consider include brucellosis, psittacosis, typhus

Plan:
Treatment
Primary: Acute infection: Doxycycline 100 mg po bid until afebrile for 5 days
Chronic infection: Ciprofloxacin 750 mg po bid and **rifampin** 300 mg po bid

Patient Education
Prevention and Hygiene: Drink pasteurized milk. Burn or bury highly infectious *Coxiella*-contaminated tissue.

Zoonotic Disease Considerations
Principal Animal Hosts: Sheep, cattle, goats, occasionally cats
Clinical Disease in Animals: Usually subclinical; anorexia, abortion.

Spriochetal Infections
ID: Leptospirosis
COL Naomi Aronson, MC, USA

Introduction: Leptospirosis is usually transmitted by contact with the urine of infected wild or domestic animals worldwide. Humans contract leptospirosis through contact with water, food or soil containing urine from these animals. *Leptospira* microorganisms cause illness after skin/mucous membrane exposure or ingestion of fresh water contaminated by the urine or tissues of infected animals. This may happens by consuming contaminated food or water or through skin contact, especially with mucous membranes or broken skin. Person to person transmission is unknown. Leptospirosis infection has an incubation period of 2-20 days. Severe leptospirosis is called Weil's disease.

Subjective: Symptoms
Fevers to 104°F which defervesce; then recurrent fever, chills, headache, conjunctival suffusion, uveitis (red, painful eye), photophobia (can last for months), jaundice, rash, myalgias; 10% of cases develop severe symptoms (Weil's disease), with jaundice, hemorrhage (skin/GI/pulmonary), renal failure, aseptic meningitis.
Focused History: *Have you had a fever that resolved after one week then recurred?* (typical pattern) *Have you recently waded, swam or bathed in freshwater?* (typical exposure in endemic area) *Do your calves or back and neck muscles ache or feel stiff?* (typical myalgias).

Objective: Signs
Using Basic Tools:
Inspection: Fever to 104°F; often biphasic (first episode lasts 3-9 days, then 3 days without fever, then recurs); day 3: conjunctival and palatal erythema; jaundice, petechiae and cutaneous hemorrhage in severe

disease,
Auscultation: Respiratory distress (in severe disease may have ARDS), tachypnea, rales
Palpation: Generalized adenopathy, tender calf muscles, tender abdominal rectus/paraspinal muscles, hepatosplenomegaly (severe disease)
Using Advanced Tools: Lab: WBC is normal to slightly elevated; urine may show proteinuria, pyuria, hematuria. CXR can show pulmonary edema like findings with diffuse whiteout, pleural effusions

Assessment:
Differential Diagnosis:
Dengue - more prominent petechial rash; jaundice unusual
Yellow Fever - endemic areas; jaundice is fulminant; bleeding manifestations; vaccine should be protective
Influenza - more nasopharyngeal symptoms, cough, sore throat; no conjunctival discharge
Hepatitis - more anorexia, malaise, jaundice; hepatitis A. E often febrile, while B, C not so likely; protective vaccine for A, B
Cytomegalovirus - more chronic fatigue; unlikely to have conjunctival redness, severe myalgias; sexual exposure is likely source
Malaria - fevers/chills; blood smears are diagnostic; cyclical fevers, but not biphasic pattern; conjunctival discharge is unusual; malarious blackwater fever or cerebral malaria could be confused with Weil's disease (severe leptospirosis).
Hantavirus pulmonary syndrome - for Weil's Disease, history of exposure to rodents versus standing freshwater may differentiate

Plan:
Treatment
Primary: Doxycycline 100 mg po bid x 7 days
Alternative: Amoxicillin 500 mg q 6 h x 7 days
Severe disease: IV **penicillin** 6-10 million units/day in q 4-6 hour doses x 7 days
NOTE: Consider pretreatment with **Tylenol** before starting antibiotics due to Jarisch-Herxheimer like reactions (fever, tachycardia, mild hypotension, chills, vasodilatation within 2 hours of treatment; peaking at 7-8 hours and resolving in one day)

Patient Education
Activity: Avoid sun exposure with **doxycycline**.
Diet: As tolerated
Medications: Photosensitivity with **doxycycline** treatment. Avoid **doxycycline** in young children and pregnant/nursing mothers.
Prevention and Hygiene: Use **doxycycline** 200 mg po q week prophylactically during period of exposure. Dispose of urine appropriately to avoid further transmission. Wear boots and avoid skin exposure to streams, standing water and mud after rainy season.
No Improvement/Deterioration: Return promptly for reevaluation.

Follow-up Actions:
Evacuation/Consultation Criteria: The cultures and serologies for definitive diagnosis are not available in the field. Acute renal failure may develop, promoting the need for peritoneal dialysis.

Zoonotic Disease Considerations:
Principal Animal Hosts: Dogs, cattle, swine, mice, rats.
Clinical Disease in Animals: Usually asymptomatic renal infection.

ID: Lyme Disease

LTC Glenn Wortmann, MC, USA & MAJ Joseph Wilde, MC, USA

Introduction: Lyme disease is a tick-borne zoonotic infection caused by *Borrelia burgdorferi*. In the U.S., it is most commonly found in the Northeast. It also occurs in Europe, Scandinavia, Russia, China, Japan and Australia and is transmitted by the bite of an ixodid tick (deer tick), primarily during the summer months when ticks are most active. The incubation period after a tick bite is 3-32 days.

Subjective: Symptoms
Early stage disease: Erythema migrans (circular, erythematous rash) at bite site in 75% of patients, with multiple lesions present in 50% of patients; fever, Bell's palsy, fatigue, malaise and cardiac abnormalities (dropped beats, chest pain, pericarditis).
Later stage disease: Joint stiffness, myalgias, pain and swelling; headache; polyneuropathy; CNS neurological problems (cerebellar ataxia, coma).
Focused History: *Do you recall being bitten by a tick?* (only 20-30 % recall bite) *Do you recall the size of the tick that bit you?* (Lyme disease is transmitted by the small deer tick) *How long ago were you bitten by a tick?* (The incubation period for erythema migrans is 3-32 days)

Objective: Signs
Using Basic Tools: Diagnostic rash of erythema migrans (EM) about 7 days after bite (see Color Plates Picture 17); peripheral cranial nerve VII palsy (facial paralysis), bradycardia, fever, generalized adenopathy, swelling in large joints (particularly knees), dropped heart beats, abnormal neurological exam.
Using Advanced Tools: EKG: Pericarditis, heart block. Serology testing is available for confirmation at most hospitals but is often negative with early infection.

Assessment:
The rash of erythema migrans (EM) is diagnostic of Lyme disease.
Differential Diagnosis:
Erythema migrans - can be confused with cellulitis, arthropod bite, contact dermatitis, pityriasis rosea, tinea corporis, drug reaction.

Plan:
Treatment: Acute infection (erythema migrans):
Primary: Amoxicillin 500 mg tid or **Doxycycline** 100 mg bid x 21 days.
Alternate: Clarithromycin 500 mg po bid x 14-21 days or **azithromycin** 500 mg po q day x 7-21 days
Neurologic infection: Penicillin 2 million units IV q 6h or **ceftriaxone** 30 mg/kg IV q d

Patient Education
General: Prevent tick bites, particularly in spring and early summer.
Activity: As tolerated. Avoid sun exposure while on **doxycycline**.
Diet: As tolerated
Medications: Occasional gastrointestinal side effects. **Doxycycline** should be avoided during pregnancy, breastfeeding and in children.
Prevention and Hygiene: Personal protective measures (insecticide, etc.) against ticks.
No Improvement/Deterioration: Return for evaluation.

Follow-up Actions
Consultation Criteria: Failure to improve.

Zoonotic Disease Considerations

Principal Animal Hosts: Deer, wild rodents
Clinical Disease in Animals: Limb or joint disease, neurologic, cardiac, renal abnormalities

ID: Relapsing Fever
COL Naomi Aronson, MC, USA

Introduction: *Borrelia recurrentis* is a spirochete transmitted by soft ticks and crushed body lice. The soft tick usually feeds on sleeping people in houses/cabins/caves/other dwellings. The risk of infection after infected tick bite is about 50%. Fevers last 2-9 days, alternating with 2-4 day afebrile periods and have many, progressively milder recurrences. The incubation period is 5-15 days.

Subjective: Symptoms
Constitutional: Acute (< 3 days): Abrupt onset fever 101-105°F, fever lasts 1-5 days then recurs multiple times, as fever ends: chills, diaphoresis, headache, arthralgias, myalgias, nonproductive cough
Specific: Sub-acute (4-14 days): Neurologic symptoms: Bell's palsy, deafness, visual changes mental status changes; myocarditis (symptoms of congestive heart failure)

Focused History: *Have you had on-and-off fevers lasting several days at a time?* (typical pattern, but may be seen with malaria and others) *Have you had a recent tick bite?* (exposure)

Objective: Signs
Using Basic Tools:
Inspection: **Acute** (< 3 days): Fever 101°F - 105°F that lyses with rigors, then recurs; with decreased fever, may see 30 minutes of tachycardia and increased blood pressure followed by diaphoresis and decreased blood pressure **Sub-acute** (4-14 days): 10% appear jaundiced, petechiae (if thrombocytopenic, esp. during fever) on neck and shoulders, conjunctival hemorrhage
Auscultation: **Sub-acute** (4-14 days): If myocarditis develops: S3 and bibasilar rales
Palpation Sub-acute (4-14 days): Splenomegaly; if neurologic involvement: cranial nerve palsies VII/VIII, bilateral or unilateral Bell's palsy
Using Advanced Tools: Lab: Giemsa or Wright's stain of a thin blood smear taken during fever may show spirochetes with darkfield exam. Examine up to 200 oil immersion fields before considering the smear negative. Blood smear may show thrombocytopenia, mild anemia, and a normal WBC count.

Assessment:
Differential Diagnosis:
Malaria - non-falciparum malaria may have periodic fevers; malaria parasite on blood smear and mosquito exposure
Lyme Disease - early in infection see skin patch/target lesion of erythema migrans; low grade temperatures usually
Leptospirosis - history of exposure to standing water/mud; usually a biphasic febrile illness with remarkable conjunctival discharge
Dengue - mosquito-borne febrile illness with prominent myalgias, headache, diffuse rash
Meningococcus - sustained febrile infection, often fulminant; petechia that progress to purpura; often with associated meningitis.
Ehrlichiosis - tick-borne fever infection; associated with pancytopenia on blood smear; usually no relapsing fever pattern
Other considerations include babesiosis, typhoid, brucellosis, rickettsial infection, Colorado tick fever.

Plan:
Treatment:
Primary: Doxycycline 100 mg po bid x 10 days
Alternative: Erythromycin 500 mg po qid x 10 days (use for pregnant women, children <8 years)

Patient Education
General: Doxycycline and **erythromycin** may cause GI upset. Relapsing fever acquired during pregnancy can lead to stillbirth, abortion.
Activity: Watch for photosensitivity while taking **doxycycline**. Avoid sun and use sunscreen.
Diet: Avoid taking milk products when taking **doxycycline**.
Medications: Start antibiotics when afebrile or near the end of a febrile period to avoid a potential Jarisch-Herxheimer reaction (fever, tachycardia, mild hypotension, chills, vasodilatation within 2 hours of treatment; peaking at 7-8 hours and resolving in one day). Give **Tylenol** 2 hours before antibiotics and 2 hours after.
Prevention and Hygiene: Prevent tick bites (spray interior of tick-infested dwelling infested with insecticide). Give **Tetracycline** 500 mg qid x 2-3 days after tick bite in endemic areas to reduce the risk of infection.
No Improvement/Deterioration: Return promptly for reevaluation.

Follow-up Actions
Evacuation/Consultation Criteria: Refer patients who develop complications.

Bacterial Infections
ID: Anthrax
COL Naomi Aronson, MC, USA

Introduction: Anthrax is an acute bacterial (spore-forming *Bacillus anthracis*) infection transmitted through broken skin or mucous membranes, inhalation, or ingestion (rare). Fatally infects herbivores (sheep, cattle, horses, pigs, goats, water buffalo, elephants, zebras or antelopes) that shed the bacilli into the environment. Spores remain viable in contaminated soil for years. Dried skins, hides, wool, bone or bone products can transmit infection. Veterinarians, farmers, tannery, wool workers are occupationally exposed. There are three forms of disease, dictated by entry site; 95% of cases are cutaneous. Less common gastrointestinal and pulmonary anthrax infections are generally fatal within days.

Subjective: Symptoms
Cutaneous anthrax: Acute (< 1 week): Local pruritus, papule with vesicles **Sub-acute (1-2 weeks):** Ulcer that dries to a black painless eschar, brawny edema, regional adenopathy **Chronic (> 2 weeks):** Without treatment, 80% resolve over 6 weeks.
Pulmonary anthrax (acute symptoms only): Fever; URI-like cough and chest discomfort; 36-48 hours after infection may observe stridor; respiratory distress; shock and death.
Gastrointestinal anthrax (acute symptoms only): Neck swelling, fever, painful swallowing, nausea, vomiting, severe abdominal pain, gastrointestinal hemorrhage and death.
Focused History: *Is anybody else in your unit sick with the same symptoms?* (If Pulmonary Sx, consider biowarfare exposure; if GI Sx, consider ingestion of infected meat.) *Do you have abdominal pain or bleeding? If so, did you eat rare meat from an animal that may have died from illness?* (GI cases usually occur in outbreaks with a history of eating raw/undercooked meat 48 hours before.) *Have you handled animal carcasses recently or local products made from animal hides or wool?* (cutaneous exposure) *How long have you been sick?* (Life threatening forms of anthrax may be fatal in 1 week; skin symptoms can last 6 weeks.) *Did you have a rash that turned into a painless sore?* (typical for cutaneous cases)

Objective: Signs
Inspection
Cutaneous anthrax: Acute(< 1 week): Papule with vesicles forms an ulcer **Sub acute (1-2 weeks):** Non-

tender ulcer with black eschar, edema (can persist) **Chronic (>2 weeks):** Eschar loosens and separates
Pulmonary anthrax: Rapid respiratory rate; neck and chest edema
Gastrointestinal anthrax: Acute(< 1 week): Edema of oropharyngeal tissues **Sub acute (1-2 weeks):** Oropharyngeal/tonsillar ulcers; may see bloody ascites
Auscultation: **Pulmonary anthrax:** Rhonchi or rales; hypotension
Palpation: **Cutaneous anthrax: Acute(< 1 week):** Regional adenopathy
Pulmonary anthrax: Neck and chest edema
Gastrointestinal anthrax: Tender abdomen
Pulmonary: Percussion: localized dullness
Using Advanced Tools: Lab: Boxcar-shaped, gram-positive rods on Gram stain of skin lesion fluid, or blood/sputum (late in course) (see Color Plates Picture 13).

Assessment:
Differential Diagnosis:
Cutaneous anthrax:
Staphylococcal boil - does not ulcerate or turn blackish; usually pus-filled and painful.
Orf - history of contact with ungulate's udders, mucous membranes; lesion can resemble anthrax but does not blacken.
Scrub typhus - "tache noire" lesion is blackish; usually patient is febrile, often with a generalized petechial rash.
Spider bite - brown recluse spider bite turns blackish with necrotic changes; usually painful lesion.
Cutaneous plague and tularemia - usually more vivid, painful skin lesions.
Pulmonary anthrax:
Influenza - initial symptoms are similar, but no hemorrhagic mediastinitis or death within 24 hours of onset.
Mediastinitis - usually seen post-operatively (thoracic surgery), with histoplasmosis or after esophageal perforation.
Gastrointestinal anthrax:
Severe gastroenteritis - early similar symptoms, but no progression to hematemesis, bloody stools, occasionally bloody ascites, shock and frequent death in 2-5 days.

Plan:
Treatment: Even if left untreated, 80% of cutaneous anthrax remains localized. However, 20% of skin cases and all other forms of anthrax may die from untreated infection.
Primary:
Cutaneous: Penicillin V 30 mg/kg in four doses/day x 5-7 days and **Cipro** 750 mg bid for 60 days
Pulmonary: Ciprofloxacin IV 400 mg q12hours x 7-10 days
Gastrointestinal: Ciprofloxacin 750 mg po bid x 60 days
Alternative:
Cutaneous: Levofloxacin 500 mg po q d x 60 days
Pulmonary/Gastrointestinal: Doxycycline 100 mg q12hour or **penicillin G** 4 MU q4hours IV for 7-10 days
Empiric (presumed exposure):
Pulmonary: Ciprofloxacin 500 mg po bid for 6 weeks, and vaccination

Patient Education
Prevention and Hygiene: A licensed vaccine is available for humans and animals. Human quarantine is unnecessary, as person-to-person transmission of any form is rare. Use body fluid precautions for duration of pulmonary and gastrointestinal illness. No viable organisms will remain in skin lesion after 24 hours of antibiotic therapy. Disinfect premises with 5% formaldehyde and incinerate wound dressings. Incinerate carcasses or bury deeply and cover with quicklime.

Follow-up Actions
Wound Care: Keep clean and dry.
Consultation Criteria: Report any case to local health authority immediately. Refer for expert care if available.
Evacuation: Evacuate when stable and body fluid precautions can be followed.

Zoonotic Disease Considerations
Principal Animal Hosts: Cattle, sheep, goats, and horses
Clinical Disease in Animals: Incubation period is 3-7 days (Range = 1 to >14 days). The peracute form in ruminants has a rapid onset and is fatal, with signs of dyspnea, ataxia, collapse, or convulsions. The acute form in ruminants begins with a rapid increase in body temperature (up to 107°F/41.5°C) and a period of excitement followed by depression, stupor, ataxia, convulsions and death. There may be bleeding from orifices. DO NOT NECROPSY any animal showing this sign. The chronic form is characterized by localized, subcutaneous, edematous swelling of the ventral neck, thorax and shoulders. In horses the disease is acute, with signs of fever, chills, colic, anorexia, and swelling of the animal's ventrum, with death in 2-3 days. Swine may die acutely without showing signs, or having a rapid swelling of the throat.
Probable Mode of Transmission: Occupational exposure (wool sorter's disease), foodborne in Africa, Asia and Russia
Known Distribution: Worldwide, associated with chalky soil.
NOTE: See also CBR: Inhalational Anthrax

ID: Bartonellosis
(Cat Scratch Disease, Trench Fever, Oroya Fever)
COL Naomi Aronson, MC, USA & MAJ Daniel Schissel, MC, USA

Introduction: There are several *Bartonella* bacterial species that cause different illnesses. *Bartonella bacilliformis* causes Oroya fever, a febrile anemia and chronic skin eruption (verruga peruana) in a limited area of mountain communities in Peru, Ecuador and Colombia. It is thought to be transmitted by the bite of a sand fly. *Bartonella quintana*, transmitted by body lice, causes trench fever, a febrile illness that which can recur years later. *Bartonella henselae* is inoculated by cat scratch or bite and causes lymphadenitis and variable fever 3-10 days after cat exposure.

Subjective: Symptoms
S. American Bartonella (Oroya fever): Fever, which can persist up to 6 weeks; pallor; weakness; chills; muscle/joint aches. After 2-20 weeks, crops of painless, red, 2-4 mm skin lesions can be seen, mainly on head and extremities.
Trench fever: Headache; fever (episodic fever of 3-5 days duration - can have fever relapses up to 10 years later); back/leg ache; shin pain; transient rash (maculopapular).
Cat scratch disease: Papule at inoculation site; regional adenopathy that can suppurate; nodes subside in 2-5 months without treatment.

Focused History: *Have you traveled in the past month to Peru, Ecuador or Colombia?* (where Bartonellosis is endemic) *Have you been around anyone with lice or poor body hygiene?* (trench fever exposure) *Have you been scratched or bitten by a cat?* (cat scratch disease exposure). *Do you have a new skin rash or sore?* (Look for the verrucous lesion of S. American Bartonellosis and the crusted papule at the inoculation site of cat scratch.) *Do you have any swollen or sore lymph nodes?* (In cat scratch disease you can see fluctuant, red, regional adenopathy.)

Objective: Signs
Using Basic Tools:
Oroya fever: Fever (up to 105°F) begins early and may persist for weeks, skin pallor and slight jaundice. After several weeks, crops of persistent, 2-4 mm, red to purple skin nodules appear on the exposed parts of the body.
Trench fever: Fever (up to 105°F) lasting 4-5 days, recurs in paroxysms for 3-6 weeks. In the immunocompro-

mised, crops of crusted red-purple papules appear.
Cat scratch disease: Variable fever >101°F is accompanied by crusted papule or pustule at inoculation site. Later: tender, fluctuant, regional adenopathy develops, which may last about 3 months (see Color Plates Picture 7). 12% have splenomegaly.
Using Advanced Tools: Lab: The Oroya fever organism can be seen on a peripheral blood smear inside red blood cells.

Assessment:
Differential Diagnosis:
Kaposi's sarcoma (KS) - trench fever skin lesions (immunocompromised patients) and Verruga peruana may resemble disseminated KS, a malignancy seen in immunocompromised hosts with HIV and in elderly men who live in the Mediterranean area.

Chronic febrile illnesses of the tropics, including Salmonella, malaria, tuberculosis, or brucellosis, may co-exist with Oroya fever or mimic it. Oroya fever is differentiated by severe anemia and the presence of bacteria in the red blood cells on blood smear.

The suppurative adenopathy associated with a skin lesion of cat scratch can mimic cutaneous plague, tularemia, toxoplasmosis, and sporotrichosis infection. Exposure to cat injury helps differentiate.

Plan:
Treatment
Primary:
Oroya fever: Chloramphenicol 2-4 g/d po in divided doses x 7 days
Verruga Peruana: Rifampin 10 mg/kg/day po x 14-21 days
Salmonella is a frequent life-threatening secondary infection, and is often suggested by splenomegaly. Infection during pregnancy can result in fetal death or abortion.
Trench fever: Aspirin prn, **doxycycline** 100 mg po qd x 15 days
Cat scratch disease: Azithromycin 500 mg po day 1, then 250 mg po qd x 4 days; analgesics
Alternate:
Oroya Fever: Doxycycline or **ampicillin**
Cat Scratch Disease: Give no treatment and infection self-resolves in 2-6 months

Patient Education
General: Oroya fever - watch for late development of skin verruca. Cat scratch disease-- apply heat locally for pain relief
Activity: As tolerated
Diet: Regular
Prevention and Hygiene: Oroya fever - use personal prevention methods during the nocturnal biting cycle of sandfly while in endemic areas. Trench fever – control body lice; use good personal hygiene.
Cat scratch disease - avoid playing with unknown cats, especially kittens.
No Improvement/Deterioration: Return if fever persists for more than one week or if (cat scratch disease) lymph nodes start to drain pus.

Follow-up Actions
Return evaluation: Cat Scratch Disease - Relieve pain in fluctuant lymph node with needle aspiration; avoid incision and drainage.
Evacuation/Consultation Criteria: Evacuation is not usually necessary, but may be indicated for some unstable patients. Serology is available on acute and convalescent serum at referral centers to confirm diagnoses.

Zoonotic Disease Considerations
Cat Scratch Fever
Principal Animal Host: Cats
Clinical Disease in Animals: Asymptomatic carriers
Known Distribution: Worldwide

ID: Brucellosis
COL Naomi Aronson, MC, USA

Introduction: Brucellosis is a common febrile illness in the Mideast, Mexico and South America. It is a bacterial infection acquired by ingesting raw milk, unpasteurized cheese or by direct contact with secretions or birth products of infected animals (cattle, goats, buffalo, camels, reindeer, caribou, yaks, coyotes, deer or swine). Average incubation period is 2 weeks, but can take up to several months.

Subjective: Symptoms
Acute (1-7days): PM fever > 100°F; profuse, malodorous sweating; flu-like symptoms and a peculiar taste in mouth **Sub-acute (1-2 weeks):** Fever, weight loss, arthralgias and myalgias **Chronic (weeks to months):** Recurrent undulant fever if not treated, arthralgias or arthritis, constipation, depression and back pain.
Focused History: Exposure– *Have you had any raw milk or cheese? Have you come in contact with cattle, goats, buffalo, camels, reindeer, caribou, yaks, coyotes, deer or swine?* **Fever–** *Do you have a fever?* (up to 104°F). *How long have you had fever?* (weeks to months). **Pain–** *Do any of your joints hurt?* (arthritis is usually in the knee or hip). *Do you have back pain?* (particularly unilateral sacro-iliac symptoms).

Objective: Signs
Inspection: Temperatures to 104°F
Palpation: **Acute (1-7 days):** Generalized adenopathy **Sub-acute (1-2 weeks):** Hepatomegaly (>50%), splenomegaly (30%)

Assessment: Travel history, animal exposure and consuming unpasteurized milk products.
Differential Diagnosis: Enteric fever, nonpulmonary tuberculosis, or non-falciparum malaria (with chronic relapses of fever), may have similar symptoms.

Plan:
Treatment
Primary: Doxycycline 100mg po bid x 6 weeks with **gentamicin** 5mg/kg q 24 hours IV x 14 days (danger of renal insufficiency)
Alternate: Doxycycline 100 mg po bid and **rifampin** 900 mg qd, but twice as high relapse rate

Patient Education
Diet: Avoid untreated milk; boil milk if pasteurization status is unknown.
Medications: Avoid **doxycycline** in children <7. **Trimethoprim-sulfamethoxazole** and **rifampin** can be used as an alternate regimen but 30% relapse rate.
Prevention and Hygiene: Handle animal carcasses carefully. Hunters should use gloves or barriers when dressing wild animals. Be careful when handling tissues and fetuses from aborted animals.
No Improvement/Deterioration: Return for persistent fever over 7 days.

Follow-up Actions
Consultation Criteria: Refer chronically febrile patients to higher level of care when available.

Zoonotic Disease Considerations
Agent: *Brucella canis, B. suis, B. abortus, B. melitensis*
Principal Animal Hosts: Dogs *(B. canis)*, swine *(B. suis)*, cattle *(B. abortus)*, goats *(B. melitensis)*
Clinical Disease in Animals: Dogs *(B. canis)* - Last trimester abortions; swine *(B. suis)* - abortion; cattle *(B. abortus)* - last half gestation abortions; goats *(B. melitensis)* - abortion.
Probable Mode of Transmission: *B. abortus* - Contact with birth products and consumption of milk products, *B. melitensis* - Consumption of milk products
Known Distribution: *B. canis* - rare, *B. suis* - northern hemisphere, *B. abortus* & *B. melitensis* - worldwide

ID: Ehrlichiosis
LTC Glenn Wortmann, MC, USA

Introduction: Ehrlichiosis is a tick-borne bacterial illness. Human infection with *Ehrlichia* has only recently been appreciated, with the first case reported in 1987. Most disease occurs in the United States, although it has been reported in Africa, Scandinavia and Western Europe. The severity of infection ranges from subclinical to fatal.

Subjective: Symptoms
Headache, fever, rash, myalgias.
Focused History: *Have you had a tick bite recently?* (typical exposure) *Do you have a headache and fever?* (suggests ehrlichiosis in late spring and early summer) *How long have you felt sick?* (often presents with the abrupt onset of fever, headache and myalgias)

Objective: Signs
Using Basic Tools: Temperature over 101°F; rash in 30% (most commonly involving the trunk; not associated with site of tick bite).

Assessment:
Differential Diagnosis: Many infections present similarly, but fever and headache (especially in a patient with exposure to ticks) should prompt the consideration of ehrlichiosis. Rocky Mountain Spotted Fever and meningitis have substantial clinical overlap.
NOTE: Serology testing of blood is available at many hospitals.

Plan:
Treatment:
Doxycycline 100 mg bid for x 7-10 days. Treatment often results in rapid clinical improvement and defervescence within one to two days.

Patient Education:
General: Avoid tick exposure.
Activity: As tolerated.
Diet: As tolerated.
Medications: Doxycycline can cause photosensitivity, and should not be given to young children or pregnant/ breastfeeding women.
Prevention and Hygiene: Avoid tick exposure.
No Improvement/Deterioration: Return for evaluation.

Follow-up Actions
Consultation Criteria: Failure to improve after initiation of antibiotics.

ID: Plague
COL Naomi Aronson, MC, USA

Introduction: Plague (*Yersinia pestis*), a highly fatal illness, is usually transmitted by a bite from a rodent flea or by contact with infected wild rodents (rats, squirrels, chipmunks, prairie dogs), rabbits, or domestic cats. Ingestion of infected animal tissues, direct handling, inoculation or inhalation of contaminated tissues are less common causes. The current plague vaccine does not protect against pneumonic plague, which can spread by droplet inhalation. The incubation period is 1-7 days. This rare disease has foci in the western USA, South America, Asia and Africa. The bubonic form (90-95% of cases) has a very rapid onset associated with a toxic state characterized by enlarged and very tender lymph nodes (buboes). The other forms of plague, pneumonic (progression of 5% of bubonic cases) and septicemic (5-10% of cases) are rapidly toxic and nearly always fatal if untreated.

Subjective: Symptoms
Constitutional: Acute (< 2 hr): Sudden onset fever, chills, headache, myalgias, lethargy
Specific: Sub-acute (24-48 hr): Cough, if pneumonic form; buboes* (85% cases), diarrhea (septicemic plague), abdominal pain **Chronic (>48 hr):** Bloody, frothy sputum; buboes suppurate
*Tender, swollen lymph nodes with red, edematous overlying skin, often in inguinal and axillary regions.
Focused History: *Do you have tender lymph nodes? If so, are they so sore that touching or movement can cause great discomfort?* (typical in bubonic form) *Can you see or recall a flea bite (a scab, pustule, ulcer) near the bubo?* (typical exposure) *Have you been handling rodents lately?* (typical exposure) *How long have symptoms been present?* (Plague is usually rapidly progressive)

Objective: Signs
Using Basic Tools:
Inspection: **Acute (< 48 hrs):** Fever to 104°F; a skin lesion at site of flea bite (25%); lethargy; tachycardia 110-140; hypotension; toxic appearance; Pneumonic: respiratory distress, tachypnea **Chronic (>2 weeks):** Can see meningeal signs: Plague meningitis is particularly associated with axillary buboes.
Auscultation: **Sub-acute (up to one week):** Rales if pneumonic
Palpation: **Acute (< 48 hrs):** Buboes **Sub-acute (up to one week):** Buboes may start to recede, also could suppurate.
Using Advanced Tools: Lab: Gram's stain of bubo needle aspirate (usually need to inject and withdraw some saline to get sample) or sputum may show gram-negative coccobacillus with a bipolar (safety pin) staining appearance. In severely septicemic patients this may even be seen with a Gram's stain of blood.
CXR: Can be normal but rapidly progress to diffuse pneumonitis (increased markings throughout lung fields).

Assessment:
Differential Diagnosis: Tularemia, enteric fever, rickettsial infection, acute lymphadenitis, dengue, typhus and Hantavirus.

Plan:
Treatment
Primary: Gentamicin 2 mg/kg load then 1.7 mg/kg q 8 hours IV (adjust lower if kidney dysfunction as noted on urinalysis)
Alternative: Doxycycline 200 mg first day then 100 mg bid x 10-14 days
NOTE: If meningitis develops, use **ciprofloxacin** 750 mg po bid or **chloramphenicol** 50 mg/kg/day x10-14 days.

Patient Education
General: Plague is a quarantinable disease subject to World Health Organization (WHO) international health regulations.
Activity/Diet: As tolerated
Prevention and Hygiene: Remove any fleas. Isolate with respiratory and secretion precautions for first 48 hours of treatment. Avoid sick and dead animals. Wear gloves when handling carcasses. Post warnings near areas of known plague. Use licensed vaccine, which protects against bubonic but not pneumonic plague.
Provide post-exposure prophylaxis: Doxycycline 100 mg po bid x 7 days.
No Improvement/Deterioration: Evacuation.

Follow-up Actions
Wound Care: Do not perform incision and drainage of suppurative nodes.
Evacuation/Consultation Criteria: Transport to a hospital level of care. Serology can be used to confirm diagnosis.

Zoonotic Disease Considerations
Principal Animal Hosts: Wild rodents (rats, squirrels, chipmunks, prairie dogs), rabbits, and domestic cats.
Clinical Disease in Animals: Fever, pneumonia, lymphadenitis

ID: Rat Bite Fever
LTC Glenn Wortmann, MC, USA

Introduction: Caused by infection with *Streptobacillus moniliformis* or *Spirillum minor*, rat-bite fever is transmitted by the bite of a rodent and has worldwide distribution. Contaminated water and milk have been implicated in outbreaks. Incubation period is 3-10 days.

Subjective: Symptoms
Fever, chills, headache, myalgias, arthralgias and rash
Focused History: *Do you recall being bitten by a rat?* (typical history) *How long ago were you bitten by a rat?* (incubation period is 3-10 days) *Do you have a rash?* (typical in most patients)

Objective: Signs
Using Basic Tools: The rash can be maculopapular, morbilliform or petechial, and erupts over the palms, soles and extremities. Approximately 50% of patients develop an asymmetric polyarthritis.
Using Advanced Tools: Lab: Culture of blood, joint fluid or pus may demonstrate organism.

Assessment:
Differential Diagnosis: Other potential causes of fever/rash are varied, and include measles, meningococcemia, Rocky Mountain Spotted Fever and secondary syphilis.

Plan:
Treatment: The presence of fever and rash after a rat-bite should suggest a diagnosis of rat-bite fever and treatment should be given empirically. For moderately to severely ill patients, use **procaine penicillin G** 600,000 units IM or IV **penicillin** every 12 hours. For mildly ill patients, **amoxicillin** 500 mg po tid x 14 days is probably adequate. For **penicillin**-allergic patients, **tetracycline** 500 mg po every 6 hours or **doxycycline** 100 mg po every 12 hours may be given.

Patient Education
General: Avoid rat bites
Activity: As tolerated
Diet: As tolerated
Medications: Occasional gastrointestinal side effects. **Tetracycline** and **doxycycline** should be avoided in children and pregnant women.
Prevention and Hygiene: Avoid rat bites.
No Improvement/Deterioration: Return for evaluation.

Follow-up Actions
Return evaluation: Referral as needed.
Consultation Criteria: Failure to improve.

Zoonotic Disease Considerations
Principal Animal Hosts: Rodents
Clinical Disease in Animals: Arthritis, pericarditis
Probable Mode of Transmission: Rodent bites, water-or food-borne
Known Distribution: Worldwide; rare

ID: Acute Rheumatic Fever
COL Naomi Aronson, MC, USA

Introduction: Acute rheumatic fever (ARF) typically occurs approximately 19 days after untreated streptococcal pharyngitis in 1-3% persons. Certain group A strep strains are more likely to cause ARF. Peak incidence is from 5-20 years of age. Recurrences are common (50%).

Subjective: Symptoms
Fever; migratory polyarthralgias in knees, ankles, elbows or wrists; rash which can wax and wane over months; subcutaneous nodules and chorea. If carditis is present, patient may have palpitations, chest pain, or SOB.
Focused History: *Have you had a sore throat in the last few weeks?* (ARF follows a strep throat infection.) *Have you taken your temperature or do you feel hot?* (expect to document a temperature >100° F) *Have you felt sick for a long time?* (ARF may present with chronic signs such as Sydenham's chorea and the rash of Erythema marginatum. Both are explained below.)

Objective: Signs
Using Basic Tools: Inspection: Fever to 102°F (up to 21 days); rapid respiratory rate (if carditis); E. marginatum rash (irregularly edged, transient, lacy, macular rash [pink rimmed with internal blanching] found on the trunk and extremities) which waxes and wanes for months; chorea (short, abrupt, non-purposeful movements, which often disappear during sleep and grimacing). Palpation: migratory large joint inflammation (arthritis); 10% have subcutaneous nodules on extensor elbows and forearms which last up to 4 weeks. Auscultation: In carditis: bibasilar rales, aortic insufficiency or mitral regurgitation murmurs, pericardial friction rub, S3 gallop.
Using Advanced Tools: In carditis, CXR may show cardiomegaly. EKG may have prolonged PR interval.

Assessment:
Diagnosis is based on clinical observation and application of the Jones criteria for diagnosis: 2 major criteria, OR 1 major and 2 minor criteria AND evidence for prior streptococcal infections (prior scarlet fever, + throat culture)
Major Criteria: Carditis, polyarthritis, subcutaneous nodules, and chorea, E. marginatum rash
Minor Criteria: Fever, arthralgias prior rheumatic fever, heart block seen on EKG
Differential Diagnosis
Polyarthritis is often the main presenting symptom. The differential diagnosis should include:
Gonococcal arthritis - sexual exposure; dysuria; urethral discharge; characteristic gun metal blue skin lesion
Subacute bacterial endocarditis - diagnose with blood cultures (if available); usually few joints involved; may see embolic skin lesions
Lyme disease - fever less frequent; fewer joints involved; history of tick bite
Reiter's syndrome - fever is unusual; prior history of sexually transmitted disease or acute diarrheal illness; characteristic skin lesions (painless, superficial erosions) in genital area/soles of feet

Carditis:
Viral myocarditis - unlikely to see joint symptoms or the E. marginatum rash
Pericarditis - may see diffuse ST segment elevations

Plan:
Treatment
Primary: **Benzathine penicillin** 1.2 MU IM (600,000 U in children) or oral **penicillin** 250 mg po qid x 10 days
Salicylates: 4-8 grams **aspirin** per day x 3-4 weeks
Alternative: **Erythromycin** x 10 days or **azithromycin** x 5 days can be used in patients allergic to **penicillin**. Other NSAIDs may be used in place of aspirin

Patient Education
General: This can be a relapsing condition.
Activity: Patients with arthritis or carditis should be on bedrest for 2 weeks (up to 8 weeks if cardiac failure present).
Diet: Regular.
Medications: Expect tinnitus with high dose aspirin.

Follow-up Actions
Secondary prophylaxis with **benzathine penicillin** 1.2 MU IM q month until age 18, or for 5 years after acute episode. If carditis/congestive heart failure then use prophylaxis until age 25 or 10 years after last ARF episode.

ID: Streptococcal Infections
COL Naomi Aronson, MC, USA

Streptococcal Pharyngitis and Scarlet fever: see Symptoms: Sore Throat and Pediatric chapter on CD-ROM
Group B Streptococcal Infection: see Symptom: OB Problems: Preterm Labor
Impetigo and Erysipelas: see Dermatology chapter

Introduction: Streptococci are gram positive bacteria. Some species are pathogenic in humans, responsible for illnesses ranging in severity from mild upper respiratory infections to life-threatening necrotizing fasciitis. Fulminant necrotizing fasciitis due to group A streptococcus may begin at a site of trivial trauma (for example a paper cut, abrasion on a cinder block), with rapid progression over 24-72 hours and 20-70% mortality even with ICU care. This is a deep-seated, fast moving infection that destroys fascia, fat, muscle and other tissue but may spare the skin. It often involves the extremities. Streptococcal infection of muscle (myositis) often is from blood infection, not associated with trauma. It may involve a single muscle group and be associated with compartment syndrome (see Procedure: Compartment Syndrome Management).

Subjective: Symptoms
Initial flu-like symptoms with fever to 104°F, extreme pain and erythema at infection site.
Focused History: *On a scale 1-10, how severe is the pain in the affected area?* (Great pain accompanying mild skin changes suggests a deeper infection.) *Have you had any trauma in this area recently?* (risk factor) *Have you had any blisters?* (suggests a deep tissue infection if no local injury such as frostbite or burns).

Objective: Signs
Inspection: Acute (< 24 hr): Ill-appearing, fever as high as 105°F, SBP <110, increased heart rate, abrupt onset red/swollen affected area, confusion/mental status changes common. **Sub-acute** (1-7 days):

Continuation of acute signs, red/swollen area spreads locally, color turns to blue/purple with bullae, may see gangrene by day 4-5. Chronic (>7 days): Earlier signs continue, clear demarcation at site of involvement
Palpation: **Acute** (< 24 hr): Tenderness >inflammatory change, area warm to touch; may see capillary leak syndrome with generalized edema (the "Michelin man" look), in myositis find changes c/w compartment syndrome of affected area. **Sub-acute/Chronic**: Continuation of earlier signs, other physical findings related to multi-organ system failure in severe cases (see Respiratory: ARDS)
Using Advanced Skills: Lab: Gram stain of fluid or tissue (blood, throat, wound and debrided tissue) for gram-positive cocci in chains (see Color Plates Picture 26). CXR to assess for early ARDS development.

Assessment:
Differential Diagnosis:
Necrotizing fasciitis - due to Clostridia; crepitus on physical exam or gas on X-ray.
Cellulitis or lymphangitis - deep streptococcal infection is much more painful and purpuric.
ARDS or multi-organ failure due to other cause - see Respiratory: ARDS

Plan:
Treatment
1. If suspect diagnosis, evacuate immediately for surgical and intensive care support.
2. If evacuation is delayed or not possible, give antibiotics. IV **Penicillin** 24 MU/day (4 MU q 4 hrs) and **Clindamycin** 900 mg q 8 hrs for 10-14 days. **Clindamycin** decreases streptococcal toxin production. Intravenous **immunoglobulin** (IVIG) 150 mg/kg/day for 5 days given early may decrease mortality.
3. Aggressively debride deep-seated infection. External findings are often the "tip of the iceberg" regarding tissue involvement. (see Procedure: Wound Debridement)
4. Oxygen, IV fluids (LR or NS) and other treatment as in ARDS section. Consider adding ceftriaxone or ciprofloxacin.

Patient Education
General: The infection control recommendation for this is contact isolation and if in the lungs respiratory droplet precautions.
Prevention: Secondary cases are rare but consider streptococcal prophylaxis (**Pen VK** 250 mg po qid) of close contacts.

Follow-up Actions
Evacuation/Consultation Criteria: Evacuate suspected cases immediately. Consult infectious disease specialist early.

Zoonotic Disease Considerations
Agent: Group A *Streptococcus* spp
Principal Animal Hosts: Cattle, swine, and horses
Clinical Disease in Animals: Meningitis and arthritis in swine; mastitis in cattle and horses; respiratory disease and strangles in horses
Probable Mode of Transmission: Direct contact; ingestion of raw milk
Known Distribution: Worldwide

ID: Tetanus
COL Naomi Aronson, MC, USA

Introduction: *Clostridium tetani* bacteria are introduced into the body through contaminated open wounds, burns, frostbite, needles, or unclean cutting/dressing of the umbilical cord. The tetanus toxin (tetanospasmin) causes acute central nervous system intoxication. Case fatality rate is 10-90%. The incubation period is 3-30 days, depending on the dose and the distance of inoculation from the central nervous system. About one million cases per year occur worldwide, especially in tropical and developing countries (50% cases are neonatal). There are four subtypes described for tetanus: generalized, localized, cephalic and neonatal.

Subjective: Symptoms
Neonatal cases: Weakness, irritability, trouble nursing, unable to suck
Specific: Acute (1-7days): Pain at wound site, local muscle spasticity **Sub-acute (7-14 days):** Trismus (lockjaw), painful tetanic spasm, glottic or respiratory muscle spasm, urinary retention, constipation, rigid abdominal wall muscles, trouble swallowing **Chronic (>2 weeks):** Slow recovery phase (4 weeks)
Focused History: *Have you received a tetanus immunization? If so, when was the last one?* (If within 5-10 years, then tetanus is very unlikely. Also if mother of baby had tetanus vaccination, then neonatal tetanus is unlikely.) *Have you recently had a potentially contaminated wound?* (typical exposure) *Does loud noise/coughing/people touching you/gusts of air trigger painful muscle spasms?* (typical stimuli for spasms)

Objective: Signs
Inspection: Acute (1-7 days): Afebrile; localized muscle spasticity, localized pain at inoculation site, neonatal cases: unable to nurse, with stiff muscles or spasming **Sub-acute (7-14 days):** Afebrile, tetanic spasm (stimulus induced) trismus (lockjaw), opisthotonos (arched back spasm), glottic/respiratory muscle spasm, cyanosis/asphyxia, profuse sweating
Palpation: **Sub-acute (7-14 days):** Abdominal muscle wall rigidity
Percussion: **Acute (1-7 days):** Brisk local deep tendon reflexes

Assessment: Diagnose from the history and physical findings/clinical observation.
Differential Diagnosis
Meningoencephalitis - usually associated with fever; true seizures and mental status changes not seen in tetanus.
Strychnine poisoning - mimics tetanus; abdominal wall muscle rigidity more often seen in tetanus; ask about an ingestion history
Hypocalcemic tetany - involves extremities; rare to see lockjaw; tapping on facial nerve (over parotid) can induce facial muscle spasm in low calcium states (Chvostek's sign)
Generalized seizures - associated with loss of consciousness, no trismus
Phenothiazine toxicity - drug history; can see torticollis (not in tetanus); relieved with **Benadryl** (not in tetanus)

Plan:
Treatment
Primary:
1. Maintain airway (ET tube can stimulate spasm so may need early tracheostomy for respiratory difficulty)
2. Medications:
    a. **Tetanus (human) immune globulin** (HTIG, Hyper-tet) 500 IU intramuscularly or injected directly into wound
    b. **Tetanus immunization** 0.5 ml IM at site away from HTIG administration. See Table 5-4.
    c. Narcotic analgesia with **codeine**
    d. **Diazepam** titrated for effect 5-10 mg q 2-4 hours to control muscle spasms (**lorazepam** or **midazolam** are also effective)
    e. **Metronidazole** 30 mg/kg/day divided in q 6 hour dosing for 7-10 days (average about 500 mg q 6 hrs IV); can also be given 1 gm per rectum q 8 hrs
3. Nursing: Keep patient in a quiet, darkened room; avoid unnecessary touching; use Foley catheter for urinary retention.

Alternate Antibiotics: Penicillin G 4 million units IV q 6 hrs for 10 days or **doxycycline** 100 mg q 12 hrs IV

Patient Education
Activity: Bedrest

Diet: High calorie, initially use tube or IV feeding.
Prevention and Hygiene: Use topical antibiotics to umbilical stump. Clean all wounds thoroughly. Maintain current Tetanus immunization status (see Table 5-4):

Table 5-4
Tetanus Immunization Chart

Tetanus Immunization Status	Minor clean wound	Major clean wound	Contaminated wound
Fully immunized recent Td* booster	—	—	—
Fully immunized Td booster 5-10 years ago	—	Td	Td
Fully immunized no booster >10 years	Td	Td	Td
Unknown, none, or incomplete immunization	Td	Td and TIG** (250 U)	Td and TIG (500 U)

*Td is tetanus toxoid; ** TIG is Tetanus Immune Globulin
NOTE: Tetanus vaccination of mother gives her protection and protects the newborn in the first few weeks of life.

Follow-up Actions
Wound Care: If needed, debride the wound to avoid secondary infection. See Procedure: Wound Debridement.
Return evaluation: Those with natural tetanus do not develop immunity (not enough toxin exposure). They need to be re-vaccinated at 4-6 weeks then one month later.
Evacuation/Consultation Criteria: The level of care requires transfer to hospital

ID: Tularemia
COL Naomi Aronson, MC, USA

Introduction: *Francisella tularensis* is a small gram-negative bacillus that can enter the body by ingestion (eating undercooked meat, drinking contaminated water), inoculation (bites, skinning/ trapping animals), inhalation or contamination. This organism can penetrate unbroken skin. It is common in rabbits, opossums, beavers, water rats, raccoons, muskrats and feral cats. It is transmitted by ticks, and less commonly by deer flies. This is primarily a disease of the Northern Hemisphere, where it often affects hunters. Symptoms start 2-5 days after contact. There are six recognized forms of tularemia: ulceroglandular (most common), typhoidal, oculoglandular, glandular, oropharyngeal and pneumonic.

Subjective: Symptoms
Constitutional: Acute (< 2 hr): Fever 101-104°F, chills, malaise, anorexia, fatigue **Sub-acute (2-48 hr):** Fever **Chronic (>48 hr):** Fever, if untreated, can last up to 30 days; chronic debilitation for months.
Specific: Acute (< 2 hr): Tender adenopathy; oculoglandular: tearing, photophobia; pharyngeal: severe sore throat; typhoidal: dry cough; pneumonic: Dry cough **Sub-acute (2-48 hr):** All: rash (35%); ulcer (skin); typhoidal: abdominal pain, diarrhea; pneumonic: pleurisy, rare hemoptysis; **Chronic (>48 hr):** Suppuration of lymph nodes
Focused History: *Have you recently had a tick exposure?* (typical transmission) *Have you been hunting and skinned/dressed/ate any small wild animals?* (typical exposure) *How long have you felt feverish?* (Fever in ulceroglandular tularemia can last 1 month if untreated.)

Objective: Signs
Inspection: **Acute (< 2 hr):** Fever to 104°F; relative bradycardia. Ulceroglandular form: papule; pharyngeal form: exudative pharyngitis; oculoglandular form: purulent conjunctivitis, eyelid edema **Sub-acute (2-48 hr):** sometimes develops membrane or oral ulcer; rash (35%): acneiform, E. Nodosum, E. multiforme, urticaria, Diffuse maculopapular **Chronic (>48 hr):** Ulcer (skin)
Auscultation: **Sub-acute (2-48 hr):** Pneumonic form: rales
Palpation: **Acute (< 2 hr):** Swollen localized lymphadenopathy **Sub acute (2-48 hr):** Fluctuant regional lymph nodes, typhoidal form: hepatosplenomegaly

Percussion: **Acute (< 2 hr); Sub acute (2-48 hr)** Typhoidal form: Enlarged liver and spleen; pneumonic form: Signs of consolidation (dullness to percussion), egophony.

Assessment:
Differential Diagnosis:
Meningococcal infection - rapidly progressive; petechial to purpuric rash; often with meningeal signs; no tick or animal exposure
Rickettsial infection - febrile illness with rash after tick bite; not likely to have suppurative lymphadenopathy or skin ulcers
Cat scratch disease - can be confused with ulceroglandular tularemia, history of cat scratch and lack of risk factors for tularemia
Sporotrichosis - history of thorn/cactus/plant related injury to body; lymphangitis seen commonly with sporotrichosis while it is less common with oculoglandular tularemia
Syphilis - sexual contact; genital ulcer, which is painless; patient often afebrile
Lymphogranuloma venereum - sexually transmitted infection; can cause regional and generalized adenopathy with genital ulcer; less likely to have ulcer elsewhere on body
Other less likely possibilities are cutaneous tuberculosis, plague and anthrax.

Plan:
Treatment
Primary: Gentamicin 5 mg/kg/day divided q 8 hrs IV or IM for 7-10 days. Due to risk of renal toxicity, ensure adequate hydration and urine output. If in doubt, use alternate drug.
Alternative: Tetracycline 500 mg q 6 hrs until fever breaks, then 250 mg qid for 5-7 days (higher relapse than **gentamicin**)
Primitive: Wet saline dressings to skin lesions or eyes for comfort.
Empiric: Cover with **ceftriaxone** 1 gm q 12 hrs and **gentamicin** as above, and transport to definitive care.
NOTE: Tetracycline 250 mg po qid for 2 weeks is an effective prophylaxis after exposure to tularemia.

Patient Education
General: This organism can live for a long time, even frozen in carcasses, water and mud.
Activity: As tolerated
Prevention and Hygiene: Avoid ticks; handle rodents/rabbits with protective clothing, including gloves; thoroughly cook wild birds and game; disinfect drinking water

Follow-up Actions
Consultation Criteria: Patients with suspected tularemia should be transported to hospital for further care and treatment. This infection usually requires consultation to make diagnosis. Serology is available and suggests diagnosis if titer >1:160 after 10 days of illness, however it cross-reacts with *Brucella*.

Zoonotic Disease Considerations
Principal Animal Hosts: Rodents, rabbits, cats, sheep
Clinical Disease in Animals: Sudden onset of high fever, lethargy, anorexia, reduced mobility, stiffness; increased pulse and respiratory rate; coughing, diarrhea and pollakiuria may develop; prostration and death in a few days.
Probable Mode of Transmission: Ingestion, inhalation, occupational exposure, insect bites
Known Distribution: Circumpolar in America, Europe, Asia

ID: Typhoid Fever
COL Naomi Aronson, MC, USA

Introduction: Typhoid fever is caused by *Salmonella typhi*. Typhoid fever is a nonspecific febrile illness common in developing countries with poor sanitation. Multidrug resistant strains of *Salmonella typhi* have been found in Asia, the Middle East and Latin America. Food (especially undercooked meat and eggs) and water contaminated by feces or urine from patients or chronic carriers is implicated in transmission. The incubation period is 1-3 weeks after exposure.

Subjective: Symptoms
Constitutional: Acute (3-7days): Fever, flu-like symptoms, chills, weakness, anorexia, myalgias **Chronic (>3 weeks):** If fever> 4 weeks, consider metastatic focus. If no Rx, 5-10% have relapsing fever pattern
Specific: Acute (3-7days): Sore throat, non-productive cough, constipation, diarrhea (5-10%), abdominal discomfort **Sub-acute (1-3 weeks):** Diarrhea **Chronic (>3 weeks):** Abdominal pain, abdominal perforation, lower GI bleed
Focused History: *How long have you felt feverish?* (fever gradually builds and lasts for 3 weeks) *Have you noticed any red to pink spots on your abdomen or chest?* (transient rose spots seen in 10-50% of patients) *Have you recently traveled in a developing country?* (very common cause of fever in endemic areas).

Objective: Signs
Using Basic Tools: Acute (3-7 days) Inspection: Stepladder temperatures to 104°F (usually in afternoon/night), relative bradycardia in 25%; moderately ill appearing; rose spots (2-3 mm pink to red papules on chest/abdomen that fade with pressure) in fair skinned persons, furry tongue; (thick white to brown coating that spares edges) Palpation: Abdominal distension; mild, diffuse abdominal tenderness. **Sub-acute** (1-3 weeks): Palpation: splenomegaly (50%), Percussion: liver can be slightly enlarged 2-3 cm below costal margin.

Using Advanced Tools: Lab: Urine culture is positive after one week, blood culture may be positive for first 2 weeks, and stool culture is positive for weeks 3-5. Other clues from stool include fecal leukocytes (may suggest an invasive gastroenteritis). Blood smear may demonstrate low white blood count and anemia.

Assessment:
Differential Diagnosis
Nontyphoidal Salmonella - infections are generally milder, without rose spots.
Tuberculosis - generally chronic, lower grade fever; night sweats; cough; hemoptysis; abnormal CXR.
Hepatitis - more often see jaundice, dark urine, gastrointestinal symptoms, malaise and fatigue
Leptospirosis - remarkable conjunctival injection; has similar fever pattern; history of exposure to contaminated fresh water
Malaria - nocturnal fever pattern typical; thick and thin blood smears will help detect the malaria parasite
Amebic liver abscess - may see more tenderness in hepatic region
Brucellosis - chronic febrile illness with relative bradycardia, splenomegaly; animal exposure, occupation may help differentiate

Plan:
Treatment: Antibiotics prolong Salmonella excretion in the stool and should not be given unless patient is febrile.
Primary: Ciprofloxacin 500 mg q 12 hrs x 10 days (some resistance has been seen)
Alternative: Ceftriaxone 1 gm q 12 hrs IV or IM x 14 days (recent study suggests that a 5 day course may be as effective), **azithromycin** 1 gm load on day 1 then 500 mg po days 2-6
NOTE: For delirium or shock, give steroids before first dose of antibiotic: **dexamethasone** load 3 mg/kg, then 1mg/kg q 6 hrs for 48 hrs

Patient Education
General: Use tepid baths/fanning to bring down temperature.
Activity: Vigorous oral rehydration, 3-4 liters first day then follow and replace losses.
Diet: Maintain nutrition/electrolytes.
Medications: Avoid laxatives and salicylates or other antipyretics.
Prevention and Hygiene: Immunize with live attenuated oral ty21a vaccine (boost q 3 years) or the **typhoid polysaccharide Vi vaccine** (boost q 2 years) pre-deployment. Practice handwashing, fly control, water treatment. *Salmonella typhi* is killed by heating food or water to 135°F, iodination or chlorination. Do not have patients infected with *salmonella* participate in general food preparation. Avoid fresh, uncooked vegetables and fruits unless you can peel or carefully wash them yourself.

Follow-up Actions
Return Evaluation: Consider non-GI source of infection such as endocarditis, visceral or renal abscesses, osteomyelitis.
Evacuation: Send to higher level of care where more diagnostic tools available when stable.

NOTES: Paratyphoid fever presents in a similar way but is generally milder, and is caused by *Salmonella enteritidis* bioserotypes.

Chapter 14: Preventive Medicine
COL (Ret) Joel Gaydos, MC, USA & SFC Jeffrey Crainich, USA

Introduction: Preventive medicine (PM) procedures minimize disease/non-battle injuries (DNBI) during war and contingency operations. PM measures should be integrated into all missions and training exercises. Responsibilities of the medic include the following: monitoring the acquisition and treatment of potable water; monitoring the acquisition, handling and preparation of food; monitoring and implementing vector (insect and rodent) control programs; and monitoring the construction and maintenance of personal hygiene (washing) facilities and solid and liquid waste disposal systems.

Table 5-5

REQUIRED IMMUNIZATIONS

Vaccine	Initial dosage/route	Booster dose	Comments
Anthrax	0.5 ml SC at 0, 2, 4 weeks, then 6, 12, and 18 months	0.5 ml SC annually	Annual booster
Hepatitis A	1.0 ml IM at 0, and 6-12 months (HAVRAX or VAQTA)	None currently required	None
Hepatitis B	1.0 ml IM at 0, 1, and 6 months	None currently required	None
Influenza	0.5 ml IM annually in October	0.5 ml IM at 6 months if still in risk area	Annual. Consider booster if in Southern Hemisphere April-September
Japanese B Encephalitis	1.0 ml SC at 0, 7, and 30 days	1.0 ml SC every three years	By geographic area only

Vaccine	Primary Series	Booster	Notes
Measles	0.5 ml SC of Attenuvax or MMR (preferred)	None currently required	Required if no serological evidence of previous exposure
Meningococcal	0.5 ml SC single dose	0.5 ml SC every 5 years	More frequent boosters for some countries
Mumps	0.5 ml SC of Mumpsvax or MMR (preferred)	None currently required	Required if no serological evidence of previous exposure
Plague	2 of the 3 shot series, 1.0 ml IM at 0 months, then 0.2 ml at 1-3 months	0.2 ml IM upon deployment to a risk area and every 6 months while there	Boost at direction of unit surgeon for high-risk areas (CDC blue sheet)
Polio	IPV (Inactivated Polio Vaccine), 0.5 ml SC at 0, 1-2, and 6-12 months.	0.5 ml SC for travel to highly endemic areas	Boost at direction of unit surgeon for high-risk areas.
Rabies	0.1 ml ID (or 1.0 ml IM) at 0, 7, and 21-28 days. Do not mix ID & IM doses within a series. Must use IM dosing if taking chloroquine or mefloquine within 14 days of any dose	None currently required	Initial series for personnel in Medical SOF, and personnel assigned, attached, or OPCON to an operational team
Rubella	0.5 ml SC of MMR (preferred) or Meruvax II	None currently required	Required if no serological evidence of previous exposure
Tetanus-Diphtheria Toxoid (absorbed)	0.5 ml IM at 0, 1-2, and 8-14 months	0.5 ml IM every 10 years or as prophylaxis after severe/dirty wounds	Required if no documentation of previous immunization.
Typhoid (2 types of vaccine used)	a. Injectable Vaccine (Typhoid Vi, Pasteur-Merieux Connaught): 0.5 ml IM b. Oral vaccine (Vivotif Berna, Berna): one capsule every other day for a total of four capsules	Boost with same vaccine if possible. a. Typhoid Vi 0.5 ml IM every two years b. Berna oral vaccine: repeat 4 dose series every five years	For capsules, do not lengthen interval. Swallow (do not chew) with cool drink 1 hr before a meal. Do not take within 24 hours of taking mefloquine or antibiotics
Varicella	0.5 ml SC at 0, and 1-2 months	None currently required	Required if no serological evidence of previous exposure
Yellow Fever	0.5 ml SC	0.5 ml SC every 10 years	None

Surveillance for Illness and Injury

While on deployments, including those involving host nation personnel, medics must compile daily sick call logs to review for any possible disease outbreaks or biological warfare exposures. This tool can be used to educate host nation personnel in preventive medicine and disease surveillance. The information gathered must be forwarded to your next higher headquarters and can be used to fill out SODARS reports. Use this form as a guideline for disease non-battle injury (DNBI) surveillance (next page):

Table 5-6: DNBI Report

Line 1   LOCATION: _____
Line 2   Strength: a. Male_____ b. Female_____ c. Total_____
Line 3   Reporting Period: a. DD/MM/YY _____ (Sunday 0001HR) through
         b. DD/MM/YY _____ (Saturday 2359HR)
Line 4   Prepared by: a. Name _____ b. Phone _____ c. E-Mail:_____

Category	Initial Visits this Period (a)	Total to Date (b)	Rate: Line_a divided by 2c (c)	Days of Light Duty (d)	Lost Work Days (e)	Admits to Hospital (f)
Line 5 Stress Reactions (Combat/Operational)						
Line 6 Dermatologic						
Line 7 GI, Infectious						
Line 8 Gynecologic						
Line 9 Heat/Cold Injuries						
Line 10 Injury, Recreational/Sports						
Line 11 Injury, MVA						
Line 12 Injury, Work /Training						
Line 13 Injury, Other						
Line 14 Ophthalmologic						
Line 15 Psychiatric, Mental Disorders						
Line 16 Respiratory						
Line 17 STDs						
Line 18 Fever, Unexplained						

Line 19 All Other, Medical/Surgical						
Line 20 TOTAL DNBI						
Line 21 Dental						

Field Sanitation

1. **General:**
    a. Factors that create a high risk for food-borne diseases: poor food inspection and sanitation, poor personal hygiene habits, inadequate refrigeration, and lack of eradication programs for food-borne diseases such as hepatitis A and brucellosis.
    b. Food transportation, storage, preparation and service have direct bearing upon the success or failure of a mission. Dining Facility sanitation is a chronic operational problem. The prospect of disease outbreaks, particularly dysentery and food poisoning, is always present and must be recognized as a constant threat to unit health.
    c. Potentially Hazardous Foods (PHFs): Any food that contains milk, milk products, eggs, meat, poultry, fish, shellfish, or other ingredients in a form capable of supporting rapid growth of infectious or toxic microorganisms. PHFs are typically high in protein and have a water content greater than 85% and a pH greater than 4.5.
    d. Factors that most often cause food-borne disease outbreaks:
        1. Failure to keep PHF cold (below 40°F internal temp.).
        2. Failure to keep PHF hot (above 145°F internal temp.).
        3. Preparing foods a day or more before being served.
        4. Allowing sick employees who practice poor personal hygiene to handle food.
    e. Food contamination can be classified into three categories:
        1. Biological - contamination by pathogenic microorganisms (protozoa, bacteria, fungus, virus) or unacceptable levels of spoilage. This category is the major threat to personnel.
        2. Chemical - contamination with chemical warfare agents, industrial chemicals, and/or other adulterating chemicals (zinc, copper, cadmium, pesticides, etc.).
        3. Physical - contamination by arthropods, debris, radioactive particles, etc.
    f. Bacteria that multiply at temperatures between 60°F and 125°F cause most food-borne illness. Maintain the internal temperature of cooked foods that will be served hot at 145°F or above. Maintain the internal temperatures of foods that will be served cold at 40°F or below to control any bacteria that may be present in the food,
    g. High food temperatures (160°F to 212°F) reached in boiling, baking, frying, and roasting will kill most bacteria that can cause food-borne illness. Prompt refrigeration to 40°F or below in containers less than 2 inches deep inhibits growth of most (but not all) of these bacteria. Freezing at 0°F or below essentially stops bacteria growth but will not kill bacteria that are already present.
    h. Thorough reheating to an internal temperature of 165°F or above will kill bacteria that may have grown during storage. However, foods that have been improperly stored or otherwise mishandled cannot be made safe by reheating.
    i. Ensure everything that touches food during preparation and serving is clean to avoid introducing illness-causing bacteria.

2. **Procurement of Food:**
    a. Order of preference for food acquisition:
        1. US Military rations brought with unit or previously cached.
        2. Local food procured from sources approved by supporting Veterinary and Environmental Science Officers.
        3. Local food procured from unapproved sources.
    b. Special Operations Forces will probably have to procure food from unapproved sources during real world contingencies, presenting a serious medical threat to the team and the mission. Use the following

guidelines:
1. Avoid local street vendors. Their personal hygiene habits tend to be poor, which results in contaminated food (i.e., fecal-oral contamination)
2. Consider all ice contaminated. It is often made from non-potable water and freezing will not kill disease-causing organisms. Anything with ice in it or on it should be considered contaminated (i.e., alcohol in the drink does not make the ice in it safe).
3. Semi-perishable rations (canned and dried products) are relatively safe and should be chosen over fresh food. Protect canned and dried foods from extreme heating and freezing. Do not use swollen or leaking cans.
4. Do not procure moldy grain or grain contaminated with insect larvae.
5. Raw fruit and vegetables may be grown in areas where "nightsoil" (human fecal matter) is used as fertilizer or where gastrointestinal or parasitic diseases are prevalent. Wash raw fresh fruits and vegetables in potable water and disinfect with one of the methods described below:
    a. Dip in boiling water for 15 seconds. Place small amounts of produce in net bags, completely submerge items for 15 seconds, remove and allow cooling. Not recommended for leafy vegetables.
    b. Disinfect with chlorine. Immerse for at least 15 minutes in a 100 ppm solution of chlorine or 30 minutes in a 50 ppm solution. Rinse the produce thoroughly with potable water before cooking or eating. Break apart "head" produce such as lettuce, cabbage or celery before disinfection.
    Bleach: (Clorox)
        4.84 oz in 32 gallons = 50 ppm
        9.68 oz in 32 gallons = 100 ppm
        1 tablespoon per gallon = 200 ppm
    70% Calcium Hypochlorite:
        0.32 oz in 32 gallons = 50 ppm
        0.64 oz in 32 gallons = 100 ppm
6. Always cook eggs to prevent salmonellosis. Blood and meat spots are acceptable, but cracked and rotten eggs are not acceptable and should be discarded.
7. Boil unpasteurized dairy products for at least 15 seconds to prevent tuberculosis, brucellosis, Q fever, etc. Avoid cheese, butter and ice cream made from unpasteurized milk, which can carry these diseases.
8. Cook all seafood to prevent hepatitis, tapeworms, flukes, cholera, etc.
    a. Avoid shellfish - cooking does not degrade some toxins (red tide).
    b. Certain saltwater fish have heat stable toxins that are not destroyed during cooking. Do not eat any species that the native population does not eat.
    c. Avoid large predatory reef fish, like barracuda, grouper, snapper, jack, mackerel, and triggerfish, which may accumulate toxins (ciguatera).
9. Eat carcass or muscle meat rather than visceral meat (liver, heart, kidney, etc.). Muscle flesh is less likely to be contaminated. Fresh meat from healthy animals is safe if cooked thoroughly.
    a. Be aware of geographic areas where toxins may occur in seafood.
    b. Perform an antemortem examination (before slaughtering), use correct field slaughter methods and perform a postmortem examination (after slaughtering). (See Vet Medicine Chapter.)
    c. Color of meat should be red to slightly-red brown. Do not consume green or brown beef if possible. Avoid meat with off odors, such as sour or sweet, fruity smells. Cook meat until it is WELL DONE- do not eat rare, medium, or bloody meat. Sausages and meat products should be well cooked.

3. Food Storage and Preservation
   a. Protect canned and dried foods from extreme heat and freezing.
   b. Store and preserve perishables such as meat, poultry, fish, etc. by refrigerating at or below 40°F. Because refrigeration or potable ice is often not available, slaughter what you need, cook thoroughly and then consume immediately. Meat can be preserved by methods other than refrigeration if time and

resources are available by smoking, curing, making jerky or pemmican, salting, and pickling. (See Vet Medicine Chapter.)
c. Semi-perishable foods such as potatoes and onions should be stored in a dry place off the ground, allowing air to circulate around them, retarding decay and spoilage.
d. Store staple products (flour, sugar etc.) in metal cans with tight-fitting lids.
e. Do not store acidic foods or beverages such as tomatoes or citric juices in galvanized cans. This will prevent zinc poisoning.

4. Preparing and Serving Food:
a. Use pesticides according to the directions on the container. Limit residual sprays to crack and crevice treatment only. Protect all foods and food contact surfaces when applying pesticides.
b. Coordinate food preparation and consumption to eliminate unnecessary lapses of time.
c. Leftover food presents a problem. Plan meals to reduce the amount of leftovers. Discard items held at unsafe temperatures (45°F to 140°F) for 3 or more hours. Never save PHF foods such as creamed beef, casseroles or gravies.
d. Meat may contain disease-producing agents that cannot be detected by inspection. Follow cooking procedures strictly to ensure that heat penetrates to the center of the meat and that all the meat is cooked to at least 165°F. This applies to poultry, pork, beef and any stuffing or other foods containing these meats.

5. Cleaning and Disinfecting:
a Cooking utensils and mess kits should be cleaned, disinfected, and properly stored after each use. (See Figure 5-8) They must be scraped free of food particles, washed in hot (120°F to 130°F) soapy water, rinsed in boiling water, sanitized for at least 10 seconds in another container of boiling water, and allowed to air dry. They must be stored in a clean, covered container that is protected from dust and vermin.
b When it is impossible to heat the water, utensils must be washed in soapy water, rinsed in two cans of clear water, then immersed in the fourth container of chlorine sanitizing solution for at least 30

Figure 5-8: Mess Wash Set-up

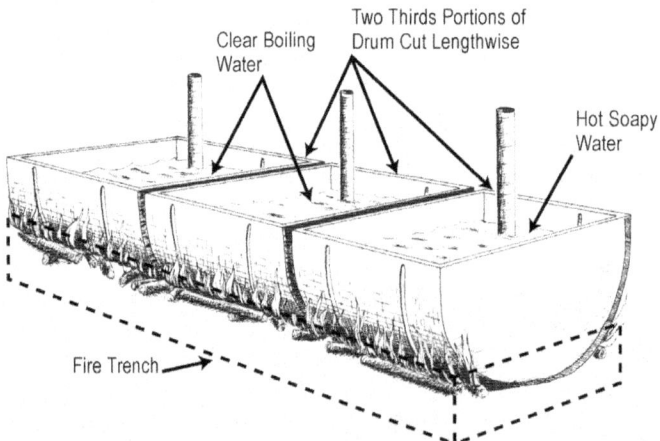

seconds. Chemical sanitizing solutions are prepared in the following order of preference:
1) Use Disinfectant, Food Service, (NSN:6840-01-035-5432) as specified on the label.
2) Use 1 level mess kit spoonful of calcium hypochlorite for every 10 gallons of water (250 ppm solution).
3) Use 1 canteen cup of 5% liquid bleach in 32 gallons of water (250 ppm solution).

c. If mess kits become soiled or contaminated between meals, they should be rewashed prior to use as described above. A pre-wash of boiling water should be available for use prior to all meals.

WASTE DISPOSAL

1. **General:**
   a. Proper waste disposal is necessary to prevent disease during real world contingencies and training exercises. Liquid and solid wastes produced under field conditions can amount to 100 pounds per person per day, especially when shower facilities are available. Without proper waste disposal methods a camp or bivouac site will soon become an ideal breeding ground for flies, mosquitoes, rats, mice and other pests, which can spread diseases such as plague, dysentery, typhoid, dengue fever, and other vector-borne diseases.
   b. There are four different types of waste: 1. Human waste; 2. Garbage (Food); 3. Liquid; 4. Rubbish (Paper/Plastic/Cans/Glass)
   c. Select disposal methods compatible with location, military situation and regulations.

2. **Human Waste Disposal:**
   a. Under field conditions, bury human waste when feasible.
   b. Devices most commonly used for various field situations are as follows:
      1. Cat Hole Latrine (for patrols) Figure 5-9
      2. Straddle Trench Latrine (1-3 day bivouac site) Figure 5-10

Figure 5-9: Cat Hole Latrine

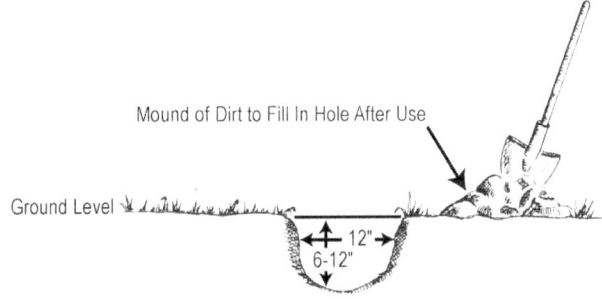

Figure 5-10: Straddle Trench Latrine

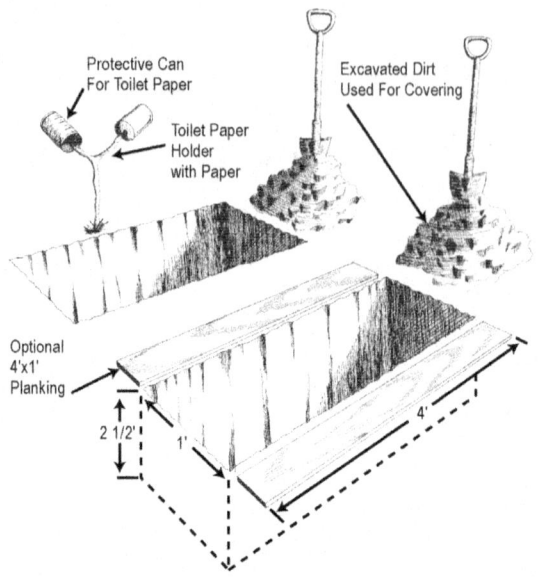

Figure 5-11: Deep Pit Latrine

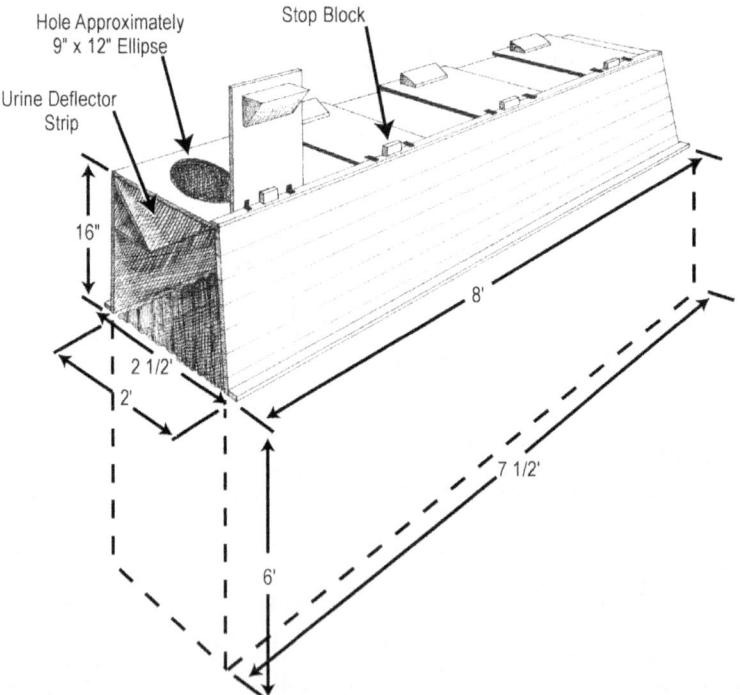

Figure 5-12: Burn Out Latrine

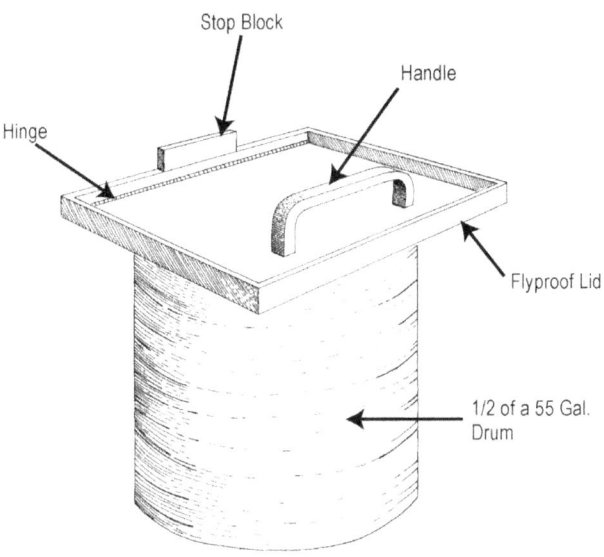

Figure 5-13: Pail Latrine

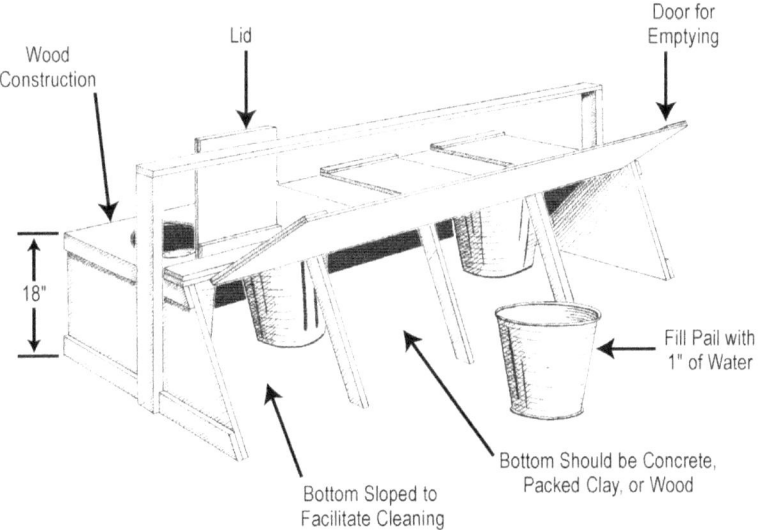

Figure 5-14: Trough Urinal

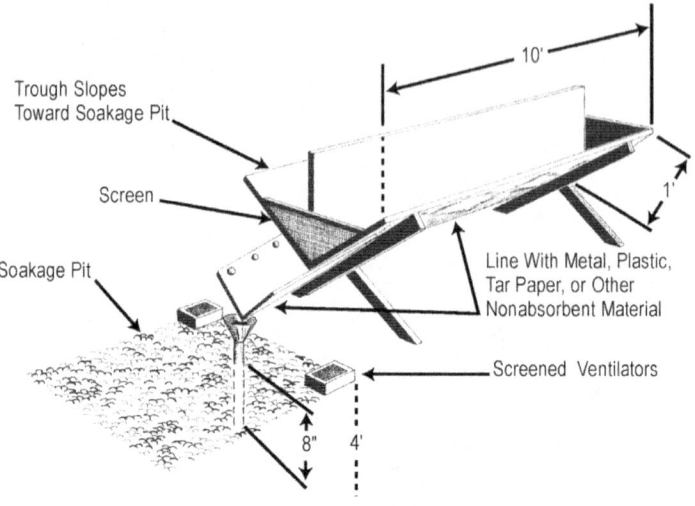

Figure 5-15: Pipe Urinal

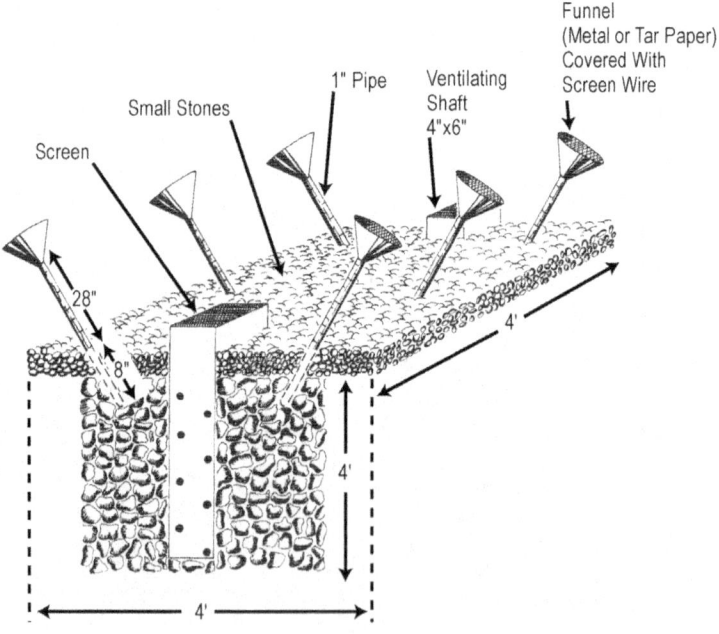

3. Deep Pit Latrine (temporary camp) Figure 5-11
4. Mound Latrine (high water table areas)
5. Burn Out Latrine (rocky, frozen, or high water table) Figure 5-12
6. Pail Latrine (rocky, frozen, or high water table) Figure 5-13
7. Urine Soakage Pits (overnight or longer)
    a) Trough urinal Figure 5-14
    b) Pipe urinal Figure 5-15
c. Toilet paper should be provided.
d. Each latrine must have an easily operated hand-washing device with soap installed outside. (Unscented soap in a nylon stocking may be tied to or near the hand-washing device for easy access. Figure 5-20)
e. Latrine Maintenance: Conduct the following procedures routinely:
    1 Latrine should be inspected for insects. Insecticide should be sprayed at least twice a week as a minimum precaution.
    2 Latrine seats and boxes should be scrubbed with soap and water and sanitized with a disinfectant on a daily basis.
    3 Ensure that the latrine box remains insect proof (fly-tight) at all times.
f. Pail Latrine: Clean daily. Waste may be disposed of by burning (using 1 part gasoline and 4 parts diesel fuel) or hauling waste to a suitable area and burying (rinse pail with water and empty rinse water in the disposal site). If available, dispose of waste at a sewage treatment plant. Emptying and hauling containers of waste must be closely supervised to prevent careless spillage. The use of plastic bag liners for pails reduces the risk of accidental spillage. The filled bags are tied at the top then are disposed of by burning or burial.
g. Constructing and Closing Urinals: Construct 1 per 20 men. Locate urine pits so the urine will not drain into the pit latrine unless the ground is porous enough to absorb the extra liquid. The urine soakage pit is the best means of urine disposal in the field. When a urine soakage pit is closed or abandoned it should be sprayed with residual pesticide and mounded over with 2 feet of compacted soil. Mark the

Figure 5-16: Barrel Filter Grease Trap

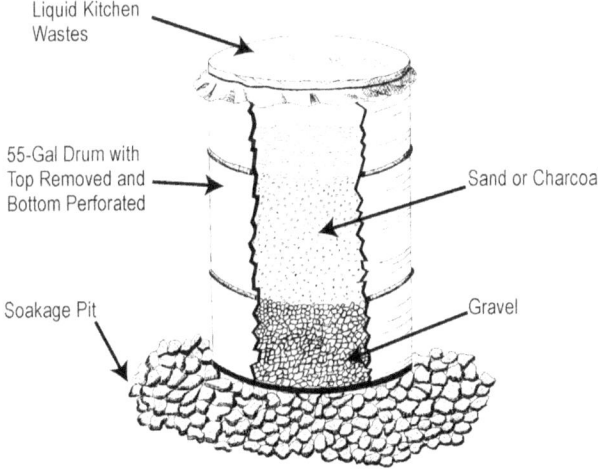

Figure 5-17: Barrel Incinerator

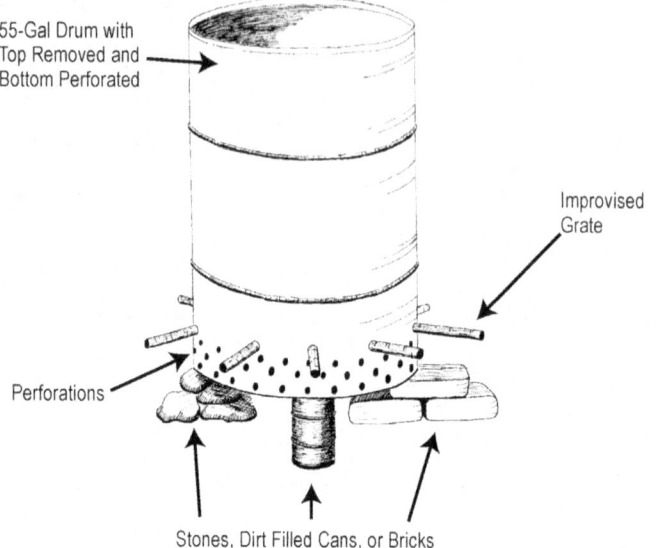

Figure 5-18: Baffle Grease Trap

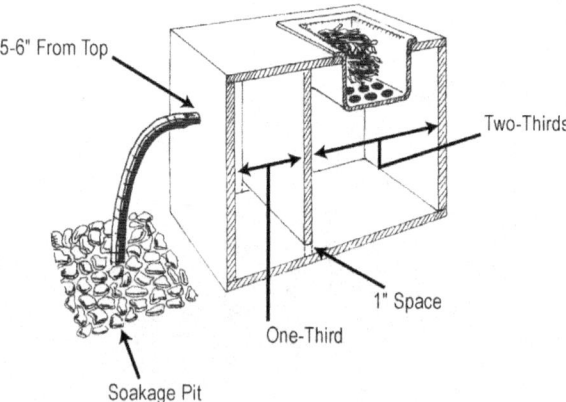

Figure 5-19: Solar Heated Shower

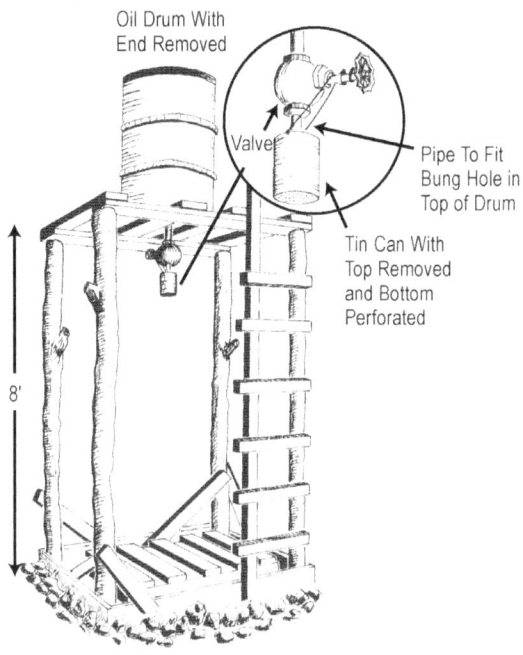

Figure 5-20: Improvised Handwashing Station

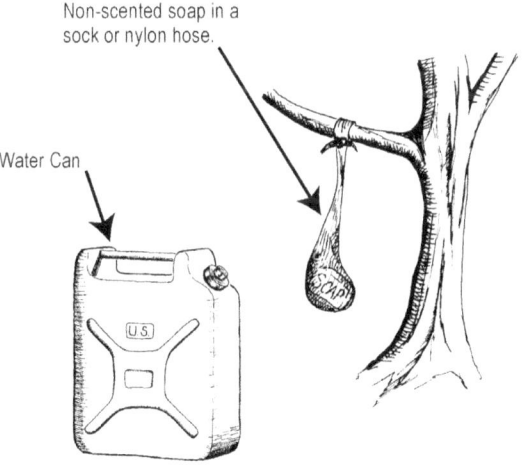

site with a sign that labeled: CLOSED SOAKAGE PIT, date and unit (security permitting).
  h. Closing Latrines: Close a latrine when it is filled to within 1 foot of the ground surface or when it is abandoned. Spray the contents of the pit, the sidewalls and the ground surface surrounding the pit with a residual pesticide. Fill the pit up to the ground surface with successive 3-inch layers of earth. Pack each layer down and spray it with pesticide before adding the next layer of dirt. Mound the latrine pit over with at least 1 foot of dirt and spray it again. Indicate the location of the latrine with a sign marked: CLOSED LATRINE, date and unit (security permitting).

3. **Liquid Waste:**
   a. There are two main types of liquid waste: kitchen and bath. The devices most commonly used to dispose of liquid waste are: soakage pits, soakage trenches, evaporation beds, and grease traps.
   b. Use soakage pits for all water facilities (i.e., under water trailer faucets, Lyster bags, and handwashing devices).
   c. Rules for construction, maintenance, and closing of liquid waste disposal devices:
      1. Do not dig pit or trench into ground water table to avoid contamination.
      2. Place one site adjacent to the mess kit laundry to avoid spillage.
      3. Dig drainage ditches around each pit/trench to prevent surface water runoff from flowing into the soakage pits or trenches.
      4. Police the soakage pit/trench area as needed.
      5. Use an approved residual pesticide on the pit/trench contents and the surrounding ground area to control insects.
      6. When the pit/trench becomes clogged, close it by covering it with 1 foot of compacted dirt and mark it with a sign indicating the type of pit, date closed, and unit (security permitting).

4. **Garbage and Rubbish (Refuse) Disposal:**
   a. Dispose of garbage and rubbish by burial or incineration.
   b. Collect refuse produced by personnel in: a 32 gallon galvanized can with cover, a 55 gallon drum with improvised cover, or in plastic bags (rubbish only).
   c. Incinerate combustible refuse when the tactical situation and local policy permits. Incinerate refuse in a barrel or inclined plane incinerator. Burial methods depend on the amount of refuse to be buried- small amounts (1 - 2 barrels) can be buried using a pit, but large amounts will need to be placed in a sanitary landfill on the local economy.

Field Water Purification

1. Treat and disinfect all sources of water other than US Military Installation or Quartermaster produced or approved water in the field.
2. Minimize the possibility of water-borne illness by selecting proper sources.
3. Water consumed by personnel will come from four possible sources. These sources are prioritized in the order in which they should be chosen for use:
   a. Fixed Facility (closed pipe system with treatment)
   b. Water Production Points (portable units, i.e., (ROWPU - Reverse Osmosis Water Purification Unit)
   c. Bottled water
   d. Emergency (raw water from the five natural sources below) from
      1. Surface water (lakes, rivers, streams)
      2. Ground water (wells, springs)
      3. Rain water
      4. Ice
      5. Snow
4. Easy access to large quantities of water will usually make surface water the best emergency source. When selecting a water source for a Special Operations unit, consider certain factors:
   a. Military Situation: Does the site provide cover and concealment? Is the site accessible to soldiers? Can water be extracted with available equipment? Can the source be used without interference from the enemy? Is the water source accessible under all weather conditions? Is the site a safe distance from targets?

b. Quantity of Water: Is there enough water in the source to sustain the troops for the desired time? Can enough water be acquired quickly?

c. Quality of the Water: A detailed site survey is critical in selecting a quality water source. Check the site for possible sources of pollution: dead fish, frogs, or other animals; excessive algae growth; oil slicks or sludge deposits; and the conditions of vegetation around the site. Dead or mottled vegetation may indicate chemical agents contamination. If possible, reconnoiter for a distance of two miles upstream of the source to locate any possible sources of contamination. Locate any bivouac site at least 100 feet downstream of the water point. Avoid using stagnant or swampy areas as water sources.

5. Water Treatment:

a. During deployments personnel will utilize the following prioritization and standards for water treatment of the four types of water sources:

(1) Fixed Facility - Chlorinate to a minimum of 2 ppm prior to consumption. If individual containers (2 quarts or less) are to be used for transport/storage of water, treat with iodine tabs (2 tabs/quart) or chlorinate to 2 ppm prior to consumption. If bulk containers (> 5 gallons) are to be used for storage/transportation of water, chlorinate to 2 ppm. This water source is preferrable to over all others.

(2) Water Production Points - Chlorinate to 2 ppm prior to consumption.

(3) Bottled Water - Carbonated bottled water needs no further treatment. If the containers are broken down and the water is placed in other containers (not the originals), treat the water with iodine (2 tabs/quart) or chlorine (2 ppm). This source is preferred when approved fixed facilities are not available.

(4) Emergency Water - Select the least contaminated raw water source available. Filter water with a KATADYN or SWEETWATER filter system (both systems are GSA approved and can be purchased through your logistics channels) or through any system with a pore size of 0.2 microns or smaller. Treat water with iodine (2 tabs/quart) or chlorine (2 ppm) prior to consumption. Seawater must not be utilized for consumption.

b. Use the following guidelines for treating water:

NOTE: Adding 250 mg of vitamin C (ascorbic acid) per quart of water after the contact time has elapsed will improve the taste of chemically purified water.

(1) Individual Canteen (1 or 2 Quart)
  (a) Iodine Tabs - NSN 6850-00-985-7166. Use 2 tabs per quart.
  (b) Chlorine - Chlorination kit: NSN 6850-00-270-6225
    1. Locate water source. Fill canteen with cleanest water available.
    2. Prepare a solution by pouring the contents of one (1) ampule of calcium hypochlorite into 1/2 canteen cup of water. Thoroughly mix the solution.
    3. Add 1 canteen capful (NBC WATER CAP) per quart.
    4. Shake the canteen to mix. Wait five (5) minutes (contact time). Loosen cap to allow water to seep around the threads of the neck and cap of the canteen. Re-tighten cap. Wait an additional twenty-five (25) minutes before using the water.
    5. In cold weather, wait 40 minutes before using the water.
  (c) Chlorine - Bleach:
    1. Locate water source. Fill canteen with cleanest water available.
    2. Use 2-3 drops of household bleach per quart.
    3. Follow directions in 4 and 5 above.
  (d) Boiling (least preferred):
    1. Locate water source. Use cleanest water available.
    2. Bring water to a rolling boil for 3-5 minutes. This will kill most organisms that are known to cause intestinal diseases.
    3. In areas where *Giardia, Entamoeba histolytica* or viral hepatitis are known to be present, boil

water for thirty (30) minutes to ensure destruction of the microorganisms.
4. In emergency situations, boil water for a minimum of 15-30 seconds.
5. High altitudes may require additional boiling.
6. Allow water to cool before dispensing or drinking.
7. Boiling provides no residual protection against recontamination and should only be used as a last resort. Water can become recontaminated if not protected properly after decontamination.
   (e) Other Chemical treatments:
   1. Povidone-iodine solution (Betadine solution, 10%, NOT THE SCRUB SOLUTION): 16 drops per liter gives 8 ppm iodine. Contact time is 20 minutes minimum or 90 minutes for cold, turbid water.
   2. Chlor-Floc tablets: 1 tablet per liter of water makes 8.4 ppm chlorine. Contact time is 15 minutes minimum or 60-90 minutes for cold, turbid water. These tablets have a flocculation material to clear turbid water. After treatment, the water must be strained before drinking, i.e., through a T-shirt.

Malaria Prevention and Control

1. Identify the type(s) of malaria, information regarding malaria drug resistance, the geographic areas at risk and the seasons of the year for risk. Consult Armed Forces Medical Intelligence Center (AFMIC) or other sources.
2. Design a prevention and control program to include the use of prophylactic drugs; personal protective measures like skin and clothing repellents, and Permethrin bed nets; and area mosquito control measures and education programs.
3. Administer prophylatic drugs when indicated to those who are not allergic, comply with pre-deployment dosing requirements and advise patients of side effects such as photosensitivity with **doxycycline**.
4. Provide alternative, effective drug prophylaxis for those unable to take the first line regimen.
5. Ensure that all infected patients are protected from biting mosquitoes to prevent transmitting malaria to others.
6. Depending on the nature and extent of the operation, determine the need to conduct area mosquito control operations, to include control or elimination of breeding sites, use of larvicides and use of sprays.

Pest Control

1. **NOTE:** Pests that require a blood meal are attracted to humans or animals by carbon dioxide emitted from the body.
2. Identify and eliminate breeding sites for mosquitoes and other insects by improving drainage, disposing of refuse properly and applying appropriate chemicals.
3. Clear dense vegetation or other harborages inhabited by pests from living areas.
4. Use chemical pesticides properly. Follow the instructions on the label.
   a. Ensure that the chemical pesticides you are using are effective against the vectors you want to control.
   b. Mix chemical pesticides in proper concentrations and dispense in sufficient density to control the desired pest.
   c. Spray approved pesticides into areas of heavy vegetation that cannot be cleared.
   d. Apply approved bait pesticides in areas where rodents are suspected to frequent. Trap rodents when feasible, since it is a safer alternative to chemical pesticides.
   e. Dispense chemical pesticides using proper personal protective measures. Wear eye protection, rubber gloves and facemask respirator when handling pesticides.
   f. Ensure that chemical pesticides used inside living areas are labeled safe for such use.
   g. Ensure all chemical pesticides are always stored in safe, secured areas.
   h. Properly dispose of all empty pesticide containers and materials contaminated with pesticides according to product labels.
      (1) Remove food sources. Avoid eating in sleeping areas. Even small amounts of food attract insects

and rodents.
(2) Do not have pets near living areas. They harbor fleas, ticks and other insects and can attract mosquitoes and other pests.
(3) Use personal protective measures:
  (a) Use the DOD Arthropod Repellent System (DEET and Permethrin).
  (b) Ensure that personnel are properly supplied with DEET and Permethrin and know how to use these products.
  (c) Ensure that soldiers sleep under Permethrin-treated bed netting.
  (d) Use personal protective measures, to include respiratory protection, when entering areas suspected of housing rodents or birds.

Rabies Control

1. Assess the rabies threat in the deployment area and initiate a control program if needed.
2. Identify personnel at risk and vaccinate them pre-deployment.
3. Maintain and review current guidelines for pre- and post-exposure rabies management.
4. Identify rabies testing laboratory (if available) and domestic and wild animal control resources in the deployment area.
5. Immunize pets and domestic animals.
6. Impound stray animals.
7. Work with animal control personnel to reduce the wild animal reservoir if necessary and feasible.
8. Identify, evaluate, treat and report human exposures.
9. Conduct surveillance for human cases and cases in domestic and wild animals.
10. Advise the command on the rabies threat and recommend preventive countermeasures.
11. Inform at-risk personnel about the transmission, prevention and clinical aspects of rabies. Stress the importance of reporting animal bites or other suspicious animal contact.

Landfill Management

NOTE: In the US, military personnel must abide by EPA standards for landfills even on deployments. Consult land managers or custodians for guidance as needed.

Suggestions for a host nation landfill operation:
Identify a large area of land that will not be used for many years after the landfill is closed.
Use a dump truck and a bucket loader if available.
Find an area close to the site to store excavated dirt while the landfill is constructed.
The pit will need to be lined will a nonporous membrane (such as clay) to prevent pollutants from leaching into the water table and contaminating the water.
Size: Use 1 acre per year per 10,000 people as an estimate. High water tables or rocky soil will limit pit depth.

Pit Operations:
1. The pit must be accessible to vehicles (dump trucks) and allow them to enter the pit. The bucket loader must cover the refuse throughout the day and at the end of the day.
2. Dispose of refuse in 10 ft. wide subcompartments in the landfill and cover them as the landfill is filled.
3. Do not fill the subcompartments with more than 6 feet of trash. Cover at the end of every day or when full.
4. Use a windscreen on the downwind side of the landfill to catch debris.

Chapter 15: Veterinary Medicine
Vet Medicine Procedure: Antemortem Exam
MAJ Joseph Williamson, VC, USA

When: Antemortem exam is the inspection of a live animal prior to slaughtering it for food purposes. Accept only those animals that are healthy, free of harmful diseases and chemicals and capable of being converted into wholesome products for consumption. This screening process only removes obviously diseased animals. A postmortem exam should be conducted prior to consuming any tissue or organ system (see section on Postmortem Exam).

What You Need: Gloves and a stethoscope

What To Do: Observe the animal at rest and in motion. You may see lameness, pain, neuromuscular deficits and/or systemic disease states in a moving animal that are not apparent in an animal at rest. Look for abnormal conditions such as continuous scratching/rubbing, emaciation or depression.

Examination Specifics:
Lameness: Reject if limbs are deformed or have gross swelling around joints. Do not use a limb if it is damaged or broken. You may consume the rest of the carcass if it is normal.
Emaciation: Reject animal if in poor state of nutrition, as evidenced by extreme thinness.

Organ Systems Analysis:
Respiratory: Reject if animal has difficulty breathing, severe coughing or excessive muco-purulent discharges.
Digestive: Reject if animal fails to eat, drink or defecate.
Urinary: Reject if posture is abnormal when urinating, if animal strains to urinate or if urine has an unnatural color (hematuria).
Reproductive: Reject animals with foul discharges from vulva, mammaries or prepuce; or with retained placentas/fetal membranes.
CNS: Reject all animals that show depression or disinterest in environment, are "downer" animals (prefer to stay down on the ground), that will not respond to stimuli, have abnormal gaits or movements, or are hypersensitive to normal stimuli.
Mucous Membranes: Reject if mucous membranes are pale, "muddy" or yellow-colored.
Skin and Hair coat: Reject if skin is yellow-colored or has diffuse discolorations (red or black) or lesions. Consider rejecting animals that have obvious hair loss indicative of systemic disease.

What Not To Do:
1. Do not accept animal if diffuse lesions are found. If lesions are localized they may be trimmed and the carcass retained for consumption.
2. Do not consume an animal from an unknown source unless the carcass passes the antemortem and postmortem examinations and is cleared for consumption.

Vet Medicine Procedure: Humane Slaughter and Field Dressing
MAJ Joseph Williamson, VC, USA

When: It may become necessary to capture, dress, and slaughter game in order to eat and continue the mission. The following guide is one of many ways to humanely slaughter and dress animals in a field environment. Perform an antemortem exam prior to slaughtering the animal, and a post mortem exam after. See Food Preservation section to process meat that is not immediately consumed.

What You Need: Knife, rope, gloves

What To Do:
1. **Figure 5-21: Humane Kill:** The following diagrams illustrate the proper position for humane kill of various livestock species:

2. **Figure 5-22.1: Procedures for Field Dressing:**
a. After killing the animal, bleed it promptly by cutting its throat at point A. If the head is to be salvaged, then insert knife at point B, cutting deeply until blood flows freely.

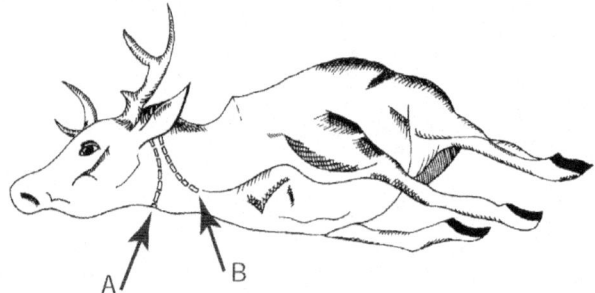

b. **Figure 5-22.2:** Remove genitals or udder. Prop the carcass belly up using rocks or brush for support. Cut circular area shown in illustration. Remove musk glands at points A and B to prevent tainting the meat.

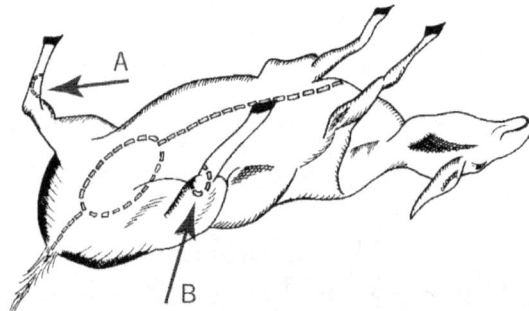

c. **Figure 5-22.3:** Split hide from tail to throat by **carefully** inserting the knife under the skin, but do not perforate the paunch or intestines. Cut around the anus and free the bung so it may be removed with the intestines. Cut around the diaphragm to free it from rib cage.

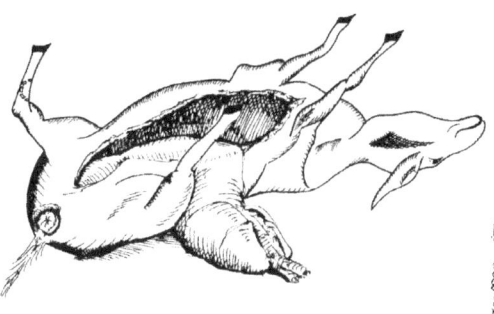

d. Reach forward to cut the vessels, gullet and windpipe. Free the gullet and viscera and remove them from the animal. If skinning the animal, it is best to peel the hide off. Using a knife can perforate and contaminate the meat.

3. Figure 5-22.4: Skinning: Make circumferential cuts around each limb above the elbow; connect them to midline cut and PEEL, DO NOT CUT OR SCRAPE away hide.

4. Preparation:
Pork: After slaughter, ensure animal is bled out completely before scalding or boiling off the hair.
Poultry: There are several methods of slaughter: wringing the neck, dislocating the neck or beheading. Bleed out animal completely before boiling. Boil the bird to remove skin contaminants and ease the removal of the feathers. Then eviscerate by opening the abdominal cavity and removing organs. The carcass is ready for cooking and consumption.

What Not To Do:
Do not slaughter sick or debilitated animals for consumption.
Do not use a blade to skin the carcass; PEEL the hide from the body.
Do not consume meat until it is well cooked.

Vet Medicine Procedure: Postmortem Exam
MAJ Joseph Williamson, VC, USA

When: Examine animal carcasses immediately after slaughter and evisceration for possible lesions that indicate unsuitability of meat for consumption. Examine all parts of the carcass. The following guidelines are for any species that may be consumed in the field environment. Avoid introducing external contamination to the carcass.

What You Need: Gloves, sharp knife

What To Do: Examine:
1. **General:** Condemn animals with gross contamination of interior surfaces or organ systems and/or discoloration of peritoneal or pleural cavities. Generalized abscesses, emaciation, and jaundiced organs or tissues are reasons for condemnation. **Localized lesions are acceptable but generalized conditions are not.** Consider acute disease processes vs. chronic processes. Lymphadenopathy indicates disease or inflammation in the area drained by the enlarged nodes. Local adenopathy may indicate a local process only (condemn only affected area), while more extensive adenopathy probably implies widespread disease process.
2. **Head:** Inspect for swelling or firm masses along jaw or face. Palpate and examine lymph nodes of the head and neck for gross swellings or lesions.
3. **Viscera:** Palpate and examine the lymph nodes. Inspect and palpate all surfaces for abnormalities, discoloration, masses and parasites; examine the heart, lungs and diaphragm as well. Slice open organs and examine for parasites, infection or disease states such as tumors. **NOTE:** Examine viscera away from the carcass to avoid contamination.
4. **Joints and Skeletal Muscles:** Bruises and localized lesions may be removed and the rest of the carcass consumed. For arthritic and swollen joints, remove affected limb, and then consume carcass if arthritis is not due to systemic disease such as septicemia or caseous (cheese-like) lymphadenitis. Do not consume broken or mangled limbs.
5. **Neoplasia, Tumors or Abnormal Growths:** Condemn organ system and/or carcass if spread throughout.
6. **Off Odors:** Condemn carcass with strong odors of urine, ketones (a fruity smell) or pungent sexual odors.

What Not To Do: DO NOT consume organs that appear discolored. DO NOT consume the liver if it appears spotty, discolored, or friable (crumbly).
REMEMBER: Systemic illness and internal disease states may not be evident on antemortem exam, therefore postmortem exam is a necessity when an animal is from an unknown or non-approved source.

Vet Medicine: Food Storage and Preservation
MAJ Joseph Williamson, VC, USA

When: The easiest way to avoid food-borne illnesses in a field environment is to immediately consume well-cleaned and cooked foodstuffs. When excess food must be stored and preserved for future use, follow these rules: preserve and store only wholesome foods that were initially safe to eat; use only potable water and spices when curing or preserving food; cold storage/freezing is the best method if available; periodic re-examination of stored products is essential to ensure wholesomeness and prevent consumption of contaminated or deteriorated food (moldy, infested, stale). Avoid food- and water-borne diseases through continuous use of these guidelines.

What You Need: Knife, meat, potable water, salt, 1% salt solution (brine), string, green hardwoods, building, saltpeter (potassium nitrate), spices, fire source, hay, salt box and/or brine pan, boiling pot.

What To Do:

I. **Curing** - Although it may be done alone, curing should be done in association with smoking. Various spices, salts, sugars and brines can be used.
1. Raw meat should be clean, edible and sliced against the grain into manageable pieces (step one of beef prep for smoking). Salt the meat in a dry, sheltered area secure from rodents and insects. Always CURE before SMOKING.
2. Use coarse salt, not table salt. Additional spices may be added to the salt for flavor. If using brine, then the solution should be 1% salt (one pound of salt to 9 pints water). Use clean plastic, glass or earthenware containers, not wood or metal containers to hold brine solutions.
3. Construct a salt box large enough to hold all of the meat. Cover the bottom of the box with salt. Rub salt into meat thoroughly and place in box. Separate pieces of meat to avoid contact. Cover with salt. Repeat this procedure in two days and again in two days. DO NOT REUSE salt or brine; discard after each use and begin with fresh salt or brine.
4. On day six, remove from salt box. Dry the meat by pacing a layer of green pine straw, hay, grasses, etc. on the floor and cover hay with salt. Place meat on salt-covered hay; cover again with salt then top with hay. Ensure that the area is free of rodents and insects.
5. Wash salt-cured meat before eating.

II. **Smoking** - There are several acceptable procedures for smoking meat and different step by step processes. The one outlined here has the elements that are common to all methods.
1. Smokehouse - Use any well-sealed building with a vented roof and a floor that can have a fire pit. Fire pit should be centered, roughly 2 feet deep. The diameter depends on the building size and how much meat is to be smoked.
2. Firewood - Use green wood from deciduous trees (ones that shed leaves in winter). Conifers such as pines and firs give an odd taste to meat and should not be used. Let fire burn down to coals and then stoke it with green wood to produce "cold smoke" (less than 85°F). Avoid flames during the smoking process.
3. Rafters - Rafters should also be green wood and run the length of the smokehouse. Suspend meat from rope or twine 4-5 ft from top of fire pit. Allow even smoking and avoid contact spoilage by ensuring that all meat hangs free.
4. Time - Smoke meat for 4-5 days, depending on size of house, size and number of pieces of meat to smoke.
5. Meat Preparation- Prepare meat following these guidelines: **BEEF -** Remove large bones and joints. Trim fat and save for pemmican (a meat and fat sausage). Cut across the grain into manageable pieces and secure with a string. The hole for string should be centralized enough to prevent meat ripping during smoking. Hang meat and prepare smoking record (see preservation records and recommendations below). **PORK -** Use hot water to remove hair from skin of animal. **DO NOT** remove layered fat or the bones, except ball and socket joints. Do not scrape the fat that oozes during the smoking process (rendered fat).
6. Smoked meat should be edible for up to one year depending on climate, condition of meat prior to smoking and insect and rodent control. Souring or the appearance of holes or moisture patches does not condemn the meat. Open up the sour area. If it clears up in 24 hours then it is still edible, if not then discard. **If in doubt, throw it out.**

III. **Jerky -** light and nutritious. Use only red meat.
1. Trim fat from meat and cut meat WITH the grain of muscle into 12-inch long strips no more than 1 inch thick and ½ inch wide.
2. Pack meat into salt for 10-12 days. Completely cover each strip with salt and do not allow strips to touch.
3. Smoke meat after salting.
4. Meat may also be dried over slow coals or sun-dried (sprinkle with pepper and hang about 20 ft into air above insect line).
5. Wash before eating if salt cured.

IV. **Pemmican -** Two basic ingredients: lean meat that is not salt cured and rendered fat.

1. Use 6 lbs. of beef to make one pound of pemmican. Dry, pound and shred the meat.
2. Prepare a casing, such as intestine, by cleaning (strip out contents and boil) and tying one end.
3. Place shredded beef lightly into casing, DO NOT PACK.
4. Render fat by boiling cut up or ground up (preferred) fat. The fat will separate into tallow, the liquefied oil from fat, and (cracklings), the fat residue. Cracklings can be eaten.
5. Pour hot tallow into casing which heats the meat and fills the casing. The mixture in the casing should be 60% tallow and 40% meat. Tie casing closed and seal it by pouring tallow on tied ends.
6. Allow pemmican to harden. Should last for approximately 5 years.

V. Salting and Pickling- Dry salt meat or immerse in a salt solution. Follow guidelines in **Curing** Section above. Use 10:1 table salt to saltpeter (potassium nitrate) for both. For pickling, mix 50 pounds of salt and 5 pounds of saltpeter with 20 gallons of water.

VI. Canning and Other Methods- These procedures are effective but require resources and equipment not readily available in a field environment.

VII. Preservation Records- Record the steps taken during meat preservation. Records should have the following information at a minimum: meat type, date, source of meat, weight and cut of meat, total time cured (preserved), wood used and/or type and amount of salt/seasoning/brine used.

What Not To Do:
Do not use meat that is unfit for consumption based on ante- or postmortem exams. Use only potable water.

Vet Medicine: Procedure:
Animal Restraint and Physical Exam
MAJ Joseph Williamson, VC, USA

When: Physical exams are an important part of animal care and ownership. If utilizing animals for the carrying of equipment, as a food source, or treating them as part of an exercise, examinations should be conducted in a thorough way using the SOAP format. Handle animals with caution when you examine or treat them. Insist that owners restrain livestock. Do not attempt restraint without assistance. Apply only the restraint necessary to perform required tasks.

What You Need: Rope, twitch, nose lead, stethoscope, pen, paper, leather gloves, exam gloves, light source, rectal thermometer (large animal style preferred)

What To Do: 1. Restraint: Allow owner and/or indigenous persons to handle and restrain the animals as much as possible. This is probably the most difficult part of the examination and may be the most dangerous. Use the following to assist the locals in restraining the animal.

Figure 5-23
a. Halter - Fasten a rope loop around the animal's neck with a bowline knot to make a temporary rope halter. Pull a bight of the standing end through the loop from rear to front and place over the animal's nose. Pull tight when in use.

Figure 5-24
b. Twitch - For horses. A twitch is a small loop of rope or smooth chain twisted around the upper lip of the horse to divert attention from work being done elsewhere on the horse. Twist the rope or chain with a stick or rod to tighten the twitch, but avoid circulatory compromise. Too much force will harm the horse's lip.

Figure 5-25
c. Casting a cow (Burley Method) - You will need approximately 40 ft of rope, with the center of the rope over the withers and wrapped as shown in the diagram.

While maintaining control of the head, pull tightly on the ends of the rope and the cow will fall. To tie the rear legs, keep both ropes taut and slide the uppermost rope along the undersurface of the rear leg to the fetlock. Then carry the end around the leg and above the hock, across the cannon bone and back around the fetlock. Secure leg with several of these figure 8s.

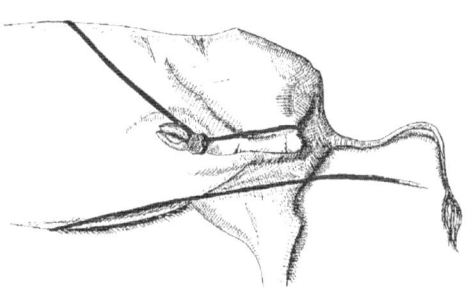

Figure 5-26
d. Restraining the Legs - Tie all four feet together to restrain the animal after it has been cast (dropped). Tie a rope to one leg below the fetlock. Tie the other legs to this one alternately, first a front leg, then a

rear one and repeat.

e. Cattle Tail Restraint - Bend the tail of the cow toward the side or back of the animal to distract the cow. Stand to the side of cow to avoid being kicked. Secure tail base with both hands to avoid damaging the tail and bend.

2. Examination: Once an animal is sufficiently and securely restrained, begin the physical exam. General observations and clinical signs are similar to those found in humans. Follow the SOAP approach, just as when examining a human patient. Remember: the diseases and injuries of animals can be similar to those in humans, but seek advice from appropriate veterinary providers or the Merck Veterinary Manual if available. One can only diagnose and treat based on his level of knowledge and understanding of veterinary medicine.

Table 5-7
Normal Physiologic Values

Species	Rectal Temp (°F)	Heart Rate	Resp. Rate	Feces (lbs./day)	Urine (ml/kg/day)	W.B.C. x1000	HCT%
Horse	100.5	23-70	12	30-50	3-18	6-12	39-52
Cow	100.5	60-70	30	30-100	17-45	4-12	24-48
Sheep	103	60-120	19	2-6.5	10-14	4-12	24-50
Goat	104	70-135	15-20		10-14	6-16	24-48
Pig	102	58-86	15-18	1-6.5	5-30	11-22	32-50
Dog	101.5	100-130	22	0-1.5	20-100	6-18	37-55
Cat	101.5	110-140	26		10-20	8-15	24-45
Rabbit	102.5	123-304	55				

3. Animal Care and Management:
a. Have enough clean, potable water available to the animals. Maintain a clean source and keep it free of feces and foreign material. Many animals will not drink contaminated or soiled water.
b. Have food or forage readily available. Allow animals to graze land and pastures. Keep feed clean, palatable and free of pests.
c. Shelter animals when possible.
d. CONTINUOUS monitoring of the animals for signs of disease and/or parasites will prevent disease transmission within the herd and to humans. A zoonosis is a disease transmissible between animals and man. Many zoonoses are threats in the field environment and precautions need to be taken to minimize them. Monitor and cull sick or debilitated animals. Review the Preventive Medicine chapter and individual infectious disease sections for specifics on zoonoses and how to prevent them.

What Not To Do:
Do not attempt to restrain animals by yourself.
Do not abuse animals. Abuse is unethical, unnecessary and may jeopardize the relationship with native personnel.
Do not alienate local medical and veterinary personnel. Work with them and assist them.

Vet Medicine: Large Animal Obstetrics
MAJ Joseph Williamson, VC, USA

Introduction: Many millions of animals have been born without assistance and forced extraction may do more harm than good. **If the cow is calving naturally, leave it alone. Do not assist.** When to intervene is dependent on the state of parturition, the presentation of the fetus, duration of labor and history of underlying disease processes. The owner will usually be more able to perform intervention if it is necessary. This outline will provide only the basics of "normal" parturition and guidelines for observation and minimal intervention. We will use the cow model throughout.

Subjective: Symptoms
Prior to parturition a normal animal will walk with difficulty, often looking back at her flanks. The udder may swell and become distended with milk, the tailhead ligament will relax and the vulva may swell and begin to discharge mucus or fluid. Restlessness and seeking a quiet isolated area is common. Some may demonstrate an aversion to food and human contact. Pasture animals such as sheep may separate from the herd and lamb on their own. NOT all animals will lie down for childbirth. Duration of labor varies considerably between species (15 minutes for horse; up to 7 hours for a pig litter), and is longer in animals giving birth for the first time. The farmer should know pregnancy status of animals and their due date. He/she may also be of assistance in controlling the animal and giving medical history.

Objective: Signs
No tools or special equipment are required for a normal birth. Use of rectal thermometer is contraindicated. Animal will present with an enlarged abdomen and a drop in body temperature (1-2° below normal 12-24 hours before birth–see Physical Exam section of this chapter). In cattle, fetal membranes filled with fluid are visible outside the cow. Forelimbs and head emerge first from vagina, followed by rest of animal. Anterior presentation with front feet first is the norm, but posterior presentation may occur and not be a cause for alarm. Placenta and afterbirth will follow. Problem/difficult births can include posterior presentation, head or limb deviations and complications arising from multiple births.

Assessment:
Differential Diagnosis: Tumor, bloat, pseudopregnancy.

Plan:
Treatment:
1. The owner/farmer should be in control of the animal at all times.
2. Minimize exposure to fetal fluids or tissues.
3. Allow fetal membranes to burst naturally during labor.
4. Occasionally the membranes will need to be ruptured after prolonged, non-progressive labor. Use a blunt, sterile instrument to make an opening in the membranes without harming the fetus.
5. If a complication arises, including labor for an extended period of time with little or no result, then the farmer may attempt intervention.
6. Veterinary assistance is needed for complicated deliveries.
7. To assist in extracting a fetus that is partially exposed with visible head, forelimbs or shoulder, GENTLY provide traction on the fetus downward and toward the hind limbs of the mother. If difficulty occurs consult with any available veterinary professionals.
8. After the delivery, ensure the young are being cleaned and cared for by the mother and that they have risen and attempted to move about.
9. Do not attempt to remove fetal membranes after birth - this may tear uterine tissue. Allow them to be expelled naturally.
10. Examine offspring for abnormalities and/or deformities. Turn over to farmer.

Vet Medicine Procedure:
Administer an Intravenous Infusion to an Animal

When: An animal needs intravenous medication or fluid resuscitation.

What You Need: IV catheters, IV administration sets, alcohol or Betadine swabs and appropriate medication/fluids.

What To Do:
1. Inspect and prepare the equipment for an IV infusion.
2. Select the IV site. Sites differ according to species: Equine: jugular vein. Avoid the upper 1/3 of the vein to minimize the risk of perforating the carotid artery. Bovine, caprine or ovine: jugular vein. Alternate sites: tail vein or the milk vein in females. Canine or feline: jugular vein. Alternate site: the cephalic vein.
3. Prep the IV site. Shave the IV site if possible and wipe with **Betadine** or alcohol swabs.
4. Administer the IV. Follow the same procedure steps as with humans.
5. Secure the IV. Ensure that the animal cannot pull out the IV and injure itself.
6. Record the procedure.

What Not To Do: Do not allow air to enter the blood stream. Do not allow blood to infiltrate the IV. Do not allow the animal to pull out the IV.

Vet Medicine: Animal Disease: Bloat in Bovine
MAJ Joseph Williamson, VC, USA

Introduction: Bloat is an over-distention of the rumen and reticulum in cattle. Frothy bloat is cause by switching rapidly from poor to rich diets or by diets high in grain or legumes. Free gas bloat is generally due to failure to eructate (belch) free gas because of a physical obstruction. Swollen, gas-filled distention of the abdominal organs may prevent normal respiration.

Subjective: Symptoms
Owner complaints about animals: difficulty breathing and frothing about the mouth, standing with legs splayed, refusing to eat or drink.

Objective: Signs
Distention of the left flank, tympanic gas-filled or froth-filled abdominal cavity, dyspnea, tachycardia. Collapse and death may result if problem persists.

Assessment:
Differential Diagnosis: Peritonitis, ascites, pregnancy.

Plan:
Treatment: Primary
Insert oral stomach tube only with appropriate veterinary supervision and gastric lavage with anti-frothing agents such as vegetable oil.

Treatment: Alternative
Trocarization: Insert a 14GA needle or trocar into gas-filled rumen (rumenotomy), expose the stomach and suture it to body wall if animal is down or condition life threatening. **These procedures should be done by trained veterinary personnel.**

Vet Medicine: Animal Disease: Colic in Equine
MAJ Joseph Williamson, VC, USA

Introduction: Colic is a nonspecific term describing sporadic abdominal pain and discomfort in the equine. Horses may suffer from a myriad of gastrointestinal problems, including intestinal impaction or strangulation, which fall under the general term of colic. Signs can vary from mild discomfort to shock and death. **The severity of clinical signs is not necessarily associated with the seriousness of the disease.**

Subjective: Symptoms

Owner complaints about animal: Restlessness, poor appetite. Inquire about duration of signs and progression (clue to severity), fecal output (indicates obstruction), history of colic in this animal and diet.

Objective: Signs
Auscultate for bowel sounds in all four quadrants of the abdomen. Check mucous membranes and hydration status. Examine feces for blood or mucus. Rectal exam should **be done only by a veterinarian.**
Mild: Animal seems uncomfortable, frequent urination, looking at flank, mild sweating.
Moderate: Increased heart and respiratory rates, restlessness (getting up and down), kicking at flank, diminished stool output, increased sweating
Severe: Increased heart and respiratory rates, extreme anxiety and restlessness (rolling on the ground), profuse sweating, muddy colored mucous membranes, signs of shock
REMEMBER: Clinical signs may not directly relate to actual severity of disease. Horses, as in humans, have varying thresholds for pain and therefore the clinical signs may be misleading. Consult a veterinarian (if available) for all colic cases.

Assessment:
Differential Diagnosis: Trauma to abdomen, gas, GI tract worm infection, pregnancy with/without complications (ask owner about animal's mating history)

Plan:
Treatment
1. Walk the horse. Walking may relieve gas and aid in the movement of obstructions or stool into colon and out of the animal. It will also keep the horse from injury by keeping it up and moving.
2. Provide fresh clean water.
3. Laxatives and wetting agents are useful. Administer orally or through a naso-gastric tube: mineral oil 2-4 liters every 12 hrs, bran mashes (mash food to soften it before giving to horse with sore gums), water. Naso-gastric intubation should only be attempted by experienced personnel.
4. Analgesics may be given if signs warrant. However, they may mask a deteriorating condition so use with caution and under veterinary supervision. Analgesics: **Flunixin meglumine:** 1.1 mg/kg every 12 hrs IV, **Xylazine:** 0.1-1.0 mg/kg as necessary IM or IV, **Butorphanol:** 0.02-0.05 mg/kg IM or IV as necessary.

Owner Education
Manage the herd properly to reduce the likelihood of colic. Offer a high-roughage diet and elevate hay bins to prevent sand impactions. Provide a good de-worming program that ensures pasture rotation and varying de-wormers. Give plenty of fresh, clean water.

Vet Medicine: Animal Disease: Foot Rot in Caprines
MAJ Joseph Williamson, VC, USA

Introduction: Foot rot is a significant problem in sheep and goats, occurring most commonly on pastures during times of persistent moisture. Susceptibility to the disease varies by species and environment. Morbidity can be up to 75% within a flock, from either the primary or a secondary infection.

Subjective: Symptoms
Owner complaints about animals: slow weight gains or weight loss, reluctance to move about, fetid odors emanating from the feet, lameness. Systemic signs may also be reported.

Objective: Signs

Benign: No clinical signs. Other animals in herd affected
Intermediate: mild to moderate signs, +/- fever (normal temp is 103-104°F), foot odor may be present, mild lameness
Virulent Form: Systemic signs of fever/anorexia, lameness, fetid odor, swelling of soft tissue, sloughing of the foot, inflammation of deep tissues, secondary infections.
Using Advanced Tools: Lab: Bacterial cultures may identify organism.

Assessment:
Differential Diagnosis: Other bacterial or fungal infections, trauma, bluetongue (inflammation of mouth and nose; sometimes accompanied by lameness), ulcerative dermatosis

Plan:
Treatment: Treat infected animals. Observe other animals in flock for signs of infection.
Trim (debride) all exposed and necrotic tissues. This is critical for proper treatment of tissues.
Apply local disinfectant (**5% formalin, 10% ZnSO4, or 5% CuSO4**).
Administer antibiotics (infected animals): IM **Penicillin** 50-70,000 u/kg and **dihydrostreptomycin** at 22,500-32,000 u/lb.
Move animal to dry pasture or ground.
Cull (remove from herd) or destroy recurrent carriers to protect rest of animals.

Vet Medicine: Equine Lameness
MAJ Joseph Williamson, VC, USA

Introduction: Lameness is a general term associated with pain in one or more legs due to injury, illness or conformational abnormalities. Lameness resulting from injury and infection will be discussed here. Conformational injuries, fractures and the majority of the soft tissue malfunctions require extensive treatment and/or diagnostic tests that are beyond the capabilities of the medic in the field. Systematically conduct a physical examination. Begin by examining the hoof and work your way up to the shoulder or pelvis. Look for lesions, injuries, swelling and bounding pulses. Use hoof testers (special pliers used to squeeze soft tissue) to apply light pressure to all parts of the hoof, to include the wall, frog and sole. Check for sores, abscesses and painful conditions not readily seen on exam.

Subjective: Symptoms
Owner complains of animal being lame (not bearing weight on the affected limb or carrying it in an unnatural or awkward position). Subtle lameness may only be detected at the trot or on hard-packed surfaces. Owner may report foul odor from hoof; obvious injury, swelling or lesions; animal constantly shifting weight or standing with hind limbs under its body to take weight off the front limbs.

Objective: Signs
Using Basic Tools: Fetid odor from hoof, bounding digital pulses, swelling of joints or leg, hot hoof walls, tenderness elicited with hoof testers (or substitute), abnormal or stilted gait, bowed or swollen tendon and tendon sheaths.

Assessment:
Differential Diagnosis: Thrush - Fetid odor with moist exudative dermatitis of the underside of the hoof. Laminitis, conformational abnormalities, bowed tendons, traumatic injury, fractures

Plan:
Treatment
Trim the affected tissue
Treat thrush with antiseptics, such as **iodine** or **copper sulfate**
Treat laminitis with pain relievers: **Phenylbutazone** 4.4-8.8 mg/kg/day divided bid or tid slowly IV or po,

tapering off over 5 days (po is treatment of choice) or **flunixin meglumine** (banamine) 1.1 mg/kg once daily up to tid, PO/IM or IV (with endotoxemia).
Consider performing a nerve block only if competent and familiar with the technique.
Prevention: Maintain proper hygiene and foot care by keeping feet trimmed and picked clean. A farrier should perform proper trimming, padding and shoeing of hoof.

Vet Medicine: Acute Mastitis (Animal)
MAJ Joseph Williamson, VC, USA

Introduction: Mastitis is the most costly disease to the dairy industry in the US and in most countries of the world. Many bacteria, including *Staphylococcus*, *Streptococcus* and *Coliforms* can cause it. These pathogens cause inflammation of the gland after traumatic injury or exposure to chemical irritants. It is usually a herd health problem. Good hygiene and sanitation are necessary, or treatment will fail.

Subjective: Symptoms
Owner complains that the animal is reluctant to be milked, has swollen milk glands, will not eat, is generally depressed, will not rise or move around (walking may cause discomfort to the gland).

Objective: Signs
Using Basic Tools: Peracute*: Swollen, hot, tender milk glands; abnormal secretions; fever **Acute:** mild systemic signs; gland changes as with peracute **Sub-acute:** no systemic changes, mild gland changes
* **Peracute -** Very acute or violent

Assessment:
Differential Diagnosis: Tumor, cellulitis, stone in milk duct, trauma
The California Mastitis Test can be used as a diagnostic test when coupled with clinical signs. Perform the test by stripping milk from each quarter (4 quarters per udder), mixing it with the reagent in the kit and observing for clumped or stringy milk.

Plan:
Treatment:
1. Strip (milk) affected quarters dry, twice daily during therapy. Continue until condition resolved.
2. DO NOT consume milk.
3. C & S is critical prior to treatment with antibiotics. Only treat with antibiotics labeled **"FOR VETERINARY USE ONLY"** and if applies, **"MAMMARY INFUSION."** Medics will have to shop for these. For Streptococcal species - **Procaine Penicillin G** Intramammary Infusion for lactating animals at 100,000 units per gland for 3 days. For Staphylococcal species - Dry treatment (not milking) is best; results disappointing if treated during lactation. Mammary infusion with pre-mixed antibiotics. For other species - base treatment on C & S. Infuse paste antibiotics into the milk glands. The paste is in a pre-measured plastic infusion syringe. Gently push the tip into teat duct and infuse antibiotics.
4. If paste antibiotics are unavailable, apply hot compresses as often as possible during peracute and acute phases.
5. Give IV antibiotics if animal has signs of systemic infection (fever, lassitude, poor appetite, etc.)

Vet Medicine: Animal Disease: Diarrhea in Porcine
MAJ Joseph Williamson, VC, USA

Introduction: Diarrhea, or scours, is a common and highly contagious problem in pigs. It may be attributed to many agents: Enterogenic *E. coli*, *Treponema hyodysenteriae*, *Salmonella*, Rotavirus or others. The disease may affect individuals or the whole herd but is not a zoonotic threat. Herd health and condition is a vital tool in assessing diarrhea in the pigs. Sporadic death or deaths only in newborns may suggest the diagnosis. Diarrhea will be found in the pens and on the ground. Diarrhea storms, with sudden deaths or high death rates, are not uncommon.

Subjective: Symptoms
Animal owner complains of unthrifty pigs (dry skin, thin, dirty/covered with feces), diarrhea, anorexia and weakness, sudden death.

Objective: Signs
Using Basic Tools: Diarrhea (pale yellow, watery to mucopurulent with flecks of mucosa), fever, lethargy, anorexia, conjunctivitis, failure to grow; weight loss and sudden death. Remember, with multiple etiological agents, signs may be varied or even subclinical.

Assessment:
Definitive diagnosis requires lab support. Tissue samples gathered postmortem should be analyzed in a competent lab. Necropsy lesions and findings may also be helpful in diagnosis.
Differential Diagnosis: Many agents associated with a variety of diseases can cause diarrhea in pigs.

Plan:
Treatment: Segregate, isolate and treat all affected pigs.
Provide fluids: Fresh water or oral rehydration salts. Antibiotics have limited value and probably should not be used. Provide high quality nutritious feed. In some cases, eradication and depopulation will be warranted. **Local government must direct this action.**
"All-in and All-out" practices when replenishing stock: remove all pigs, sanitize stalls and premises and replenish with new animals from one source all at the same time.
Prevention: Report large outbreaks to appropriate agencies. Improve the sanitation of the farm. Vaccines exist for some of the disease agents. Coordinate vaccination programs through the local veterinarian and the appropriate ministries.

Vet Medicine: Conjunctivitis (Pinkeye)
MAJ Joseph Williamson, VC, USA

Introduction: Pinkeye is a common ocular disturbance that may be associated with irritation or trauma to the eye caused by *Moraxella bovis*. Transmission through a herd is usually by dust, droplets, or by flies or other insects.

Subjective: Symptoms
Owner complains that animals have conjunctivitis and discharge from the affected eye.

Objective: Signs
Conjunctivitis, central corneal ulceration-opacity (opacity begins on the periphery and migrates centrally), mucopurulent discharge (yellow-green color, viscous discharge), edematous eyelids, periorbital edema. May have underlying trauma or irritation.

Assessment:
Differential Diagnosis: Corneal ulcer or traumatic injury

Plan:
Treatment: Administer antibiotics (**penicillin, nitrofurazone, tetracycline** or **gentamicin**) as ophthalmic ointments or by subconjunctival injection. Apply topical ocular anesthetic (**tetracaine** will suffice). Direct a 25ga needle into subconjunctiva and slowly administer treatment.
Prevention: Isolate affected animals. Maintain hygiene and insect control.

Chapter 16: Human Nutrional Deficiencies

Vitamin and Mineral Facts

MAJ Victor Yu, SP, USA & LCDR (sel) Edward T. Moldenhauer, MSC, USN

Table 5-8

Vitamin Name	Dietary Sources	Deficiency	Treatment
B1 Thiamine	Whole and enriched grains, dried beans, peas, sunflower seeds, pork	Dry beriberi-peripheral neuropathy in legs. Cerebral beriberi-dementia, Wernicke-Korsakoff syndrome. Wet beriberi-heart failure	The usual treatment daily dose for adults is 50 to 100 mg IV or IM x 7-14 days. Subsequently, a dose of 10 mg per day po until the patient is fully recovered. Adjust dosage for children. Improvement should be apparent within 6-24 hours by reduced restlessness; the disappearance of cyanosis, reduction in heart rate, respiratory rate, and cardiac size, and clearing of pulmonary congestion.
B2 Riboflavin	Milk, mushrooms, spinach, enriched grains, liver	Inflammation of mouth and tongue, cracks at corners of the mouth	RDA is 1.1 to 1.3 mg
B3 Niacin	Bran, enriched grains, peanuts, mushrooms, tuna, salmon, chicken, beef	Pellagra-thickened skin, swollen tongue, bloody diarrhea, psychosis	Administration of niacin will reverse the neurologic symptoms and signs. 10-20 mg per day in the presence of adequate amounts of dietary protein is sufficient.
B6 Pyridoxine	Animal protein foods, sunflower seeds, spinach, broccoli, bananas	Headache, anemia, convulsions, nausea, vomiting, flaky skin, sore tongue	RDA is 1.3 - 1.7 mg
B12 Cyanocobalamin	Animal foods, especially organ meats, oysters, clams. Note: Not found in plants!	Macrocytic anemia, poor nerve function	RDA is 2-3 mcg
C Ascorbic acid	Citrus fruits such as oranges and strawberries; broccoli; greens	Scurvy-bleeding and bruising around hair follicles on the lower legs, splinter hemorrhages, bleeding gums, loose teeth	Infantile scurvy - 50 mg of ascorbic acid qid x 1 week then 50 mg tid x 1 month, supplemented fruits and vegetables such as orange, strawberry, Brussels sprouts, and broccoli. Adult scurvy - 250 mg qid until asymptomatic. Ascorbic acid of 300-500 mg per day in divided doses po for several months in chronic scurvy.
Folate Folic acid	Green leafy vegetables, fortified cereals, organ meats, sunflower seeds, oranges	Diarrhea, poor growth, depression	RDA is 400 mcg

Table 5-8 Continued

Vitamin Name	Dietary Sources	Deficiency	Treatment
A Retinoids and Carotenoids	Fortified milk, fortified breakfast cereals. Yellow-orange pigmented fruit such as cantaloupe and papaya; sweet potatoes and carrots. Dark leafy vegetables (spinach, broccoli, etc.). Organ meats such as liver.	Night blindness, xerophthalmia, poor growth, dry skin	Oleovitamin A, 15-25 thousand units once or twice a day po. If absorption defect is present, give same dosage IM. Be aware that the minimum toxic dose in adults is about 75-100 thousand units daily.
D Calciferol	Fortified milk, fish oils, sardines and salmon. **NOTE**: Most people get vitamin D from exposure to sunlight.	Rickets (children)-bowed bones, painful walking, tetany; Osteomalacia (adults)-bowed bones, fractures	Treatment can only protect against further deformities. Diet high in calcium and phosphorus, 25-100 thousand units vitamin D daily. Treat contributing disease if present.
E Tocopherol	Vegetable oils, margarine, green vegetables, nuts, wheat germ and whole grains	Hemolysis of red blood cells, nerve destruction	RDA is 8-10 mg of tocopherol equivalents
K Phytonadione	Liver, green leafy vegetables, GI flora can produce from diet	Hemorrhage, bruising	RDA is 60-80 micrograms

Table 5-8 Continued

Mineral Name	Dietary Sources	Deficiency	Treatment
Calcium	Milk and dairy products, canned fish such as sardine, tofu	High risk for osteoporosis	RDA is 1000-1300 mg
Chromium	Egg yolks, whole grains, pork, nuts, mushrooms	Peripheral neuropathy	RDA is 50-200 mcg
Copper	Organ meats, whole grains, beans, nuts	Anemia, low white blood cell count, poor growth	RDA is 1-3 mg
Fluoride	Seaweed, tea, toothpaste, fluoridated water	Increased risk of dental caries	RDA is 3-4 mg
Iodine	Iodized salt, saltwater fish and shellfish	Goiter; poor growth in infancy when mother is iodine deficient during pregnancy	Iodine therapy 5 gtts daily saturated solution of potassium iodine or 5-10 gtts of a strong iodine solution in a glass of water. Continue until gland returns to normal size, then place patient on maintenance dose 1-2 gtts daily or use iodized table salt. Encourage local government to iodize salt.
Iron	Red meats, seafood, broccoli, bran enriched products. Note: Plant foods are not good sources of iron!	Low blood iron; small, pale red blood cells; low blood hemoglobin values	RDA is 10 mg for men and 15 for women
Magnesium	Wheat bran, nuts, legumes, green vegetables	Weakness, muscle pain, poor heart function	RDA is 300-500 mg
Phosphorus	Dairy products, meats, poultry, fish	Possibly poor bone maintenance	RDA is 700 mg for adults, and 1300 for ages 9-18
Zinc	Seafoods, meats, greens, whole grains	Skin rash, diarrhea, decreased appetite and sense of taste, hair loss, poor growth and development, poor wound healing	RDA is 15 mg for men and 12 for women

Chapter 17: Toxicology
Toxicology: Poisoning: General
COL Clifford Cloonan, MC, USA

Introduction: Almost anything in sufficient quantity can be toxic, even substances that are essential to life such as water and oxygen. A poison is any substance that even in small quantities produces harmful physiologic or psychological effects. Poisonings are responsible for 10% of all emergency department visits, 9% of all ambulance patient transports, and 5%-10% of all medical admissions to hospitals. They are the third leading cause of accidental death in the U.S. Risk Factors: Approximately 80% of all accidental poisonings occur in children ages 1-4, who typically ingest household products. Few of these incidents are fatal. Adolescents and young adults are at highest risk for intentional poisonings (drug abuse/suicide). The majority of poisoning deaths occur in individuals age 20-49, and are usually intentional. Geographic Associations: Local health care providers can describe what toxins/drugs are commonly used in a specific culture for the purpose of committing suicide and to achieve altered mental status.

Poisons enter the body through a variety of different routes - ingestion, inhalation, injection, and surface or dermal absorption. The toxic effects of ingested poisons may be immediate when inhaled or injected, or delayed when absorbed through the skin or ingested. Because most substances are absorbed through the small intestine, it may take several hours for the poison to enter the bloodstream. Alcohol, which is absorbed in the stomach, is a notable exception. Early management of ingested poisons focuses on removing the toxin from the stomach and chemically binding the toxin to prevent absorption in the small intestine.

Alterations in mental status are common in poisonings but there are many other causes of altered mental status that should ALWAYS be considered. In particular, do NOT assume that altered mental status is due to alcohol or drug intoxication even when the patient has clearly been drinking. Use the mnemonic AEIOUTIPS to recall other causes of altered mental status: A - Alcohol and other toxins/drugs, E - Endocrine (hypothyroidism); I - Insulin, too much (hypoglycemia), or Insulin, too little (hyperglycemia); O - Opiates (heroin, morphine, etc.) and Oxygen, too little (hypoxia); U - Uremia (kidney failure); T - Trauma (head injury, shock)/Temperature (hyper/hypothermia); I - Infection (meningitis, encephalitis); P - Psychiatric (pseudocoma); S - Space-occupying lesion (epidural/subdural hematoma), Stroke, Subarachnoid hemorrhage, Shock.

When treating a poisoned/intoxicated patient, the medic should protect himself. If the patient has been poisoned by a hazardous material, this substance may also pose a risk to the medic. Patients who are intoxicated may behave irrationally or violently.

Subjective: Symptoms

An accurate history is the most important component of the workup. If poisoning was suicidal in nature or involved the use of illicit drugs, history from patient is often inaccurate or intentionally misleading. Obtain history from family members and obtain description of the scene from persons who initially found the patient. Determine which drugs (legal and illicit) or toxins to which the patient may have had access. In the event of ingestion, determine what was ingested, when it was ingested, and whether the patient vomited. In cases of possible occupational exposure, identify the patient's job and the types of toxins to which he/she may have been exposed.

Symptoms:	Acute (< 2 hr)	Sub-acute (2-48 hr)	Chronic (>48 hr)
Constitutional	Nausea/vomiting are common	Signs/symptoms of organ failure	Death or recovery +/- symptoms of chronic organ system damage
Location: Three organ systems are most likely to produce immediate morbidity and mortality.			
Respiratory	Difficulty breathing, shortness of breath	Shortness of breath on exertion	Recovery or chronic shortness of breath, chronic cough, etc...

Cardiovascular	Fainting/near fainting, palpitations, chest pain	Postural hypotension, shortness of breath on exertion	Recovery or symptoms of CHF
CNS	Hallucinations, difficulty concentrating, headaches, visual disturbances	Numbness/tingling/painful sensations, visual disturbances	Recovery or symptoms of learning disabilities, chronic pain, long term visual disturbances

Objective: Signs

Focus of the initial physical examination should be on ruling out life/limb/sight threatening conditions. In poisonings these involve the respiratory, cardiovascular, and central nervous systems.

Using Basic Tools: Inspection: Should reveal spontaneous conversation, gait, posture, general appearance, affect (depressed, agitated, happy) and appearance of the skin (needle track marks or other evidence of drug use, evidence of trauma, discoloration). Vital signs can indicate type and severity of systemic effects. Monitor cardiac function. Perform a basic neuro examination with a focus on mental status (assess for agitation, mania, depression, etc., as well as basic orientation) and eyes (pupil size, equality, and reactivity, nystagmus, visual acuity, and extraocular muscles). Observe gait if possible and perform tests of cerebellar function (i.e., finger-to-nose, rapidly alternating hand movements, heel-to-shin, Romberg).

Signs	Acute(< 2 hr)	Sub acute (2-48 hr)	Chronic (>48 hr)
Respiratory	Dyspnea, wheezing, stridor, apnea, hypo/hyperventilation,	Shortness of breath on exertion	Recovery or COPD, restrictive lung disease, emphysema, etc...
Cardiovascular	Hypo/hypertension, tachy/bradyarrhytmias	Signs of ischemia/infarction, postural hypotension, early CHF	Recovery or signs of chronic CV disease, i.e. CHF, cardiomyopathy, recurrent tachy/bradyarrhythmias
CNS	Stroke, seizures, altered mental status to include coma, agitation/somnolence/depression	Paralysis, seizures persistent altered mental status	Recovery or persisting paralysis, recurrent seizures, mental retardation, persistent vegetative state

Using Advanced Tools: Pulse oximetry (**WARNING:** Pulse oximetry may be normal in carbon monoxide poisoning, cyanide poisoning initially, methemoglobinemia, and other conditions causing inadequate oxygenation of the tissues). The presence of a normal pulse oximetry reading does not always indicate adequate oxygenation. EKG (arrhythmias). Lab: urinalysis, blood glucose. Drug testing is generally not available in field environments.

Assessment:
Differential Diagnosis: See AEIOU-TIPS discussion in Introduction.

Plan:
Treatment
1. Secure airway. If patient is hypoxic and/or hypoventilating apply oxygen and assist respirations.
2. Start an IV in all presumed poisoned patients for drug access. Fluid resuscitate as needed to support blood pressure.
3. Treat arrhythmias per ACLS.
4. Decontamination procedures: Decontaminate skin and mucous membranes as required with mild soap and water. Remove patient from any further exposure to toxic vapors/fumes. Give **syrup of Ipecac** (if within

20 minutes of ingestion), gastric lavage, and/or **activated charcoal** to minimize toxins in the GI tract (see below).
NOTE: Inducing diarrhea is NOT effective and is likely to make the patient worse (dehydration, fluid/electrolyte imbalance).

Syrup of ipecac - effective in some cases if administered within 20 minutes of ingestion.
1. Administration:
    a. In patients 1-12 years old, give 15 ml followed by 2-3 glasses of water.
    b. In patients over 12 years, give 30 ml followed by 2-3 glasses of water.
    b. May be repeated in 20 minutes if vomiting does not occur.
2. Complications: Mallory-Weiss tear of the esophagus, causing bleeding; pneumomediastinum (air trapped in chest cavity outside the lungs); diaphragmatic or gastric rupture; and/or aspiration pneumonitis
3. Contraindications: Patient < 1 year old, altered level of consciousness (aspiration), ingestion of caustic substances, loss of gag reflex, seizures, pregnancy, acute myocardial infarction, ingestion of: acids, alkalis, ammonia, petroleum distillates, non-toxic agents, rapidly acting central nervous system agents, or hydrocarbons.

Gastric lavage - may provide opportunity for immediate recovery of a portion of gastric contents.
1. Administration:
    A. Use large-bore orogastric tube rather than a smaller nasogastric tube (Size 36-40 French for adults, size 24-28 French for children).
    B. Never insert large orogastric tubes nasally (may fracture/amputate nasal turbinate and/or cause serious bleeding).
2. Complications: Agitation, tracheal intubation, esophageal perforation, aspiration pneumonitis, pediatric fluid and electrolyte imbalances.
3. Contraindications: Altered levels of consciousness (relative contraindication if the airway is protected), low-viscosity hydrocarbons or caustic agent ingestion.

Activated charcoal
1. Administration
    a. Administering 20-30 minutes before gastric lavage may double the effectiveness of lavage.
    b. Do not administer until after vomiting if **ipecac** has already been given.
    c. Form slurry of 1-2 g/kg body weight (30-100 g for adults, 15-30 g for children), and administered orally or by gastric tube.
2. INDICATIONS/ CONTRAINDICATIONS:
    a. Safe and effective treatment in most toxic ingestions.
    b. Do not use for strong acid, strong alkali.
    c. Not effective for cyanide, iron or alcohol

Patient Education
Prevention and Hygiene: In cases of toddler poisonings educate mother/father regarding "poison proofing" of home. Remove all cleaning products and other toxins from child's reach; apply locks to cabinet doors, etc.

Follow-up Actions
Evacuation/Consultation Criteria: Evacuate if patient unstable. If there is any question as to the severity of the poisoning or whether the patient may have been committing an act of self harm, consult emergency medicine/toxicology if patient unstable or serious poisoning suspected and psychiatry if poisoning is felt to have been an act of self-harm.

Toxicology: Venomous Snake Bite
COL Clifford Cloonan, MC, USA

Introduction: Venomous snakes cause injuries and deaths worldwide in all temperate and tropical climates, but they are a particular problem in Australia, which has 40% of the world's neurotoxic snakes and about 23% of all venomous snakes. In North America, poisonous snakes cause only 14-20 deaths per year. The risk of death is greatest in the very young, the very old, those with medical problems involving the cardiovascular and respiratory systems and those who sustain multiple bites. Only about 1/5 of snake bites in the U.S. are inflicted by venomous snakes and not all bites by poisonous snakes result in envenomation. Rattlesnakes fail to inject venom in up to 20% of bites. The typical victim of a pit viper is a young male 11 - 19 years of age who is bitten on the hand while trying to handle the snake. Alcohol use is often a contributing factor. Because snakes either hibernate or are inactive during winter, the peak snakebite season in temperate climates is April-October. In the United States the great majority of poisonous snakebites are caused by pit vipers (*Crotalidae*), specifically rattlesnakes, copperheads and cottonmouth snakes. Eastern and western diamondback rattlesnakes, although causing only about 10% of all snakebites in the U.S., are responsible for 95% of all snakebite deaths in the U.S. The other poisonous species of snakes in North America (not pit vipers) are the Eastern and Texas coral snakes. They are members of the *Elapidae* family, along with cobras, kraits, and mambas. Sea snakes belong to the *Hydrophidae* family.

Subjective: Symptoms
Variable depending on type of snake, amount of venom injected, age of victim and other factors.

	Acute (2 hr)	Sub-acute (2-48 hr)	Chronic (>48 hr)
Constitutional	*Crotalidae*: Rapid onset of severe pain at bite site, severe HA, marked thirst. *Elapidae/Hydrophidae*: Little/no immediate pain at bite site	*Crotalidae*: Persistent severe pain, HA, thirst, dizziness, chills, nausea. *Elapidae*: Excessive perspiration *Hydrophidae*: Muscle aches/ pains/stiffness and pain on passive movement of arm, thigh, neck, trunk muscles	Either improving or organ system failure (renal, respiratory, cardiovascular), disseminated hemorrhage, Pruritus, fever, myalgia, arthritis suggests serum sickness secondary to antivenin admin.
Respiratory	If anaphylaxis: Difficulty breathing, shortness of breath. *Elapidae/ Hydrophidae*: Severe envenomation may cause respiratory paralysis/arrest	Onset of anaphylaxis may be delayed > 2 hr. so consider if SOB/ bronchospasm occur. *Elapidae/Hydrophidae*: Respiratory paralysis/arrest possible	*Elapidae/Hydrophidae*: Respiratory paralysis may be prolonged (up to 7 days)
Cardiovascular	If anaphylaxis: Fainting/ near fainting, shock symptoms, severe envenomation may cause arrest	Palpitations, shock symptoms	Usually no long term effects
GI	Nausea/Vomiting	Nausea/Vomiting	Usually no long term effects
Neuro	*Elapidae/Hydrophidae*: Blurred vision	*Elapidae/Hydrophidae*: Paresthesias (numbness of lips/soles of feet)	*Elapidae/Hydrophidae*: Recovery or possible long-term numbness, burning/ tingling sensation

Focused History: *Can you identify the snake?* (give appropriate antivenin) *When was your last tetanus immunization?* (need for current tetanus protection) *Do you have allergies to horses/horse serum?* (check before giving serum derived from horses)

Objective: Signs
Rapid onset suggests a more severe envenomation.
Using Basic Tools:

	Acute (2 hr)	Sub-acute (2-48 hr)	Chronic (>48 hr)

Respiratory	If anaphylaxis: Bronchospasm/respiratory arrest *Elapidae/ Hydrophidae*: May produce early respiratory paralysis/arrest but usually delayed	*Elapidae/Hydrophidae*: Respiratory paralysis/arrest, death	*Elapidae/Hydrophidae*: Respiratory paralysis/arrest can last up to a week. Death, if it occurs, tends to occur early
Cardiovascular	Anaphylaxis may cause hypotension/shock. *Crotalidae*: Diffuse bleeding, 3rd spacing, hypotension, shock. *Elapidae*: Arrhythmias, cardiac arrest	Hypotension, shock, diffuse ecchymosis, significant swelling	Usually no long term complications
GI	All: Vomiting	*Elapidae/Hydrophidae*: Diarrhea	Usually no long term complications
Neuro	*Elapidae/Hydrophidae*: Difficulty focusing; paralysis of the eye muscles, eyelids; difficulty opening mouth, speaking, swallowing, paralysis of the jaw and tongue	*Elapidae/Hydrophidae*: Muscular incoordination, twitching, muscle paralysis (include respiratory muscles); altered mental status, coma	↓Range of motion (ROM), weakness, numbness, burning/tingling sensation
Renal	Usually no early renal problems	*Crotalidae*: Gross hematuria. *Hydrophidae*: Reddish-brown urine	*Crotalidae*: Renal failure or recovery
Soft Tissue	*Crotalidae*: Usually two fang punctures at site of bite, rapid onset of swelling	*Crotalidae*: Significant swelling, tissue necrosis, petechiae, ecchymosis, bullae– local & poss. diffuse	*Crotalidae*: Usually no long term morbidity, but compartment syndrome, tissue necrosis, ↓ROM may occur

Using Advanced Tools: Lab: Hematocrit, urinalysis and 12-lead EKG to assess renal and cardiac complications. Blood transfusion: type and crossmatch as required.

Assessment
Differential Diagnosis - Non-venomous snakebite; venomous bite from animal other than a snake; other sources of intoxication.

Plan
Treatment Goals: rapid transport to hospital-level care, delay progress of envenomation and alleviate early symptoms.
Primary
1. Ensure airway is patent and adequate -- if not, secure airway. If hypoxic and/or hypoventilating, apply O_2 and assist respirations, prevent aspiration (lay the patient on their side), intubate as required.
2. Start an IV in all snakebitten patients (in an unbitten extremity). Fluid resuscitation to support blood pressure and maintain urine output (see below). Drink water as tolerated but otherwise NPO. NO ALCOHOL! Be prepared for shock.
3. Monitor vital signs with pulse oximetry and cardiac monitoring if available. Treat arrhythmias per ACLS guidelines – **NOTE:** Muscle breakdown may release significant potassium, so consider hyperkalemia if

arrhythmias occur.
4. Limit the systemic spread of the venom thru methods described below:
   a. Keep the patient as calm and inactive as possible. Reassure. Give benzodiazepam (e.g., **Valium** 5 mg) po as needed.
   b. Gently clean around the bite site to remove any venom from the skin.
   c. Immobilize the bitten limb in a dependant position.
   d. Suctioning the bite site (NOT with mouth) within minutes after bite is reasonable if remote from hospital care. Do not incise over the puncture site. The use of suction is controversial but all agree: never use the mouth to apply suction.
   e. Do not apply tourniquets, ligatures, or constricting bands **unless** the snake is primarily neurotoxic (Australian elapid, sea snake, krait, cobra or other neurotoxic species). Neurotoxic bites only: apply a constricting band approximately 1 in. wide 2-4 inches above the bite and loose enough to admit a finger. Alternatively, wrap the bitten extremity with an elastic bandage or place it in an air splint. Another method: Place a thick pad over the area of the bite and hold it in place with a tight wrap, wrapping from distal to proximal. If more than 30 minutes after the bite, do not apply a constricting band. Do NOT treat pit viper bites with these methods. Always check for a pulse after applying – this is not a tourniquet!
   f. Measure the circumference of a bitten extremity 10 cm proximal to the bite. Track this measurement and pulses over time.
5. Remove all jewelry from bitten extremity.
6. Insert Foley catheter, record urine output and monitor fluid balance. Check urine for myoglobin (positive for blood on urine dipstick but no RBCs on microscopic exam) and blood. Avoid overhydration (rales, wheezing, orthopnea, respiratory distress, and distended jugular veins). Cautiously hydrate to maintain urine output > 30-50 cc/hr (adults). Administer **furosemide (Lasix)** up to 100 mg to promote urine output as needed. Give low dose **dopamine** (2.5 kg/minute) by continuous infusion if necessary to maintain urine output. Give adult victims with myoglobinuria and decreased urine output 25 grams of **Mannitol** and 100 mEq (generally two ampules) **sodium bicarbonate** added to 1 liter 5% dextrose and infused over 4 hours to prevent myoglobinuric nephropathy.
7. Treat pain with **acetaminophen** and opiates as required. Avoid NSAIDs, which interfere with platelet function.
8. Treat nausea/vomiting with **Compazine** (give slowly if administering by IV - this can cause/worsen hypotension).
9. Give tetanus toxoid as required.
10. **DO NOT** cauterize, incise, or amputate the bite site. **DO NOT** apply electric shock or pack bitten limb in ice.
11. If the snake can be **SAFELY** killed, bring in for identification (see below). Avoid handling the snake. Be sure it is dead. **WARNING – dead snakes can still reflexively bite!**
12. Give antivenin, which is the only proven therapy for snakebite, only if it is specific for the snake involved (monovalent), or if the envenomation is severe (polyvalent). **See (and follow) package insert for antivenin-specific instructions**. The administration of any type of antivenin has a risk of allergic reaction and serum sickness that can be **life-threatening**. DO NOT administer antivenin unless the specific criteria for administration are met. Remember: death from snakebite is rare and snakebite without envenomation is common. Inappropriate administration of antivenin can kill a patient who would otherwise have survived without permanent sequelae. In the U.S. even when the offending snake is venomous, and envenomation has occurred (e.g., copperhead), antivenin administration is often not necessary.
13. Be prepared to treat anaphylaxis after giving antivenin with **epinephrine** 0.3 cc 1:1000 IM and **diphenhydramine** 50 mg IM. If patient rapidly becomes hypotensive and/or develops acute severe respiratory distress it may necessary to give 1mg (10cc) 1:10000 **epinephrine** slowly by IV.

NOTES: 1. Snakebites on the extremities can produce extensive swelling that may (but rarely does) lead to the development of a compartment syndrome (pain on passive stretching and active flexing of the involved muscle groups, distal paresthesias, pulselessness, tense overlying tissues). Doing a fasciotomy in a patient with a venom-induced bleeding disorder and local tissue necrosis may cause significant, even life-threatening, bleeding and/or infection. An aggressive surgical approach is more likely to cause harm than good, so delay fasciotomy (see Procedures: Compartment Syndrome Management) as long as feasible.

2. Early, prophylactic, broad-spectrum antibiotic therapy (second-generation cephalosporin – **Keflex/Ancef**) is reasonable but not generally recommended. If suction by mouth has been done, prophylax with **erythromycin**. Treat infection, if it develops as appropriate. NOTE: Redness, swelling, pain, and increased warmth in the surrounding tissue occur in both envenomation and infection.

Alternate: Support the airway, maintain adequate oxygenation, ventilation, urine output and blood pressure until specific, neutralizing, antivenin can be administered. There is no good evidence supporting that any first aid measures aside from those describe herein.

Primitive: Maintaining airway and urine output will save most patients. An overly aggressive surgical approach and resorting to various unproven therapies will cause more harm than good.

Empiric: In the proper circumstances, assume snakebite with envenomation and observe patient for 4-6 hours for development of signs/symptoms. If no signs/symptoms after 6 hours, consider bite w/o envenomation.

Patient Education

General: Do not handle snakes, especially after drinking alcohol.
Activity: Limit activity after bite. May return to activity as tolerated after resolution of symptoms.
Diet: Initially, NPO except for water. Regular diet as tolerated later.
Wound Care: Cleanse wound gently, remove any venom that may be present on the skin. Debride wound in 48-72 hrs if indicated – avoid early, excessive debridement. Watch for infection, which is difficult to distinguish from envenomation in the early stages.
No Improvement/Deterioration: Return for reassessment for cold or pulseless limb, changes in mental status, or development of blood in urine and/or decreasing urinary output.

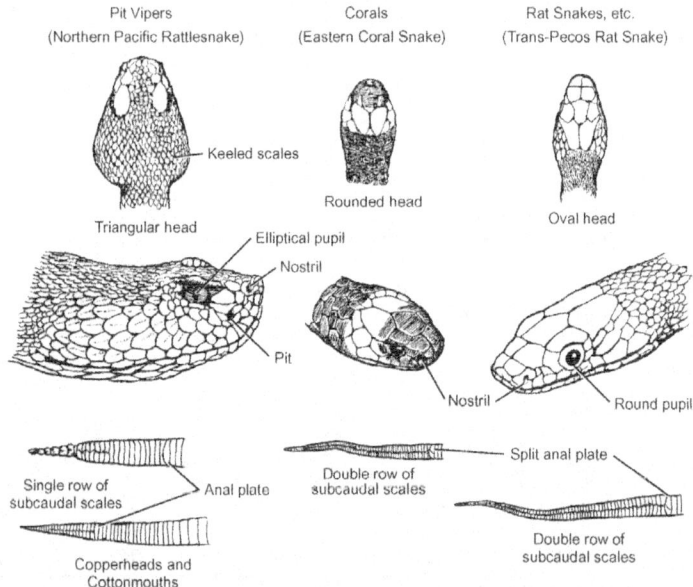

General Characteristics of Venomous and Non-venomous Snakes

Follow-up Actions
Return evaluation: If antivenin has already been given, give more. Administering an inadequate amount of antivenin is a common mistake. Watch closely for sudden anaphylactic shock.
Evacuation/Consultation Criteria: Evacuate snakebite victims for intensive care if possible. Consult with emergency medicine physician if available.

NOTES: Snake identification–Pit vipers: "pit" located below each nostril; triangular-shaped head; elliptical, not round, pupils; hollow fangs; single, not double row of scales on the ventral (belly) side distal to the anal plate; rattlesnakes usually have a rattle.
Coral snakes in U.S. - encircling colored bands of black, red and yellow/white, with the latter bands touching ("red on yellow, kill a fellow; red on black, venom lack"); no long fangs; small mouth makes it difficult for them to bite anything larger than finger.

Chapter 18: Mental Health

Anxiety Disorders: See Symptom: Anxiety

Mental Health: Operational Stress

MAJ Michael Doyle, MC, USA

Introduction: Battle fatigue is a normal response to the abnormal stress of combat, and is the term applied to any combat-related stress reaction requiring treatment. Combat and Operational Stress Reaction is the term applied to service members who present psychologically or emotionally disturbed in non-combat situations. Most service members presenting with signs and symptoms of an emotional or psychological disturbance do not have a mental disorder, but rather, are struggling with the abnormal stress of military operations. Sorting those from the relatively small number that have actual mental disorders is a process called Neuropsychiatric Triage. In a deployed/operational setting, service members who present for evaluation of emotional or psychological symptoms (or are brought in by the chain of command) do so because of an impairment in duty performance, concerns for safety, or both. Always think SAFETY. Have the chain-of-command secure the service member's weapon and send the service member with an escort if there is any concern for safety. Battle Fatigue is classified as either **Light** or **Heavy**. This classification guides treatment planning depending on the tactical situation as well as the severity of symptoms. Light: minimal to mild impairment in functioning; symptoms are present but do not significantly impact duty performance; duty performance complaints are more subjective than objective. Tactical situation allows for forward treatment. Heavy: obvious impairment in duty performance or tactical situation precludes treatment at unit or forward aid station.

Subjective: Symptoms
Anxiety, nervousness, fear, panic, terror, sadness, guilt, depression, anger, insomnia, hallucinations, delusions, hyper-alertness, agitation, inattention, carelessness, erratic actions, outbursts, or physical exhaustion, immobility, panic running, loss of skills, or loss of memory; loss of confidence, hope, or faith; somatic complaints: muteness, blindness, deafness, paralysis or weakness; thoughts of hurting oneself or someone else.
Focused History: *How long have you had these feelings?* (longer than few days suggest mental disorder) *Can you fight? Can you do your job?* (ask the same of a supervisor – Light Battle Fatigue if affirmative) *Do you have any thoughts of wanting to hurt yourself of anyone else? Have you ever had those thoughts in the past?* (**Always ask -** suggests instability and a danger to the unit)

Objective: Signs
Using Basic Tools: Usually none; tachycardia. Repeat the exam to ensure nothing was missed.

Assessment:
Differential Diagnosis: Undetected physical trauma - Always be concerned with a hidden injury missed

in the primary or even secondary survey!
Almost any mental disorder; consider the following:
Major depressive, bipolar, or schizophreniform disorders - severe, long-lasting conditions that will not resolve without definitive psychiatric care, unlike battle fatigue, which will largely resolve within 72 hours with restoration of confidence, reassurance, replenishment, and rest.
Brief psychotic, acute stress, or conversion disorders - temporary disturbances that may last weeks to months. Sometimes difficult to differentiate from battle fatigue. Degree of impairment/severity of symptoms coupled with prompt resolution suggests battle fatigue. Management principles for battle fatigue can provide support and symptom relief.
Post-Traumatic Stress Disorder (PTSD) - also a long-lasting disorder that appears after some trauma; not likely to occur in acute settings or immediately post-event unless there is some history of prior trauma. PTSD has symptoms similar to those of battle fatigue; history is crucial, as is time course of illness. Someone with PTSD may also suffer battle fatigue in an acutely stressful event.
Substance Intoxication - history will distinguish this from battle fatigue. Assess the whole unit: substance abuse may indicate increased susceptibility to battle fatigue for individuals and units.

Plan:
Treatment guided by the acronym *PIES*:
Proximity: Treat service member as close to the unit as tactically and symptomatically possible. A violent, out-of-control patient cannot likely be treated at the battalion aid station if the battalion is actively engaged in combat, whereas one who is physically exhausted may be treated in his platoon area during a lull in the fighting.
Immediacy: Do not delay initiation of treatment; treat as soon as symptoms are identified and tactically feasible.
Expectancy: "This will get better. You will return to your unit." Positively convey the expectation that this condition will improve and that the service member will **not** be evacuated.
Simplicity: Keep treatment simple. Provide the "Four Rs":
  1. Rest: Provide place and time for adequate rest. Consider **Valium** 5 mg po or **Ativan** 1-2 mg po or IM.
  2. Reassurance: Reassure service member that this condition will improve with rest and he will soon return to his unit.
  3. Replenishment: Food, water, hygiene
  4. Restoration (of confidence): Often service members will have lost confidence in themselves, their equipment or their leadership. Work to restore confidence by keeping the service member in his military role—do not emphasize a "patient" role. Assign simple tasks and duties such as rehearsing battle drills, checking weapons, etc.

Prevention: Combat resiliency, or the ability to ward off the impairing features of combat and operational stress, is best attained through tough, realistic training, physical stamina, high morale and esprit, strong unit cohesion, and unity of effort. This is a leadership and command responsibility, but medics play an important role in assessing a unit's health along these lines.

Follow-up Actions
Evacuation/Consultation Criteria: At any level of care, the provider must make a determination to **HOLD**, return to **DUTY**, or **REFER** battle fatigue casualties. This decision is based on the severity of symptoms and the tactical situation. Using the treatment principles above, 80% of service members presenting with impairment from psychological or emotional disturbances can be returned to duty within 72 hours. Another 10% may benefit from holding and treating for up to a week. The remaining 10% likely have mental disorders and will need referral higher for definitive care. The SOF medic's role in those cases is to ensure safety and stability for transport.
NOTES: Although specific to battle fatigue, these assessments and treatment principles can effectively be applied to non-combat situations such as Peace and Peace Enforcement operations, Stability and Support Operations, routine deployments and exercises.

Mental Health: Suicide Prevention

MAJ Michael Doyle, MC, USA

Introduction: Suicide is the third leading cause of death in the U.S. Armed Forces. Suicidal service members present a challenge to command and medical personnel. Suicide attempts and completed suicides are very disruptive to units that experience them. Many suicides can be prevented through awareness of warning signs and early intervention. **The vast majority of Military suicides are committed by 17-25 year old white males, E-1 to E-4, with relationship problems.** Suicide may be attempted by healthy personnel with an abnormal reaction to operational or relationship stress, or by someone with a mental disorder.

Subjective: Symptoms
Overly stressed, sad, anxious, frustrated, worthless, hopeless, helpless, or guilty; thoughts of self-harm, harm to others, death or being better off dead; relationship problems.
Focused History: *Have you had thoughts of hurting yourself or anyone else? Are you having thoughts now? Have you ever attempted suicide in the past or have you ever intentionally injured yourself? Have you been using alcohol or drugs? Do you have access to a gun? How is your relationship with your wife/girlfriend/husband/boyfriend/significant other? Have you been giving away personal effects?* (ask peers or supervisors if they have observed this).

Objective: Signs
Using Basic Tools: Increasing agitation, interpersonal conflicts, anger, frustration, irritability; change in mood to sadness, new appearance of depression, social isolation and withdrawal; giving away personal effects; increased impulsivity or a history of impulsive or violent behaviors.

Assessment:
Differential Diagnosis: Presence of a mental disorder or personality disorder, self-mutilation and self-injurious behavior.

Plan:
Treatment - Suicidal Ideation and Attempted Suicide
1. Secure the individual's weapons and ammo. Protect patient and others, including health care staff.
2. Monitor/accompany the suicidal individual at all times.
3. Treat injuries or medical conditions.
4. If imminently dangerous to self or others, hospitalize or place under 24-hour watch. Otherwise, manage with a "buddy watch" until the crisis has settled.
5. Identify the stressor that has precipitated this event. A chaplain can often be very helpful in settling down home-front crises.
6. Involve the chain-of-command in disposition plans.
7. If discharging a service member to his unit's custody, always have the service member contract verbally or in writing to return immediately if the thoughts of harm recur.
8. After a serious attempt or completed suicide, ask a chaplain or mental health professional to meet with the patient's ship, squad or section mates to address feelings of guilt, remorse or anger. Do not forget the medical personnel involved.
9. Remain vigilant for suicide clusters in units known for low morale, frequent or high rates of AWOL/UA, poor leadership or other discipline problems.

Prevention: Closely follow service members who report suicidal ideation, even if ideation is not accompanied by intent. Let the service member know that you care and can help. Enlist the support of the chain of command.

Follow-up Actions
Evacuation/Consultation Criteria: Personnel who attempt suicide usually require evacuation. Consult mental health professionals at any point in your evaluation of a service member who presents with indications of increased risk for dangerous behavior or acts.

Mental Health: Substance Abuse
MAJ Michael Doyle, MC, USA

Introduction: Opiates, including codeine and oxycodone preparations and benzodiazepines are the two categories of drugs for which life-threatening intoxication can be reversed. Barbiturate overdose is frequently fatal. Alcohol can infrequently be fatal. Alcohol, benzodiazepines, and barbiturates (found in medications like **Fiorinal** and **Fioricet** used in the treatment of migraine headaches) can cause life-threatening withdrawal after chronic use. Opiate withdrawal is uncomfortable, but not life threatening.

Subjective, Objective, Assessment and Plan
Differential Diagnosis: Intoxicated patients should always be monitored for overt and covert overdose. At times, severe withdrawal states may present as delirium or as psychosis (primarily in alcohol withdrawal, and this is rare). It is not uncommon to see signs of withdrawal from a substance (alcohol, illicit or prescribed drugs) in service-members early in the course of an operation, once access to the substance is denied.
Table 5-9

Intoxication	T	P	R	BP	Motor	Eyes	Complaints	Mental Status
Opiates	↓	↓	↓↓	↓	slowed	miotic pupils pinpoint in overdose	euphoria when high, N/V and constipation later	slurred speech, "nodding off", unresponsive in OD
	Rx: **Naloxone**, 0.8 mg/70 kg slowly administered by IV.							
Benzo-diazepine	NC	NC	↓	↓ (-)	ataxia uncoordination	NC to sluggish, nystagmus	talkative sedation with ↑dose	irritabilty, emotional dis-inhibition, confusion and stupor; in severe OD, coma and death
	Rx: 1. Intoxication: **Flumazenil**, 0.2 mg IV over 15 seconds; repeated after 45 secs and again each subsequent minute until sedation reversed/relieved to a maximum of 1.0 mg total dose given. 2. Overdose: **Flumazenil**, 0.2 mg IV over 30 seconds. 0.3 mg given after 30 secs if no response and up to 0.5 mg each subsequent 30 second interval to a cumulative dose of 3.0 mg.							
Barbiturates	NC	NC	↓↓	↓	ataxia uncoordination	NC to sluggish, nystagmus	talkative, slow speech and thinking	dis-inhibition, confusion, inattention, slurred speech; in OD, coma and death
	Rx: Very little beyond supportive measures of ventilation, hydration and nutrition support can be provided.							
Withdrawal								
Alcohol 12-72 hours after last drink	↑	↑↑	↑	↑↑	fine tremor, restless.	↓visual acuity	N/V, fatigue, anxiety, insomnia	agitated, irritable, hallucinations and delusions, illusions, confusion, seizures
	Rx: Treat those with autonomic evidence of withdrawal (↑ pulse, temp, BP or visible tremors) with **Valium** 5-10 mg po 3-4 times a day on the first day of withdrawal. Do not give a dose once the patient begins to feel groggy or sleepy. Continue to assess and monitor vitals and treat on days 2 and 3 if T, P, BP are still elevated.							

Intoxication	T	P	R	BP	Motor	Eyes	Complaints	Mental Status
Benzo-diazepine 24-48 hours after last dose	↑	↑↑	↑	↑↑	fine tremor, restless, seizure	↓visual acuity	N/V, fatigue, anxiety, insomnia	agitated, irritable, hallucinations and delusions, illusions, confusion, seizures
	Rx: Treat those with autonomic evidence of withdrawal (↑ pulse, temp, BP or visible tremors) with **Valium** 5-10 mg po 3-4 times a day on the first day of withdrawal. Do not give a dose once the patient begins to feel groggy or sleepy. Continue to assess and monitor vitals and treat on days 2 and 3 if T, P, BP are still elevated.							
Barbiturate onset at 24 hours, may last up to 14 daus	↑	↑↑	↑	↑↑	fine tremor, restless,	↓visual acuity	N/V, fatigue, anxiety, insomnia	agitated, irritable, hallucinations and delusions, illusions, confusion, seizures
	Rx: Give **phenobarbital** 120 mg po every 1 to 2 hours until 3 of the following 5 signs are present: 1) nystagmus, 2) drowsiness, 3) ataxia, 4) slurred speech, and 5) emotional lability. Then give no more. Phenobarbital has a long half-life and will self-taper.							
Opiates Symptoms peak at 48 hours	↑	↑↑	↑	↑↑	restless	mydriasis, twitching and kicking. "the habit"	drug craving, bone, back & muscle pain, isnomnia, N/V and darrhea	anxious, sad, irritable, with yawning, rhinorrhea, lacrimation, "gooseflesh" ("cold turkey")
	Rx: Not life-threatening, only uncomfortable. Treat with **Motrin** 800 mg po tid and **clonidine** 0.1-0.3 mg po tid or qid, not to exceed more than 1 mg total dose/day. Follow for signs of hypotension after each dose/day.							

Similarly, indigenous people and host nation personnel may present for care with signs and symptoms of withdrawal or intoxication.

Procedures
Essential: Supportive measures, IV hydration
Recommended: Monitoring (Vitals at a minimum) (see Table 5-9)

Mood Disorders: See Symptom: Depression

Mental Health: Psychosis versus Delirium
MAJ Michael Doyle, MC, USA

Introduction: Delirium represents a disturbance in consciousness accompanied by a change in cognitive function that develops acutely. Someone who is delirious has impairments in awareness, alertness, memory and executive functioning (i.e., difficulty buttoning a shirt) and may be hallucinating. Psychosis is not a specific disorder, but rather describes a degree of severity in certain mental disorders. Someone with psychosis or a psychotic disorder has gross or obvious impairment in perceiving reality. The individual misperceives external cues and responds often to internal stimuli. He appears cognitively impaired, behaviorally disturbed, or both. Psychotic disorders are generally not amenable to treatment in a theater of operations. The most important consideration here is distinguishing psychosis (which is largely idiopathic) from delirium (which is a manifestation of a life-threatening medical condition that may be reversible). Always maintain a high index of suspicion for a physical or CNS injury!

Subjective: Symptoms
Patient often unaware of symptoms; in delirium, history of recent injury, illness
Paranoia: Unreasonable suspiciousness; feelings of persecution, being single out or watched; includes feelings of grandiosity
Hallucinations: False sensory perceptions not associated with real external stimuli; delirious patients often have tactile and visual hallucinations; psychotic patients more often have auditory hallucinations
Delusions: False belief, based on incorrect inference about external reality; not consistent with patient's intelligence and cultural background; cannot be corrected with reasoning
Focused History: *How long has this been going on?* (Delirium has a rapid onset of hours to days; psychosis takes days to weeks.) *Do the symptoms change at night?* (Delirium is often worse at night; psychosis is not so variable.)

Objective: Signs
Using Basic Tools: Autonomic instability (delirium) versus normal vital signs (psychosis), assess for head trauma or occult injury (differential); diffuse hyperreflexia (delirium).
Mental Status Exam:
1. Alertness: Diminished (delirium); normal or increased (psychosis); not responsive to external stimuli (both)
2. Orientation: Disoriented to person, place, time, situation or all (delirium); oriented (psychosis) but answers may be contrived and bizarre
3. Activity: Agitated, especially in evenings (delirium); catatonia—purposeless movements or rigid posturing with waxy flexibility (psychosis)
4. Speech: Slurred words or difficult to comprehend (delirium); disorganized and uses made up words called neologisms (psychosis)
5. Thought Content: Delusions, paranoid ideation, simplified thinking (psychosis)
6. Thought Processes: Difficult to follow because of loose associations or flight of ideas; thoughts often derail or stop abruptly (psychosis)
7. Mood: Disorganized (psychosis)
8. Affect: Inappropriate to situation or stated mood; often blunted or flat (psychosis)
9. Cognition: Impaired memory, attention tasks (both)
10. Judgment: Impaired (both)

Using Advanced Tools: Evidence of medical illness in delirium; ↑WBC, ↓HCT, etc.

Assessment:
Differential Diagnosis
Delirium - orientation is generally impaired; identify underlying medical problem and treat it.
Psychosis - orientation generally preserved; identify underlying medical problem and treat it.
Substance Abuse - alcohol withdrawal, PCP, amphetamine, cocaine intoxication appear psychotic
Seizure Disorder - temporal lobe epilepsy (often with herpes encephalitis) can appear psychotic.
Head Injury - may cause delirium; obtain history from other unit members.
Mental Disorders principally associated with psychosis:
    Schizophreniform disorder and schizophrenia - ages 15-25 men, 20-35 women
    Bipolar Disorder, manic with psychotic features - 3rd and 4th decade, sometimes earlier
    Major Depressive Disorder, severe with psychotic features - more common in an older population
    Brief Psychotic Disorder - may or may not have an identifiable precipitant; begins and resolves within 30 days, often with supportive measures alone.
(Identify the presence of a mental disorder first; do not worry too much about what type it is.)

Plan:
Treatment
1. Calm the patient to protect him and others around him. Psychotic and delirious patients may pose a danger to self or others simply through agitation, reckless behavior or inappropriate activities. For both use:
    a. Benzodiazepines (**diazepam** 5-10 mg po or IM, **lorazepam** 2 mg po, IM or IV).
    b. Neuroleptics (**haloperidol** 2-5 mg IM or po) can settle an agitated patient. If **haloperdol**

is unavailable, use **chlorpromazine** 50-100 mg po or IM. Also consider giving **diphenhydramine** 25-50 mg po or IM coincident with the **haloperidol** or **chlorpromazine**.
   c. AVOID USE OF DIPHENHYDRAMINE, CHLOPROMAZINE OR OTHER MEDICATIONS WITH ANTI-CHOLINERGIC SIDE EFFECTS IF YOU SUSPECT DELIRIUM. THESE WILL WORSEN THE DELRIUM!
   d. If medication fails to settle down an agitated patient, see Psychiatric Restraint Procedures (on CD-ROM). If leather restraints are unavailable, consider restraint with sheets, wrapped around patient on litter. Pharmacological or physical restraint may be necessary to better evaluate and treat a delirious patient.
2. Consider IV hydration for those unable to care for self.
3. Place on watch.

Follow-up Actions:
Evacuation/Consultation Criteria: Evacuate US service members suffering from a psychotic or delirious disorder. Host nation service members and persons should be given behavioral redirection and managed with a goal of maintaining safety for all parties.

Mental Health: Recovering Human Remains
How to Prepare Yourself, Your Buddies, and the Unit
MAJ Michael Doyle, MC, USA

The Mission: May include collecting the bodies of fellow service members so that the Mortuary Affairs specialists can return them to the United States for identification and burial. It may include gathering and possibly burying the bodies of enemy or civilian dead to safeguard public health. The numbers of dead may be small and very personal, or they may be vast. The dead may include young men and women, elderly people, small children or infants, for whom we feel an innate empathy. Being exposed to children who have died can be especially distressing, particularly for individuals who have children of their own.

What To Expect: Seeing mutilated bodies evokes horror in most human beings, although most people quickly form a tough, protective mental "shell". The dead bodies may be wasted by starvation, dehydration and disease (e.g., Rwanda refugees or some POW and concentration camp victims). They may have been crushed and dug out from under rubble, (e.g., the Beirut barracks bombing or earthquake victims). They may be badly mutilated by fire, impact, blast or projectiles (e.g., the victims of the air crashes at Gander, Newfoundland; the civilians killed by collateral damage and fire near the Commandancia in Panama City, or the Iraqi army dead north of Kuwait). They may be victims of deliberate atrocity (e.g., the Shiites of south Iraq or any side in Bosnia). Survivor reactions may include grief, anger, shock, gratitude or ingratitude, numbness or indifference. Such reactions may seem appropriate or inappropriate to you, and may affect your own reactions to the dead. In situations where the cause of death leaves few signs on the bodies (e.g., the mass suicide with cyanide at Jonestown, Guyana) caregivers often have more difficulty adapting because it is harder to form the "shell." The degree of decomposition of the bodies will be determined by the temperature, climate and length of time since death. Bodies will emit a strong odor of decomposition. Workers may have to touch the remains, move them and perhaps hear the sounds of autopsies being performed or other burial activities. These sensations may interfere with work, and create disturbing memories. In body handling situations, many personnel naturally tend towards what is aptly called "graveyard humor." This is a normal human reaction or "safety valve" for very uncomfortable feelings. Other feelings may occur, including sorrow, regret, repulsion, disgust, anger and futility.

When: Personnel may have to perform these services after any death, natural or traumatic.

What You Need: Body bags, shovels, reporting forms, pens or pencils, bags for personal effects, labels, gloves, visual barriers/screens, deodorants

What To Do:
Guidelines for How to Work with Human Remains
BEFORE
1. Learn as much as possible about the history, cultural background and circumstances of the disaster or tragedy. How did it come to happen? Try to understand it the way a historian or neutral investigating commission would.
2. Look at videos and photographs of the area of operation and of the victims. The television news networks and magazines may be sources. If pictures of the current situation are not available, look up ones from previous similar tragedies in the library archives. Share them as a team, and talk about them.
3. Understand the importance and value of the mission. Giving the deceased a respectful burial (even if in some cases it must be a hasty and mass burial), saving their remains the indignity of simply being left on the ground to decay, helping survivors know their loved ones have died rather than remaining uncertain for years and providing a safer environment for the living are all difficult but important. Concentrate on the overall mission, not on each individual, to maintain effectiveness when seeing or working with bodies.

DURING
1. Personnel who examine personal effects for identification and other purposes must not be those who have handled or seen the body.
2. Do **not** desecrate or take souvenirs from the bodies. These are criminal acts.
3. Conduct funeral ceremonies consistent with your own beliefs and background. Unit chaplains and/or local clergy may also conduct rites or ceremonies.
4. Limit exposure to the bodies. Have screens, partitions, covers, body bags or barriers so that people do not see the bodies unless it is necessary. Wear gloves if the job calls for touching the bodies. Mask odors with disinfectants, air-fresheners or deodorants. (Using other scents such as perfume or aftershave lotions are of limited value in the presence of the bodies, and are perhaps better saved for when taking breaks away from the work area.)
5. When the mission allows, schedule frequent short breaks away from working with or around bodies.
6. Drink plenty of fluids, continue to eat well, and maintain good hygiene. To the extent possible, the command should ensure facilities for washing hands, clothing and taking hot showers after each shift. (If water is rationed, the command should make clear what can be provided and how it should be used and conserved.)
7. Hold team debriefings frequently to share thoughts and feelings with teammates.
8. Have a mental health/stress control team or chaplain lead a Critical Event Debriefing after a particularly bad event or at the end of the operation.
9. Plan team, as well as individual, activities to relax and think about things other than the tragedy. Do not abide feelings of guilt, or frustration about not being able to fix the situation. DO WHAT CAN BE DONE WITH RESOURCES AVAILABLE, ONE STEP AT A TIME.
10. Stay physically fit.
11. Keep the unit Family Readiness Group fully informed about what is happening, and make sure family members and significant others are included in and supported by it.
12. Take special care of new unit members, and those with recent changes or special problems back home. If the stress caused by working with dead bodies begins to interfere with performance or ability to relax, TAKE ACTION. Do not ignore the stress. Seek out a buddy or someone to talk with about new or unusual feelings. Other people are likely to be feeling the same things. Do NOT withdraw. The unit chaplain, medic or a combat stress control/mental health team member can often help.
13. Help your buddy, coworkers, subordinate or superior if he or she shows signs of distress. Give support and encouragement, and try to get the other person to talk through the problems or feelings they are having. This will improve each person's ability to cope with the situation.

AFTER
1. Take an active part in an end-of-tour debriefing and pre-homecoming information briefing in the unit prior to leaving the operational area.
2. Follow through with Family Support Group activities which recognize and honor what the unit has done

and share the experience (and the praise for a hard job well done) with the families.
3. Do not be surprised if being at home brings back upsetting memories from the operation. It may be hard to talk about the memories from the operation, especially with those who were not there. This is very common, but try to talk to them anyway, and talk with teammates from the operation (best option). Do not hesitate to talk with a chaplain or with the community mental health or stress control team.

WHAT NOT TO DO
Do not keep these emotions inside. Do not withdraw. Do not desecrate or take souvenirs from the bodies.

NOTE: Source: *When the Mission Requires Recovering Dead Bodies: How to prepare yourself, your buddies, and the unit.* Combat Stress Action Office; HSHA-PO, Department of Preventive Health Services, AMEDDC&S, Fort Sam Houston, Texas.

Chapter 19: Anesthesia

Anesthesia: Total Intravenous Anesthesia (TIVA)
LTC Howard Burtnett, AN, USA & MAJ Steven Hendrix, AN, USA

What:
Induction and maintenance of general anesthesia using only the infusion of intravenous anesthetic medications.

Fundamental tenets of TIVA are:
1. The combination of medications selected for infusion must provide all the components of anesthesia: amnesia (hypnosis), analgesia, autonomic stability, and if required, areflexia (complete muscle relaxation). Table 5-12 lists examples of combinations that have been successfully used in the field environment.
2. Continuous infusion techniques provide more precise control over the pharmacological effects of the medications being administered, avoiding the autonomic "peaks and valleys" seen with intermittent bolus administration.
3. Vigilant titration, based on observed and anticipated patient response, is essential. Successful use of this technique allows for small increases in the anesthetic effect when necessary, as well as aggressive, but methodical downward titration of the infusion rates throughout the course of the anesthetic, resulting in a smooth emergence.

When:
As an alternative to inhalation anesthetics when:
1. Inhalation agents or their delivery devices (vaporizers) are unavailable
2. A patient requires high concentrations of oxygen and minimal drug-induced cardiovascular depression.

What You Need:
1. Normal monitoring capability (BP cuff, pulse oximeter, EKG and stethoscope). Peripheral nerve stimulator is desirable if infusing muscle relaxants.
2. Bag-Valve-Mask or other device capable of delivering positive-pressure ventilations
3. Endotracheal tube or other devices to assist in maintaining or securing the airway.
4. Established IV access and infuse maintenance fluids. Medications will be infused through this line, so it is imperative to have a patent line.
5. Infusion assisting devices, such as a Dial-A-Flow IV rate control clamp. Electronic infusion pumps are ideal but likely not available.
6. Intravenous anesthetic medications (Table 5-10). These recommended dosages are designed for healthy adult patients requiring acute surgical intervention. Do not confuse these dosages with those used for long term sedation/analgesia in critical care/intensive care situations. These dosages are not recommended

for the pediatric population.

Table 5-10: Dosing Guidlines for IV Agents

Drug	Loading Dose	Incremental Dose	Infusion Dose
Amnestics			
Propofol	1.0 - 2.5 mg/kg	0.25 - 1.0 mg/kg	75 - 200 mcg/kg/min
Midazolam	50 - 150 mcg/kg	5 - 10 mcg/kg	0.25 - 1 mcg/kg/min
Analgesics			
Fentanyl	2.0 - 4.0 mcg/kg	0.25 - 1.0 mcg/kg	0.02 - 0.1 mcg/kg/min
Sufentanil	0.2 - 0.6 mcg/kg	0.1 - 0.15 mcg/kg	0.003 - 0.01 mcg/kg/min
Ketamine	0.5 - 2.0 mg/kg	0.5 - 1.0 mg/kg	10 - 40 mcg/kg/min
Muscle Relaxants	*Intubating Dose*		
Vecuronium	0.08 - 0.1 mg/kg	0.01 - 0.015 mg/kg	0.8 - 1.2 mcg/kg/min
Succinylcholine	1 - 2 mg/kg	20 mg boluses	0.25 - 0.5 mg/kg/min

What To Do:
1. Conduct pre-operative system check (Table 5-11).
2. Induce general anesthesia with selected medications (loading doses listed on Table 5-10).
3. Secure the airway and ensure adequate oxygenation and ventilation of the patient.
4. Initiate maintenance infusions of selected medications (Table 5-10).
5. Monitor the patient vigilantly and titrate infusions. If a patient has not responded to surgical stimulation during the previous 10-15 minutes, and a substantial increase in the level of surgical stress is not imminent:
   a. Reduce the infusion rate by 20%.
   b. If the patient subsequently begins to respond to surgical stimulation:
      - Increase the infusion rate to a setting between the original rate and the reduced rate (or approximately a 10% reduction from the original setting) and
      - Administer a bolus equivalent to the amount of drug the infusion will provide during the next 5-minute period.
6. General considerations when employing continuous infusion techniques:
   a. Using a powered infusion pump will ensure continuous infusion, providing optimal control and ideal effect-site concentration.
   b. Use small-bore tubing for the medication infusion and place as close to the IV cannulation site as possible.
   c. Check carrier fluid and connections often. Empty carrier IVs will result in emergence from anesthesia or overdose.
   d. Limit the use of muscle relaxants: determine titration of infusion rates based on the usual clinical signs of anesthetic depth (lack of movement, stable blood pressure and heart rate, regular respiratory rate and rhythm). Overzealously using muscle relaxants blinds the practitioner to some of these signs, necessitating the delivery of positive-pressure ventilations throughout the procedure.
   e. Force downward titration: This is the most critical consideration when rapid post-surgical recovery is required. Review section 5.b.

What Not To Do:
Contraindications: Other than allergic reaction to the selected medication, there are no absolute contraindications to this technique.

Complications:
1. Compatibility concerns: All drugs being infused through a single access line MUST be compatible. This compatibility applies to not only the anesthetic agents, but also other medications (antibiotics) that might be administered during the course of an anesthetic.
2. False sense of security: Practice vigilant downward titration to avoid overdosing the patient but provide adequate levels of anesthesia at the same time. Remain constantly aware of your patient's vital signs.
3. Awareness: Avoid overuse of muscle relaxants, which can mask the purposeful movement usually

associated with inadequate depth of anesthesia. With inadequate levels of anesthesia, patients may recall later what happened in surgery.

Table 5-11: Pre-And Intraoperative System Checks For Continuous Infusion

Preoperative checks
1. Medication infusion lines "piggybacked" into established IV line.
2. Medication infusion lines are primed and purged of air.
3. Clips and clamps have been removed from medication infusion lines.
4. If electrical infusion devices are available:
   - Functional and in the "ON" position.
   - Dosage settings are correct.
   - Connected to a reliable electrical supply.

Intraoperative checks
1. Medication lines remain connected to carrier fluid.
2. Carrier fluid is present and flowing.
3. Pump alarms remain active.
4. Drug supply in reservoir is adequate.

Table 5-12: Examples of Techniques Used In The Field:
Example 1.

Continuous infusion of **midazolam**, **ketamine**, and **vecuronium**.
Induction: **midazolam** (1 mg/kg), followed by **ketamine** (2 mg/kg) and **vecuronium** 0.1 mg/kg. Following intubation, initiate infusion of the same medications at the following rates:
Midazolam   (50 mcg/kg/hr),
Ketamine   (1 mg/kg/hr)
Vecuronium (60 mcg/kg/hr)

Obtain this infusion mixture by adding **midazolam** 5 mg, **ketamine** 100 mg and **vecuronium** 6 mg to a 50 ml bag of 0.9% normal saline and infuse at the rates mentioned above. Determine the infusion rate to deliver this concentration by using the formula:

$$\frac{\text{Patient's weight (kg)}}{2} = \text{ml to be infused per hour}$$

So that a patient weighing 70 kg would require an infusion rate of 35 ml/hr.

Sample Calculations:
1. **Midazolam** infusion of 50 mcg/kg/hr, means 50 mcg x 70 kg or 3500 mcg/hr is required.
2. Using the formula:

$\frac{\text{Patient's weight (kg)}}{2}$ = ml to be infused per hour **or**   70/2 = 35 ml/hr to be infused for this patient

Therefore, 3500 mcg/hr required divided by 35 ml/hr (determined infusion rate) yields a concentration requirement of 100 mcg/ml.

3. Adding 5 mg of **midazolam** to 50 ml of 0.9% normal saline yields a concentration of 100 mcg/ml Similar calculations yield requirements for 100 mg of **ketamine** and 6 mg of **vecuronium**.

Example 2.
Continuous infusion of **propofol** and **ketamine**
Induction: **Propofol** (2 mg/kg) followed by **ketamine** 1 mg/kg and **vecuronium** 0.1 mg/kg.
Following intubation, initiate an infusion of **propofol** and **ketamine**. This combination requires two separate infusion bags, as the **propofol** infusion is deliberately titrated downward more aggressively than the **ketamine**.
Infusion rates are as follows:
Ketamine (1.5 mg/kg/hr)
Propofol (12 mg/kg/hr) for the first 30 minutes, then (9 mg/kg/hr) for 30 minutes, lastly (6 mg/kg/hr).

Prepare the two infusions as follows:
Ketamine 150 mg in a 50 ml bag of 0.9% normal saline, yielding a concentration of 3 mg/ml. Infuse at the rates mentioned above. Determine the infusion rate to deliver this concentration by using the formula:

$$\frac{\text{Patient's weight (kg)}}{2} = \text{ml to be infused per hour}$$

Sample Calculations: For a patient weighing 80 kg
1. **Ketamine** infusion of 1.5 mg/kg/hr, means 1.5 mg x 80 kg or 120 mg/hr is required.
2. Using the formula:

$\frac{\text{Patient's weight (kg)}}{2}$ = ml to be infused per hour **or** 80/2 = 40 ml/hr to be infused for this patient

Therefore, 120 mg/hr required divided by 40 ml/hr (determined infusion rate) yields a 3 mg/ml concentration required.

3. Adding 150 mg of **ketamine** to 50 ml of 0.9% normal saline yields a concentration of 3 mg/ml

Propofol can be mixed in a 50:50 solution using 0.9% normal saline. The resulting concentration (5 mg/ml) allows for easy titration. To prepare the mixture, add 1000 mg of **propofol** to a 100 ml bag of 0.9% normal saline. This results in 1000 mg in 200 ml and creates enough **propofol** for most procedures under one hour in duration. With the 80 kg patient, the first 30 minutes would require approximately 100 ml of the solution:
   12 mg/kg/hr X 80 kg = 960 mg/hr
   960 mg/hr divided by 5 mg/ml (concentration) = 192 mg/hr
   192 mg/hr divided by .5 (first 30 minutes) = 96 ml.

The second 30 minutes would required approximately 75 ml of the solution, with the subsequent infusion rate being approximately 50 ml/hr. Discontinuing both infusions 15-20 minutes prior to the end of surgery results in a smooth emergence.

Example 3.
Continuous infusion of **propofol** with incremental boluses of **fentanyl**

Induction: **Propofol** (2 mg/kg) followed by **fentanyl** 3 mcg/kg and **vecuronium** 0.1 mg/kg.
Following intubation, initiate an infusion of **propofol** at the following rates:
1. **Propofol** (12 mg/kg/hr) for the first 30 minutes, then (9 mg/kg/hr) for 30 minutes, then (6 mg/kg/hr) thereafter.
2. **Propofol** can be mixed in a 50:50 solution using 0.9% normal saline. The infusion technique is described in Example 2.
3. Administer **fentanyl** boluses of 0.25 mcg/kg as required, not to exceed 2 mcg/kg/hr. Toward the end of surgery, titrate **fentanyl** boluses to the patient's respiratory effort.

Anesthesia: Local Anesthetics / Regional Anesthesia
LTC Howard Burtnett, AN, USA & LTC Eugene Acosta, AN, USA

Introduction: Regional techniques utilizing local anesthetics may be used for surgical, diagnostic or therapeutic procedures or for providing relief of acute or chronic pain. Techniques range from subcutaneous infiltration of a small, specified area to major regional plexus blockade.

I. Local Infiltration of an Area

When: Whenever good operative conditions can be obtained with a moderate volume of local anesthetic or to perform a minor surgical procedure such as suturing a wound or excising a small tumor.

What You Need: Sterile barrier (towels or drapes), an antiseptic prepping solution (**Betadine** or alcohol), appropriate sized syringe, 25 or 27 gauge needle, local anesthetic (Table 5-13). Note: A 50:50 ratio of **lidocaine** and **bupivacaine** offers the luxury of a fast onset and long duration.

What to Do:
1. Assemble equipment.
2. Prep area and establish a sterile field with the chosen barrier.
3. Use the smallest possible gauge needle to minimize pain. With the bevel of the needle facing down and parallel to the skin, quickly insert the needle up to the hub. Begin infiltrating the desired area while withdrawing the needle. Slowly inject the medication. Perform subsequent needle insertions from this anesthetized area.
4. Injections should be systematic and delivered in a triangular geometric pattern to ensure an adequate block. Should deeper tissue levels need anesthetizing, utilize a systematic approach again, anesthetizing progressively deeper layers.
5. Adding the vasoconstrictor **epinephrine** delays the systemic absorption of the anesthetic, which prolongs the duration of the block and allows for the safe administration of a larger dose of local anesthetic. (Table 5-13).

What Not To Do:
Contraindications: There are no specific contraindications to this technique.
1. Avoid local anesthetics in patients reporting allergies to them.
2. Avoid using vasoconstrictors with these blocks in the fingers or toes.
    Complications: tissue ischemia and necrosis-- large volumes of local anesthetics and high concentrations of **epinephrine** can lead to ischemia and necrosis of wound edges. Should **epinephrine** be utilized, a concentration of 1:200,000 affords the maximum vasoconstrictive property sought for these procedures.

Table 5-13: Infiltration Anesthetics

Drug	Concentration (%)	Plain Solutions		Epinephrine Containing Solutions	
		Maximum Dose (milligrams)	Duration (minutes)	Maximum Dose (milligrams)	Duration (minutes)
Lidocaine	0.5 - 1.0	300	30 - 60	500	120 - 360
Prilocaine	0.5 - 1.0	500	30 - 90	600	60 - 120
Bupivacaine	0.25 - 0.5	175	120 - 240	225	180 - 240

II. Digital Block of the Finger or Toe (Figures 5-27 through 5-32)

When: Whenever good operative conditions can be obtained with a moderate volume of local anesthetic. To perform a surgical procedure on the area.

What You Need: An antiseptic prepping solution (**Betadine** or alcohol), appropriate sized syringe, 25 or 27 gauge needle, local anesthetic (Table 5-14).

What To Do:
1. Assemble equipment:
2. Prep area.
3. Each digit is supplied with two pairs of nerves: dorsal and palmar in the hand, and dorsal and plantar in the foot.
4. Insert the needle in the dorsal aspect of the digit at the level of the metacarpal (metatarsal) head.
5. Advancing the needle in the palmar (plantar) direction, inject 1-2 mls of local anesthetic on each side

Figure 5-27

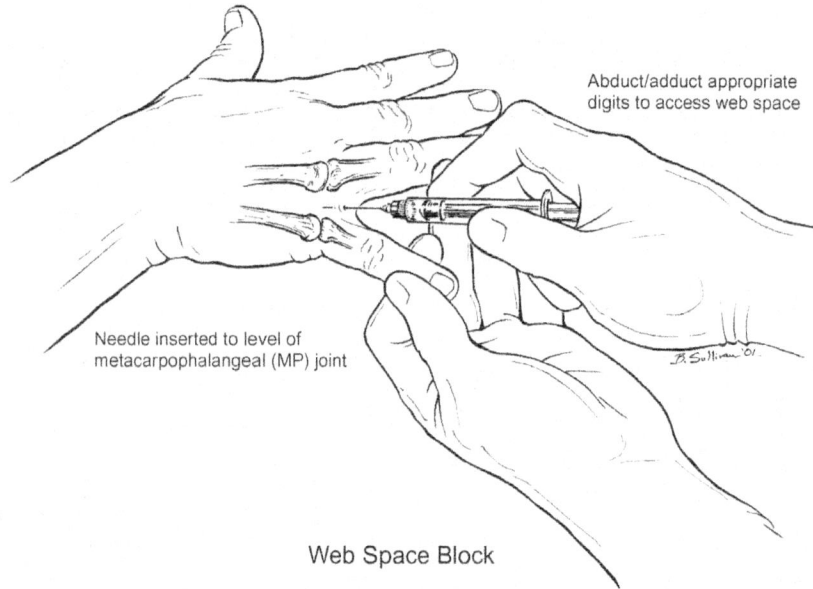

Abduct/adduct appropriate digits to access web space

Needle inserted to level of metacarpophalangeal (MP) joint

Web Space Block

Figure 5-28

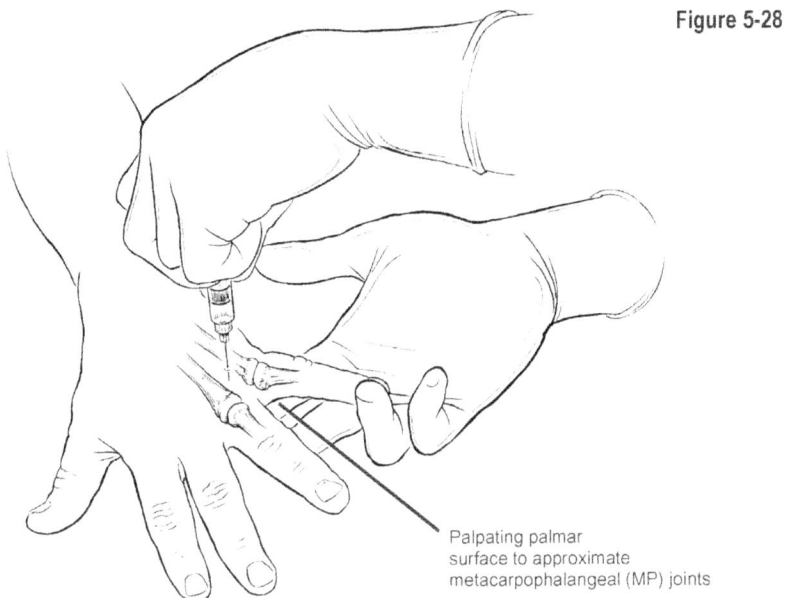

Palpating palmar surface to approximate metacarpophalangeal (MP) joints

Digital Nerve Block, Classical Approach

Figure 5-29

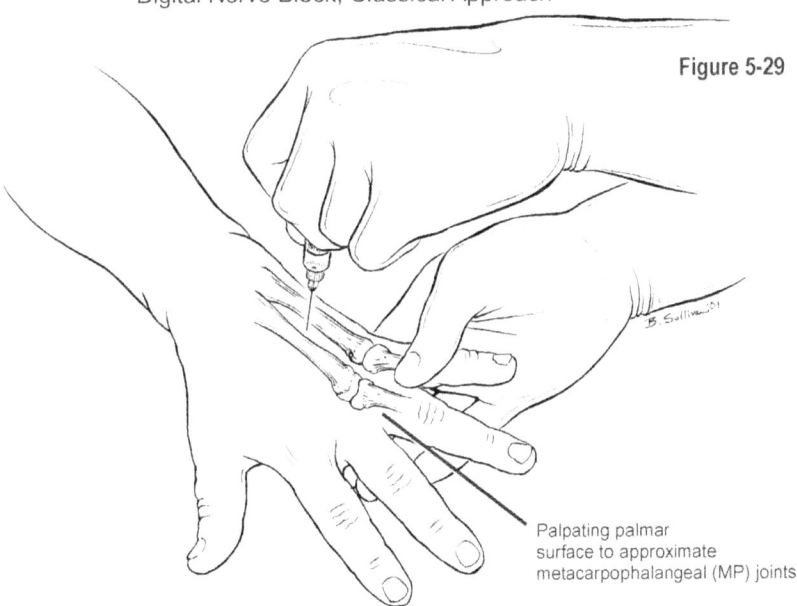

Palpating palmar surface to approximate metacarpophalangeal (MP) joints

Digital Nerve Block, Metacarpal Approach

Figure 5-30

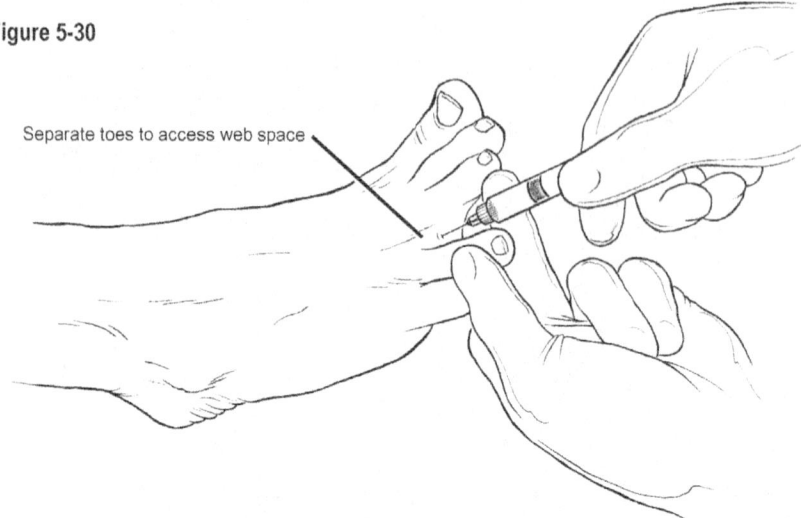

Separate toes to access web space

Web Space Block

Figure 5-31

Digital block

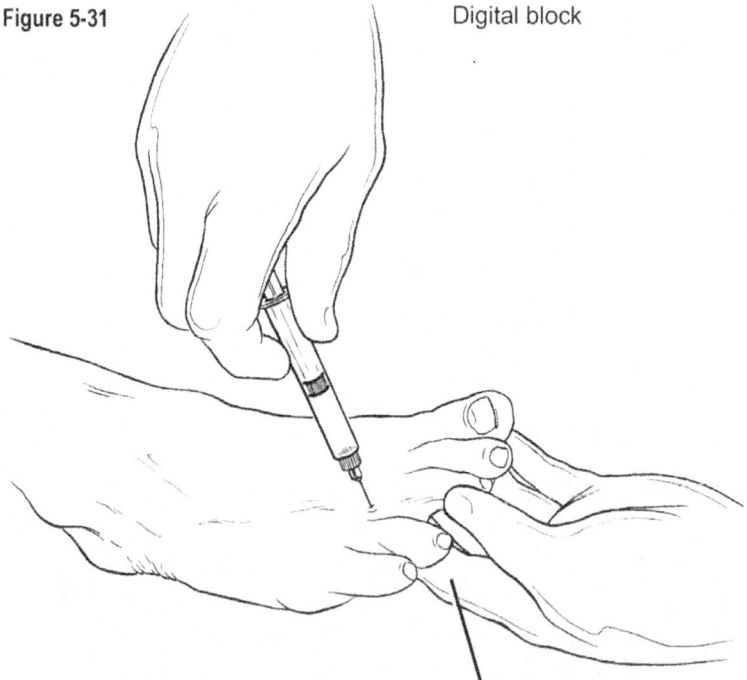

Palpation of plantar surface to approximate metatarsophalangeal (MP) joints

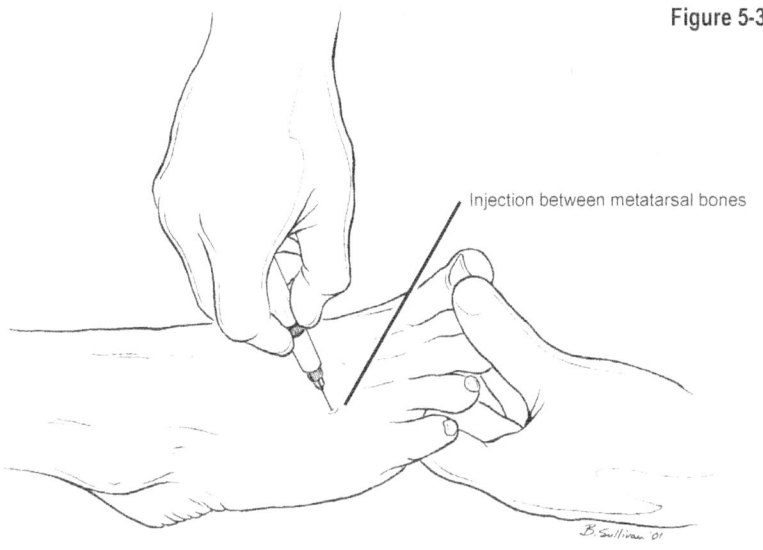

Figure 5-32

Metatarsal Block

of the digit.
6. Slowly inject the medication with the smallest possible gauge needle to minimize pain.
7. Repeat steps 4-7 for the palm

What Not To Do:
Contraindications:
1. Avoid local anesthetics in patients reporting allergies to them.
2. Avoid using vasoconstrictors with these blocks.

Complications: Tissue ischemia and necrosis. Large volumes of local anesthetics can lead to mechanical compression and ischemia. Deposit the anesthetic only on the lateral aspects of the digits to be anesthetized. Avoid total circumferential administration.

Table 5-14: Anesthetics for Digital Blocks

Drug	Concentration (%)	Plain Solutions Maximum Dose (milligrams)	Duration (minutes)
Lidocaine	0.5 - 1.0	300	30 - 60
Prilocaine	0.5 - 1.0	500	30 - 90
Bupivacaine	0.25 - 0.5	175	120 - 240

III. Standard Approach To Blocks

a. For the rest of this section, the following equipment will be referred to as **Standard Equipment** in the **What you Need** paragraph: An antiseptic prepping solution (**Betadine** or alcohol), appropriate sized syringe, 25 or 27-gauge needle, local anesthetic.

b. For the rest of this section, the following rationale will be referred to as **Standard Contraindications** in the **What Not To Do paragraph**, unless otherwise indicated: Local anesthetics should be avoided in patients reporting allergies to them.

c. For the rest of this section, the following rationale will be referred to as **Standard Complications** in the **What Not To Do** paragraph, unless otherwise indicated: Tissue ischemia and necrosis. Large volumes of local anesthetics can lead to mechanical compression and ischemia.

IV. Nerve Blockade of the Hand and Wrist

When: This blockade is particularly valuable when you need to maintain some motor function during surgery. Blockade will produce sensory loss in their hand, and motor loss in the intrinsic muscles of the hand, but not loss of the extension or flexion of the hand or wrist. If complete sensory and motor blockade is required, brachial plexus blockade is a more acceptable alternative.

What You Need: Standard Equipment: In order to achieve analgesia of the entire hand and wrist the ulnar, both the median and radial nerves must be blocked. **NOTE:** Additional techniques are discussed on the CD-ROM.

a. Ulnar Nerve Block at the Wrist (Figures 5-33, 5-34)

What To Do:
1. Assemble equipment.
2. Prep area.
3. Block the palmar branch by inserting a 25 or 27 gauge needle at 90° to the skin, lateral to the flexor

Figure 5-33

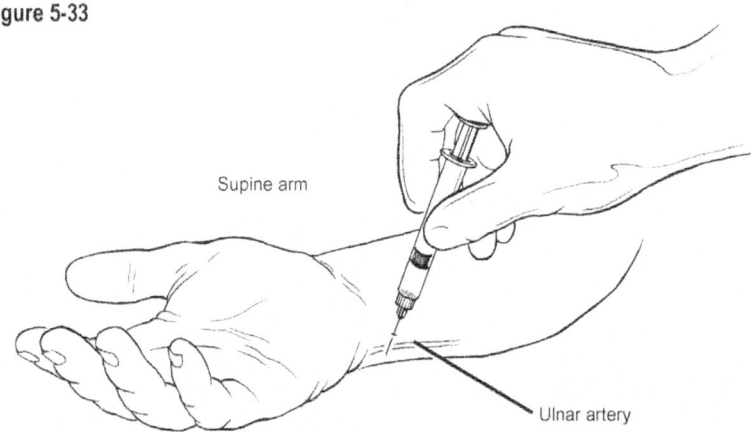

Ulnar Nerve Block, Ventral Approach

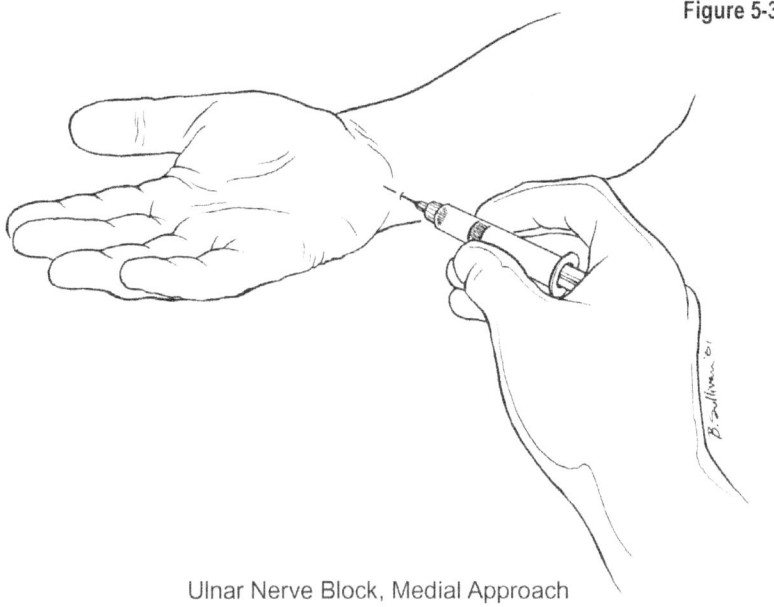

Figure 5-34

Ulnar Nerve Block, Medial Approach

carpi ulnaris tendon and medial to the ulnar artery. This artery can be palpated with the wrist in flexion.
4. At 1-1.5 cm in depth, slowly inject up to 5 mls of local anesthetic solution.
5. Withdraw the needle to the subcutaneous tissue and redirect the needle laterally around the ulnar aspect of the flexor carpi ulnaris, injecting an additional 5 mls of local anesthetic solution as you proceed. This will block the dorsal branch of the ulnar nerve laterally.

What Not To Do: See IIIb & IIIc
Contraindications: Ulnar nerve blocks should be avoided when significant paresthesia can be elicited easily by palpating the nerve.

b. **Median Nerve Block at the Wrist (Figure 5-35)**

What To Do:
1. Assemble equipment.
2. Prep area.
3. Have the patient flex their wrist against resistance to identify the palmaris longus tendon. If present, insert a 25 or 27 gauge needle just lateral to the tendon. If the palmaris longus is absent, insert the needle 1 cm medial to the ulnar border of flexor carpi radialis tendon. Insert the needle past the flexor reticulum, which is indicated by increased

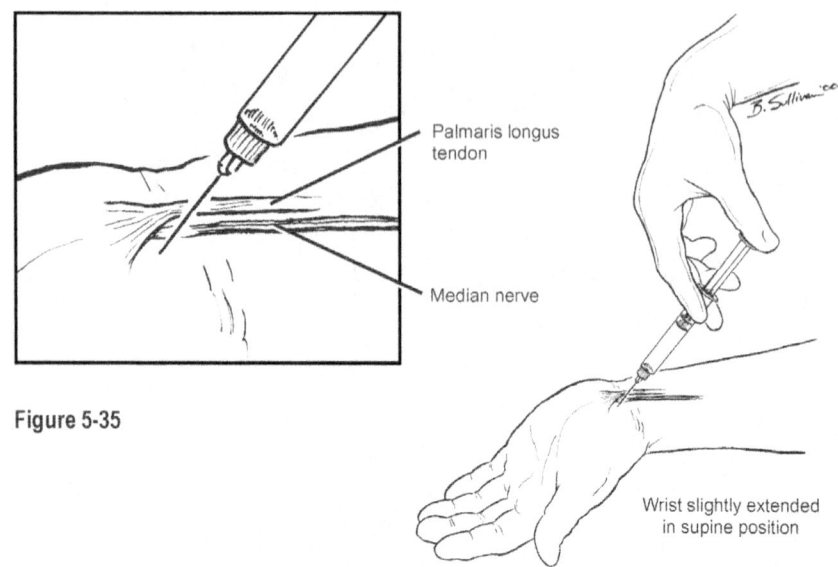

Figure 5-35

Median Nerve Block at the Wrist

resistance. Then slowly advanced the needle an additional 2-3 millimeters.
4. If paresthesia is elicited, then slowly inject 5 mls of local anesthetic solution. In the event of resistance to injection or pain (could be due to intraneural injection) stop and withdraw the needle 2 millimeters before continuing
5. Withdraw the needle to the subcutaneous level while injecting an additional 2-3 mls of local anesthetic solution.

What Not To Do: See IIIb & IIIc
Contraindications: Do not use a median nerve block in the presence of carpal tunnel syndrome.

c. **Radial Nerve Block at the Wrist (Figure 5-36)**

What To Do:
1. Assemble equipment.
2. Prep area
3. Extend the thumb against resistance, revealing the "anatomical snuff box," which is the area just above the styloid process of the radius.
4. Insert a 25 or 27 gauge needle close to the tendon of extensor pollicis longus over the styloid process of the radius. Direct it subcutaneously across the dorsum of the wrist towards the ulnar border. Inject 5-7 mls of local anesthetic solution as the needle is advanced.
5. Withdraw the needle to the insertion point and redirect it across the tendon of the flexor pollicis brevis and inject and additional 2-3 mls of solution subcutaneously

Figure 5-36

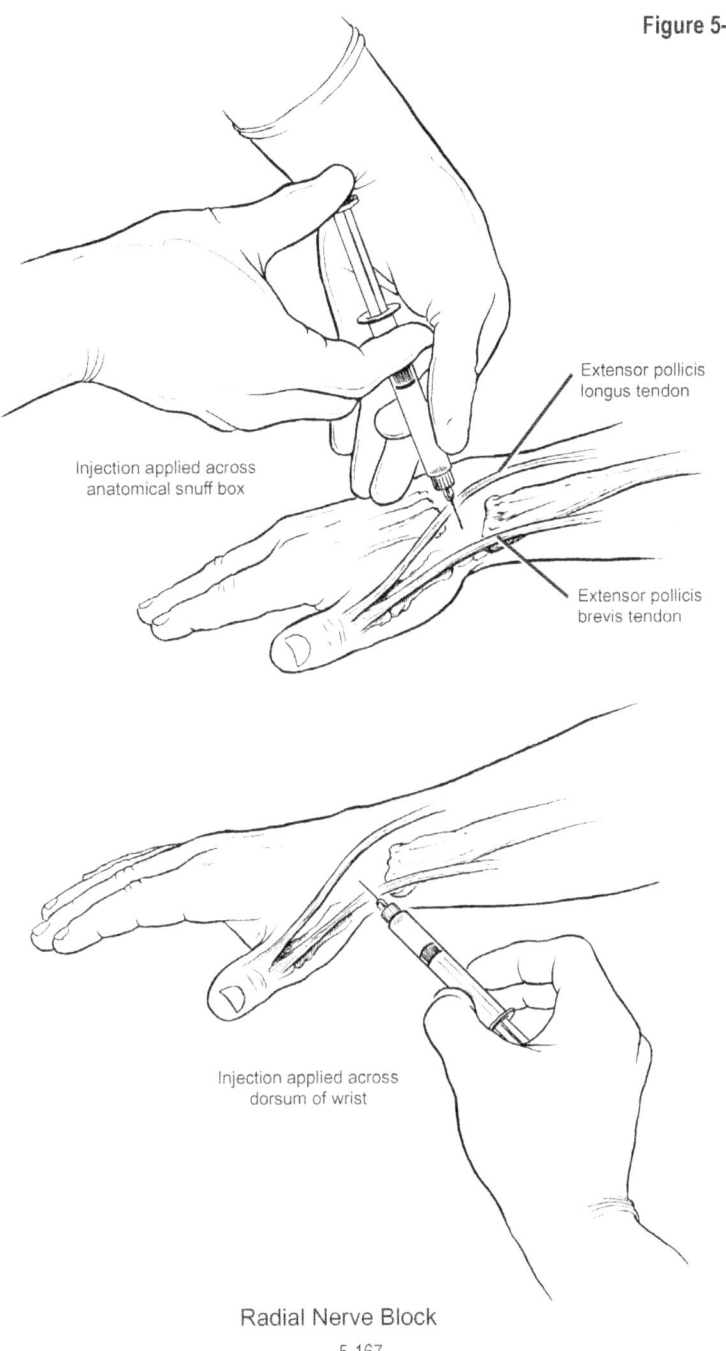

Radial Nerve Block

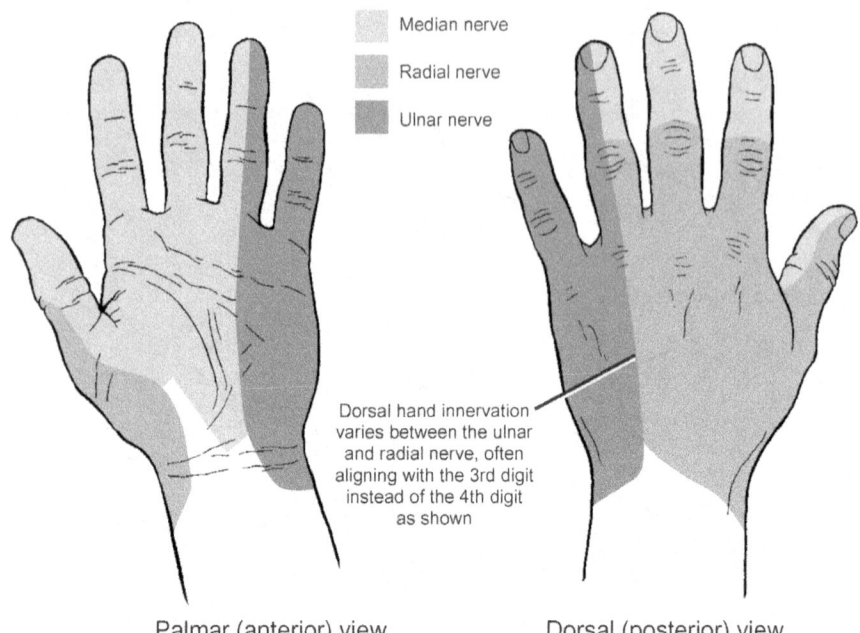

Figure 5-37
Dermatomes of Cutaneous Innervation of the Hand

What Not To Do: See IIIb & IIIc

V. Nerve Blockade of the Foot and Ankle:

When: Surgery of the toes and distal foot. The term "ankle block" denotes the location at which the local anesthetic solution is applied. This block is not suitable for surgery of the ankle. Rarely will all five nerves require blockade. Surgery of the medial foot requires blockade of the superficial, deep peroneal, saphenous, and tibial nerves, but not the sural nerve. Surgery of the lateral aspect of the foot requires blockade of all but the saphenous nerve.

What You Need: Standard Equipment: Complete ankle blockade is accomplished by injecting and blocking all five nerves that supply the foot. The nervous supply to the foot is made of five terminals: superficial n., deep peroneal n., saphenous n., sural n., and tibial n.
NOTE: Additional techniques are discussed on the CD-ROM.

What Not To Do: See IIIb & IIIc

a. Superficial and Saphenous Nerve Block (Figure 5-38)

What To Do:
1. Assemble equipment.
2. Prep area.

Figure 5-38

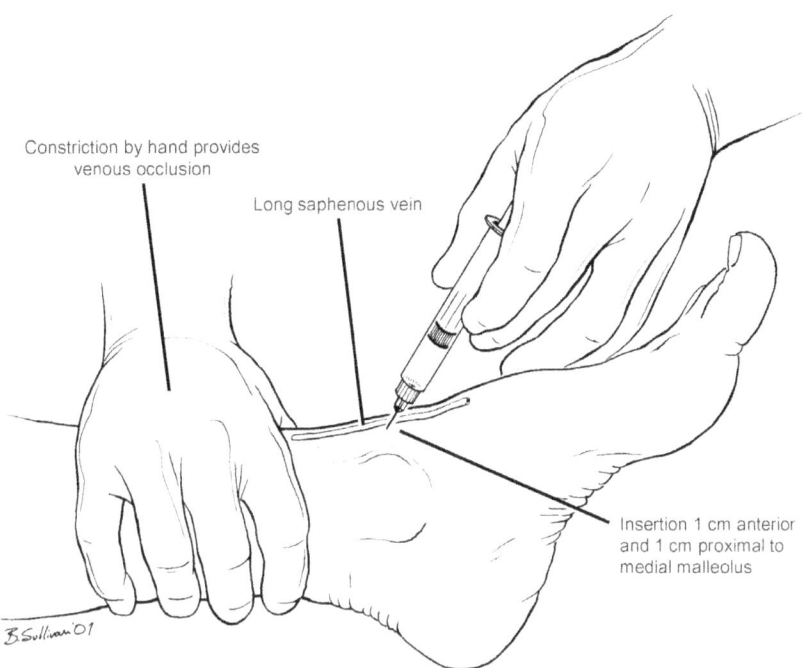

Saphenous Nerve Block

3. These two nerves can be anesthetized by injecting a "ring" of local anesthetic solution in the subcutaneous tissue.
4. Locate the upper aspect of the medial malleolus and inject a 10 ml ring of local anesthetic solution while advancing the needle toward the medial border of the lateral malleolus.

b. **Sural Nerve Block (Figure 5-39)**

   What To Do:
   1. Assemble equipment.

Figure 5-39

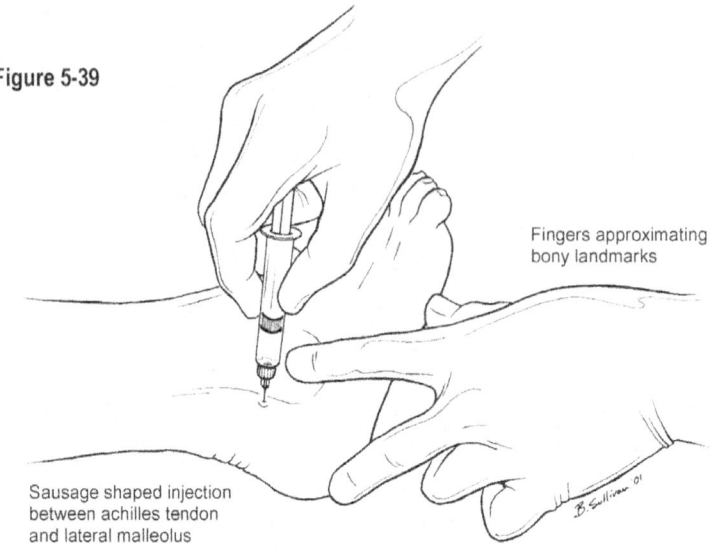

Fingers approximating bony landmarks

Sausage shaped injection between achilles tendon and lateral malleolus

Sural Nerve Block

2. Prep area.
3. Rotate the leg medially to gain access to the lateral posterior aspect of the foot.
4. Inject 5 mls of local anesthetic solution on the lateral side of the foot between the Achilles tendon and the posterior lateral malleolus.

c. Deep Peroneal Nerve Block (Figure 5-40)

What To Do:
1. Assemble equipment.
2. Prep area.
3. On the dorsal surface of the foot, locate the dorsalis pedis artery and the extensor hallucis longus tendon (tendon to the great toe).
4. Insert the needle between the dorsalis pedis a. and the extensor hallucis longus. If a paresthesia of the great or second toe is obtained, inject 5 mls of local anesthetic solution.
5. If a paresthesia is not obtained, advance the needle until bone is contacted and inject the 5 mls of local anesthetic solution in a fan-like manner while withdrawing the needle.

d. Tibial Nerve Block (Figure 5-41)

What To Do:
1. Assemble equipment.
2. Prep area.
3. Rotate the leg laterally to gain access to the medial posterior aspect of the foot.

Figure 5-40

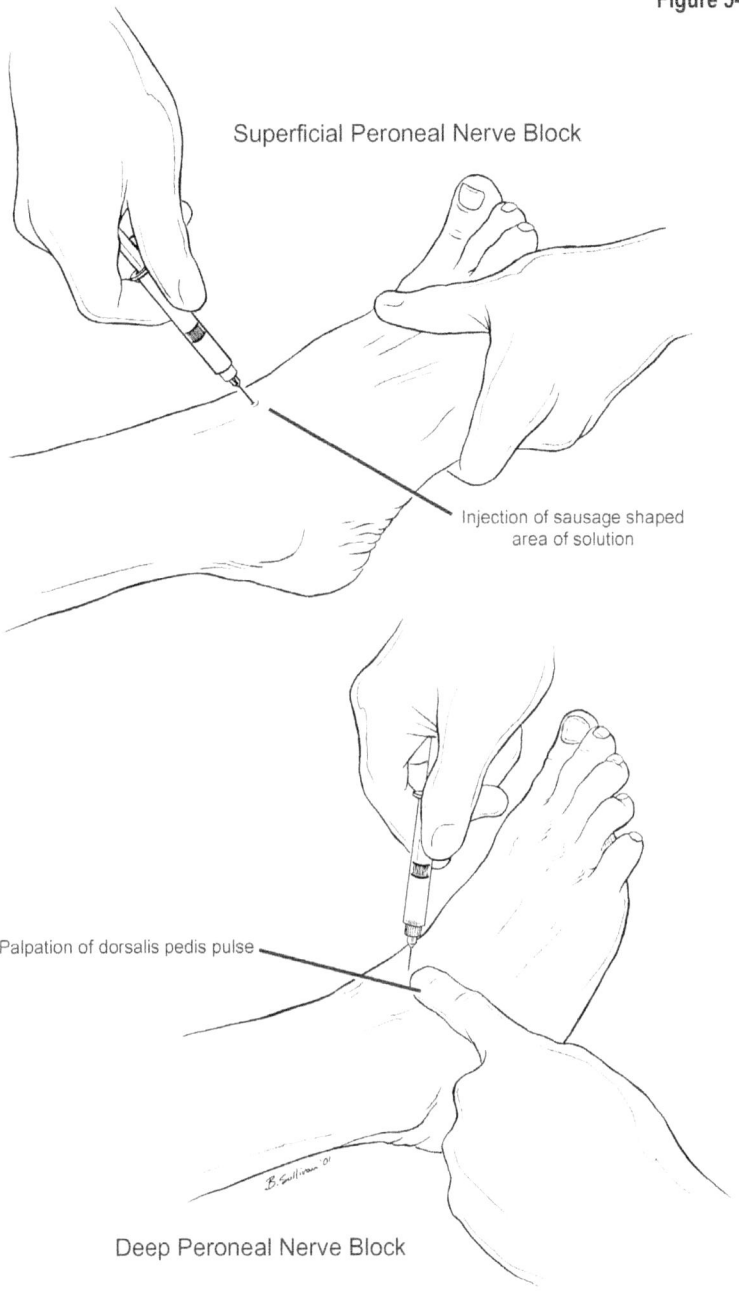

Figure 5-41

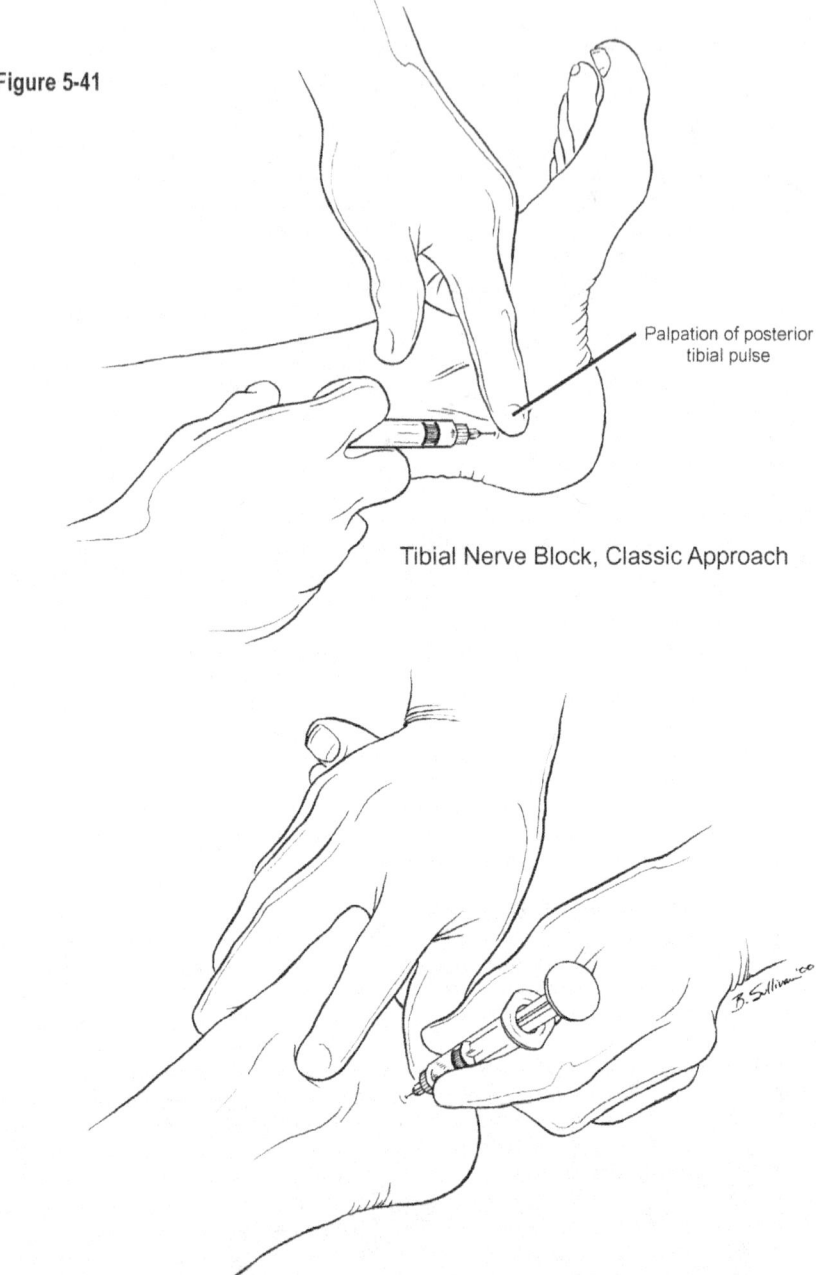

Palpation of posterior tibial pulse

Tibial Nerve Block, Classic Approach

Tibial Nerve Block, Sustenaculum Tali Approach

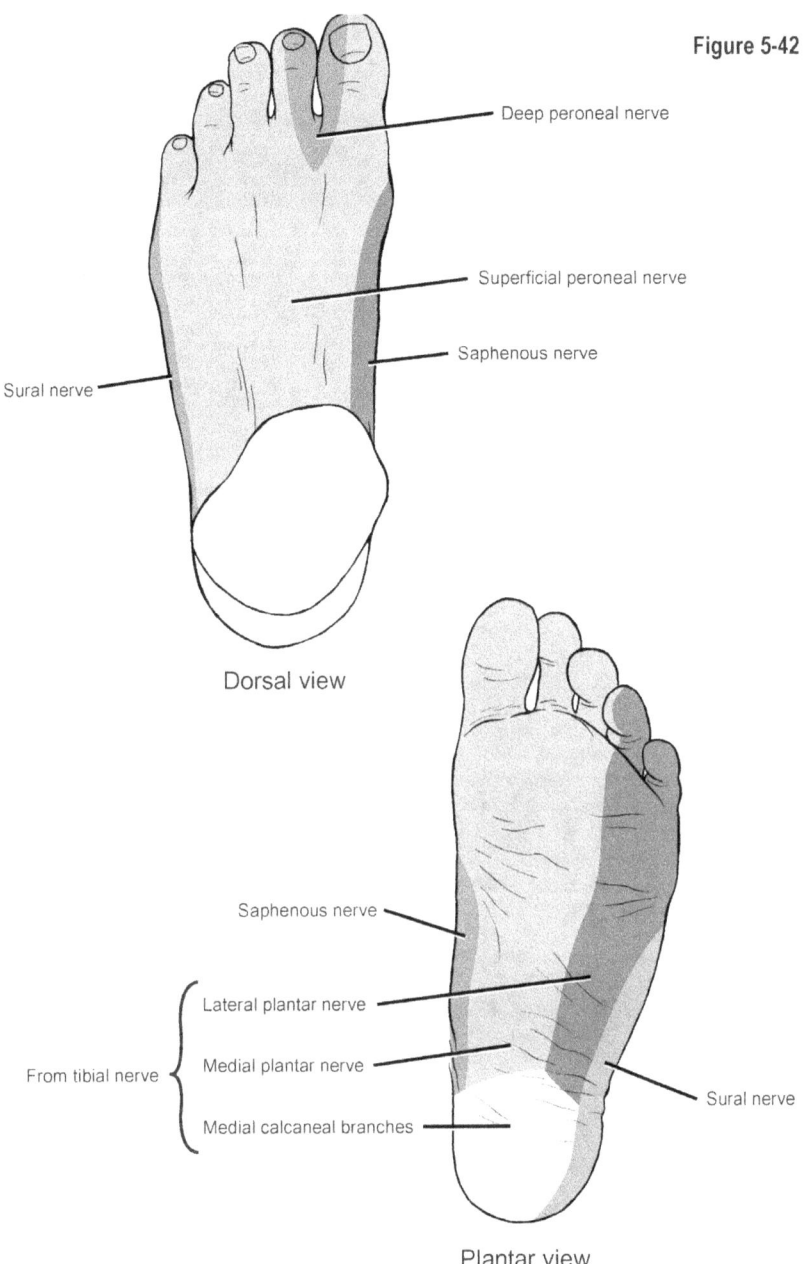

Figure 5-42

Cutaneous Innervation of the Foot
5-173

4. Locate the posterior tibial artery. The tibial nerve lies posterior and lateral to the tibial artery.
5. Insert the needle toward the artery, anterior to the Achilles tendon and posterior to the medial malleolus, injecting 5 mls of local anesthetic solution in a fan-like manner while withdrawing the needle.

VI. Axillary Blockade

Achieve regional anesthesia for surgery below the elbow by brachial plexus blockade using the axillary approach. The nervous and vascular structures of the upper arm are encased within a perivascular sheath, a tubular structure surrounding the nerves and vessels. Introducing local anesthetic solution into this sheath at the axillary level affords excellent blockade below the elbow, and many times, satisfactory anesthesia of the elbow itself.

When: For any surgical procedure of the forearm or hand; upon surgeon request or in high-risk patients where a general anesthetic would be deleterious.

What You Need:
Equipment: 22 gauge needle, 60cc syringe, tourniquet, sterile gloves, anesthetic agent(s), **epinephrine**, bicarb, razor, prep solution and resuscitation equipment (including oxygen).
Local Anesthetic (Table 5-15)
1. Total of 50-60 ml needed: 40 ml to fill the plexus sheath and an additional 10 - 20 ml are needed for ring block and musculocutaneous nerve.
2. Add 1:200,000 **epinephrine** as a marker to assist in detecting accidental intravascular injection (add 0.25cc of 1:1000 **epinephrine** to each 50 ml of local for a 1:200,000 solution).
3. Combinations: Combining a fast onset local anesthetic with another of long duration is done at times to take advantage of both of these desirable characteristics. Remember to not exceed maximum dosage levels.
4. Add NaHCO3, 1 mEq per 10 ml of local anesthetic solution for all agent **(except bupivacaine, where 0.1 mEq is added)** to accelerate the onset of block.
Choice of local anesthetic solution and the appropriate concentration depends on nerves to be blocked, desired onset time and desired duration of action.

Table 5-15

Drug	Concentration	Onset (min)	Duration (w/epi)	Max Dose-w/epi
Chloroprocaine	2-3%	10-20 min	60-120 min	1000mg-14mg/kg
Lidocaine	0.5-1.5%	15-30 min	150-180 min	500mg -7mg/kg
Mepivacaine	1-1.5%	15-30 min	180-240 min	500mg-7mg/kg
Bupivacaine	0.25-0.5%	20-30 min	180-360 min	225mg-3mg/kg
Tetracaine	0.2-0.25%	10-20 min	240-360 min	200mg

NOTE: Add epinephrine 1:200,000 to all mixtures (see above)

What To Do:
1. Basic set up for general anesthesia (see Total Intravenous Anesthesia section).
2. Prepare patient with appropriate premedications and monitors (see Total Intravenous Anesthesia section).
3. Patient position: Supine, with head turned away from side to be blocked and the arm abducted ~90°. The forearm is flexed to 90° and externally rotated so the dorsum of the hand lies on the table and the forearm is parallel to the long axis of the patient's body. Do not place the hand of the arm to be blocked under the head. This hyperabduction obliterates the brachial artery pulse.
4. Paresthesia technique (Optional: Transarterial and Peripheral Nerve Stimulator Techniques found on CD-ROM).
   a. Use the index finger of the non-dominant hand to palpate the axillary artery as high up into the axilla as possible.
   b. Prep the area with razor if needed and with povidone-iodine swabs. Place a skin wheal over the artery proximal to the index finger.

    c. Look for paresthesias by inserting the needle through the skin wheal and directing it slightly above or below the arterial pulsation (attempting to stimulate the median, ulnar or radial nerve).
    d. Angle the needle to almost the same plane as the sheath. When the patient reports a paresthesia, aspirate, then inject 2-3 ml of local anesthesia as a test. The paresthesia should fade. If so, continue to inject, aspirating the syringe every 5 ml to check for intravascular injection, until a volume of 40cc has been injected.
    e. Intense burning associated with the injection of a small volume indicates possible intraneural injection (STOP INJECTING).
    f. Use the remaining 10 ml are to block the musculocutaneous, intercostobrachialis and medial cutaneous nerves as below. Prep the areas as in 4b above.
    g. Musculocutaneous nerve: insert the needle into the body of the coracobrachialis muscle until it touches the humerus, withdraw 2-3 mm, then inject 5 ml of the remaining anesthetic solution into the muscle.
    h. Intercostobrachialis and medial brachial cutaneous n.: subcutaneously direct the needle inferiorly and superiorly to the insertion site for the musculocutaneous block and inject the remaining 5-10 ml solution as a **ring block**.
5. After placing the block, bring the arm to the side and massage the axilla for a few minutes to spread the anesthetic. Maintain pressure over auxiliary injection site to decrease bleeding and keep agent high in axilla.
6. Continually assess the patient for signs of systemic absorption of the anesthetic and possible toxic reaction. Elevated vital signs of excitement are seen at lower blood levels of anesthetic, progressing to CNS depression as blood levels rise.
Warning signs include: Patient reporting metallic taste in mouth or circumoral paresthesias, tinnitus, drowsiness/dizziness/disorientation, visual disturbance, slurred speech, generalized twitching and tremors.
7. Always be prepared to provide airway and cardiovascular support whenever administering brachial plexus blockade.
8. Assess the block - "Push, pull, pinch, pinch" (support patient's arm during these maneuvers!)
PUSH: The patient attempts to extend the arm. Inability to do so implies block of the radial n.
PULL: Attempt to flex the arm at the elbow; inability to do so implies musculocutaneous n. block.
PINCH: The hand at the base of the thenar eminence (prominence at the base of the thumb); sensory loss implies median n. block.
PINCH: The base of the little finger; sensory loss implies ulnar n. block. Inability to spread fingers against resistance implies ulnar n. block.
9. Apply tourniquet above level of surgery prior to the operation.

What Not To Do:

Contraindications: Uncooperative patient/refusal, bleeding disorders, infection at injection site or allergies to local anesthetics.
Complications: Toxic systemic absorption/intravascular injection that can lead to cardiovascular collapse. Hematoma at injection site from axillary artery puncture. Accidental injection of agent into the nerves.

PART 6: OPERATIONAL ENVIRONMENTS
Chapter 20: Dive Medicine
Dive Medicine: Barotrauma to Ears
CPT Jeffrey Morgan, MC, USA

Introduction: The volume of a gas changes inversely to air pressure-gas expands as pressure drops and compresses as pressure rises. An enclosed, gas-filled cavity in the body is susceptible to injury from the expansion or compression of the gas if it is not able to equalize to the outside pressure. The middle ear is the most frequently injured body part. There are three types of ear barotraumas: external ear barotrauma (pinna to tympanic membrane). middle ear barotrauma (tympanic membrane to cochlea) and inner ear barotrauma (round or oval window rupture, perilymph fistula). **NOTES:** All tables in this chapter refer to the Treatment Tables in the United States Navy Dive Manual and are included at the end of this chapter. All references to paragraphs within the Treatment Tables refer to paragraphs in the Navy Dive Manual, Revision 4.

Subjective: Symptoms
External ear barotrauma: Diving with ear plugs, tight fitting diving hood, cerumen impaction or some other object in the external ear canal (EAC) creates a confined space between the object and tympanic membrane (TM). This space is compressed with increasing depth, creating a vacuum, tugging the sensitive TM and creating pain. Divers feel pain during descent, typically at shallow depths.
Middle ear barotrauma: Pain, usually on descent (squeeze) but may be on ascent (reverse squeeze); caused by the eustachian tube (ET) failing to equalize pressure. ET dysfunction is often secondary to upper respiratory infection or allergies. If a diver continues deeper despite the pain, blood may fill the middle ear cavity and cause temporary conductive hearing loss and give a feeling of fullness in the ear. Continued pressure may result in TM rupture. If the TM ruptures, the pain will stop but nausea and vomiting may ensue as cold water enters the middle ear and causes vertigo. The dive should be immediately aborted after a TM rupture. Alternatively, a prolonged vacuum in the middle ear will be relieved with a serous effusion seeping from lining tissues. Symptoms of a serous effusion are mild pain, popping sensations in the ear and temporary conductive hearing loss.
Inner ear barotrauma: Often associated with, and usually secondary to middle ear barotrauma. Following a forceful Valsalva, the diver may have roaring tinnitus and sensorineural hearing loss. If the vestibular symptoms are present for less than one minute, the vertigo is considered transient. Vertigo underwater is a life-threatening situation and the injured diver needs immediate assistance to the surface. If the vertigo lasts for more than one minute, the vertigo is considered persistent.

Objective: Signs
Using Basic Tools: 512 Hz Tuning fork
Using Advanced Tools: Otoscope with insufflation bulb
Sudden loss of balance, nausea and vomiting, tinnitus, and hearing loss are seen in all 3 conditions.
External ear barotrauma: Inflammation of the TM unrelieved by Valsalva, edema or blood in the EAC, very tender in EAC and on tragus, severe cases have TM irritation and/or rupture, cerumen impaction is sometimes seen.
Middle ear barotrauma: TM redness, hemorrhage or rupture; blood behind TM; middle ear barotrauma graded on Teed Classification (listed below) serous effusions are typically amber-colored fluid and sometimes have bubbles behind the TM.
Inner ear barotrauma: Nausea, vomiting, ataxia, vertigo, tinnitus, sensorineural hearing loss (high frequency loss more common than low), positive fistula test*. TM rupture due to middle ear barotrauma may also be seen.
*Fistula test: Pressurize inner ear using insufflation bulb on otoscope, and evaluate for vertigo to test for round window or oval window rupture. The insufflation bulb can also be used to test for an intact TM.

Assessment:
Differential Diagnosis
External ear barotrauma - swimmer's ear
Middle ear barotrauma - viral/bacterial ear infection, URI
Inner ear barotrauma - tertiary syphilis, Meniere's disease, Arterial Gas Embolism, inner ear decompression sickness.

Plan:
Treatment
External ear barotrauma: Avoid plugs and hoods while diving. Remove from dive status until ear is healed. Keep ear dry. Clean ear of any obstructions. If infection sets in, treat as external otitis (swimmer's ear) with **Cortisporin** otic drops.
Middle ear barotrauma: Remove from dive status until asymptomatic. Use nasal decongestants (i.e., **Afrin**) for no more than 5 days or systemic decongestants. Do not use any eardrops if TM is ruptured. Avoid exertion and all swimming while TM is healing.
Inner ear barotrauma: Remove patient from water and continuously monitor until evaluated by Diving Medical Officer (DMO) or ENT physician (within 72 hrs). Maintain strict bed rest with head slightly elevated. Avoid straining or Valsalva. Do not allow flying until injury is fully healed. Obtain immediate consult with ENT if diver has persistent vertigo and ataxia. A round window or oval window rupture causes permanent disqualification from military diving.

Patient Education
General: Pressure can greatly affect the sensitive tissues in the ear. Time away from pressure is usually enough to allow the ears to heal.
Diet: Normal
Medications: Topical decongestants like **Afrin** should be not taken for more than 5 days continuously due to rebound effect. Do not use topical eardrops when there is a chance the TM is ruptured.
Prevention and Hygiene: Avoid diving hoods, equalize ears often, avoid forceful Valsalva.
No Improvement/Deterioration: Consult ENT or DMO.

Follow-up Actions
Return evaluation: Weekly until healed.
Consultation Criteria: Patients with persistent vertigo, loss or significant decrease in hearing, non-healing TM rupture, large TM rupture, or foreign body stuck in EAC should be referred to ENT.

TEED CLASSIFICATION

Grade	Clinical description	Expected Healing Time
TEED 0	Symptoms without otoscopic signs	May dive same day
TEED 1	Diffuse redness and retractions of the tympanic membrane	1-2 days
TEED 2	Grade 1 plus slight hemorrhage within the tympanic membrane	1-4 days
TEED 3	Grade 1 plus gross hemorrhage within the tympanic membrane	3-7 days
TEED 4	Dark and slightly bulging tympanic membrane due to free blood in the middle ear; a fluid level may be visualized behind the tympanic membrane	5-12 days
TEED 5	Free hemorrhage into the middle ear with tympanic membrane perforation	7-15 days. Perforation must be totally healed before diving again.

References: Bove, Alfred A., Bove and Davis' Diving Medicine, 3rd Edition, WB Sauders Company, Philadelphia, 1997, pp. 241.

Dive Medicine: Barotrauma, Other
(Dental, Sinus, GI, Skin, and Face Mask)
CPT Jeffrey Morgan, MC, USA

Introduction: The volume of a gas changes inversely to air pressure-gas expands as pressure drops and compresses as pressure rises. An enclosed, gas-filled cavity in the body is susceptible to injury from the expansion or compression of the gas if it cannot equalize to the outside pressure. Middle ear injuries are the most frequent, but several other areas are subject to barotrauma, primarily on descent. Dental fillings or caries that have air trapped in the teeth can cause pain called dental barotrauma. The lining of the sinuses can be pulled off the bone allowing the cavity to fill with blood. In very rare circumstances, the bone forming the sinus may fracture. Large amounts of gas in the intestines may expand on ascent and cause intestinal pain (GI barotrauma). Poorly fitting dry suits can have air trapped in folds and wrinkles that will compress at depth, sucking skin in to fill the vacuum (skin barotraumas/suit squeeze). If a diver does not equalize the vacuum in the facemask during descent, a mask squeeze may result.

Subjective: Symptoms
Dental barotrauma (barodontalgia): Pain in teeth usually presents on descent. Once pain is felt, patient normally ascends and relieves pressure. Tooth may implode if diver continues to descend through initial pain. Often patient cannot point out exact tooth. History typically includes recent fillings or poor dental health.
Sinus barotrauma: Usually occurs in maxillary and frontal sinuses and presents with pain over affected sinus or in maxillary teeth. Blood may be seen in the mask. Pressure on ascent can occur (reverse squeeze), usually secondary to mucosal polyps or plugs. Ascending or descending to decrease sinus pressure relieves pain. Predisposing conditions include sinusitis, upper respiratory infections, mucosal polyps, deviated septum, nasal polyps, smoking, persistent use of topical decongestants, and allergies.
GI barotrauma: Flatulence and diffuse pain in abdomen (usually mild) present on ascent. History often includes pre-dive carbonated drinks and gas-producing foods.
Skin barotrauma: Pain and bruising on skin in areas of fold from a poorly fitting dry suit.
Facemask squeeze: Pain in area of mask, headache, transient blurring vision. History may include diving with goggles instead of mask.

Objective: Signs
Using Basic Tools:
Sinus barotrauma: Pain over sinus with percussion. Transient, bloody nasal discharge, or blood in mask, transillumination of sinus can occasionally reveal fluid in the sinus.
GI barotrauma: Tympanic sound on abdominal percussion possible.
Skin barotrauma: Bruising on skin typically at folds in the ill-fitting dry suit.
Facemask squeeze: Bruising in area of mask seal, conjunctival and scleral hemorrhages.
Using Advanced Tools: Dental barotrauma: air in tooth chamber on XR, **Facemask squeeze:** funduscopic exam for retinal damage

Assessment:
Differential Diagnosis
Dental barotrauma - sinus barotrauma, tooth abscess
Sinus barotrauma - sinus infection, dental barotrauma
GI barotrauma - gastritis, acute abdominal disorders
Skin barotrauma - rash, bleeding disorders, jelly fish stings
Facemask squeeze - facial trauma, jellyfish stings

Plan:
Treatment
Primary:

Dental barotrauma: Analgesics. Re-fill the tooth ensuring no air remains in chamber.
Sinus barotrauma: Stop dive and surface slowly, topical and systemic pain relief as necessary. If bleeding does not stop, pack turbinates with gauze coated in antibiotic ointment in the opening of the affected sinus. If packing material is left in more than one hour, patient must be prescribed a prophylactic antibiotic (i.e., **Augmentin**) to prevent toxic shock syndrome.
GI barotrauma: Ascend slowly from dive
Skin barotrauma: Analgesics as necessary
Facemask squeeze: Analgesics as necessary.

Patient Education
General: Educate patient on the cause of injury. Prevention is possible with all of these injuries.
Dental barotrauma: Practice good oral hygiene. Ensure fillings have no trapped gas pocket.
Sinus barotrauma: Explain how sinuses can become blocked and force blood vessels to leak while under increasing pressure. Avoid diving with URIs, mucosal polyps or allergic symptoms.
GI barotrauma: Encourage flatulence prn on ascent.
Skin barotrauma: Ensure dry suit fits properly.
Facemask squeeze: Equalize mask.
Diet: Normal
Medications: Over-the-counter analgesics prn. **Tylenol** is usually enough for pain relief (avoid aspirin and first generation NSAIDS if persistent bleeding is present), topical and systemic decongestants prn.
Prevention and hygiene: Do not dive until all symptoms have resolved.
No Improvement/Deterioration: Consult Diving Medical Officer.

Follow-up Actions
Return evaluation: As needed.
Consultation Criteria: Dental barotrauma requires dental consult prior to returning to diving.

Decompression Injuries
Dive Medicine: Decompression Sickness
(the Bends, Caisson Disease)
CPT Jeffrey Morgan, MC, USA

Introduction: While breathing air under pressure (i.e., underwater), nitrogen is dissolved into body tissues. Increased pressure and prolonged exposure saturates these tissues with nitrogen. When the body is decompressed, the nitrogen comes out of solution to form bubbles in the vasculature or tissues. When the body is decompressed too fast in relation to the nitrogen load (based on time and depth of dive; listed extensively in the Navy Dive Tables), too many bubbles can form. These bubbles may precipitate out into the vascular system, skin, lungs, ears and other tissues causing a variety of symptoms that are categorized into Type I (mild) and Type II (severe) Decompression Sickness (DCS). Any neurological symptom is classified as Type II DCS. Spinal symptoms (neurological) are common in DCS associated with diving. Cerebral symptoms are common from rapid decompression at high altitudes (i.e., pilot ejecting at altitude). Some authors estimate that 30% of people with typical Type I pains can have accompanying Type II symptoms. Most symptoms of DCS occur within the first 24 hours of surfacing. When the caregiver is not sure if the DCS is Type I or II, treat as Type II DCS.

Subjective: Symptoms
DCS (Type I): Skin itching and rashes; skin marbling; red/purple patches; limb swelling; peau d'orange (skin is dimpled like an orange); most joint and/or muscle pain (classically a deep dull ache)
DCS (Type II): Any Type I symptom can be present plus: various **neurological symptoms**: numbness, tingling, tremors, paralysis, paresthesia, mental status changes, fatigue, amnesia or bizarre behavior, light-headedness, poor coordination or incontinence; **inner ear symptoms** (staggers): ringing in ears, vertigo, hearing loss, dizziness, nausea, or vomiting; **cardiopulmonary symptoms** (chokes): chest pain (worse with

inspiration), tachypnea, cough, irritability, or loss of consciousness; **pain** in the hip, abdomen, thorax, pelvis or spine.

Objective: Signs
Using Basic Tools: Perform a complete dive history and neurological exam per the example in the appendix. Numbness, pain, weakness or paralysis of limbs, diminished or absent reflexes, decreased cognitive function, poor coordination, ataxia, hearing loss, vomiting, tachypnea, coughing; unconsciousness. Auscultation of crackles in lung fields.
Using Advanced Tools: CXR to rule out pneumothorax.

Assessment:
Differential Diagnosis: Arterial gas embolism, myocardial infarction, trauma, pulmonary embolus and many others.

Plan:
Treatment
Primary
1. 100% oxygen immediately.
2. Hyperbaric oxygen (HBO) recompression therapy as soon as possible. Refer to Figure 6-1 and treatment tables in this chapter.
3. Complete neurological exam prior to recompression if symptoms allow. Otherwise, perform exam at depth.
4. Pregnancy test for all women of childbearing age. If patient is pregnant, benefit of recompression treatment needs to outweigh risks to unborn child (possible: retrolental fibroplasia, in utero death, birth defects).
5. If transportation to HBO chamber is necessary, transport supine and fly below 1000 ft or pressurize the cabin to below 1000 ft. If transporting on land, avoid mountain areas if possible (exacerbates DCS symptoms). Use fluid instead of air in bulbs on Foley, ET tube, etc. to reduce risk of rupture with pressure changes.
6. Ensure patient remains well hydrated. Fully conscious patients may be given fluids by mouth. 1-2 L of water, juice or non-carbonated drink, over the course of a Treatment Table 5 or 6, is usually sufficient. Stuporous or unconscious patients should always be given IV fluids (normal saline at a rate of 75-125 cc/hr). If the patient is obviously dehydrated, an initial 1L bolus of normal saline may be given (for an otherwise healthy patient). No Ringer's lactate until patient is urinating. Keep fluids running so urine is clear and > 30 cc/hour. Catheterize patients unable to urinate. Keep patient warm and have them avoid exertion during recompression.
7. Surface if ACLS is needed and can be administered. Operational chambers are not usually equipped or approved for ACLS treatment. Follow algorithm at end of this section (Figure 6-1) for no improvement or deteriortion. If it appears that the patient has died in the chamber, a qualified medical person who may examine and pronounce someone dead must be consulted prior to aborting the recompression treatment. If treatment is aborted, chamber should surface as early as possible ensuring the inside tender does not get DCS.

Alternate: Submarine escape pod may be used if no hyperbaric chamber is available. 100% oxygen.
Primitive: 100% oxygen. In-water recompression (extremely risky; follow "In Water Recompression" instructions in Navy Dive Manual, Revision 4).

Patient Education
General: Recompression and hyperbaric oxygen is the treatment of choice for DCS.
It is strongly recommended that pregnant women should not dive. Patient should not fly within 72 hours of recompression treatment.
Diet: Continue taking in clear fluids. Urine should be clear and of adequate volume (30 cc/hr). Eat solid

Figure 6-1: Treatment of Arterial Gas Embolism or Decompression Sickness

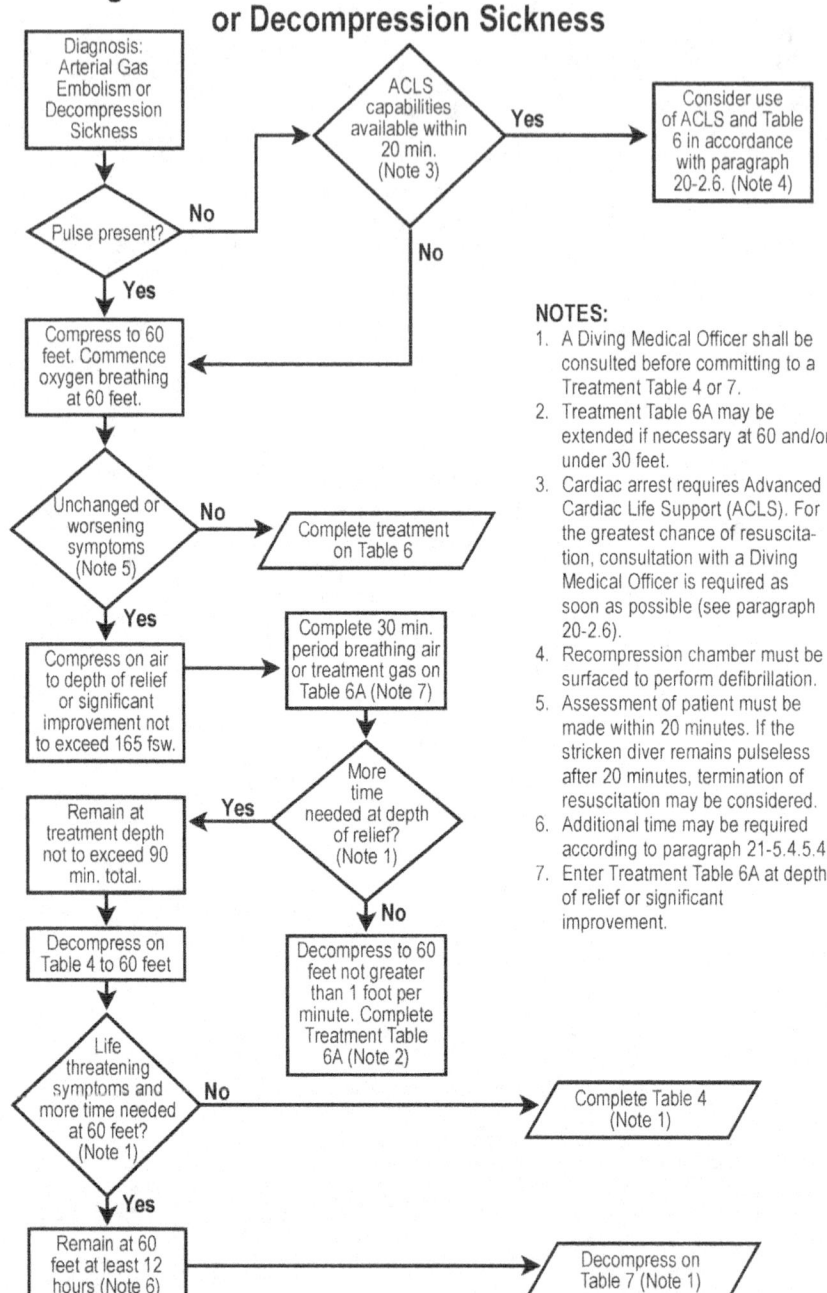

foods as tolerated.
Prevention and Hygiene: Follow decompression tables, stay fit, avoid alcohol, avoid trauma, limit medications. Patient should remain at recompression chamber facility for 6 hours, and be within one hour of chamber for 24 hours following recompression treatment in case further treatment is needed.
No Improvement/Deterioration: Return for immediate re-evaluation.

Follow-up Actions
Return evaluation: If residual symptoms are present, additional recompression treatments may be performed AFTER consult and approval from a Diving Medical Officer (DMO).
Consultation Criteria: DMO should be consulted immediately after a dive injury is identified. Any residual symptoms should be discussed with a DMO and appropriate medical specialist (i.e., neurologist).
Evacuation: As soon as patient is stable. Comply with warnings about altitude.

Dive Medicine: Pulmonary Over Inflation Syndrome
(Including Arterial Gas Embolism)
CPT Jeffrey Morgan, MC, USA

Introduction: Pulmonary Over Inflation Syndrome (POIS) results typically from gas expanding in the lung as a diver ascends from depth without exhaling. The expanding gas ruptures lung and vascular tissues. Gas can then enter the pleural space (pneumothorax), mediastinum (mediastinal emphysema), pulmonary venous system (creating emboli) and other tissues. The pulmonary venous emboli return to the left heart, enter the arterial system (arterial gas embolism [AGE]), travel anywhere in the body, block blood flow and cause ischemia. The central nervous system (CNS) and the heart are most susceptible to serious injury from localized hypoxia due to an AGE. As a basic rule, any diver who has obtained a breath of compressed gas at depth from any source, whether diving apparatus or diving bell, and who surfaces unconscious or loses consciousness within 10 minutes of surfacing, must be assumed to be suffering from an AGE.

Subjective: Symptoms
Symptoms: Sudden onset CNS symptoms are most common and include: dizziness, limb weakness or paralysis, hemiparesis, numbness, mental status changes, loss of consciousness, confusion, tingling, poor coordination, ataxia, difficulty speaking, visual disturbances, convulsions, personality changes, urinary retention, abdominal complaints of nausea and/or vomiting. Virtually any CNS symptom may be associated with an AGE. Symptoms of other POIS conditions may also be seen.
Focused History: *How long after surfacing did the symptoms start?* (AGE symptoms usually manifest while surfacing or within 10 minutes of surfacing from a dive.)

Objective: Signs
Using Basic Tools:
Follow the neurological examination checklist from the US Navy Dive Manual in Handbook Appendix.
Neurological: Numb to: Light/deep touch, pain/temperature, proprioception, vibration, two-point tactile discrimination; decreased strength or paralysis, including hemiparesis or hemiplegia; diminished reflexes; decreased mental functioning.
Cardio-pulmonary symptoms: Tracheal deviation, respiratory distress; blood in sputum, EKG changes (4-5%), cardiac arrest, crackles in lungs, decreased breath sounds, tympanic areas in thorax.
Using Advanced Tools: CXR if available to assess for POIS (pneumothorax, air in mediastinum or other tissues).

Assessment:
Differential Diagnosis: Decompression sickness ('the bends'), hypoxia, carbon monoxide poisoning and other diving causes, as well as non-diving causes, including hypoglycemia, seizure, near drowning and myocardial infarction.

Plan:
Treatment: See Treatment Tables at the end of this chapter.
Primary:
1. 100% oxygen immediately. Do not let the patient sit up prior to recompression
2. Hyperbaric oxygen (HBO) recompression therapy as soon as possible.
3. If transportation to HBO chamber is necessary, transport supine, and fly below 1000 ft or pressurize the cabin to below 1000 ft. Use fluid, as opposed to air, in bulbs on Foley, ET tube, etc. to reduce risk of rupture with pressure changes.
4. Hydrate patient. Fully conscious patients may be given po fluids: two liters of water, juice or non-carbonated drink, over the course of a Treatment Table 6, is usually sufficient. Stuporous or unconscious patients should always be given IV fluids (normal saline at a rate of 75-125 cc/hour). If the patient is dehydrated, give a one-liter bolus of normal saline. Keep fluids running so urine is clear and at least 30cc per hour. Ringer's Lactate can be used after patient is producing urine. Catheterize patients unable to urinate.
5. Be prepared to immediately treat clinically diagnosed pneumothorax with 14-16-gauge needle (needle thoracentesis) and chest tube while at depth or while surfacing (See Procedure: Thoracostomy).

Alternative: Submarine escape pod may be used for decompression if no chamber is around, 100% Oxygen
Primitive: In-water recompression (extremely risky; follow "In Water Recompression" instructions in Navy Dive Manual Revision 4), 100% Oxygen

Patient Education
General: An AGE stops blood flow to tissues and organs distal to the blockage. By going back down to pressure, the gas bubbles causing the blockage will shrink to alleviate the AGE. Rapid recompression is essential to minimize neurological damage.
Activity: Lay supine during travel and recompression to minimize neurological damage.
Diet: Drink fluids to remain hydrated (if able).
Prevention and Hygiene: Avoid diving until cleared by a Diving Medical Officer.
No Improvement/Deterioration: Neurological symptoms can get worse and AGE can lead to death. Recompression as soon as possible is the best treatment to prevent permanent damage. Return daily for 3 days for assessment of possible residual symptoms.

Follow-up Actions
Return evaluation: Assess possible residual symptoms daily for 3 days and follow the algorithm (Figure 6-1) from USN Dive Manual. If residual symptoms are present, additional HBO therapy may be indicated (contact Diving Medical Officer). Physical rehabilitation may be beneficial and neurological follow-up is required.
Evacuation Consultation Criteria: A Diving Medical Officer should be consulted as soon as possible. Evacuation should be considered as soon as patient is stable.

NOTES: If a limb is paralyzed, a deep vein thrombosis may form, increasing the risk of pulmonary embolism. Prophylactic low molecular weight heparin (LMWH) may be beneficial if HBO therapy does not alleviate the paralysis. LMWH is dangerous because it greatly increases the risk of uncontrollable hemorrhage. It must be used very cautiously.

Dive Medicine: Hypoxia
(Including Shallow Water Blackout)
CPT Jeffrey Morgan, MC, USA

Introduction: Hypoxia is the most common cause of unconsciousness in diving operations. As a diver depletes the residual gas from his tanks or from his lungs (breath hold diving), the partial pressure of oxygen in the diver drops insidiously, causing hypoxia and unconsciousness. The increased ambient pressure during descent and at depth also increases the partial pressure of oxygen and other gases. At depth, the elevated partial pressure of oxygen delays the onset of the physiological, hypercapnia-induced drive to breathe. By the time that drive induces him to surface, the diver may have stayed too long at depth. He may then have insufficient oxygen to sustain him during ascent as the partial pressure of oxygen decreases quickly. The diver can become hypoxic and unconscious under the surface. This phenomenon is called a "shallow water blackout" and is seen more frequently in breath hold diving. Divers with an underwater breathing apparatus are trained to surface with residual oxygen in their tanks to avoid this danger.

Subjective: Symptoms
Light-headedness, confusion, tingling, or numbness

Objective: Signs
Using Basic Tools: Brief period of confusion and ataxia preceding unconsciousness, which is often the first sign.
Using Advanced Tools: Hypoxia can be assessed with a pulse oximeter. CXR can help determine whether or not there is a Pulmonary Over Inflation Syndrome (see adjacent section in this chapter).

Assessment:
Differential Diagnosis - arterial gas embolism (AGE), carbon monoxide poisoning, decompression sickness, trauma, shock and other causes.

Plan:
Treatment: Treat all unconscious divers (who were breathing compressed gas) for an AGE until proven otherwise.
Primary: Perform ABCs of resuscitation. Place patient on 100% oxygen, maintain oxygen until full recovery, and then slowly wean. Keep patient prone until full recovery. Monitor for 24 hours for residual symptoms. If patient was breathing compressed gas, treat for an AGE until proven otherwise (i.e., diver is witnessed passing out secondary to holding breath too long and responds quickly and completely to oxygen treatment). Secure diver's breathing source and tanks for testing.
Primitive: Perform artificial respiration if patient in respiratory arrest.

Patient Education
General: Educate divers on hypoxia and shallow water blackout
Diet: Normal
Prevention and Hygiene: Education and training in diving practices. Do not hyperventilate before a breath hold dive - hyperventilating drives carbon dioxide level even lower
No Improvement/Deterioration: Consult Diving Medical Officer (DMO); treat as AGE

Follow-up Actions
Return Evaluation: Monitor for 24 hours for full improvement.
Consultation Criteria: Any residual symptoms need to be evaluated by a DMO.

Dive Medicine: Oxygen Toxicity
CPT Jeffrey Morgan, MC, USA

Introduction: Depth provides circumstances in which oxygen may become toxic to the body. As a diver descends in water, the partial pressure of oxygen (ppO_2) increases. The US Navy Dive Manual limits the ppO_2 during diving operations based the depth and duration of dives. Exceeding these limits risks an oxygen toxicity injury for the diver. Oxygen toxicity affects various tissues in the body, most notably the pulmonary system and the central nervous system (CNS).

Subjective: Symptoms
Pulmonary: Non-productive cough and difficulty breathing, usually exacerbated on inspiration, progressing to substernal burning and severe pain on inspiration.

Objective: Signs
Using Basic Tools: Pulmonary: Dyspnea, non-productive cough, diminished air exchange (cannot blow out a match or candle 12-14 inches away).
CNS: "VENTID-C" is the mnemonic for oxygen toxicity. There is no specific order in which these signs and symptoms appear. The first sign may be convulsions. Only one sign may present or several of the signs may present: V- **V**isual disturbances (blurred or tunnel vision), E- **E**ars (tinnitus), N- **N**ausea, T- **T**witching/tingling (often seen around the eyes and mouth), I- **I**rritability, D- **D**izziness, C- **C**onvulsions (tonic-clonic), often without warning.
Using Advanced Tools: CXR may reveal thickening of alveolar and interlobular septa and lung edema.

Assessment:
Differential Diagnosis - decompression sickness (DCS), Arterial Gas Embolism, Pulmonary DCS ("Chokes"), Pulmonary Over Inflation Syndrome

Plan:
Treatment
Pulmonary: Wean patient from oxygen source while maintaining normal respiratory function (pulse-ox above 92%.)
CNS: Remove patient from oxygen source and return to room air at sea level. If convulsions occur at depth, slowly bring patient to surface with regulator in mouth while compressing the abdomen.

Patient Education
General: Oxygen can be toxic for prolonged exposure and at high partial pressures. When diving with oxygen rigs a diver needs to stay within the limits established in the Navy Dive Manual.
Diet: Normal
Prevention and Hygiene: Do not return to diving for at least 24 hours and until DMO has cleared the diver.
No Improvement/Deterioration: Consult Diving Medical Officer, pulmonolgist or neurologist.

Follow-up Actions
Return evaluation: Follow patient daily to ensure improvement.
Consultation Criteria: If breathing or neurological problems persist, consult DMO, pulmonologist or neurologist.

Single Depth Oxygen Exposure Limits (from US Navy Dive Manual Revision 4)

Depth	Maximum Oxygen Time
25 FSW	240 minutes
30 FSW	80 minutes
35 FSW	25 minutes
40 FSW	15 minutes
50 FSW	10 minutes

A diver should never be on 100% oxygen for more than 4 hours at any depth in any 24-hour period during normal diving operations.

Dive Medicine: Carbon Monoxide Poisoning
CPT Jeffrey Morgan, MC, USA

Introduction: Carbon monoxide is a colorless, odorless and tasteless gas that binds very strongly to hemoglobin. It is typically produced by incomplete combustion of fuels. When filling SCUBA tanks, there is a risk that the air intake of the compressor is located near a source of carbon monoxide (exhaust pipe of engine, etc.). Carbon monoxide (CO) has 210-250 times the affinity for hemoglobin that oxygen does. If hemoglobin is bound to CO it cannot bind to and transport oxygen to tissues. The severity of symptoms is not directly correlated to the percent of hemoglobin bound to carbon monoxide. Carbon monoxide poisoning causes hypoxia and other intra-cellular problems stemming from the production of free radicals. Factors that affect carbon monoxide absorption are concentration of inhaled CO, duration of CO exposure, and respiratory rate. Symptoms usually begin while ascending from a dive (due to dropping partial pressure of oxygen). Significant exposure may cause the symptoms to begin anytime during the dive. Hyperbaric oxygen therapy reduces the half-life of carbon monoxide saturation considerably.

Subjective: Symptoms
Headache, nausea, vomiting, dizziness, dyspnea, tinnitus, weakness, irritability, memory loss, confusion, collapse, stupor, unconsciousness, coma and death. Unconsciousness can occur with very few prior symptoms or warning signs.
NOTE: Delayed CO symptoms can occur up to 21 days after exposure. Up to 23% of patients who have mild initial symptoms and are only treated with 100% oxygen will have delayed symptoms.
NOTE: Pregnant women who have been exposed to CO and have no symptoms may have a fetus in distress. The fetus must be evaluated also. Fetal hemoglobin binds with CO at approximately 15 times that of the mother.

Objective: Signs
Using Basic Tools: Abnormal mini-mental status exam; weakness; vomiting; increased respiratory rate but lungs clear to auscultation; bright, cherry-red lips are a rare and late sign of CO poisoning.
Using Advanced Tools: Depressed S-T segments or arrhythmias on EKG. Cannot assess hypoxia with pulse oximeter.

Assessment:
Differential Diagnosis - cyanide poisoning, hypoxia (both can be treated with HBO also)

Plan:
Treatment: Treat a diver with suspicious clinical symptoms, including any CNS symptoms beyond a mild headache.
Primary: Remove diver from source of carbon monoxide. Administer 100% oxygen as soon as possible pending hyperbaric oxygen (HBO) therapy. Maintain hydration to ensure urine output remains at least 30 cc per hour (urinating every 2-3 hours). Perform pregnancy test for all women of childbearing age. If patient is pregnant, benefit of recompression treatment needs to outweigh risk to unborn child (possible: retrolental fibroplasias, in utero death, birth defects).
Alternative: if HBO chamber is not available, treat with 100% oxygen until symptoms have resolved.

Patient Education
General: High pressures of oxygen are needed to offset the CO. If mild symptoms are treated only with 100% oxygen, monitor the patient for about 24 hours. Return daily for follow-up for three days. CO poisoning symptoms may begin over the next 21 days. If symptoms occur, patient needs to return for HBO treatment.
Diet: Keep patient hydrated.

Prevention and Hygiene: Use only approved USN air sources for diving.
No Improvement/Deterioration: Return to a medical care provider or emergency room immediately.

Follow-up Action
Return evaluation: Follow up daily for three days after symptoms have resolved to evaluate for recurrence. Avoid diving for at least four weeks after symptoms resolve, and until cleared by a Diving Medical Officer (DMO).
Consultation Criteria: DMO consultation as soon as possible.
Evacuation: Avoid altitude over 1000 feet unless in aircraft with pressurized cabin.

Dive Medicine: Carbon Dioxide Poisoning
CPT Jeffrey Morgan, MC, USA

Introduction: Carbon dioxide (CO_2), a colorless, odorless, tasteless gas, is a normal component of the atmosphere that can be toxic in high concentrations. A diver may experience CO_2 poisoning (hypercapnia) even without a deficiency of oxygen. Hypercapnia often results from improperly venting expired CO_2 in chamber or hard hat diving operations. Difficulties with rebreather (closed circuit or semi closed circuit) rigs like the MARK XVI and LAR V (MARK XXV), which use chemicals to remove CO_2 from the breathing supply can also lead to hypercapnia. Skip breathing (voluntary hypoventilation) while diving also causes CO_2 to build up in the blood stream. Increased work rates and shivering due to cold water increase CO_2 generation and the chances of CO_2 poisoning. Patients usually recover within 15 minutes by breathing fresh air, but headache, nausea, and dizziness may persist after treatment.

Subjective: Symptoms
Headache, dizziness, confusion, euphoria, unconsciousness. Note: a diver may experience no signs or symptoms other than sudden unconsciousness.
Focused History: *Have you been diving with a rebreather? In a chamber/hard hat?* (common causes)

Objective: Signs
Using Basic Tools: Increased rate and depth of respirations, shortness of breath, increased pulse rate. Neurological exam including mini-mental status exam (see Appendices): decreased mental status (usually obvious during history questions), decreased balance, decreased strength, numbness in the extremities and unconsciousness; diver may become unconscious underwater and drown.

Assessment:
Differential Diagnosis
Hypoxia - more pronounced fatigue and confusion, more frequent cyanosis, more visual changes.
Oxygen toxicity - convulsions, tunnel vision, twitching of facial muscles
Pulmonary Over Inflation Syndrome (POIS), including Arterial Gas Embolism (AGE) - often specific and/or catastrophic neurological findings within 10 minutes of surfacing; pneumothorax, subcutaneous emphysema. See POIS section in this chapter.
Decompression sickness (DCS) - focal or general neurological findings, which typically begin hours after a dive

Plan:
Treatment
1. Treat for AGE or hypoxia if these conditions are suspected or cannot be ruled out. See respective sections in this chapter.
2. Otherwise, have patient rest and breathe fresh air with deep inspirations for about 30 minutes. Significant recovery should quickly occur with the possibility of some residual headache, nausea and dizziness.
3. If patient has not made significant improvement in 30 minutes, consider immediate evacuation and reconsider differential diagnoses.

Patient Education
General: Symptoms should resolve quickly by taking deep breaths of fresh air.
Activity: Rest until all symptoms resolve.
Diet: Drink plenty of replacement fluids.
Prevention and Hygiene: Change CO_2 absorbent material in LAR V prior to each dive. Work slowly in water, especially if water is very cold. Avoid skip breathing. Ventilate frequently if in a chamber or diving surface-supplied rigs.
No Improvement/Deterioration: Return promptly for reevaluation.

Follow-up Actions
Return evaluation: See one day after treatment to ensure there are no residual symptoms.
Evacuation/Consultation Criteria: Evacuate if patient does not respond to treatment within 30 minutes. Consult a Diving Medical Officer at the earliest opportunity after beginning therapy to confirm diagnosis and treatment plan.

Dive Medicine: Dangerous Marine Life - Venomous Animals
CPT Jeffrey Morgan, MC, USA

Introduction: There are a variety of creatures encountered in the water that injure through venom. Some of the venoms are mild and easily treated. Others are very toxic and extremely life threatening. Divers are often unaware of the type of animal that bit or injured them. The medical caregiver must be prepared for a variety of signs and symptoms. This section does not address all the many creatures that can envenomate humans, but does include the noteworthy ones. Medical personnel routinely covering dives should have a text on dangerous marine life that includes animals that can envenomate.

Subjective Symptoms/Objective Signs: Discussed under each animal species.

Highly Toxic Fish (Stonefish, Zebrafish, and Scorpionfish): Stings by these fish can kill, and are usually accidental-- a diver steps on the fish or handles it. These fish carry venom in their spines much like other venomous fish, but their venom is much more toxic. Initial local symptoms: severe pain followed by numbness and/or hypersensitivity around the wound site lasting for days. Generalized reactions include respiratory failure and cardiovascular collapse. There is an antivenin for stonefish toxin that seems to be somewhat effective for zebrafish and scorpionfish toxins (see below for ordering information).

Other Venomous Fish: Most fish envenomate through fin spines, while a diver is stepping on or handling the fish. Venom will continue to flow into the injury while the spine sheath is still in the patient. The venom is usually heat labile and may decompose in hot water (about 115°F). Initial local symptoms: severe pain followed by numbness and/or hypersensitivity around the wound site lasting for hours. General symptoms may include nausea, vomiting, sweating, mild fever, respiratory distress and collapse. Serious anaphylactic reactions are possible.

Stingrays: Stingrays are common in tropical and temperate areas. They hide in the ocean floor sand with eyes and tail exposed. Most attacks from stingrays occur from swimmers or divers stepping on them. The tail will whip up in self-defense and impale the diver/swimmer's leg with a barbed spine. The wound area has a blue rim and is typically swollen, painful, and pale. Generalized reactions can include fainting, nausea, vomiting, frequent urination and salivation, sweating, respiratory difficulty, and cardiovascular collapse. Symptoms may take months to resolve. Secondary infections and necrotic lesions often develop. The toxin is heat labile (113°F). No antivenin is available.

Coelenterates: Hazardous coelenterates include Portuguese man-of-war, sea wasp or box jellyfish, sea nettle, sea blubber, sea anemone, and rosy anemone. The most common stinging injury is the jellyfish sting, occurring worldwide. Most jellyfish stings result only in local skin irritation and modest pain. However, a box jellyfish (sea wasp) sting can result in death within 10 minutes from cardiovascular collapse, respiratory failure and paralysis. Antivenin is available for box jellyfish toxin (see below for address). The Portuguese man-of-war is rarely fatal but does cause similar generalized symptoms as the box jellyfish that resolve after about a day. Stings from the man-of-war look like a red string of beads.

Coral: This porous, rock-like formation found in water often has sharp edges. Usually cuts from coral are self-limiting and have only a mild skin reaction. Unfortunately, they usually take a long while to heal. Some coral can sting (coelenterate family). One of the deadliest poisons known was found recently in coral (genus *Palythoa*). If it is introduced into a deep cut in the body, it may be fatal. No antidote is known. Divers should wear dive suits to protect them from coral cuts, especially in surging waves.

Octopi: The octopus is an underwater chameleon, changing colors often in the water trying to conceal itself from its enemies. Most species of octopi found in the U.S. are harmless to humans. Octopi can envenomate by biting and injecting venom from salivary glands. An octopus bite consists of two small punctures, surrounded by swollen, red and painful tissue. Bleeding may be severe due to anticoagulant effects of the venom. The blue ringed octopus found in Australian and Indo-Pacific waters is often deadly. It injects a neuromuscular blocker called maculotoxin that may cause paralysis, vomiting, respiratory difficulty, visual disturbances and cardiovascular collapse. No antivenin is available. Paralysis may last 4-12 hours (with mechanical ventilation).

Sea Urchins: These round, spiny creatures carry venom in the long spines. Divers and bathers step on these animals, impaling their feet on the long spines, which typically break off. The venom may cause numbness, generalized weakness, paresthesias, nausea, and vomiting. Cardiac dysrhythmias have been reported. The toxin is heat labile. Immersion of affected area in water above 115°F is recommended. Allergic reactions can accompany these injuries.

Cone Shells: The cone shell is widely distributed throughout the world. The shell is a symmetrical spiral with a distinctive head. Venom is contained in darts inside the proboscis, which extrudes out of the narrow end but can reach most of the shell. A stinging or burning sensation begins at the site of the sting; followed by numbness and tingling that spread from the wound to the rest of the body. Involvement of the mouth and lips is severe. Generalized symptoms include muscular paralysis, difficulty swallowing or talking, visual disturbances and respiratory distress. A cone shell sting should be viewed as severe as snakebite. Cone shell victims will probably experience paralysis or paresis of skeletal muscle with our without myalgia. Symptoms develop within minutes and can last up to 24 hours. No antivenin is available and mortality reaches 25%.

Sea Snakes: Sea snakes are air-breathing reptiles that swim underwater for great distances (over 100 miles from land). They inhabit the Pacific and Indian Oceans and the Red Sea. There are some unsubstantiated reports of sea snakes in the Caribbean Sea (reportedly coming through the Panama Canal). The neurotoxin venom of a sea snake is 2-10 times more potent than that of a cobra. Bites are usually not painful, and only about 25% cause envenomization. There is a latent period of 10 minutes to several hours after the bite before generalized symptoms develop: muscle aches and stiffness, thick tongue sensation, progressive paralysis, nausea, vomiting, difficulty with speech and swallowing, respiratory distress and failure, and smoky colored urine from myoglobinuria (which may progress to kidney failure). The venom is heat-stable and antivenin is available (see information below).

Assessment:
Differential Diagnosis - see different species of animals that envenomate

Plan:
Treatment/Procedures
Venomous Fish, Highly Toxic Fish, Stingrays
1. Lay patient supine, reassure and observe for shock.
2. Irrigate wound with cold saline or salt water to rinse remaining toxin. Minor surgery may be required

to open wound.
3. Evacuate urgently to medical treatment facility for administration of antivenin if available and intensive care support if needed. Envenomizations should not be treated in the field if the patient can be evacuated.
4. Soak wound in water as hot as patient can tolerate (<122°F) for 30-90 minutes. Use hot compresses if wound is on the face.
5. Give **diazepam** IM 5 mg for muscle spasms.
6. Manage pain (see Procedure: Pain Assessment and Control). Do NOT use narcotics in cases of RESPIRATORY DISTRESS or FAILURE. **Lidocaine** may be used in the wound for pain relief, but NEVER use **epinephrine**.
7. Explore the wound and clean out any remaining spines or barbs.
8. Immobilize the affected extremity.
9. Administer **tetanus** and antibiotic prophylaxis (**tetracycline** 250 mg po qid, and **neomycin** or **bacitracin** topically).
10. Consider x-ray to look for any remaining sheaths or barbs.

Coelenterates
1. Apply vinegar or a 3-10% solution of **acetic acid** (or carbonated beverage) to sting site to neutralize stingers.
2. Gently remove any remaining tentacles with a towel or cloth.
3. For box jellyfish stings, administer antivenin slowly: one container (vial) IV and three containers IM route. Treat sensitivity reactions to the antivenin with a SC injection of **epinephrine**, along with corticosteroids and antihistamines (see Shock: Anaphylactic). Treat hypotension with volume expanders and pressor medication if available.
4. Evacuate victims of box jellyfish and Portuguese man-of-war stings immediately.
5. Use topical and/or local anesthetic agents (e.g., **lidocaine**).

Coral
1. Control bleeding.
2. Promptly clean with **hydrogen peroxide** or 10% **povidone-iodine** solution and debride the wound. Remove all foreign particles.
3. Cover with a clean dressing.
4. Administer **tetanus** prophylaxis.
5. Topical antibiotic ointment.
6. Manage pain (see Procedure: Pain Assessment and Control). Do NOT use narcotics in cases of RESPIRTORY DISTRESS or FAILURE.
7. Evacuate to medical treatment facility if symptoms are severe.

Octopi
1. Control local bleeding with pressure
2. Clean and debride wound. Cover with clean dressing.
3. If blue-ringed octopus is suspected:
   a. Apply pressure bandage and immobilize the bitten extremity. Place extremity lower than the heart.
   b. Be prepared to administer CPR and ventilate patient
   c. Immediately evacuate patient to medical treatment facility for intensive care.
4. Administer **tetanus** prophylaxis.

Sea Urchins
1. Remove all protruding spine fragments from wound if possible. Do not break large pieces off in the wound.
2. Bathe wound in vinegar, then soak wound in as hot water as can be tolerated (no more than 122°F).
3. Clean and debride wound and apply topical antibiotic ointment.
4. Surgical removal is often necessary for deep spines. If necessary, evacuate patient to medical treatment facility to have this performed. X-rays can identify broken spines.
5. Treat allergic reactions and bronchospasms with SC **epinephrine** and antihistamines (see Shock:

Anaphylactic).
6. Administer **tetanus** prophylaxis.

Cone Shells
1. Place patient supine.
2. Apply pressure bandage to wound and place injury site at a level below the heart. Keep the patient from moving.
3. Immediately transport patient to medical treatment facility for intensive care. Be prepared to ventilate patient and administer CPR. Treat symptoms with supportive care as they present.
4. Avoid any analgesics that cause respiratory depression (narcotics).
5. Administer **tetanus** prophylaxis

Sea Snakes
1. Keep victim still.
2. Apply pressure dressing to bite site and place bite in a position below the heart.
3. Incise wound and apply suction if within 2 minutes of time of bite.
4. Transport patient immediately to medical treatment facility for antivenin treatment and intensive care.
5. Place IV and administer a bolus of 1 L normal saline. Continue IV at a rate of 125 ml/hr to keep urine output at least 30 cc/hour. Watch patient's urine looking for smoky color (indicating myoglobinuria, renal failure).
6. Be prepared to ventilate patient and perform CPR.
7. Observe patient for at least 12 hours after a bite due to possible latent effects.
8. Administer **tetanus** prophylaxis.

Patient Education
General: Keep the patient calm and reassure him that he has probably been envenomated and will receive further treatment.
Activity: Ensure the patient rests and remains calm.
Diet: Keep patient hydrated with IV NS and keep him NPO until you are sure surgery will not be required.
Medications: Antivenins have a high incidence of serum sickness. Treat this with **epinephrine** and **antihistamines**. If at all possible, administer antivenin in a medical treatment facility in order to treat the serum sickness appropriately with intensive care support.
Prevention and Hygiene: Avoid these types of marine animals.
Wound Care: Keep dressing clean and dry. Antibiotic ointment will help prevent secondary wound infections

Follow-up Actions
Return evaluation: Monitor patients until all symptoms resolve.
Evacuation/Consultation Criteria: Refer to each individual treatment plan above for evacuation guidance. Consult an emergency medicine specialist, an internist or a Diving Medical Officer (DMO) for unstable patients after envenomization. Refer recovered divers to a DMO for clearance to return to diving duty.

NOTE: Antivenins are available from Commonwealth Serum Lab; 45 Poplar Rd, Parkville; Melbourne, Victoria, Australia. Telephone: 011-61-3-389-1911. Telex: AA-32789.

Dive Medicine: Dangerous Marine Life - Biting Animals
CPT Jeffrey Morgan, MC, USA

Introduction: Diving in open water puts a diver in an environment with numerous dangerous sea creatures. Many of these creatures are predatory and can cause significant harm to humans. This section will only deal with the marine predators that injure by biting. Other marine life may injure humans through venom (see preceding section), electrical shock, pinching and other means. Preparedness and avoidance is the best way to prevent encounters with dangerous marine life. Signs, symptoms, assessment and treatment are all very similar for the various animals listed.

Sharks: Attacks are rare but severe. Some species (great white, tiger, whitetip and others) are more aggressive towards divers. Injuries range from bumps or scratches caused by contact with rough skin to large bite wounds. Treat all sharks with respect and avoid them, because they are fast, strong and potentially aggressive.

Killer Whales (Orcas): Killer whales are mammals that usually travel in pods of 3-40 whales. These carnivores are at the very top of the food chain in the ocean. They have great intelligence, size, speed, interlocking teeth, and powerful jaws. Even though there are no recorded attacks on humans, there is potential that any animal this big could strike or bite an irritating diver.

Barracudas: Barracudas are predatory fish found in the tropics that can grow to 10 feet in length. Most are much smaller (3-5 feet). The barracuda is a fast swimmer with extremely sharp teeth, but attacks are usually less severe than those of sharks. It rarely attacks divers, but is known to attack surface swimmers or limbs dangling from boats, and is attracted to bright objects.

Moray Eels: Commonly found in holes and crevices near the ocean floor in tropical and subtropical waters, the moray eel resembles a large sea snake (up to 10 feet long) in shape and movements. Its face appears more like a dog with sharp teeth. Once it bites, it may be very difficult to dislodge. The moray eel will usually bite when a diver comes too close or sticks a hand into its hole or crevice. Mild envenomization may occur with certain species. Supportive care is all that is needed for this venom.

Sea Lions: These mammals can be very aggressive with divers. Large male sea lions have a reputation for nipping divers during the mating season. Divers often mistakenly assume these animals as friendly and try to approach sea lions, provoking a bite. These bites resemble dog bites and are not usually severe.

Crocodiles: Many experts feel crocodiles are more dangerous than sharks. The saltwater and Nile crocodiles can grow up to 30 feet long. Crocodiles are fast, strong and aggressive reptiles. They are territorial and often found near river estuaries and brackish water. Treat any crocodile over three feet long as dangerous. Divers should immediately exit any area inhabited by a crocodile.

Subjective: Symptoms
Severe pain in area of bite or injury; confusion and shock may set in very quickly. Infection may set in 3-7 days after wounding, with pain, redness, swelling, warmth and fever. **Focused History:** *What bit you?*

Objective: Signs
Evaluate as a trauma patient.
Using Basic Tools: Airway damage extending from mouth to lungs: tachypnea, labored breathing, spitting blood, sucking chest wound. Hemorrhagic shock: fast heart rate, pallor, hypotension. Tissue loss, deformed body parts at bite site.

Assessment:
Differential Diagnosis - other animal bites, blast trauma, other lacerating or penetrating trauma.

Plan:
Treatment - Treat per trauma protocol in this Handbook.
Airway, Breathing, Circulation
1. Control bleeding: apply compression dressings, elevate and apply direct pressure. Use pressure points, tourniquet as needed.
2. Ensure airway is well established and administer 100% O_2
3. Infuse 1-2 L Ringer's Lactate or Normal saline by IV for severe blood loss. Monitor vital signs.

4. Treat severe pain with **morphine** 5mg IV. Add 1-2 mg doses IV. Do not exceed 10 mg of **morphine** in 24 hours.
5. Evacuate to medical treatment facility as soon as possible if trauma is severe.
6. Send any severed limbs with the patient in saline-moistened gauze on ice.
7. Cleanse and debride wounds in a clean environment. Explore wounds thoroughly (shark teeth may not appear on x-ray).
8. Splint any extremities that look deformed or that have possible fractures. X-ray when possible.
9. Monitor pulses and urine function for possible compartment syndrome and/or myoglobinurea from crush injuries.
10. Administer tetanus prophylaxis: **Tetanus toxoid** 0.5 ml IM.
11. Culture wounds for aerobes and anaerobes before starting broad-spectrum antibiotics.
12. If patient has unexplained neurological symptoms, consider decompression sickness and arterial gas embolism.

Patient Education
General: Reassure patient
Activity: Rest until all injuries are identified and treated.
Diet: Keep patient NPO since he will possibly undergo surgery very shortly
Medications: Watch for respiratory suppression with **morphine**.
Prevention and Hygiene: Do not swim with or challenge dangerous marine life. Avoid crocodiles on land. Do not wear shiny objects while diving. Do not panic in the water. Only as a last resort if bitten, hit snout, gills or eyes to drive attacker away.
Wound Care: Keep wounds clean, dry and covered.

Follow-up Actions
Return evaluation: Patients should not be allowed back into the water until cleared by a Diving Medical Officer (DMO).
Evacuation/Consultation Criteria: Evacuate those with severe wounds, which require surgery. Consult a general surgeon in these cases. Consult a DMO for other diving injuries (e.g., decompression sickness).

Dive Medicine: Underwater Blast/Explosion/Sound Injury
CPT Jeffrey Morgan, MC, USA

Introduction: Underwater explosions create shock waves that move out in all directions at the speed of sound. Water is non-compressible, therefore it transmits the shock wave a greater distance and with more intensity than an explosion in air (this is the concept used by depth charges against submarines). Unlike air explosions, shrapnel and debris injuries are minimal because water quickly dampens any particles propelled away from the explosion. The shock wave generated by the explosions can bounce off of the surface, ocean floor, boat hulls or sea walls and strike divers in the water with a compounding impact. Factors affecting the intensity of an underwater explosion: size of the charge, distance from the blast, protective clothing, depth of the diver (deeper is worse), depth of ocean floor (shallow depths reflect blast more intensely) and firmness of reflective surfaces near explosion (firmer surfaces reflect blast waves better, therefore are worse for the diver). Shock waves cause implosions of gas-filled cavities in the body, tearing of tissues at gas-liquid interfaces, and emboli.

Subjective: Symptoms
May be unable to talk; complaints may include chest pain, cough, pain with breathing, spitting blood, severe abdominal pain, confusion, headache, seeing "floaters" or "stars," ear pain, loss of hearing, and ringing of ears.

Objective: Signs
Decompression sickness or arterial gas embolism (see appropriate sections of this chapter) may occur secondary to a rapidly aborted dive, or because the trauma of the explosion may suddenly move the diver into more shallow water.
Using Basic Tools: General: Fracture or dislocation of any body part; swollen extremities (possible heart failure).
HEENT: Subcutaneous emphysema ('crackling' with palpation of the skin) over the neck (leak in the respiratory tract, such as pneumothorax); ruptured tympanic membrane (TM), tender or collapsed sinuses, blood in nose and ears, nystagmus, dizziness, vertigo, and loss of balance; distended neck veins (heart failure or cardiac tamponade).
Chest: Hemoptysis, obvious discomfort with deep inspiration, rales and friction rubs upon auscultation, cough, and respiratory failure; decreased or absent breath sounds (simple, or tension pneumothorax, hemothorax); subcutaneous emphysema.
GI tract: Blood in stool; rigid abdomen; rebound tenderness (probably due to intestinal rupture/hemorrhage).
CNS: Mental status changes (delirium, confusion, unresponsiveness), and paralysis of any part of the body (stroke from emboli).
Using Advanced Tools: X-rays: CXR: Pneumothorax; patchy or diffuse infiltrates (pulmonary contusion) a few hours after the blast injury. Cervical spine: assess for fractures, dislocations and other abnormalities.

Assessment:
Differential Diagnosis - arterial gas embolism (AGE) and decompression sickness (DCS) (see appropriate sections in this chapter), as well as these possible blast-associated injuries:
Lungs - pulmonary contusions, pneumothorax, tension pneumothorax, hemothorax, alveolar rupture.
Gastrointestinal - intestinal hemorrhage, intestinal perforation, paralytic ileus, acute abdomen
Heart - cardiac contusion, congestive heart failure, tamponade
Brain and nervous system - brain injury/contusion, paralysis, stroke and C-spine injury.
Ears and sinuses - ruptured TMs, conductive hearing loss, ruptured sinuses and inner ear barotraumas including round window and oval window ruptures.
Body - fractured bones, joint dislocations.

Plan:
Treatment
1. Treat as major trauma patient (see Trauma on CD-ROM). Immobilize the head and neck with a cervical collar until head/neck trauma is ruled out.
2. Secure the airway. Intubate if there is a doubt that the patient will be able to maintain his own airway. Place on 100% O_2 initially and then wean down over the next 12 hours (except CNS injury).
3. Perform needle decompression in 2nd intercostal space in the mid-clavicular line if any type of pneumothorax develops (see Procedure: Thoracostomy). Follow this later with a chest tube in the 5th - 6th intercostal space in the mid-axillary line (also for hemothorax).
4. Keep patient hydrated. Adjust IV infusion rate to ensure a urine output of at least 30 cc/hour. Do not over-hydrate patient. Do not use any fluids with a dextrose component, since the sugars may cause increased swelling in the neurological tissues.
5. Perform serial neurological examinations. Use Glasgow coma scale (See Appendices).
6. For CNS injuries, hyperventilate with 100% O_2. Immobilize trunk for any suspected spinal cord injury (SCI). For SCI with NO BRAIN INJURY, give high-dose **methylprednisolone** within the first 8 hours of injury ONLY: one dose of 30 mg/kg IV, followed by 5.4 mg/kg/hour IV for the next 23 hours.
7. Splint any deformed extremities.
8. If TM is ruptured, keep ear canal clean and dry to allow the TM to heal. If the rupture is large, an ENT surgeon may need to surgically repair the TM. Do not insert any drops or topical antibiotics in the ear canal. If there is any loss of balance, keep the patient supine with the head up about 30°. Start broad-spectrum systemic antibiotics: **imipenem** 500 mg IV q 6 hrs or **ciprofloxacin** 250-500 mg po bid.

9. Control any bleeding in sinuses (see Dive Medicine: Barotrauma, Other).
10. Transport to a medical treatment facility as soon as possible.
11. Monitor patient for at least 24 hours after injury. Some of the injuries listed above may present after a latent period. Treat for AGE and DCS as needed. Any signs of acute abdomen must be evaluated by a general surgeon.

Patient Education
General: Reassure patient.
Activity: Bed rest for at least 24 hours; immobilize until spine injury is ruled out. As tolerated after the first 24 hours.
Diet: Keep patient NPO until surgery is ruled out. Keep patient hydrated with IV fluids (normal saline or Ringer's Lactate). Do not over hydrate.
Medications: Methylprednisolone side effects may include: peptic ulcers, masking of infections, increased intracranial pressure, glaucoma, hypokalemia, hypocalcemia, and hyponatremia.
Wound Care: Keep wounds clean, dry and dressed.

Follow-up Actions
Return evaluation: Observe patient for at least 24 hours. Once discharged from constant (hospital) care, follow-up daily or every other day until all injuries resolve. Treat for decompression.
Evacuation/Consultation Criteria: Evacuate all unstable patients, including those with CNS effects. Consult a general surgeon for any complicated patient, as well as any ENT or orthopedic surgeon as required.

Calculating safe distance from underwater detonation

$$\text{PSI (lbs/in)} = \frac{13{,}000 \times 3(\text{Weight of the explosive}) \times \text{RE Factor}}{\text{Distance in feet}}$$

RE (Relative Effectiveness) Factor: Based on the type of explosive used. Examples include:
Dynamite (M1) = 0.92
TNT = 1.0
C4 (M112) = 1.34
HBX-1, HBX-3, and H-6 = 1.68

< 50 PSI: Safe area with no expected injuries
200-300 PSI: Lung and Gastrointestinal injuries expected
>500 PSI: Possible fatality
>2000 PSI: Certain death.

Dive Medicine: 'Caustic Cocktail' Chemical Burn
CPT Jeffrey Morgan, MC, USA

Introduction: Closed-circuit (no respiratory gases escape the breathing apparatus) and semi closed-circuit diving rigs use carbon dioxide (CO_2) absorbent chemicals to remove or scrub CO_2 from breathing gases. If water mixes with this solid substance (i.e., water may leak into the chemical canister), an alkaline solution forms. Inhaling or swallowing this solution causes chemical burns of the pharynx and trachea.

Subjective: Symptoms
Burning in mouth and throat, possible headache, choking and gagging with a foul taste in mouth; in severe cases burning may extend all the way down into the lungs. History may include water getting into the rebreather rig. Disassembling the rig may find water in the CO_2 canister.
Focused History: *Did your mouth start burning after you inhaled water from your rebreather rig?* (typical history)

Objective: Signs
Using Basic Tools: Rapid respiratory rate; choking and gagging; red, irritated and burned mucosa; pharynx may swell, compromising airway; lungs may develop rales.

Assessment:
Differential Diagnosis - laryngospasms secondary to inhalation of saltwater. There will not be any burning or redness with this event.

Plan:
Treatment
1. Place patient in an upright position in the water and remove mouth from caustic source.
2. Repeatedly rinse mouth with fresh water. When foul or sour taste is gone, have the diver swallow several mouthfuls. If only seawater is available, only rinse mouth. Do not have the patient swallow any of the seawater.
3. If the injury is severe and extends down to the lungs, secure the airway and apply 100% O_2. Patient may require intubation and mechanical ventilation, as well as immediate evacuation.

Patient Education
General: Remain calm and do not hyperventilate, which may worsen symptoms.
Activity: Rest until symptoms are gone and medic ensures there is no pulmonary or airway involvement.
Diet: Avoid eating until foul or sour taste is gone and pulmonary involvement is ruled out.
Prevention and Hygiene: Avoid getting water into CO_2 scrubber canister.
Wound Care: Rinse burned areas with water.

Follow-up Actions
Return Evaluation: If no hospitalization is needed, follow-up daily until symptoms are completely resolved.
Evacuation/Consultation Criteria: If respiration becomes labored or airway starts to swell, secure airway immediately and apply 100% O_2. Evacuate immediately and consult a Diving Medical Officer or pulmonologist.

Dive Medicine: Disqualifying Conditions for Military Diving
CPT Jeffrey Morgan, MC, USA

Introduction: All the conditions that make diving dangerous cannot be adequately discussed in this short section. Consequently, diving medical personnel must evaluate each disease/injury process and the limitations it may produce in the context of diving physiology. Each branch of service has different standards for disqualifying individuals from diving duty, but much authority rests in the hands of the Diving Medical Officer (DMO). Standards for qualified divers are different from those who are entering training. The following list includes medical conditions that are generally considered disqualifying (temporary or permanent) for diving duty in all branches of the U.S. military.

Ocular: Visual acuity that does not correct to 20/20. Each service has different regulations for uncorrected acuity. The main consideration is that poor visual acuity must be correctable to 20/20. All services prohibit radial keratotomy surgery, but photo radial keratotomy is acceptable as a surgical correction for acuity. Color and night vision must be normal.

Pulmonary: Spontaneous pneumothorax, traumatic pneumothorax (6 month disqualification), history of sarcoidosis; **isoniazid** (INH) treatment for positive PPD, recurrent pulmonary barotrauma, any exercise-limiting pulmonary diseases (including pnuemonia), respiratory airway diseases or asthma after the age of 12; COPD.

Psychiatric: Any diseases with potential to hinder performance, judgment or reliability; any psychotic disorder; any depressive or anxiety disorder that required hospitalization, involved work loss, involved suicidal gestures or required medication; alcohol dependence w/o 1 year of sobriety and aftercare; panic prone individual; claustrophobia.

Neurological: Headache requiring medication or with neurological deficits; history of penetrating head injury; history of closed head injury with CSF leak that lasted more than 7 days, intracranial bleed, depressed skull fracture, or loss of consciousness; seizure disorders except toxic, febrile or immediately post traumatic; recurrent syncope or vertigo; history of decompression sickness or arterial gas embolism with residual impairment; cerebral vascular disease; history of heat stroke with residual neurological impairment; stammering or stuttering; symptomatic disk disease; spine surgery in past 6 months (must be asymptomatic for the 6 months); neurosurgery for cancer

Ears, Nose and Throat: Eustachian tube dysfunction, or ineffectiveness; middle ear surgery except tympanoplasty; any inner ear surgery; any inner ear pathology; any inner ear hearing dysfunction below basic diving qualification standards; any laryngeal or tracheal framework surgery.

Cardiac: Congestive heart failure; coronary artery disease; history of myocardial infarction; patent foramen ovale; any heart surgery; Wolff-Parkinson-White syndrome (symptomatic, not just EKG findings); any electroconductive disorder in the heart; blood pressure in excess of 140 systolic or 90 diastolic (in a qualified diver, hypertension requiring only 1 medication to control may be acceptable).

Endocrine: Diabetes (insulin dependent or requiring oral medication); uncontrolled thyroid dysfunction.

Gastrointestinal: Any condition that causes air trapping in the GI tract; severe gastroesophageal reflux; achalasia; paraesophageal and incarcerated sliding hiatal hernias; chronic partial gastric-outlet obstruction; severe postgastrectomy dumping syndrome; chronic, recurrent or acute small bowel obstruction; symptomatic gallstones; any uncorrected abdominal wall hernia.

Obstetrics and Gynecology: Pregnancy

Oral, Maxillofacial, and Dental: Any disorder that prevents holding a regulator in the mouth or the ability to clear the ears; poor dental health or care (below category 2) (broken or badly carious teeth).

Orthopedic: Any disorder that prevents the diver from performing essential emergency procedures in the water.

Medications: There is no list of disqualifying medications. The disease process for which the diver is taking the medication is the main concern and must be evaluated for its limitations on diving. Side effects of the medications in question must fully be understood by the diving medical personnel prior to allowing a diver to enter the water. **Diving medical personnel are expected to use their judgment to prevent unsafe/unhealthy divers from entering the water.**

NOTES: The US Navy is the lead agent for all US military diving. BUMED-21 addresses diving medicine issues at (202) 762-3444. The Naval Experimental Diving Unit (NEDU) is available for consultations on diving emergencies (24 hrs a day) at (850) 230-3100.

The lead civilian diving resource is the Diver's Alert Network (DAN), located at Duke University in North Carolina. DAN may be reached at (919) 684-8111 (24 hrs a day) for emergencies, and at (919) 684-2948, Ext. 222, for non-emergency diving questions.

Dive Medicine: Treatment Tables

Treatment Table 4

1. Descent rate - 20 ft/min.

2. Ascent rate - 1 ft/min.

3. Time at 165 feet includes compression.

4. If only air is available, decompress on air. If oxygen is available, patient begins oxygen breathing upon arrival at 60 feet with appropriate air breaks. Both tender and patient breathe oxygen beginning 2 hours before leaving 30 feet. (see paragraph 21-5.4.4.2).

5. Ensure life-support considerations can be met before committing to a Table 4. (see paragraph 21-5.6) Internal chamber temperature should be below 85°F.

6. If oxygen breathing is interrupted, no compensatory lengthening of the table is required.

7. If switching from Treatment Table 6A or 3 at 165 feet, stay a maximum of 2 hours at 165 feet before decompressing.

8. If the chamber is equipped with a high - O_2 treatment gas, it may be administered at 165 fsw, not to exceed 2.8 ata O_2. Treatment gas is administered for 25 minutes interupted by 5 minutes of air.

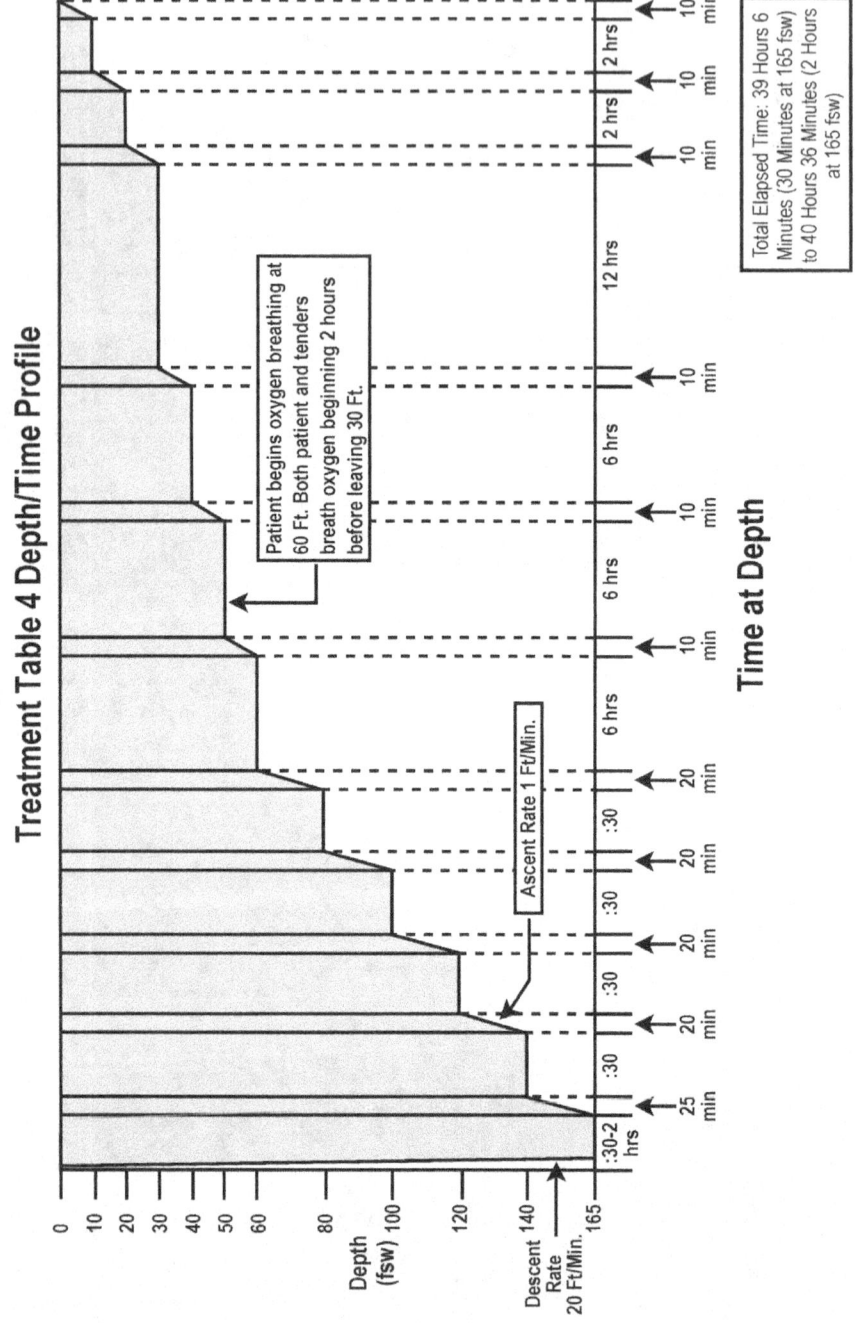

Treatment Table 5 Depth/Time Profile

1. Descent rate - 20 ft/min.
2. Ascent rate - Not to exceed 1 ft/min. Do not compensate for slower ascent rates. Compensate for faster rates by halting the ascent.
3. Time on oxygen begins on arrival at 60 feet.
4. If oxygen breathing must be interrupted because of CNS Oxygen Toxicity, allow 15 minutes after the reaction has entirely subsided and resume schedule at point of interruption. (see paragraph 21-5.5.6.1)
5. Treatment Table may be extended two oxygen-breathing periods at the 30-foot stop. No air break required between oxygen-breathing periods or prior to ascent.
6. Tender breathes 100 percent O_2 during ascent from the 30-foot stop to the surface. If the tender had a previous hyperbaric exposure in the previous 12 hours, an additional 20 minutes of oxygen breathing is required prior to ascent.

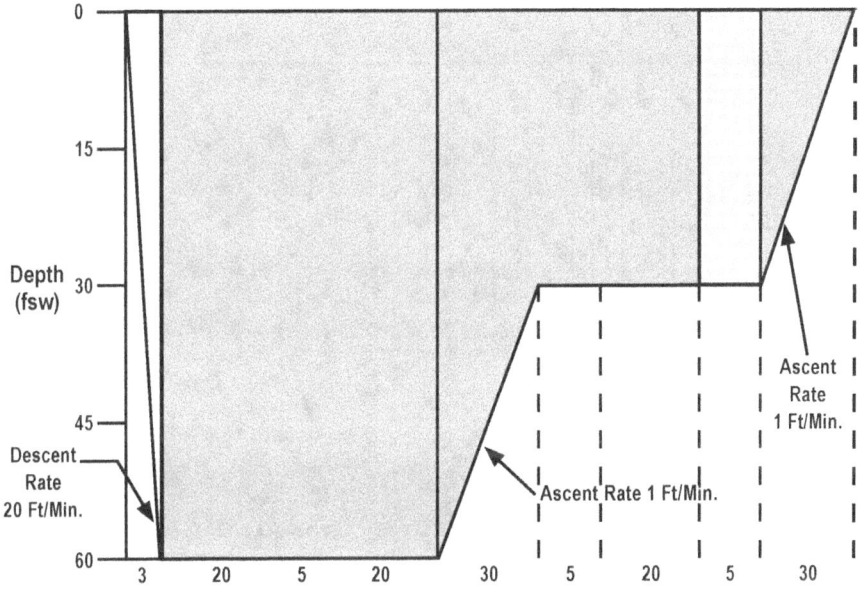

Total Elapsed Time:
135 Minutes
2 Hours 15 Minutes
(Not Including Descent Time)

Treatment Table 6 Depth/Time Profile

1. Descent rate - 20 ft/min.
2. Ascent rate - Not to exceed 1 ft/min. Do not compensate for slower ascent rates. Compensate for faster rates by halting the ascent.
3. Time on oxygen begins on arrival at 60 feet.
4. If oxygen breathing must be interrupted because of CNS Oxygen Toxicity, allow 15 minutes after the reaction has entirely subsided and resume schedule at point of interruption. (see paragraph 21-5.5.6.1)
5. Table 6 can be lengthened up to 2 additional 25-minute periods at 60 feet (20 minutes on oxygen and 5 minutes on air), or up to 2 additional 75-minute periods at 30 feet (15 minutes on air and 60 minutes on oxygen), or both.
6. Tender breathes 100 percent O_2 during the last 30 min. at 30 fsw and during ascent to the surface for an unmodified table or where there has been only a single extension at 30 or 60 feet. If there has been more than one extension, the O_2 breathing at 30 feet is increased to 60 minutes. If the tender had a hyperbaric exposure within the past 12 hours an additional 60-minute O_2 period is taken at 30 feet.

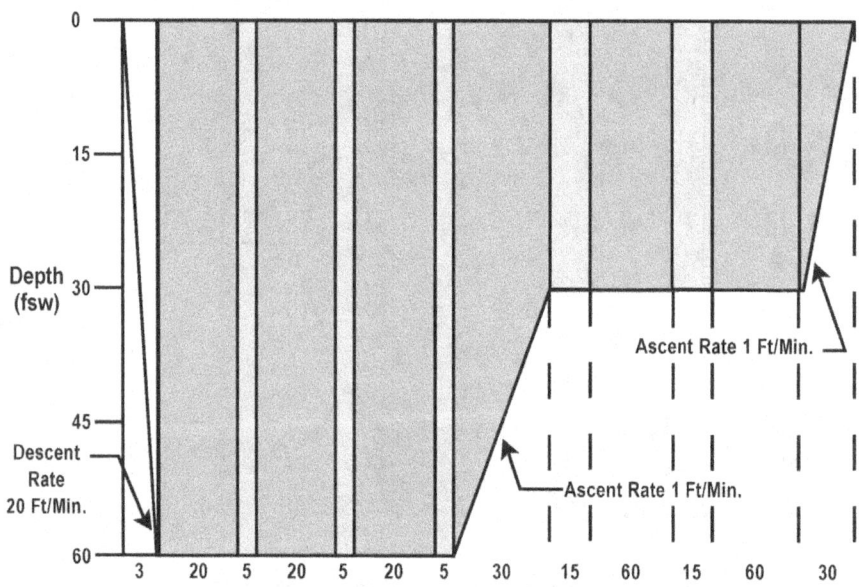

Time at Depth (minutes)

Total Elapsed Time:
285 Minutes
4 Hours 45 Minutes
(Not Including Descent Time)

Treatment Table 6A

1. Descent rate - 20 ft/min.

2. Ascent rate - 165 fsw to 60 fsw not to exceed 3 ft/min, 60 fsw and shallower, not to exceed 1 ft/min. Do not compensate for slower ascent rates. Compensate for faster rates by halting the ascent.

3. Time at treatment depth does not include compression time.

4. Table begins with initial compression to depth of 60 fsw. If initial treatment was at 60 feet, up to 20 minutes may be spent at 60 feet before compression to 165 fsw. Contact a Diving Medical Officer.

5. If a chamber is equipped with a high-O_2 treatment gas, it may be administered at 165 fsw and shallower, not to exceed 2.8 ata O_2 in accordance with paragraph 21-5-7. Treatment gas is administered for 25 minutes interrupted by 5 minutes of air. Treatment gas is breathed during ascent from the treatment depth to 60 fsw.

6. Deeper than 60 feet, if treatment gas must be interrupted because of CNS oxygen toxicity, allow 15 minutes after the reaction has entirely subsided before resuming treatment gas. The time off treatment gas is counted as part of the time at treatment depth. If at 60 feet or shallower and oxygen breathing must be interrupted because of CNS oxygen toxicity, allow 15 minutes after the reaction has entirely subsided and resume schedule at point of interruption. (see paragraph 21-5.5.6.1.1).

7. Table 6A can be lengthened up to 2 additional 25-minute periods at 60 feet (20 minutes on oxygen and 5 minutes on air), or up to 2 additional 75-minute periods at 30 feet (60-minutes on oxygen and 15 minutes on air), or both.

8. Tenders breathes 100 percent O_2 during the last 60 minutes at 30 fsw and during ascent to the surface for an unmodified table or where there has been only a single extension at 30 or 60 fsw. If there has been more than one extension, the O_2 breathing at 30 fsw is increased to 90 minutes. If the tender had a hyperbaric exposure within the past 12 hours, an additional 60 minute O_2 breathing period is taken at 30 fsw.

9. If significant improvement is not obtained within 30 minutes at 165 feet, consult with a Diving Medical Officer before switching to Treatment Table 4.

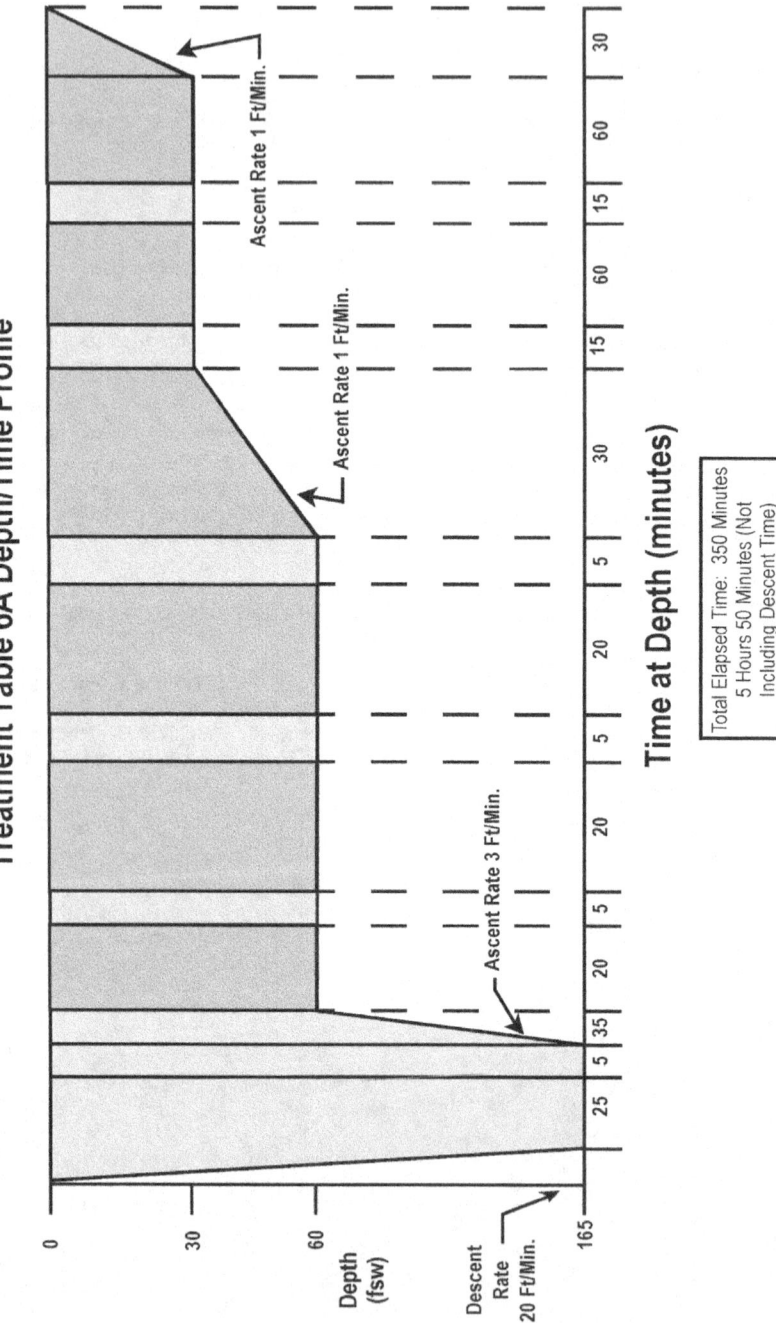

Treatment Table 7

1. Table begins upon arrival at 60 feet. Arrival at 60 feet is accomplished by initial treatment on Table 6, 6A or 4. If initial treatment has progressed to a depth shallower than 60 feet, compress to 60 feet at 20 ft/min to begin Table 7.

2. Maximum duration at 60 feet is unlimited. Remain at 60 feet a minimum of 12 hours unless overriding circumstances dictate earlier decompression.

3. Patient begins oxygen breathing periods at 60 feet. Tender need breathe only chamber atmosphere throughout. If oxygen breathing is interrupted, no lengthening of the table is required.

4. Minimum chamber O_2 concentration is 19 percent. Maximum CO_2 concentration is 1.5 percent SEV (11.4mmHg). Maximum chamber internal temperature is 85°F (see paragraph 21-5.6.5).

5. Decompression starts with a 2-foot upward excursion from 60 to 58 feet. Decompress with stops every 2 feet for times shown in profile below. Ascent time between stops is approximately 30 seconds. Stop time begins with ascent from deeper to next shallower step. Stop at 4 feet for 4 hours and then ascend to the surface at 1 ft/min.

6. Ensure chamber life-support requirements can be met before committing to a Treatment Table 7.

7. A Diving Medical Officer shall be consulted before committing to this treatment table.

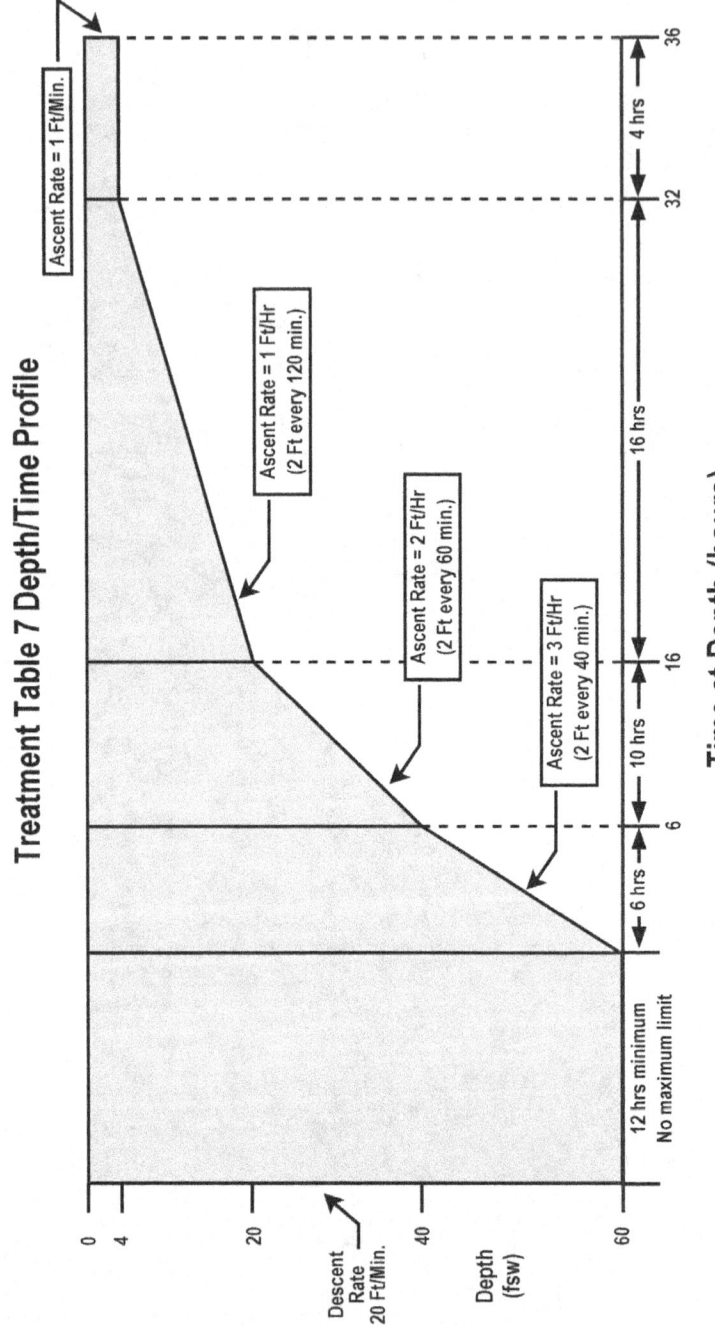

Chapter 21: Aerospace Medicine

Aerospace Medicine: Hypoxia
LTC Brian Campbell, MC, USA

Introduction: Oxygen makes up approximately 20% of the air we breathe. As ambient atmospheric pressure decreases with increased altitude, partial pressures of oxygen also drop. This results in decreased oxygenation of the blood and tissue hypoxia. Healthy individuals can easily tolerate ambient pressures of 10,000 feet above mean sea level (MSL) for prolonged periods without hypoxic effects. However, everyone will eventually become impaired above 10,000 feet. Rapid decompression at extreme altitudes may result in death. See also related injuries such as acute mountain sickness, high altitude pulmonary edema, high altitude cerebral edema hypoxia associated with diving. Risk Factor: Underlying pulmonary disease (acute or chronic).

Subjective: Symptoms
Variable from person to person, but consistent in an individual across exposures; impaired judgement, thinking, and vision (particularly color vision); paresthesias, especially of the face and hands, are common.
Focused History: *When did symptoms start?* (acute onset with change in altitude is typical) *Have you had a fever or other signs of illness prior to this flight?* (pulmonary disease is risk factor, or may suggest alternate diagnosis) *Have you been diving or breathing compressed air in the last 48 hours?* (may suggest decompression sickness [DCS]) *Have you been flying? If so, do you know the cabin altitude of the flight?* (>10,000 feet is typical) *Are you taking any drugs or medications of any kind?* (consider substance abuse, drug reaction or allergic reaction)

Objective: Signs
Using Basic Tools: Lack of fine motor control; personality changes (e.g., withdrawn, violent, or highly flamboyant activity; inattentiveness or absence periods; respiratory rate and pulse may increase; seizure activity is possible.

Assessment: Resolution of signs and symptoms with return to lower altitudes or treatment with oxygen confirms the diagnosis.
Differential Diagnosis:
DCS - may rule out if not > 18,000 feet MSL and negative history of breathing compressed air within last 48 hours
Substance abuse - may rule out if no history or other evidence/suspicion of taking drugs, medications, "nutritional supplements"
Allergic or idiopathic reaction to prescription or OTC medications (including "nutritional supplements") - may rule out if not taking any of these substances)
Atypical seizure activity - must consider if positive history of seizure disorder or history of head trauma involving loss of consciousness within the last 10 years

Plan:
Treatment
Primary: Supplemental oxygen under pressure via fitted mask.
Alternative: Increase oxygen concentration of inspired air by any other means (different masks, etc.).
Primitive: Increase ambient pressure by decreasing altitude or adjusting cabin altitude

Patient Education
General: Effects of hypoxia are experienced only while exposed to a hypoxic environment. No residual effects should present. Patient will likely be affected in similar manner by exposure to similar altitudes.
Prevention: Early recognition of the symptoms of hypoxia allows early intervention and avoids performance decrement. Use pre-mission altitude chamber testing to determine individual team member's response to the hypobaric environment.

Follow-up Actions
Return evaluation: If any sequelae present, must consider neurologic DCS. Evaluation by neurologist is indicated at earliest opportunity.
Consultation Criteria: Consultation with Aerospace Medicine specialist, Diving Medicine specialist, Flight Surgeon, or Aeromedical PA is helpful but not necessary in uncomplicated cases.

NOTES: Hypoxia is a particular danger for HALO and HAHO missions. There is tremendous opportunity for impairment of judgement and thinking, which can quickly lead to death or serious injury. Hypoxia may also manifest less acutely in association with living and working at high altitude. An acute hypoxic event without complications or sequelae is not grounds for restriction of special duty status.

Aerospace Medicine: Barodontalgia
LTC Brian Campbell, MC, USA

Introduction: Dental decay can produce small pockets of gas in or around teeth. When the ambient pressure changes, the pressure differential in the trapped pocket of gas cannot be equalized and severe pain results. This rare condition almost always occurs on ascent. Since barodontalgia may develop during diving or flight operations, see also Dive Medicine chapter.
Risk Factors: Dental pathology (e.g., active caries, pulpitis, periapical abscess), recently placed amalgam restorations or crowns.

Subjective: Symptoms
Acute onset of sharp, stabbing pain in a single tooth on ascent. This pain can be quite severe and will increase in severity with continued ascent. Symptoms abate with return to sea level.
Focused History: *When did symptoms start?* (acute onset associated with change in altitude [i.e. pressure] is typical) *Did you have a fever or other signs of illness prior to this flight?* (may suggest barodontalgia or other alternate diagnosis) *Have you had any dental problems or dental treatments recently?* (typical exposure history; body slowly absorbs free gas, so older treatment less problematic)

Objective: Signs
Using Basic Tools: Obvious dental lesions help make the diagnosis. Percussion of the involved tooth should exacerbate symptoms.

Assessment: History makes the diagnosis, especially in a patient with risk factors.
Differential Diagnosis
Barosinusitis - may be indicated with maxillary tooth symptoms-look for signs and symptoms of maxillary sinus involvement
Acute pulpitis or periapical abscess of the tooth - unrelated to pressure changes
Acute infectious sinusitis - may be indicated if systemic symptoms or purulent nasal discharge is present-see ENT chapter

Plan:
Treatment
Primary: Increase ambient pressure by descent to sea level if possible.
Alternative: Analgesics as required. Consider fracture of tooth or removal/replacement of restoration if

symptoms are unrelenting in an operational setting that precludes descent.

Patient Education
General: Whenever possible do not fly while suffering from acute dental disease.
Prevention: Good dental hygiene and regular check-ups with dentist.

Follow-up Actions
Return evaluation: Dental exam for diagnosis and treatment of underlying pathology.
Consultation Criteria: For severe cases, when symptoms not relieved upon descent, or when underlying pathology is the cause.

NOTES: Barodontalgia should temporarily restrict personnel from special duty involving flight operations until cleared by a Flight Surgeon.

Aerospace Medicine: Barosinusitis
LTC Brian Campbell, MC, USA

Introduction: The paranasal sinuses are bony structures with a fixed volume, containing air, mucus, and water vapor. The pressure within the sinuses is normally equal to ambient air pressure and equalizes as ambient pressure changes. When the pressure cannot be equalized due to malformations or swelling of the sinus outflow tract, pain results. This is more likely to occur when descent from altitude results in a sinus "squeeze." Upper respiratory tract infections (URIs), smoking, and untreated seasonal allergic conditions greatly increase the risk of developing barosinusitis during flight. Barosinusitis is also seen with diving. See Dive Medicine chapter.

Subjective: Symptoms
Acute onset of sharp, stabbing sinus pain on descent is classic. This pain can be quite severe and will increase in severity with continued descent. Symptoms will abate with decreasing ambient pressure resulting from return to altitude. Patients will bleed into the sinus in severe cases, thereby relieving the pressure differential and significantly reducing the pain. Symptoms of sinus congestion and facial pressure akin to that with a URI will then result.
Focused History: *When did symptoms start?* (Acute onset associated with change in altitude is a key factor in differentiating from infectious etiology.) *Have you had a fever or other signs of illness prior to this flight?* (History of recent illness can illuminate risk factors and/or provide clues to make differential diagnosis.) *Have you had any dental problems or dental treatments recently?* (Affirmative answer may lead toward barodontalgia as diagnosis, although maxillary tooth pain may occur with any sinusitis.)

Objective: Signs
Using Basic Tools: Patient is generally in obvious distress. Some redness and swelling of the face in the affected area may be noted though not to the degree one would expect with purulent sinusitis. Increased lacrimation is also possible. Bloody/mucoid nasal discharge may be seen from affected sinus in severe cases. Transillumination of the sinus should reveal this.
Using Advanced Tools: X-ray: Sinus series X-rays (looking for air/fluid levels) are helpful if available.

Assessment:
Differential Diagnosis
Barodontalgia - caused by air trapped under a crown or amalgam or within a tooth cavity; may be suggested by maxillary sinus symptoms present on ascent
Pulpitis or periapical abscess of the tooth - if maxillary sinus symptoms associated with a diseased tooth are present on ascent

Acute infectious sinusitis - may be indicated if systemic symptoms and/or purulent discharge is present

Plan
Treatment
Primary: Increase cabin altitude to decrease pressure differential in sinus and alleviate pain. Spray nasopharynx with decongestant spray (e.g., **Afrin**). Slowly descend while frequently using modified Valsalva maneuver (i.e., pinch nose and exhale against closed nostrils) as needed to equalize pressure in sinuses.
Alternative: Use of po decongestants (e.g., pseudoephedrine) prior to and during flight may decrease the risk of barosinusitis in a patient who must fly with a URI.
Primitive: Modified Valsalva maneuver.

Patient Education
General: It is easier to prevent a sinus "squeeze" than to treat one. Use modified Valsalva maneuver to equalize pressure in sinuses frequently during descent; do not wait until pain develops to attempt to equalize.
Prevention: Whenever possible, do not fly while suffering from a URI.

Follow-up Actions
Return evaluation: Patient should be followed and nasal decongestants (po or topical) should be used for several days following an episode of barosinusitis. If sinus bleeding occurs, treat with **amoxicillin** (250 mg po tid) and manage like acute sinusitis.
Consultation Criteria: No consultation is necessary in uncomplicated cases. Consult dentist if dental condition is suspected; consult ENT specialist, if possible, for sinus bleeds.

NOTES: An uncomplicated sinus squeeze may not require restriction of special duty status. An underlying URI or other sinus problem or barosinusitis with secondary sinus bleed, however, should temporarily restrict affected personnel from special duty involving flying or diving operations until cleared by a Flight Surgeon or Diving Medical Officer.

Aerospace Medicine: Barotitis
LTC Brian Campbell, MC, USA

Introduction: The middle ear is a bony structure with a fixed volume containing air and water vapor. The pressure within the middle ear is normally equal to the ambient pressure and equalizes through the Eustachian tube as ambient pressure changes. Pain results when the pressure cannot be equalized due to malformations or swelling of the Eustachian tube. This is barotitis media. Barotitis externa can result if a foreign body (e.g., ear plug, cerumen plug) blocks the external auditory canal (EAC). Both conditions are more likely to occur with descent, but a reverse "squeeze" can occur on ascent with similar symptoms. As barotitis may develop during both diving or flight operations, see Dive Medicine chapter also. Upper respiratory tract infections (URIs) and untreated seasonal allergic conditions greatly increase the risk of developing an ear "block" during flight. Smoking may also be a contributory factor.

Subjective: Symptoms
Acute onset of sharp, stabbing pain in the ear on descent is classic. This pain can be quite severe and will increase in severity with continued descent. Symptoms will abate with return to altitude. In severe cases, patients will rupture the tympanic membrane (TM), thereby relieving the pressure differential and significantly reducing the pain. Hearing loss will then result. Vertigo may be associated with barotitis media or with TM rupture from either condition.
Focused History: *When did symptoms start?* (Acute onset associated with change in altitude is a key factor in differentiating from infectious etiology.) *Have you had a fever or other signs of illness prior to this flight?* (History of recent illness can illuminate risk factors and/or provide clues to make differential diagnosis.)

Objective: Signs
Using Basic Tools: Patient is generally in obvious distress. Otoscopic exam will reveal blocked EAC in barotitis externa and retracted TM in barotitis media. Blood may be visualized behind the TM in severe cases of barotitis media. Symptoms of URI may also be present given the association of barotitis with URI and seasonal allergies.

Assessment:
Differential Diagnosis
Acute otitis media of infectious etiology - bulging, red TM with poor temporal correlation of symptoms to altitude changes.
Acute trauma to the EAC or TM - e.g., insect or arthropod bite, usually visualized on otoscopic exam.
Other barotrauma to the ear (inner or middle), including cochlear tear and/or round window rupture - may be present if blood behind TM, vertigo, or no relief of symptoms upon return to original altitude.

Plan:
Treatment
Primary: Change cabin altitude to decrease pressure differential and alleviate pain (increase altitude if symptoms presented on descent, or vice versa). Spray nasopharynx with decongestant spray (e.g., **Afrin**) to decrease swelling of Eustachian tube in barotitis media. Remove EAC obstruction in barotitis externa. Slowly descend while using modified Valsalva maneuver (i.e., pinch nose and exhale against closed nostrils) to equalize pressure in sinuses. Use a Politzer bag, a device similar in appearance to an Ambu bag (and often carried on medical evacuation aircraft), to force air into the nasopharynx while the patient swallows. This will probably be more effective than the modified Valsalva maneuver. Myringotomy may be necessary in extreme cases of barotitis media.
Alternative: PO decongestants (e.g., pseudoephedrine) prior to and during flight decrease risk of barotitis in patients with URI.
Primitive: Modified Valsalva maneuver (as above)

Patient Education
General: It is easier to prevent an ear "block" than to treat one. Use modified Valsalva maneuver to equalize pressure in middle ear frequently during descent; do not wait until pain develops to attempt to equalize.
Prevention: Whenever possible, do not fly while suffering from an upper respiratory infection.

Follow-up Actions
Return evaluation: Patient should be followed and decongestants (po or nasal) should be used for several days following an episode of barotitis. If large hemorrhagic bullae are present in the EAC in barotitis externa, evacuation of blood with a syringe and sterile needle should be performed; small hemorrhages do not require treatment. Manage TM rupture like that from any other cause; empiric antibiotic treatment is not recommended.
Consultation Criteria: Consult ENT specialist for TM rupture, suspected middle/inner ear barotrauma, or after myringotomy.

*NOTES: An uncomplicated ear squeeze may not require restriction of special duty status. An underlying URI or other ear problem or barotitis with secondary bleed or inner/middle ear trauma, however, should temporarily restrict affected personnel from special duty involving flying or diving operations until cleared by a Flight Surgeon or Diving Medical Officer.

Aerospace Medicine: Decompression Sickness
LTC Brian Campbell, MC, USA

Introduction: Nitrogen makes up approximately 80% of the atmosphere. An inert gas, it saturates all tissues of the body. Nitrogen will expand because of decreased atmospheric pressure at altitudes above 18,000 feet above mean sea level (MSL), forming bubbles in body tissues. The tissue affected by these bubbles determines the severity of decompression syndrome (DCS). As DCS may be associated with ascent from diving operations or with exposure to low ambient pressures during flight operations, see Diving Medicine section also.

Subjective: Symptoms
"Bends" – musculoskeletal (primarily joint) pain
"Chokes" – shortness of breath, chest pain, non-productive cough
"Creeps" or "itches" – pruritus or feeling of insects crawling on the skin
Neurologic DCS – Any neurologic symptom is possible, including paralysis, paresis, paresthesia, loss of consciousness, headache, fatigue, seizure, and personality changes.
Focused History: *Do you have* [each of the symptoms listed above]? (Answers provide insight to which body tissues are affected, thus how severe DCS may be.) *Are you taking any drugs, medications or "nutritional supplements" of any kind?* (Affirmative answer opens the possibility of allergic or idiopathic drug reaction.) *Have you flown over 18,000 feet, or been diving in the past 24 hours?* (typical exposure)

Objective: Signs
Using Basic Tools: Conduct a normal physical exam since symptoms vary according to the area of the body affected. Diffuse mottling of skin or central neurologic signs may indicate arterial gas embolism and are ominous. Involvement of more than one joint is indicative of more serious DCS. Examiner must perform a complete neurologic examination, to include mental status exam (see Appendix).

Assessment:
Differential Diagnosis
Acute hypoxia - indicated by immediate, complete resolution of symptoms with supplemental oxygen
Allergic or idiopathic reaction to prescription or OTC medications, including "nutritional supplements" - may rule this out if not taking any of these substances
Seizure disorder - may be indicated by history of seizure disorder or head trauma involving loss of consciousness within last 10 years
Arterial gas embolism (see Dive Medicine chapter) - may be indicated in Aerospace Medicine as a consequence of catastrophic decompression of aircraft with loss of consciousness or sudden death less than 10 minutes following event

Plan:
Treatment
Primary: 100% oxygen until hyperbaric oxygen recompression can be accomplished
Alternative: Gamow bag or Hyperlite transport chamber
Primitive: 100% oxygen
Empiric: Return to sea level

Patient Education
General: DCS is a very serious condition. The threat of death or permanent neurologic injury is out of proportion to the usually mild symptoms. Pre-existing acute musculoskeletal injuries, heavy intercurrent exercise and alcohol use increase the risk of developing DCS.
Prevention and Hygiene: Allow at least 24 hours between diving operations and flying or other high altitude operations. Avoid strenuous physical activity before high altitude operations.

Follow-up Actions
Return evaluation: Consultation with neurologist is indicated for any suspected neurologic DCS case.
Consultation Criteria: Consultation with Diving Medical Officer (DMO) is recommended. Aerospace medical or Flight Surgeon consult may be helpful if DMO is not available.

NOTES: DCS is a danger in any flying operation that requires ascent to 18,000 feet MSL or above. Rapid return to lower altitudes in HALO missions mitigates the threat to some degree, but compression in a hyperbaric chamber is required in any DCS case. Neurologic DCS is life threatening and should be treated as a true medical emergency. Consider any episode of DCS as grounds for temporary restriction of special duty status until cleared by competent medical authority.

Chapter 22: High Altitude Illnesses
High Altitude Illnesses: Acute Mountain Sickness
COL Paul Rock, MC, USA & LTC Brian Campbell, MC, USA

Introduction: Acute mountain sickness (AMS) is a short-lived (days to a week) illness that occurs in people from low altitude (less than 5000 ft) who travel rapidly to higher areas (usually more than 8000 - 9000 ft) and remain there for more than several hours. It is caused by the decreased amount of oxygen available at high altitude (see Aerospace Medicine: Hypoxia). Symptoms of AMS usually go away as a person's body adapts to lower oxygen levels over a week to 10 days (altitude acclimatization). It is impossible to predict who will be more susceptible to AMS. Although not life-threatening, AMS can degrade physical and mental performance. Additionally, AMS can progress to more serious altitude illness such as high altitude cerebral edema (HACE) and high altitude pulmonary edema (HAPE).

Subjective: Symptoms
Similar to a alcoholic "hangover"- headache (often severe), nausea (with or without vomiting), fatigue, decreased appetite, disturbed sleep. Symptoms begin within 3 to 24 hours after ascending to a higher elevation and are most severe in the first 24 to 48 hours.
Focused History: *When did symptoms begin?* (typically start within 24 hours **after** traveling to higher altitude) *Did you have these symptoms or an illness* **before** *going to higher altitude?* (affirmative answer suggests viral illness or other preexisting condition) *Have you taken any medications, drugs or alcohol?* (Intoxication with these substances can cause symptoms similar to AMS.) *Have you been in a tent, cave or vehicle with a stove or motor running?* (Carbon monoxide poisoning can cause similar symptoms.) *Do you have a cough or difficulty breathing?* (suggests HAPE) *How is your coordination? Have you been stumbling or falling?* (suggests HACE)

Objective: Signs
Using Basic Tools: Patient appears 'sick'; decreased urination; poor balance (truncal ataxia) when carrying backpack; mild swelling (edema) in the hands, feet and/or face ('puffy' around the eyes.)
Using Advanced Tools: Pulse oxygen levels will be lower with AMS or other altitude illnesses.

Assessment: Diagnosis is made on basis of history.
Differential Diagnosis
Early HACE - significant ataxia (cannot do 'heel-to-toe walk'); swelling of optic nerve (papilledema).
Coexistent HAPE - cough; rales; frothy, pink, or blood-tinged sputum.
Other causes of headache (migraine, cluster, or tension headache; viral syndrome; meningitis; head trauma; etc.) Symptoms before ascending to altitude - stiff neck, fever or increased white cell count; history of head trauma.
Intoxication - history of ingesting medications, recreational drugs, alcohol. (see Toxicology chapter)
Carbon monoxide poisoning - history of exposure to combustion fumes, occasional cherry-red skin color. (see Toxicology chapter)
Hypothermia - lack of headache and nausea, decreased body temperature. (see Cold Illnesses: Hyporthermia)
Hyperthermia and/or dehydration - history of decreased fluid intake, elevated body temperature or tenting of skin. (see Heat Related Illnesses chapter)

Plan:
Treatment
Primary: Stop ascent! Descend to lower altitude until symptoms resolve. Once symptoms resolve, continue ascent slowly (See Preventive below). Treat with **acetazolamide*** 125-150 mg po tid/qid. **Aspirin, acetaminophen, ibuprofen, indomethacin,** or **naproxen** in usual doses can be used to treat headache pain. Nausea and vomiting can be treated with **prochlorperazine** 10 mg po every six hours or 25 mg by rectal suppository every 12 hours.
Alternative: Stay at higher altitude during day, but sleep at lower altitude (See Preventive below). **Dexamethasone*** 4 mg po 4 qid (Save for people allergic to **acetazolamide** or other sulfa drugs).
Primitive: Descent is the best treatment for all altitude illnesses (e.g., AMS, HACE, HAPE). Native people of the Andes Mountains in South America chew coca leaves or drink coca tea to prevent and treat AMS.

* These medications can be stopped 1-2 days after symptoms resolve. AMS symptoms may recur after stopping dexamethasone, but do not recur after stopping acetazolamide.

Patient Education
General: AMS is caused by ascending too rapidly, before body has chance to adjust to altitude. Symptoms will improve over several days as body adjusts to altitude.
Activity: Avoid strenuous activity until acclimatized to altitude.
Diet: Stay hydrated. High carbohydrate (starches and sugars) diet to decreas symptoms.
Medications: Acetazolamide causes tingling sensations in lips, nose and fingertips and makes carbonated beverages to taste funny. Do not stop taking **acetazolamide** because of these side effects.
Prevention and Hygiene: Ascend slowly (1000 - 2000 feet/day above 8000 ft) with a rest day (no ascent) every 3-4 days. Sleep at least 1000-2000 ft lower than working altitude. **Acetazolamide** 125-250 mg po tid/qid beginning 12-24 hours before starting ascent and continuing for 48 hours after reach destination altitude.
No Improvement/Deterioration: Seek medical aid if headache worsens, develop difficulty with walking, coordination, cough, cough up frothy, pink or bloody sputum, 'gurgling' sounds in chest when breathing.

Follow-up Action
Reevaluation: No follow up is necessary unless symptoms return.
Evacuation/Consultation Criteria: Evacuate to lower altitude as discussed above.
There is no way to predict which person is more susceptible to AMS. Consider medical profile limiting deployment to altitude for those with recurrent or prolonged AMS.

High Altitude Illnesses: High Altitude Cerebral Edema
COL Paul Rock, MC, USA & LTC Brian Campbell, MC, USA

Introduction: High altitude cerebral edema (HACE) is a potentially fatal accumulation of fluid (edema) in brain tissue which sometimes occurs in people from low altitude (less than 5000 feet) who ascend rapidly to high altitude (greater than 8,000 feet; but rare below 11,500 feet) and remain there for several days. It is caused by the decreased amount of oxygen available to the body in the low pressure atmosphere at high altitude (see Aerospace Medicine: Hypoxia). HACE is a severe form of acute mountain sickness (AMS) (see Acute Mountain Sickness section) and most often occurs in people who have AMS symptoms and continue to ascend. Although rare (usually less than 1-2% of persons going to high altitude), if left untreated, HACE can progress to coma and death in 12 hours or less. High altitude pulmonary edema (HAPE), which can also be rapidly fatal, often occurs with HACE. (see High Altitude Pulmonary Edema in following section.)

Subjective: Symptoms
Early: Symptoms of AMS (severe headache, nausea with vomiting, decreased appetite and fatigue); later: progressive weakness, fatigue and clumsiness; confusion and disorientation; vivid hallucinations (visual and/or auditory).

Focused History: *When did the symptoms begin?* (typically begin 3-10 days after ascent; later than AMS) *Have you had symptoms of AMS?* (risk factor; worsening AMS symptoms after 48-72 hours are likely due to HACE.) *How is your coordination? Have you been stumbling or falling?* (Ataxia and clumsiness are typical.) *Are you seeing or hearing unexpected or unusual things (having hallucinations)? Do you have a cough or difficulty breathing?* (typical of HAPE; accompanies 1/3 of HACE cases) *Have you taken any medications, drugs or alcohol?* (Intoxication could cause similar symptoms.) *Have you been in a tent, cave or vehicle with a stove or motor running?* (Carbon monoxide poisoning could cause similar symptoms.)

Objective: Signs
Using Basic Tools: Early: Behavioral changes (agitated or quiet and withdrawn); later: disorientation, confusion, ataxia (cannot do 'heel-to-toe-walk'), incoordination and often hallucinations (visual and auditory); abnormal deep tendon reflexes, decreased consciousness, coma and death; may have rales, cough, and frothy, pink or bloody sputum (concomitant HAPE).
Using Advanced Tools: Ophthalmoscope: Retinal hemorrhages and swelling of optic nerve in the back of the eye (papilledema). Pulse oxygen levels will be lower with HACE or other altitude illnesses.

Assessment:
Differential Diagnosis
Other causes of headache - migraine, cluster, or tension headache; infection. History of headache at low altitude. Check for stiff neck, fever or increased white cell count (see Neurology: Meningitis).
Head trauma
Intoxication - history of ingesting medications, recreational drugs, alcohol (see Toxicology: Poisoning).
Carbon monoxide poisoning - history of exposure to combustion fumes (see Toxicology: Poisoning).
Hypothermia - lack of headache and nausea. Decreased body temperature (see Cold Illnesses: Hypothermia.)
Hyperthermia and/or dehydration - history of decreased fluid intake, elevated body temperature (see Heat-Related Illnesses).
AMS - HACE patients have papilledema and/or ataxia, and may have deteriorating mental status.

Plan:
Treatment
Primary:
1. Evacuate to lower altitude immediately (1000 to 2000 feet change may be lifesaving).
2. Oxygen: 6 L/minute or more by oxygen mask. Insert endotracheal tube if comatose.
3. **Dexamethasone** 8mg initially, then 4mg every six hours po or IV.
4. Evacuate patient to advanced medical care.

Alternative: Oxygen, **dexamethasone** and portable hyperbaric ('pressure') chamber*. (see Procedures: Portable Pressure Chamber)
Primitive: Descent is the best treatment for all altitude illnesses (e.g., AMS, HACE, HAPE.)

Patient Education
General: HACE is caused by rapid ascent before the body can adjust to altitude. Symptoms will worsen with further ascent.
Activity: Bed rest or very limited activity - can descend under own power in emergency *if* accompanied
Medications: Dexamethasone can cause psychosis, puffy face, and increase appetite. Taper the dose after taking for >3-4 days.
Prevention and Hygiene: Ascend slowly (less than 1000 ft/day) with rest day every 3-4 days. Do not continue ascending with symptoms of altitude illness. Sleep at as low an altitude as possible (1000-2000 ft lower than working altitude) until body adjusts to altitude (7-10 days). **Acetazolamide** 125-250 mg po tid/qid beginning 12-24 hours before ascending may help prevent HACE. **NOTE:** Individuals with allergic reactions to sulfa-containing substances should not be given **acetazolamide**.
No Improvement/Deterioration: HACE is rapidly fatal if not treated. Seek medical attention if have headache and difficulty with balance or have hallucinations. HACE is often accompanied by HAPE. Seek

medical attention for cough or frothy, pink or bloody sputum or gurgling sounds in your chest or throat when breathing.

Follow-up Actions
Return evaluation: Evaluate individuals who survive for neurologic deficits that might affect their performance of military duties. Individuals who have had one episode of HACE are at increased risk of future episodes and should be referred for possible medical profile to restrict exposure to altitudes greater than 8,000 feet.
Evacuation/Consultation Criteria: Evacuate all patients with HACE to higher echelon of medical care, preferably to hospital facility. CT scan or MRI imaging of the brain may show cerebral edema. If not evacuated rapidly, even patients who survive may have prolonged or permanent neurologic damage.

*NOTES: Portable hyperbaric chambers (e.g., Gamow bag, CERTEC bag, Hyperlite chamber, Portable Altitude Chamber [PAC]) are not normally available in the military medical supply inventory. They are available in the civilian sector in the USA and many European countries. These lightweight, highly portable cloth chambers are extremely useful in treating altitude illnesses (including AMS). When deploying rapidly to high altitude terrain, consider procuring such a chamber. Given that the incidence of altitude illness diminishes greatly after acclimatization to altitude (7-10 days), the portable chamber could be stored (or discarded) after that time.

High Altitude Illnesses: High Altitude Pulmonary Edema
COL Paul Rock, MC, USA & LTC Brian Campbell, MC, USA

Introduction: High Altitude Pulmonary Edema (HAPE) is a potentially fatal accumulation of fluid (edema) in the lungs that occurs in people from low altitude (less than 5000 ft) who ascend rapidly to high altitudes (usually greater than 9000 ft) and remain there for several days. It is caused by the decreased amount of oxygen available in the low-pressure atmosphere at high altitude (see Aerospace Medicine: Hypoxia). It often begins after the first or second night spent at high altitude and is most common during first week. Young men who do heavy physical exertion upon arrival at high altitude are very susceptible. Although not common (usually less than 10% of persons going to altitudes above 12,000 ft), once HAPE develops, it can be rapidly fatal (6-12 hours) if not treated. Half of individuals with HAPE will also have symptoms of acute mountain sickness (AMS) (see Acute Mountain Sickness section), and some may develop high altitude cerebral edema (HACE) (see High Altitude Cerebral Edema section), which can also be fatal.

Subjective: Symptoms
Early: shortness of breath, dry cough, dyspnea on exertion; later: dyspnea at rest, symptoms of AMS (headache, nausea and vomiting, decreased appetite and fatigue), clear and watery sputum; still later: frothy, blood-streaked or pink sputum; feel or hear 'gurgling' in chest with breathing.

Focused History: *Do you have difficulty breathing or a cough? Are you coughing up frothy, pink or bloody sputum?* (typical symptoms) *When did symptoms begin?* (typically after exercise or after sleeping during first week at altitude) *Did you have these symptoms or any illness before going to altitude?* (rule out preexisting condition) *How is your coordination? Have you been stumbling or falling? Are you seeing or hearing unexpected or unusual things (hallucinations)?* (suggest coexisting HACE) *Have you had fever and chills? Have you coughed up thick, greenish or yellow-colored sputum?* (suggest bronchopneumonia) *Do you have any pain or swelling in your legs? Does your chest hurt when you breathe?* (typical of possible deep venous blood clot in legs with subsequent blood clot in lungs) *Have you ever had asthma or hay fever in the past?* (rule out asthma)

Objective: Signs
Using Basic Tools: Early signs: Tachypnea and tachycardia during physical activity (compared to companions at same altitude), dry cough, crackling sounds (rales) in lungs (mid-lung area); progressive signs: tachypnea and tachycardia at rest (compared to unaffected companions); 'gurgling' breath sounds; excessive (compared to companions at same altitude) bluish color of lips, fingernail bed, tip of nose and ears (cyanosis); cough productive of frothy and/or pink, blood streaked sputum; low-grade fever. Late signs: coma, respiratory

failure, death.
Using Advanced Tools: Pulse oximeter: Oxygen levels will be lower with HAPE or other altitude illnesses. CXR: Fluffy infiltrates in mid-lung fields or spreading throughout lungs.

Assessment:
Diagnosis is made on basis of clinical presentation and exposure history.
Differential Diagnosis
AMS or HACE in addition to HAPE - see Acute Mountain Sickness and High Altitude Cerebral Edema sections)
Pneumonia - fever greater than 101°F; infected (purulent) sputum; symptoms *before* traveling to high altitude (see Respiratory: Pneumonia)
Pulmonary embolus - chest pain with breathing, blood clot in leg veins (pain and swelling) (see Respiratory: Pulmonary Embolus)
High-altitude cough - chronic, dry cough can occur at very high altitude (usually greater than 15,000 ft) due to irritation of throat by breathing cold, dry air. Not associated with rales, sputum production, or other signs or symptoms of HAPE.
Asthma - history of asthma; breathing cold air or allergen exposure; wheezing (see Respiratory: Asthma)

Plan:
Treatment
Primary:
1. Immediately evacuate to lower altitude (1000 to 2000 feet lower may be lifesaving) by litter (walking will worsen HAPE.).
2. Oxygen : 6 liters/minute or greater by mask.
3. **Nifedipine:** break 20 mg capsule and hold under tongue, then 20 mg sustained release tablet every six hours swallowed.
4. If HACE also present, treat it (see High Altitude Cerebral Edema section).
5. If comatose, endotracheal tube intubation to protect airway. (see Procedure: Intubation a Patient)
NOTE: **Nifedipine** is not on SOF drug list, but is available through military medical supply channels and civilian sources.
Alternative:
1. If evacuation to lower altitude not possible, bed rest, high-flow oxygen by mask, **nifedipine**, and treatment in portable hyperbaric ('pressure') chamber*. (see Procedure: Portable Pressure Chamber)
2. If nifedipine not available, use **acetazolamide** 250 mg (do not give if allergic to sulfa) po q 6-8 hours.
3. End-positive-airway-pressure (EPAP) mask may be helpful, if available.
Primitive: Descent is the best treatment for all altitude illnesses (e.g., AMS, HACE, HAPE). 'Pursed-lip' breathing may help increase oxygenation. Patient in prone, slightly head-down position for brief period (10-20 min) may help drain lung fluid through the mouth temporarily (if patient can tolerate that position).

Patient Education
General: HAPE is caused by rapid ascent before the body has a chance to adjust to high altitude. More likely to occur if exercise vigorously during first 3 to 5 days after ascent, or keep ascending while symptomatic.
Activity: Bed rest (physical activity makes HAPE worse).
Medications: Nifedipine can lower the blood pressure and cause dizziness when sitting up or standing rapidly from a prone position. It can cause swelling (edema) of the hands, lower legs and feet.
Prevention and Hygiene: Ascend slowly (less than 1000 ft/day above 8000) with rest day (no ascent) every 3-4 days. Do not continue ascent with symptoms of altitude illness or difficulty breathing. Sleep at as low an altitude as possible (1000-2000 ft lower than working altitude) until body adjusts (7-10 days). Avoid vigorous physical activity for first 3 to 5 days after ascent. If you have had HAPE during previous trips to high mountains, you should take **nifedipine** when going to altitude in the future.
No Improvement/Deterioration: HAPE can be rapidly fatal if not treated. Seek medical attention for more difficulty than companions breathing during exercise or at rest. Seek medical attention for a cough, or cough up blood or pink, or frothy sputum. Seek medical attention if hear or feel 'gurgling' in chest when breathing.

HAPE is often accompanied by HACE, so seek medical attention if have severe headache, difficulty keeping your balance or have hallucinations.

Follow-up Actions
Return evaluation: If evacuated promptly, patient may recover rapidly (hours to days) and completely. Individuals who have had one episode of HAPE are at increased risk of recurrence.
Evacuation/Consultation Criteria: Evacuate all patients with more than mild HAPE to a hospital facility. They should be evaluated for possible medical profile to restrict exposure to altitudes above 8000 ft.

*NOTES: Portable hyperbaric chambers (e.g., Gamow bag, CERTEC bag, Hyperlite chamber, Portable Altitude Chamber [PAC]) are not normally available in the military medical supply inventory. They are available in the civilian sector in the USA and many European countries. These light weight, highly portable cloth chambers are extremely useful in treating altitude illnesses (including AMS). When deploying rapidly to high altitude terrain, consider procuring such a chamber. Given that the incidence of altitude illness diminishes greatly after acclimatization to altitude (7-10 days), portable chamber could be stored (or discarded) after that time.

Chapter 23: Cold Illnesses and Injuries
Cold Illnesses and Injuries: Freezing Injury (Frostbite)
Murray Hamlet, DVM

Introduction: Frostbite occurs at temperatures below freezing (32°F), most often in exhausted, wet, discouraged soldiers who are poorly dressed and inattentive to prevention. Although there are four basic degree categories of frostbite, it is important to differentiate merely superficial from deep or severe frostbite since they are managed differently (see below). Superficial involves just the surface of the skin but no blisters form. Deep or severe frostbite involves partial or full-thickness skin injury, causes blisters and demarcates over a period of days or weeks. Extremities and exposed skin are at an increased risk of injury.

Subjective: Symptoms
Progression from cool, to cold and uncomfortable, to numb and painless tissue; injury is often concealed by mittens, gloves or boots; slow, stiff movements. Injured tissue becomes extremely painful upon rewarming. Deep injury can produce acidosis, rhabdomyolysis, fever and coagulopathies.

Objective: Signs
Using Basic Tools: Frozen tissue is blanched white or pale yellow, completely ischemic and hard to the touch. The skin is immobile over joints. Upon rewarming the skin becomes red, swollen and may turn gray or deeply red to purple-blue. Large blisters containing either clear or hemorrhagic fluid form in severe frostbite.

Assessment:
Differential Diagnosis - gangrene from other sources (ischemia, burns, and severe infections) may present similarly but the history of cold exposure should clarify the diagnosis.

Plan:
Treatment
Primary:
1. **Do not thaw tissue** if there is any threat of re-freezing during evacuation.
2. Warm superficial frostbite gradually in the axilla, groin or in warm water.
3. Deep frostbite is best managed with:
    a. Moving water immersion (whirlpool) at 104°F or 40°C for thirty minutes. This produces significant pain but affords the best tissue salvage.
    b. Apply a loose, dry dressing for transport.

c. Low molecular weight **Dextran** (1 L of 6% solution IV, followed by 500 ml/day for 5 days) or **heparin** management (15 units/kg IV stat, then a total of 70 units/kg in the first 24 hours) helps to prevent sludging of blood.
d. Vasodilator **phenoxybenzamine hydrochloride** (**Dibenzyline**) 10 mg bid po. Increase dose 10 mg each day to max of 60-80mg qd.
e. Surgically debride dead tissue* (see Procedure: Wound Debridement)
4. Manage pain with appropriate medications: Aspirin, NSAIDs or narcotics if indicated (deep, long standing injury may require morphine sedation).
5. All medications have more benefit if given before thawing.
*NOTES: Debriding frostbitten tissue too early is the most common error in frostbite management. Debriding too early results in retraction, infections, graft failures and removal of viable tissue. Wait to debride mummified tissue for 4 to 8 weeks unless fever and coagulopathies mandate earlier intervention. There is a 2 mm liquefaction line between viable tissue and distal mummifying tissue. Late surgical debridement of indurated tissue should only be to this line.

Patient Education
General: Use personal protective measures to prevent cold injury.
Diet: Eat 5% to 10% more calories. Include more fat in the diet.
Medications: Give NSAIDs with food. Monitor respiratory status if morphine is used.
Wound Care: Manage wounds with warm water baths bid, pat dry and bandage with a loose, dry dressing.

Follow-up Actions
Evacuation/Consultant Criteria: Minor frostbite can be managed quite successfully in the field, but deep frostbite will require evacuation. Consult a general surgeon as needed for management.

Cold Illnesses and Injuries: Hypothermia
Murray Hamlet, DVM

Introduction: Hypothermia is a reduction in body temperature below the normal range. Risk factors for hypothermia include trauma, wind and wetness, physical and mental exhaustion, poor clothing and cold, particularly during rapid changes in weather. Freezing temperatures are not necessary to cause hypothermia. Water immersion can produce extremely rapid cooling. Use core temperature taken via rectum or esophagus. Simply feeling between the shoulder blades to determine if the victim is cold is a field expedient diagnostic test.

Subjective: Symptoms (see Table 6-1, Classification of Level of Hypothermia)
Mild hypothermia (core temperature between 90° and 95°F): Poor coordination, stumbling and shivering.
Moderate hypothermia (81° - 90°F): Muscle and joint stiffness, poor coordination, slurred speech, extreme disorientation and confusion.
Severe hypothermia (below 81°F): Asleep or unconscious. Below 77°F: Spontaneous ventricular fibrillation.

Objective: Signs (Use rectal thermometer to take core temperature. Do not place in stool.)
Using Basic Tools: Mild: Lethargic, diminished fine motor control, shuffling, stumbling gait, shivering; **Moderate:** Lack of shivering, slow to react, disoriented, makes major errors in judgment, loses consciousness; **Severe:** Heart and respiration rates slow, difficult to perceive a pulse, muscles become too stiff to move, cardiac arrhythmias often develop. If unconscious, the victim may have spontaneous ventricular fibrillation or cardiac arrest if handled roughly. Below 77°F spontaneous cardiac arrest is likely. There are often no obvious vital signs and the victim may appear clinically dead, but in fact is not.

Assessment:
Differential Diagnosis - head trauma, hypoglycemia and alcoholic stupor. Differentiate by history of cold exposure.

Plan:
Treatment (see Figure 6-2, Hypothermia: EMS Prehospital Care)
Primary:
1. Handle individuals carefully to prevent ventricular arrhythmias.
2. Ventilate by mouth or by mask, O_2 if available. Ventilation changes the fibrillatory threshold and allows safer transport.
3. If pulse and respiration are absent while in the field, do not initiate chest compression; merely ventilate. If in a vehicle and no cardiac activity is evident, initiate CPR and defibrillate.
4. If more than thirty minutes from definitive care, and warming will compromise the evacuation effort, do not initiate warming procedures. However, any warmth applied inside the wrapped insulation layers is useful.
5. Remove wet clothes and insulate, particularly the torso, head and neck. Apply a vapor barrier over the insulating layers.
6. Apply sweet fluids orally if awake. Otherwise, give 250-500 cc IV bolus of warmed normal saline followed by rapid drip.
7. Patients may appear cold, stiff, blue and may appear to be dead, but this diagnosis cannot be made until they have been rewarmed in a treatment facility.

Patient Education
General: Follow preventive measures, including proper use of cold weather clothing, staying dry, getting out of the wind, and monitoring buddies.
Diet: Eat a high calorie, high fat diet to improve performance in the cold.
Medications: Avoid medications that compromise judgment and shivering, including tranquilizers, alcohol, and some anti-depressants.
Prevention and Hygiene: Stay dry, well fed and rested.
Revaluation: Core temperature may continue to decrease (after drop) after the patient is removed from the cold. This can be life threatening if a two or three degree drop occurs at a core temperature of 88°F or less.

Follow-up Actions
Evacuation/Consultant Criteria: Depending on the timeliness of evacuation, patients with severe hypothermia should be transported. Lesser degrees of hypothermia can usually be treated locally.
NOTES: Many trauma victims become slowly hypothermic, which may be as life threatening as the trauma itself. Do not overlook it. A short period of successful ventilation oxygenates the patient perceived to be dead, and allows them to be handled, insulated, packaged and transported, while minimizing the likelihood of ventricular fibrillation during this process.

Cold Illnesses and Injuries: Non-freezing Cold Injury (Trenchfoot and Immersion Foot)
Murray Hamlet, DVM

Introduction: Having cold, wet feet for an extended period (2 days or more) will produce trenchfoot. Sitting in a life raft with wet extremities produces immersion foot. Tissue death occurs as a result of long-term vasospasm from cold, usually above freezing. The colder it is, the less time it takes to produce damage and vice versa. This injury is common in POWs, escape and evasion victims, and life-raft survivors. It is also common in combat soldiers exposed to water-filled trenches. Chilblains follow cold, wet exposure of the hands or feet of less than 12 hours. They will be swollen, pink, mildly tender, and pruritic, but will recover in 24 hours. A longer exposure (12 hours or more) produces pernio, resulting in thin, partial thickness, necrotic plaques on the dorsum of the hands or feet. These will slough without scarring in a few days, but the area may remain very painful for months or years.

Table 6-1

CLASSIFICATION OF LEVEL OF HYPOTHERMIA

Core Temperature	Thermoregulatory Status	Signs and Symptoms		Classification
98.6°F				Normal
98.6°F		- Cold Sensation - Shivering		Mild
95 - 89.6°F	Control and Responses Fully Active	Physical Impairment - Fine Motor - Gross Motor	Mental Impairment - Complex - Simple	Mild
89.6 - 82.4°F	Responses Attenuated/ Extinguished	~86°F - Shivering Stops - Loss of Consciousness		Moderate
<82.4°F	Responses Absent	- Rigidity - Vital Signs Reduced or Absent - Risk of VF/Cardiac Arrest (Rough Handling)		Severe
<77°F	- Spontaneous Ventricular Fibrillation - Cardiac Arrest			Severe

HYPOTHERMIA: EMS PREHOSPITAL CARE

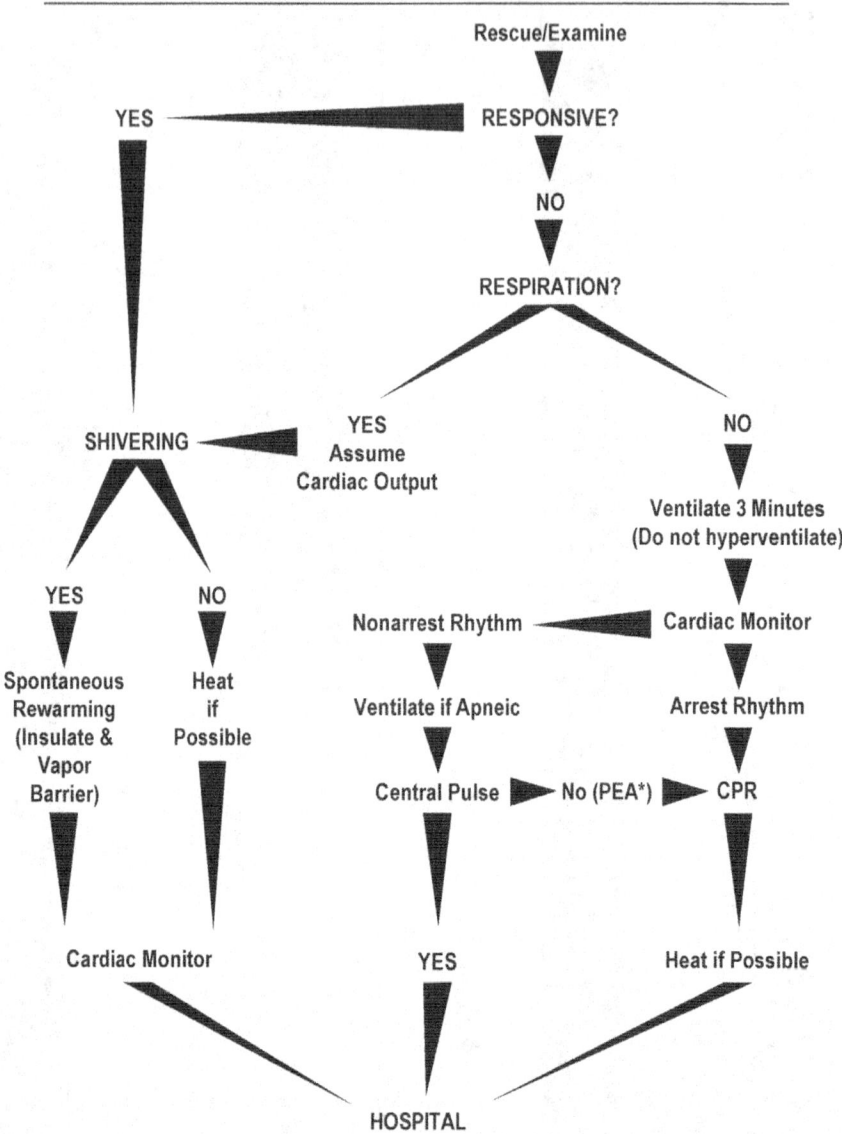

Figure 6-2: *Pulseless Electrical Activity (new term for Electromechanical Dissociation)*

Subjective: Symptoms
Initially cold, wet feet progressing to numbness. Upon warming the torso and feet, they become hot and very painful. The patient is unable to wear boots or walk. Those that can have a shuffling gait and will describe walking as if on "wooden limbs."

Objective: Signs
Initially pale, pulseless, and numb tissue that has slow capillary refill. Upon warming, they are edematous, bright red or purple, hot and painful. Digital pressure produces pitting and slow rebound. After a number of days, liquefaction necrosis or mummification of distal parts occurs. Fever develops and debridement is necessary.

Assessment: The history of being cold and wet along with visualization of the limbs is diagnostic.
Differential Diagnosis:
Frostbite requires below freezing temperatures and produces a dry, mummifying gangrene (see Frostbite). Chilblains are the early stages of trenchfoot and results in just swelling and itching of the extremity, which subsides in 24 hours.
Pernio are thin, necrotic plaques on the dorsum of the hands or feet. They may be proximal to a more serious distal trenchfoot injury.

Plan:
Treatment
Primary: Pat dry, DO NOT RUB. Elevate the feet, warm the torso and hydrate orally. Use urine output (color, frequency) as hydration gauge. Pain meds help to some degree. NSAIDs may help. Most patients need litters, and may need sleep meds.

Patient Education
General: Keep feet dry. Wounds will have a long, slow healing process. Patients may have residual symptoms for years. Fanning the feet at night might help sleep.
Diet: High calorie, high protein diet, and adequate hydration.
Medications: NSAIDs and sleep meds.
Prevention and Hygiene: Change to dry socks daily in cold, wet conditions. Gently massage feet at night to improve blood flow. Treat and/or evacuate those with early symptoms to avoid serious injury.
Wound Care: If necrotic, auto amputation occurs, clean the wound and use loose dry dressings. Evacuate.

Follow-up Actions
Evacuation/Consultant Criteria: Patients who are unable to ambulate or perform their mission, have recurrent injury, have auto amputation of digits or develop osteomyelitis should be evacuated.

Chapter 24: Heat-Related Illnesses
Heat-Related Illnesses: Introduction
LTC Richard Kramp, MC, USA

Introduction: The body sweats to maintain a constant body temperature when stressed by heat. Adequate intake of both water and sodium are essential to replace losses from sweating. Insufficient water intake leads to dehydration while inadequate sodium intake or excessive water intake can lead to hyponatremia.

Acclimatization: It takes about 2 weeks to fully acclimate to a hotter environment. During this time the member should gradually increase his heat exposure and activity. This reduces the likelihood of becoming a heat casualty, but does not prevent it - caution is always needed. The member's need for drinking water will increase with corresponding increases in activity and heat exposure. Acclimated members will sweat earlier and more profusely but with lower salt loss. Heat injuries range in severity from heat cramps to heat exhaustion to heat stroke. While the mechanism of heat cramps is not understood, there is convincing evidence to suggest it is the result of sodium depletion or over-hydration. Heat exhaustion and heat stroke probably represent a continuum of disease, varying in intensity and severity of tissue damage. Both are characterized by water and sodium loss.

Heat-Related Illnesses: Heat Cramps
LTC Richard Kramp, MC, USA

Risk Factors: More likely to occur in sodium depleted or over-hydrated soldiers after strenuous physical activity in a hot, humid environment. Troops that are not fully acclimatized are at increased risk.

Subjective: Symptoms
Painful, tonic contractions of skeletal muscles frequently preceded by palpable or visible fasciculation.
Focused History: *Which muscles are involved?* (usually all or part of large skeletal muscles) *What does the pain feel like?* (usually severe) *Do you have frequent leg cramps at night?* (suggest vascular problems, not heat cramps) *What makes the symptoms worse?* (Manipulation of the muscle may precipitate cramping)

Objective: Signs
Inspection/ Palpation: Muscle cramping with possible fasciculations, rapid resolution after oral or IV salt solution.

Assessment:
Differential Diagnosis - tetany due to alkalosis (hyperventilation, severe gastroenteritis, cholera) or hypocalcemia; strychnine poisoning; black widow spider envenomation or abdominal colic. These are usually easily to eliminate with a short history.

Plan:
Treatment
Primary: Re-hydrate with 0.1% salt solution po (salt tablets are not recommended) or normal saline solution IV if more rapid treatment needed. If a sports drink is used, dilute it 50/50 with water.

Patient Education
General: Patients with heat cramps usually have sodium deficits or over-hydration. Eating the entire MRE and adding salt to tray pack meals should replenish salt stores over several days.
Activity: Allow 2 to 3 days to replenish salt and water deficits before resuming work in the heat.
Diet: Increase salt application to food and water intake.
Prevention and Hygiene: Consume adequate quantities of salt and water as part of the normal diet.
No improvement/Deterioration: If recovery is not rapid (within 1-2 hours with oral fluids, within 15-30 minutes with normal saline), return for reevaluation.

Follow-up Actions
Consultation Criteria: If recover is not rapid (within 1-2 hours with oral fluids, within 15-30 minutes with normal saline).
NOTE: An attempt should be made to determine a reason for the episode so that appropriate advice can be given to the command and the operator.

Heat-Related Illnesses: Heat Exhaustion

LTC Richard Kramp, MC, USA

Introduction: Heat exhaustion is the most common heat illness. Heat exhaustion may develop over several days and is a manifestation of strain on the cardiovascular system. It occurs when the demands for blood flow (to the skin for temperature control through convection and sweating, to the muscles for work, and other vital organs) exceed the cardiac output. Risk Factors: Dehydrated and sodium-deficient members are at risk after strenuous physical activity in the heat. Operators that are not fully acclimatized are at increased risk.

Subjective: Symptoms
Profound fatigue, thirst, nausea/vomiting, tingling of the lips, shortness of breath, orthostatic dizziness, headache, and syncope
Focused History: *What have you had to eat and drink in the last 48 hours?* (Water and salt intake may be too low. If they have consumed several gallons of fluid in the past 2 hours consider hyponatremia- water intoxication.) *Do you know who you are, where you are and what day it is?* (While heat exhaustion patients may be confused, it is a common sign of heat stroke. A patient with mental status changes should be treated as a heat stroke patient until it is proven otherwise.)

Objective: Signs
Inspection: Pale skin; anxiety and agitation; muscle spasms; vomiting; orthostatic hypotension and shortness of breath. If the patient is confused, assume heat stroke until proven otherwise. *IF THE RECTAL TEMPERATURE CONTINUES TO RISE, TREAT AS A HEAT STROKE.*
Auscultation: Hyperventilation
Palpation: Skin cool and moist to touch. It may be dry in desert environments where evaporation is rapid.

Assessment:
Differential Diagnosis - heat stroke, simple dehydration, febrile illness

Plan:
Treatment
1. Reduce the load on the heart with rest and cooling. Place casualty in shade and remove heavy clothing. Apply cool water to the skin if available.
2. Correct water and electrolyte depletion by administering oral or IV fluids. IV fluids replenish the volume and correct symptoms quickly. Patients with resting tachycardia or orthostatic hypotension should initially receive 200-250 cc boluses of normal saline (NS) repeatedly until these vital signs are corrected. No more than 2 liters of NS should be administered without laboratory surveillance. Subsequent IV fluid replacement should be D5/0.5 NS or D5/0.2 NS. Since this is seldom available, alternating D5W with NS or Ringers Lactate may be the best alternative. If patient can tolerate oral fluids use a 0.1% salt solution.

Patient Education
General: Maintain adequate fluid and water intake and work/rest cycles in heat. Avoid direct sunlight and other risk factors.
Activity: Heat exhaustion patients have rapid clinical recovery. However, they all need at least 24 hours of rest and re-hydration under first echelon or unit level medical supervision to reverse water-electrolyte depletion.
Diet: Regular diet augmented with salted food and increased water intake.
Prevention and Hygiene: Acclimatize gradually with adequate water and dietary salt. Forced drinking may help to avoid dehydration.
No improvement/Deterioration: Return quickly for reevaluation.

Follow-up Actions
Return Evaluation: If the patient fails to improve rapidly, assume the patient is a heat stroke casualty and treat as such with rapid cooling and evacuation. A single episode of heat exhaustion does not imply a predisposition to heat injury and no continuing follow up or profile evaluation is required.
Consultation Criteria: Repeated episodes of heat exhaustion require a temporary profile against heat exposure, evacuation and referral for a thorough evaluation.

NOTES: A case of heat exhaustion or heat stroke should alert the command to instigate work/rest cycles, increased water and electrolyte intake and reduced workload. The difference between heat exhaustion and heat stroke is usually impossible to determine. Soldiers who do not respond dramatically to rest and fluid/electrolyte repletion should be observed for 24 hours for delayed complications of heat stroke.

Heat-Related Illnesses: Heat Stroke
LTC Richard Kramp, MC, USA

Introduction: Heat stroke is a medical emergency, distinguished from heat exhaustion by the presence of neurological symptoms. If heat stroke is suspected and body temperature is elevated, start cooling immediately! Do not delay for a diagnostic evaluation. Cooling and evaluation should proceed simultaneously.
Risk Factors: A history of previous heat stroke, poor physical conditioning, dehydration, high work loads in a hot environment, illness with fever, medications that interfere with sweating or contribute to dehydration such as caffeine, alcohol and diuretics.

Subjective: Symptoms
Dizziness, exhaustion, weakness, nausea, possible involuntary urination, confusion, delirium and other mental status changes.
Focused History: *What have you had to eat and drink in the last 48 hours?* (Water and salt intake may be too low. If they have consumed several gallons of fluid in the past 2 hours consider hyponatremia- water intoxication.) *Do you know who you are, where you are and what day it is?* (While heat exhaustion patients may be confused, it is a common sign of heat stroke. A patient with mental status changes should be treated as a heat stroke patient until it is proven otherwise.) *How did your skin get wet?* (Sweating is rare but happens in heat stroke. Cooling liquids may have been applied.) *What were you doing today?* (typical exposure of work in hot environment).

Objective: Signs
Using Basic Tools: Inspection: Sudden collapse and unconsciousness; diminished or absent sweating with hot, red skin; markedly elevated rectal temperature to 106-110°F (not universal); convulsions; seizures; diminished urination.
Auscultation: Elevated blood pressure; rapid, deep respirations dropping off to shallow and irregular respirations.
Palpation: Diminished or absent sweating with hot, red skin; rapid, thready pulse.

Assessment:
Differential Diagnosis - infection (particularly meningococcemia and *P. falciparum* malaria), pontene or hypothalamic hemorrhage, drug intoxication (cocaine, amphetamines, phencyclidine, theophylline, tricyclic antidepressants), alcohol or sedative withdrawal, severe hypertonic dehydration and thyroid storm.

Plan:
Treatment
1. Reduce body temperature rapidly: Use any means available. Ice water is preferred but seldom available in the field. Field expedient baths, which will keep the water cool, can be constructed by digging plastic-lined, shaded pits. Discontinue active cooling when the rectal temperature reaches 101°F in order to avoid

overcooling. Constantly monitor the patient's body temperature and alternate heating and cooling until the temperature stabilizes. Continue monitoring the patient's temperature every 10 minutes for the next 48 hours.
2. Hydrate with 1 1/2 liters of D5NS over the first few hours. Over-hydration can increase the likelihood of complications.
3. Control the airway to prevent vomiting. Intubate if patient is unconscious. Consider NG tube.
4. Give **diazepam** 5-10 mg IV or IM to control seizures.
NOTE: **Epinephrine**, **sodium amytal** and **morphine** are contraindicated. **Atropine** and other drugs that interfere with sweating are also contraindicated.

Patient Education
General: Avoid heat exposure until clinical recovery and a thorough medical evaluation are complete. Recovery is primarily a function of the magnitude and duration of the temperature elevation. There is an increased risk for future heat stroke.
Activity: Patients should receive profiles restricting heat exposure (a permanent profile may be issued later) until clinical recovery is complete and their heat tolerance is evaluated.
Diet: None during initial symptoms, then as tolerated.
Medications: Avoid alcohol, caffeine and other diuretics during convalescence.
Prevention and Hygiene: Avoid heat exposure for several weeks until the body can regulate heat correctly again.
No improvement/Deterioration: Evacuate for additional testing and treatment with continued cooling en route.

Follow-up Actions
Reevaluation: Hypotensive patients who do not respond to saline may benefit from carefully titrated dopamine.
Evacuation/Consultation Criteria: All heat stroke patients need mandatory evacuation and referral. Evaluation of the potential complications of heat stroke (encephalopathy, coagulopathy, hepatic injury, renal failure and rhabdomyolysis) requires laboratory tests not available in the basic or advanced management tools.

Chapter 25: Chemical, Biological, and Radiation (CBR) Injuries
CBR: Chemical Weapons of Mass Destruction
Lt Col John McAtee, USAF, BSC

Introduction: Weapons of Mass Destruction (WMD) offer unique challenges to the SOF community. Generally, SOF forces will be operating independently, without medical support or decontamination capability and in non-permissive areas. Any individual who suddenly becomes a casualty without being wounded, or is suffering a greater degree of incapacitation than is compatible with his injury should be considered a possible chemical victim. It is unlikely that a chemical agent would produce only a single casualty under field conditions, and a chemical attack should be considered with any sudden increase in numbers of unexplained casualities. Report any use or suspicion of use of a WMD to higher command ASAP. Chemical weapons come in three basic types: nerve, blood and blister agents.
Symptom onset and severity will vary depending on the following:
1. Whether the agent is vapor or liquid form.
2. Temperature, wind conditions, terrain and humidity.
3. Route of absorption.
4. Specific agent, quantity and duration of exposure.
5. Pre- and Post-exposure treatment and protection.

Nerve Agents

Nerve agents inhibit CNS function and are highly lethal. Common nerve agents include GA (tabun), GB (sarin), GD (soman), GF and VX. Use of **pyridostigmine** as a preventive measure for GA and GD is a command decision.

Subjective: Symptoms
Eyes, nose and throat: Eye pain, dim vision, photophobia, nasal congestion, hoarseness.
Respiratory: SOB, tightness of chest, dyspnea
GI: Nausea, anorexia, epigastric tightness, heartburn, abdominal cramping
CNS: Apprehension, giddiness, insomnia, headache, drowsiness, difficulty concentrating, confusion, poor memory, weakness

Objective: Signs
Using Basic Tools:
Eyes, nose and throat: miosis (pupillary contraction), lacrimation, conjunctivitis, rhinorrhea, nasal hyperemia
Respiratory: Tachypnea, wheezing, increased bronchial secretions, cough, Cheyne-Stokes respirations (5-30 seconds of apnea)
Cardiac: Occasional early tachycardia followed by bradycardia and hypotension
GI: Salivation, vomiting, diarrhea, involuntary defecation, extreme urgency
GU: Frequent urination, incontinence
CNS: Coma, seizures, ataxia, areflexia
Other: Muscle fasciculation, sweating, pallor, cyanosis

Assessment: Diagnosis based on clinical signs and symptoms, environment and probability.

Plan:
Treatment
Mask self, mask patient and inject **atropine** as follows:
Mild Exposure (miosis, headache, rhinorrhea, salivation, dyspnea): 1 **Mark 1 auto-injector** IM
Severe Exposure (all of the above plus SOB, apnea, generalized twitching, convulsions, urinary and stool incontinence and paralysis): 3 **Mark 1 auto-injectors** IM and **diazepam** IM by auto-injector or 10 IM/IV if by Carpuject (use of more than 1 Mark 1 increases the risk of heat illness)
Evacuation Plan: Evacuate if unstable after decontamination.

Blood Agents

High concentrations of blood agents such as cyanide exert their effect rapidly, causing unconsciousness and death in a matter of minutes. However, if the patient is still alive after the cloud has passed (more than 5 minutes after presumed exposure), he will probably recover spontaneously.

Subjective: Symptoms
Eye and skin irritation. Low levels can cause weakness, headache, disorientation and nausea.

Objective: Signs
Using Basic Tools: Violent convulsions; increased deep respirations followed by cessation of respiration within one minute; slowing of the heart rate until death.

Assessment: Diagnosis based on clinical signs and symptoms, environment and probability

Plan:
Treatment
Mask self and mask patient. Advanced Cardiac Life Support (ACLS) medications are usually not available in the field, so it is useless to initiate ACLS without them.
Replace C_2 canisters on the protective mask after initial exposure.

Evacuation Plan: Evacuate if unstable after decontamination.

Blister Agents

Blister agents such as phosgene, mustard gas or Lewisite attack exposed skin and mucous membranes. They penetrate clothing and force troops to wear full protective equipment, degrading fighting efficiency. The mask protects against eye and lung damage but provides only limited protection against systemic effects. Extensive, slow-healing skin lesions will place a heavy burden on the medic. No drug is available for the prevention of the effects on the skin and mucous membranes. Phosgene penetrates garments and rubber easier than other chemical agents and produces a rapid onset of severe and prolonged effects. When mixed with other chemicals, the rapid skin damage caused by phosgene will make the skin more susceptible to the second agent. If an unmasked victim were exposed to phosgene before donning his mask, the pain caused by the agent will prompt him to unmask again.

Subjective: Symptoms
Burns and blisters, itching, pain, conjunctivitis, coughing, shortness of breath, vomiting and diarrhea.

Objective: Signs
Using Basic Tools:
General: Shock after large exposure to Lewisite, resulting from protein and plasma leakage from capillaries and subsequent hemoconcentration and hypotension.
Eyes: Significant conjunctivitis with possible later scarring. Phosgene in the eye results in immediate pain, conjunctivitis and keratitis.
Skin: Reddened and extremely pruritic; progresses in 4-24 hours to blistering, which may be severe depending on agent and exposure. Phosgene may cause immediate pain followed by skin necrosis. Apart from mucous membranes, the face, neck and skin-on-skin areas (armpits, genitalia, webs of the digits, etc.) are most susceptible.
Respiratory tract: Swelling impeding the airway, tissue sloughing, hyperactive airways, tracheobronchial stenosis, pulmonary edema.
GI tract: Nausea and vomiting which may worsen.

Assessment: Diagnosis based on clinical signs and symptoms, environment and probability

Plan:
Treatment
Mask self. Mask patient.
1. Burns: Treat burns similarly to second-degree thermal burns (clean, prevent infection). Irrigate copiously with water to remove any persistent agent. Apply **Calamine lotion** or suitable substitute to relieve pruritus. Small blisters (<1cm) should be left alone. Irrigate surrounding area daily and apply **Silvadene Cream**. Larger blisters should be unroofed and the area irrigated 4 times daily with soapy water and covered liberally with **Silvadene Cream** or suitable substitute. Give systemic pain relief and fluids.
2. Prevent infection and scarring of eyes: Irrigate copiously with water. Treat mild exposure as conjuctivitis More severe injuries require daily irrigation, topical antibiotics and a topical mydriatic. Apply **Vaseline** to lid edges to prevent adherence, reduce scar formation and allow for a path for infection to drain if present.
3. Control pain with systemic analgesics.
4. Topical steroid may be helpful in the first 48 hrs but of no benefit after that period.
5. Maintain adequate oxygenation. Intubate patients with severe pulmonary involvement early to allow for assisted ventilation and suction of necrotic and inflammatory debris. Oxygen may be necessary for prolonged period.
6. Antitussives and demulcents help the severe non-productive cough. Bronchodilators may be necessary to control airway irritability and systemic steroids may also be of benefit.
7. Establish IV accesses in an unaffected area. Perform fluid resuscitation with Ringer's lactate or normal saline. With severe blistering or if Lewisite is suspected, treat as a significant thermal injury. (See burn management above).

NOTE: Dimercaprol (BAL) is no longer available for skin or eye application. However it is available for IM use and will reduce systemic effects of Lewisite exposure.

Evacuation Plan: Evacuate if unstable and after decontamination completed.

CBR: Set Up a Casualty Decontamination Station
18D Skills and Training Manual, Reviewed by LTC Richard Broadhurst, MC, ARNG

What: Establish an area in which to decontaminate and treat a casualty. It consists of a **decontamination area**: triage, emergency treatment (may be co-located with triage) and skin decontamination; a **treatment area**: clean holding area pending treatment, advanced treatment facility; a clean **patient holding** area for those pending evacuation (can be located inside the treatment area); and a **hot-line** separating the decontamination and treatment areas.

When: Chemical agents used against your unit, or against personnel you support. Similar procedures can be used to remove nuclear fallout and biological agents.

What You Need: Required: water source, supertropical bleach (STB), shovels, personnel MOPP4 ensemble for decontamination crews, protective shelter (tents, buildings, tree cover, caves, etc.). Optional: medical equipment sets (MES) for patient decontamination and patient treatment (contains many of these items), tentage, plastic sheeting, chemical agent alarms, chemical agent monitors, engineer tape or wire, field radio or telephone, windsock, camouflage netting, brushes and/or sponges, plastic bags, litters, litter stands, and contaminated disposal containers.

What To Do:
1. Select primary and alternate sites.
 a. Select primary and alternate sites in advance of operations. If the prevailing winds change direction, use of the primary site may no longer be possible.
 b. Site selection factors:
  1) Direction of the prevailing winds.
  2) The location of friendly facilities downwind from the chemical hazard released at the decontamination station.
  3) Availability of protective shelters or buildings to house clean treatment facilities.
  4) Terrain
  5) Availability of cover and concealment. The protective shelter may have visual, audible and infrared signatures that can compromise concealment.
  6) General tactical situation
  7) Availability of evacuation routes (contaminated and clean).
  8) Location of the supported unit's vehicle decontamination point, personnel decontamination point and MOPP exchange point. It is sometimes best to collocate with these unit decontamination sites. The arrangement of the operational areas must be kept flexible and adaptable to both the medical and tactical situations.

2. Set up the decontamination area.
 a. Triage area. Do not decontaminate expectant patients. Personal equipment, including weapons, should be returned to the patient's unit if possible, for decontamination and management. Patient equipment can be decontaminated in the decontamination area as an option.
 b. Emergency treatment area. **NOTE:** Sometimes triage and emergency treatment are conducted in the same area.
 c. Skin decontamination area. Mix the STB with water in buckets and apply to garments and skin with brushes, sponges or rags. Sequentially decontaminate the chemical agent protective ensemble, remove the components as well as any clothing, decontaminate the underlying skin, and pass the clean patient to the shuffle pit (see below).

d. Overhead cover: erect an overhead cover, at least 20 x 50 feet, to cover the decontamination, clean waiting and triage/ emergency treatment areas. If the protective shelter is used, the overhead cover should overlap the air lock entrance. If plastic sheeting is not available, alternate materials such as trailer covers, ponchos or tarpaulins may be used.

3. Set up the clean side of the decontamination station on the upwind side of the contaminated areas.

4. Set up the shuffle pit on the hot line as the only point of access between the decontamination area and the clean waiting and treatment area.
    a. Turn over the soil in an area that is 1-2 inches deep, and of sufficient length and width to accommodate a litter stand. Can also use mulch, sawdust or similar material if available. The shuffle pit should be wide enough to force the litter bearers to stand in the pit also.
    b. Mix supertropical bleach (STB) with the soil in a ratio of 2 parts STB to 3 parts soil.
    c. Transfer newly decontaminated patients into a patient decon bag on a litter in the shuffle pit (if available). If these items are not available, logroll patients from the arms of the decon team to those of the litter team. The patients should then be transported by the litter team to the treatment area.

5. Set up the treatment area on the upwind side of the decontamination area.
    a. Set up a protective shelter over the treatment area attached to the air lock adjoining the clean side of the decontamination station.
    b. When a protective shelter or air lock is not available for use, set up a covered medical treatment facility (use tents, fixed facility, etc.) 30-50 meters upwind from the shuffle pit.

6. Set up the evacuation holding area (can be part of the treatment area).
    a. Set up an overhead cover of plastic sheeting at least 20x25 feet.
    b. Make sure the cover overlaps part of the clean treatment area and part of the protective shelter.
    c. Avoid setting the protective shelter up near the generator.

7. **Mark the hot line.** Ensure that the entire hot line is clearly marked. Use wire, engineer's tape or other similar material to mark the entire perimeter of the hot line.

8. **Establish ambulance points on both the "clean" and "dirty" evacuation routes.**
    a. Establish a "dirty" ambulance point downwind from the triage area in the decontamination station.
    b. Establish a "clean" ambulance point upwind from the evacuation holding area on the clean side of the decontamination station.

9. **Set up a contaminated (dirty) dump.**
    a. Establish the contaminated dump 75-100 meters downwind from the decontamination station.
    b. Clearly mark the dump with NATO chemical warning markers.

10. **Emplace chemical agent alarms.** Set these alarms around the area, particularly between the decontamination and treatment areas.

11. Camouflage areas IAW tactical directives

What Not To Do:

Do not select only one decontamination site. Have an alternate site in case the wind direction or the operational situation changes.

Do not fail to determine the prevailing wind direction. It may be different at the primary and alternate sites.

Chapter 26: CBR: Biological Warfare
COL Theodore Cieslak, MC, USA & COL Edward Eitzen, MC, USA

Introduction: Dozens of biological organisms and toxins are potential weapons of war or terrorism. Detailed discussion of each of these agents is beyond the scope of this manual. This guide focuses on six agents discerned to represent the most significant or problematic threats by a panel of military and civilian experts. The motives of terrorists are often difficult to ascertain, and since belligerents may employ weapons of opportunity, these six agents are not necessarily those most likely to be employed, but rather those that, IF employed, might pose the greatest threat to health and to operations. Because agents other than those discussed here might be encountered, and because patients may present with a nonspecific febrile illness, empiric therapy might often be necessary. This is especially true on the battlefield where sophisticated diagnostic tools and expert consultation are less likely to be available.

Subjective: Symptoms
Using Basic Tools: Predominantly pulmonary: neuromuscular impairment of muscles of respiration (botulism), shortness of breath and cough; non-specific febrile illness: fever, headache, myalgias, fatigue, malaise, and weakness.
Using Advanced Tools: Lab: CBC, nasal and throat swabs, blood and sputum for Gram stain and culture; CXR; if possible obtain and save serum for future serologic studies, blood and/or urine for future toxin analysis and throat swabs for viral culture.

Objective: Signs
Depend on agent but may include fever, cough, tachypnea, tachycardia, rales, dyspnea, cyanosis, diaphoresis, hypotension, and muscular weakness.

Assessment:
Use the following epidemiological clues to differentiate a potential biological attack from naturally occurring illnesses and infectious diseases: tight clusters (in time and location) of casualties; unusually high infection rates; unusual geography (presence of a presumed disease in an area where it does not naturally occur); unusual or unexpected clinical presentations, unusual munitions, evidence of a point-source for outbreak; dead or dying animals; lower illness rates in protected personnel.

Plan:
Treatment: This section deals with empiric therapy provided when biological attack is suspected but the identity of the specific agent is unknown. In cases where a specific agent is identified or strongly suspected the medic should refer to the appropriate section of this manual. Remember that empiric therapy is not a substitute for the continued pursuit of a definitive diagnosis and the consequent provision of definitive therapy.
Primary: When dealing with casualties exhibiting pulmonary symptoms, or when dealing with large numbers of casualties exhibiting significant but nonspecific febrile illness, empiric antibiotic therapy might be warranted. This would be the case if patients were deteriorating and lives were in jeopardy, and it might also be the case if the tactical situation would be compromised by large numbers of casualites. In this setting, oral **doxycycline** 100 mg q 12 hours can be prescribed. Supportive care (oxygen, intravenous fluids, antipyretics) should also be provided as needed.
Alternative: Ciprofloxacin 500 mg po q 12 hours (primary if anthrax suspected)
Consider **tetracycline** or other fluoroquinolones as alternatives.
NOTE: See individual discussions below concerning vaccines, barrier protection, quarantine, evacuation and other issues.

Biological Agents: Inhalational Anthrax

Introduction: The inhalational form of the disease would likely result from intentional aerosol delivery. It is one of the most viable potential biological weapons due to the stability of anthrax spores, the relative ease of their dissemination and the high lethality of the disease (see ID: Anthrax).

Subjective: Symptoms
Classic inhalational anthrax follows a "biphasic" course, with a brief period of early non-specific flu-like symptoms: fever, malaise, fatigue, muscle aches, headaches, mild chest discomfort and non-productive cough. Following these symptoms patients may experience a partial recovery. However, 1-2 days later, in the final stage of the disease, patients complain of high fever and significant shortness of breath.

Objective: Signs
Using Basic Tools: Early: Fever, tachypnea; Late: fever, tachycardia, tachypnea, dyspnea, cyanosis, diaphoresis, hypotension, chest wall edema, meningismus, hemorrhagic mediastinitis, sepsis.
Using Advanced Tools: CXR: Chest x-rays will often demonstrate pleural effusions and widening of the mediastinum late in the course of the disease. The lung fields themselves may be relatively clear, allowing differentiation of anthrax from most forms of pneumonia. Early in the disease, chest x-rays may be normal.
Lab: Gram-stain of blood; blood cultures

Assessment:
Differential diagnosis - pneumonia (from either conventional etiologies or other potential biological weapons: plague, tularemia, staphylococcal enterotoxins), gram-negative sepsis.

Plan:
Treatment: The prognosis for symptomatic anthrax victims is very poor; in all likelihood, 85% or more of symptomatic victims will succumb even in the face of appropriate therapy. However, EARLY treatment may be lifesaving.
Primary: Ciprofloxacin 400 mg IV every 12 hours, oxygen, intravenous fluids, antipyretics.
Alternative: IV **doxycycline, tetracycline,** or **penicillin** G if **ciprofloxacin** is unavailable.
Primitive: Use oral **ciprofloxacin** if IV therapy is not possible.

Patient Education
General: Caregivers need only use standard precautions when dealing with patients since inhalational anthrax is not contagious.
Prevention: Immunization is an effective preventive measure against anthrax. Consider empiric therapy as soon as anthrax is suspected. Start asymptomatic persons thought to have been exposed to aerosolized anthrax on oral **ciprofloxacin** (500 mg po q 12 hours). Oral **doxycycline** (100 mg 12 hours) is an acceptable substitute. Other fluoroquinolones, **tetracycline,** or **penicillin** V are alternates. In addition, asymptomatic, exposed persons who have not received anthrax vaccine should be immunized with at least three doses of vaccine: at "time zero", and at 2 and 4 weeks after the first dose.

Follow-up Actions:
Evacuation/Consultation Criteria: Consult Preventive Medicine early for suspected cases. Evacuate if stable and likely to tolerate travel.

Biological Agents: Botulism

Introduction: Botulism is caused by exposure to one of seven neurotoxins produced by *Clostridium botulinum* and related anaerobic bacteria. It is NOT due to infection with the bacteria, **so it cannot be treated with antibiotics and is not contagious.** While botulism might be acquired in a number of ways (consuming contaminated canned foods, inhalation, and rarely, percutaneous inoculation), it is likely to be encountered in aerosol form if weaponized. Botulism presents similarly regardless of route of inoculation.

Subjective: Symptoms
Following a latent period of several hours to several days, descending, symmetrical, flaccid paralysis; blurry vision, difficulty swallowing and speaking, dry or sore throat, and dizziness. As the paralysis proceeds downward, weakness and difficulty breathing become significant problems.

Objective: Signs
Using Basic Tools: Mydriasis (abnormal pupillary dilation), ptosis (sagging eyelid), difficulty speaking and swallowing, postural hypotension, absent gag reflex, extraocular muscle palsies, cyanosis, and progressive, descending, symmetrical, muscle weakness. Fever is generally absent.

Assessment:
Differential Diagnosis - myasthenia gravis, Guillan-Barre syndrome, Eaton-Lambert syndrome, poliomyelitis, tick paralysis. Nerve agent exposure can also cause paralysis on the battlefield, but the paralysis would be spastic. Moreover, miosis (abnormal papillary dilation), copious secretions and immediate onset of symptoms should differentiate nerve agent intoxication botulism.

Plan:
Treatment
Primary: Supportive (oxygen, assisted ventilation if necessary, IV fluids).

Patient Education:
General: Even those with access to antiserum may have a very prolonged course, requiring months of recovery. Provide adequate ventilatory support throughout this prolonged course.
Prevention: Asymptomatic persons thought to have been exposed to botulism may be salvaged by prompt administration of antitoxin. A licensed antitoxin is available through the CDC but this product is only effective against 3 of the 7 types of botulinum toxin (types A, B, E).

Follow-up Actions:
Evacuation/Consultation Criteria: Evacuate all victims. Consult preventive medicine or infectious disease experts early.

Biological Agents: Pneumonic Plague

Introduction: Plague is caused by infection with *Yersinia pestis*, a Gram-negative bacillus. Although bubonic and primary septicemic forms are known, the pneumonic form of the disease would likely occur after intentional aerosol delivery. A large percentage of symptomatic victims will succumb even in the face of appropriate therapy. Incubation period is 1-7 days. See also ID: Plague.

Subjective: Symptoms
Fever, malaise, fatigue, cough and shortness of breath.

Objective: Signs
Using Basic Tools: Fever, tachycardia, tachypnea, dyspnea, cyanosis, diaphoresis, rales, and hypotension. The classic finding in pneumonic plague is the production of bloody sputum in a previously healthy patient, although this is not present in every case.
Using Advanced Tools: CXR: Increased markings and consolidation of pneumonia; Lab: Gram-stain of sputum will demonstrate short, bipolar Gram-negative rods, often with a "safety-pin" appearance; blood cultures.

Assessment:
Differential Diagnosis - other forms of pneumonia (both conventional etiologies and other potential biological weapons: tularemia, staphylococcal enterotoxins), sepsis caused by other gram-negative bacteria, anthrax.

Plan:
Treatment
Primary: IV **gentamicin** (1.5 mg/kg every 8 hours), oxygen, intravenous fluids, antipyretics.
Alternative: IV **streptomycin, doxycycline, tetracycline,** or **ciprofloxacin** if **gentamicin** is unavailable.
Primitive: Oral **doxycycline** or **ciprofloxacin** if IV therapy is not possible.

Patient Education
General: Pneumonic plague is contagious; caregivers should employ droplet precautions when dealing with patients. At a minimum, mask either the casualty or the health-care team and close contacts.
Prevention: Contacts exposed to aerosolized plague should take prophylactic oral **doxycycline** (100 mg every 12 hours). Oral **ciprofloxacin** (500 mg every 12 hours) is an alternate. Other fluoroquinolones and **tetracycline** are other options.

Follow-up Actions
Evacuation/Consultant Criteria: Evacuate patients promptly, maintaining droplet protection for care providers and aircrew. Consult Preventive Medicine or infectious disease experts early.

Biological Agents: Smallpox

Introduction: Smallpox is caused by infection with Variola virus. Naturally occurring smallpox has been globally eradicated since the last case occurred in Somalia in 1977. Authorized stockpiles of virus exist in only two high-security laboratories. The categorization of smallpox virus as a viable weapon stems from the fear that belligerent groups may possess clandestine stocks. Moreover, fear exists that other closely related orthopoxviruses (such as monkeypox or cowpox) might be genetically manipulated to produce variola-like disease. The incubation period is 7-17 days.

Subjective: Symptoms
Begin abruptly with malaise, fever, rigors, headache, backache, and vomiting.

Objective: Signs
Using Basic Tools: Characteristic rash appears 2-3 days after the onset of symptoms; all lesions progress synchronously from macules to papules to pustules, and are concentrated on the hands, face and trunk; fever and mental status changes; complications include viral "sepsis", hepatic insufficiency, encephalopathy, skin hemorrhage.

Assessment:
Differential Diagnosis - chickenpox (lesions in various stages of progression and not concentrated on the trunk), monkeypox, enteroviral exanthems (such as hand-foot-mouth disease).

Plan:
Treatment
Primary: Supportive (oxygen, intravenous fluids, and antipyretics).

Patient Education
General: Smallpox is contagious. Caregivers should employ airborne and contact precautions when dealing

with patients. At a minimum, mask either the casualty or the health-care team and close contacts. Wear gloves when touching the patient.
Prevention: Report any suspected case immediately to public health officials. Vaccinia vaccine is effective at preventing smallpox. Immunize those exposed to smallpox promptly. Those immunized within the first several days after exposure may be protected against the development of smallpox. Vaccinia Immune Globulin (VIG) is not useful in smallpox victims.

Follow-up Actions
Evacuation/Consultant Criteria: Consult early with preventive medicine experts. Isolate and do not evacuate patient. Quarantine contacts for 17 days (incubation period) to ensure they will not be secondary cases.

Biological Agents: Tularemia

Introduction: Tularemia is caused by infection with *Francisella tularensis*, a gram-negative coccobacillary organism. Although several forms are known, the pneumonic or typhoidal forms of the disease would likely occur after intentional aerosol delivery. See also ID: Tularemia.

Subjective: Symptoms
Fever, malaise, fatigue, cough, shortness of breath and abdominal pain.

Objective: Signs
Using Basic Tools: Fever, tachycardia, tachypnea, dyspnea, cyanosis, diaphoresis, rales, hypotension, and abdominal tenderness and pneumonia and sepsis later.
Using Advanced Tools: CXR: Increased markings and consolidation of pneumonia Lab: Gram-stain of sputum may demonstrate short gram-negative rods; blood cultures.

Assessment:
Differential Diagnosis - other forms of pneumonia (both conventional etiologies and other potential biological weapons: plague, staphylococcal enterotoxins); sepsis caused by other gram-negative bacteria, typhoid fever, anthrax.

Plan:

Treatment
Primary: Intravenous **gentamicin** (1.5 mg/kg every 8 hours), oxygen, intravenous fluids, antipyretics.
Alternative: IV **streptomycin, doxycycline, tetracycline,** or **ciprofloxacin** if **gentamicin** is unavailable.
Primitive: Oral **doxycycline** or **ciprofloxacin** if IV therapy is not possible.

Patient Education
General: Tularemia is not typically contagious; caregivers need only employ standard precautions when dealing with patients.
Prevention: Start asymptomatic persons thought to have been exposed to tularemia via aerosol on oral **doxycycline** (100 mg every 12 hours). Oral **ciprofloxacin** (500 mg every 12 hours) is an acceptable substitute. Try other fluoroquinolones or **tetracycline** if **doxycycline** or **ciprofloxacin** is unavailable.

Follow-up Actions
Evacuation/Consultant Criteria: Evacuate patients promptly. Consult preventive medicine or infectious disease experts early.

Biological Agents: Viral Hemorrhagic Fevers

Introduction: The viral hemorrhagic fevers (VHFs) are a diverse group of diseases caused by viruses of at least four families. They share a propensity to cause bleeding but otherwise vary considerably in their clinical manifestations and severity. Among the VHFs are Ebola and Marburg, certain Hantavirus infections, Argentinian and Bolivian hemorrhagic fevers, Lassa fever, Crimean-Congo hemorrhagic fever, and yellow fever. Incubation periods vary from several days to as long as several weeks.

Subjective: Symptoms
Fever, malaise, myalgias, headache, photophobia, vomiting, diarrhea, abdominal pain, cough and dizziness.

Objective: Signs
Using Basic Tools: Commonly seen: fever, GI bleeding, pulmonary hemorrhage, facial flushing, conjunctival injection, petechiae, purpura, bleeding from the mucous membranes, and skin ecchymoses. Other symptoms: hematuria, hypotension, shock, edema, hepatic tenderness (hepatic failure), pharyngitis, hyperesthesias.
Using Advanced Tools: Lab: Blood culture to rule out meningococcemia and typhoid fever.

Assessment:
Differential diagnosis - any cause of a bleeding, diathesis or disseminated intravascular coagulation (both conventional causes as well as plague): dengue (which can cause hemorrhagic fever but is not transmissible by aerosol), malaria, typhoid fever, meningococcemia, rickettsial diseases, leptospirosis, shigellosis, fulminant hepatitis, leukemia, lupus, hemolytic-uremic syndrome, and thrombocytopenic purpuras. Most of these conditions are discussed in this book and can be differentiated based on differences in presentation and laboratory findings.

Plan:
Treatment
Primary: Supportive (oxygen, intravenous fluids, and antipyretics). Avoid **aspirin** and IM injections to avoid additional bleeding.

Patient Education
General: Many VHFs are contagious; caregivers should employ contact precautions when dealing with patients. At a minimum, this entails wearing gloves when touching the patient and disinfecting medical equipment (such as stethoscopes) between patient encounters.
Prevention: A licensed vaccine is available for yellow fever. In the presence of endemic yellow fever personnel should be immunized.

Follow-up Actions
Evacuation/Consultant Criteria: Consult early with preventive medicine experts. Isolate and do not evacuate patient. Quarantine contacts for 21 days (incubation period) to ensure they will not be secondary cases. Shorten the quarantine period to reflect the appropriate incubation period when a definitive diagnosis is available.
NOTES: Advanced treatments and diagnostic tests will not be available in a field setting. Intravenous **ribavirin** may be beneficial in certain VHFs (Argentinian & Bolivian hemorrhagic fevers, Lassa fever, Hantavirus infection). The platelet count, PT, PTT, and serum protein levels are likely to be abnormal in VHF patients, but such findings are not diagnostic. Lumbar punctures may be necessary to rule out meningitis in patients with meningeal signs and/or altered mental status.

Chapter 27: CBR: Radiation Injury
Lt Col Aimee Hawley, USAF, MC

Introduction: Acute radiation injury results from high doses of radiation associated with a nuclear explosion, leak of radioactive material, or detonation of a radiation dispersal device made from highly radioactive material. The two most radiosensitive organ systems in the human body are the hematopoietic (blood forming) and gastrointestinal (GI).

Subjective: Symptoms
Three phases: 1. Prodrome: (within hours of exposure) nausea, vomiting, diarrhea, fatigue, weakness, fever and headache; time to onset, duration and severity of these symptoms varies with radiation dose received. 2. Relatively symptom-free latent phase, lasting 2-6 weeks depending on dose received. 3. Clinical symptoms in the affected major organ system (hematopoietic, gastrointestinal, neurovascular).

Objective: Signs
Using Basic Tools: Dose dependent: Survivable dose- no objective physical signs; higher dose- signs of bone marrow suppression (infection, bleeding) or gastrointestinal syndrome (GI bleeding, diarrhea) follow days to weeks after exposure; concomitant trauma.
Using Advanced Tools: Lab: WBC count immediately and q 4 h (50% drop in lymphocytes within the first 24 hours, with total lymphocyte count less than 1×10^9 indicates significant radiation injury);
If patient not evacuated: Draw 15 cc of peripheral blood in a heparinized tube, refrigerate and forward it with patient for analysis in cytogenetic lab for biodosimetry (estimate of dose received). Collect baseline urine and stool samples for evaluation of internal contamination (first 24 hours following injury);
Radiation Monitoring (if available): Whole-body monitoring with RADIACS meter or assessment of inhalation injury with nasal swabs (forward for analysis) for anyone not wearing respiratory protection (protective mask).

Assessment:
Differential Diagnosis - radiogenic vomiting may be confused with psychogenic vomiting that often results from stress and fear reactions.

Plan:
Treatment
Primary:
Decontaminate following stabilization. Soap and water are 95% effective.
Control radiogenic emesis with antiemetics: **Granisetron (Kytril)** 1 mg IV over 30 seconds or **ondansetron (Zofran)** 8 mg IM, or IV over 30 seconds are best.

Patient Education
General: A patient who receives a minimal dose should be reassured and returned to duty. A patient with any lymphocyte depletion within the first 24 hours should be evacuated as quickly as possible for definitive care of subsequent infectious and gastrointestinal complications.
Diet: No special diet is indicated.
Medications: Antiemetic drugs may cause drowsiness.
Prevention and Hygiene: Definitive surgical management of associated wounds and trauma must be completed within 36 hours in a patient with significant radiologic injury so as to avoid infection and increased morbidity associated with poor wound healing.
No Improvement/Deterioration: Evaluate for signs of infection.
Wound Care: As above, definitive wound care within 36 hours of injury is mandated.

Follow-up Actions
Evacuation/Consultation Criteria: Evacuate a patient with suspected significant radiologic injury as soon as possible. **NOTE:** A variety of antidotes are available for specialized treatment of internal contamination. Common ones include **potassium** or **sodium iodide (KI** or **NaI)** to prevent thyroid uptake of radioactive iodine, which occurs after reactor accidents. Significant internal contamination by plutonium, the primary radioactive contaminant in nuclear weapons accidents, is treated with chelating agents such as **calcium diethylenetriaminepentaacetic acid (DPTA)**.

Part 7: TRAUMA
Chapter 28
Trauma: Primary and Secondary Survey

LTC John Holcomb, MC, USA & CPT Robert Mabry, MC, USA

Primary Survey

The standard ABC approach as outlined in civilian models provides an excellent methodology for addressing life-threatening injuries in systematic fashion. The "ABC" pneumonic prioritizes the search for injuries in accordance with their potential to kill the patient; it is simple to remember and it provides an anchor point from which patients can re-assessed if they deteriorate. This system may require some modification in a tactical setting. For example, in a mass casualty situation, the SOF medic may need to address the ABCs of several patients at once. Simply asking the casualties where they are injured can do this. Those casualties who answer the question appropriately have an intact airway, are breathing and are conscious. The medic should then focus his attention on those casualties who are unconscious or in obvious distress. Meanwhile, the medic can direct the lightly injured casualties or non-medical team members to assist in controlling the bleeding of those patients with active hemorrhage, thus addressing the circulation step.

During combat, moving the patient to a safe location takes priority over the Primary and Secondary Survey unless a rapid maneuver can be performed for an obvious life-threatening injury, i.e., the application of a tourniquet. Rapid control of hemorrhage is a mainstay of combat casualty care.

Airway: A conscious spontaneously breathing patient requires no immediate airway intervention. If the patient is able to talk normally his airway is intact. If the patient is semi-conscious or unconscious, the flaccid tongue is the most common source of airway obstruction. The chin lift or jaw thrust maneuver should be attempted and should readily relieve any obstruction created by the tongue. Once the airway is opened or if further difficulty is encountered, a nasopharyngeal or oropharyngeal airway should be inserted. The nasopharyngeal airway is better tolerated in the semi-conscious patient and the patient with an intact gag reflex. If the above measures fail to provide an adequate airway or if the patient is unconscious, unresponsive and apneic, orotracheal intubation should be considered. Orotracheal intubation done on a trauma patient with an intact gag reflex without the use of pharmacological sedation and paralysis will be difficult and may cause additional complications such as vomiting, airway trauma and increased intracranial pressure, and thus should be avoided except as a last resort. If the patient is breathing and definitive airway control if needed, blind nasotracheal intubation (BNTI) may be attempted. Severe facial fractures and basilar skull fractures are relative contraindications to BNTI.

Other adjuncts to airway management can and should be used if available and if the medic is skilled in their use. Other possible adjuncts to airway management include the Laryngeal Mask Airway (LMA) the Intubating LMA, the Combitube, and the Lighted Stylet.

If the patient has obvious maxillofacial trauma with signs of airway compromise or if orotracheal intubation fails, then a surgical cricothyroidotomy may be a necessary and lifesaving maneuver (see Procedure: Cricothyroidotomy). The most common mistake when performing a surgical airway is delaying too long before starting the procedure.

Civilian models of trauma care include cervical spine control and immobilization with airway management. Few if any battlefield casualties with penetrating trauma will have associated injury to the cervical spine unless they have combined blunt injuries from vehicle or aircraft crashes, falls or crush injuries, or penetrating injury to the spinal cord. Meticulous attention to presumed cervical spine injury on the battlefield is not warranted if penetrating trauma is the obvious mechanism. Furthermore, the medic or the casualty may sustain additional injury if evacuation from the battlefield, and/or treatment of other injuries such as hemorrhage is delayed while the cervical spine is immobilized.

Breathing: In the conscious patient, who is alert and breathing normally, no interventions are required. If the patient has signs of respiratory distress such as tachypnea, dyspnea, or cyanosis, which may be associated with agitation or decreasing mental status, an aggressive search for an etiology is required. Injuries that may result in significant respiratory compromise include tension pneumothorax, open pneumothorax (sucking chest wound), flail chest, and massive hemothorax. The patient's chest and back should be quickly exposed and inspected for obvious signs of trauma, asymmetrical or paradoxical movement of the chest wall, accessory muscle use and jugular venous distention. If possible, auscultation should be performed listening for abnormal or decreased breath sounds. The chest wall should be palpated to identify areas of tenderness, crepitus, subcutaneous emphysema or deformity.

Open pneumothorax should be treated with a three-sided occlusive dressing and a tension pneumothorax with needle decompression (see Procedure: Thoracostomy, Needle).

The field management of a flail chest centers on controlling the patient's pain and augmentation of the patient's respiratory efforts with bag valve mask ventilation. Chest wall splinting with tape, sandbags and the like has been advocated in the past but should no longer be performed as it decreases the movement of the chest wall and will further compromise the patient's ability to ventilate. These casualties may have significant injury to the underlying lung and may deteriorate rapidly requiring endotracheal intubation and positive pressure ventilation.

Management of a massive hemothorax in the field should be directed at maintaining adequate ventilation with a BVM. If evacuation is delayed and the patient continues to deteriorate, consideration may be given for the placement of a chest tube. If more than 1000cc of blood is immediately drained by the chest tube or if the output is more than 200cc per hour for 4 hours, the patient likely has injury to the great vessels, hilum, heart or vessels in the chest wall that will require surgical repair.

Flail chest and massive hemothorax are difficult injuries to treat in the field and should be evacuated are rapidly as possible.

Circulation: Uncontrolled hemorrhage is the leading cause of preventable battlefield deaths. Rapid identification and effective management of bleeding is perhaps the single most important aspect of the primary survey while caring for the combat casualty.

Obvious external sources of bleeding should be controlled with direct pressure initially followed by a field dressing or pressure dressing. If bleeding is not controlled by the previous measures or if gross arterial bleeding is present, an effective tourniquet should immediately be applied. Clamping of injured vessels is not indicated unless the bleeding vessel can be directly visualized. Blind clamping of vessels may result in additional injury to neurovascular structures and should not be done.

NOTE: The current ATLS manual discourages the use of tourniquets in the pre-hospital setting because of distal tissue ischemia, tissue crush injury at the tourniquet site, which may necessitate subsequent amputation. This admonition is based on the civilian model of trauma care where most penetrating injuries are low velocity in nature and rapid evacuation to a trauma center is available. Withholding the use of tourniquets on the battlefield for patients with severe extremity hemorrhage may result in additional death or injury that might have otherwise been prevented.

Sources of internal hemorrhage should be identified. A significant amount of blood can be lost into the chest and abdominal cavities, the retroperitoneal space and the soft tissues surrounding fractures of the pelvis and lower extremities. Significant bleeding into the thoracic and abdominal cavities following trauma will require surgical exploration. In the absence of a head injury, hypotensive resuscitation will help prevent more bleeding. Bleeding from injuries to the pelvis and groin or from fractures of the lower extremities not otherwise amenable to treatment with a tourniquet and not associated with thoracic injuries may be controlled with the application of Pneumatic Anti-Shock Garment (PASG), AKA Military Anti-Shock Trousers (MAST).

After sources of hemorrhage are identified and controlled, the need for intravenous access should be considered. If the patient has an isolated extremity wound, the bleeding has been controlled and there are no signs of shock, there is no need for immediate intravenous fluid resuscitation. Intravenous access with a saline lock should be considered for all casualties with significant injuries. If there is a truncal injury and if signs of shock are present, or if blood pressure continues to drop, intravenous access should be obtained with a 16 or 18-gauge catheter followed by a 1-2 liter bolus of normal saline or lactated Ringers, or 500 milliliters of Hespan. If the patient has improvement of the clinical signs of shock following the initial bolus, subsequent intravenous fluids should be titrated to achieve only a good peripheral pulse and an improvement in sensorium rather than to normalize blood pressure. If there is no clinical improvement following the initial IV fluid bolus, the possibility of severe uncontrolled intra-abdominal or intrathoracic bleeding should be considered. Further fluid resuscitation in uncontrolled hemorrhage is not indicated, may be harmful, and may waste the limited fluids available to the SOF medic.

Cardiopulmonary arrest from hemorrhage has a very high mortality in the hospital setting. Attempting to resuscitate patients who are in cardiac arrest secondary to hemorrhage while in the field will almost certainly be futile.

Disability: A brief neurological assessment should be performed using the AVPU scale:
A-Alert
V-Responds to verbal stimuli
P-Responds to painful stimuli
U-Unresponsive

Exposure: Clothing and protective equipment such as helmets and body armor should only be removed as required to evaluate and treat specific injuries. If the patient is conscious with a single extremity wound, only the area surrounding the injury should be exposed. Unconscious patients may require more extensive exposure in order to discover potentially serious injures but must subsequently be protected from the elements and the environment. Hypothermia is to be avoided in trauma patients.

Vital Signs: Vital signs should be assessed frequently, especially after specific therapeutic interventions, and before and after moving patients. The SOF medic should be sensitive to subtle changes in vital signs in wounded SOF operators. As a group these patients are in excellent physical condition and may have tremendous physiological reserves. They may not manifest significant changes in vital signs until they are in severe shock. The vital signs include:
  Pulse: The rate and character of the pulse should be evaluated. A weak, rapid, barely palpable radial pulse indicates the presence of hemorrhagic shock.
  Respiration: Respiratory rate can be an extremely sensitive indicator of physiologic stress. Resting tachypnea should be considered abnormal and should prompt investigation if there is no obvious cause.
  Blood Pressure: The SOF medic is not expected to carry a sphygmomanometer during combat operations. Palpation of distal and central pulses provides a rough guide to systolic blood pressure.
  Radial- at least 70 mmHg
  Femoral- at least 60 mmHg
  Carotid- at least 50 mmHg
  Temperature: Only if hypo or hyperthermia are suspected. Hypothermia is an often unrecognized and yet significant contributor to traumatic death.

Secondary Survey
During the Secondary Survey, a more methodical search for non-life threatening injuries is conducted. These injuries should be treated as they are encountered. Like the Primary Survey above, the Secondary Survey may need to be modified and adapted according to the tactical situation and the number and type of casualties encountered. The vast majority (75%) of casualties who are wounded in action (WIA) will have isolated penetrating trauma to the extremities. These patients do not require a detailed head to toe exam

in the Secondary Survey. They will need to have a bandage and/or splint applied with evaluation of their neurovascular status distal to the injury before and after treatment. They then need to be frequently reassessed for signs of deterioration as the tactical situation permits. Patients who are severely injured or unconscious will require a more detailed Secondary Survey as outlined below. Evacuation should not be delayed to perform a Secondary Survey or for the treatment of non-life threatening injuries.

The Secondary Survey should be conducted in a systematic head to toe, front to back fashion using visual Inspection, Auscultation, Palpation and Percussion (IAPP) where applicable.

HEENT: The head and face should be inspected for obvious laceration, burns, contusion, asymmetry or hemorrhage. The bones of the face and head should then be palpated to identify crepitus, bony step-off, depressions or abnormal mobility of the mandible and mid-face. The eyes should be opened and examined for signs of trauma, globe rupture, or hyphema. The orbits and zygomatic arches should be palpated for signs of fractures. Pupils should be checked for reactivity and symmetry. If the patient is awake, extra-ocular movements can be assessed along with gross visual acuity. The ears should be inspected for obvious trauma and the ear canals for blood or cerebral spinal fluid (CSF). Battle's sign indicating possible basilar skull fracture may be observed over the mastoid processes. The nares should be inspected for blood or CSF. The mouth and oropharynx should be inspected for trauma or bleeding. Loose teeth, dental appliances or other potential airway obstructions should be removed. Any previous airway interventions should be reassessed.
Neck: The neck should be visually inspected searching for obvious trauma or deformity, tracheal deviation, jugular venous distention (JVD), or signs of respiratory accessory muscle use. The cervical spine should be palpated for step-off, tenderness or deformity.
Chest: The chest wall should be observed for penetrating injury or blunt injury, asymmetrical breathing movements or retractions. Auscultation over the anterior lung fields, posterior lung bases and heart should follow. The entire rib cage, sternum and chest wall should be palpated for tenderness, flail segments, subcutaneous emphysema or crepitus. Percussion may be performed looking for hyperresonance or dullness.
Abdomen: The abdomen should be observed for signs of blunt or penetrating injury. The presence or absence of bowel sounds should evaluated. Palpation searching for tenderness, guarding or rigidity should follow. Percussion may elicit subtle rebound tenderness.
Pelvis: The pelvis should be inspected for signs of penetrating trauma or deformity. Pelvic instability and fracture should be suspected with movement of the anterior iliac crests when lateral and anterior pressure is applied. The perineum and genitals are inspected next for signs of injury. Scrotal, vulvar and perineal hematomas or blood at the urethral meatus may indicate pelvic fracture. Likewise the rectal exam will yield information about the location of the prostate, and presence or absence of gross blood in the rectum.
Extremities: The extremities are inspected and palpated proximally to distally. Each bone and joint distal to the pelvis and clavicle should be assessed for crepitus, tenderness, deformity and abnormal joint motion. Distal pulses and capillary refill are then examined. Asking the patient if he can feel the examiner lightly touching his hands and feet tests gross sensation. Gross motor strength is tested by having the patient squeeze the examiner's fingers and by moving his toes up and down against the resistance of the examiner's hands.
Neurological: A field neurological exam should consist of observation of the pupils for reactivity and asymmetry (done during HEENT exam), the level of consciousness, gross sensory and motor function (assessed during examination of the extremities) and calculation of the Glasgow Coma Scale (GCS). The GCS is a useful tool that can be used to monitor the clinical status of seriously injured patients. A declining GCS score over time indicates further neurological deterioration. A GCS less than 9 indicates severe neurological injury (see Appendices: Glasgow Coma Scale).
Pain Management: Intravenous **morphine sulfate** is an excellent analgesic for traumatic injuries. **Morphine** has a rapid onset, is easily titratable and can be readily reversed by **naloxone** if the patient becomes obtunded or experiences respiratory depression. It should be used with caution in patients with injuries that may compromise respiratory function and it is contraindicated in patients with head injuries or altered levels of consciousness. Doses should be given in 5-mg increments every 10-15 minutes until adequate levels of analgesia are obtained. The practice of withholding narcotic analgesics or more frequently, giving inadequate doses because of concerns of abuse or respiratory depression in otherwise healthy SOF operators are based on unrealistic concerns and should be avoided. Casualties with combat wounds require treatment for their

pain, preferably through the intravenous route (see Procedure: Pain Assessment and Control).
Antibiotics: There are many antibiotic regimens. A reasonable approach is to use **cefotetan** 1-2 grams IV q12 hrs, IV or IM. It is cheap, readily available, easily stored and rarely causes an allergic reaction. Minor injuries in the field often become infected so pre-hospital antibiotic use should be liberal. Of course all open fractures, abdominal wounds and extensive soft tissue injury should be cleaned, dressed and given antibiotics (see Appendices: Organism/Antibiotic Chart)
Reassessment: The most valuable diagnostic tool available to the SOF medic while caring for combat casualties is a repeated physical exam. Patients should be frequently reassessed with attention given to potential complications of their particular injuries and treatments. For example, distal pulses and sensation should be re-examined in those patients with extremity injuries looking for the development of compartment syndrome. Wounds treated with bandages and tourniquets should be reassessed for further bleeding. Patients with head injuries should have frequent neurological examinations looking for signs of deterioration. Those with chest wounds will require repeated auscultation to rule development or re-accumulation of a pneumothorax.

Trauma: Primary/Secondary Survey Checklist
Furnished by JSOMTC

Scene Size-Up
Tactical Situation/Security/Time on Site
Body Substance Isolation (BSI) Precautions
Mechanism of Injury (MOI)/History of Events
Determine # of Patients
Request Help if needed
Direct/Provide C-spine Stabilization if indicated (always if multisystem trauma, altered LOC or blunt injury above clavicle)

Primary Survey
INITIAL ASSESSMENT
    AVPU/GCS (GCS of 8 or less requires intubation)
Chief Complaint if Conscious
    Determine Apparent Life Threats (Massive Hemorrhage)/Stop Gaps
ASSESS AIRWAY
    Assess Airway for 5-10 Seconds
    Open Airway/Modified Jaw Thrust/Chin Lift
    Inspect Mouth and Clear (Suction/Heimlich/Laryngoscope)
    Insert Indicated Airway Adjunct (NPA/OPA/ET/CRIC)
    Reassess Airway/BLS as Required
ASSESS BREATHING
    Inspect Anterior Chest (bilateral rise and fall)
    Occlude Chest Wounds
    Auscultate Anterior Chest X1 bilateral
    Palpate Anterior Chest
    Percuss Anterior Chest X1 bilateral
    Palpate Posterior Chest
    Identify/Stop Gap Treat Posterior Wound
    Manage Injuries that Could Complicate Breathing
ASSESS CIRCULATION
    Identify and Control Major Bleeding Pack, Clamp, Dress PRN
    Apply Tourniquet PRN (Amputation, Major Bleeding)
    Assess Pulse (Radial/Femoral/Carotid)
    Assess Peripheral Perfusion (Skin Color/Temp)
    State General Impression of Patient Based on Injuries/Findings

Log Roll, Identify/Definitively Treat Downside Wounds (tension pneumo after movement if not treated)
Expose as needed
Initiate Movement within 10 minutes

Secondary Survey
Reassess all Previous Treatments Following Movement
Move Patient Safely
Mental Status (AVPU/GCS)
Reassess ABC/TX
Airway (ET if unconscious) (reassess placement PRN)
Neck
    Assess for Jugular Vein Distention
    Assess for Tracheal Deviation
    Palpate C-spine
    Treat all Injuries to Neck
Anterior Chest
    Reassess Prior Treatment/Reinforce PRN
    Auscultate x 3 Bilateral
    Auscultate Heart
    Percuss x 3 Bilateral
    Palpate
Abdomen/Pelvis
    Inspect the Abdomen for Obvious Injury
    Assess the Pelvis/Shoulders
Extremities (Hemorrhage/Fracture)
    Inspect & Palpate Arms
    Inspect & Palpate Legs
L-Spine
    Palpate L-spine/Treat Injuries
Log Roll
    Maintain C/L-Spine Cntrl
Assess/TX Posterior
    Head to Foot
Reassess Airway
    Check Placement of ET Tube
Reassess Circulation
    Reinforce Treatments, Apply Stump Dressing PRN, etc.
Obtain Baseline Vital Signs
Obtain Patent Percutaneous IV
PERRL
AVPU/GCS
Identify and Treat All Life Threatening Injuries
Full Set of Vital Signs
Fracture Immobilization
    Splint all Long Bone Fractures

EVACUATE
Move Patient Safely
Reassess ABCs w/placement Check
Resuscitative Phase
    Vital Signs
    Treat for Hypothermia
    Patient Positioning

Perform Detailed Physical Exam
   HEENT
   Thorax/Heart
   Abdomen/GU
   Musculoskeletal
   Neurological
   Identify and Treat all Wounds
   Cricothyroidotomy, Tube Thoracostomy, Venous Cutdown
   Splint Cutdown Site
   9-Line CASEVAC (ASAP)
   Field Medical Cord (FMC)

Chapter 29: Trauma: Human and Animal Bites
COL Roland J. Weisser, Jr., MC, USA

Introduction: Human bite wounds have a notorious reputation, based primarily on the risks associated with a single injury, the "clenched fist" injury. Human bites in other areas of the body pose no greater risks than animal bites. Institutionalized patients with poor impulse control create an especially high-risk environment for human bite wounds. Three types of human "bite" wounds may lead to complications:
1. Clenched fist injury: This wound most often results when a clenched fist strikes the mouth/teeth of an adversary, and the force of the punch breaks the skin. The hand is flexed when the injury is sustained, inoculating bacteria directly into the wound. Then, when the hand is subsequently relaxed, the tendon retracts into its sheath, carrying the inoculum into the tendon sheath, making normal irrigation and cleansing techniques difficult and less effective. This type of bite wound has the highest risk of infection.
2. Bite to a finger: Fingers are enveloped in only a thin layer of overlying skin that constrains the underlying tendons and their sheaths, only a few millimeters beneath the surface. Hence, when a finger is bitten, even though the wound may appear to be only a superficial abrasion there is potential for inoculation of the tendon sheaths through an unnoticed skin defect.
3. Puncture wounds about the head: This type of injury is usually sustained during "horseplay" among children of all ages. The tooth impacts the head, producing the wound. Although the wound may appear innocuous on the surface, deep contamination may occur.

Animal Bites: There are an estimated 1-3 million animal bites in the U.S. per year. 80-90% inflicted by dogs, 5-15% by cats, 2-5% by rodents, and the balance by rabbits, ferrets, farm animals, monkeys, reptiles and other species (see Toxicology: Venomous Snake Bites). Monkey bites have a notorious reputation, but the reputation is primarily based on anecdotal reports. Dog bites cause a crushing-type wound due to their rounded teeth and strong jaws. The pressure may damage deeper structures such as bones, vessels, tendons, muscle and nerves. Cats, due to their sharp, pointed teeth, usually cause puncture wounds and lacerations with the inoculation of bacteria into deep tissues. Cats more frequently bite women. Dogs more frequently bite men. The hand is the anatomical part most frequently bitten.

Bites on the hand have a risk of infection due to the relatively poor blood supply, and anatomic considerations that make adequate cleansing of the wound difficult. In general, the better the blood supply, the easier the wound is to clean (e.g., laceration vs. puncture), thus lowering the risk of infection. Nearly any group of pathogens, including bacteria, viruses, rickettsia, spirochetes, and fungi may cause infection. Dog and cats are more likely to host *pasteurella* and *staph. aureus*, among others. However, many infected bite wounds are mixed infections, with any of the organisms ultimately having the potential to produce sepsis, meningitis, osteomyelitis, or septic arthritis.

Subjective: Symptoms
A puncture, laceration or abrasion possibly with contusions, erythema, edema, pain, throbbing or itching.

Because human bite wounds may be intentional injuries, always consider the potential for alcohol, child, and spouse abuse.

Focused History: *Who or what bit you? When were you bitten? Where were you when you were bitten?* (aids in assessment and treatment, but the history provided concerning human bites is known to be notoriously unreliable, unlike animal bites. Animal bites: Determine the kind of animal and its status [i.e., general health, rabies vaccination status, behavior]; the time and location of the incident; the circumstances surrounding the bite [i.e., defensive vs. unprovoked]; and the whereabouts of the animal [loose in the wild or observable in quarantine]. Animal bites, particularly in developing countries, carry a high risk of rabies infection [see ID: Rabies]. Human bite wounds: Often infected when patients present for the first time because the wounds appeared so innocuous and the patients delayed seeking care. Other aspects that influence presentation include the patient's health status: HIV status of biter, tetanus immunization status, time delay since receipt of the injury, amount of disability encountered, and any underlying disease). *Have you been immunized against tetanus?* (determines status of tetanus prophylaxis.)

Objective: Signs

Using Basic Tools: Adequate visualization may require superficial surgical extension of the wound to determine involvement of tendon sheaths. Maintain high index of suspicion for infection.
1. Clenched fist injury: Evaluate integrity of extensor tendons; inspect for signs of infection (hot, swollen, red), palpate for crepitus; inspect for loss of knuckle height, or penetration into the joint capsule.
2. Bite to a finger: Carefully inspect and palpate all bite wounds of the fingers for deeper penetration into underlying structures. Evaluate for integrity of the extensor and flexor tendons; inspect for evidence of flexor tenosynovitis.
3. Bites about the head: Ear and nose bites: Inspect for loss of tissue; palpate for cartilage tears and depth of penetration into adjacent structures.
4. Consider when appropriate: Distal neurovascular status, tendon or tendon sheath involvement, bony injury (particularly in the skull of infants and children), joint space violation, visceral injuries, and foreign bodies (e.g., teeth in the wound).
5. Consider possible cervical spine injuries inflicted from shaking by large animals.

Using Advanced Tools: X-Rays: Clenched fist injuries have an associated risk of metacarpal head fracture that may require surgical treatment.
"Old" infected bites may reveal cortical erosion, periosteal new bone formation or bone loss seen with osteomyelitis.

Assessment:
Diagnose based on history of a bite, wound appearance and captured specimens (teeth).
Differential Diagnosis - other animal bites, insect bites; cellulitis or deep hand infections due to other causes; foreign bodies from other trauma

Plan:

Treatment
1. Bite site and appearance:
    A. Superficial human bites (mixture of abrasions and contusions) can be managed adequately with only local cleansing and tetanus immunization.
    B. Bites to the ear and nose: When associated with tissue loss or violation of cartilage require consultation with surgery (plastic surgery or ENT if possible). Penetrating bites in cartilage are slow to heal due to the limited blood supply and difficulty in treating chondritis.
    C. Consider primary closure in relatively clean wounds or wounds that can be effectively cleansed. Facial wounds, because of the excellent blood supply, are at a low risk for infection even with primary closure.
    D. Deep bite wounds, animal bites, those to the lower extremities, those with a delay in presentation, or those in compromised hosts generally should be left for closure by secondary intention.
2. Irrigate wounds with water or sterile saline (preferred) using a 19 gauge blunt needle and a 35ml syringe to provide adequate pressure (7psi) and volume. Flush individual punctures with approximately 200 cc of irrigation solution. Heavily contaminated wounds may require more.
3. Debride devitalized tissue, particulate matter and clots to provide clean wound edges that will result in

smaller scars and promote faster healing.
4. Use prophylactic antibiotics in all bites. If antibiotics are used for an active infection, the duration of therapy should be 7-14 days depending on the severity of the infection and the clinical response. Most likely organisms for human bites: *Streptococcus viridans* 100%; *Bacteroides* 82%.
    a. Early (not yet clinically infected): **Amoxacillin/Clavulanate (Augmentin)** 875/125 mg bid po x 5 days
    b. Later (Signs of infection 3-24 hours): **Ampicillin/Sulbactam** 1.5gm q 6h IV or **cephotoxin** 2.0gm q 8h IV or **ticarcillin/clauvulanate** 3.1 gm q 6h IV or **piperacillin/tazobactam** 3.375 gm q 6h IV or 4.5 gm q 8h IV. If penicillin allergic, use **clindamycin** 300 mg qid po (+) either **ciprofloxacin** 500 mg bid po or **trimethoprim/sulfamethoxazole DS** po bid x 7 days.
    c. Bat, raccoon, and skunk bites (very high infection risks): **Amoxacillin/Clavulanate (Augmentin)** 875/125 mg bid or 500/125 mg tid po x 7 days: Alternates: **Doxycycline** 100 mg bid po x 7 days.
    d. Cat bites (80% become infected): **Amoxacillin/Clavulanate (Augmentin)** 875/125 mg bid or 500/125 mg tid po x 7 days; Alternates: **Cefuroxime axetil** 0.5 gm q 12 hours po or **doxycycline** 100 mg bid po. Resistant organisms seen, so non-healing infections may respond to **penicillin G** IV or **penicillin VK** po. Observe for osteomyelitis.
    e. Dog bites (5% become infected): **Amoxacillin/Clavuanate** 875/125 mg bid or 500/125 mg tid po x 7 days. Alternates: **Clindamycin** 300 mg qid po (+) **ciprofloxacin** 500 mg bid po x 7 days. Observe for osteomyelitis.
    f. Bites from hospitalized patients: Consider including coverage for aerobic gram negative bacilli.
5. Give tetanus antitoxin if >5 years since last dose.
6. Use narcotics or benzodiazepines judiciously for agitation (see Pain Assessment and Control).
7. Consider anti-rabies therapy (see ID: Rabies)
    a. Infiltrate around inoculation site with ½ dose of **human rabies immune globulin** (HRIG 20 IU/kg), give remaining ½ IM into gluteal region.
    b. If patient is not immunized against rabies, give **human diploid cell rabies vaccine** (1 ml IM in deltoid x 5 as detailed on Immunization Chart in Preventive Medicine Chapter) beginning immediately. For individuals who have been fully vaccinated against rabies previously (including ID and IM protocols given to most SOF personnel), give 1 ml IM booster dose in deltoid immediately at presentation and again 3 days later. Pre-exposure vaccination does not guarantee protection against rabies, but it does buy time to get to definitive treatment if bitten, and it does decrease the number of post-exposure boosters required.
    c. If possible, isolate suspected animal source and observe 10 days for signs of rabies.

Patient Education:
General: This wound has a high risk of infection so close follow-up is needed. Return promptly for fever and hot, red, or swollen wound, particularly if accompanied by swollen nodes or streaks (blood poisoning) traveling away from the wound.

Follow-Up Actions:
Wound Care/Return Evaluation: Recheck patient in 24-48 hours if not infected at first visit, and followed daily if infected.
Evacuation/Consultation Criteria: Bites with extensive tissue loss, involvement of complex/deep structures, penetration of the skull, and infection failing to respond to above antibiotic regimens should be evacuated ASAP. Consult general surgery or infectious disease specialist in these cases, and others as needed.

Chapter 30: Shock
Shock: Anaphylactic
COL Clifford Cloonan, MC, USA

Introduction:
Anaphylaxis is an acute, generalized allergic reaction affecting the cardiovascular, respiratory, cutaneous, and gastrointestinal systems. This is a severe immune-mediated reaction that occurs when a previously sensitized patient is again exposed to an allergen. Common causes: bee/wasp stings, penicillin or other drug allergies (esp. when given IM/SC/IV), seafood (esp. shrimp/shellfish) and nuts of various types. Allergens may produce an allergic reaction by being ingested, inhaled, injected, or absorbed through the skin/mucous membranes. Shock is produced by the release of histamine that causes bronchospasm/wheezing. Histamine also causes "leaky" vessels resulting in hives/edema and hypotension. This produces both a volume problem and a vascular resistance problem. Anaphylactic shock differs from less severe allergic reactions in that it is characterized by hypotension and obstructed airflow (upper and/or lower) that can be life-threatening. History of allergies is a risk factor.

Subjective: Symptoms
General malaise/weakness, lightheadedness, anxiety/feeling of impending doom diffuse itching/"scratchy" sensation in the back of the throat, chest tightness/difficulty breathing. If allergen was ingested, there may be associated nausea/vomiting and diarrhea.

Objective: Signs
Using Basic Tools: Vital Signs: Pulse: tachycardia; B/P: hypotension and orthostatic hypotension; +/- narrowing pulse pressure (systolic-diastolic pressure); Respirations: tachypnea/hyperpnea
NOTE: Children and physically fit young adults may maintain near normal vital signs until significant shock is present and death is imminent
Inspection: Anxious appearance with increased respiratory rate +/- audible stridor, +/- altered mental status, flushed/red skin with hives (urticaria), and edema (esp., periorbital/perioral). Nasal mucosa may be congested, swollen and inflamed. Profuse watery rhinorrhea, itchy eyes, followed by wheezing are characteristic. Generalized itching with hives (pruritus and urticaria), and occasionally angioedema of the face (swelling of the eyelids, lips, cheeks) Capillary refill: Delayed in shock, longer than 3 seconds.
Auscultation: Inspiratory and expiratory wheezing (bronchospasm)
Palpation: Warm skin (until severe hypotension develops at which point skin becomes cool/moist)
GI: Bowel edema, causing cramps and water diarrhea.
Using Advanced Tools: Pulse oximetry

Assessment:
Differential Diagnosis
Allergic reaction without hypotension and/or airway obstruction
Vasovagal reaction after injection/immunization (common)
Cardiogenic shock
Angioedema

Plan:
Diagnostic Tests
Essential: Clinical observation is the only diagnostic test. Use the rapidity and constellation of symptoms to suggest the diagnosis. A prior history of similar symptoms may be the only other clue.
Close observation with frequent assessment/reassessment of mental status, vital signs, and pulse oximetry
Recommended: Continuously monitor urinary output. If patient is intubated and ventilations are being supported, frequently reassess the pressures needed to ventilate.

Procedures
Essential: Reduce/eliminate all further exposure to allergen. If due to an injected drug or venom, apply

loose tourniquet proximal to injection/bite/sting site and place injection site in a dependant position to reduce venous/lymphatic circulation. Apply ice to, and consider injecting small dose of **epinephrine** (0.1-0.2 ml 1:1,000) into, the injection site, unless contraindicated. If due to bee/wasp sting(s), carefully remove all stingers. Avoid applying pressure to venom sac while stinger is inserted in patient. If due to an ingested food/drug, give **activated charcoal** (50 gm for adults).

Recommended: Evaluate for common causes of anaplylactic shock.
A. Any food can contain an allergen that could cause anaphylaxis
B. Any drug is capable of causing anaphylaxis
C. Insect bites and stings - biting flies, and stinging insects - most types
D. Latex allergy - contact reactions that get progressively worse among medical personnel

Treatment
NOTE: IV administration of **epinephrine (adrenaline)** is DANGEROUS if it is given in too large a dose, if given too quickly IV, and/or if given unnecessarily. DO NOT USE IV **epinephrine** to treat a simple allergic reaction without signs of shock and/or severe bronchospasm and/or stridor - give either subcutaneously, or intramuscularly.
Epinephrine: **WARNING:** Use only 1:10,000 concentration for IV - if only 1:1,000 concentration is available dilute to 1:10,000 before using.
Dose: 0.1ml/kg 1:10,000 slow IV (up to 5 ml total dose for adults) given over 15-20 minutes in 1 ml increments every 3-5 minutes. Give **epinephrine** endotracheally if necessary to treat severe hypotension and bronchospasm. If unable to obtain an IV and patient is not intubated, **epinephrine** may be given intraosseously or deep IM (avoid SC administration in hypotensive patients due to poor absorption).
Bronchospasm: Treat with inhaled beta agonists (i.e., **albuterol** or **epinephrine**)
Benadryl 50-100 mg IV over 3 minutes. Consider adding **cimetidine** 300 mg IV q 6 hr
Solu-Medrol (methylprednisolone 125-250 mg q 6 hr)
Trendelenburg position
Oxygen
Crystalloid (saline) fluid bolus IV, titrated to restore and maintain blood pressure
Apply PASG (MAST) if available and if hypotension is unresponsive to epinephrine and fluids

Patient Education
General: Activity: Following resuscitation, inform patient that he/she has experienced a life threatening allergic reaction. If the triggering allergen is known, warn patient to avoid any future exposure.
Diet: Avoidance of allergen if known
Medications: Anaphylaxis kit (Epi-Pen autoinjector; Ana-Kit) for use in the event of recurrence. **Albuterol** (or other beta-agonist) inhaler if bronchospasm was a prominent symptom (be sure to properly teach patient how to use inhaler with spacer).
Prevention and Hygiene: Avoid circumstances in which recurrent exposure is possible/likely
No Improvement/Deterioration: Return immediately for any recurrence of symptoms after first self-administering anaphylaxis kit.

Follow-up Actions
Return evaluation: Return for repeat evaluation/treatment if fainting/near fainting occur; if difficulty breathing does not resolve with treatment
Consultation Criteria: If cause of allergic reaction is unknown, patient should be advised to seek allergist in an effort to isolate cause

Shock: Hypovolemic
COL Clifford Cloonan, MC, USA

Introduction: Hypovolemic shock is usually caused by hemorrhage but may also be caused by burns, severe or prolonged diarrhea (cholera), prolonged vomiting, internal third space loss (as in peritonitis), or crush injury. Tissue injury from trauma may worsen shock by causing microemboli that further activate the inflammatory and coagulation systems. Hemorrhage sufficient to cause shock usually happens in the torso, in the thigh(s) (femur fracture), or externally. Fractures of the femur, pelvis, and/or traumatic amputation are associated with substantial blood loss. From a clinical perspective, attempts to quantify blood loss in order to determine a shock category is of little value because even external blood loss is notoriously difficult to quantify and quite often trauma patients have significant internal as well as external hemorrhage. Treat the patient- not the evident blood loss.

Subjective: Symptoms
Constitutional: Diffuse weakness, anxiety/feelings of impending doom, difficulty concentrating, c/o being chilled to the bone; progressive thirst; shortness of breath. Consider thirst progressing in severity and breathing that becomes progressively deeper and more rapid to be evidence of worsening shock until proven otherwise.

Objective: Signs
Using Basic Tools
Vital Signs: **WARNING**: Children and physically fit young adults may maintain near normal vital signs until significant shock is present and death is imminent!
Pulse: Tachycardia except in some cases where an unexpected bradycardia is found (penetrating abdominal trauma, ruptured ectopic pregnancy or other pelvic bleeding); B/P: Progressive hypotension and orthostatic hypotension; narrowing pulse pressure (systolic - diastolic pressure); Respirations: Tachypnea/hyperpnea; measurement of orthostatic vital signs may be helpful when significant postural hypotension is documented but this test is neither sensitive nor specific for shock.
NOTES:
1. Most useful of all the vital signs in assessing hypovolemic/hemorrhagic shock is the pulse pressure (systolic - diastolic pressure), which becomes progressively narrowed as shock proceeds. The normal pulse pressure for an adult is between 30 and 40 mm Hg. If a blood pressure cuff is not available, estimate the pulse pressure by the strength of the pulse. A weakening pulse implies a narrowing pulse pressure.
2. More important than the absolute value of any of the vital signs at a given point is their trend over time. Do not overlook falling blood pressure, a narrowing pulse pressure, and a rising heart rate- these are signs of progressing shock.
3. Continuously measuring the hourly urinary output is a readily obtainable, objective means of determining the adequacy of intravascular fluids.

Inspection: Pale, diaphoretic, anxious appearing. If degree of shock/hypotension is severe (i.e., systolic B/P < 60-70 mm Hg) then there may be evidence of altered mental status ranging from confusion to unconsciousness. If shock is due to trauma there is often external evidence of traumatic injury.
Auscultation: Clear lungs with deep, rapid respirations unless there is intrathoracic trauma
Palpation: Cool, moist skin. In non-hypothermic patients an ascending palpation of the skin from feet to chest to note the point at which the skin becomes warm is a useful, rapid, method for estimating the degree of shock. The more severe the shock, the more proximal the level of warmth.
Mental status: Often described as being altered in moderate to severe shock but because cerebral blood flow is preserved to the last, a patient's mental status may be normal or near normal until right before death (when MS is abnormal look for other causes, especially closed head injury in a traumatized patient)
Capillary refill: Normally prolonged beyond 3 seconds in shock BUT interpretation is difficult in elderly patients (normal is up to 4.5 sec), cold environment or poor lighting.
Using Advanced Tools: Continuous monitoring of urinary output and pulse oximetry.

Assessment:
Differential Diagnosis
Vaso-vagal faint - transient hypotension due to bradycardia caused by parasympathetic stimulation
Hemorrhagic shock - intravascular volume depletion through blood loss. Body responses intact (narrow pulse pressure).

Septic shock - loss of vascular tone due to release of infectious toxins in the circulatory system. (Body responses impaired)
Cardiac shock - loss of cardiac output to sustain blood pressure. Body responses usually intact unless on B/P medications.
Anaphylactic shock - loss of vascular tone due to an allergic reaction. Body responses impaired.
Neurogenic shock - loss of vascular tone due to impaired neural or spinal function. Body responses impaired.
Distributive shock - distribution of blood flow is impaired (pulmonary embolus - blocks blood from entering into lungs, cardiac tamponade - mechanical impairment of the pumping action of the heart and tension pneumothorax- loss of blood return to the thorax and heart due to pressures building in the chest cavity).

Plan:
Procedures
Essential: CONTROL ALL HEMORRHAGE. Eliminate all possible sources of ongoing intravascular volume loss.
Once hemorrhage is controlled, initiate blood administration as soon as it can safely be accomplished.
If unable to control hemorrhage (i.e., intrathoracic/intraabdominal bleeding), urgent evacuation to a medical facility with general surgical capability is indicated ASAP - do not delay evacuation to initiate any procedures other than to secure the airway. Further efforts at hemorrhage control should be performed en route and should not delay evacuation. Initiate two large bore IV infusions with appropriate fluid (saline, RL, Hespan, etc.). If hemorrhage has been controlled, administer fluid to obtain systolic BP > 90 mm Hg (see Shock: Fluid Resuscitation).
Recommended: For lower abdominal/pelvic bleeding and/or large thigh/buttocks wounds or femur fractures PASG (MAST) application is reasonable and appropriate. Applying PASG when patient has uncontrolled intrathoracic bleeding is contra-indicated (See Procedure: PASG).

Treatment
Primary:
1. Control hemorrhage first! If volume loss is due to other causes, i.e., burns, diarrhea, vomiting, etc., the same principle applies - try to prevent further loss but in these cases early aggressive fluid resuscitation is appropriate. Measure urine output and all blood and fluid loss to insure replacement and use guidelines such as those used to direct fluid resuscitation of burned patients to estimate volume losses.
2. Volume and IV fluid resuscitation is a temporizing measure, not a treatment for uncontrolled hemorrhage. When hemorrhage has been controlled, administer sufficient fluids (po and/or IV) to maintain an hourly urinary output of > 50 cc/hr. Attempting to maintain urinary output (and systolic pressure) of patients with on-going, uncontrollable, blood loss may temporarily maintain their urinary output at the expense of increasing the rate of red blood cell loss. There is no simple answer to this dilemma and the most appropriate response probably depends upon a variety of factors, such as prior hydration and health status, anticipated time to surgery, availability of IV fluids.
3. Preserve body heat by passive rewarming (by blankets, etc.). External heating (active rewarming) before volume has been restored may produce undesired vasodilation, worsening hypotension.
4. Place all badly wounded patients in a position with their feet elevated about 12 inches with the head and heart low. Utilize the head-down position unless it causes obvious distress, labored respiration or cyanosis, even in patients with chest wounds and with head wounds as long as their systolic blood pressure remains below 80 mm Hg. When the systolic blood pressure has risen above 80 mm Hg, gradually start slow elevation of the head.
5. When hemorrhage cannot be controlled, the priority is rapid transport to a surgeon for surgery ASAP. Attempt to control pelvic bleeding and bleeding into the thigh(s) with PASG. When bleeding is into the chest/abdomen, limit fluid administration (see Table 7-2).
6. Both normal saline and Ringer's Lactate are appropriate fluids, as is Hespan. The only resuscitation fluid for hemorrhagic shock that is significantly better is blood- it is the only oxygen-carrying fluid available in the field today.

NOTE: Do **NOT** treat bradycardia in hypovolemic patients with atropine. The appropriate treatment is to stop further blood loss and rapidly restore intravascular volume, preferably with blood.

Primitive: When IV fluids cannot be administered to restore intravascular volume, rehydrate orally. Start with small volumes and increase as tolerated (see Table 7-1).

Patient Education
General: Activity: Bedrest
Diet: If surgery is imminent, keep patient NPO but clear liquids as tolerated is appropriate in most circumstances
Prevention and Hygiene: If cause is penetrating injury emphasize importance of use of body armor when patient recovers.
No Improvement/Deterioration: Search for other causes of symptoms. Evacuate for surgery if cause is on-going, uncontrollable, hemorrhagic
Wound Care: Patients in shock are more susceptible to infection than those with normal perfusion. Keep wounds as clean as possible and watch closely for developing infection
Evacuation/Consultation Criteria: Uncontrollable hemorrhage requires rapid evacuation to a general surgeon

Shock: Fluid Resuscitation

This text accompanies Table 7-1 on the next page.
1. Maintenance rate calculation when NPO: (Weight in Kg) + 40 = ml per hour of infusion rate
2. End Points of resuscitation:
    a. Controlled hemorrhage, dehydration, burns: Normal BP, pulse, urine output (0.5-1 ml/Kg/Hr), normal capillary refill, good mentation.
    b. Uncontrolled hemorrhage of the trunk: Accept lower BP. Begin fluids if radial pulse not palpable, or pulse is > 120. Run ASAP. Stop when radial pulse palpable < 120. Monitor. May need to restart.

Table 7-1: Fluid Resuscitation
COL David Burris, MC, USA & Mark Calkins, MD

Fluid	Indication	O_2	Potential Benefit	Cautions
Crystalloids Saline Ringer's Lactate	Hypovolemia Dehydration Hemorrhage Shock, Burns	No	Easy to store Inexpensive Effective	Weight ratio- requires 3:1 for lost blood Dilution. Edema, Coagulopathy
Hypertonic Saline 3% 7.5% Hypertonic Saline-Colloid Combinations HTS Dextran HTS Hetastarch	Hemorrhagic Shock Burns-only one dose initially	No	Lighter weight Small volume = large effect Longer duration of effect than plain HTS?	Hypernatremia Do not use for dehydration from vomiting, diarrhea or sweating Do not repeat without addition of other fluids See Colloids
Colloids Albumin Artificial Colloids Dextran Hetastarch *Other commercial names depending on country	Hemorrhagic Shock Burns? Third day.	No	Longer duration 1:1 replacement for blood	Overuse may lead to "leak" into tissue Artificial: Coagulopathy Allergic reaction
Oral Rehydration Fluids	Dehydration Controlled hemorrhage Burns	No	Fluids of opportunity. Non-sterile. Ingredients: 4 tsp sugar, 1/2-1 tsp salt, 1 liter water	Austere option in abdominal wounds and unconscious patients, but use with caution
Blood: Red Cells	Hemorrhage	Yes	Ideal	Storage, type & cross-match
Artificial blood Hemoglobin based Fluorocarbon based	Hemorrhage	Yes	Easy storage	Experimental only Not yet available for use Future option?

Table 7-2: Routes of Fluid Administration

Route	Indication	Potential Benefit	Caution
By mouth	Dehydration, hemorrhage	Normal, may use any fluid	Aspiration risk in unconscious. Avoid in abdominal wounds if possible
Gastric tube	Unconscious. Avoid in abdominal wounds.	May use any fluids	Aspiration risk. Limit infusion rate to 200 hour/hour and check residuals each hour - stop if more than 200 hr
Intravenous	Unable to take oral. Need large volumes	High volumes	Use care in increasing blood pressure for uncontrolled hemorrhage
Interosseous	Unable to obtain IV	High volumes	Same as IV, watch for leak into tissues
Rectal	No other route available, or no sterile fluids available.	Any fluid	Limit flow rates to 200 hr per hour

Chapter 31: Burns, Blast, Lightning, & Electrical Injuries

Burns
LTC Lee Cancio, MC, USA & 1LT Harold Becker, SP, USAR

Introduction: Burns can be classified by their cause, as thermal (i.e., heat), flame, flash, contact (with a hot radiator, etc.), scald (hot water, oil, or other liquid), chemical, electric (covered in Lightning and Electrical Injuries section), or radiation (sunburn, x-rays, nuclear). Inhalation burn injury may occur with or without skin injury, and may be life threatening. The depth of burn is often classified as first, second, or third degree (see below). "Partial-thickness burns" refers to first- and second-degree burns, whereas "full-thickness burns" refers to third degree burns. A burn of 20% of the total body surface area (20% TBSA) or greater is a life-threatening burn (in the very young, very old, or those with serious medical diseases, 10% TBSA or more can be life-threatening). These patients need IV fluid resuscitation and aeromedical evacuation, if possible. Eschar (a layer of burned skin) will form at the site of injury. With time, the eschar will slough off and be replaced with epithelium (new skin) if the burn is partial thickness, or with granulation tissue if the burn is full thickness. Full thickness burns may eventually heal, particularly across the joints, by wound contracture. The best definitive treatment for large open wounds is skin grafting.

Subjective: Symptoms
Painful, red skin without (1st degree) or with (2nd degree) blisters; or dry, charred, non-painful skin (3rd degree), or a combination. May complain of hoarseness or coughing.

Objective: Signs

	First-Degree	Second-Degree	Third-Degree
Typical causes	Sun, hot liquids, brief flash burns	Hot liquids, flash or flame, chemical	Flame, prolonged contact with hot liquid or hot object, electricity, chemical
Color	Pink or red	Pink or mottled red	Dark brown, charred, pearly white, translucent with visible, thrombosed veins
Surface	Dry	Moist, weeping, blisters	Dry and inelastic
Sensation	Painful	Very painful	Anesthetic
Depth	Epidermis	Epidermis and portions of the dermis	Epidermis, dermis, and possibly deep structures
Healing	Few days	Few weeks	Skin grafting or slow inward contraction of edges

Warning: Airway obstruction may present suddenly or gradually over a period of hours with: stridor, hoarseness, coughing, carbon in the sputum or in the mouth, rapid or labored breathing, and finally respiratory distress.

Assessment:
Differential Diagnosis - dermatitis with erythematous or bullous features, abrasions, blast, other trauma.

Plan:
Treatment
Directed toward burns of 20% TBSA or greater, and those with inhalation injury. (For small burns, focus

on wound care.)
Primary:
1. Stop the burning process. Decontaminate chemical burns at the scene. Remove any hot synthetic clothing. For patients with tar burns, immerse the injured areas in cold water until the hot tar has cooled down.
2. Protect the C-spine. Cervical injury is common following high-speed motor vehicle accidents, explosions, high-voltage electrical injury or falls/jumps.
3. Airway. Secure the airway. Prophylactically intubate patients with mild symptoms of airway obstruction (swelling of the face, upper airway or larynx) or smoke inhalation injury, and before a prolonged aeromedical or ground evacuation. See Procedure: Intabate a Patient
4. Breathing. Give 100% O_2 by non-rebreather mask for burn, shock or carbon monoxide poisoning. Intubated patients can be bag-ventilated for prolonged periods (up to 12 hours) during evacuation.
5. Circulation. IV circulatory support: insert 2 large-bore cannulas through (in order of preference): unburned skin, eschar, cut-down or intraosseous cannula. Start Lactated Ringer's (LR) at 500 cc/hr for adults or 250 cc/hr for children age 5-15. Do not give an initial fluid bolus (contraindicated in burn patients), unless the patient has low blood pressure or major mechanical trauma (i.e., bleeding). Secure the lines with suture: tape does not stick well to burned skin. This IV fluid rate will need to be adjusted based on burn size and weight (see paragraph 8 below). Insert Foley catheter to monitor fluid output.
6. Disability. Neurological exam and treat neuro injuries. Even patients with massive injuries should initially be alert, unless they have received drugs, have sustained a head injury, are in shock, or have ingested a toxic substance (carbon monoxide, drugs, alcohol).
7. Exposure and environment. Keep the patient warm by all available means (aluminum combat casualty blanket, warm IV fluids, sleeping bag, etc.). Burn patients lose heat through the damaged skin, and severe hypothermia can result if the environment is not kept hot. Cool only the smallest burns. Never soak a burn patient in wet linens unless he also has heat stroke. Monitor core temperature at least hourly if possible.
8. Fluid Resuscitation. Carefully measure burn size and preburn weight, and estimate fluid resuscitation needs according to the formula below (see Notes).
9. Do a careful secondary survey. Remove all the clothing, roll the patient to inspect the back, and remove all jewelry (especially rings, since fingers can swell causing damage beneath rings). Examine the corneas with fluorescein and Wood's lamp, looking for corneal defects in all patients with facial burns and those who complain of eye problems. Treat corneal abrasions with ophthalmic antibiotics such as **erythromycin** and **gentamicin**. Look for non-thermal trauma. Burns can make it more difficult to detect spinal or extremity fractures, or intraabdominal injury. A diagnostic peritoneal lavage can be done through burned skin. Check the tympanic membranes for rupture in blast injuries.
10. Open fractures in burn patients are at high risk for developing osteomyelitis. Immobilize the fracture with splints. A plaster cast can be used over a burn, but should be immediately bivalved to permit wound care and to allow for post-burn swelling. Definitive care is external fixation.
11. Use frequent, low-dose IV narcotics (e.g., **morphine, fentanyl, methadone**) for pain control. Avoid IM narcotics. IV **ketamine** is useful for painful procedures. Give **Phenergan** 25 mg IV, IM, or po to potentiate the effects of narcotics and treat nausea.
12. Place a nasogastric tube to prevent gastric ileus, vomiting and aspiration. To prevent stress ulceration of the stomach and duodenum, give 30 cc of **magnesium-** or **aluminum-containing antacids** q 2 hrs, preferably via nasogastric tube. Clamp the tube for 30 minutes after each dose, or give an IV H_2-blocker such as **cimetidine**.
13. Immunize against tetanus as needed.
14. If the patient will be evacuated within 24 hrs of injury, then no specific wound care of the burn is needed. Otherwise, cleanse the burns with an antimicrobial solution and daily shower. (Use normal saline or similar to cleanse the face.) In general, do not debride small (< 2 cm diameter) blisters if they are intact, but unroof them if they rupture, lie across major joints, or are large. If the burns are in the scalp, shave the hair. Apply an antimicrobial burn cream such as **silver sulfadiazine** (**Silvadene, Flamazine,** etc.) or **mafenide acetate** (**Sulfamylon**) bid. Following application, the wounds can be left open, but in the field it is best to cover the wounds with sterile gauze dressings or clean linen. IV **ketamine** or narcotics are useful for pain control during dressing changes. Alternate topical treatment: apply

Bacitracin to the face and to burn areas. If burn creams are not available, a 0.5 percent solution of **silver nitrate** in water is also very effective. Apply this solution to a thick layer of gauze dressings at least once every 6 hrs (must be kept moist). Do not put silver nitrate on the face (stains the corneas black). Other options: any antibiotic ointment, **Betadine** ointment or even honey.

15. Do not use prophylactic IV or oral antibiotics. A centimeter of redness surrounding a burn wound is common, and results from local inflammation rather than true infection. If redness spreads and other symptoms of infection appear, the patient has cellulitis and needs anti-streptococcal antibiotics (**penicillin, vancomycin**, or 1st generation cephalosporin). When effective burn creams are not used, the patient may develop invasive gram-negative burn wound infection. Look for systemic signs of sepsis and changes in the color and odor of the burn wound. This is a life-threatening problem– give aggressive fluid resuscitation, and two broad-spectrum IV antipseudomonal antibiotics (**piperacillin, ticarcillin** or a 3rd generation cephalosporin; and an aminoglycoside).

16. Burn patients need more calories and protein. Supplement patients with burns over 30 % TBSA with milkshakes or any similar high-calorie, high-protein food source.

17. Burn patients with deep burns across most or all of the anterior and lateral chest may develop a "chest-eschar syndrome" during the first 24 h post-burn. Full-thickness burned skin (leathery, tight, and inelastic) may act like a straightjacket, inhibiting chest movement during inspiration or bag ventilation. Using a scalpel or electrocautery, cut through the eschar on the chest from mid-clavicular line to anterior axillary line down past the costal margin. Then, connect right and left across the epigastrium (see Figure 7-2, chest escharotomy). Do this procedure immediately when it is needed.

18. Burn patients with circumferential deep burns of the extremities are at risk for an extremity eschar syndrome, in which swelling beneath the inelastic eschar causes gradual constriction of the blood vessels. This can result in nerve and muscle damage, and eventually life-threatening infection of dead muscle and/or limb loss. This syndrome is diagnosed by loss of distal pulses in a patient with deep (full-thickness or deep partial thickness) burns of an extremity. (Note also that low blood pressure due to severe shock may also cause loss of peripheral pulses in burn patients.) Treat with escharotomy: incise the tight, inelastic eschar with a scalpel or electrocautery. Place the incision in the mid-lateral and/or mid-medial line of the extremity (see Figure 7-2). Cut all the way through the skin, but no deeper into the subcutaneous fat than is necessary to release the tension. Low-dose IV narcotics or **ketamine** will help control pain. Check distal pulses after the procedure to make sure it was successful.

Alternate: Chemical Injuries

1. Decontaminate the patient. Decontaminate at the site of injury as thoroughly as possible. Determine exactly what compound caused the injury. Following decontamination, treat in the same manner as thermal injuries.
2. Acids or bases: brush off any solid material, then flush with copious amounts of water—at least 30 minutes for acids, hours for bases. Test the skin with pH paper to determine when it is safe to stop decontamination. Never attempt to neutralize a chemical by applying a basic compound to an acid burn. Burns of the eyes should be continuously irrigated (can use IV line) at the inner canthus. Alkali burns of the eyes may require irrigation for 8-12 hours.
3. White phosphorus (WP): an incendiary compound that ignites on contact with air at $32°C$ ($89.6°F$). To prevent this, wounds containing WP fragments must be continuously immersed in water, saline solution, or similar liquid. Remove the fragments in an operating room and place them in a container of water. A Wood's lamp (UV light) can be used in a dark room to identify these fluorescent fragments.
4. Hydrofluoric acid (HF): HF absorption can cause deep tissue damage and can deplete circulating calcium and magnesium, resulting in lethal dysrhythmias. Topical application of **calcium gluconate** in a gel such as **Surgilube** will chelate the fluoride anion and prevent systemic absorption. This mixture can be placed inside a surgical glove for those patients with HF hand burns.
5. Tar and asphalt. Hot tar and asphalt cause a deep thermal injury. Cool the injured areas in water. Then, apply white petrolatum (**Vaseline**, etc.), mineral oil, or vegetable oil to the area in order to dissolve and soften the material. Do not apply gasoline or other petroleum-based solvents.

Alternate: Fluid Resuscitation

1. If LR is not available, you may use normal saline or alternate between normal saline and sodium bicarbonate solution (mix 2 1/2 50-meq ampules of **sodium bicarbonate** per liter of D5W to make a

Figure 7-1
The Rule of Nines. The numbers give you the percentage of the total body surface area (TBSA) of each of the body parts shown.

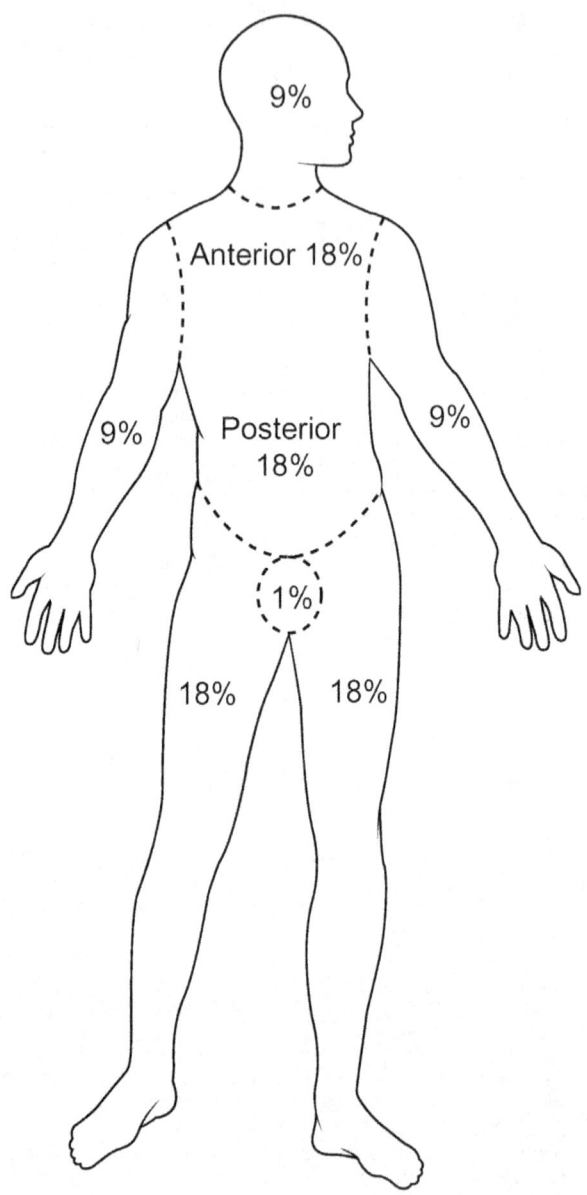

Rule of Nines

Figure 7-2
Perform escharotomies along the lines shown. Avoid making incisions across any involved joints, as indicated by bold lines. Otherwise repair will be very difficult.

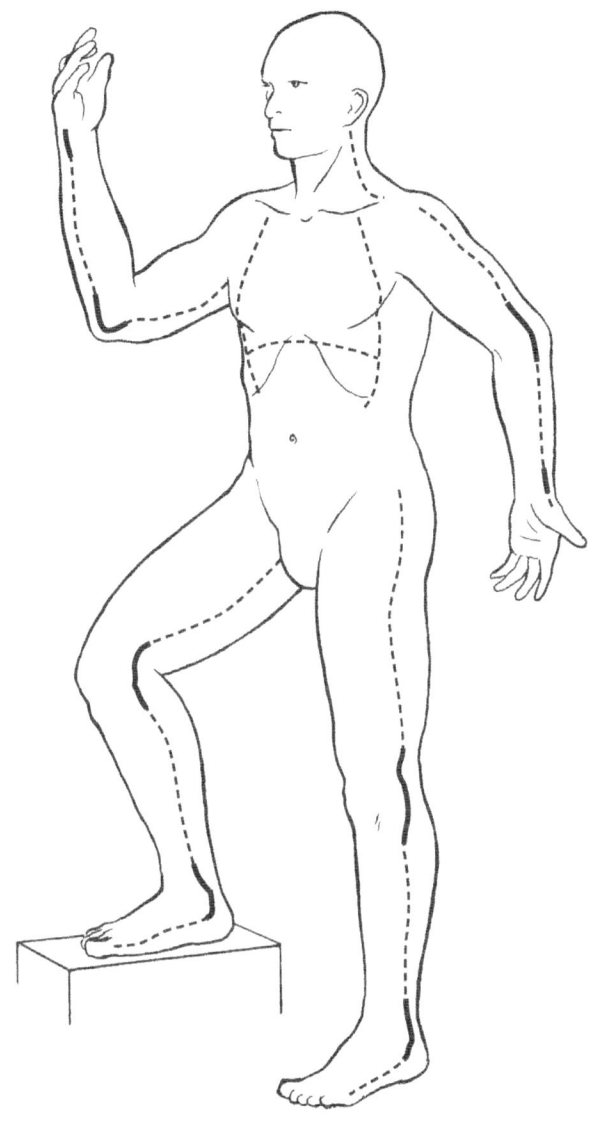

Regions of Escharotomy

solution of 125 meq/L). If you give normal saline alone, you should also start D5W at the rate given below (1cc/kg/TBSA burned/24 h), because of the high sodium concentration in normal saline. You can also use 5% albumin (or fresh frozen plasma) during the first 24 hours post-burn in combination with crystalloid, according to a formula such as albumin (0.5 cc/kg/TBSA burned/24 h) plus LR or saline (1.5cc/kg/TBSA burned/24 h).
2. If unable to provide IV fluids, start oral resuscitation using the following oral formula: 1 liter of water, ¼ tsp of salt, ¼ tsp of sodium bicarbonate (if no bicarbonate, use total of ½ tsp salt), 2 tbsp of sugar or honey and a little orange or lemon juice.

Patient Education
Wound care: Change the dressings bid. Soak the dressings off to decrease pain and prevent disruption of the healing process, wash with soap and water, apply topical burn cream and new dressings. Keep the dressings dry and clean. After the burns are healed, apply a moisturizing cream to the burned area bid. Use sunscreen and sun precautions, since burned skin is more susceptible to sunburn.
Activity: Ambulate, and do range of motion exercises 5-10 minutes every hour to prevent edema and contractures.
Diet: Eat a diet high in calories and proteins during the healing process.
Prevention: Avoid burn agents to prevent further accidents.
No Improvement/Deterioration: Return to the clinic for any difficulty breathing, hoarseness, or worsening cough, and for any increased pain in the burn area, increased redness and blanching to the burn area, red streaks from the burn area, or fever.

Follow-up Actions
Return evaluation: If the patient is reliable, the burn encompasses < 10% TBSA partial thickness burn, and there are no associated injuries such as inhalation burns, then the patient may be able to perform wound care at home. See the patient at least twice weekly until healed. If the patient is unreliable or if burns are $\geq$ 10% TBSA, then daily follow-up or hospitalization is recommended.
Evacuation/Consultation Criteria: Evacuate those with $\geq$ 20% TBSA burned or those with serious associated injuries, such as inhalation injury or fractures. Consult a trauma surgeon or emergency medicine specialist for all burn cases other than sunburns.

NOTES: After initiating IV fluid therapy, calculate the total body surface area (TBSA) burned using the **"Rule of Nines"** (Figure 7-1). First-degree burns are not significant and need not be included. The hand, including the 5 fingers, represents approximately 1% of the total body surface. Find out or estimate the patient's pre-burn weight in kilograms (kg).
The total fluid (Lactated Ringers) required during the first 24 h after injury is estimated using the modified Brooke formula for adults: **2 cc/ kg/%TBSA**. Plan to give ½ of the estimated fluid in the first 8 hrs. In children weighing less than 30 kg the infusion rate is estimated at 3 cc/kg/%TBSA. Plan to give ½ of the estimated fluid over the first 8 h. Children will also need maintenance fluids of 5% dextrose in ½ normal saline. This should be given using a rule such as the 4-2-1 rule: 4cc/kg/h for the first 10 kg, 2cc/kg/h for the next 10 kg, and 1cc/kg/h for the next 10 kg. If a patient's resuscitation is delayed by a few hours, then give fluid more rapidly.
Adjust the initial fluid infusion rate to the urine output. Failure to monitor and record the urine output (catheter or bedpan) and adjust the fluid rate hourly may result in death or in severe complications. Adequate urine output is 30-50cc/h in an adult and 1cc/kg/h in a child who weighs less than 30 kg. If the output is greater, or less, than the target for 2 consecutive hours, decrease, or increase, the IV rate by 20% respectively until the rate is satisfactory.
After the first 24 hours, switch resuscitation fluids to albumin (diluted to 5% in normal saline or similar crystalloid) at 0.5 cc/kg/%TBSA burned over 24 h at a constant rate. If albumin is unavailable, use fresh frozen plasma, or continue the LR while adjusting the rate as before. At the 24 hour point after injury also start D5W at 1cc/kg/%TBSA burned each 24 h at a constant rate (subtract po amount from the total needed per day).
After 48 hours, patients with large burns continue to need fluid (i.e., water or D5W) at 1cc/kg/%TBSA/day, but may be very thirsty, so watch their intake carefully to avoid over-drinking.

Blast Injuries
Lt Col John Wightman, USAF, MC

Introduction: High explosive (HE), thermobaric, and nuclear detonations cause extreme compression of surrounding air or water molecules, creating a blast shock wave. The force of this shock wave can be transmitted into the human body, causing tissue tearing through pressure differentials and forces at air-tissue interfaces. The following identifies the most common consequences of blast injury:

Lung: *Hemorrhage* – Pulmonary contusion, hemoptysis (may threaten airway) hemothorax; *Escape of Air* – pneumothorax, pulmonary pseudocyst, arterial gas embolism (AGE).
Gastrointestinal (GI) Tract: *Hemorrhage* – Hematoma leading to obstruction, upper or lower GI bleeding, hemoperitoneum; *Escape of Contents* – Mediastinitis, peritonitis.
Ear: *Middle ear* – Ruptured tympanic membrane (TM), temporary conductive hearing loss; *Inner ear* – Temporary or permanent sensory hearing loss.

See the appropriate sections in the Trauma chapter on CD-ROM for management of specific trauma.

Subjective: Symptoms
Focused History: Pulmonary: *Are you short of breath?* (Pulmonary contusion inhibits oxygen diffusion and requires more effort to inhale. Pneumothorax and hemothorax decrease volume of air that can be inspired. Shock will cause sensation of dyspnea due to poor tissue perfusion.) *Do you have chest pain?* (Chest pain indicates possibility of penetrating or blunt trauma, pneumothorax, or myocardial ischemia due to coronary AGE.) *What does your pain feel like?* (Chest pain of pulmonary contusion is often described as dull and diffuse, but breathing may also feel tight due to difficulty expanding the chest. Pain may wax and wane with respirations. The pain of pneumothorax will often be sharp and focal, but may be lateral or central. It is usually pleuritic until the lung is completely collapsed. Chest pain that seems like it would be consistent with myocardial ischemia may be due to AGE.) *How much effort is required breathe?* (Dyspnea at rest indicates shock due to external or internal hemorrhage, pneumothorax, or serious pulmonary contusion. The more exertion required to elicit dyspnea, the less lung injury is likely.)
Abdominal: *Do you have abdominal or testicular pain, nausea, urge to defecate, or blood in your stools?* (Penetrating and blunt abdominal trauma cause pain, but primary blast injury of air-containing structures in the GI tract may cause any of the listed symptoms.) *What does your pain feel like?* (The pain of stretched bowel will feel like a persistent gas bubble, though it may have sharp and crampy waves as it is affected by peristalsis. Once the bowel ruptures, pain will decrease until peritonitis begins. The pain of peritonitis is usually diffuse and severe).
Special Senses: *Do you have pain or problems with your eyes or ears?* (evaluate for penetrating or blunt trauma) *What does your pain feel like?* (Any eye pain associated with decreased vision is a penetrating foreign body until proven otherwise. Ear pain caused by a ruptured TM is often initially sharp but wanes over time.)

Objective: Signs
Using Basic Tools: Perform neurological examination as described in appendix.
General: External hemorrhage, dyspnea, altered mental status, seizures.
Vital Signs: Tachycardia (probable hemorrhage); bradycardia (blast-induced vasovagal reaction); irregular heart rhythm (shock or AGE); hypotension (hemorrhage, other causes of shock, or vasovagal reaction).
Inspection: Penetrating wounds, traumatic amputations, cyanosis (indicating hypoxia or cyanide); mottling or blanching of skin (AGE); otorrhea or bleeding from ears (TM rupture or basilar skull fracture).
Auscultation: Unilateral absent breath sounds (pneumothorax or hemothorax).
Palpation: Subcutaneous emphysema (AGE, pneumothorax); abdominal tenderness (penetrating, blunt, or blast trauma)
Pulse Oximeter: SPO2 < 95% on room air following blast indicates some degree of lung injury, inadequate

respirations, shock, or exposure to chemical agent such as cyanide. Persistent SPO2 75-95% on room air indicates mild lung injury.
Using Advanced Tools:
Stool Guaiac: Gross hematochezia (bowel injury), guaiac-positive stool (occult penetrating, blunt, or blast trauma). Ophthalmoscope: Penetrating anterior-eye trauma, lack of red reflex (indicates posterior-eye trauma). Otoscope: Ruptured TM.

Assessment:
Differential Diagnosis
Rapid unconsciousness - penetrating or blunt brain or cardiac trauma, vasovagal syncope, cerebral or cardiac AGE, chemical nerve-agent or cyanide inhalation.
Airway compromise - altered mental status, penetrating or blunt face or neck trauma, inhalation injury, massive hemoptysis, foreign-body aspiration.
Ventilatory insufficiency - pulmonary contusion, pneumothorax (all types), rib fractures, bronchopleural fistula, chemical-agent or biological-toxin exposure.
Shock - external or internal hemorrhage, tension pneumothorax, hypoxia from pulmonary injury, GI bleed (more often lower), coronary AGE.
Focal neurological deficits - head injury, spinal injury, peripheral nerve injury, cerebral or spinal AGE.

Plan:
Treatment: Evaluate for necessity of tetanus immunization booster.
1. **Shock:** Resuscitate as per Shock chapter. Bolus with one quarter the usual amount (crystalloid or hetastarch) and reevaluate to avoid exacerbating lung or brain injury. Repeat boluses as necessary to restore mental status, as the endpoint of resuscitation.
2. **AGE: Primary:** Evacuate URGENT to hyperbaric oxygen chamber. Administer 100% oxygen.
   Alternate: Place casualty in coma position with left side down (halfway between left-lateral decubitus and prone) and head at same level as heart (Figure 1-1). Perform in-water recompression as a last resort in divers exposed to blast.
3. **Massive hemoptysis compromising airway:** Selective intubation of mainstem bronchus on least injured side. Use lumen of tube to facilitate gas exchange in and out of lung with lighter bleeding. Use cuff to prevent blood from side of heavier bleeding crossing into mainstem bronchus of better lung. See algorithm (see Figure 7-5) below. Numbers in boxes of algorithm indicate order of preference of interventions listed.
4. **Pneumothorax, tension pneumothorax, hemothorax, pulmonary contusion:** Consult appropriate section of this book. Do not use morphine if bradycardic. Predict need for positive-pressure ventilation (PPV) or positive end-expiratory pressure (PEEP) using chart below. PEEP will not be available in the field, but PPV can be done with a mouth-to-mask or bag-valve-mask/tube with slower and less forceful deliveries. AGE is most common cause of death in immediate survivors and often occurs when PPV is initiated. Contusion usually requires temporary supplemental oxygen. Pharyngeal petechiae (but not TM rupture) predict higher likelihood of pulmonary contusion.
   Prediction of Respiratory Problems
   Insignificant pulmonary injury may be defined as no dyspnea with exertion after 1 hour of rest post-blast. Significant pulmonary blast injuries may be classified based on pulse oximetry. This may predict likelihood of complications, and requirement for PPV and PEEP.
   Mild: SPO2 > 75% on room air, unlikely to need PPV, normal PEEP if PPV initiated, pneumothoraces occur, bronchopleural fistulae rare
   Moderate: SPO2 > 90% on 100% oxygen, likely to need conventional PPV, PEEP of 5-10 cm H_2O usually needed, pneumothoraces common
   Severe: SPO2 < 90% on 100% oxygen, likely to need unconventional PPV, PEEP > 10 cm H_2O usually needed, pneumothoraces almost universal, bronchopleural fistulae common

5. **GI bleeding, GI tract rupture:** See Abdominal Trauma section on CD-ROM for additional guidance.

Figure 7-5

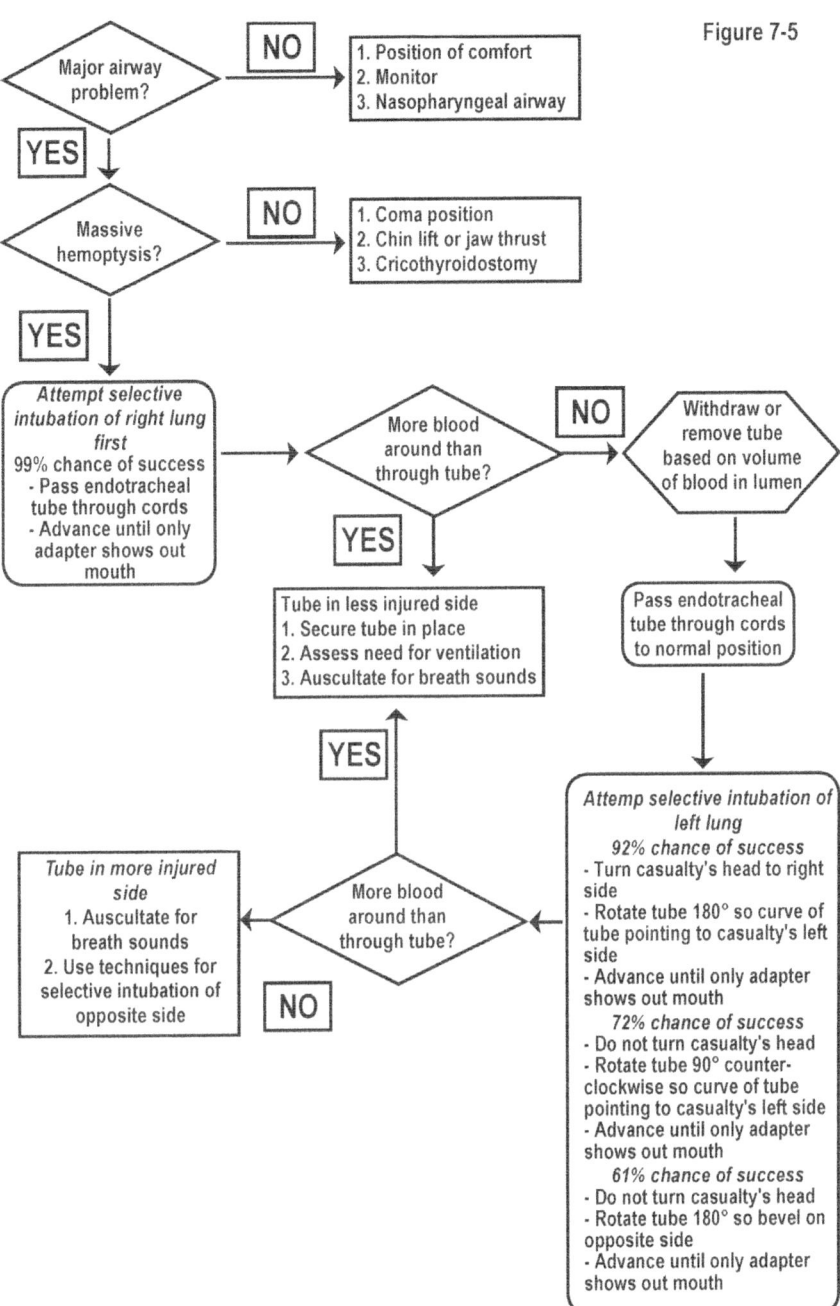

7-25

Primary: NPO. Maintenance IV fluid at 125 ml/hr. **Cefoxitin** 1 gm q 8 hrs or **ceftriaxone** 2 gm q day IV or IM. Evacuate PRIORITY for surgical care within 4 hours. Monitor for peritonitis and sepsis. **Prochlorperazine** 25 mg or **promethazine** 25 mg up to qid IV or IM, if needed to prevent recurrent vomiting.
Alternative: Maintenance po water, if no IV and evacuation time > 4 hours. **Ciprofloxacin** 500 mg q 12 hrs and **metronidazole** 500 mg q 8 hrs po, if parenteral cephalosporins not carried or casualty is allergic to them. Virtually any broad-spectrum antibiotic coverage is better than nothing when time to definitive care is prolonged.

6. **TM rupture:** Patient may not dive, swim or shower until TM heals. Risk of otitis progressing to encephalitis, permanent hearing loss, vertigo, tinnitus and other complications.
   Primary: Do not attempt removal of foreign debris. Prevent water and other non-sterile material from entering ear canal.
   Empiric: Prophylactic antibiotics not indicated. If infection of TM (myringitis) develops, instill ophthalmological (eye) **gentamicin** 2 drops (not ointment) q 6 hrs for 10 days. Otological (ear) suspensions for otitis externa contraindicated when TM ruptured.
   Alternative: Amoxicillin/clavulanate 875 mg q 12 hrs (500 mg q 8 hrs) or **ciprofloxacin** 500 mg q 12 hrs po, if ophthalmological antibiotic drops not available.
   Return Evaluation: Inspect area surrounding ear, external ear, ear canal, and TM daily for redness, swelling, or purulent drainage. Pain when gently pulling up and back on pinna or pressing on cartilage just in front of canal also indicates otitis externa.
   Consultation Criteria: ENT consult within 3 days; sooner if significant debris in canal; up to 2 weeks acceptable, if no infection.

7. **Head trauma from blast:** See Trauma on CD-ROM. Transient amnesia is common after loss of consciousness caused by explosions. Vertigo is usually due to head injury, not blast effects on the ear.
   Meclizine 25 mg q 8 hrs as needed po can improve symptoms but can also sedate, thus impairing ability to function and making assessment of mental status more difficult

Evacuation: Pneumothorax, AGE, and bowel-wall stretching injuries are more likely to be initiated or exacerbated by decreased ambient external pressure (associated with ascent from dive, travel to altitude in ground vehicle or aircraft, or combination) on the casualty. Reassess frequently! Monitor cardiac (EKG monitor if available) and pulmonary status (including pulse oximetry) throughout trip. If there is a possibility the patient has a pneumothorax, place a chest tube before any ascent to altitude. Tension pneumoperitoneum affecting respiration is rare, but may require 14-gauge needle paracentesis in midline just above umbilicus for decompression. Remove air from cuffs on Foley catheters and endotracheal tubes and replace with liquid. Notify receiving facility of urgent need for appropriate special services: neurosurgeon, general surgeon, hyperbaric chamber, intensive care, etc.

Electrical and Lightning Injuries
Maj Michael Curriston, USAF, MC, LTC Lee Cancio, MC, USA & 1LT Harold Becker, SP, PA, USAR

Introduction: Electrical and lightning injuries span a wide spectrum of potential injuries. Low voltage AC tends to result in tetanic muscle contractions "freezing" the victim to the source. High voltage DC exposures tend to be brief and explosive in nature, often throwing the victim from the source. Lightning strikes rarely cause the injuries associated with "man-made" forms of electrocution due to the extremely brief duration. In mass casualty triage following lightning strike, employ a form of **"reverse triage":** Give immediate care first to those victims in cardiopulmonary arrest. The heart often resumes spontaneous rhythmic contractions and patients may only require airway control and ventilatory assistance. Conscious victims of a lightning strike are unlikely to develop imminent demise. **NOTE:** Victims of electrocution or lightning are not electrically "charged" and may be touched immediately after injury provided they are no longer in contact with the "live" electrical source.

Subjective: Symptoms
Cardiopulmonary arrest, loss of consciousness, entry/exit burns and wounds, delirium, pain, numbness and tingling.

Focused History: *Were there any witnesses to your injury? If so, what did they see?* (may confirm electrocution and suggest possible associated injuries.) *Do you have any pain, numbness, tingling, or other abnormal sensations or feelings?* (directs you to areas of injury for closer evaluation and monitoring on serial exams.) *Have you had any change in urine color or frequency?* (decreased quantity and/or frequency of urine, and Coke or tea-colored urine may suggest renal failure due to rhabdomyolysis).

Objective: Signs
Using Basic Tools: Approach patient as any other trauma victim, performing primary and secondary surveys. Assess for evidence of blunt trauma to head, neck and spinal column, chest, abdomen, and musculoskeletal system (fractures, dislocations). Include thorough eye and ear exams. Assess for Compartment Syndrome*. Superficial "fern-like" burns of the skin are indicative of lightning injury. Severe electrical burns may occur with "entrance" and "exit" wounds and extensive tissue damage along the path of the electrical current. Injured muscle releases large amounts of myoglobin that may damage the kidneys. This may be seen clinically as dark tea or coke-colored urine.

*Compartment syndrome may jeopardize the survivability of a limb and results from swelling within tight fascial compartments of the extremities causing compression of nerves and blood vessels. Compartment syndrome may be characterized by the "7 Ps", the last two of which are ominous findings:
1. Pain out of proportion to visible injury
2. Pain worsened with passive stretch of the involved muscles
3. Pressure palpable over the compartment
4. Paralysis or weakness of affected muscles
5. Paresthesias in distribution of affected compartmental nerves
6. Pallor
7. Pulselessness

See Procedure: Compartment Syndrome Management
Using Advanced Tools: Lab: Presumptively diagnose myoglobinuria if positive for blood on urine dipstick and no red blood cells visible on microscopic urine exam. Monitor EKG and cardiac status. RBC count and hemoglobin &/or hematocrit as trauma baseline labs; repeat if patient's condition deteriorates or fails to improve (look for evidence of occult hemorrhage &/or hemolysis).

Assessment:
Differential Diagnosis
Consider traumatic head injury &/or stroke as causes of prolonged altered level of consciousness or coma. Consider internal bleeding (chest, abdomen, GI) as cause of unexplained hypotension.

Plan:
Treatment
Primary: Resuscitate patient following ABCs and cervical-spine control as per BLS/ACLS & ATLS protocols. Establish IV access. Give oxygen and assist ventilation as needed. Perform meticulous wound and burn care. Give antibiotics (e.g., **ampicillin/sulbactam** 1.5-3 gm IV q 6 hrs, or **ceftriaxone** 1-2 gm IV q 24 hrs, or **ciprofloxacin** 200-400 mg IV q 12 hrs) if fever or other sign of infection develops. **Tetanus prophylaxis** is essential. Due to variable extent and depth of burns, burn formulas are not helpful. Closely monitor I & O and titrate IV fluids (normal saline or lactated Ringers) to maintain urine output at 0.5-1 ml/kg/hour (consider Foley catheter and/or NG tube).
Empiric: If patient develops evidence of myoglobinuria, increase IV fluids to maintain urine output twice normal at 1-2 ml/kg/hour (~60-150 ml/hr). Alkalinizing the urine aids in solubility and excretion of myoglobin. Add 1-3 ampules of sodium bicarbonate to each liter of IV normal saline or lactated Ringers. Monitor urine pH with urine dipsticks and adjust amount of bicarb to keep pH greater than 6.5 (add more bicarb to increase urine pH; less to decrease). Monitor lung sounds and jugular veins for evidence of induced pulmonary edema. Consider also giving **furosemide** (**Lasix**) 20-60 mg IV titrated to achieve and maintain diuresis, and/or **mannitol** 12.5 gm IV bolus followed by infusion at 12.5 gm/hour. Do not allow patient to become

hypotensive from resultant diuresis and hypovolemia - adjust IV fluids accordingly. Monitor blood pressure, urinary output.

Primitive: If sterile IV solutions are not available or are in limited supply, and the patient has bowel sounds present and is not vomiting, consider substituting an oral rehydration solution (ORS). You may use any of the commercially available solutions (e.g., Ricelyte, Cera-Lyte) or make your own ORS**, which can be taken orally given through a nasogastric tube "drip" or "infusion". If the patient develops myoglobinuria, attempt to alkalinize the urine by increasing the amount of baking soda (bicarbonate) to 1-3 tbsp/L and adjusting the drip rate to achieve urine output of 1-2 ml/kg/hour. Adjust the amount of baking soda added to the ORS based on a target urine pH of 6.5 or greater. The solution can be placed into a clean, used IV fluid bag that has been cut open just enough at the top to add the ORS. Then hang the bag with IV tubing attached and the other end of the IV tubing attached to the NG tube. Adjust the flow rate in the same manner as any other IV fluid infusion. Adjust the continuous NG "drip" to maintain adequate urine output. The bag and tubing may be used repeatedly in the same patient. Add more ORS solution to the bag as needed. **NOTE: If patient will be going to surgery within 4-6 hours, discontinue oral/NG tube intake unless told otherwise by the surgical team.**

** **ORS Recipe:** 1 tsp of salt, 1 tsp of salt substitute (potassium chloride), 1 tsp baking soda (bicarbonate), 2-3 tbsp of table sugar or 2 tbsp of honey or Karo syrup, all mixed in 1L of clean or disinfected water.

Patient Education
Medications: Furosemide and **mannitol** are potent diuretic agents.
Prevention and Hygiene: Remember scene safety. Turn off the electrical source before making physical contact with the victim. Do not become a victim yourself. Remember that lightning does strike twice in the same area. Avoid being the tallest object in an open area, and being near one. Shelter in a grove of trees if possible. If caught in the open, crouch low and/or seek low ground.

Follow-up Actions
Return Evaluation: If myoglobinuria does not clear within 24-36 hours of adequate urine diuresis and alkalinization, then a source of undetected myonecrosis or muscle ischemia should be sought. Look for areas of swelling and tenderness
Consultation Criteria: Except for the most trivial of electric shocks, all victims of electrical injury should be evaluated by a physician as soon as tactically &/or operationally feasible. Patients with suspected compartment syndrome might require "limb-saving" fasciotomy (see Procedure: Compartment Syndrome Management). **NOTE:** Fasciotomy is not the same procedure as escharotomy, which is done on 3rd degree burn victims to relieve constricting overlying burn eschar.

Chapter 32: Non-Lethal Weapon Injuries
Non-Lethal Weapon Injuries: Laser Eye Injuries
Lt Col John McAtee, USAF, BSC

Introduction: Even low levels of laser energy can burn the cornea or retina of the eye. The retina is particularly vulnerable because the optics of the eye focus the damaging energy of laser light on the retina. The severity of injury depends on duration of exposure, laser wavelength, area of retina damaged and type of lenses or personal protection used. Due to the importance of vision for mission execution and success, as well as the need to protect others from similar burns, laser injuries must be promptly identified, personnel must be quickly moved from the threat environment and the command (and intelligence personnel) must be immediately notified.

Subjective: Symptoms
Range from mild eye irritation to extreme pain and photophobia, immediate partial or complete loss of vision (may be temporary), or loss of peripheral vision

Objective: Signs
Using Basic Tools: Loss of visual acuity-assess with newsprint, or if available, a Snellen Chart or Vision Screener. Loss of visual fields-assess peripheral vision in all quadrants (confrontation test with fingers). If available use Amsler Grid Chart. Corneal or periorbital burn-corneal ulcer or inflammation (fluorescein exam), skin burns. Unequal, unreactive or abnormal pupils.
Using Advanced Tools: Ophthalmoscope: Hemorrhagic debris in the vitreous humor from retinal damage (inability to focus on the retina); disrupted macula.

Assessment:
Diagnose based on clinical signs and symptoms, environment and probability.
Differential Diagnosis
Traumatic eye injury (abrasion, blunt trauma, penetrating trauma, etc)
Infection (iritis, conjunctivitis, blepharitis, etc.)

Plan:
Treatment
Corneal Injury: Treat as an ultraviolet keratitis. If painful, apply topical anesthetic drops, a short acting cycloplegic medication, topical ophthalmic antibiotic and patch. Evacuate if medically or operationally indicated.
Vitreoretinal Injury: Maintain at bedrest if possible, with head elevated and eye(s) patched to facilitate blood settling down and away from the macula. Immediate evacuation is recommended. Do not use steroids to reduce intraocular inflammation without obtaining approval from a physician consultant.

Patient Education
Activity: Bedrest if operationally possible.
Prevention and Hygiene: Use laser protective eyewear in recognized threat environment.

Follow-up Actions
Wound Care: Maintain eye patch for 24 hours for corneal injury, and for the duration of evacuation in the case of vitreoretinal injury.
Evacuation/Consultation Criteria: Evacuate as discussed above. Use chart below for additional guidance. Consult ophthalmology or emergency medicine specialist for all cases of laser eye injuries.
Laser Exposure Evacuation Criteria:

Visual Acuity	Normal	Minor Defect	Major Defect
Macular Damage			
20/63 or worse in one/both eyes	Evacuate	Evacuate	Evacuate
20/50 or better in both eyes	Return to duty	Reevaluate in 15 min.	Evacuate

PART 8: PROCEDURES
Chapter 33 Basic Medical Skills
Procedure: Airway Management

18D Skills and Training Manual, Adapted by COL Warren Whitlock, MC, USA

What: How to assess and control the patient's airway. This guideline does NOT address an entire trauma assessment.

When: You have a casualty in respiratory distress. **Airway:** Check for airway patency. Open the casualty's airway and establish the least invasive but most effective airway. **Breathing**: Determine if the casualty is exchanging air sufficiently to maintain oxygen saturation, or requires assisted ventilations. **Monitor**: After checking and correcting the airway and breathing status, monitor to insure no deterioration. Perform these procedures without causing further injury to the patient.

What You Need: Various sizes of nasal and oropharyngeal airways (see below), gloves, gauze pads, tongue blades, bag-valve-mask (BVM) system, water-soluble lubricant, 10cc syringe to inflate the cuff, stylette, laryngoscope with blades, endotracheal tubes (rough size of little finger diameter; 7-9 for adult, 6-7 for adolescents, 4-6 for children [uncuffed], 3.5-4 for infants [uncuffed]), and oxygen/suction (if available), and emergency drugs.

What To Do:
1. Assess consciousness: does casualty respond to shake and shout, or painful stimuli?
    a. If patient is conscious, go to step 2.
    b. If patient is unconscious, go to step 3.
2. Assess airway and respirations in a conscious casualty.
NOTE: Assessing the airway and respirations are two different steps in the trauma sequence, but every time the airway is assessed, the respiratory effort can also be partially assessed. However, a clear airway with respiratory effort detected does not fully clear the respiratory system. After assessing the airway, assess respiratory effort bilaterally to ensure that both lungs are working and air movement is adequate.
    a. Ask casualty simple questions to determine status of airway.
        (1) If casualty can talk to you without difficulty, airway is clear.
        (2) If the patient answers with difficulty, coughing, pain, hoarseness or other difficulty, manage the airway using the same procedure as if the casualty were unconscious (Step 3).
    b. Auscultate both lungs to ensure that air is being exchanged equally bilaterally.
    c. If history does not point to respiratory/airway involvement and there are no signs of respiratory distress present, continue primary assessment.
        (1) Monitor the patient's airway and respirations.
        (2) Monitor for signs and symptoms of hypoxia.
    d. If signs of respiratory distress develop:
        (1) Initiate appropriate treatment immediately.
        NOTE: Do not attempt to insert oropharyngeal airways or endotracheal tubes in conscious casualties unless they have a history or signs of inhalation burns/injuries.
        (2) Give supplemental oxygen, if available.
        NOTE: Failure to notice signs and symptoms of hypoxia or respiratory distress early may have catastrophic effect on the patient.
    e. If casualty becomes unconscious, manage casualty IAW step 3.
3. Assess airway and respirations in an unconscious casualty.
NOTE: If patient is in a position that makes assessing the airway impossible, move the patient as little as

possible to assess the airway. Be aware of C-Spine control and other possible injuries when moving patient. Remember life has precedent over limb.

    a. Look, listen and feel for respiratory effort.
        (1) Look for bilateral rise and fall of the chest.
        (2) Listen for air escaping during exhalation.
        (3) Feel for breath exhaling from the casualty's mouth on the side of your face.
    b. If respiratory effort is detected, assess the respiratory effort for at least 6 seconds.
        (1) Assess the quality of the respiratory effort as strong, moderate, or weak.
        (2) Assess the rhythm of the respiratory effort as regular or irregular.
        (3) Assess the rate of the respiratory effort: < 10 respirations per minute or >20 respirations per minute are indicators for assisted ventilations.

NOTE: Multiply the number of respirations detected in a 6 second period x 10 to get the number of respirations per minute.

    c. If no respiratory effort is detected, check pulse.
        (1) If the casualty is pulseless:
            (a) In a combat situation, an unresponsive, non-breathing, pulseless casualty is a fatality. End of this task.
            (b) In a noncombat situation, initiate CPR (see Cardiac Resuscitation).
        (2) If the casualty has a pulse, establish an airway immediately.

4. Open and inspect the airway of an unconscious casualty.
    a. Inspect head, face, and throat for signs of trauma and inhalation injuries. Signs of inhalation injuries may include reddened face or singed eyebrows and nasal hair.
    b. Open the airway using the appropriate technique.
        (1) If working on a trauma casualty, use the jaw thrust technique:
            (a) Kneel at the top of the casualty's head.
            (b) Grasp the angles of the casualty's lower jaw.
            (c) Rest your elbows on the surface on which the casualty is lying.
            (d) Lift with both hands, displacing the lower jaw forward.
        (2) If working on a non-trauma casualty, use the head-tilt/chin-lift method.
            (a) Kneel at the level of the casualty's shoulders.
            (b) Place one hand on the casualty's forehead and apply firm, backward pressure with the palm of the hand to tilt the head back.
            (c) Place the fingertips of the other hand under the bony part of the casualty's lower jaw, bringing the chin forward.

CAUTIONS: (1) Do not use the thumb to lift the lower jaw. (2) Do not press deeply into the soft tissue under the chin with the fingers. (3) Do not completely close the casualty's mouth.

    c. Inspect the oral cavity for foreign material, blood, vomitus, avulsed teeth, and signs of inhalation injuries. If the casualty has signs of trauma, foreign objects, and/or complications, continue with this step.
        (1) If casualty is breathing with adequate respiratory effort/air exchange and has no signs of trauma, foreign objects, or complications of the upper airway, proceed to step 5.
        (2) If airway is clear but no respiratory effort is detected, see step 6.
        (3) If airway is not clear, regardless of respiratory effort, see step 7.

5. Insert an oropharyngeal airway (J tube) if the casualty is breathing, has no history of apnea, and no trauma or complications of the upper airway. Have suction available before attempting.
    a. The oropharyngeal airway should be approximately the same length as the distance from the corner of the casualty's mouth to tip of his ear lobe.
    b. Insert the airway inverted until past the tongue and then rotate 180°.

WARNING: It is more traumatic (and contraindicated in children) to use this "corkscrew" technique. If a tongue depressor is available, it is preferable to use it to depress the tongue and insert the oral airway under direct vision.

    c. Check for respiratory effort after J tube is inserted. Respiratory effort should be the same or improved after insertion of J tube. If decreased, remove tube, re-inspect airway, reinsert J tube and reassess.
    d. Have assistant provide ventilations and administer oxygen if available.

    e. Continue with the survey.
6. If the casualty has no respiratory effort and no apparent obstruction of the airway, attempt to give 2 breaths using the rescue breathing technique.
    a. If the breaths go in, intubate and ventilate the casualty (see Procedure: Intubate a Patient).
    b. If the breaths do not go in, attempt to reopen the airway again and give 2 more breaths.
       (1) If the breaths go in, intubate and ventilate the casualty.
       (2) If breaths still do not go in, insert laryngoscope and inspect the oropharynx for foreign body, blood, vomitus, swelling or other causes of obstruction.
       (3) Using forceps, attempt to remove any foreign objects seen.
          (a) If able to clear airway, attempt 2 breaths and assess for return of spontaneous respirations.

NOTE: If at any time spontaneous respirations return after clearing an airway, the casualty requires assisted ventilations with an oropharyngeal airway or ET tube. Casualties who were apnic for any length of time will have an elevated CO_2 level. Traumatized casualties who were apneic will have difficulty regaining O_2 saturation. They may start off breathing adequately, but their CO_2 deficit will cause them to destabilize over time. Failure to assist ventilations in a formerly apneic casualty WILL cause harm and possible death.

          (b) If unable to clear airway, perform surgical cricothyroidotomy (see Procedure: Cricothyroidotomy).
          (c) If the situation makes it impossible to perform an immediate surgical cricothyroidotomy, perform a needle cricothyroidotomy (see Procedure: Cricothyroidotomy).
       (4) If no obstruction is seen but vocal cords are visualized, attempt to intubate casualty
          (a) Successful intubation: ventilate casualty (see Procedure: Intubation).
          (b) Unsuccessful intubation: perform surgical or needle cricothyroidotomy (see Procedure: Cricothyroidotomy).
       (5) If no obstruction of airway is seen but vocal cords are not visualized, perform surgical cricothyroidotomy.
7. Clear the airway of a casualty who may or may not be breathing.
    a. Clear any foreign material or vomitus from the mouth as quickly as possible using forceps or the finger sweep method.
    b. If casualty is vomiting, turn head to the side or roll casualty on side to prevent aspiration.

CAUTION: Be aware of C-spine and other injuries.

    c. Stem bleeding into the oral cavity with packed gauze, but only after a secure airway is in place.
    d. After clearing the obstruction, assess the respirations and determine the type of airway required based on the cause of the obstruction and the situation.

NOTE: Casualties who are vomiting or bleeding into their naso-oropharynx need a secured airway, i.e., ET tube, to protect against aspiration. In a combat situation, the medic may have to settle for a J tube until time and circumstances permit him to intubate the casualty.

       (1) If the casualty is breathing on his own with little or no chance of aspiration, insert J tube.
       (2) If the casualty is not breathing or has minimal respiratory effort, or there is a chance for aspiration, intubation is preferred.
    e. Secure airway with an oropharyngeal airway or an ET tube.
    f. If blockage cannot be removed or injuries make obtaining a secure oral airway improbable, give casualty a cricothyroidotomy immediately (see Procedure: Cricothyroidotomy).
    g. Assist ventilations with Bag-valve Mask and oxygen if available.
8. Monitor airway and respiratory effort for at least q 5 min while you continue the primary survey.
    a. After Primary Survey is complete, reassess casualty's LOC, airway, and respiratory status to determine if additional management is required to further control and protect the airway.
    b. Unconscious casualties require intubation to further control and protect airway (see Procedure: Intubation)
    c. If the casualty is in severe respiratory distress or arrest and cannot be intubated, you must perform a cricothyroidotomy (see Procedure: Cricothyroidotomy)
9. Monitor and assess casualty on regular basis to determine if ABCs are improving or worsening.
10. Adjust treatment to compensate for improving or worsening status of the casualty.
11. Evacuate casualty to nearest appropriate medical treatment facility.

NOTES: Providing oxygen allows time to treat the underlying respiratory problem.
1. The nasal cannula is the simplest method suitable for a spontaneously breathing patient. Each additional liter/min of flow adds approximately 4% to the 21% O_2 available normally at sea level.
2. Facemasks provide higher and more precise levels of inspired oxygen—up to .35-.60
    a. Venturi mask delivers 24%-50% Fi O_2
    b. Non-rebreather delivers 60%-90% Fi O_2
    c. A continuous positive pressure device (CPAP) can deliver up to 100%
3. Use a BVM device to assist or control ventilation until a more secure airway can be obtained. If used correctly, 100% oxygen can be delivered this way.

What Not To Do:
If it takes 2 additional people to hold down a casualty to intubate them, re-evaluate the need for intubation since they have to be exchanging oxygen to maintain muscle strength and resist.

Do not proceed to directly to intubation in a patient with respiratory disease. Evaluate ways to improve their airway, then assist with respiratory effort. Ambu or bag-valve mask ventilation, timed with a patient's efforts can help relax and improve their respiratory status, and potentially avoid the risk of intubation.

Procedure: Intubation
MAJ John Hlavnicka, AN, USA

What: Establish a temporary emergency airway through the mouth or nose, and pharynx.

When: To control the airway during cardiopulmonary resuscitation or respiratory failure, prior to the onset of expected complications (e.g., laryngeal edema from inhalation burns), during complications from surgical anesthesia or other complications.

What You Need: Oxygen source and tubing, tonsil-tip suction and source, bag-valve-mask (BVM) device with self-inflating reservoir and oxygen coupling, face masks of different sizes, oral and nasopharyngeal airways (different sizes), water-soluble lubricant, straight and curved blade laryngoscopes, endotracheal tubes of different sizes, a syringe to inflate the cuff, stylets, tongue blades, nasogastric tube, and emergency drugs.

What To Do:
First: Patient Evaluation
Evaluate the airway during the initial injury assessment, and administer supplemental oxygen during this time if possible. Continual airway assessment is crucial since subtle changes in mental or respiratory status can occur at any time. Airway characteristics that can make fitting the mask and tracheal intubation difficult include:
1. Short, thick, muscular or fat neck with full set of teeth;
2. Full beard, facial burns, or facial injuries;
3. Receding or malformed jaw;
4. Protruding maxillary incisors; and
5. Poor mandibular (lower jaw) mobility.

Co-existing injuries such as known or suspected cervical spine injury, thoracic trauma, skull fractures, scalp lacerations, ocular injuries and airway trauma must be included when planning airway management.

Second: Technique
Endotracheal intubation indications include anatomic traits making mask management difficult or impossible, need for frequent suctioning, prevention of aspiration of gastric contents, respiratory failure or insufficiency, disease or trauma to airway, type of surgery or position of patient during surgery, need for postoperative ventilatory support, and traumatic injuries or musculoskeletal malformations making ventilation difficult.
1. Gather and check all previously listed equipment for proper function. Check light on laryngoscope, inflate ET cuff with 5-10cc air and check for leaks, then deflate and leave syringe attached, insert lubricated stylet so it does not protrude beyond distal end of ET and bend into hockey stick form, and have suction on.

2. Hyperventilate with 100% O_2 for several minutes using BVM.
3. Have assistant hold cricoid pressure if aspiration is a risk.
4. If orotracheal intubation is planned, hold the laryngoscope in left hand and insert the blade on right side of mouth pushing the tongue to the left and avoiding the lips, teeth and tongue. Holding the left wrist rigid (to avoid using the scope as a fulcrum and damaging the teeth), visualize the epiglottis.
5. If a straight (Miller) blade is used, pass the blade tip beneath the laryngeal surface of the epiglottis and lift forward and upward to expose the glottic opening. If a curved (Macintosh) blade is used, advance the tip of the blade into the space between the base of the tongue and the pharyngeal surface of the epiglottis (the vallecula) to expose the glottic opening.
6. Insert the ET with the right hand through the vocal cords until the cuff disappears. Remove the stylet and advance the tube slightly further. Inflate the cuff with air until no leak is heard when ventilated with bag. Adult women use a 7.0mm; men use an 8.0mm ET.
7. Verify correct placement by listening over both lungs for bilateral, equal breath sounds and observe the chest for symmetric, bilateral movements. Listen over the stomach, where you should not hear breath sounds. Note depth of insertion by centimeter markings on the tube at the lips, and tape the tube in place.
8. For **nasotracheal intubation** when the mouth cannot be opened or the patient cannot be ventilated by another means, or if the patient is conscious and requiring intubation, follow steps 1-3 using a lubricated (water-soluble), size 7-7.5 ET without the stylet. Insert the ET tube straight down into the larger nares until it reaches the posterior pharyngeal wall. If doing a blind nasal intubation, listen for the patient to inhale and insert the ET quickly into the trachea with a single smooth motion. If intubating under direct visualization, now insert the blade as previously described and pass the ET through the cords. Inflate the cuff and verify placement as above.

What Not To Do:
1. Do not mishandle laryngoscope blade and handle. Teeth can be broken and aspirated, or lips or gums lacerated with resultant bleeding. In addition, cardiac arrhythmias can occur with manipulation of the trachea and esophagus.
2. Do not allow the ET tube to be moved or removed accidentally. It must be adequately secured after successful placement to avoid compromising respiratory status in order to replace it.
3. Never perform a nasal intubation in a patient with a known or suspected basilar skull fracture or cribriform plate fracture. The ET can end up in the brain! Never force the ET against tissue resistance. Bleeding and inflammation can result, making future attempts at intubating difficult or impossible.

Procedure: Cricothyroidotomy, Needle and Surgical
18D Skills and Training Manual, Adapted by COL Warren Whitlock, MC, USA

What: Methods to establish a temporary emergency airway through the neck.

When: Consider cricothyroidotomy to establish an airway in casualties having a total upper airway obstruction or inhalation burns preventing intubation. Two methods are available:
1. Needle penetration of the cricothyroid membrane
2. Surgical placement of an airway tube through the cricothyroid membrane - when a cricothyroidotomy needle is unavailable or performing a needle cricothyroidotomy is not effective.

What You Need: Gather pre-assembled cricothyroidotomy kit (every medic should have an easily accessible 'Cric Kit' that contains all required items) or minimum essential equipment as below:
Cutting instrument: #10 or 11 scalpel, knife blade, 12-14 Gauge catheter-over-needle (e.g., Angiocath) with 10cc syringe attached for needle cricothyroidotomy (below). Syringe can also be used to inflate cuff on ET tube. Airway tube: IV catheter 12-14 gauge (from above), ET tube, cannula, or any noncollapsible tube that will allow sufficient airflow to maintain O_2 saturation. In a field setting, an ET tube is preferred because it is easy to secure. Use a size 6 -7 and insure that the cuff will hold air. Other instruments: 2 Hemostats, needle holder, tissue forceps, scissors. Other supplies: Oxygen source and tubing, Ambu bag, suctioning apparatus,

povidone-iodine prep, gauze, (sterile) gloves, blanket, silk free ties (for bleeders; size 3-0), 3-0 silk suture material on a cutting needle, and tape.

What To Do: Needle and Surgical Cricothyroidotomy

1. Preparation.
    a. Place the casualty in the supine position.
    b. Place a blanket or poncho rolled up under the casualty's neck or between the shoulder blades to hyperextend the casualty's neck and straighten the airway. WARNING: Do not hyperextend the casualty's neck if a cervical injury is suspected.
    c. Assemble needle/syringe set if not already done.
2. Locate and prep the cricothyroid membrane.
    a. Place a finger of the nondominant hand on the thyroid cartilage (Adam's apple) and slide the finger down to find the cricoid cartilage.
    b. Palpate for the "V" notch of the thyroid cartilage.
    c. Slide the index finger down into the depression between the thyroid and cricoid cartilage, the cricothyroid membrane.
    d. Prep the skin over the membrane with povidone-iodine.
3. Put on gloves (sterile if available) after assembling equipment and supplies.
4. **Needle Cricothyroidotomy**
    a. Make a small nick in the skin with a #11 blade to open a hole for the IV catheter to slide through the skin
    b. Using the needle/catheter/syringe, penetrate the skin and fascia over the cricothyroid membrane at a 90° angle to the trachea while applying suction on the syringe. Advance the catheter through the cricothyroid membrane.
    c. Once air freely returns into the syringe, STOP advancement, and direct the needle toward the feet at a 45° angle.
    d. Hold the syringe in one hand, and use the other hand to advance the catheter off the needle towards the lower trachea.
    e. Slide the catheter in up to the hub—**CAUTION:** Do not release the catheter until it is adequately secured into place.
    f. Check for air movement through the catheter by using the syringe to inject air through it and confirm free airflow. If air does not flow freely, straighten the tube and try again or withdraw the catheter and begin again at step 4b above.
    g. If air flows freely and the patient is breathing on his own, use the 3-0 suture to make a stitch through the skin beside the catheter. Secure the catheter to the stitch with several knots. Connect catheter to an oxygen source at a flow rate of 50 psi or 15 L/min. See Step 6 and 7 below for wound care and on-going management.
    h. If the patient is NOT breathing on his own, attach the syringe to the catheter, remove the plunger and deliver artificial respirations through the syringe and catheter. If the patient does not recover spontaneous respirations after several minutes, or if oxygen source is not available, proceed to Surgical Cricothyroidotomy below.
5. **Surgical Cricothyroidotomy** (If Needle cricothyroidotomy is not possible or is insufficient)
    a. Proceed through steps 1-3 if not already done. Test ET cuff to ensure it holds air.
    b. Raise the skin to form a tent-like appearance over the cricothyroid space, using the index finger and thumb.
    c. With a cutting instrument in the dominant hand, make a 1 inch horizontal incision through the raised skin to the cricothyroid space.
    CAUTION: Do not cut the cricothyroid membrane with this incision.
    d. Relocate the cricothyroid space by touch and sight.
    e. Stabilize the larynx with one hand and cut or poke a 1 inch incision through the cricothyroid membrane with the scalpel blade. **NOTE:** A rush of air may be felt through the opening. Look for bilateral rise and fall of the chest.
    f. Insert the ET tube or other airway tube through the opening into the trachea at a 90° angle to the trachea. Once in the trachea, direct the tube toward the feet at a 45° angle. Do NOT

insert an ET tube, or other long airway more than 3-4 inches to avoid intubating a single bronchus. Inflate the ET cuff if applicable. Do NOT release the airway tube until it is secured (see below).

g. Connect the Ambu bag to the tube and inflate the lungs, or have someone perform mouth to tube respirations. Auscultate the abdomen and both lung fields while observing for bilateral rise and fall of the chest. If there are bilateral breath sounds and bilateral rise and fall of the chest, the tube is properly placed and may be secured (see below). If not, reposition the tube as follows until adequate placement is obtained: (1) Unilateral breath sounds and unilateral rise or fall of the chest indicate that the tube is past the carina. Deflate the cuff on an ET tube, retract the tube 1-2 inches, inflate the ET cuff and recheck air exchange and placement. (2) Air coming out of the casualty's mouth indicates that the tube is pointed away from the lungs. Deflate the cuff on an ET tube, remove the tube, reinsert, inflate the cuff and recheck for air exchange and placement. (3) Any other problem indicates tube is not in the trachea. Follow the preceding step.

h. If air flows freely, and the patient is breathing on his own, proceed to next step. If the patient is **NOT** breathing on his own, continue providing respirations via the Ambu bag with oxygen if available, or via mouth to tube assistance at the rate of about 20/min.

i. Secure the airway tube using tape (temporary), or use the 3-0 suture to make a stitch through the skin beside the tube. Secure the tube to the stitch with several knots.

j. Suction the casualty's airway, as necessary. Insert the suction catheter 4 to 5 inches into the tube. Apply suction only while withdrawing the catheter. Administer 1 cc of saline solution into the airway to loosen secretions and help facilitate suctioning. **NOTE:** Ventilate the casualty several times or allow him to take several breaths between suctionings.

6. Apply a dressing to further protect the tube or catheter and incision using one of the techniques below.
   a. Cut two 4 X 4s or 4 X 8s halfway through. Place them on opposite sides of the tube so that the tube comes up through the cut and the gauze overlaps. Tape securely.
   b. Apply a sterile dressing under the casualty's tube by making a V-shaped fold in a 4 X 8 gauze pad and placing it under the edge of the catheter to prevent irritation to the casualty. Tape securely.

7. Monitor casualty's respirations on a regular basis.
   a. Reassess air exchange and placement every time the casualty is moved.
   b. Assist respirations if respiratory rate falls below 12 or rises above 20 per minute.

What Not To Do:
Do not remove needle before advancing the catheter into trachea. (NEEDLE Cricothyroidotomy)
Do not forget to insure that the tube is correctly placed, and secured. (SURGICAL Cricothyroidotomy)
Do not fail to monitor.

Procedure: Thoracostomy, Needle and Chest Tube
COL Warren Whitlock, MC, USA

What: Mechanisms to treat pneumothorax and hemothorax.

When: A needle thoracostomy can be performed faster than a tube thoracostomy in a rapidly deteriorating patient having signs of a tension pneumothorax. This can be life saving and gives enough relief to provide time for the medic to insert a chest tube. Once the chest tube is properly inserted, remove the needle.

What You Need: 18 gauge needle, 16-18 gauge Intracath, 10-20 cc syringe, sterile saline, alcohol pads, **Betadine**, latex sterile gloves, assorted chest tubes (sizes 28-32 French for adult pending air evacuation, 36-40°F for adult with hemothorax, 12-14°F for children), water seal drainage system (e.g., Pleur-Evac) and connection tubing for suction (alternate: one-way valve made from finger of latex glove), instruments: scalpel, forceps, gauze (may be in prepared tray), **Lidocaine** 1-2 % without epinephrine, petrolatum gauze, external dressing (4x4), adhesive tape, pulse oximetry

What To Do:
Figure out which lung has the pneumothorax! Insure that the procedure is performed on the side suspected

of having a pneumothorax (tension pneumothorax, simple pneumothorax, hemothorax), which will be the lung without breath sounds. Hyper-resonance is also a helpful sign, but the lack of breath sounds after penetrating or blunt trauma is a definitive sign.

Needle Thoracostomy:
1. Prep the chest wall by pouring **Betadine** over the intended site or swab with an alcohol wipe.
2. Insert an 18 Ga (or larger) 1.5 inch needle or IV catheter into the 2nd intercostal space, along the midclavicular line (an imaginary line from the middle of the collarbone, or clavicle; the interspace immediately below the clavicle is the 1st interspace). Run your finger down the midclavicular line, over the 2nd rib, to the 2nd intercostal space. Insert the IV catheter immediately above the 3rd rib.
3. This will release a rush of air from the pressure built up in the pleural space. Advance the catheter up to the hub, then remove the needle stylette and discard. The patient's ability to spontaneously breathe usually improves immediately. Leave the catheter in place, and attach a three-way stopcock, which can be used to drain air as it accumulates
4. This can improve the patient's symptoms and be life saving. Primarily, it is fast and easy to perform, providing enough time for the medic to set up for inserting a chest tube. The life-threatening emergency is the tension pneumothorax, not the simple pneumothorax that remains.
5. Once the chest tube is properly inserted, the catheter can be removed.

Alternative Technique: Remove the plunger from a 10-20 cc syringe filled with sterile saline, attach an 18 Ga needle/catheter (or larger) and use it to perform the thoracostomy. This allows handling of the needle/IV catheter more precisely and provides visual "bubbles" when the trapped air is released into the syringe. This is helpful in a noisy environment. Once the catheter is placed and the needle removed, setup for chest tube can begin. If the location is not safe for the second procedure, leave the catheter in place, attach a three-way stopcock to drain air as it accumulates, cover the catheter with gauze and tape, and move to a secure location for the procedure.

Tube Thoracostomy: Setup for a tube thoracostomy is more labor intensive than for a needle thoracostomy. Perform a tube thoracostomy after or in lieu of a needle thoracostomy to treat a simple pneumothorax (required prior to air evacuation).

1. Prep the chest wall by pouring **Betadine** over the intended site.
2. Site of insertion: along the mid-axillary line (a line running straight down from the middle of the armpit), always above the level of nipples in males (5th intercostal space since below this level there is a risk of puncturing the diaphragm).
3. Infiltrate 1% **lidocaine** along the track to be used. Generally, the tube is placed in the 3rd to 5th intercostal space on the mid-axillary line.
4. Cut a 3 cm long incision on top of the rib, NOT in the intercostal space. Cut along the axis of the rib, down to the bone, crossing the mid-axillary line.
5. Insert a large curved hemostat (Kelly Clamp) with the curve pointed toward the ribs and create a tunnel over the top of the rib. This tunnel helps to stabilize and seal the chest tube after placement. Curve the clamp over the top of the rib. Advance it slowly, opening and closing the jaws of the hemostat to clear a path and then puncture into the thoracic cavity. Do not advance straight in.
6. Digitally explore the pleural space to remove any pleural adhesions and insure the lung is free to fall away from the chest wall.
7. Have chest tube ready. Use size 28-32 French for adult pending air evacuation, 36-40°F for adult with hemothorax, 12-14°F for children. When in doubt, use a larger size because it will allow drainage of either air or blood from the chest cavity. Grasp the tube between the jaws of the clamp and insert into the pleural space. Direct the clamp tip posterior, towards the apex and the spine. Make sure that the tube is completely inserted so that no holes are left outside the chest.
8. Connect the free end of the chest tube to an underwater seal drainage system (Pleur-Evac), and then suture into place with Nylon 2/0. If an underwater seal drainage system is unavailable, make a field expedient version by securing the free end of the tube in a container of water that is lower than the level of the inserted end of the tube. This system prevents the patient drawing air back into the chest cavity. Bubbles coming out of the free end of the tube are a positive sign, indicating that the patient

is expelling free air.
9. In emergency conditions, use a one-way Heimlich valve instead. Cut a finger off of a latex glove. Fasten it as air-tightly as possible over the end of the tube: insert the free end of the chest tube inside the open end of the glove finger and tape the glove finger around the tube. Cut a 2 cm slit in the closed end of the glove finger. This will allow air to escape, but the glove finger will collapse on inspiration and prevent air from entering the lung. This will also collect blood draining from the chest tube.
10. Place **Vaseline**-impregnated gauze around the tube at the incision site, cover over that with 4x4 gauze and tape in place. The Vaseline gauze will prevent air leaks.
11. If the patient's condition does not improve, or deteriorates, the placement of the tube is suspect and it should be checked thoroughly for proper placement, or repositioned.
12. **Other considerations:** Antibiotics- load with 1 gm **ceftriaxone (Rocephin)** IM or IV or similar broad-spectrum antibiotic for all chest tubes inserted in the field.

What Not To Do:
Do not insert a chest tube if a tension pneumothorax is suspected and the patient is rapidly deteriorating—perform a needle thoracostomy instead for rapid relief.
Do not reposition or remove and replace a suspect tube if the patient shows signs of a repeat tension pneumothorax. Perform a second needle thoracentesis, then insert a second chest tube.

Procedure: Pulse Oximetry Monitoring
18D Skills and Training Manual, Reviewed by COL Warren Whitlock, MC, USA

What: Assess oxygen saturation using a pulse oximetry device.

When: There are three major reasons to use pulse oximetry:
a. To assess the current status of oxygenation of the blood and the need for supplemental oxygen.
b. To monitor patient's oxygenation continuously when there is a risk of respiratory failure.
c. To monitor pulse rate if there is no better monitoring capability available.

What You Need: A commercial pulse oximeter.

What To Do:
1. Understand the principles that make pulse oximetry possible: Pulse oximeters represent the percent of oxygen bound to hemoglobin (oxygenated blood – bright red; poorly oxygenated – dark red), by measuring the transmission of red and near infrared light through arterial beds. Hemoglobin absorbs red and infrared light waves differently when it is bound with oxygen (oxyhemoglobin) versus when it is not (reduced hemoglobin).
Oxyhemoglobin absorbs more infrared than red light and reduced hemoglobin absorbs more red than infrared light. Pulse oximetry reveals arterial saturation by measuring this difference. The pulse oximeter probe contains a sensor and a light source, and is usually packaged in a clip or flat wrap that can be attached to a source of good capillary perfusion. One side (light source) emits wavelengths of light into the arterial bed and the other side (sensor) detects the presence of red or infrared light.
2. Select a site: The probe is normally placed on the finger or toe of an uninjured limb in an adult, or on the ear of an infant or small child.
3. Understand appropriate and inappropriate readings for pulse oximetry:
   a. Normal SaO_2 (oxygen saturation) of hemoglobin is 93% or greater.
   b. Levels below 90% usually indicate insufficient oxygenation of the blood.
   NOTE: These patients need supplemental oxygen or evaluation of their ventilation or both.
4. Recognize that certain poisons can displace oxygen, saturate hemoglobin and produce FALSE normal pulse oximetry readings. True oxygen saturation is LOW with carbon monoxide (carboxyhemoglobin) or

nitrite (methemoglobin) poisonings, although pulse oximetry will be normal.
5. Record all findings in the patient's medical record.

What Not To Do:
Do not fail to recognize explanations (below) for false readings, which are common:
Excessive ambient light on the oximeter's probe.
Hypotension - causes vasoconstriction of capillary beds
Hypothermia – causes vasoconstriction of capillary beds
Patient's use of vasoconstrictive drugs.
Patient's use of nail polish.
Jaundice.
Very dark pigmented skin (choose area that has less pigment – finger tips, toes, etc.)

Procedure: 3-Lead Electrocardiography

18D Skills and Training Manual, reviewed by COL Warren Whitlock, MC, USA and LTC Richard Broadhurst, MC, ARNG

What: Guidelines on preparing for and conducting an EKG using only three leads.

When: EKG is used for diagnostic and monitoring purposes.
Diagnose: Metabolic and toxic disorders of the heart. (example: high potassium, low calcium, toxic quinidine or digoxin), chamber enlargement (example: ventricular and atrial enlargement), acute myocardial infarction and myocardial ischemia (myocardial infarction or angina), or arrythmias and conduction system abnormalities (example: bundle branch blocks seen in Chagas' cardiomyopathy).
Monitor: Therapeutic changes made in any of the above can be monitored by EKG changes.

What You Need: A cardiac monitor/defibrillator, EKG paper, electrodes, EKG paste (conductive gel), heart rate calculator ruler, alcohol prep pads, surgical lubricant, drapes, and tape.

What To Do:
NOTE: A 3-lead EKG is used to monitor the heart solely for dysrhythmia. The following information can be obtained from using Lead II:
Rate: How fast the heart is beating (electrically).
Rhythm: Life-threatening dysrhythmias.
Intervals: How long it is taking to conduct an impulse through the parts of the heart.

The following information **cannot** be obtained from using Lead II:
The presence or location of an infarct.
Axis deviation or chamber enlargement.
Right-to-left differences in the conduction impulse formation.
The quality or presence of pumping action.

1. Prepare the equipment.
   a. Read the manufacturer's instructions for proper use of the equipment on hand if not familiar with it.
   b. Plug the machine into a wall outlet that is grounded.
   c. Turn the power switch ON and allow the machine to perform its self-checks and warm up for 5 minutes.
   d. Check the machine's graph paper supply.
   e. Verify that the machine is set on the standard settings: paper speed at 25 mm/sec; amplitude at 10 mm/mv.
   f. Verify that all other equipment is on hand: EKG paste, 3 clean electrodes, alcohol pads, drapes or towels.

2. Prepare the area.
    a. Free from electrical interference.
    b. Comfortable and private for the patient.
    c. Convenient for the person obtaining the EKG.
    d. Free from distractions (noise, traffic).
3. Prepare the patient.
    a. Explain the procedure to the patient and answer any questions.
    NOTE: Many patients will be apprehensive about being connected to an electrical instrument. Reassure the patient that there is no danger and they will feel no pain.
    b. Ask (or assist) the patient to remove all clothing from the waist up. Provide a chest drape for female patients.
    c. Ask (or assist) the patient to lie supine on the bed or examination table.
    NOTE: Patients with respiratory problems may be unable to tolerate the supine position. If necessary, elevate the head of the bed to 45°.
    CAUTION: When necessary, ensure the patient's IV tubing and/or urinary catheter tubing is handled with care and positioned properly to avoid discomfort to the patient.
    d. Ensure the patient's body is not in contact with the bed frame or any metal objects and that all limbs are firmly supported.
    e. Ask the patient to relax and breathe normally throughout the procedure.
4. Apply the chest electrodes.
    a. Clean the sites for electrode placement by rubbing with an alcohol prep pad to remove dead skin, oils, and traces of soap or dirt.
    b. Apply a small amount of EKG paste to the sites.
    c. Attach the electrodes, being careful to place them over the intercostal spaces and not directly over the ribs. Lead II (bipolar) is the most commonly used lead.
        (1) Attach the negative electrode (white) to the right arm, right shoulder, or upper right anterior chest wall.
        NOTE: Place the electrodes on "meaty" places NOT directly over bone because bone is not a good conductor of electricity.
        (2) Attach the positive electrode (red) to the left leg, left thigh, or lower left anterior chest wall at the intersection of the fourth intercostal space and the mid-clavicular line.
        NOTE: The latter is least preferred since it may interfere with the defibrillator paddles if a dysrhythmia should occur.
        (3) Attach the electrically neutral electrode (black) to the left arm or right anterior chest.
    d. Ensure these 3 leads form a triangle around the heart referred to as Einthoven's Triangle. Placement of Lead-II electrodes.
    e. Obtain a readable EKG strip.
        (1) Interpret the current flow for Lead II.
            (a) When an electrical current flows toward the positive electrode, a positive (upward) deflection is recorded on the EKG.
            (b) When an electrical current flows away from the positive electrode, a negative (downward) deflection is recorded on the EKG.
            (c) The absence of any electrical impulse, results in the recording of an isoelectric line (flat line).
5. Interpret whether the rhythm strip is normal or abnormal.
    a. Step 1: Determine whether the rate is normal. Compare to pulse found on patient.
        (1) Six-second method.
            (a) Count the number of R-waves in a 6-second interval by noting two 3-second marks at the top of the EKG paper.
            (b) Multiply the number of R-waves within the 6-second strip by 10, which will give the heart rate per minute.
        (2) Triplicate method
            (a) Locate two R-waves that fall on dark lines on the graph paper.
            (b) Assign numbers corresponding to the heart rate to successive dark lines starting with the

dark line on which the first R-wave fell.
- (c) The order is 300, 150, 100, 75, 60, 50, and so forth. The assigned numbers are a result of dividing 300 by 1, 2, 3, 4, 5, 6, and so forth.
- (d) The number corresponding to the line where the second R-wave falls is the pulse. Normal rate is considered 60-100 beats/ minute, although some fit individuals will have resting rates down into the 40s.
b. Step 2: Determine if the rhythm is regular (P-wave and QRS complexes occur at regular intervals). R-waves fall on top of each other when the EKG paper is folded in half and held it up to the light. Not regular is abnormal.
c. Step 3: Analyze the P-wave. Determine whether there is a P-wave for every QRS complex, the P-waves are upright, and they are regular and similar in appearance. Any NO is abnormal.
d. Step 4: Determine if there is any ST segment elevation or depression of 2 small squares or more.
e. Step 5: If any heart abnormalities are discovered, refer to Acute MI or Cardiac Resuscitation sections and evacuate the patient. Many arrhythmias are not addressed in those sections, but medics are not trained or equipped in the field to treat them. Tachycardias can be treated by diving reflex, carotid massage (rule out bruit first). Ventricular tachycardia and fibrillation should be treated as soon as possible.

What Not To Do:
Never treat an EKG. Always treat the patient. Does the information from the EKG correlate to the patient? Example: When a rhythm looks like atrial fibrillation, feel the patient's pulse. If it is regular and full, then the information from the EKG machine "does not correlate," and is in error.

Procedure: Pericardiocentesis
COL Warren Whitlock, MC, USA

What: Mechanism to relieve fluid or air inside the sac surrounding the heart.

When: The patient has sustained a penetrating wound in the chest that may have entered the heart covering (pericardium), and is showing signs of shock - hypotension, tachycardia, and tachypnea with narrowed pulse pressure, muffled heart sounds, pulsus paradoxicus (heart rate increasing with expiration, decreasing with inspiration—greater than normally seen). The medic must be aware that this procedure is dangerous and should not be attempted without prior training, and only as a last resort in life threatening emergencies when vital signs deteriorate: (narrow pulse pressure) low mean arterial pressure, +/- muffled heart sounds. If the vital signs are stable, the medic should continue IV fluids and monitor the patient only. There are also more rare etiologies for developing fluid (viral pericarditis), as well as air (pneumopericardium in diving) in the pericardial sac. These conditions can be relieved with the same procedure outlined below to relieve blood in the pericardial sac.

What You Need: 18 gauge spinal needle or Pericardiocentesis kit with Mansfield catheter, a 60cc syringe, sterile preparation kit (alcohol wipe may be adequate in emergencies), local anesthetic, sterile needles, 3-way stopcock, alligator clips, EKG with defibrillator/monitor, gauze pads/bandage. Emergency drugs (atropine, lidocaine, epinephrine, oxygen)

What To Do:
Preparation:
1. Have the patient lie supine.
2. Administer oxygen (nasal cannula or face mask) and pulse oximeter if available.
3. Start an IV – volume loading may help attenuate the effects of cardiac tamponade.
4. Set up EKG machine to monitor the cardiac rhythm.
5. Clean subxiphoid area with antiseptic

Procedure:
Option A – Emergency Technique to Withdraw Fluid Once
1. Connect 18 gauge spinal needle and 60 cc syringe

2. Place your finger 0.5 cm below the costal margin of the patient's xiphoid to mark the point of needle insertion. Raise the needle to a 30° angle from parallel to the patient's chest. Aim the needle at the tip of the ipsilateral (same side) scapula (shoulder blade)
   3. Insert the needle, maintaining slight suction and advanced until blood flow is obtained, and then stop advancement.
   4. Withdraw as much a blood as possible and then withdraw the needle. Only a small amount (5-10 cc) removed can have a marked improvement in vital signs.

Option B - Technique to Withdraw Fluid Multiple Times
   1. Attach a central line needle and catheter to a 60 cc syringe.
   2. Follow the procedure as above.
   3. When fluid is obtained, hold the syringe and needle in one hand, and gently advance (slide) the catheter into the pericardial space. Withdraw the needle from the catheter.
   4. Attach the 3-way stopcock (closed position) to the hub of the catheter
   5. Remove the needle from the syringe and discard. Connect one end of IV tubing to one port of the 3 way stopcock, the other end of the IV tubing attach to a 60 cc syringe (optional to connect another IV line to the third port of the stopcock for ejecting blood from the syringe).
   6. Open the 3 way stopcock to the syringe to withdraw fluid from the pericardial space, then turn open to ejection port IV line to eject the fluid out.
   7. When no further fluid/blood return, turn the 3-way stopcock to closed or in-between position.

Option C - Technique to Monitor Needle Approach with EKG Lead (use with both A & B above) to Withdraw Fluid Multiple Times
   1. Connect V1 lead of 12 lead EKG with an alligator clip to the metallic hub of the needle or needle stylet of the catheter, then insert needle as directed above.
   2. During the advancement of the needle, monitor the EKG for ST segment elevation. This indicates that the needle tip is in contact with myocardium and should be withdrawn.
   3. This monitoring improves the safety of the procedure, but is not practical in immediate life and death field scenarios.

Post-Procedure:
   1. Monitor for pneumothorax and arrhythmias
   2. Collect samples of pericardial fluid in cases not related to penetrating trauma for later analysis.

What Not To Do:
Care must be taken not to insert the needle more than 1/8 of an inch once blood is obtained.
The catheter can be left in if the medic has a catheter line that can be switched between closed and open.
Remember, small movements of the syringe can have large effects on movement of the tip of the needle causing lacerations of the myocardium or coronary arteries.

Procedure: Pneumatic Anti-Shock Garment
COL Clifford Cloonan, MC, USA

What: Apply pneumatic antishock garments

When: Otherwise uncontrollable, on-going hemorrhage in the lower abdomen/pelvis and/or buttocks/thigh(s); neurogenic shock (especially when unable to avoid a head-up attitude, i.e. during extraction/ evacuation); anaphylactic shock; pelvic fracture, femur/tibia fracture when traction splint not available (used as a pneumatic splint).

What You Need: Pneumatic Anti-Shock Garment (a.k.a. – M.A.S.T.)

What To Do:
1. Apply the garment – there are various methods to do this (follow manufacturer recommendations/ATLS

guidelines).
2. Inflate garment, legs first, then abdominal compartment, until either the patient's blood pressure becomes adequate (around 80 mm systolic), the Velcro on the suit begins to crackle (indicating that separation is imminent), and/or bleeding is controlled and/or fracture(s) is/are adequately stabilized.
3. If applying a PASG that has a pressure gauge that measures the pressure inside the PASG it is ESSENTIAL that this pressure NOT be used as an end-point for inflation. This pressure gauge should ONLY be used to allow the care provider to maintain a constant PASG inflation pressure despite changes in altitude and/or temperature.
4. PASG should ONLY be applied as a temporizing measure and NOT as a substitute for other interventions, particularly surgery.
5. If the patient has on going bleeding rapid evacuation for surgical stabilization is indicated.

What Not To Do:
Absolute Contraindication to application of PASG/MAST
Pulmonary edema/congestive heart failure /cardiogenic shock - The increased peripheral vascular resistance which PASG/MAST produces increases the work of the heart and will worsen these conditions

Relative Contraindications to application of PASG/MAST
Head injury/cerebrovascular accident - use of PASG/MAST will increase intracranial pressure.
Severe and uncontrollable bleeding above the diaphragm - the possibility exists of increasing intrathoracic bleeding as the blood pressure increases.
Ruptured diaphragm - inflation of the abdominal compartment of the PASG/MAST will force abdominal contents into the chest cavity
Third trimester pregnancy – do not inflate abdominal compartment.
Impaled object in the abdomen - do not inflate abdominal compartment.

DO NOT remove PASG/MAST until patient is in a location where his/her underlying problem can be fully addressed (i.e., surgery). Premature removal of the PASG/MAST can lead to severe hypotension. When the decision is made to deflate the PASG/MAST, slowly deflate the garment, abdominal compartment first, then legs, until the systolic blood pressure drops by 5 mm Hg. Then, discontinue deflation, and provide fluid replacement until the pressure is restored. Repeat these steps until deflation is complete. Circumstances may require a more rapid deflation but this should ONLY be done in the OR when the surgeon and the anesthetist/ anesthesiologist are fully prepared to deal with the consequences.

NOTE: Treatment of hypovolemic/hemorrhagic shock by applying PASG has fallen into such disfavor that it is rarely recommended for this application any more. There are good reasons, however for continuing to recommend their use in combat situations – specifically for use in stabilizing pelvic fractures and tamponading bleeding in the pelvis, buttocks, and/or groin/upper thigh where a tourniquet cannot be applied to control hemorrhage.

Procedure: Blood Transfusion
18D Skills and Training Manual, reviewed by COL Warren Whitlock, MC, USA

When: You have a trauma patient who may require a transfusion in a medical facility with blood replacement capability. You must correctly assess the trauma patient to determine whether or not he requires blood replacement and if he does, what those requirements are.

What You Need:
1. A thermometer, blood pressure cuff, stethoscope, IV stand, tourniquet, large bore IV catheters, tape, alcohol and Betadine prep pads, vacutainers, needle and syringe, gloves, crystalloids, and the patient's clinical record.
2. 2 large bore IVs already established and the following materials, blood transfusion recipient set ("Y" type), 500ml or 1000ml bag of 0.9% normal saline, blood, IV stand, needle and syringe.

What To Do:
1. Perform a survey of the casualty to ensure airway stabilization, adequate respirations, and hemorrhage control.
   a. Establish two large bore (18 gauge or larger) IV lines.
      (1) Draw blood.
      (2) Request and/or perform labs, H/H with type and crossmatch.
   b. Initiate IV fluids for resuscitation. Give Ringer's Lactate as a first choice, normal saline as a second choice.
   NOTE: The usual initial volume for resuscitation is 1-2 liters in an adult and 20ml/kg in pediatric cases. Other diagnostic decisions are based on the observed response.
   c. Establish a set of baseline vitals.
   d. Perform other resuscitative procedures as required: ABCs and secondary survey.
2. Monitor patient and determine if patient requires blood transfusion:
   a. Indications that patient does not require a blood transfusion:
      (1) Stable vital signs within normal limits, Class I or Class II shock.
      (2) Patients who have lost less than 20% of their blood volume and are no longer hemorrhaging require no further fluid bolus or immediate blood administration.
      (3) H/H within normal limits.
   CAUTION: With rapid blood loss, H/H will lag behind actual blood volume.
   b. Indications for blood transfusion:
      (1) Vital signs not stable, patient in Class III shock.
      (2) Patients who have lost 20% to 40% of their blood volume and are still hemorrhaging will show marked deterioration. Continued fluid therapy and blood replacement are indicated.
      (3) Patients with little or no response to the initial fluid therapy.
      (4) H/H below normal.
   NOTE: Isovolemic patients can have adequate O_2 carrying capacity with Hb levels as low as 7 gm/dL.
3. Determine the type of blood to give to the patient. (Type Specific or Universal Donor - O Negative)
   NOTE: Response to blood administration should identify patients that are still bleeding and require rapid surgical intervention.
   a. Select the appropriate blood type based on the type and crossmatch and the types of blood available.
   b. Ideally, the patient should receive the same type of blood that they have.
   c. In urgent situations, type O RBCs (not whole blood) may be used for patients of other blood types, and either A or B RBCs may be used for AB recipients (but not both together).
   d. Rh-negative patients should always receive Rh-negative blood except in life-threatening emergencies when Rh-negative blood may be unavailable.
   e. Rh-positive patients may receive either Rh-positive or Rh-negative blood.
4. Select the appropriate blood component, if available:
   a. Whole blood: used for rapid massive blood loss and exchange transfusions. May be necessary if a component is unavailable. (Higher incidence of transfusion reaction due to plasma)
   b. Packed RBCs: also used for rapid massive blood loss and exchange transfusions. Packed RBCs are preferred due to the lower chance of complications.
      (1) Transfused to replace Hb or O_2 carrying capacity, including blood lost at surgery or as a result of trauma.
      (2) Consider the patient's age, cause, degree of anemia, circulatory stability and the condition of heart, lungs, and blood vessels.
      (3) When volume expansion is required, other fluids can be used concurrently or separately.
   c. Fresh frozen plasma is an unconcentrated source of all clotting factors except platelets.
      (1) Used to correct a bleeding tendency of unknown cause or one associated with liver failure.
      (2) Can supplement RBCs when whole blood is unavailable for exchange transfusion.
      (3) Except when prepared from autologous donations, not ordinarily used as a volume expander.
   d. Washed RBCs (by continuous-flow washing) are free of almost all traces of plasma, most WBCs, and platelets. Used for patients who have severe reactions to plasma (e.g., severe allergies or IgA

immunization) or for those who have rare WBC antibodies and repeated febrile transfusion reactions.
5. Determine the amount of blood to administer:
    a. Blood is usually replaced at a 1:1 ratio.
    b. Stabilization of adequate vital signs is primary indicator of sufficient blood volume.
6. Order type specific or type O blood or blood component based on the type and crossmatch.
7. Verify and inspect the blood pack received from the laboratory for abnormalities such as gas bubbles or black or gray colored sediment (indicative of bacterial growth):
    a. Note the time the blood pack was received and record the time.
    b. Check the label for blood components, type, and expiration date.
    NOTES: Two people, if possible, should independently verify the information match of type and cross match patient to label. Infusion of a blood pack should be started within 30 minutes of being issued.
8. Establish baseline data:
    a. Reconfirm data from the patient's history regarding allergies or previous reactions to blood or blood products.
    b. Measure and evaluate the vital signs.
    c. Record the vital signs on the chart.
9. Reduce the chances of reactions by applying prophylactic measures:
    a. Meticulous identification and clear labeling at the bedside of casualty's blood samples intended for compatibility testing.
    b. Ensure that blood has been screened for hepatitis, AIDS, malaria, and syphilis.
    c. Carefully identify casualty and donor blood at the time of transfusion.
        (1) Identify casualty by ID tags, ID bracelet, or other means.
        (2) Identify the donor blood and has a second person check the blood and the casualty if possible.
    d. Handle blood products with care to prevent hemolysis or destruction of RBCs by over-warming stored blood.
        (1) Warming devices applied to the blood container itself (e.g., microwave warmers) are contraindicated because:
            (a) A high incidence of hemolysis occurs.
            (b) Any interruption in transfusion may encourage bacterial growth in unused warmed blood.
            (c) The blood bank will not know if unused blood has been warmed and re-chilled.
        (2) Use an IV set that includes a heat exchange device, if available, to warm blood gently (but not > 37° C) during delivery.
    e. Prevent contact with inappropriate IV solutions such as injections of distilled water or non-isotonic solutions.
    f. In a patient with a history of allergies or an allergic transfusion reaction, give an antihistamine prophylactically just before or at the beginning of the transfusion (e.g., diphenhydramine 50 mg orally or IM). It must never be mixed with the blood.
    g. In cases where volume overload is possible, whole blood is contraindicated. A rise in venous pressure can be avoided by infusing RBCs at a slow-to-moderate rate. The patient should be observed for signs of increased venous pressure or pulmonary congestion. If possible, direct observation of venous pressure during the infusion is a useful precaution.
    h. Administer the first 10 to 30 ml over 15 min, while observing the patient for reactions.
    i. Use transfusion sets that include a filter to trap the clots and fibrin shreds present in stored blood units.
    j. For patients likely to receive large amounts of blood that has been stored > 5 to 6 days use microaggregate filters to remove particles as small as 20-40µm.
    k. Guard against air in tubing with any pressure infusion and when changing IV sets to prevent air embolism.
    CAUTION: Consult manufacturer's instructions. Microaggregates can be detected in the lungs after massive transfusions and have been implicated as a cause of the syndrome of posttraumatic pulmonary insufficiency, though direct evidence is lacking.
10. Prepare the blood and the blood recipient set:
    NOTE: Use only tubing that is designed for the administration of blood products. It is equipped with a filter designed for the fine filtration required for blood products. If "Y" type recipient tubing is not available, use regular infusion tubing for the normal saline (0.9% Normal Saline only) and the available blood

recipient tubing for the blood pack. Prime each set. Attach a sterile, large bore (16 or 18 gauge) needle to the end of the blood tubing and "piggyback" the blood into the normal saline line below the level of the roller clamp. Hang the blood pack at least 6 inches higher than the normal saline.
   a. Close all three clamps on the "Y" tubing.
   b. Aseptically insert one of the tubing spikes into the container of normal saline. Invert and hang this container about 3 feet above the level of the patient.
   c. Open the clamp on the normal saline line and prime the upper line and the blood filter.
   d. Open the clamp on the empty line on which you will eventually hang the blood. Normal saline will flow up the empty line to prime that portion of the tubing.
   e. Once the blood line is primed with saline, close the clamp on the blood line.
   f. Leave the clamp on the normal saline line open.
   g. Open the main roller clamp to prime the lower infusion tubing.
   h. Close the main roller clamp.
   i. Aseptically expose the blood port on the blood pack.
   j. Aseptically insert the remaining spike into the blood port and hang the blood at the same level as the normal saline container.
11. Connect the bloodline:
    a. Patients receiving blood should have two patent IV sites in the event of complications or emergencies. Establish one or two new IV sites as needed.
       (1) Use a large gauge IV catheter (14, 16, or 18) to enhance the flow of blood and prevent hemolysis of the cells.
    b. If the patient already has 2 IV sites, aseptically switch the blood line with one of the existing IV lines or piggyback the blood line into an existing IV line.
12. Begin the infusion of blood:
    a. Attach the primed infusion set to the catheter, tape it securely, and open the main roller clamp.
    b. Close the roller clamp to the normal saline and open the roller clamp to the blood.
    c. Adjust the flow rate with the main roller clamp.
       (1) Set the flow rate to deliver approximately 10 to 30 cc of blood over the first 15 minutes.
       (2) Monitor the vital signs every 5 minutes for the first 15 minutes and observe for indications of an adverse reaction to the blood.
       (3) If after the first 15 minutes, no adverse reaction is suspected and the vital signs are stable, set the main roller clamp to deliver the prescribed flow rate.
       NOTE: Prolonged transfusions pose a hazard of bacterial growth because blood quickly reaches room temperature.
13. Monitor and evaluate the patient throughout the procedure.
    a. Monitor vital signs every 15 minutes or IAW local SOP.
    b. Compare the vital signs with previous and baseline vital signs.
    c. Observe for changes that indicate an adverse reaction to the blood.
    d. If a reaction is suspected, stop the blood, infuse normal saline, and identify and treat the reaction IAW. Procedure: Blood Transfusion Reactions.
       CAUTION: When a transfusion reaction occurs or is suspected, stop the blood immediately and infuse normal saline. The unused blood and recipient tubing must be sent to the laboratory along with a 10 ml specimen of the patient's venous blood and a post-transfusion urine specimen.
14. Discontinue the infusion of blood when the patient's vital signs have stabilized or the transfusion is finished.
    a. Close the clamp to the blood and open the clamp to the normal saline.
    b. Flush the tubing and filter with approximately 50 cc of normal saline to deliver the residual blood.
    c. After the residual blood has been delivered, run the normal saline at a TKO rate or hang another solution, as needed.
    d. Take and record the vital signs at the completion of the transfusion and 1 hour later.
    NOTE: As a rule, a unit of blood should be infused within 2 to 4 hours unless contraindicated by risk of circulatory overload.
15. Dispose of the used blood pack IAW local SOP.
    a. Return it to the laboratory blood bank.

b. Discard it in a container for contaminated waste.
16. Document the procedure and significant observations on the appropriate forms IAW local SOP.

What Not To Do:
Do not transfuse blood products when ANY doubt exists as to the crossmatching, or type of blood. Transfusion reactions can convert a critical situation into a "fatal" situation if safe medical procedures are not followed.
Do not withhold blood products in a patient that is hypotension, tachycardic and actively bleeding, with a normal hematocrit. Remember, the hematocrit will take several hours to fall in a bleeding patient, so that a normal percentage is not unusual in acute trauma.

Procedure: Field Transfusion
COL Richard Tenglin, MC, USA

Introduction: Transfusion of red blood cells in a hospital is a complicated, risky procedure. Obtaining fresh whole blood from one individual and transfusing it to another under austere field conditions is even more risky, so there is no reason to transfuse red cells to a patient if bleeding can be stopped. Bleeding from an extremity can be stopped with a tourniquet. Continuous bleeding from neck, chest or abdomen requires surgical intervention and blood transfusion can only support a patient for a short time while accessing resuscitative surgery. Aggressive transfusion can increase blood loss by increasing intravascular pressure and diluting coagulation factors. Within these very significant limitations, field transfusion can be done as follows.

What You Need: An established large bore IV in the patient that will receive the transfusion, a suitable blood donor, a blood collection system, clamps, alcohol prep pads, and a blood pressure cuff

What To Do:
1. The donor and recipient must be males, or females who have never been pregnant (pregnancy can induce transfusion-significant antibodies) of known ABO and Rh compatibility (donor: same ABO and Rh type as recipient, or Type O positive for RH positive recipient, or Type O negative). The medic should document blood types for all team members prior to deployment–attempts to type and cross in the field in an emergency only adds to the already considerable risk of field transfusion.

2. Any sterile, closed blood collection system may be used, but the current system used to collect donated blood in the US should be considered the standard. Clamp or tie off all tubes from the collection bag except two–one that will be used to collect the blood from the donor and one that will be used to infuse the blood into the recipient.

3. Draw blood from the donor first. Position a blood pressure cup on the donor's arm above the elbow. Inflate the blood pressure cuff so that it is between systolic and diastolic pressure. Find the antecubital vein, prep the area with alcohol and puncture it with the needle. Collected blood into the largest of the collection bags, (the one with the anticoagulant), mixing it frequently and gently.

4. When the collection bag is full, deflate the BP cuff, clamp the collecting tube and remove the needle from the donor.

5. Immediately infuse the blood into the recipient through the remaining line. It is best to piggyback this through an established large bore IV.

Procedure: Blood Transfusion Reaction

18D Skills and Training Manual
Reviewed by COL Warren Whitlock, MC, USA

When: You have a patient who has received or is receiving a transfusion of whole blood or a blood component. The patient is having a reaction to the transfusion. You must correctly identify the cause of the blood transfusion reaction and manage it without causing further injury to the patient.

What You Need: Thermometer, blood pressure cuff, stethoscope, gloves, tourniquet, alcohol and Betadine prep pads, IV catheters, tape, blood transfusion recipient sets ("Y" type), 500ml or 1000ml bag of 0.9% normal saline, blood, IV stand, needles and syringes, Ringer's lactate, **Benadryl**, epinephrine 1:1000 solution, oxygen and related equipment, **mannitol**, **bicarbonate**, **Tylenol** or **ibuprofen** and the patient's clinical record.

What To Do:

NOTE: Patient has received or is receiving transfusions.
1. Monitor the patient during and after transfusion.
2. Compare baseline vital signs to current vital signs.
3. Monitor the site of the infusion for edema, warmth, and urticaria.
4. Examine the patient for systemic signs and symptoms of reaction.
5. Identify the type of reaction based on the signs and symptoms.
   a. Hemolytic reactions result in the lysis (destruction) of recipient or donor RBCs (usually the latter) during or after administration of solutions, plasma, blood, or blood components.
   NOTES: Hemolytic reaction is the rarest and most severe transfusion reaction. It usually starts within the first 10 minutes of the transfusion. The most severe reaction occurs when donor RBCs are hemolyzed by antibody (Ab) in the recipient's plasma.
   STOP THE TRANSFUSION IF A HEMOLYTIC REACTION IS SUSPECTED
   Onset is usually acute, within 1 hour; it may occur during or immediately after a transfusion. Patient complains of discomfort and anxiety, or may have no symptoms. They may have difficulty breathing, precordial oppression, a bursting sensation in the head, facial flushing, and severe pain in the neck, chest, or especially the lumbar area. They may have chills, fever, evidence of shock, a rapid feeble pulse, cold clammy skin, dyspnea, drop in BP, nausea, and vomiting. Dark urine, free Hb may be found in the plasma or urine, followed by elevated serum bilirubin and clinical jaundice.
   NOTES: Group O whole blood can cause hemolytic reactions due to plasma anti-A and anti-B hemolytic antibodies or IgG immunogloblins. It is dangerous when given in emergencies to a recipient with another blood group. Remove the plasma, which contains most of the antibodies, first. Group O packed RBCs should be used rather than whole blood.
   b. Febrile reactions: signs and symptoms within 30 minutes. Chills & fever with a rise of at least 1° C in body temperature, headache and back pain, stable vital signs, rarely progressing to cyanosis and shock.
   c. Allergic reactions are most common with multiple transfusions and in people with a history of allergies. Signs and symptoms of mild reactions include rashes, urticaria, edema, occasional dizziness, fever, and headache during or immediately after the transfusion. Signs and symptoms of severe reactions include dyspnea, wheezing, and tracheal edema. Incontinence may be present, indicating generalized smooth muscle spasms. Rarely, anaphylaxis may occur.
   d. Circulatory overload: dyspnea, sudden anxiety, neck vein distention, crackles at the base of the lungs (signs of pulmonary edema).
   e. Transmission of large amounts of air into a vein can cause foaming of blood in the heart with consequent inefficient pumping, leading to heart failure.
   f. Rapid transfusion of cold blood can cause arrhythmia or cardiac arrest.
   NOTE: Air embolism is largely a complication of pressure infusion of blood from rigid glass bottles, but it can also happen when an IV set is changed or a plastic blood bag is erroneously vented.
6. Initiate the appropriate treatment for the transfusion reaction.

a. Treatment for hemolytic reactions.
   (1) Stop the transfusion immediately and change the IV tubing.
   (2) Leave the needle in place and reconnect the new tubing to the needle.
   (3) Send a sample of the patient's fresh blood and urine to the laboratory for analysis, if possible.
   (4) Start Ringers lactate or normal saline IV to correct hypotension.
   (5) Monitor vital signs and urine output.
   (6) Administer oxygen.
   (7) Administer medications. To establish osmotic diuresis, start an infusion of **mannitol** 20gm IV (e.g., 100 ml of 20% solution) at once and continue it at 10 to 15 ml/min until 1000ml (200 gm) have been given. If diuresis ensues, mannitol should be continued to a maximum of 100gm/day, or volume may be maintained with other IV fluids until hemoglobinemia and hemoglobinuria have cleared. **Bicarb** IV, one ampule per liter of fluid given. **Benadryl** 50mg IM.
   (8) Use blankets to relieve chills.
   NOTE: **Lasix** may be substituted for **mannitol** to maintain urine output.
b. Treatment for febrile reactions: Febrile transfusion reactions can usually be managed with antipyretics and gentle patient cooling. Change tubing, but keep the venous access open with normal saline. Check the patient's temperature every 30 minutes. Give **Tylenol** or **ibuprofen** for fever. Document the episode, time, and IV fluid given. When symptoms recur with blood that is otherwise compatible, further transfusions should consist of RBCs that have been washed or specially filtered to remove WBCs.
c. Treatment for allergic reactions: Stop the transfusion and change the tubing. An antihistamine usually controls mild cases (e.g., **diphenhydramine** 50mg IM or IV). Administer fluids such as Ringer's Lactate or Normal Saline IV, to support BP. Give **Tylenol** or **ibuprofen** for pain. For more severe reactions: **Epinephrine** 0.5 to 1ml of 1:1000 solution sq (or, in extreme emergencies, 0.05 to 0.2ml of 1:1000 solution diluted to 1:10,000 and injected slowly IV) should be given. Give **Benadryl** 50mg IV STAT then 50mg po q4h. Start IV normal saline at a rate to support BP. A corticosteroid (e.g., **dexamethasone sodium phosphate** 4 to 20mg IV) may occasionally be required.
d. Treatment for circulatory overload. The transfusion should be discontinued immediately. Place the patient in an upright position. Keep the IV line open with a slow infusion of normal saline. Use diuretics (**Lasix**) and **morphine** if necessary.
e. Treatment for air embolism: stop the source of the air and bleed or replace the line. Turn the patient on the left side, head down, to allow the air to escape a little at a time from the right atrium. Monitor the patient for pulmonary or cerebral embolisms.
f. Treatment for arrhythmia or arrest due to infusion of cold blood: Stop the infusion. Manage the arrhythmia or arrest. Warm the blood before resuming infusion.

7. Explain to the patient that there are possible late complications of blood transfusions and that they should notify medical personnel immediately if the develop signs or symptoms of late complications.
   a. Serum hepatitis.
   b. Malaria: Can be transmitted by asymptomatic donors. Patients may develop high fever and headaches weeks after the transfusion.
   c. Syphilis.
   d. AIDS virus.
   NOTE: All donated blood should be tested for antibodies to the AIDS virus.
   NOTE: Storing blood for more than 96 hours at 4°C inactivates the spirochetes.
   e. Delayed hemolytic reactions can occur from 1 to 2 weeks after transfusion. Signs are fever, mild jaundice, gradual fall in hemoglobin level, positive Coombs' test.
   f. Bacterial infection: A few contaminating bacteria, particularly gram-negative, can grow in refrigerated blood and may cause severe reactions and sepsis if transfused. Procedures that allow blood to reach room temperature (prolonged transfusions or warming blood) may accelerate bacterial growth and are potentially hazardous.
8. Record all treatment given.
9. Recommend evaluation by a physician.

What Not To Do:
Do not continue a transfusion when a patient complains of difficulty breathing and feeling bad during a transfusion. Stop the transfusion, continue normal saline, and evaluate for possible transfusion reaction. Do not reassure a patient that all transfusions are completely safe and without risk. There is always risk with transfusion of blood products. Transfusion is preferred when the benefit of the transfusion, out weighs the risk of the complications.

Procedure: Intraosseous Infusion

18D Manual Skills and Training Manual, Reviewed by LTC (Ret) John Zotter, VC, USA

When: Intravenous access is unavailable or impossible to obtain. Infusion of fluids, blood or medications into the tibia of a child or the sternum of an adult is the only route available (rare in adults since, central venous access is usually possible).

What You Need: Alcohol swabs, intraosseous needles, an IV administration set, and IV solution.

What To Do:
1. Intraosseous (IO) infusion should only be used on unconscious or gravely ill children under 6 years of age when peripheral cannulation is unobtainable.
2. Apply gloves for personal and patient protection.
3. Clean the site as for an IV infusion.
4. NOTE: The site of choice for intraosseous infusion is the superior portion of the tibia, 1-2 finger widths below the tibial tuberosity in children under age 6. The single recommended site of insertion for adults is the manubrium (top 1/3 of the sternum), or on the midline and 1.5 cm (5/8 inch) below the sternal notch.
5. Select an intraosseous (bone marrow) needle or 16 or 18 ga. spinal needle.
6. Put the child in a supine position. Use an uninjured lower leg and use padding to put the knee into a 30° flexion.
7. Use a boring or screwing motion to advance the needle distally (away from the epiphysial plate) at a 45° angle until it penetrates into the marrow. There should be a decrease in the resistance to advancement.
8. To check your needle placement, aspirate bone marrow into 10 cc syringe filled halfway with sterile saline. It will not always be possible to aspirate marrow.
9. Cannot aspirate: Apply just enough pressure to plunger to clear needle tip of possible clot/bone particle, then aspirate again.
10. Infuse 5-10ml of NS by syringe to ensure correct placement and to clear clots. Swelling around injection site: indicates penetration of bone has not occurred and saline is being injected into fleshy tissue; reassess and try again.
11. Secure the needle with tape if necessary.
12. Attach an IV infusion set as with normal IV infusion.
13. **NOTE:** Crystalloid solutions, blood, and most ALS drugs can be administered by this route.
14. Recognize the contraindications to intraosseous infusion.
    a. Fracture at or proximal to the infusion site.
    b. Traumatized extremity.
    c. Cellulitis at the infusion site.
    d. Congenital bone disease.
15. Record all treatment in the patient's medical record.
16. FAST1 adult intraosseous infusion system is currently the **only** FDA kit approved for adults. Web site for additional information **www.pyng.com**

What Not To Do:
Do not forget that medications can be delivered through this route at the same dose that would be given IV. Do not use the sternal technique in small adults or children.

Do not the use the sternal technique in patients with previous sternotomy (heart by-pass), evidence of sternal skin infection or burns over the insertion site, or fracture or vascular injury that could compromise the integrity sternum.

Procedure: Suturing

Adapted from 18D Skills and Training Manual with additional information from COL Glen Reside, DC, USA

When: You must stitch a wound closed without contamination.

What You Need: A large Kelly clamp, a needle holder, tissue forceps, a mosquito clamp, appropriate sutures and needles or pre-package suture/needle combinations (preferred), four towel clamps, sterile gloves, several 4x8 inch gauze pads, four hand towels, shaving razor, antiseptic solution and sponges, irrigation syringe (may use attached catheter for increased pressure), sterile saline for irrigation and suture wash, surgical bowl(s), a 22-23 gauge needle with 5-10 cc syringe, and **lidocaine** anesthesia (with or without epinephrine as appropriate).

What To Do:
1. Gather the appropriate equipment. Inspect for damage or tampering.
2. Prepare the wound site.
   a. Protect the wound with sterile gauze.
   b. Shave the skin 3-5 inches around the wound.
   c. Perform a surgical scrub.
      (1) Clean the wound area with circular motions, making sure not to let antiseptic solution wash into the wound.
      (2) Clean from the wound edges out.
      (3) Dispose of the sponges.
      (4) Repeat as needed.
   d. Irrigate the wound with sterile solution.
      (1) Irrigate from one end toward the other, usually from the proximal to distal.
      (2) Avoid "suck back" with the syringe.
3. Drape the patient appropriately with hand towels.
4. Check the patient for allergy to medications.
5. Administer the anesthesia.
   a. Inject anesthetic into the subcutaneous layer.
   b. Insert the needle full length.
   c. Aspirate for blood.
   d. Inject anesthesia upon withdrawal of the needle.
   e. Wait for 3 minutes, then test for pain.

WARNING: Do not use **Lidocaine** with **epinephrine** to anesthetize the fingers, nose, ears, toes, or penis.

6. Select the proper suture.
   a. Use 5-0 to 6-0 for the face, 3-0 to 4-0 for the arms, legs, and trunk. 2-0 may be used to secure chest tubes and other high-stress applications. Tend to use smaller sizes in children and in lower stress areas.
      (1) Chromic suture: Use for the bowel, muscle, and peritoneum. Resorbs within 14-21 days. Packaged in isopropyl alcohol, which is an irritant. Rinse suture with saline prior to use. Causes less inflammation than plain suture.
      (2) Plain suture (gut): Use for subcutaneous tissue and ligation of small vessels. Resorbs within 7-14 days. Packaged in isopropyl alcohol, which is an irritant. Rinse suture with saline prior to use.
      (3) Vicryl, Maxon, and Dexon: Strongest absorbable sutures. They are easier to handle and tie than plain suture, have higher tensile strength, and cause less tissue reaction. Resorb in 4-6 weeks. Vicryl comes dyed and undyed - avoid using dyed on the face.
      (4) Nylon: Most popular skin suture. Not absorbable. Good tensile strength, minimal tissue reaction
      (5) Polyester or Polypropylene: Not absorbable. Easier to tie than nylon. High tensile strength, moderate/minimal tissue reaction.

    (6) Silk: Not absorbable. Fair tensile strength, excellent handling/knot tying. Moderate tissue reaction.
  b. Use wire sutures on bone and tendons. Stainless steel is the strongest suture material, with the most secure knots, and is well tolerance by tissue unless corroded. Stiffness of the metal can cause irritation and tissue damage.
  c. Tissue reaction to suture: localized acute, aseptic inflammation. Suture, especially braided suture, can provide a wick through the skin allowing pathogens access to a wound.
 7. Suture the patient. Thread needles with desired suture if not using pre-packaged needle/suture combinations.

HINTS: Align the edges of the wound, and stitch the middle of the wound first, if possible. Use the needle holder to clamp on the back of the needle near but not ON the suture material, with the needle perpendicular to the holder. Some recommend grasping the needle half way around the curve. Insert the point of the needle perpendicular to the skin and then follow the curve of the needle through when piercing tissue. Suture an equal width of tissue on each side of the wound. Go deeper rather than wider with the stitches if need to achieve greater wound closure. Do not have stitches too tense - make sure tissue is not blanched by the stitch. Keep stitches uniform approximately 5-10mm apart and 5-10mm from the wound edge, with knots away from the wound edge.

  a. Simple interrupted suture: Puncture the skin with the needle and exit into the wound, traversing the skin only. Pull the needle out through the wound, and enter the opposite side of the wound at the same depth. Curve the needle up through the skin, positioning it as described above. Tie a square knot plus an additional throw, then clip excess suture material. This is the most common suture used. Advantages: strength; successive sutures can be placed following the path of the laceration; distance, depth and tissue eversion can vary from stitch to stitch (see Figure 8-1).
  b. Vertical mattress suture: Puncture the skin with the needle at least 1 cm from the edge of the wound, and exit into the wound, traversing the skin and subcutaneous tissue (at least 1 cm deep). Pull the needle out through the wound, and enter the opposite side of the wound at the same depth (subcutaneous tissue). Curve the needle up through the skin at least 1 cm from the edge of the wound, positioning it as on the other side. Reverse the orientation of the needle, as if to sew with the opposite hand. Point the sharp tip away from yourself and insert the needle approximately 5mm from the wound. Make the return suture either subcutaneous or a skin closure. Tie a knot as above and clip the excessive suture material. Advantages: more tissue eversion, broad wound contact, watertight. Useful for preventing broad scar formation (see Figure 8-1).
   Disadvantage: constricts blood supply at wound edges, possibly causing necrosis and dehiscence.
  c. Continuous suture: Insert the needle and exit through the subcutaneous tissue. Tie a knot as for a simple interrupted stitch, but do not clip the suture end. Suture continuously the entire length of the wound without tying any additional knots until the end. This method can be modified to 'lock' each stitch by bringing the suture back across the wound after the stitch and passing it under the piece of suture coming from the previous stitch. All the locks should be aligned on the same side of the wound. Tie the final knot on the opposite side of the wound. Clip the excessive material. Advantages: aligns perpendicular to the wound, distributing tension evenly; allows watertight, rapid closure; locking feature prevents continuous tightening of the stitches as suturing progresses. Disadvantages: not able to adjust to tension from edema; should not be used on areas of existing tension (see Figure 8-1).
 8. Eliminate the dead space by rolling the wound proximally to distally with a rolled gauze pad.
 9. Apply **bacitracin** or other topical antibiotic as appropriate and then bandage the wound.
 10. Tell the patient when to return to have the sutures removed or to return earlier if the wound shows signs of infection (red, hot, swollen, wound draining pus; fever; red streaks from wound). Remove stitches from eyelid in 3 days; cheek in 3-5 days; nose, forehead and scalp in 5-7 days; arm, leg, hand, foot in 7-10+ days; and chest, back, and abdomen in 7-10+ days.

What Not To Do:

Do not suture opposite sides of the wound at different depths or distances from the wound edge. This will create uneven skin alignment, overriding edges and poor or delayed healing, as well as a poor cosmetic result.
Do not tie sutures too tight, so as to compromise blood flow to the wound edges, which need it most.

Figure 8-1

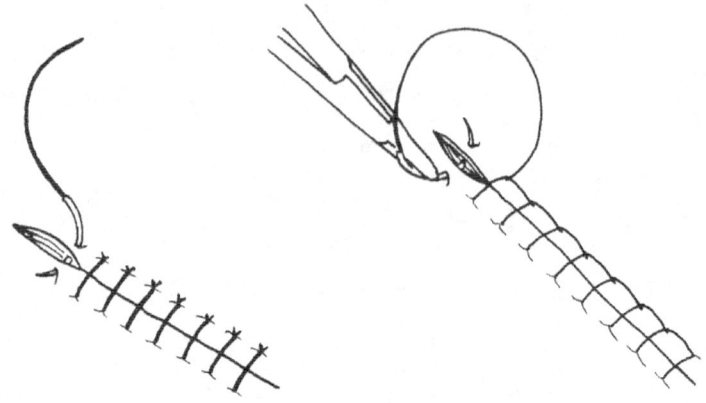

Simple Interrupted Suture   Continuous Locking Suture

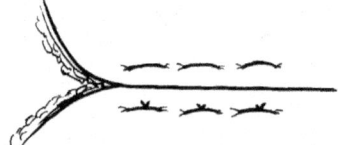

Horizontal Interrupted Mattress Suture

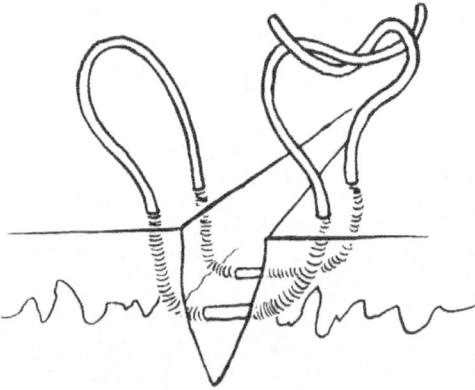

Suture Techniques

Do not suture a dirty or contaminated wound. If it is still dirty after irrigation, or if irrigation is not possible, allow the wound to heal by granulation. If there is danger that the skin may close prior to the deeper tissues granulating, then pack the wound with iodoform gauze (or **Betadine** soaked gauze) daily until the wound heals up to the skin.

Procedure: Wound Debridement

1LT Harold Becker, SP, PA, USAR & LTC Lee Cancio, MC, USA

What: Remove dead or devitalized tissue to decrease infection and improve healing.

When: It is essential to debride a traumatic wound to prepare it for closure. Devitalized tissue inhibits leukocyte phagocytosis, acts as a culture medium for bacteria growth, and provides an anaerobic environment that limits leukocyte function. Debridement relieves excess tension, provides drainage, and removes bacteria and devitalized tissue that impair the wound's ability to ward off infection. Whether the wound is secondary to an abrasion, laceration, burn, frostbite or gunshot, debridement should be rational, not radical.

What You Need: Recommended: Skin hooks, Iris scissors, Metzenbaum or Mayo scissors, scalpel with #10, #11, #15 blades, tissue forceps, 35cc syringe, 16 or 18 gauge needle or plastic cannula, toothbrush or a surgical scrub brush, NaCl for irrigation, retractors: Sims or Army-Navy
Improvised: Any type of scissors, scalpel with any type of blade or pocketknife, any type of tissue forceps or hemostats, IV with catheter (any type of clean fluid), any type of brush, tap water (boiled if possible), any type of retractor or pliable object (i.e., SAMS splint)
NOTE: Items should be sterilized with cold sterilization or boiling water if possible.

What To Do:
1. Determine the margin between devitalized and viable tissue. Use clinical judgment. Within 24 hours there is usually a sharp demarcation between devitalized skin and viable skin. Longer time is usually recommended in frostbite and gangrene. It is more difficult to differentiate nonviable muscle from muscle that is injured but will heal. Use color, contraction, consistency, and circulation (the 4Cs) as guidelines when excising muscle. Identify devitalized muscle by its dark color, mushy consistency, inability to contract when grasped with forceps and a lack of brisk bleeding when cut.
2. Prep and drape the wound.
3. Irrigate the wound. Use a 35cc syringe with a 16 or 18-gauge cannula and NaCl for irrigation. (This will provide approximately 8 psi, which will dislodge most particles). Irrigate the wound copiously.

Minor wounds not involving muscle
   1. Stabilize the skin edges with the skin hooks by retracting the wound at both ends. Use your fingers pull the skin being debrided perpendicular to the laceration (this will prevent the skin from rolling in, providing an even, clean edge).
   2. Using the scalpel with a #11 or #15 blade, hold it angled away from the wound edge and excise the devitalized skin. Holding it at an angle will ensure that eversion is achieved when the edges are approximated.
   3. After excising the skin edges, inspect subcutaneous tissue. Excise any devitalized tissue with iris scissors.
   4. Irrigate the wound once again with copious amounts of NaCl.
   5. Close the wound either primarily or secondarily depending on the location, initial debridement, and the level of contamination.

Pearl: A technique that helps distinguish devitalized tissue: apply fluorescein dye to a gauze pack and pack the wound. The fluorescien will stain devitalized tissue, which can then easily be debrided. If unsure, excise the skin until active bleeding starts.

Abrasions

Particular attention should be given to abrasions in order to prevent a traumatic tattoo. These occur when fine particles become embedded and are incorporated into the epithelium.
1. Ensure adequate anesthetic has been administered (either locally or via block).
2. **Xylocaine** gel can be applied to the wound for 5-10 minutes. This may assist in providing adequate anesthesia.
3. Irrigate the abrasion copiously with NaCl.
4. Use a sterile toothbrush or surgical scrub brush soaked with NaCl or surgical soap to help remove the debris.
5. Use the tip of a #11 blade to remove large or deeply embedded particles.
6. Use mineral oil, **Vaseline**, peanut butter, or mayonnaise (or some other oil-based product) to help remove tar.
7. Leave the wound open and clean it daily.

Penetrating wounds
1. Use a scalpel with a #11 or #15 blade to excise the entry and exit wounds. The incisions should be sufficient to allow optimal surgical exposure and drainage. The excised skin should include the underlying subcutaneous tissue, and be incised oriented parallel to the underlying muscle fiber.
2. Incise the fascia parallel to the muscle fiber with Mayo or Metzenbaum scissors in both directions. Open the muscle surrounding the missile tract in the direction of the fibers to allow adequate exposure for inspecting the tract.
3. Inspect the wound tract. Remove any foreign bodies. Excise any muscle that is compromised and nonviable with a scalpel or scissors.
4. Utilize the retractors at this time to help with visualization and debridement. Be careful when using retractors in order to avoid damaging vessels, nerves, and healthy tissue.
5. Perform this procedure at both the entry and exit wounds. Debride the mid-track through extended entry and exit wounds. This prevents cutting across muscle groups to connect the two wounds.
6. Appropriate drainage of the wound may be difficult to achieve. Liberal incisions tend to facilitate drainage from the deep recesses. Remember to excise skin, fascia, vessels, nerves, and bone conservatively, and muscle more liberally. Try to save periosteum and tendons unless severely contaminated or compromised.
7. Irrigate the wound copiously again as above.
8. Do not pack the wound. The additional pressure can cause tissue necrosis, due to its already compromised blood supply. Lightly lay dry sterile gauze in the wound.
9. Leave the wound open with delayed primary closure in 4-10 days.

What Not To Do:
Do not debride good, viable tissue. Wait until the tissue declares itself, or makes it apparent that it is dead.
Do not close the debrided wounds, but let them drain. They may be closed later (delayed primary closure) if not infected.
Do not pack debrided wounds tightly, but allow them space to expand.

Procedure: Skin Mass Removal
Maj. Frederick Shuler, USAF, MC

What: Surgical procedures for treating various masses and conditions of the skin including abscess, epidermal inclusion cyst (EIC), lipoma, mole, etc. These masses may be inflamed (abscess) or non-inflamed (lipoma).

When: The patient complains of a mass in the skin that is either infected or a hindrance to activity and mission performance. Manage lesions that do not fit into these categories conservatively until return from the mission. The medic should not remove vascular masses.
Non-inflamed: These are best treated electively with excision (the removal of the entire lesion) and submis-

sion for pathologic evaluation. Differential diagnosis includes lipoma, fibroma, neuroma, and fibrohistiocytoma (potentially malignant), hence the need for pathology review.

Inflamed: Although antibiotics can control and sometimes reverse the inflammation of an abscess, those that appear to be infected and unresponsive to conservative therapy should be incised and drained (I&D). This however, creates an open wound that requires dressing changes for 1-2 weeks. An EIC can be excised as opposed to incised as a non-inflamed mass if it is not actively inflamed. An attempt should be made to remove the entire cyst wall. If the wound remains sterile (the cyst is not accidentally opened during the procedure), it can be closed at the end of the procedure. If the EIC is actively infected, treat it as an abscess.

What You Need: Sterile prep and drape, needles: 18 and 24-27 ga, 10 cc syringe, alcohol prep pads, local anesthetic, preferably with **epinephrine**, scalpel: #15 blade, irrigation: sterile NS/LR/water or hydrogen peroxide, sterile gloves, sterile 4x4s, sterile hemostat, Adson pickups, Metzenbaum or Iris scissors, Allis clamp (if available), specimen container and label (store/send in a watertight container [e.g., urine cup] filled with formaldehyde), saline, IV fluid or sterile water, tape and dressings.
Excision: 3-0 dissolvable suture (taper needle), back lesion: 2-0 nylon (cutting needle), extremity or scalp lesion: 3-0 nylon (cutting needle)
I&D: 2x2 or 4x4 gauze (or iodoform) for packing, tape

What To Do:

Prep: For inflamed and non-inflamed lesions, scrub and prep the area around the lesion with **Betadine** and drape with sterile towels. Infiltrate local anesthetic in a field block at 2-4 sites around the area of the lesion. This is a much more tolerable approach to anesthetizing the inflamed lesion, but works well in providing pain control for either lesion. Try not to inject the EIC, as the distention of the capsule can cause increased pain or spray the contents of the lesion out through the EIC orifice back at the surgeon. Allow several minutes for the anesthesia to take effect. Subcutaneous lesions (lipomas, etc.) require deeper anesthesia, but they should likewise NOT be injected. Plan an incision along the Lines of Langer (natural lines of tension) to minimize the scar formation and promote efficient healing. Make the incision: use an elliptical incision for excision of an epidermal mass or EIC, but use a straight incision for an abscess or subepidermal lesion. Include the entire epidermal lesion, as well as the EIC (skin overlying cyst plus punctate, follicular orifice), in the excised tissue to prevent recurrence. Similarly, remove deeper masses in their entirety.

Non-inflamed superficial mass:

Do not remove these lesions unless they fit the criteria above. Make an elliptical incision around any superficial mass. Grasp the tissue to be excised with a clamp to allow retraction and demonstration of the lines of tension of the surrounding tissue. Dissect under the mass, remaining in the dermis if the mass is indeed superficial. Remove the tissue plug. Control hemostasis by gentle pressure. Irrigate the wound then close it in one layer with nylon suture.

Standard guidelines: The specimen should be sent to the pathologist for evaluation. Use mattress suture technique rather than simple interrupted technique in areas of higher tension (i.e., back, joints) to prevent dehiscence of the wound. A dry bandage should be kept in place for 36-48 hours to allow re-epithelialization of the wound. Sutures should be left in 5 days on the face, 7-10 days elsewhere, and 10-14 days on high-tension areas. No antibiotics are needed. Profile of the soldier/patient should include limited movement of the surgical wound for 2-3 weeks.

Non-inflamed subcutaneous mass:

Do not remove these lesions unless they fit the criteria above. Unless the lesion is an EIC, make a single incision through the skin over the mass, large enough to allow visualization and dissection in the wound. If it is an EIC that is not actively inflamed, make an elliptical incision as above for a superficial mass. Be certain the ellipse will include both the EIC and its epidermal opening. Gently spread the subcutaneous tissue to locate the mass, and use scissors when needed to dissect the mass out of the wound intact. The capsule of the EIC is usually adjacent to the dermis. Dissection should proceed carefully in order to prevent rupture of the EIC capsule and spillage of the foul smelling contents. If the mass has a capsule that ruptures, attempt to remove the mass and capsule "piecemeal." Inspect the wound for retained fragments of wall. If the rupture was large, or the capsule cannot be entirely removed, manage the mass as an inflamed subcutaneous mass (see below). Otherwise, send the specimen to the pathologist for evaluation. Control hemostasis by gentle pressure. Irrigate the wound, then close it in one or two layers. Close the dermal layer using inverted, interrupted stitches with dissolvable suture. Include some fat in the stitch if the dissection extended into the fat tissue. Follow the standard guidance above. Use this surgical approach for the removal of all non-infected, subcutaneous masses such as lipomas or fibromas. It is important to send lesions for pathologic evaluation, as further radical surgery may be necessary for the rare malignancy.

Inflamed superficial or subcutaneous mass:
Treat with appropriate antibiotics. If the lesion is an EIC that responded to therapy, treat as in the previous paragraph. If the lesion is an abscess or unresponsive EIC (or similar lesion, such as furuncle), interferes with the mission performance and cannot be safely managed until the end of the mission, perform the following surgery. Mark any sinus tract by placing a needle or other object into the tract. Incise the abscess/cyst (avoid spraying the contents on any person) and evacuate its contents. Explore the cavity with a hemostat and spread the jaws to break down walls and adhesions in the abscess. If the abscess had a sinus tract communicating with the epidermis, open the tract and expose it to therapy. Obtain hemostasis with pressure. Irrigate with hydrogen peroxide or saline and pack with damp (sterile saline) 2x2s or 4x4s (or iodoform), and apply a dry dressing. Do not close this infected wound. Do not continue antibiotics unless cellulitis is severe, the infection does not resolve with I&D or the patient is immunosuppressed (i.e., diabetic, HIV, malnourished, on chemotherapy). Then continue antibiotics only until the wound demonstrates that it is healing. Start wet-to-dry dressings.

Wet-to-dry dressings: This requires moist packing that "dries" during the interim between dressing changes. When the packing is removed, it debrides the wound by removing the dead cells that stick to it. Do not allow the packing to dry completely. If it does, then change the dressing more frequently. Remove the packing daily with non-sterile gloves, irrigate the wound, and replace the packing until the wound closes (1-3 weeks). The irrigation does not need to be "sterile" as potable water can be used (the wound is already colonized with skin flora and is by definition not sterile). **NOTE:** Only the wound should be in contact with the moist dressing, because exposure of the surrounding skin to the continuous moisture can denude the skin and lead to further infection. The patient can even remove the packing, take a shower and wash the wound with a soap and water, before repacking the wound.

What Not To Do:
DO NOT leave obvious cyst wall behind in the wound.
DO NOT close grossly infected wounds.
DO NOT forget to send the lesion to a pathologist.
DO NOT attempt to remove a lump over a joint at the wrist or fingers- this is often a ganglion cyst and connects directly to a tendon sheath and the joint space.

Procedure: Joint Aspiration
COL Warren Whitlock, MC, USA

When: To analyze joint fluid for suspected infection or for therapeutic reasons; relieve pain by draining an effusion from a swollen synovial space; instill medications/steroids.

What You Need: Alcohol swabs, Povidone-iodine prep solution, sterile gloves and towels, gauze, forceps, local anesthesia with **ethyl chloride vinyl spray** and/or **lidocaine** 1%, appropriate syringes, needles, and chocolate (Thayer-Martin) media if gonococcal arthritis is suspected.

What To Do:
General Procedure
1. Identify landmarks and mark the entry point with a scratch or indentation on the skin.
2. Sterilize the skin in a wide field around the puncture site.
3. Anesthetize the skin with the 1% **lidocaine** with the 10-ml syringe and 22 to 27-gauge needle; continue

down to the joint capsule.
4. Select an appropriate needle (usually 20-gauge for knee, shoulder, elbow, or ankle; 25 or 27-gauge for small hand joints).
5. Stabilize the joint to be aspirated. Advance the needle into the joint space while continually applying suction. Generally, a sudden "give" will be felt when the needle passes through the synovium into the joint space.
6. Withdraw as much fluid as possible. Then withdraw the needle and apply firm pressure over the site for 1-2 minutes.
7. Cover the site with an adhesive dressing.

Specific Joint Techniques

1. Shoulder: Have patient sit with arm in lap (this positions the shoulder in mild internal rotation and adduction). Identify insertion site inferior and slightly lateral to tip of coracoid. Direct the needle (20- or 22-gauge 1/2-in needle) to joint space medial to the head of the humerus and just below the palpable tip of the coracoid process.

2. Wrist: Position the wrist in the prone position with about 20° of flexion. Identify insertion site by marking the distal ends of the ulna and radius. Enter a bulging, inflamed joint space at the wrist dorsally at prominent areas of swelling; such areas are invariably found on the radial or ulnar sides of the wrist during examination. If possible, avoid inserting needles in the palmar or dorsal aspects of the wrist to prevent damaging nerves or blood vessels over the joint. Use a 20- or 22-gauge 1/2-in needle.

3. Elbow: Have patient sit with the arm supported horizontal to the ground and the elbow bent at 30°. Identify insertion site on the lateral aspect of the elbow in the shallow depression immediately anterior and inferior to the lateral epicondyle of the humerus. Advance the needle medially and slightly proximally into the joint space. With significant effusion, the bulging synovium should be evident laterally. **CAUTION:** Be sure that it is not an olecronan bursitis, which would not require a joint procedure. Use a 20- or 22-gauge 1/2-in needle.

4. Knee: Place patient supine with quadriceps muscle relaxed (patella should be freely movable). Identify the insertion site immediately beneath the lateral or medial edge of the patella. Enter the joint space of the knee either medially or laterally. Pressure on the opposite side of the joint will make the synovium bulge more prominently and toward the needle. Direct the needle parallel to the plane of the underside of the patella. From the lateral aspect, the entrance site is at the intersection of lines extended from the upper and lateral margins of the patella. Fluid should be obtained before the needle tip reaches midline. Use a 20- or 22-gauge 1/2-in needle.

Synovial Fluid Analysis
Record the physical characteristics or the fluid:
1. Total volume
2. Color and clarity (cloudy if you cannot read print held up behind it)
3. Viscosity (joint fluid usually will stretch 1-2 inches)

Laboratory Studies:
1. Cell count and differential
2. Special stains: All fluids should have Gram stain. Special stains for fungi and acid-fast bacilli should also be performed with chronic joint problems.
3. Examine for crystals: Examine the specimen immediately under the microscope.
4. Culture: Sterile container holding 1-2 ml (GC requires special media)
5. Fluid glucose: Low fluid glucose suggests infection (< 20 mg/dl).
6. Protein content: High fluid protein indicates inflammation (Usually 1/3 of serum).
7. Fat stain: Sudan stain (a few drops of glacial acetic acid and Sudan III in ethanol) positive indicates presence of free fat that suggests intra-articular fracture

What Not To Do:
Do not perform under the following circumstances:
  Local infection along the proposed needle entrance tract (e.g., overlying cellulitis, periarticular infection)
  Uncooperative patient, especially if unable to keep the joint immobile throughout procedure
  Difficult to identify bony landmarks
  A poorly accessible joint space, as in hip aspiration in the obese patient
  Inability to demonstrate a joint effusion on physical examination, except when septic arthritis is strongly suspected.

Procedure: Compartment Syndrome Management
LTC Winston Warme, MC, USA

What: Relieve pressure of compartment syndrome by fasciotomy.

When: Compartment Syndrome (CS) is a condition in which structures such as muscles, vessels and nerves are constricted within a tight fascial compartment. It results when internal or external pressure reduces capillary perfusion below the level necessary for tissue viability in a closed fascial space or muscle compartment. This is most common in the leg secondary to blunt trauma (but may occur in the arms) and may be due to crush injury, muscle rupture and burns. On exam, the compartments will be tense in comparison to the other side. The patient will complain of pain, especially with passive movement joints distal to the injury. The other Ps (pallor, paresthesia and pulselessness) are late findings and only strengthen the diagnosis already made.

What You Need: IV fluids, an IV infusion set, splinting material, medications for pain, antibiotics, tetanus prophylaxis, bandages for open wounds, a minor surgical set (scalpel, blades, tissue forceps, retractors, scissors), +/- a urinary catheter set.

What To Do:
1. Identify a suspected compartment syndrome (CS) based on patient history.

NOTE: There are three main physiological reactions that can cause compartment syndromes: increased accumulation of fluid (bleeding, e.g., a closed tibia fracture), decreased volume (constriction of compartment), or external compression (crush injury).
   a. Fractures: closed or open.
   b. Severe contusion with no fracture, (e.g., s/p kicked in the leg in a soccer game).
   c. Recently applied splint, cast or bandage.
   d. Injury to a major blood vessel may produce CS: bleeding in the compartment, partial occlusion of the artery secondary to spasm/initial tear with inadequate collateral circulation, or postischemic swelling without restoration of circulation (delayed more than 6 hours).
   e. Extreme exertion - long distance running with improper training. Different from the chronic condition of exertional compartment syndrome that may require elective release, but is not an emergency. This presents with recurrent mild pain in the anterior or lateral compartments of lower leg, sometimes with a foot drop or neurologic signs.
   f. Burns: Decrease compartment size with massive edema; coalesces the skin, subcutaneous tissue & fascia into one tight, constricting eschar; underlying compression of nerves/muscles. May require escharotomy (see Burns chapter).
   g. Other causes: anticoagulant therapy, arterial puncture, hemophilia, infiltration of IV fluid, and snake bites.
2. Identify a suspected compartment syndrome based on signs and symptoms.
   a. The most important symptom if an impending CS is pain that is out of proportion to the primary problem or injury.
   b. The 6 Ps: Pressure - the earliest finding; swollen, palpably tense compartment. Pain on stretch -

passive movement of the digits may produce pain in the involved ischemic muscles. Paresis - muscle weakness due to primary nerve involvement, muscle ischemia, or guarding secondary to pain. Paresthesia or anesthesia - a late physical finding in a conscious and cooperative patient is a sensory deficit. Must perform a nerve (sensory) exam to find affected compartment. Pulses present - peripheral pulses are palpable in 90% patients. Pallor - capillary refill is routinely present

3. Rule out other probable causes because of morbidity of treatment (fasciotomy).
   a. Diagnosis of compartment syndrome: Peripheral pulses intact R/O arterial injury. Nerve injuries cause little pain.
   b. Diagnosis for arterial injury - usually absent pulses, poor skin color and decreased skin temperature.
   c. Diagnosis for nerve damage (neurapraxia): Remarkable paresis or paresthesia, nerve damage (neurapraxia) associated with a fracture or contusion.
4. Assess the patient
   a. Monitor and maintain vital signs and urinary output to minimize hypotension.
      NOTE: Insert a urinary catheter as necessary.
   b. Administer suitable medications: to alleviate pain and anxiety (see Procedure: Pain Assessment and Control), antibiotic therapy and tetanus prophylaxis for open wounds (see Burns).
   c. Monitor the patient for crush syndrome (similar to Compartment Syndrome, but also suffer distal pulse and neurological damage) and manage accordingly. May need kidney support and fasciotomy.
5. Decompress the affected compartments. This is a skill that should be developed in medlab or during ATLS training. Trying this out when you have never done it in a lab or in the OR is not recommended. Do your training beforehand and be ready for this contingency.
   a. Loosen/remove volume-restricting plaster casts and circular dressings. Splitting and spreading a plaster cast may result in a 65% decrease in intra-compartmental pressure. Keep the limb at heart level. Do not elevate it as this can decrease perfusion. If symptoms of neurologic deficit persist more than 1 hour after cast splitting, the cast and all circular dressings must be removed and the limb re-examined.
   b. There are no satisfactory nonsurgical methods for treating compartment syndromes. Surgical decompression, which allows the volume of the compartments to increase, is the primary means of relieving pressure.
   c. Evacuate all cases of suspected Compartment Syndrome. If evacuation is not possible, proceed as below.
   d. Treatment: Adequate decompression of the muscular compartments with scissors (fasciotomies). Incise skin 12 centimeters along the anterolateral and posteromedial sides of the leg to allow for release of the anterior and lateral compartments and superficial and deep compartments of the leg, respectively. In the arm, two 5 cm vertical incisions are placed on the dorsum of the hand between the index and middle metacarpals and the ring and small metacarpals. The fascia is then split. 20 cm incisions are placed on the volar and dorsal aspects of the forearm to release the underlying fascia under direct visualization. Release of the carpal tunnel is also needed on the volar aspect of the wrist, but due to possible damage to the median nerve, should not be attempted without seeing it done before.
   e. Do not close the skin. May need to perform additional surgery.
6. Perform postoperative care
   a. Leg and arm wounds are cared for in the routine manner (see Procedure: Skin Mass Removal).
   b. Without fractures: Closure in a week (delayed primary closure, if possible) with or without skin grafting. Necrotic muscle is debrided once or twice a week until a satisfactory granulation bed is present (see Procedure: Wound Debridement).
      CAUTION: Skin grafting or closure prior to this may lead to infection and the need for subsequent amputation. Prevent development of contractures: the ankle is splinted in the neutral position and the forearm is splinted in the position of function (holding a beer can).
   c. Fracture present: Closed fractures require referral for fixation. Open fractures require referral for fixation.
   d. If renal insufficiency develops, reduce fluid administration and evacuate the patient.
7. Record all treatment given.

What Not To Do:
Do not cut too deeply or too early when performing fasciotomy.
Do not close fasciotomy incisions until one week post-op.

Procedure: Splint Application
LTC Winston Warme, MC, USA

When: A casualty has a fractured or dislocated bone that requires stabilization. Apply a splint to relieve pain and prevent further harm by immobilizing the underlying bone, the joint above the injury and the joint below the injury.

What You Need: Water (NOT HOT), padding material (rolled cotton or Webril), a knife, scissors or cast saw, plaster or fiberglass casting material. Alternatives for splints: thin boards, sticks, or adjacent body parts for fingers/toes/legs or to splint an arm against the body.

What To Do:
1. Before applying a splint, inspect skin carefully to ensure that there are no sores or breaks in the skin that should be cleaned and dressed with Telfa before casting. Small puncture wounds might be open fractures and should be treated emergently to decrease the incidence of infection. Patients with open fractures should receive tetanus prophylaxis and antibiotics if available (e.g., 1gm of **Rocephin** IM or IV) before being evacuated.
2. Immobilize the injury in a position of function (as described below) extending to the joint above and below the fracture. Position the patient and encourage them to relax.
   a. Arm: Wrist–in a natural position, 15° extension (as if holding a can). Elbow–at a 90° angle
   b. Leg: Knee–5° to 15° flexion, ankle–90°. There should be no inversion or eversion of the foot.
3. Pad the fracture and joint area with sheet cotton or Webril in acute injuries and postoperative cases to provide comfort and lessen the possibility of pressure sores. Wrap the padding smoothly with the turns overlapping about ½ the width of the previous layer. Pad bony prominences with pieces of felt, or use several additional layers of Webril, or cotton.
4. Use 5X30 inch splints or rolls torn/cut to size. Dip and squeeze the casting material. Use a splint of 10 thicknesses (plies) of casting material on the posterior lower leg and continuing onto the plantar surface of the foot for ankle injuries. Extend out beyond the toes on the foot. Similarly, use 5-ply casting material to make medial and lateral splints for the arm. Extend just to the proximal palmar crease on the hand. Rub the plaster/fiberglass smooth as it is applied so that the layers blend. Rub and mold the splint with your hands over the contour of the body part until it is firmly set. Continue molding until reaching the setting point of the plaster/fiberglass.
5. Make sure the splint is not circumferential so that there is room for some swelling to occur. Elevate the extremity above heart level to minimize the swelling and maximize comfort. Encourage the patient to wiggle toes and fingers
6. Cover the splint with an ace bandage.
7. Check peripheral neurovascular status to insure that the splint is not too tight.
8. Evacuate the patient to orthopedics for definitive management.

What Not To Do:
Do not forget to reassess neurovascular status later. Remember that inflammation and swelling can continue and result in loss of neurovascular function 8 to 12 hours later. Inform the patient and arrange for re-examination.
DO NOT CAST AN ACUTELY INJURED PATIENT. Allow for tissue swelling with a non-circumferential splint.

Procedure: Common External Traction Devices
18D Skills and Training Manual, Reviewed by LTC Winston Warme, MC, USA

When: When a casualty has a fracture of the lower limb that needs traction to reduce the fracture, stabilize and align it. Apply the correct amount of traction to the extremity without causing further injury to the patient.

What You Need: Moleskin, elastic bandage, felt pads or cotton, stockingette, rope or cord, a spreader bar, soap, water, a razor and blades.

What To Do:
1. Inspect the skin of the lower extremity to determine the scope of injury.
2. Check the pulse in the lower extremity to ensure blood circulation.
NOTE: If none, assess the neurovascular status before, after, and at intervals during the procedure. If pulses continue to be absent, continue with this task and evacuate as soon as possible.
3. Wash the lower extremity.
4. Shave the leg.
5. Pad the bony prominences of the leg (medial/lateral malleolli and fibular head) to prevent injury.
6. Apply moleskin to the medial and lateral aspects of the leg to protect the skin from breakdown.
7. Apply a long piece of stockingete from the medial roximal tibia to the lateral proximal tibia, leaving a loop below the foot. Attach the spreader bar to the stockingette, in line with the long axix of the extremity. Secure the stockingette to the limb with an elastic bandage.
   a. If stockingette is not available, improvise with rope, torn clothing or belts.
   b. The spreader bar can be a stiff branch, a canteen cup with a hole drilled in the center, or an unopened abdominal bandage taped to the device.
8. Align the rope over a pulley or pivot point with tension in line with the long axis of the extremity.
9. Attach five pounds of weight to the construct. More may be needed.
CAUTION: The device used to attach the weight to the extremity must not be so tight that it restricts the blood flow. The force applied by the weight must NOT exert uneven pressure that would cause skin necrosis.
NOTE: Add additional weight until the patient experiences relief of pain. The relief will often be dramatic by adding as little as two or three pounds. Reduce the weight over the following days as the muscle spasm (that the traction device is designed to overcome) diminishes.
10. Perform a postreduction examination of the extremity to evaluate the following:
    a. Reduction efficacy.
    b. Neurovascular status.
CAUTION: Pain may indicate the possibility of compartment syndrome. Re-examine the patient frequently for this complication (see Procedure: Compartment Syndrome Management).
11. Check the pulse in the extremity hourly.
12. Record the procedure.

What Not To Do:
Do not add too much weight.
Do not constrict the limb, which restricts blood flow and causes necrosis.

Procedure: Bladder Catheterization
CAPT Leo Kusuda, MC, USN

What: Straight catheterization through urethra to drain urinary bladder.

When: When patient is unable to void or is having frequent (q 15-20 minute) voiding, suggestive of poor emptying of the bladder.

What You Need:
1. Catheter (options):
   a. Foley catheter: Has balloon at end of straight tip. Sizes generally used are 16 or 18 Fr. Balloon sizes are 5 and 30 cc. 5cc balloon is most commonly used.
   b. Red rubber Robinson: Red colored straight catheters without balloons.
   c. Short female straight catheters: 14 or 16 Fr.
   d. Coudé catheter: Has curved tip with or without balloon. If balloon is attached, check to see if the balloon port is on the same or opposite side as the curved tip. If no balloon, there is usually a small raised bump or ridge at the opposite end to orient to the tip.
2. Antiseptic solution. If none is available, use soap and water.
3. (Sterile) Water soluble lubricant or **lidocaine** gel. If none is available, use petroleum gel or even patient's saliva. Can make **lidocaine** gel by mixing 10 cc of 1% **lidocaine** solution with 10 cc of lubricant.
4. 5-30 cc syringe to inflate Foley cuff.

When You Don't Have Everything (options):
1. If no Foley or Coudé, use feeding tube, nasogastric tube, small diameter chest tube or IV tubing (try to round the edges). Standard catheter size is about 3/8 inch diameter.
2. If no sterile tube, can boil tube. A clean non-sterile tube can be safely used as long as the bladder is emptied regularly and there is no evidence of break in the mucosal lining (gross bleeding). Risk of infection is low with clean intermittent catheterization.
3. If no sterile lubricant, try to find water-soluble lubricant. Can use petroleum gel such as vaseline. If nothing available: either use water, passing the catheter very slowly; or use saliva (preferably patient's).

What To Do:
1. Using sterile technique, wash the urethral meatus with antiseptic soap. Lubricate the catheter with water soluble lubricant and gently pass the catheter into the bladder. In males, it is necessary to pass the catheter until the hub or flange is in contact with the urethra.
2. In young males, the pain caused by the catheter may cause the patient to clamp down with his sphincter preventing passage of the catheter. In such cases, use 10-20 cc of **lidocaine** gel (1-2%) which can be made by mixing 10 cc of 1% **lidocaine** solution with water soluble lubricant to make a viscous solution. Squirt the **lidocaine** gel into the urethra and squeeze the meatus shut and hold for 1-2 minutes.
3. In older males, the prostate may make it difficult to pass a catheter. A Coudé catheter with the tip pointing up toward the patient's head may be easier to pass. Also a larger, stiffer Coudé in the 20-22 Fr size may also be easier. If only a small catheter can be passed, a urethral stricture may be present.
4. In females, the meatus may be tucked under a fold in the vaginal opening or obscured by the labia. "Frog leg" the patient as much as possible and elevate the buttocks. This is best accomplished with the patient in stirrups at the end of an exam table. An assistant may be required to spread the labia apart. A strong light is also very useful.
5. Have patient empty the bladder. Measure the residual, dipstick the urine and send for a culture if possible. Record the amount emptied from the bladder and the results of the testing. Normal post-void residuals should be generally under 100 ml. In patients with large bladder capacities, a residual representing 20-25% of the total initial bladder volume is considered normal. Example: patient voids 450cc and has 150 cc left. The initial bladder volume was 600 cc with 25% remaining. This is acceptable.
6. If the bladder is palpable, place a Foley catheter. Empty the bladder as much as possible. Patients that have very large volumes (>1000 ml) when suddenly emptied may experience a profound vasovagal reaction. It would be prudent to empty the bladder with the patient lying down. To avoid a vasovagal reaction, one can empty the bladder by letting out 400 ml every few minutes. Extremely large bladders, when drained often will cause rupture of bladder mucosal veins that had been compressed by the urine. Do not be surprised if the urine turns bloody. Irrigate out clots, as needed using sterile saline or water with a catheter tip syringe.
7. If the catheter is to be left more than a day, daily cleaning of the meatus with antiseptic soap can decrease infections. Prophylactic antibiotics are not necessary; however, if the catheter is left in place

for more than several days, pre-treat the patient with cystitis antibiotic regimen (**Septra DS** bid) the day before catheter removal and monitor the patient for signs of infection within the first week after catheter removal.

What Not To Do:
In men, do not inflate the balloon unless urine flows through the catheter, and the catheter flange is at the meatus. If this is not done, the balloon may be inflated in the prostatic urethra.
Do not make more than two attempts to pass the catheter if there is injury to the urethra. Allow the patient to try to void. Consider a suprapubic catheter and evacuation.
Do not use saline to inflate the balloon if the catheter is to be left more than a day. Use sterile waterto inflate the balloon. Saline may crystallize in the balloon port resulting in inability to deflate the balloon.
Do not leave catheters in for more than one month without changing them.
Do not forget to always pull the foreskin of an uncircumcised male back down over the glans to prevent development of paraphimosis.

Procedure: Suprapubic Bladder Aspiration (Tap)
CAPT Leo Kusuda, MC, USN

What: Insert a needle into the bladder to drain the bladder.

When: Bladder is palpable above pubic bone, patient cannot void and the urethra is injured so Foley catheterization not indicated. Indicated for anuric patients prior to transport.

What You Need: Long spinal needle, if available (if not, use a 2 inch or longer needle for adults; 1 ½ inch for children), antisetic skin prep, 10cc syringe, 60 cc syringe, stopcock and IV tubing (desirable), 10 cc of 1% **lidocaine** local anesthetic (optional), Kelly clamp (optional).

What To Do:
1. Confirm bladder is full:
    a. Palpate abdomen above pubic bone. If there is a midline mass above the pubic bone, confirm that this is the bladder by pressing on the mass. If there is increased urge to void, this suggests that the mass is the bladder. Then percuss abdomen above pubis. If the mass is dull and firm, this also suggests the mass is the bladder.
    b. Do a bimanual rectal or vaginal exam to feel for an enlarged bladder
WARNING: IF YOU CANNOT PALPATE THE BLADDER, DO NOT ATTEMPT A TAP. IF THE BLADDER IS NOT FULL, YOU MAY INJURE THE BOWEL.
2. Prep the skin, an area about the size of your hand, centered around a spot in the midline 1-2 finger-widths above the pubic bone with alcohol prep pads or Betadine.
3. If you have **lidocaine**, fill the 10 cc syringe and anesthetize the skin as described below. This is about as painful as the tap will be, but the patient will be less likely to move during the procedure.
4. Attach the spinal needle (or appropriate length needle) to the lidocaine syringe and infiltrate the skin in the midline about 1-2 finger-widths above the pubic bone. Infiltrate the tough fascia below the skin and subcutaneous fat. Direct the needle straight down, perpendicular to the long axis of the body. Do not angle up toward the head or down toward the feet. You may infiltrate below the fascia.
5. If you are using a large 18 gauge needle, now insert the supplied obturator (wire that goes inside the needle) with the needle. If the needle is smaller, you may use a syringe to aspirate in place of the obturator. Attach the stopcock, tubing and 60 cc syringe (if available) to siphon out the urine.
6. Pass the needle in the area that was anesthetized. Again direct the needle straight down.
7. The bladder should be encountered within 1 inch below after going through the fascia (tough layer below the skin). If you do not get urine back, slowly withdraw the needle while aspirating.
8. Once urine is encountered, advance the needle about ½ inch. A Kelly clamp or other large clamp may be

used to stabilize the needle. Use the 60 cc (or the 10 cc) syringe to aspirate urine.
9. Withdraw the needle once urine aspiration stops.

What Not To Do:
Do not make repeated passes in different directions.
Do not get the needle far from the bone.
Do not advance the needle too deep. In most average size adults, it is difficult to go too deep. However, do not press the hub of the needle into the skin. In children, the bladder is higher and much closer to the skin.

What to look for when emptying the bladder:
1. Suddenly emptying a large bladder (volumes >1000 ml) may precipitate a profound vasovagal reaction, so empty the bladder with the patient lying down. To avoid a vasovagal reaction empty the bladder by letting out 400 ml every few minutes. Drainage of extremely large bladders can cause rupture of bladder mucosal veins that had been compressed by the urine, so do not be surprised if the urine turns bloody.
2. Some patients will have a marked increase in urine output after the bladder is decompressed. This signifies a significant obstruction to the bladder that has resulted in some renal injury. The urine output can be very high (e.g., 800 ml/hour). In most cases, the patient's thirst mechanism may allow for adequate hydration. If urine output is very high (>200 cc/hr), monitor the vital signs every hour and ask the patient if they feel lightheaded. If the pulse rate becomes elevated (>100) or there are orthostatic symptoms (lightheadedness) or decreased blood pressure and increased pulse from supine to standing position, insert an IV catheter and give D5 ½ NS. Infuse the crystalloid over the next 2 hours at a rate to replace half the urine volume of the previous 2 hours.

Procedure: Portable Pressure Chamber
COL Paul Rock, MC, USA

What: Portable hyperbaric ('pressure') chambers are fabric bags that can be internally pressurized to create a hyperbaric (higher pressure) condition inside the bag. The higher pressure inside is generated and maintained by a small hand-, foot- or battery-powered air pump. Hyperbaric environments are useful for treating altitude illnesses because high pressure increases the oxygen available to the body. For a patient inside the pressure chamber bag, increased air pressure has the same effect as a descent in elevation. The difference between the air pressure outside and inside the chamber bag at high altitude determines the amount of simulated descent that can be achieved for the patient. The higher the actual altitude, the greater the pressure difference, and the greater the simulated descent. For most chamber bags, the difference in pressure at 15,000 feet is the equivalent of a descent to an elevation of about 8,000 feet. Because the pressure difference between the inside of the bag and the atmosphere is less at lower elevations, the degree of simulated descent is less there. However, because descent of even a few thousand feet can be a life saving treatment for some altitude illnesses, pressure bags are still useful at altitudes as low as 9,000 feet.

Portable pressure chambers are relatively lightweight (generally less than 15 lbs) and can be folded for easy carrying in backpacks. They are commercially available in the civilian sector in several countries around the world. Available chambers include the Gamow Bag and the Hyperlite chamber (USA), the CERTEC (France), the PAC (Portable Altitude Chamber, Australia). They can be purchased or rented through many outdoor/travel medical suppliers. The US military maintains no such chamber in the supply chain.

When: For temporary treatment of altitude illnesses (acute mountain sickness [AMS], high altitude cerebral edema [HACE], and high altitude pulmonary edema [HAPE]) when descent or evacuation is not available, or while waiting for evacuation. This treatment can also be used to treat decompression sickness (DCS) and arterial gas embolism/pulmonary overinflation syndrome (AGE/POIS) in diving (see Dive Medicine). The chamber's weight allows the patient to be transported in the chamber, even when evacuated by plane.

What You Need: Required: Portable hyperbaric chamber (fabric 'pressure chamber' bag) with pressure

gauge, ventilation pump (hand-, foot- or battery powered), connecting hoses
Useful/Optional: Carbon dioxide absorption device ('CO_2 scrubber'; placed inside chamber), mat or blanket for patient to lie upon (for increased comfort), supplemental oxygen (can be placed inside chamber or added to ventilation air at pump)

What To Do:
NOTE: Follow specific instructions for the type/brand of pressure chamber being used.
Begin Treatment
1. Attach ventilation pump to chamber using hose.
2. Place patient in chamber. (Be sure patient's head is visible in viewing window.)
3. Close chamber. (If patient is alert, instruct them to hold the fabric away from their face.)
4. **IMMEDIATELY** begin pressurizing/ventilating the chamber with ventilation pump.
5. Watch pressure gauge. When chamber is pressurized enough, ventilation valve will open automatically to exhaust CO_2 and prevent over-pressurization. Listen for sound of air exhaust to confirm ventilation.
6. Continue pumping at slower rate to maintain pressure and ventilation. **Do** not stop pumping while patient is inside chamber.

Length of Treatment
1. No absolute guidelines. Treat until symptoms resolve (usually takes more than 2-3 hours) or until patient can be evacuated to lower altitude.
2. Continuously observe patient through chamber window. Talk to patient to reassure—chambers transmit voices easily.
3. Consult US Navy Treatment Tables in this book for guidelines on the treatment of DCS and AGE/POIS with hyperbaric oxygen.

End Treatment
1. Instruct patient to equalize ears as necessary during chamber decompression by swallowing or Valsalva maneuver.
2. Open exhaust ('dump') valve. Continue pumping at low rate to maintain ventilation.
3. Watch pressure gauge. When pressure inside chamber is equal to pressure outside (gauge reads "0"), open chamber and remove patient. Opening chamber before pressures are equal causes sudden decompression and can damage patient's ears or sinuses. Theoretically, sudden decompression could also cause decompression sickness ('bends').

What Not To Do:
Do not stop ventilation pump while patient is inside chamber. (Constant pumping is required to insure continued oxygen supply and to exhaust carbon dioxide.)
Do not decompress chamber at rapid rate or open it before pressure inside is equal to pressure outside. (Necessary to avoid sudden decompression problems.)

Procedure: Pain Assessment and Control
LTC Michael Matthews, MC, USA & COL Warren Whitlock, MC, USA

What: Pain control is an integral part of the overall management of the sick or injured patient and often the most neglected. Many studies have shown that health care providers are poor at recognizing and treating pain and that untreated acute pain can lead to chronic pain syndromes. The medic should not allow his own feelings toward pain influence how he perceives his patient's reaction, but simply recognize that each person handles pain in their own way and try to alleviate the pain to the maximum extent possible given the circumstances. The SOF medic should teach the patient that it is easier to prevent pain than to "chase" or treat it once it has become established, and that communication of unrelieved pain is essential to its relief.

When: Pain should be assessed at its onset and reassessed frequently.

What You Need: Assessment of pain (based on history [see mnemonic below] and exam), reference list of medications and related information (including side effects), availability of medications and delivery mechanisms (needle, syringe, etc.).

What To Do:
1. Assess patient's pain using history and exam. This mnemonic will be helpful.
   OPQRST Mnemonic for Pain Assessment
   O = Onset - When did the pain start? Was it sudden or gradual?
   P = Palliate/Provoke - What palliates or provokes the pain? What makes it better or worse?
   Q = Quality - What does the pain feel like? (sharp, dull, tearing, etc.)
   R = Radiate - Does the pain radiate? If so, from where to where?
   S = Severity - How bad is the pain on a scale of 1 to 10? (with 1 being the least & 10 the most)
   T = Time - How long have you been in pain? Is it continuous or intermittent? Has it gotten worse, better, or stayed the same?
2. Consult reference material and use a stepped approach for the control of pain:
   a. *Mild to Moderate Pain:* Begin, unless there is a contraindication, with a non-steroidal anti-inflammatory drug (**NSAID**) or acetaminophen. NSAIDs or single injections of local anesthetics alone may control mild to moderate pain after relatively minor surgical procedures. NSAIDs decrease levels of inflammatory mediators generated at the site of tissue injury. At present, one NSAID (ketorolac) is approved by the Food and Drug Administration for parenteral use. All but two NSAIDs (salsalate and choline magnesium trisalicylate) appear to produce a risk of platelet dysfunction that may impair blood clotting and carry a small risk of gastrointestinal bleeding. Acetaminophen does not affect platelet aggregation, but neither does it provide peripheral anti-inflammatory activity. Benzodiazepines (anti-anxiety drugs), other muscle relaxants and other classes of drugs are also used for pain control due to their secondary effects.
   b. *Moderately Severe to Severe Pain:* Normally treat initially with an opioid analgesic, especially for more extensive surgical procedures that cause moderate to severe pain. The concurrent use of opioids and NSAIDs often provides more effective analgesia than either of the drug classes alone. Even when insufficient alone to control pain, NSAIDs or single injections of local anesthetics have a significant opioid dose-sparing effect upon postoperative pain and can be useful in reducing opioid dosages and side effects. Although it is likely that NSAIDs also act within the central nervous system, in contrast to opioids, they do not cause sedation or respiratory depression, nor do they interfere with bowel or bladder function.
   c. *Opioid Tolerance or Physiological Dependence:* These complications are very unusual in short-term use by opioid naive patients. Likewise, psychological dependence and addiction are extremely unlikely to develop when patients without prior drug abuse histories use opioids for acute pain. Proper use of opioids involves selecting: 1) appropriate drug, initial dose, and route and frequency of administration; 2) optimal drug/route/dose of non-opioid analgesic, if desired; 3) acceptable incidence and severity of side effects; and 4) an inpatient or ambulatory setting for pain management. Titrate opioids to achieve the desired therapeutic effect and to maintain that effect over time.
   d. *Other Pain:* When increasing doses of opioids are ineffective in controlling postoperative pain, search for another source of pain, and consider other unusual diagnoses such as neuropathic pain (i.e., burning, tingling or electrical shock sensation triggered by very light touch, and accompanied by a sensory deficit in the area innervated by the damaged nerve).
   e. *Other Therapies:* Remember that interventions such as relaxation, distraction, and massage can reduce pain, anxiety, and the amount of drugs needed for pain control.

3. Some analgesics available to the SOF medic are listed below.
   a. Opioids:
   Morphine - Indications: Moderate to severe pain (> 5 on 10 point scale). Example: Cardiac ischemia. Contraindications: Relative: Untreated hypotension, mild COPD or asthma, full stomach; Absolute: allergy, biliary disease, severe COPD or asthma. Route: IV preferred, IM, Oral (poorly absorbed); Dose: IV 2 to 4 mg then 2 mg every 10 to 15 minutes titrated to effect; IM 10 mg every 1 hour titrated to effect. Side Effects: Respiratory depression, muscle relaxation, sedation, histamine release.

Codeine - Indications: Mild to moderate pain (> 2 and < 8). Intermediate onset of action in 30 min. Also used for cough suppression. Contraindications: Allergy. Routes: Oral. Dose: 7.5 to 30 mg (i.e., Tylenol with Codeine #1-3) every 3 to 4 hours, 30 mg every 6 hours for cough suppression. Side Effects: Respiratory depression, sedation, constipation, diaphoresis.
Oxycodone - Indications: Moderate pain (>5). Contraindications: Allergy, known drug abuser. Routes: Oral. Dose: 5 mg every 3 hours. Available combined with acetaminophen in 2.5mg/325mg; 5mg/325mg; 7.5mg/500mg and 10mg/650mg. Side Effects: Respiratory depression, sedation, bradycardia, nausea, vomiting.
Meperidine - Indications: Moderate pain (>5 needing rapid onset in minutes). Contraindications: Allergy, MAO inhibitors. Routes: IM, PO. Dose: 75 to 150 mg, PO/IM every 2 to 4 hours. Side Effects: Reduced cardiac contractility, CNS toxicity, respiratory depression.
Fentanyl - Indications: Short term severe pain relief (>8). Short surgical procedures. Contraindications: Need for long-term pain control. Routes: IV, IM. Dose: 0.05 to 0.1 mg/kg IM or IV (slow). Usually given with midazolam for muscle relaxation. Side Effects: Muscle rigidity if given rapidly IV, arrhythmias, CNS toxicity, respiratory depression.
Nalbuphine - Indications: Moderate to severe pain (between 5-10). Contraindications: Cannot be used for surgical procedures. Not approved for use in children. Routes: IV, IM. Dose: IM/IV 10 to 15 mg every 3 to 4 hours. Side Effects: Acute withdrawal in persons addicted to opiates, respiratory depression, sedation, bradycardia.
Butorphanol - Indications: Moderate to severe pain (5-10). Contraindications: Same as Nalbuphine. Routes: IM, IV, Nasal. Dose: 1 to 2 mg every 2 to 4 hours IV/IM. 1 spray (1mg) every 4 hours for migraine. Side Effects: Same as nalbuphine

b. NSAIDs:
Ibuprofen - Indications: Mild to moderate pain (1-4). Fever reduction. Contraindications: Allergy. Gastritis, Renal disease. Routes: Oral. Dose: Adult: 400 to 800 mg every 6 to 8 hours with food, Children: 5 to 10 mg/kg every 6 to 8 hours with food. Side Effects: Gastric irritation or bleeding.
Aspirin - Indications: Mild to moderate pain (1-4). Fever reducer. Antiplatelet activity in cardiac ischemia. Contraindications: Use in children and teenagers. Allergy. Bleeding disorders. Patients on Digoxin, lithium, phenytoin. Routes: Oral, Rectal. Dose: Adults: 325 to 650 mg every 4 to 6 hours. Side Effects: Acute asthma attacks in aspirin allergic patients, Reyes Syndrome in children and teenagers with certain viral illnesses. GI upset.
Ketorolac - Indications: Mild to moderate pain (1-5). Nephrolithiasis. Contraindications: Allergy, GI disease, hepatic/renal disease. Patients on lithium, phenytoin, Digoxin. Routes: IV, IM. Dose: IV: 30 mg every 4 to 6 hours IM: 60 mg every 4 to 6 hours. Side Effects: Gastric irritation, arrhythmias.

c. Benzodiazepines:
Diazepam - Indications: Muscle relaxation, Sedation. Contraindications: Respiratory depression. Routes: Oral, IV, IM, Rectal. Dose: Oral: 5 to 10 mg every 8 hours for sedation, IV: 5 to 10 mg. IM: 10 mg, Rectal: 10mg (1 day only). Side Effects: Respiratory depression, cardiovascular collapse, paradoxic excitation.
Midazolam - Indications: Same as diazepam. Contraindications: Same as diazepam. Routes: IV, IM. Dose: IV: 1 to 2 mg; IM: 5 to 10 mg, Side Effects: Respiratory depression.
Lorazepam - Indications: Sedation in acute psychotic agitation. Contraindications: Same as diazepam. Routes: IV, IM. Dose: 0.05 to 0.1 mg/kg over 2 to 4 minutes. Maximum dose 8 mg in 12 hours. Usually mixed with haloperidol 10 mg IM for the treatment of acute psychotic agitation. Side Effects: Same as diazepam.

d. Local Anesthetics - Indications: Local or regional anesthesia. Contraindications: Allergy. Routes: Local infiltration subcutaneously. Dose: Lidocaine: 3 mg /kg, Mepivacaine: 8 mg/kg, Bupivacaine: 1.5 mg/kg. Side Effects: CNS toxicity if injected intravascularly. If mixed with epinephrine, intravascular injection can cause severe hypertension, tachycardia, distal vasoconstriction.

e. Other Agents:
Ketamine - Indications: Painful invasive procedures. Contraindications: Hypertension. Head or eye injury. Chest or abdominal surgery. Routes: IV, IM. Dose: IV 1 to 2 mg/kg. IM 4 mg/kg. IV infusion of 0.5 to 1 mg/kg per hour approved for adult use only for prolonged procedures. Atropine at 0.1 mg/kg for children to 1 to 2mg for adults should be added to control secretions. Side Effects: Emergence hallucinations (mostly in adults). Hypertension, respiratory depression if given rapidly IV.
Acetaminophen - Indications: Mild to moderate pain (1-4). Fever reduction, Contraindications: Allergy, Liver disease. Routes: Oral, rectal. Dose: Adult: 650 to 975 mg every 4 to 6 hours. (Often used in combination form with codeine or oxycodone) Children: 10 to 15 mg/kg every 4 to 6 hours. Side Effects: Gastric irritation or bleeding. Liver toxicity in long-term use or in overdose.

f. Reversing Agents:
Naloxone - Indications: Reversal of side effects of opiates. Contraindications: No absolute. Will trigger opiate withdrawal. Routes: IV,IM,SC. Dose: 0.1 to 2.0 mg q 2-3 min to max of 10 mg. Side effects: Acute withdrawal may exhibit combativeness.
Flumazenil - Indications: Reversal of benzodiazepam overdose. Contraindications: Benzo dependence or if used for seizure control as it will trigger intractable seizures, concomitant ingestion of tricyclic antidepressants (TCAs), presence of increased intracranial pressure such as in head injury. Routes: IV. Dose: 0.2 mg IV every minute until desired response to a maximum dose of 3 mg. Side effects: See contraindications.

What Not To Do:
Do not give insufficient pain medication to achieve relief.
Do not give pain medication only after the pain has returned. Anticipate the onset of pain, and give the medication 30 minutes BEFORE the pain returns to provide effective relief.
Do not fail to consider all classes of pain medications and their few side effects before prescribing. A patient that gets good pain control from morphine, but starts to vomit repeatedly as a side effect, might have benefited from a non-opioid medication.

Chapter 34: Lab Procedures
Lab Procedure: Prepare Specimens for Transport
18D Skills and Training Manual

When: To ensure the safe, intact transfer of specimens for later examination.

What You Need: Appropriate shipping containers and packing materials.

What To Do:
1. Prepare specimens to be shipped.
2. Pack specimens in appropriate containers and at the proper temperature.
3. Package specimens in accordance with federal regulations.
4. Provide specimen information to the receiving lab.
   a. Patient identification.
   b. Suspected disease or condition.
   c. Type of specimen.
   d. Body area where specimen was collected.
   e. Sending facility identification.
   f. Type of test requested.
   g. Name of the person requesting the test.
5. Place appropriate warning labels on packages containing infectious material.
6. Ship specimens in accordance with federal regulations.
7. Follow all precautions when shipping infectious material.

8. Ship two of each specimen at a minimum.
9. Maintain a log of all specimens sent to include:
   a. Date specimen was prepared.
   b. Patient identification.
   c. Specimen identification.
   d. Date specimen was shipped.
   e. Test requested.
   f. Name of the person requesting the test.

Lab Procedure: Urinalysis
18D Skills and Training Manual

When: You have to diagnose genito-urinary diseases, detect microorganisms and/or metabolites resulting from other pathologic conditions.

What You Need: A properly collected specimen a clock or watch, a refractometer, applicator sticks, reagent strips (N-Multistix), sulfosalicylic acid test materials, Clinitest materials, Acetest materials, Ictotest materials, a microscope, glass slides and coverslips, a centrifuge, distilled water, test tubes (13 x 100 mm and 15 x 85 mm), disposable transfer pipets, a laboratory request form and a logbook.

What To Do:
1. Record the appearance and color of the urine on the laboratory request form.
   a. Appearance:
      (1) If the specimen appears clear, write "CLEAR" on the form.
      (2) If the specimen is not clear but can still be seen through, write "HAZY" on the form.
      (3) If the specimen cannot be seen through at all, write "CLOUDY" on the form.
   b. Report the color as shades of yellow and report any color change that occurs on standing.
   NOTE: Normal color differs due to varying amounts of a pigment called urochrome. Normal color is straw-yellow amber. Abnormal color can result from diet, medication, and/or disease. Red in males indicates fresh blood from the lower urinary tract. Orange is caused by medication such as pyridium, which is used to treat urinary tract infections. Brown indicates hemoglobin; black, malaria (blackwater fever); colorless, polyuria (absence of urochrome).
2. Determine and record the specific gravity, pH, protein, glucose, ketones, bilirubin, nitrite, urobilinogen, blood, and leukocyte esterase. Dip the reagent strip in the specimen, remove it, and then compare each reagent area with the corresponding color chart on the bottle label at the number of seconds specified in the instructions accompanying the reagent strips.
3. Perform the following confirmation tests for positive reactions on the reagent strip or when indicated by patient's condition (if available).
   a. Specific gravity–refractometer method.
      (1) Place 1 to 2 drops of urine on the refractometer.
      (2) Point the refractometer toward a uniform, bright light source.
      (3) Look through the refractometer and read the scale.
      (4) Report the value from the left-hand scale, at the juncture of the dark and light areas.
      NOTE: Normal values are 1.003 to 1.033.
   b. Protein–sulfosalicylic acid test procedure.
      (1) Add 10 drops of centrifuged urine to a 13 X 100 mm test tube.
      (2) Layer 10 drops of sulfosalicylic acid (3 percent) over the surface of the urine. (Equal amounts of urine and SSA can also be used.)
      (3) Observe for turbidity, which confirms a positive result.
      (4) Grade the degree of turbidity.
         (a) Negative–no turbidity.
         (b) Trace–faintly visible turbidity.

                (c) 1+—definite turbidity.
                (d) 2+—heavy turbidity but no floccules.
                (e) 3+—heavy cloud with floccules.
                (f) 4+—heavy cloud with heavy floccules.
         c. Glucose–Clinitest procedure.
            (1) Place 5 drops of urine into a 15 x 85 mm test tube.
            (2) Add 10 drops of water to the test tube.
            (3) Drop in a Clinitest tablet. Boil.
            (4) Read the results 15 seconds after boiling stops. Observe during the test for pass through.
            CAUTION: Do not shake the tube while it is boiling.
            (5) Mix the tube by gentle shaking it.
            (6) Compare the color of the solution with the color chart.
         d. Ketones–Acetest procedure.
            (1) Place one Acetest tablet onto a clean, dry piece of white paper.
            (2) Dispense 1 drop of urine on top of the tablet.
            (3) After 30 seconds, compare the color of the tablet to the Acetest chart.
                (a) Positive--purple color.
                (b) Negative--tan color.
         e. Bilirubin–Ictotest procedure.
            (1) Place the special Ictotest mat on a paper towel.
            (2) Drop 10 drops of urine onto the test mat.
            (3) Place one Ictotest tablet in the center of the moistened mat.
            CAUTION: Do not touch the tablet with your fingers.
            (4) Place 1 drop of water onto the tablet and wait 5 seconds.
            (5) Place a second drop of water on the tablet so that both drops run off the tablet onto the mat.
            (6) Look for a blue to purple color of the mat at 60 seconds, which indicates the presence of bilirubin.
            NOTE: Normal urine is a slightly pink to red color on the mat.
            (7) Read the results by comparing the color of the mat to the color chart.
4. Perform microscopic examination.
     a. Centrifuge the urine specimen at 1500 to 2000 rpm for 5 minutes.
     b. Pour off the supernatant, mix, and place a drop of the remaining sediment on a glass slide covering it with a coverslip.
     c. Examine the entire cover-slipped area of the slide under low power magnification with subdued light to locate any casts.
     NOTE: Confirm the casts under high power with subdued light.
     d. Record and report the number of casts per low power field (#/LPF).
     e. Scan 10 to 15 fields under high dry (40x) magnification to identify the specific types of cells present.
     f. Record and report the count of cells as the number seen per high power field (#/HPF).
     g. Record and report elements such as mucus threads, parasites, crystals, and/or yeast as OCCASIONAL, FEW, or MANY.
     h. Grade the amount of bacteria seen from "-" to "4+".
     NOTE: Sperm is reported verbally.
5. Identify the clinical significance of abnormal results.
     a. Proteinuria.
        (1) Renal damage/ disease.
        (2) Multiple myeloma (specifically Bence Jones protein).
        (3) Strenuous exercise.
     b. Glycosuria.
        (1) Diabetes mellitus.
        (2) Central nervous system damage.
        (3) Pregnancy with undiagnosed diabetes mellitus.
        (4) Metabolism disorders.
     c. Ketonuria.

                (1) Starvation.
                (2) Diabetic acidosis.
        d. Blood.
            (1) Hematuria - intact RBCs.
                (a) Renal disease or calculi.
                (b) Exposure to toxic drugs or chemicals.
                (c) Trauma.
                (d) Strenuous exercise or menstruation.
            (2) Hemoglobinuria - lysed RBCs.
                (a) Severe burns.
                (b) Transfusion reactions.
                (c) Infections.
                (d) Strenuous exercise.
            (3) Myoglobinuria - protein found in muscle tissue.
                (a) Muscular trauma.
                (b) Lengthy coma.
                (c) Muscle-wasting diseases.
                (d) Extreme muscular exertion.
        e. Bilirubinuria.
            (1) Cirrhosis.
            (2) Hepatitis.
        f. Increased urobilinogen.
            (1) Liver disease.
            (2) Hemolytic disorders.
        g. Nitrites. Some bacteria reduce nitrate to nitrite, which is not a normal constituent of urine.
            (1) Cystitis.
            (2) Antibiotic therapy.
            (3) Urinary tract infections.
            (4) Specimen contamination from improper preservation.
        h. Specific gravity.
            (1) Diabetes insipidus (SG < 1.003).
            (2) Hydration (normal SG) and dehydration (SG > 1.033).
        i. Leukocytes–urinary tract infections.
    6. Record and report the correct results for a given specimen.

Lab Procedures: Wet Mount and KOH prep

MAJ Ann Friedmann, MC, USA

When: To assess sample (discharge, scraping, biopsy) for presence of Candida or Tinea.

What You Need: Sample, slide, coverslip, normal saline in dropper or vial, 10% KOH (potassium hydroxide), small cotton swab, and a microscope.

What To Do:
1. Obtain specimen (see Pelvic Exam procedure guide, Candidiasis section, Tinea section, Vaginal Discharge section, etc.). Collect fluid or discharge on cotton swab. Put a scraping sample or biopsy directly on slide.
2. For fluid or discharge: place 1 drop each of discharge on right and left ends of slide.
3. Place one small drop of KOH on one drop of sample. Perform whiff test (smell for amine odor).
4. Place one small drop of normal saline on other spot.
5. Cover each sample with glass cover slip.
6. Observe under microscope. Microscopically examine the KOH portion for hyphae and spores under the

100X objective. Examine the saline portion for trichomonads and clue cells under the 400X objective.
7. For dry specimens of skin scrapings or biopsies place a drop of KOH over the specimen already on the slide, cover with a coverslip (probably not practical with biopsy), and observe under microscope as above.
8. What to Look For:
   Candida (Yeast) and Fungi (tinea): KOH causes the epithelial cells to swell and break open which will allow a clearer view of any filamentous or branching yeast or fungi.
   Clue Cells: Vaginal epithelial cells to which bacteria are attached, obscuring the cell border. The edges of vaginal epithelial cells are normally sharp and clear. The attached bacteria make the edges of the cell appear fuzzy on low power. Small bacteria will be visible on higher power.
   Trichomonas: Motile flagellated protozoa. They look like sperm with balloon-shaped heads and are about the size of a normal epithelial cell.

What Not To Do:
Do not add too much KOH or saline allowing the fluids to mix or run off the slide.
Do not forget to perform the whiff test.

Lab Procedure: Gram Stain
18D Skills and Training Manual

When: You have a culture with one or more organisms growing in it. Prepare and stain a smear so that gram-negative organisms appear pink to red and gram-positive organisms appear blue to purple with 100% accuracy.

What You Need: A sample for testing, a Bunsen burner, Gram stain reagents, an inoculating loop, an inoculating needle, glass slides, water, a staining rack, a diamond point pen or lead pencil, needle, syringe, disposable transfer pipets, a timer, and a logbook.

What To Do:
1. Label a slide with a lead pencil or diamond point pen.
2. Prepare the smear.
   a. From a liquid specimen or medium, (if applicable).
      (1) Place 1 drop in the center of a labeled slide using a sterile pipet, syringe, or loop.
      (2) Spread inoculum to about the size of a dime.
      (3) Allow the smear to completely air-dry.
   b. From a colony on an agar plate (if applicable):
      (1) Place a small drop of water in the center of the labeled slide.
      (2) Touch the top center of the colony with a sterile inoculating needle and transfer a small amount of bacteria to the drop of water.
      (3) Mix the bacteria and water with the needle and spread it to about the size of a dime.
      (4) Allow the smear to completely air-dry.
   c. From a swab (if applicable):
      (1) Roll the swab across the center of a dry labeled slide.
      (2) Allow the smear to completely air-dry.
3. Use either of the following methods to fix the smear:
   a. Methanol.
      (1) Flood the smear with methanol and allow it to stand for 1 minute.
      (2) Tilt the slide to drain off excess methanol and allow it to completely air-dry.
   b. Heat. Gently heat the smear by passing it through the burner flame two or three times.
4. Stain the slide.
   a. Flood the slide with crystal violet for 1 minute.
   b. Rinse the smear with tap water.
   c. Flood the slide with Gram's iodine for 1 minute.

d. Rinse the smear with tap water.
   e. Rinse the slide with decolorizer for 2-5 seconds.
   f. Rinse the smear with tap water.
   g. Flood the slide with safranin for 30-60 seconds.
   h. Rinse the smear with tap water.
   i. Blot the smear dry by placing it between paper towels or allow the smear to air-dry.
   NOTE: Test each new batch of reagents with known gram-positive and gram-negative organisms. Prepare quality control (QC) slides by adding a drop of a 24-hour culture to a glass slide and allowing the broth to dry. Fix each slide and store them in a box until ready for QC testing. A slide containing separate drops of gram-positive and gram-negative organisms should be stained each day to test the quality of the gram stain reagent. Alternatively, a mixture containing both gram-negative and gram-positive organisms can be used to prepare the control slides.
5. Examine the smears under oil immersion for:
   a. Gram stain reaction.
   b. Cell shape.
   c. Arrangements.
6. Record the results in the logbook and on the laboratory request form.

What Not To Do: Do not overheat the slide. Do not dislodge the smear when rinsing or drying the slide.

Lab Procedure: Brucellosis Test

18D Skills and Training Manual

When: To test a serum specimen for brucella microorganisms.

What You Need: A properly prepared serum specimen, a glass plate, a ruler, a wax pencil, Rose Bengal Serum Agglutination brucellosis antigen (or other serum agglutination antigen as available), a 0.2ml serological pipette, an applicator stick, a lab request form, and a logbook.

What To Do:
1. Prepare the glass plate.
   a. Make a series of 1 1/2-inch squares with a ruler and wax pencil.
   b. Use 5 squares to test 1 antigen against the sera diluted.
   c. Clean and dry the glass plate after each use.
2. Prepare the serum dilutions.
   NOTE: 0.08 = 1:20 dilution, 0.04 = 1:40 dilution, 0.02 = 1:80 dilution, 0.001 = 1:160 dilution, and 0.005 = 1:320 dilution.
3. Pipet the specimen.
   a. Use a 0.2 ml serological pipette.
   b. Pipet the serum onto a row of squares on the plate.
   c. Pipet the serum in this order, starting with 0.08, 0.04, 0.02, 0.001, and 0.005 ml.
4. Shake the antigen well before using and place a drop of antigen on each drop of serum.
5. Mix the serum and antigen with an applicator stick.
6. Hold the glass plate in both hands.
7. Rotate the plate 15-20 times in an 8-inch circular motion.
8. Observe agglutination within 1 minute.
9. Obtain the correct results for the specimen given.
   a. Agglutination occurs when clumping of the antigen occurs with a clearing of the background fluid.
   b. Negative reactions are characterized by a homogenous suspension with no clumping or clearing.
10. Report the results.
    a. Positive reaction and the highest dilution at which it occurred.
    b. Negative reaction occurring at a 1:20 dilution.

Lab Procedure: Wright's Stain Using Cameco Quik Stain
18D Skills and Training Manual

When: You are performing a complete blood count (CBC).

What You Need: A properly collected blood sample, Wright's stain (Cameco Quik Stain), glass slides, capillary tubes, a staining rack, distilled water, a microscope, immersion oil, 95% methanol and a laboratory request form.

What To Do:
1. Prepare the glass slides by cleaning them in 95% methanol.
2. Prepare the blood smear.
   a. Place a drop at a point midway between the sides of the slide and a short distance from one end.
   b. Place the edge of the spreader slide on the specimen slide at a 30° angle.
   c. Pull the spreader slide toward the drop of blood until contact is made.
   d. Push the spreader slide toward the opposite end of the specimen slide, drawing the blood behind it into a thin film.
   e. Allow the smear to air-dry. The smear should have three areas represented.
      (1) The thick area has:
         (a) An increase of smaller cells with rouleaux formation.
         (b) Poor cellular characteristics.
      (2) The thin area has:
         (a) An area of choice for reading a differential.
         (b) Cellular characteristics which are at their best.
         (c) RBCs that are touching without overlapping.
      (3) The feathered edge has:
         (a) An increase in larger cells.
         (b) Cellular distortion with gaps between the RBCs.
3. Write the name or identification number of the patient with a lead pencil in the thick area of the smear or on the frosted end of the slide.
4. Stain the slide.
NOTE: For best results, blood smears should be stained within 2 hours after they are prepared.
   a. Fix the smear and let it air-dry.
   b. Cover the slide with Wright's stain and allow it to remain on the smear for 5 to 10 seconds. The Wright's stain is polychromatic; dyes in the stain will produce multiple colors when applied to cells. The stain is composed of two main parts, Methylene-blue and eosin.
   (1) Methylene-blue is an alkaline dye that has an affinity for the acidic portions of the cell--DNA in the nucleus and RNA in the cytoplasm. It stains varying shades of blue to purple.
   (2) Eosin is the acidic portion of the dye that has an affinity for the basic portions of the cell. It stains in varying shades from orange to pink.
   NOTE: The stain should cover the slide but should not be allowed to overflow the edges. A blue-green metallic sheen should be on the surface. The exact time that the stain is applied may vary with each batch of stain.
   c. Rinse the slide thoroughly with distilled water for 15 to 20 seconds.
   d. Wipe the stain from under the slide and allow it to air-dry.
5. Examine the smear under the oil immersion objective.
6. Recognize discrepancies with staining and the way to correct those discrepancies.
   a. For darker stained leukocytes, dip the slide in distilled water for 1 minute or more after staining.
   b. For lighter stained leukocytes, decrease the Wright's stain time.
7. Insure that the staining reactions are sufficient.
   a. RBCs appear buff pink to orange.
   b. WBCs have a blue nucleus with a lighter stained cytoplasm.

8. Report and record the results on the laboratory request form.

Lab Procedure: Ziehl-Neilson Stain
18D Skills and Training Manual

When: You have to test sputum for acid-fast bacilli (AFB) such as tuberculosis.

What You Need: A properly collected sputum sample, a glass slide, Bunsen burner, pipets, Kinyoun's carbol-fuchsin stain, Kinyoun's acid-alcohol reagent, Kinyoun's methylene blue reagent, microscope, lab request form, and a logbook.

What To Do:
1. Prepare a smear of the specimen on a glass slide. Air-dry and fix the smear by passing it briefly through a flame.
   CAUTION: Acid-fast bacteria are highly infectious. Specimens and smears must be handled with care. Wear facemask and gloves and wash hands immediately after performing the AFB examination. Perform these procedures in a biosafety cabinet, if available.
2. Stain the smear.
   a. Flood the fixed smear with Kinyoun's carbol-fuchsin stain for 3 minutes.
   b. Wash the smear gently with running water.
   c. Decolorize it with Kinyoun's acid-alcohol reagent until no color appears in the washing. This should take about 2 minutes.
   d. Wash the smear gently with running water.
   e. Counterstain with Kinyoun's methylene blue reagent for 30 seconds.
   f. Wash the smear gently with running water.
   g. Allow the smear to air-dry.
3. Examine the smear using the oil immersion objective. Scan the length of the slide at least three times in the stained area (about 300 fields).

NOTE: Acid-fast organisms will appear as red coccobacilli, rods, or filaments. Some may appear beaded. The background will be bluish.

Lab Procedure: Giemsa Stain for the Presence of Blood Parasites
18D Skills and Training Manual

When: You have to test for blood-borne parasites such as malaria.

What You Need: A properly collected blood specimen, a Cameco Wright-Giemsa stain, distilled water, glass slides, coverslips, staining dishes, water, lint-free cloth, 95% ethanol, rubber gloves, methanol, timer, microscope, immersion oil, mounting medium, capillary tubes, the laboratory request form and a logbook.

What To Do:
1. Prepare the glass slides by cleaning them in 95% ethanol and wiping them until completely dry.
2. Prepare the blood smears.
   a. Prepare the thin smear.
      (1) Place a small drop of blood at a point midway between the sides of the slide and a short distance from one end.

(2) Place the edge of another slide (spreader) on the specimen slide at a 30° angle.
(3) Slide the spreader slide toward the drop of blood until contact is made.
(4) Immediately push the spreader slide toward the opposite end of the specimen slide, drawing the blood behind it. The smear should cover 1/2 to 2/3 of the slide.
  b. Prepare the thick smear.
    (1) Place 3 or 4 drops of the blood specimen at the end of a slide.
    (2) Use a corner of another slide to stir the drop over an area about the size of a dime. It is approximately the proper thickness if ordinary newsprint is just legible when looking through the freshly prepared smear. At least two thick and two thin smears should be prepared for each specimen.
  NOTE: A positive control slide should be stained with all malaria specimens when available.
  c. Allow the smear to air-dry in a flat position.
3. Write the name or identification number of the patient in the thick area of the thin smear with a lead pencil.
4. Stain the smear within 24 hours after air-drying and examine for uniform staining qualities.
  a. Fix the thin smear in methanol for 3 to 5 seconds.
  CAUTION: Do not allow the thick smear to come in contact with the methanol or near the methanol vapors because this will fix the cells.
  b. Allow the smear to air-dry.
  c. Dip the thick smear in distilled water 3 to 5 times so that the stain will be adequately absorbed, and to lyse intact RBCs.
  d. Allow the smear to air-dry.
  e. Place the entire slide in staining solution for 10 minutes.
  f. Dip the entire slide in distilled water for 20 seconds or more (for desired color balance).
  NOTE: Time will depend on the age, thickness, and density of the smear.
  g. Remove the slides from the water and allow them to air-dry, standing on end on absorbent paper.
  CAUTION: Do not blot the slides dry. This will damage the blood smears.
5. Examine the entire smear under the oil immersion objective.
  a. If malarial parasites are present in the thick smear, report "Plasmodium species seen". Confirm by reading the thin smear and report by genus and species.
  b. If no malarial parasites are present, report "No Plasmodium species seen."
  c. If other blood parasites are present, report by genus and species.
6. Record the results in the logbook and on the laboratory request form.

Lab Procedure: Tzanck Stain

COL Warren Whitlock, MC, USA

When: To diagnose herpes lesions from herpes simplex or herpes zoster.

What You Need: Giemsa stain, Pap stain, Wright's stain, slides/coverslip, microscope, #15 blade

What To Do:
1. Obtain specimen. Gently rupture fresh vesicle with #15 scalpel blade. Gently scrape mushy debris from vesicle base using curved belly of blade. Do NOT use cotton swab. Smear debris onto microscope slide. Sample 3-4 vesicles if possible.
2. Prepare specimen. Gently heat slide over match. Air dry (fix in methyl alcohol if Giemsa is used). Place drops of stain (Wright's stain). Allow stain to sit on slide for over 1 minute. Rinse off stain under gently running water. Put on a drop of immersion oil or mineral oil. Apply cover slip
3. Examine specimen. Examine under 10x power. Scan for smear bands. Background of lightly stained debris. Look for darkly stained cells. Examine cells under 43x power. Positive test: multinucleated giant cells: Herpes Simplex Virus (HSV), Varicella Zoster Virus (VZV)

What Not To Do: Do not dig too deeply in obtaining superficial cells for examination. Do not overheat slide.

Lab Procedure: Culture Interpretation
18D Skills and Training Manual

When: To microscopically differentiate between species of microorganisms.

What You Need: Glass slides, coverslips, a microscope, a lab request form and a logbook.

What To Do:
1. Identify colonies by shape
    a. Circular
    b. Undulate
    c. Lobate
    d. Radiated lobated
    e. Crenated
    f. Dentated
    g. Concentric
    h. Regular myceloid
    i. Irregular (curled undulated edges)
    j. Irregular (filamentous)
    k. Irregular (rhizoid)
2. Identify colonies by size
    a. Dwarf = smaller than 0.5 mm
    b. Pinpoint = 0.5 mm
    c. Small = 1 mm
    d. Medium = 2 mm
    e. Large = 3 mm
    f. Spreading = 6 mm
3. Identify colonies by texture
    a. Smooth
    b. Rough
    c. Mucoid
4. Identify colonies by color. Colonies vary in color from clear white to violet.
5. Identify colonies by elevation
    a. Flat
    b. Raised
    c. Convex
    d. Umbilicated
    e. Umbonate
    f. Papillary
    g. Convex Rugose
6. Identify colonies by biochemical reaction
    a. Hemolytic
    b. Non-Hemolytic
7. Identify specific colony types based on their morphology
    a. *Staph. aureus* colonies are usually opaque, circular, smooth, raised and 1-3 mm in diameter. They may or may not be pigmented. They vary in color from white to golden. On blood agar, Beta hemolysis (clear halos) is seen around growing colonies (see Color Plates Picture 24).

b. Streptococci (general category) colonies are usually small, slightly granular, circular, convex and translucent. The most important characteristic of streptococci is their hemolytic property. Hemolytic streptococci will leave clear halos around growing colonies (see Color Plates Pictures 26, 27).
c. Pneumococci (general category) colonies are usually small round colonies surrounded by a zone of alpha hemolysis.
d. *N. gonorrhoeae* colonies are grown on Thayer-Martin media. Colonies are smooth, glistening, translucent and soft. No pigmentation is usually present, however, some strains may show and iridescent bluish-gray coloration.
e. *B. anthracis* (anthrax) produces non-hemolytic colonies having a "cut-glass" appearance in transmitted light with dull, irregular edges (see Color Plates Picture 13).
f. *Clostridium perfringens* (associated with "gas gangrene") is strictly anaerobic, and will not grow in the presence of O_2. They may show hemolysis. They present a characteristic "double zone" or target appearance
g. *E. coli* is a facultative anaerobe with 2-3 mm colonies, typically low, convex, smooth, and colorless or opaque. The colonies tend to bunch together (see Color Plates Picture 28).
h. Shigella (general category) produces colonies that are usually 2 mm, circular, convex and whitish-clear with a smooth surface.
i. *Vibrio cholera* colonies are usually 1-2 mm, round, slightly convex, translucent and grayish-yellow in appearance. On blood agar, they will have a surrounding greenish zone.
j. *H. influenzae* colonies are usually smooth, raised, slimy, mucoid, confluent and non-hemolytic.

Lab Procedure: Macroscopic Examination of Feces and Test for Occult Blood

18D Skills and Training Manual

When: To examine a freshly collected fecal specimen for visible parasites, color, consistency, mucus, and gross or occult blood.

What You Need: An occult blood developer kit with manufacturer's instructions, disposable transfer pipets, applicator sticks, a watch that indicates seconds and a laboratory request form.

What To Do:
1. Determine the consistency and color of the specimen and record the results.
2. Examine the specimen for mucus (white patches) on the surface and record the results.
3. Perform a visual inspection for red and/or tarry spots and patches (gross blood) then record the results.
4. Perform a survey for visible parasites.
5. Perform the test for occult blood. This is useful in mass screening for colorectal cancer.

NOTE: Refer to the manufacturer's instructions for the particular brand of test in the inventory. Slight variations in the following procedure are to be found due to different source companies.
   a. Label the test envelope with the patient's identification.
   b. Collect a small portion of the stool sample on one end of an applicator stick.
   c. Apply a thin smear inside the first box (labeled "A" or "1" depending on the kit).
   d. If indicated, apply a second thin smear inside the second box (labeled "B" or "2").
   e. Close the cover.
   f. Open the flap on the backside of the kit and apply 2 drops of developer to the paper directly over each smear.
   g. Read the results within 60 seconds.

NOTE: After reading the results, add developer to the positive and negative performance monitors. A blue color should appear in the positive monitor within 10 seconds. No blue color will appear in the negative monitor. If the positive and negative controls are as expected, the test envelope and reagent

are good and the results of the patient's specimen are presumed valid.
(1) Positive - any trace of blue on or at the edge of the smears.
(2) Negative - no detectable blue on or at the edge of the smears.
NOTE: Refer to the manufacturer's instructions for specific times and quantitation.
The principle of this test is the general reaction:
Hemoglobin reacts with pseudo peroxide, yielding H_2O_2, which reacts with benzidine orthotoluidine guaiac in the presence of oxygen to yield the color blue.
6. Record all results on the laboratory request form.

Lab Procedure: Feces for Ova and Parasites

18D Skills and Training Manual

When: To examine a stool sample for the presence of parasites and their eggs.

What You Need: A properly collected stool specimen: 10% formalin, ethyl acetate, Lugol's iodine or Merthiolate-Iodine-Formalin (MIF), toluene or xylene, paper cups, water, transfer pipets, graduated cylinders, applicator sticks, centrifuge, centrifuge tubes, laboratory funnel, 4" x 4" gauze, slides, coverslips, logbook, and a lab request form.

What To Do:
1. Concentrate the sample.
   a. Strain about 3 ml of the specimen suspension through either the disposable filtering device or 2 layers of wet gauze placed in the funnel into a 15 ml conical centrifuge tube.
   b. Fill the tube to 12 ml with 10% formalin.
   c. Centrifuge at 2000 rpm for 5 minutes.
   d. Remove the tubes and pour off the supernatant.
   e. Prepare slides for acid-fast bacteria (AFB) staining by placing 1 drop of the sediment on a glass slide and allowing it to air dry for recovery of cryptosporidium.
   f. Add formalin to the 8 ml mark on the tube and resuspend the sediment.
   WARNING: Formalin may be fatal if swallowed or harmful if inhaled. Exposure may create a cancer risk and may cause blindness. Keep away from heat, sparks, or flame. Avoid breathing vapor. Use only with adequate ventilation. Wash thoroughly after handling.
   g. Add ethyl acetate to the 12 ml tube, stopper the tube, and vortex.
   WARNING: Ethyl acetate is flammable. Keep away from heat, sparks, or flame. It is harmful if swallowed or inhaled. Use only with adequate ventilation. Avoid contact with eyes. Wash thoroughly after handling.
   h. Allow the tube to stand for 3 minutes.
   i. Centrifuge at 2000 rpm for 5 minutes.
   j. Ream the plug with an applicator stick and decant the supernatant.
   CAUTION: Do not allow debris or ethyl acetate to contaminate the sediment.
2. Perform the microscopic examination.
   a. Transfer 1 drop of the sediment to a glass slide.
   b. Add 1 drop of Lugol's iodine and cover with a coverslip.
   c. Scan the whole slide on low (10x) power.
   d. Identify using high-dry (40x) objective. Scan at least 75 fields under high dry.
3. Report the results.
   a. If organisms are recovered, report by genus and species name.
   b. If none are seen, report "No ova/ parasites seen".

The following steps relate to examination for pinworms, "Scotch tape test".
4. Prepare the collection slide.
   a. Anchor one end of the cellulose tape to the underside of the microscope slide.

b. Fold the tape over the near end of the slide and smooth the adhesive down along the full length of the slide.
   c. Attach a paper tab to the free end of the tape for labeling.
   d. Store in the refrigerator.
5. Collect the specimen
   a. Using the paper tab, pull the tape back from the slide leaving the fold over the underside attached.
   b. Loop the tape over a tongue depressor to hold it steady.
   c. Press the sticky tape surface against the perianal skin. Press open the perianal folds to gain access to eggs in the crevices.
   d. Use a limited area of the tape surface to decrease the area for microscopic examination.
   e. Smooth the tape back into place using a piece of cotton.
   CAUTION: The eggs will stick to you and are infectious for several hours after being passed.
6. Perform a microscopic examination of the slide.
   a. Lift the tape from the slide and place a drop of toluene or xylene on the slide. Smooth the tape back into position.
   NOTE: Slides may be stored up to several weeks in the refrigerator before being examined; however, do not put the toluene or xylene on the slide until ready for examination because the chemicals will distort the eggs.
   b. Search the slide for eggs using low power magnification and reduced light.
7. Record and report the results.

Lab Procedure: Concentration Techniques for Ova and Parasites (Con-Trate Method)

18D Skills and Training Manual

When: To microscopically examine a fecal specimen for the presence of parasites and parasite ova.

What You Need: A properly collected fecal specimen, a Feka Con-Trate System containing: filtering devices, disposable centrifuge tubes with caps, Muco Pen X (reagent A) and Ethyl Acetate (reagent B), manufacturer's instructions for Con-Trate Fecal Concentration; cotton tipped applicator sticks; microscope slides; 22 x 40 mm coverslips; 10% formalin; a centrifuge; disposable transfer pipets; a tube rack; a squirt bottle; Lugol's iodine; a vortex mixer and a laboratory request form.

What To Do:
1. Perform the concentration procedure.
   a. Strain about 3 ml of the specimen suspension through the disposable filtering device into a 15 ml conical centrifuge tube.
   b. Fill the tube to 12 ml with 10% formalin.
   c. Centrifuge at 2000 rpm for 5 minutes.
   d. Remove the tubes and pour off the supernatant.
   e. Prepare slides for acid-fast bacteria (AFB) staining by placing 1 drop of the sediment on a glass slide and allowing it to air dry.
   f. Add formalin to the 8 ml mark on the tube.
   WARNING: Formalin may be fatal if swallowed and harmful if inhaled. Exposure may create a cancer risk and may cause blindness. Keep away from heat, sparks, or flame. Avoid breathing vapor. Use only with adequate ventilation. Wash thoroughly after handling.
   g. Add ethyl acetate to the 12 ml tube, stopper the tube, and vortex.
   WARNING: Ethyl acetate is flammable. Keep away from heat, sparks, or flame. It is harmful if swallowed or inhaled. Use only with adequate ventilation. Avoid contact with eyes. Wash thoroughly after handling.
   h. Allow the tube to stand for 3 minutes.
   i. Centrifuge at 2000 rpm for 5 minutes.

j. Ream the plug with an applicator stick and decant the supernatant.
CAUTION: Do not allow debris or ethyl acetate to contaminate the sediment.
2. Perform the microscopic examination.
   a. Transfer 1 drop of the sediment to a glass slide.
   b. Add 1 drop of Lugol's iodine and cover with a coverslip.
   c. Scan the whole slide on low (10x) power.
   d. Identify using high-dry (40x) objective. Scan at least 75 fields under high dry.
3. Record and report the results on the laboratory request form.
   a. If organisms are recovered, report by genus and species.
   b. If no organisms are seen, report "No ova parasites seen."

Lab Procedure: Microhematocrit Determination

18D Skills and Training Manual

When: To measure the percentage of red blood cells in total blood volume.

What You Need: A control; a logbook; either a microhematocrit centrifuge (calibrated to 10,000 rpm), with a microhematocrit reader, sealing putty, and capillary tubes; or a Compur M1100 minicentrifuge with capillary tubes and operator's manual; a laboratory request form.

What To Do:
NOTE: The hematocrit is the ratio of red blood cells to plasma expressed as a percent of whole blood.
1. Microhematocrit Centrifuge/Reader Method, if applicable.
   a. Prepare the blood specimen collected in EDTA tubes.
   NOTE: If blood was collected in capillary tubes, remove them from the reverse side of the laboratory request form and proceed to step 1b.
      (1) Mix the control or specimen by gently inverting the tube several times before removing the stopper.
      (2) Fill two capillary tubes 3/4 full without bubbles.
      NOTE: The test should be performed in duplicate to correlate the results and to balance the centrifuge.
      (3) Seal the tubes by plunging the dry ends into a putty pad.
   b. Centrifuge the capillary tubes.
      (1) Put one tube opposite the other with the sealed ends toward the rubber gasket.
      (2) Centrifuge the tubes for 5 minutes
   c. Read the results by using the microhematocrit reader.
      (1) Place the tube in the groove of the plastic holder with the sealed end toward the center and the black index line intersecting the red blood cell-clay interface.
      (2) Match the red vertical line on the plastic holder with the numeral "100" on the outer circle of the microhematocrit reader.
      (3) Rotate the inner circle while holding the outer circle stationary until the black spiral indicator line matches the plasma/air interface line within the tube.
      (4) Rotate both the inner and outer circles of the reader clockwise simultaneously until the black spiral indicator line matches the RBC plasma interface line excluding the buffy coat.
      (5) Run the control and unknown in duplicate.
      (6) Report and record the results in percentage on the laboratory request form and in the logbook.
2. Compur M1100 method, if applicable.
   a. Prepare the blood specimen collected in the EDTA tubes.
      (1) Mix the control/specimen by gently inverting the tube several times before removing the stopper.
      (2) Completely fill two M1100 capillary pipets with blood.

b. Centrifuge the capillary tubes.
   (1) Raise the minicentrifuge rotor center post to the upper position by pressing in the lock release levers.
   (2) Place the tubes in the rotor with their lower, outer ends against the seals and the inner ends resting against the center post.
   (3) Seal the tubes into place by pressing the center post down until it locks into position.
   (4) Close the cover.
   (5) Start the rotor by sliding the ON/OFF switch to the right; it will automatically shut off when complete.
   (6) Allow the rotor to stop.
c. Read the results.
   (1) Read the hematocrit values through the clear window in the cover directly on the percent scales marked on the rotor next to the capillary tubes.
   (2) Slide the ON/OFF switch to the left.
   (3) Press the cover, release the button, and raise the cover.
   (4) Raise the rotor center post.
   (5) Remove the capillary tubes and discard.
   (6) Report and record the results in percentages on the laboratory request form and in the logbook.

Normal Values:
Males: 40% to 54% Volume Packed Red Cells (mean: 47% VPRC) Females: 37% to 47% Volume Packed Red Cells (mean: 42% VPRC)
Elderly: Values may be decreased Children: 31% to 43% Infants: 30% to 40% Newborn: 30% to 40%
NOTE: Critical values are less than 18% or greater than 61%.

Lab Procedure: RBC Count/Morphology

18D Skills and Training Manual

When: As part of a complete blood count (CBC), or to test for anemia and other red blood cell disorders.

What You Need: A properly collected blood sample, a microscope, a hemacytometer with coverslip, diluting fluid, a tally counter, an RBC pipet, aspirator tubing, wipes, a clock, a Petri dish, a logbook, and a lab request form.

What To Do:
1. Inspect the equipment.
   a. Check the hemacytometer, coverslip, and pipet for nicks and scratches.
   b. Ensure that the tally counter is operable.
2. Prepare the blood for counting using either the Unopette system or the pipet method.
   a. Prepare the blood for count using the Unopette system.
      (1) Using the protective shield on the capillary pipette, puncture the diaphragm of the reservoir.
      (2) Remove the shield from the pipette assembly by twisting.
      (3) Holding the pipette almost horizontally, touch the tip of the pipette to the blood.
      NOTE: The pipette will fill by capillary action. Filling will cease automatically when the blood reaches the end of the capillary bore in the neck of the pipet.
      (4) Wipe the outside of the capillary pipet to remove excess blood that would interfere with the dilution factor.
      (5) Squeeze the reservoir slightly to purge some air while simultaneously maintaining pressure on the reservoir.
      (6) Cover the opening of the overflow chamber of the pipet with the index finger and seat the pipet securely in the reservoir neck.
      (7) Release the pressure on the reservoir, then remove the finger from the pipet opening.

NOTE: Negative pressure will draw blood into the reservoir.
(8) Squeeze the reservoir gently two to three times to rinse the capillary bore, forcing the diluent up into (but not out of) the overflow chamber, releasing pressure each time to return the mixture to the reservoir.
(9) Place the index finger over the upper portion of the opening and gently invert several times to mix the blood thoroughly with the diluent.
(10) Let the mixture stand for 10 minutes before charging the hemacytometer.
(11) To charge the hemacytometer, convert the Unopette to a dropper assembly by withdrawing the pipet from the reservoir and reseating it securely in the reverse position.
(12) Invert the reservoir and discard the first 3-4 drops of mixture before charging the hemacytometer.
  b. Prepare the blood for count using the pipet method.
(1) Draw blood to the 0.5 mark on the RBC pipet.
NOTE: The blood column must be free of bubbles.
(2) Wipe off any excess blood on the outside of the pipet tip.
(3) Do not touch the opening of the pipet with the wipe.
(4) Draw diluting fluid to the 101 mark.
(5) Wipe the outside of the pipet as in steps a (2) and a (3) above.
(6) Place the pipet between the thumb and forefinger and rotate it in a figure-of-eight motion for 3 minutes.
(7) Discard the first 2-3 drops of mixture before charging the hemacytometer.
3. Prepare the hemacytometer.
  a. Charge both sides of the hemacytometer.
  b. Place the hemacytometer in a Petri dish.
  c. Allow the cells to settle for 10 minutes.
4. Perform the RBC count.
  a. Using low-power focus on the ruled section of the counting chamber, locate the center 1 square millimeter area for counting RBCs.
  b. Switch to high-power and locate the square in the upper left-hand corner.
  c. Count all the RBCs lying within the square and those touching the centerline of the upper and right-hand triple lines.
  NOTE: RBCs touching the left-hand and bottom-center lines are not counted.
  d. Count the remaining three corners and center.
  e. If there is a variation of more than 25 cells between any of the five areas, repeat the test.
  f. Count the second chamber in the same manner.
5. Calculate the RBC count.
  a. Add up the total number of RBCs in both counting chambers.
  b. Divide the sum by two.
  c. Multiply by the K factor (constant) of 10,000.
  d. Record the results as # RBCs X 10^{12} RBC/L.
6. Perform a Wright's stain of a blood sample (See Page 8-46).
7. Recognize the significance of abnormal RBC morphology.
  a. Anisocytosis: Variation in the size of the RBCs
    (1) Macrocyte:
      (a) Diameter: 9 microns or greater.
      (b) May be found in liver disease.
    (2) Microcyte:
      (a) Diameter: 6 microns or less.
      (b) Found in thalassemia and other anemias.
  b. Poikilocytosis: Variation in the shape of the RBCs.
    (1) Sickle shaped: associated with sickle cell disease.
    (2) Ovalocytes/elliptocytes: associated with hereditary elliptocytosis.
    (3) Spherocytes:

(a) These will have a smaller surface and thus reduced ability to carry O_2.
(b) Associated with hemolytic anemia and hereditary spherocytosis.
(4) Bizarre forms:
(a) Irregularly contracted cells.
(b) Indicative of disease process.
(5) Crenated cells:
(a) Mechanically produced irregularities due to improper slide preparation or an old blood sample.
(b) Not reported.
c. Hemoglobin variations:
(1) Hypochromasia:
(a) Decreased hemoglobin concentration.
(b) Found in iron deficiency anemia
(2) Target cells:
(a) An abnormally thin erythrocyte with a centrally stained area.
(b) They are associated with liver disease and certain hemoglobinopathies.
(3) Polychromasia: immature cytoplasm (RNA present).
d. Inclusions:
(1) Basophilic stippling:
(a) Heavy metal poisoning.
(b) Certain anemias and alcoholism.
(2) Howell-Jolly bodies:
(a) Nuclear remnants less than 1 micron in diameter.
(b) Stain dark violet with Wright's stain.
(c) Are implicated in certain anemic conditions and other forms of nuclear maturation defects.
(3) Cabot's rings:
(a) Nuclear remnants.
(b) Severe pernicious anemia (B-12 deficiency).
e. Differences in coloration:
(1) Hypochromic:
(a) A decrease in the color of the RBC indicating a decrease in their hemoglobin content.
(b) May be found in iron deficient anemia and other anemias.
(2) Hyperchromic:
(a) Deep coloration of RBCs.
(b) Probably a problem with the stain.
f. Rouleaux formations: A phenomenon where the erythrocytes adhere to one another presenting a "stack of coins" appearance. It occurs in conditions characterized by increased amounts of fibrinogen and globulin.
8. Record findings as:
a. Normal: (N/N)
(1) Normocytic: normal cell size and shape.
(2) Normochromic: normal hemoglobin content and coloration.
b. Abnormal:
(1) Use the specific terminology for the noted abnormality.
(2) Grade the degree of all abnormalities.
(a) Slight: abnormal forms seen intermittently.
(b) Moderate: few abnormal forms seen in most fields.
(c) Marked: several abnormal forms seen in each field.

Lab Procedure: WBC Count on Whole Blood

18D Skills and Training Manual

When: You need to determine the number of white blood cells (WBC) in a blood sample.

What You Need: A WBC Unopette, a hemacytometer with coverslip, a microscope, a tally counter, a moisture chamber, a clock, a laboratory request form, and a logbook.

What To Do:
1. Inspect the equipment.
    a. Check the hemacytometer and coverslip for nicks, scratches, and cleanliness. Replace if necessary.
    b. Verify that the WBC Unopette reservoir and capillary pipet are in usable condition.
    c. Verify that the tally counter is operable; if not, replace it.
2. Prepare the blood for the count.
    a. Puncture the Unopette reservoir with the protective shield over the capillary pipet.
    b. Touch the tip of the capillary pipet to the blood specimen and allow the pipet to fill by capillary action.
    NOTE: The filled capillary pipet must be free of bubbles.
    c. Wipe excess blood from the outside of the pipet without touching the tip.
    d. Squeeze the reservoir slightly, cover the overflow chamber of the pipet with the index finger, and insert the capillary pipet into the reservoir.
    e. Simultaneously, remove the finger from the overflow chamber and release pressure from the reservoir to draw the blood into the diluent.
    f. Squeeze the reservoir several times to rinse the pipet and to thoroughly mix the blood with the diluent.
    g. Let the reservoir stand for at least 10 minutes to allow the red cells to hemolyze.
    NOTE: The leukocyte count should be performed within 3 hours of collection.
3. Prepare the diluted specimen for count.
    a. Mix diluted blood by inverting the reservoir to resuspend the cells.
    b. Convert to a dropper assembly by withdrawing the pipet from the reservoir and reseating it securely with the capillary tube exposed.
    c. Clean the capillary bore by inverting the reservoir and gently squeezing the sides to discard 3 to 4 drops.
4. Prepare the hemacytometer.
    a. Charge both sides of the hemacytometer by gently squeezing the sides of the reservoir to expel the contents until the chambers are properly filled.
    b. Place the hemacytometer in a moist chamber (Petri dish containing a damp gauze pad).
    c. Allow the cells to settle for 3 to 5 minutes.
5. Perform a WBC count.
    a. Check under low power, low light for even distribution of cells.
    b. Count all 9 square millimeters of the counting chamber. Count the WBCs lying within the square and those touching the extreme upper and far right lines.
    NOTE: There should be no more than a 10 cell difference among the 9 squares counted.
    c. Count the second chamber in the same manner.
    NOTE: There should be no more than a 15 cell difference between the two sides of the chamber.
6. Calculate the WBC count
    a. Add up the total number of WBCs in both counting chambers.
    b. Divide the sum by two to get the average number of cells counted.
    c. Use the following formula to obtain the WBCs/L:

FORMULA: Avg # cells counted x "K" factor x 1,000,000 (10 to the 6th power) = WBCs/L .

K FACTOR:
    Dilution factor = 100
    Area factor = 0.111
    Depth factor = 10

K factor = dilution factor x area factor x depth factor
K factor = 100 x 0.111 x 10
K factor = 111
7. Report and record the results as WBCs/L in the logbook and on the laboratory request form.
Normal values:
Adult 4.0 to 11.0 x 10 (to the 9th power) WBCs/L
Newborn 10.0 to 30.0 x 10 (to the 9th power) WBCs/L

Lab Procedure: WBC Differential Count

18D Skills and Training Manual

When: You need to count and classify the different types of WBCs; assay the morphology of RBCs and WBCs; and estimate the number of platelets in the blood smear.

What You Need: A properly Wright's stained blood smear, a tally counter, a microscope, immersion oil, and a laboratory request form.

What To Do:
1. Inspect the smear under low-power magnification and locate the thin area on the smear where there is no overlapping of RBCs.
2. Switch to oil immersion (100x) for the count.
   a. Identify and count 100 consecutive WBCs.
   NOTE: If the WBC count is between 2.0×10^9 and 5.0×10^9 WBCs/L, classify 300 WBCs. When the count is greater than 5.0×10^9 WBCs/L, classify 500 WBCs.
   b. Record each cell type separately on the differential tally counter.
      (1) Lymphocyte
      (2) Monocyte
      (3) Neutrophilic band
      (4) Segmented neutrophil
      (5) Eosinophil
      (6) Basophil
   c. Express the number of each cell counted as a percentage.
3. Identify and report morphological variations of WBCs/RBCs/platelets.
   a. Rouleaux formation
   b. Signs of immaturity
   c. Hemoglobin variation
   d. Size and shape
   e. Staining characteristics and inclusions
   f. Nucleated RBCs (NRBCs) are reported as number per 100 WBCs counted.
   NOTE: The WBC count may have to be corrected if the number of NRBCs is 5 or more per 100 WBCs.
   FORMULA: (#WBC/L x 100) divided by (100 + #NRBCs counted) = Corrected WBC/L.
4. Perform a qualitative platelet estimate.
   a. Adequate–8 to 20 per oil immersion field.
   b. Increased–greater than 20 per oil immersion field.
   c. Decreased–less than 8 per oil immersion field.
5. Report and record the results on the laboratory request form.

Lab Procedure: ABO Grouping and Confirmation Tests
18D Skills and Training Manual

When: You need to determine a patient's blood type.

What You Need: 10 x 75 mm test tubes, a test tube rack, an indelible marker, disposable transfer pipettes, reagents, a processing worksheet, a logbook, a centrifuge (calibrated to 1000 RCF @ 3175 rpm), and an AABB Technical Manual (TM 8-227-3).

What To Do:
1. Prepare the equipment and the specimen.
    a. Obtain a test tube rack.
    b. Place the patient/donor blood tube and serum tube in the rack.
    c. Prepare a 2 to 5% cell suspension from the blood tube.
    d. Obtain five test tubes, label them properly, and place them in the rack.
        (1) Label tube # 1, "A" with the patient ID number.
        (2) Label tube # 2, "B" with the patient ID number.
        (3) Label tube # 3, "A,B" with the patient ID number.
        (4) Label tube # 4, "AC" with the patient ID number.
        (5) Label tube # 5, "BC" with the patient ID number.
2. Prepare the specimen for group testing.
    a. Add 1 drop of "Anti-A" reagent and 1 drop of the patient's cell suspension to the tube marked "A".
    b. Add 1 drop of "Anti-B" reagent and 1 drop of the patient's cell suspension to the tube marked "B".
    c. Add 1 drop of "Anti-A,B" reagent and 1 drop of the patient's cell suspension to the tube marked "A,B".
    d. Add 2 drops of the patient's serum and 1 drop of "A" cells to the tube marked "AC".
    e. Add 2 drops of the patient's serum and 1 drop of "B" cells to tube marked "BC".
3. Perform the ABO group test.
    a. Mix each tube by gentle shaking.
    b. Centrifuge the test tubes for 15 seconds.
    c. Examine each tube for hemolysis.
    d. Examine each tube for agglutination.
        (1) Write "+"on the worksheet for agglutination.
        (2) Write "O" on the worksheet for no agglutination.
    NOTE: Grade the degree of agglutination: +/-, 1+, 2+, 3+, or 4+; and annotate hemolysis.
4. Interpret the results of the tests and determine the blood group.
5. Record the results in the logbook and on the laboratory request form.

Lab Procedure: Rh Typing
18D Skills and Training Manual

When: You need to determine a patient's Rh type.

What You Need: 10 x 75 mm test tubes, an indelible marker, a test tube rack, disposable transfer pipets, prepared reagents: anti-Rh (D) typing serum and 22% bovine albumin (control), a processing worksheet, a logbook, a centrifuge (calibrated to 1000 RCF @ 3175 rpm), and an AABB Technical Manual (TM 8-227-3).

What To Do:
1. Prepare the equipment and the specimen.
    a. Obtain a test tube rack.
    b. Place the patient's blood specimen in the rack.

c. Prepare a 2 to 5% cell suspension from the patient's blood.
  d. Obtain two test tubes, properly label the tubes, and place them in the rack.
     (1) Label tube #1, "D" with the patient's ID number.
     (2) Label tube #2, "DC" with the patient's ID number.
2. Prepare the sample for Rh typing.
  a. Add 1 drop of "Anti-D" and 1 drop of the patient's cell suspension to the tube marked "D".
  b. Add 1 drop of Rh control (22% bovine albumin) and 1 drop of the patient's cell suspension to the tube marked "DC".
3. Perform the Rh (D) typing test.
  a. Mix the test tubes by gentle shaking.
  b. Centrifuge the test tubes for 15 seconds.
  c. Examine each test tube for hemolysis.
  d. Examine each test tube for agglutination.
     (1) Write "+" on the worksheet for agglutination.
     (2) Write "0" on the worksheet for no agglutination.
4. Interpret the results of the test and determine the Rh type.
   NOTE: Grade the degree of agglutination: +/-, 1+, 2+, 3+, or 4+; and annotate hemolysis. If the patient specimen is <2+ AND the anti-D reagent is polyclonal or chemically modified, OR the control tube has agglutination, proceed with the test for Rh variant (weak D).
5. Perform the test for the Rh variant (weak D antigen) if no agglutination is seen in BOTH the Anti-D and control tube.
   NOTE: The terms Rh positive and Rh negative refer to the presence or absence of the red cell antigen "D".
6. Record the results in the logbook and on the laboratory request form.

Lab Procedure: Crossmatching

18D Skills and Training Manual

When: To determine the compatibility or incompatibility of donor blood units with a recipient and evaluate the incompatibilities.

What You Need: Test tubes, gloves, an indelible marker, a test tube rack, 5 1/4-inch pipettes, gauze, saline, a rubber bulb, reagents, a work sheet, a centrifuge (calibrated to 1000 RCF @ 3175 rpm), a logbook, a completed SF 518 (Medical Record--Blood or Blood Component Transfusion), and a 37° Celsius incubator.

What To Do:
1. Verify the need to perform a crossmatch procedure.
   a. Check that the transfusion request (SF 518) is properly filled out.
   b. Verify the number of units needed for the transfusion.
   NOTE: There must be one SF 518 for each unit requested.
   c. Check that the name and ID number on the blood tubes match the request form.
   d. Check to ensure that the phlebotomist's initials are on the tube with appropriate date and time.
2. Determine the recipient's ABO group.
3. Determine the recipient's Rh type.
4. Select the donor blood to perform the testing.
   a. Obtain the donor units that are the same blood group and type as that of the recipient.
   b. Select the order of preference of donor blood, if the group and type for the recipient are not available.
      (1) If recipient is Group O, give only Group O blood.
      (2) If recipient is Group A, give Group A blood first and then Group O packed cells.
      (3) If recipient is Group B, give Group B blood first and then Group O packed cells.
      (4) If recipient is Group AB, give Group AB blood first, and then Group A packed cells, or Group B packed cells, or Group O packed cells as a last resort, in that order.

(5) Rh-positive blood should be selected for Rh-positive recipients.
(6) Rh-negative blood should be saved for Rh-negative recipients (this includes individuals that are D-positive).
NOTE: If a person is Du positive, his blood will be considered Rh-positive if he is a donor and considered Rh-negative if he is a recipient.

5. Prepare the equipment.
   a. Obtain the test tube rack.
   b. Place the patient's blood specimen tube and serum tube in the rack.
   c. Obtain two test tubes; label and place them in the rack.
      (1) Label tube #1, "XM" (Crossmatch) with the patient's ID number.
      (2) Label tube #2, "AUTO" (Auto Control) with the patient's ID number.
   d. Prepare a 2% to 5% cell suspension from a segment of the donor unit and the recipient's cells.
6. Prepare the tubes for testing.
   a. Add 2 drops of recipient serum and 1 drop of donor cell suspension to the tube labeled "XM".
   b. Add 2 drops of recipient serum and 1 drop of recipient cell suspension to the tube labeled "AUTO".
7. Prepare the test tubes for an immediate spin (saline phase).
   a. Mix the test tubes.
   b. Centrifuge the test tubes for 15 seconds.
   c. Examine each test tube for agglutination.
   NOTE: If the recipient has a negative antibody screen, the current standards suggest that an immediate spin crossmatch be performed ONLY for the detection of ABO incompatibilities. The antiglobulin phase of testing rarely uncovers clinically significant antibodies in a recipient whose antibody screening test is negative.
      (1) Write on the work sheet "+" for agglutination.
      NOTE: Grade the degree of agglutination: +/=, 1+, 2+, 3+, or 4+.
      (2) Write on the work sheet "=" for no agglutination.
   WARNING: If a positive reaction occurs in the crossmatch, **STOP** the crossmatch procedure. The blood is not compatible.
8. Perform the albumin phase.
   a. Add 2 drops of 22% bovine albumin to tubes labeled "XM" (only if no agglutination appeared after saline phase) and "AUTO".
   b. Mix the tubes.
   c. Centrifuge the test tubes for 15 seconds.
   d. Examine the test tubes for agglutination.
      (1) Write on the work sheet "+" for agglutination.
      NOTE: Grade the degree of agglutination: +/=, 1+, 2+, 3+, or 4+.
      (2) Write on the work sheet "=" for no agglutination.
   WARNING: If a positive reaction occurs in the crossmatch, STOP the crossmatch procedure: the blood is not compatible.
9. Continue testing tubes "XM" and "AUTO".
   a. Incubate the test tubes for 15 to 30 minutes at 37ø C.
   b. Centrifuge the test tubes for 15 seconds.
   c. Examine each test tube for agglutination.
      (1) Write on the work sheet "+" for agglutination.
      NOTE: Grade the degree of agglutination: +/=, 1+, 2+, 3+, or 4+.
      (2) Write on the work sheet "=" for no agglutination.
10. Perform the final testing of tubes "XM" and "AUTO".
    a. Wash the tubes four times with saline.
    b. Remove all saline from the last wash.
    c. Add 2 drops of antihuman globulin to both tubes.
    d. Mix the tubes.
    e. Centrifuge the test tubes for 15 seconds.

f. Examine each test tube for agglutination.
   (1) Write on the work sheet "+" for agglutination.
   NOTE: Grade the degree of agglutination: +/=, 1+, 2+, 3+, or 4+.
   (2) Write on the work sheet "=" for no agglutination.
   NOTE: Current standards do not require the use of optical aids before a crossmatch may be considered compatible, but their use may enhance sensitivity and consistency.
g. Add check cells to all negative tubes to confirm the reactivity of the antiglobulin reagent.
   NOTE: If the negative tubes do not show agglutination after adding check cells, repeat the bovine albumin phase (steps 8 through 10g).

11. Obtain correct results for the specimen given.

Appendices: Anatomical Plates
Figure A-1

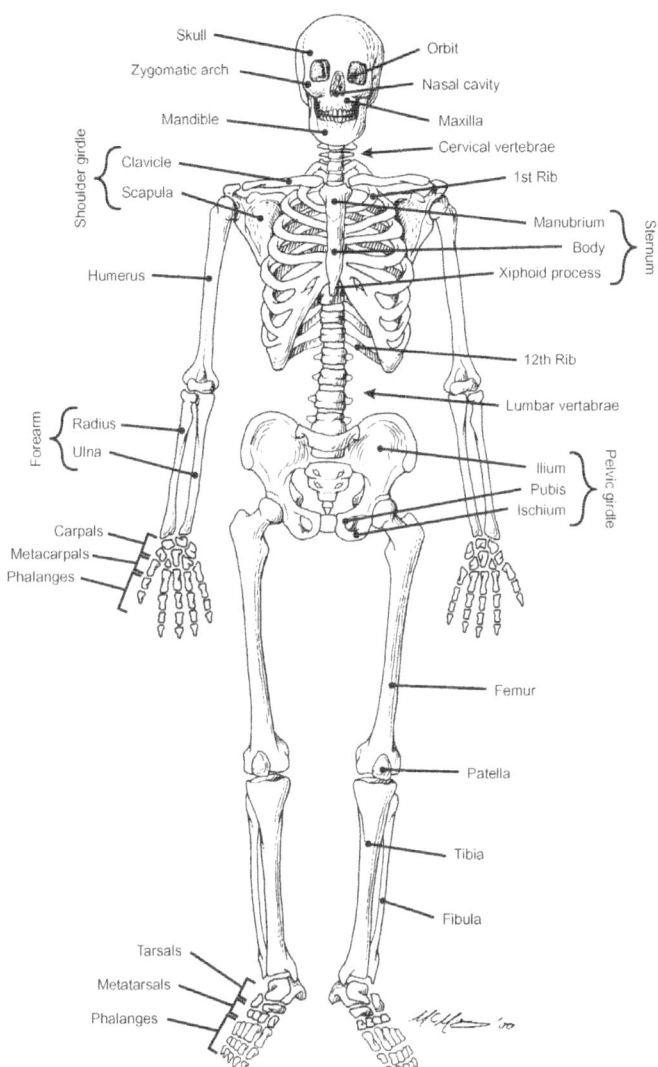

Skeletal System, Anterior View

Appendices: Anatomical Plates
Figure A-2

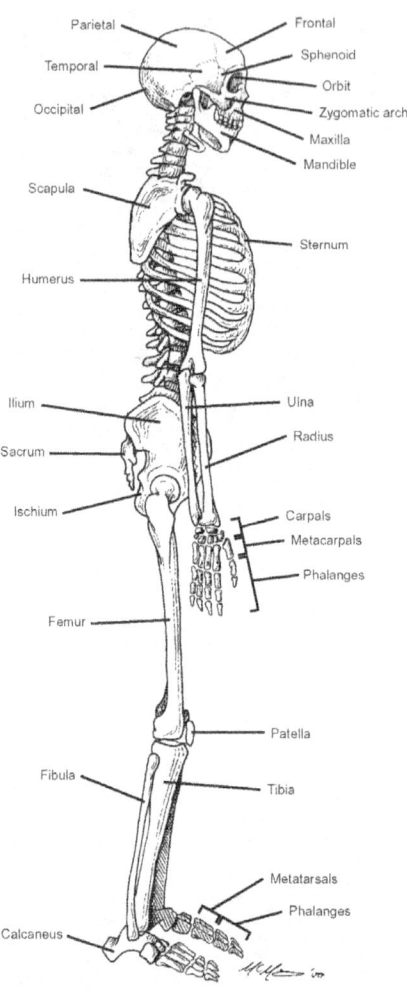

Skeletal System, Lateral View

Appendices: Anatomical Plates
Figure A-3

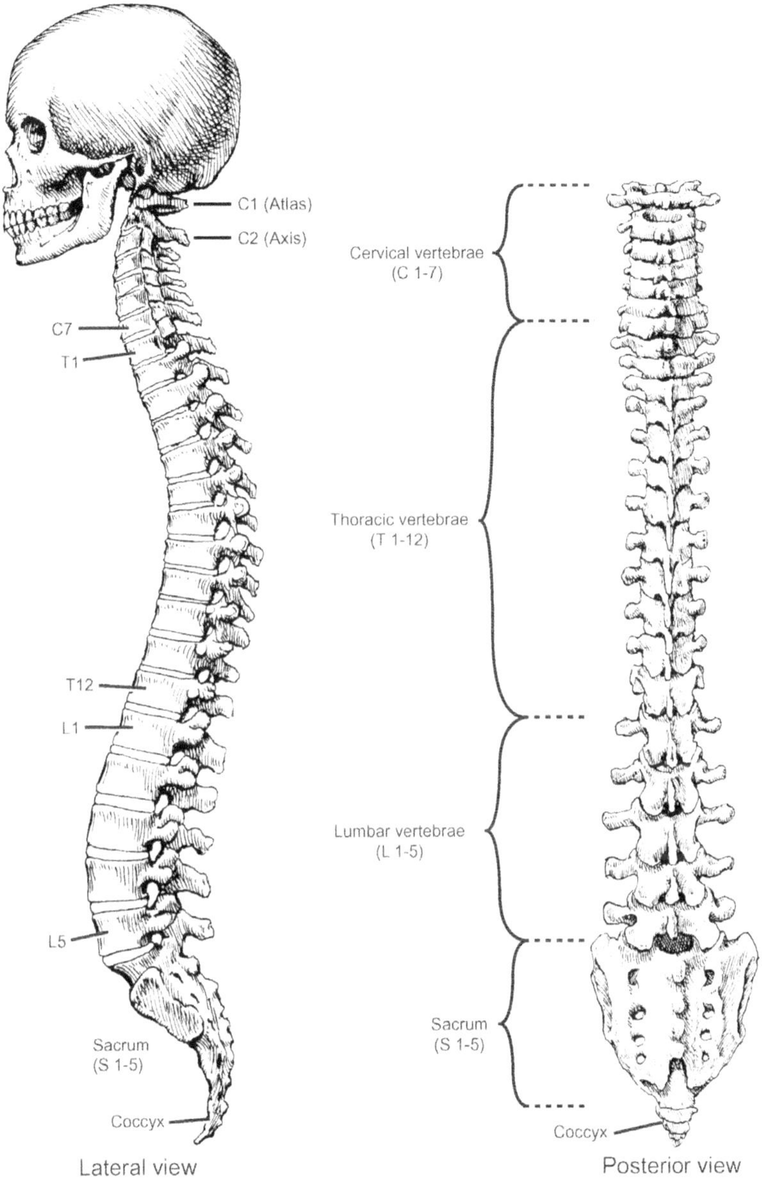

Lateral view

Posterior view

Vertebral Column

Appendices: Anatomical Plates
Figure A-4

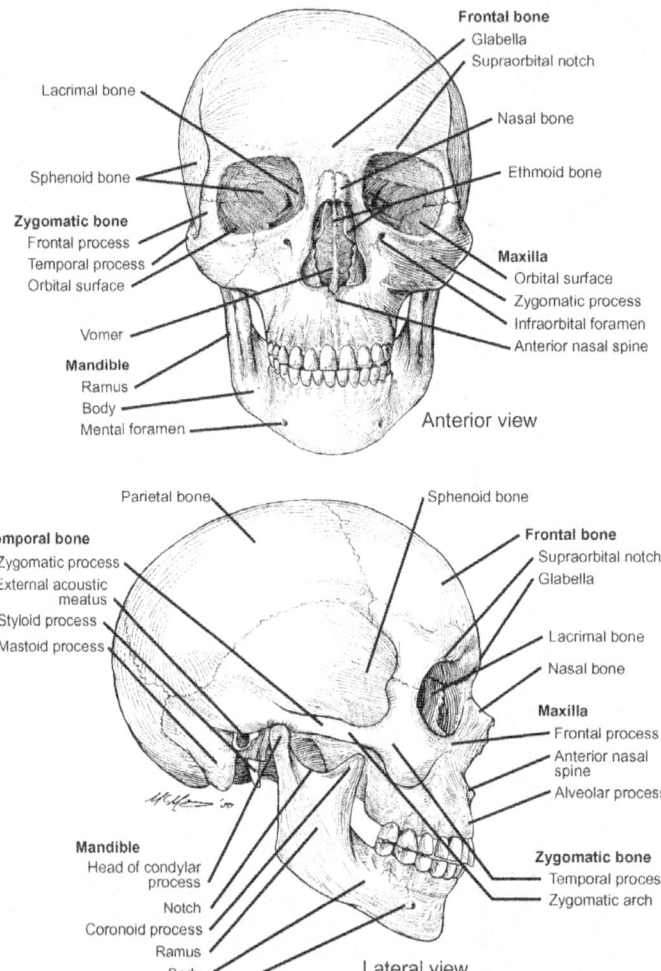

External Anatomy of the Skull

Appendices: Anatomical Plates
Figure A-5

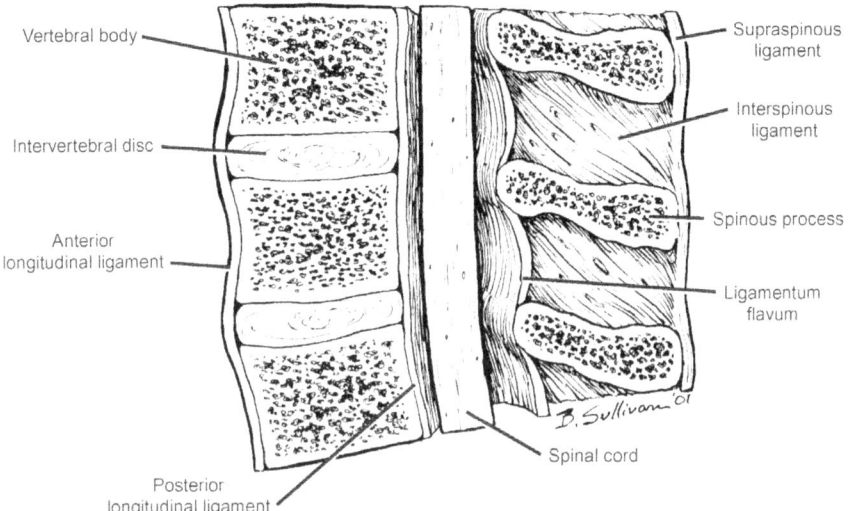

Vertebral Ligaments of the Lumbar Region
(mid-sagittal section, left lateral view)

Appendices: Anatomical Plates
Figure A-6

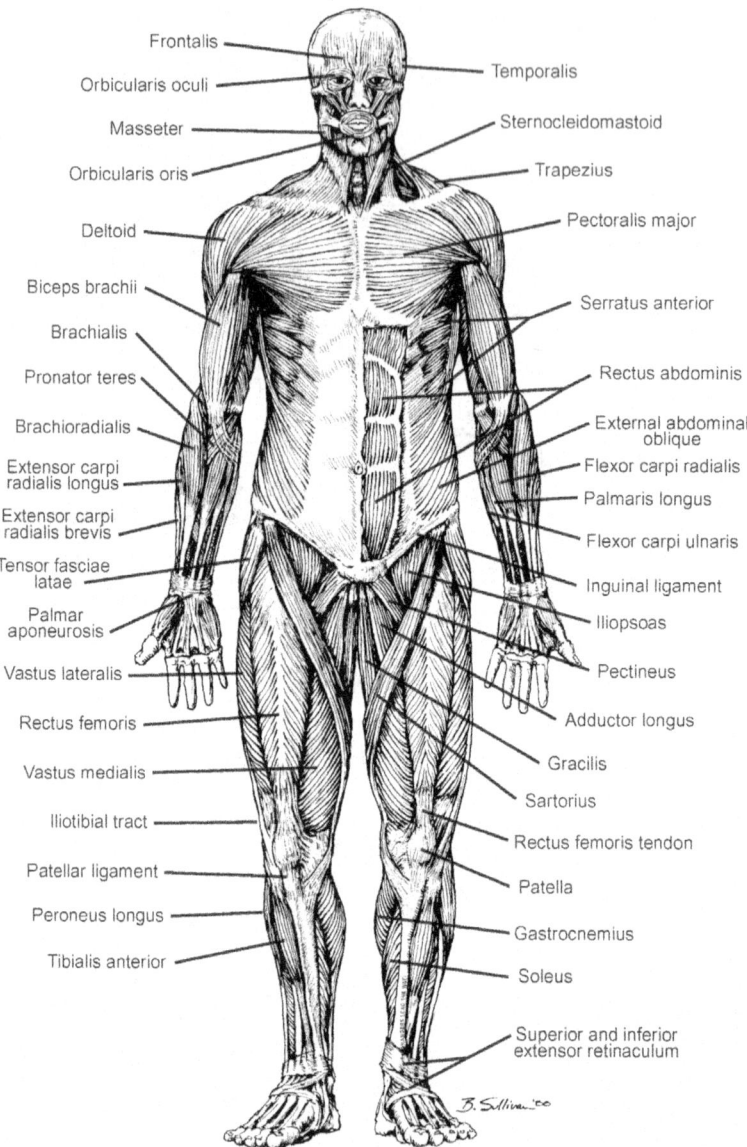

Superficial Muscles, Anterior View

Appendices: Anatomical Plates
Figure A-7

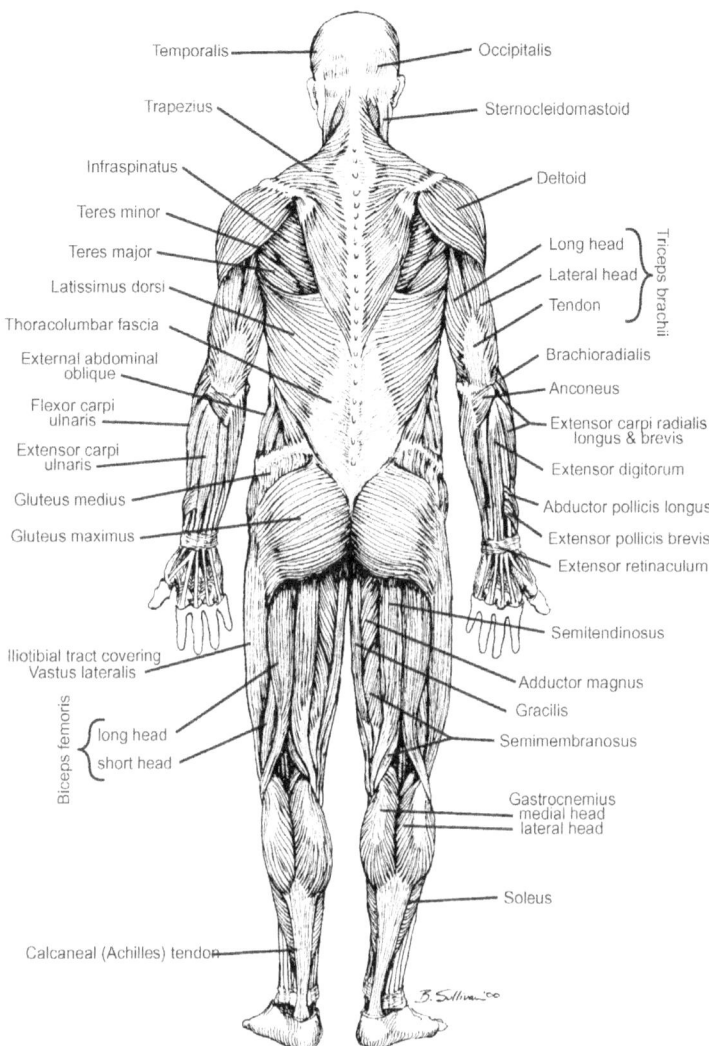

Superficial Muscles, Posteror View

Appendices: Anatomical Plates
Figure A-8

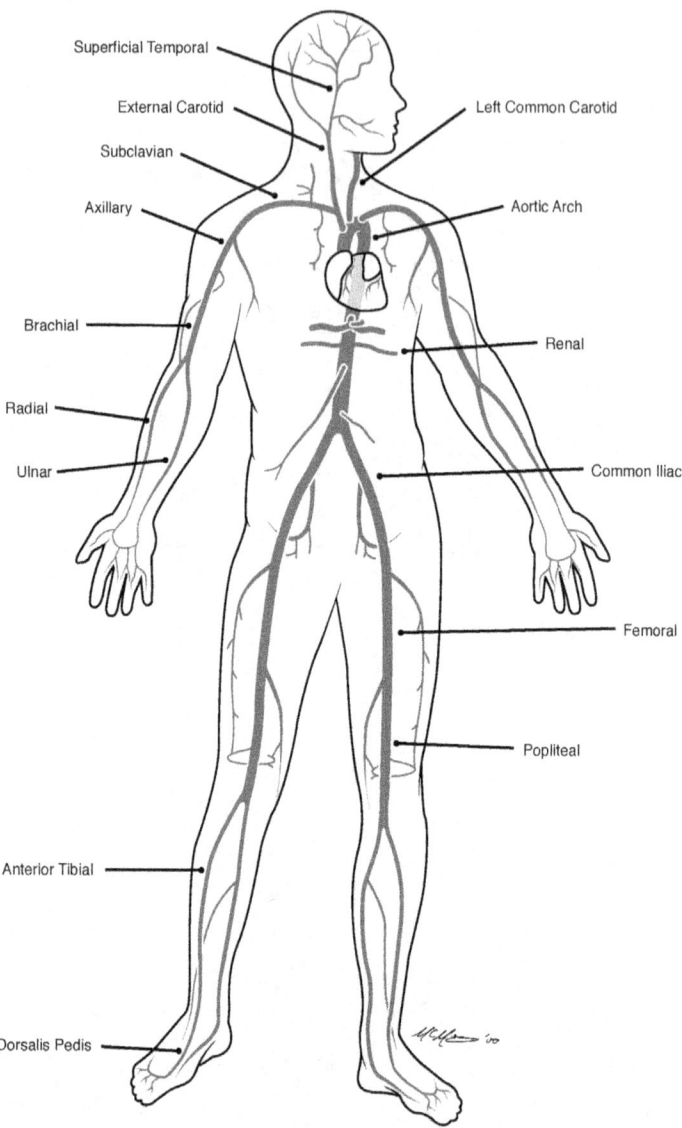

Arterial System

Appendices: Anatomical Plates
Figure A-9

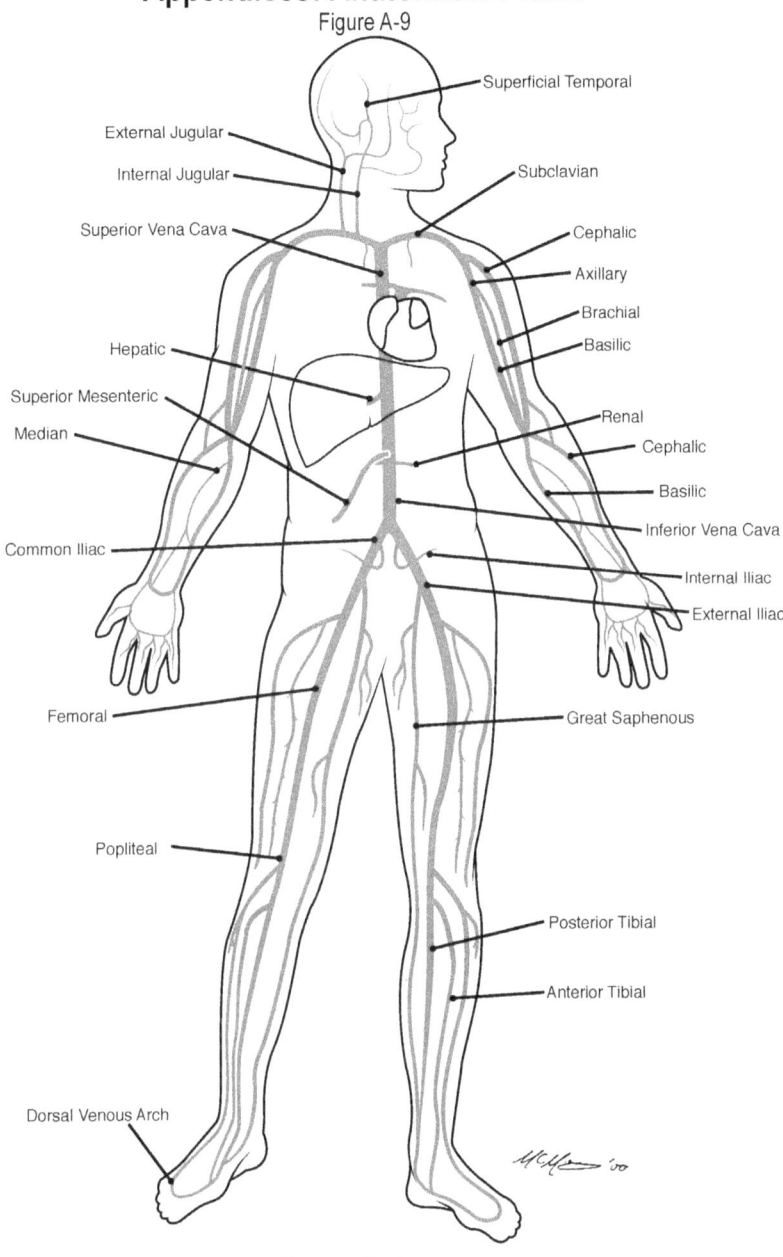

Venous System

A-9

Appendices: Anatomical Plates
Figure A-10

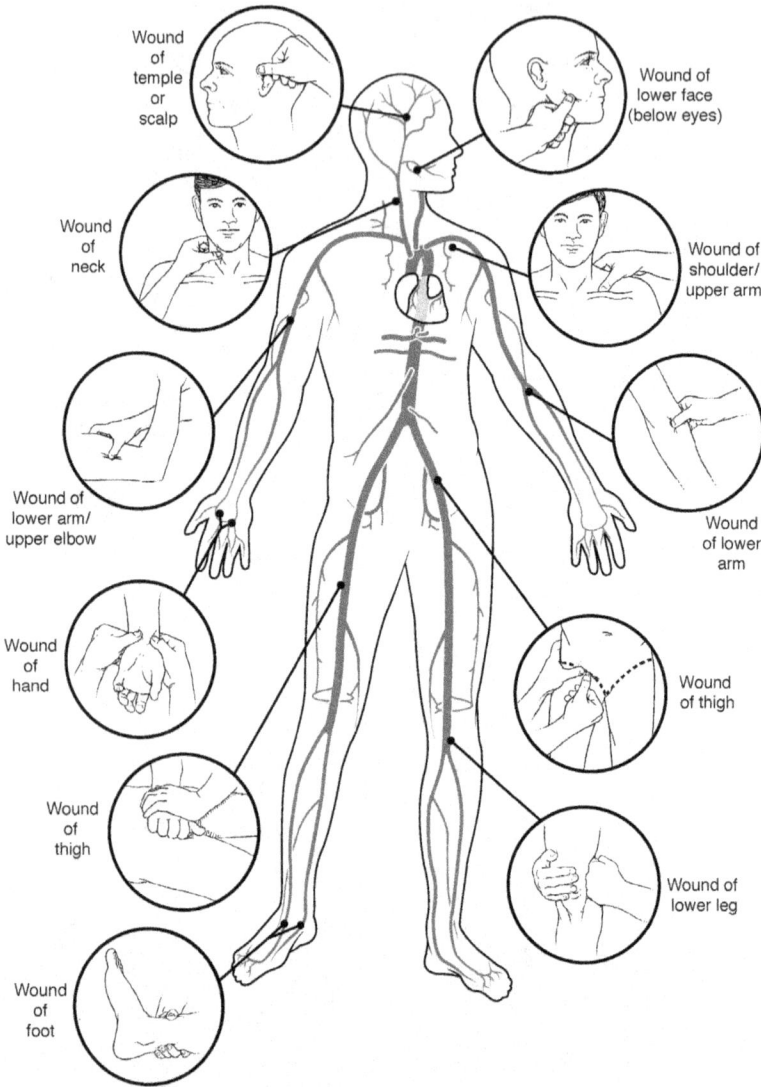

Pressure Points to Control Bleeding

Appendices: Anatomical Plates
Figure A-11

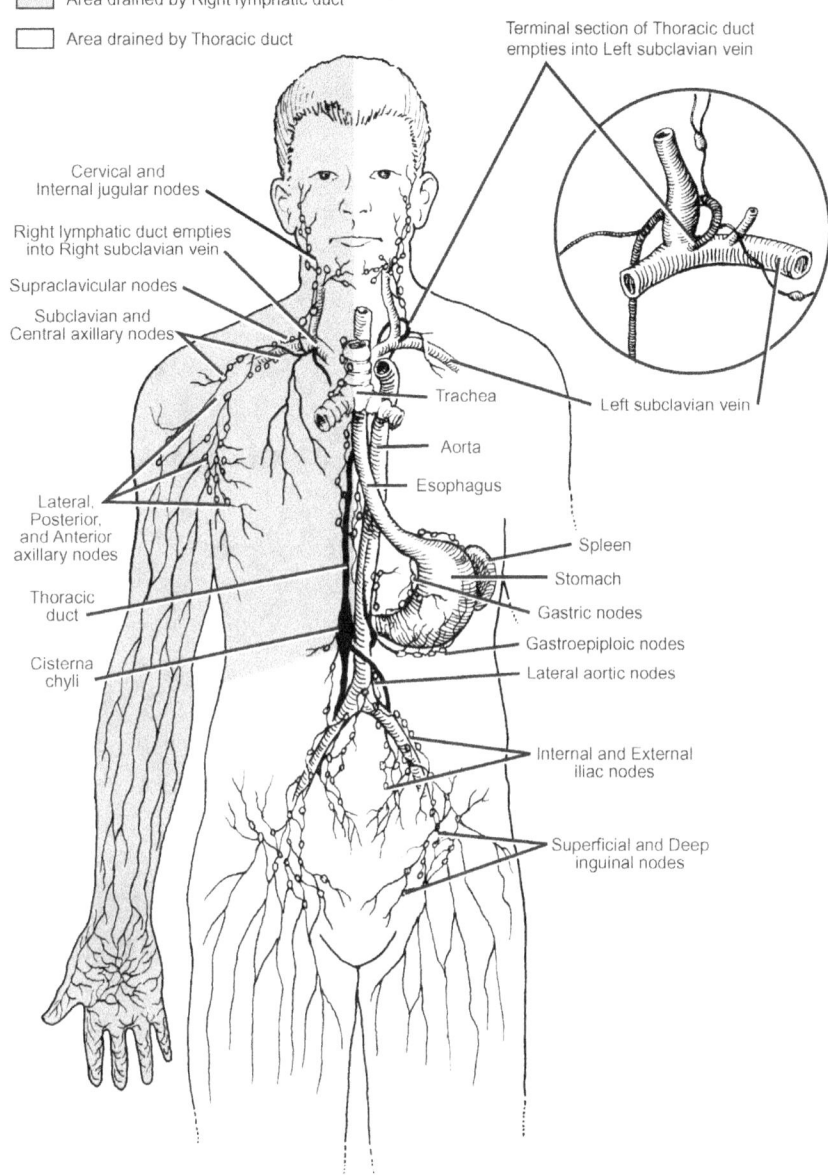

Lymphatic System (major components)

A-11

Appendices: Anatomical Plates
Figure A-12

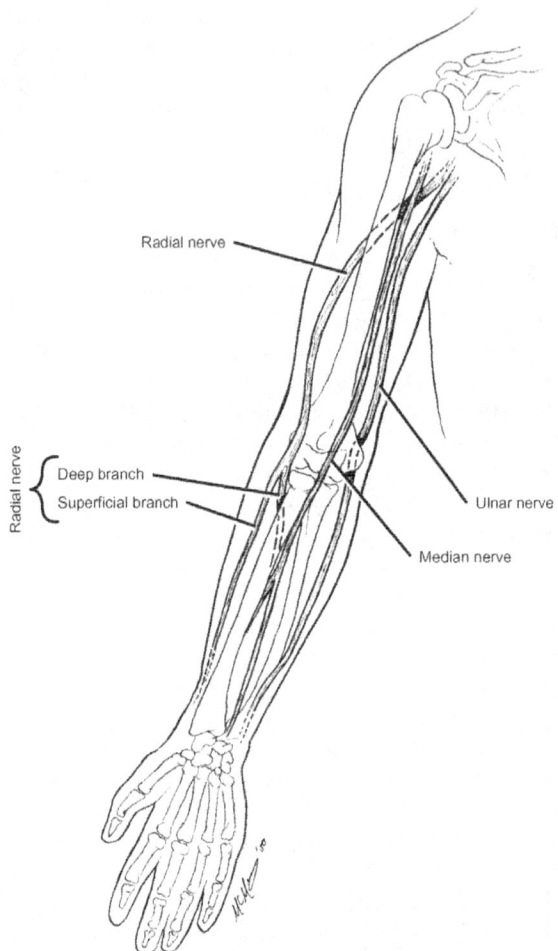

Nerves of the Arm

Appendices: Anatomical Plates
Figure A-13

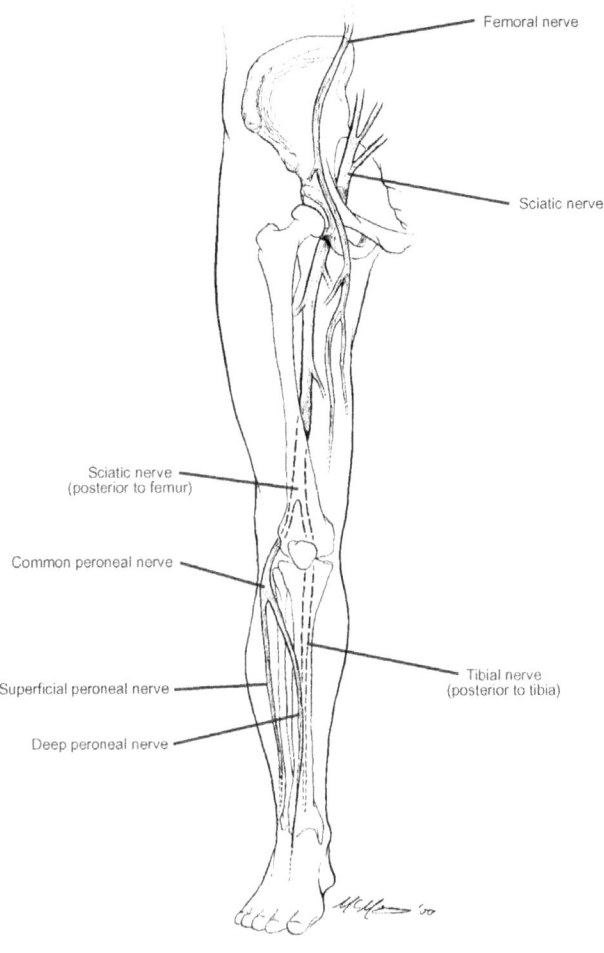

Nerves of the Leg

Appendices: Anatomical Plates
Figure A-14

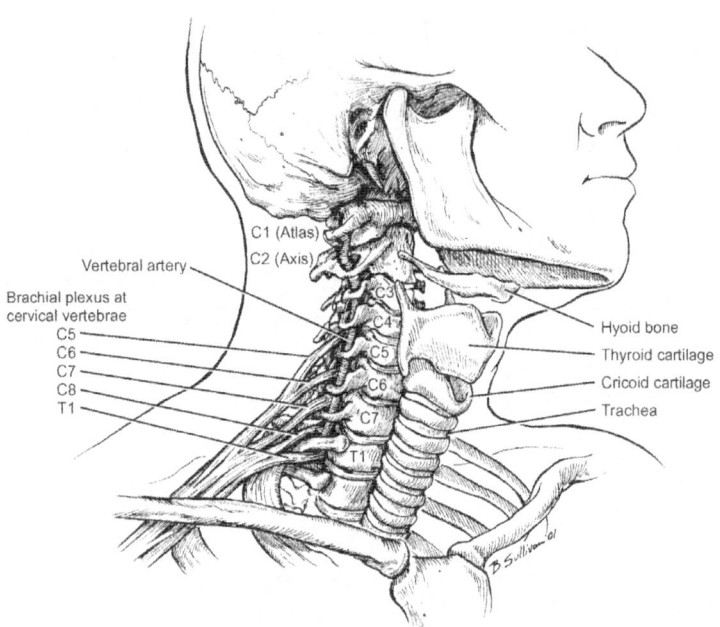

Cervical Vertebrae and Related Structures

Appendices: Anatomical Plates
Figure A-15

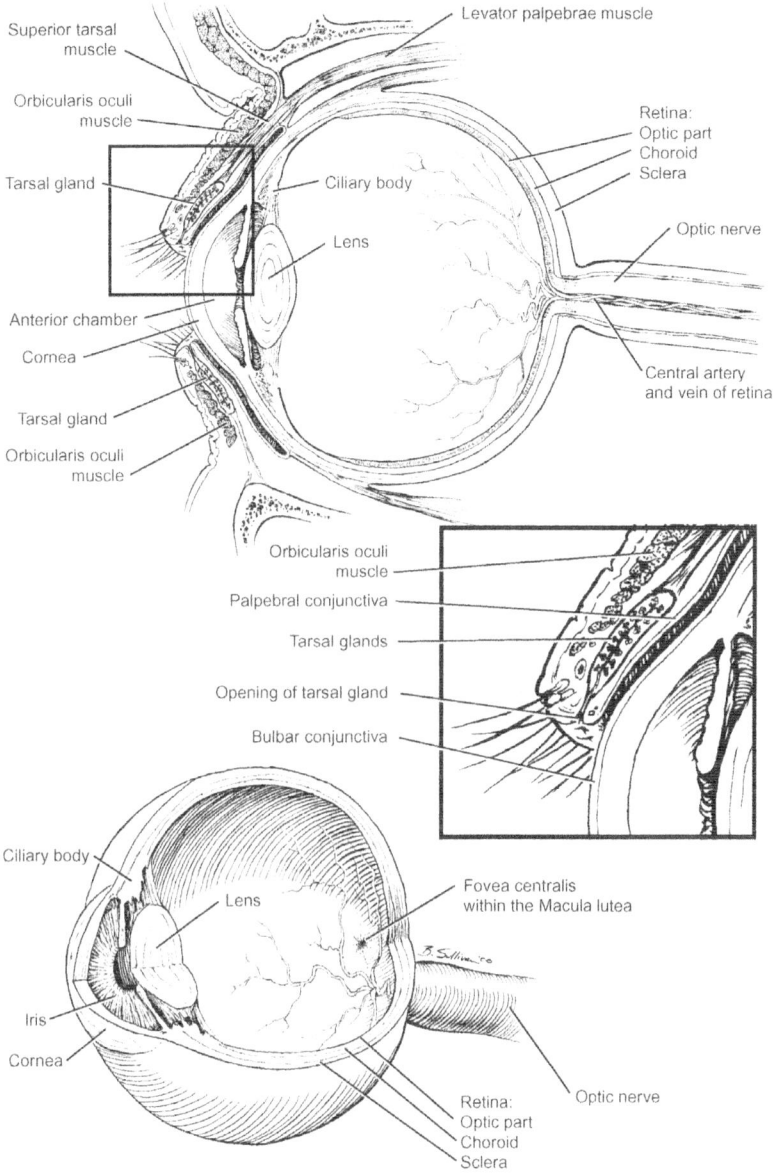

Eye anatomy

Appendices: Anatomical Plates
Figure A-16

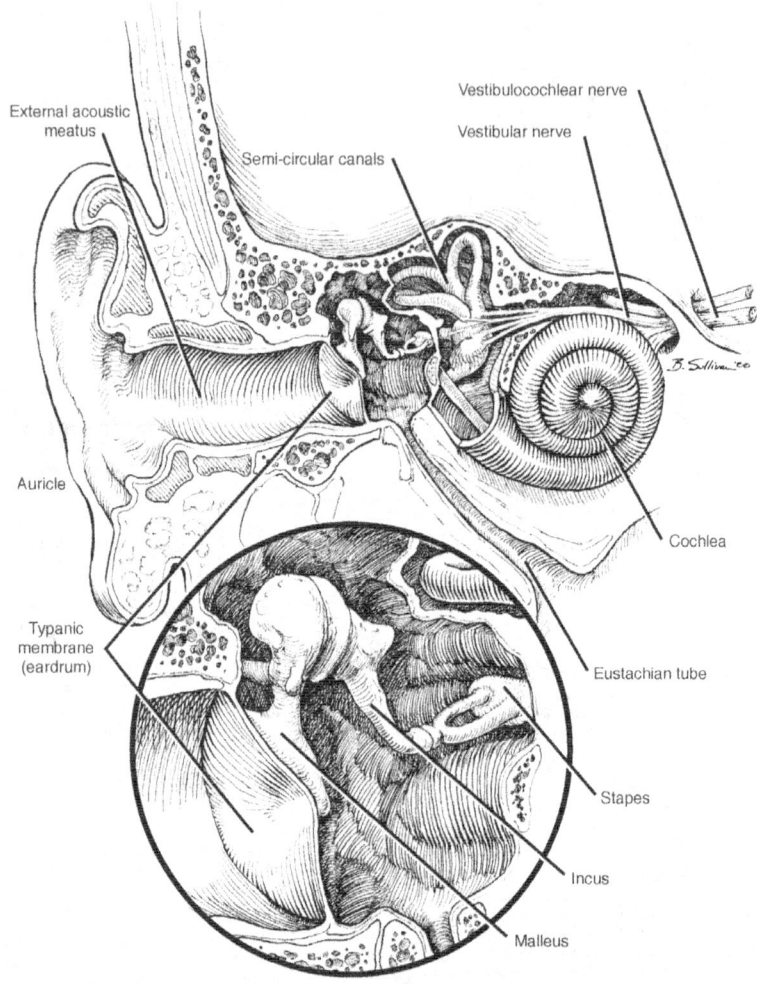

External, Middle, and Inner Ear

Appendices: Anatomical Plates
Figure A-17

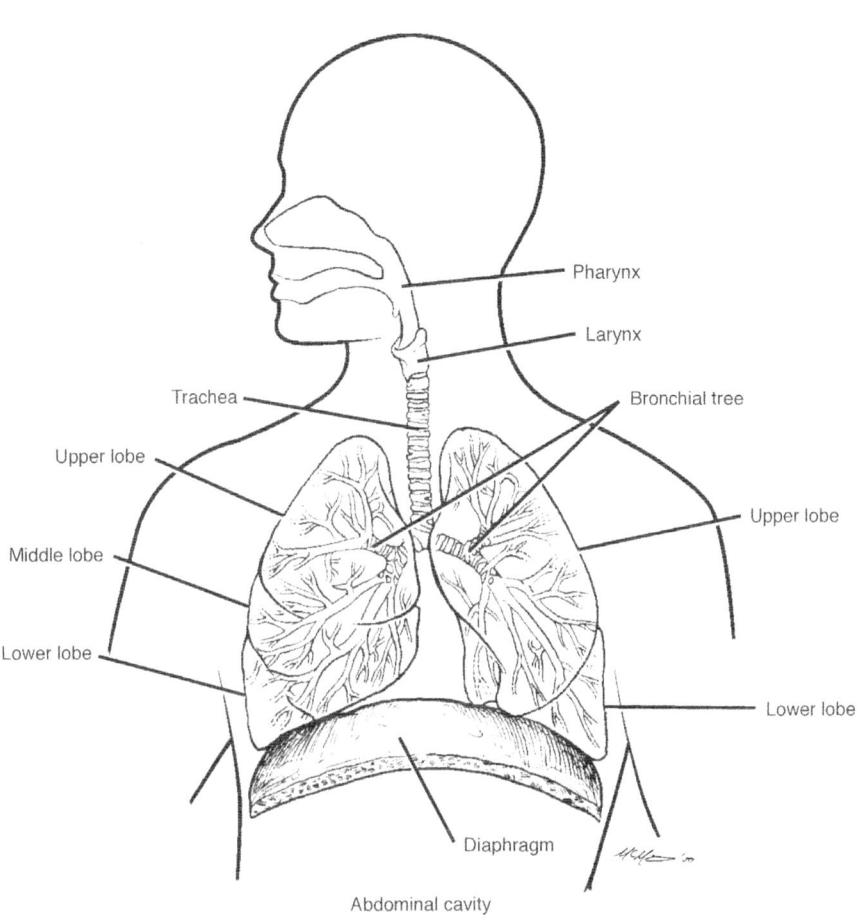

Respiratory System

Appendices: Anatomical Plates
Figure A-18

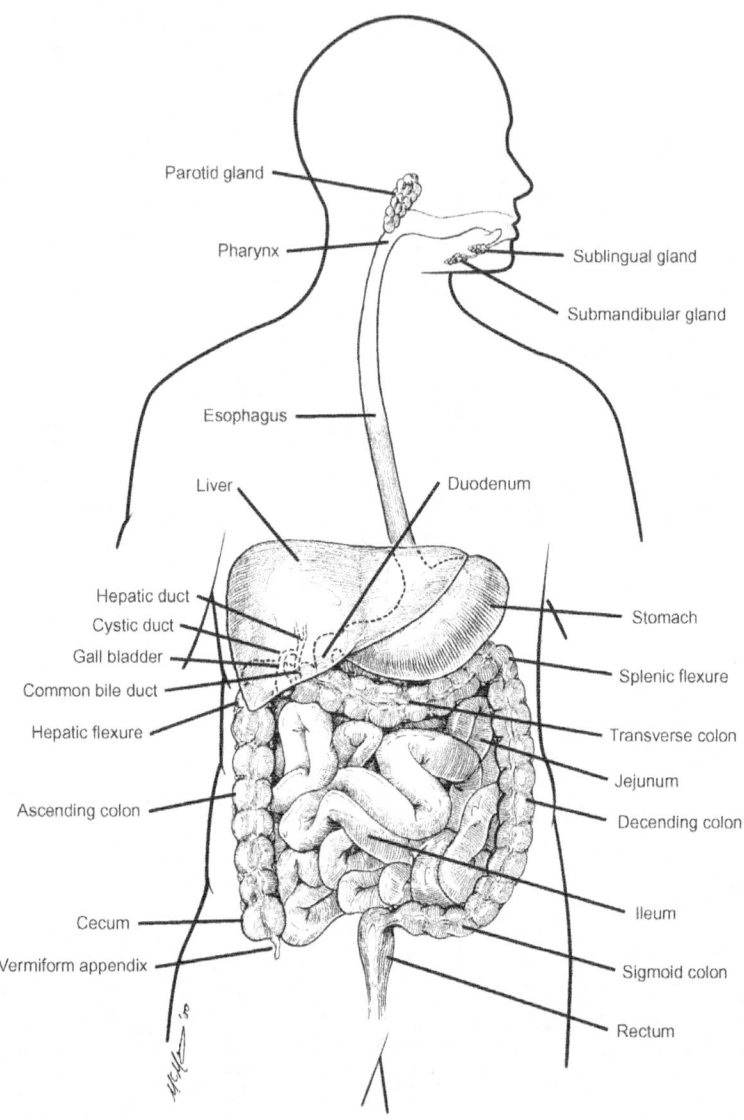

Digestive System

Appendices: Anatomical Plates
Figure A-19

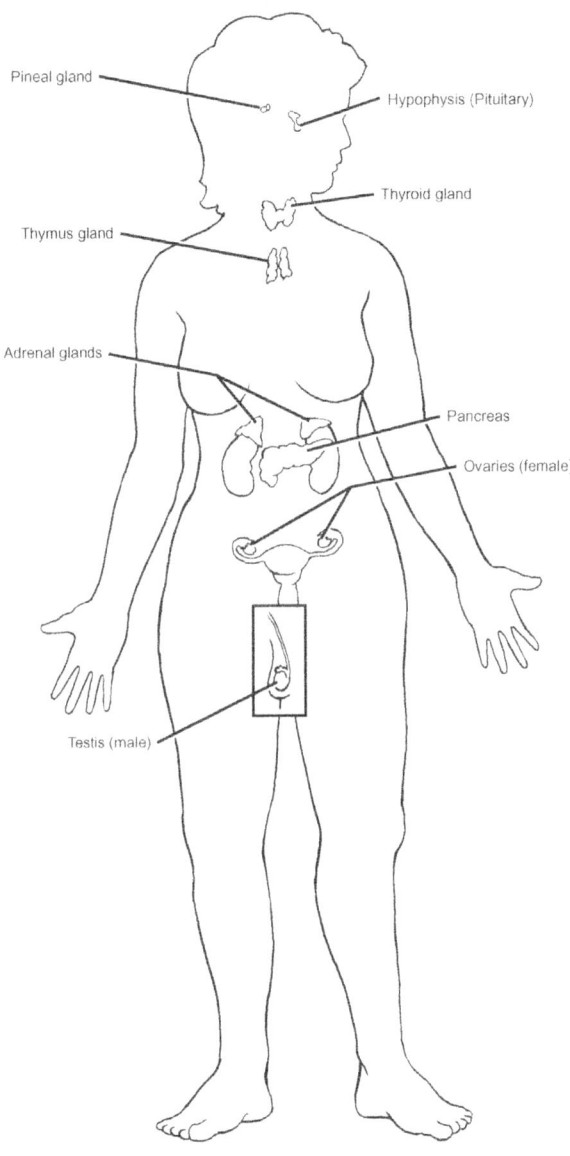

Endocrine System
A-19

Appendices: Anatomical Plates
Figure A-20

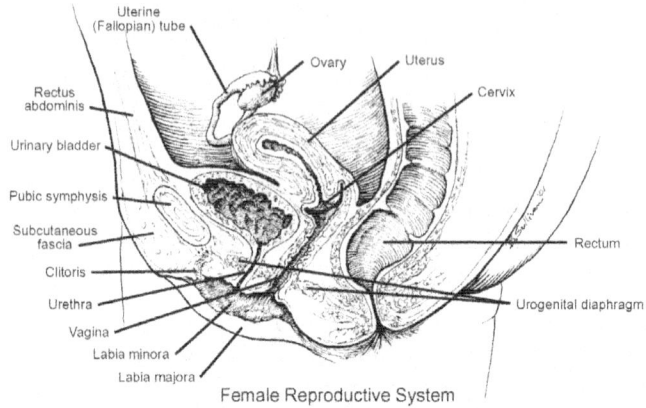

Female Reproductive System

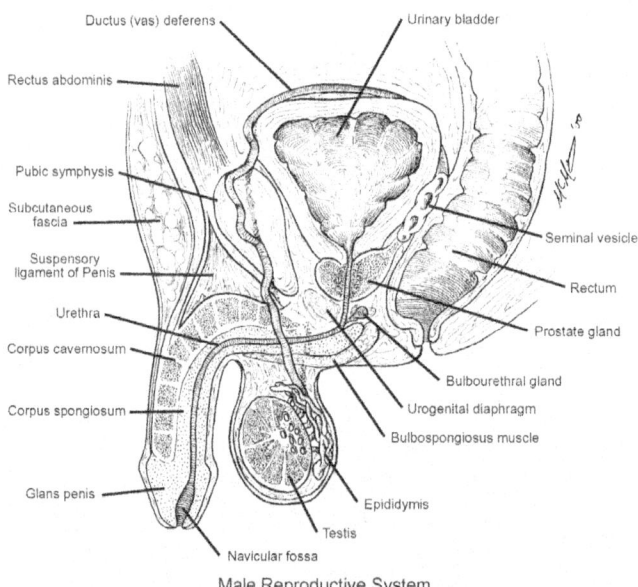

Male Reproductive System

Appendices: Color Plates

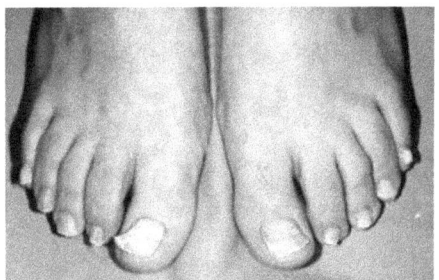

1-Dermatophyte Onychomycosis
Slide Courtesy of MAJ Dan Schissel

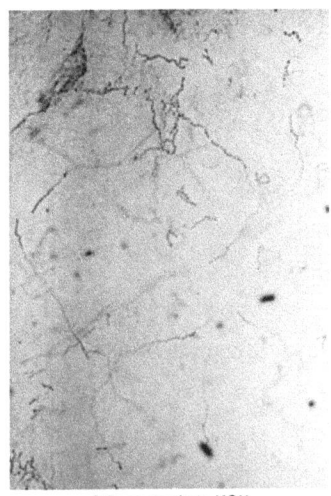

3-Dermatophyte KOH
Slide Courtesy of MAJ Dan Schissel

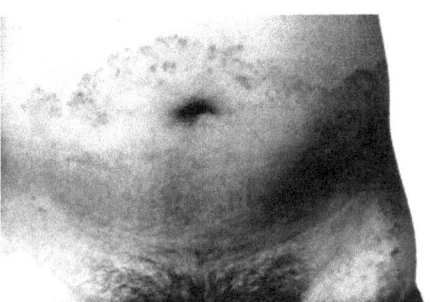

2-Dermatophyte Ringworm
Slide Courtesy of MAJ Dan Schissel

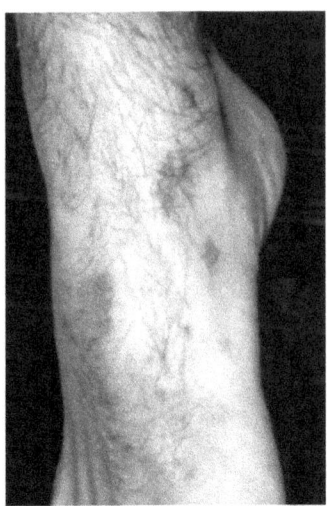

4-Disseminated Gonococcal Infection
Slide Courtesy of MAJ Dan Schissel

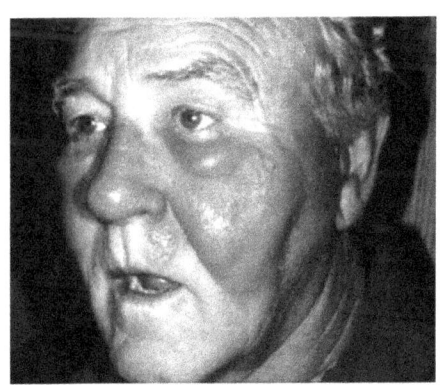

5-Erysipelas
Slide Courtesy of MAJ Dan Schissel

Appendices: Color Plates Continued

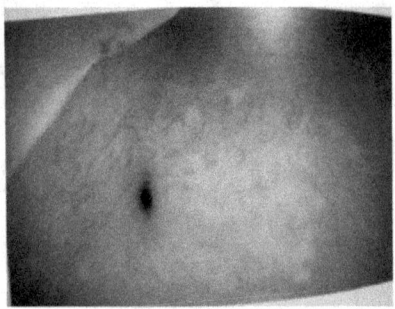

6-Tache Noire Lesion of Scrub Typhus
Slide Courtesy of Textbook of Military Medicine

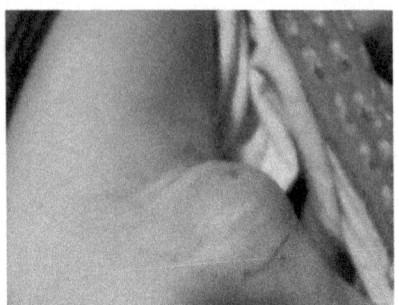

7-Enlarged Lymph Node of Cat-Scratch Disease
Slide Courtesy of COL Naomi Aronson

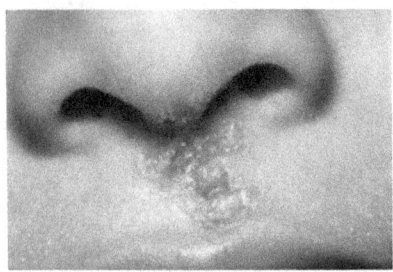

8-Impetigo Contagiosa
Slide Courtesy of MAJ Dan Schissel

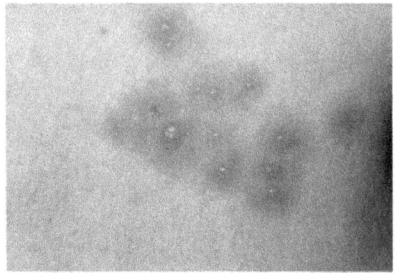

9-Herpes Zoster: Back Close-up
Slide Courtesy of MAJ Dan Schissel

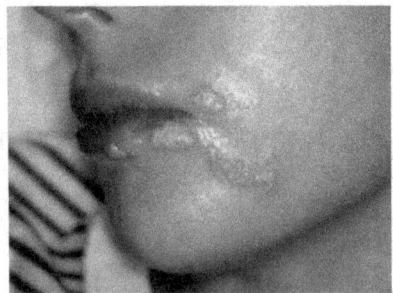

10-Herpes Simplex
Slide Courtesy of MAJ Dan Schissel

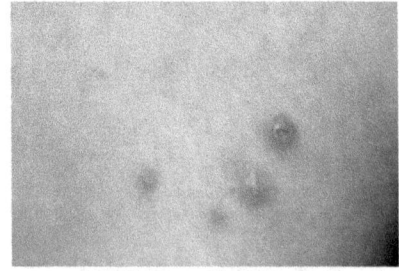

11-Molluscum Contagiosum
Slide Courtesy of MAJ Dan Schissel

Appendices: Color Plates

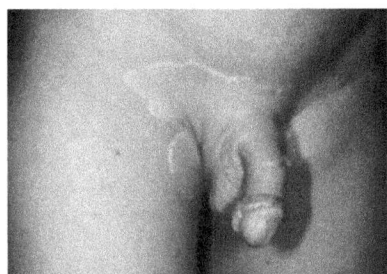

12-Staphylococcal Scalded Skin Syndrome
Slide Courtesy of MAJ Dan Schissel

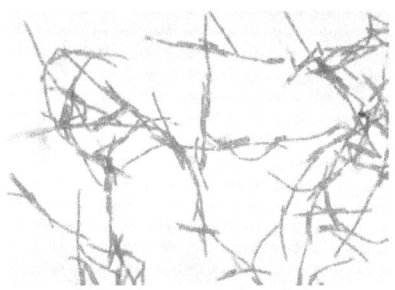

13-*Bacillus anthracis* Gram Stain
Slide Courtesy of CDC/Dr. William A. Clark

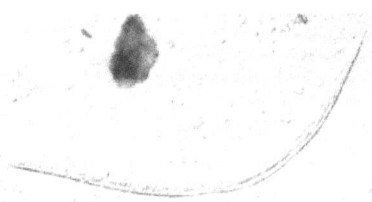

14-*Strongyloides stercoralis* First-Stage Larva
Preserved in 10% Formalin
Slide Courtesy of CDC Parasite Image Library

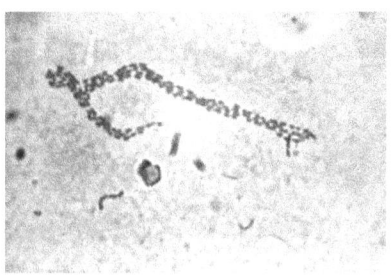

15-*Haemophilus ducreyi*
Slide Courtesy of COL Naomi Aronson

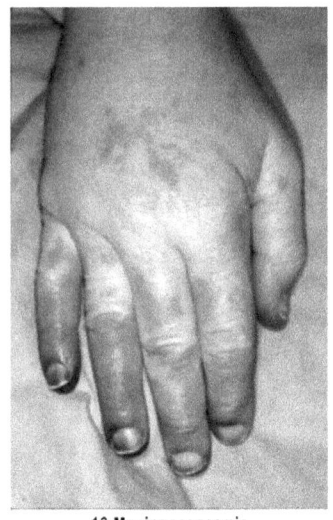

16-Meningococcemia
Slide Courtesy of MAJ Dan Schissel

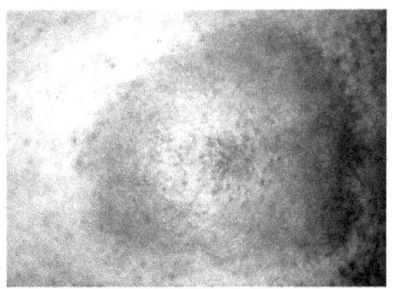

17-Erythema Migrans of Lyme Disease
Slide Courtesy of MAJ Joseph Wilde

Appendices: Color Plates

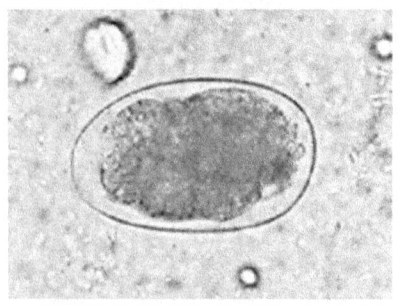

18-Hookworm eggs examined on wet mount
Slide Courtesy of CDC Parasite Image Library

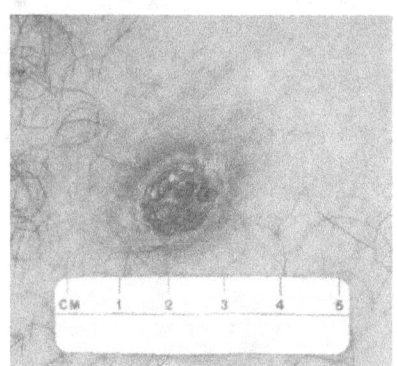

19-Leishmaniasis Skin Ulcer
Slide Courtesy of LTC Glenn Wortmann

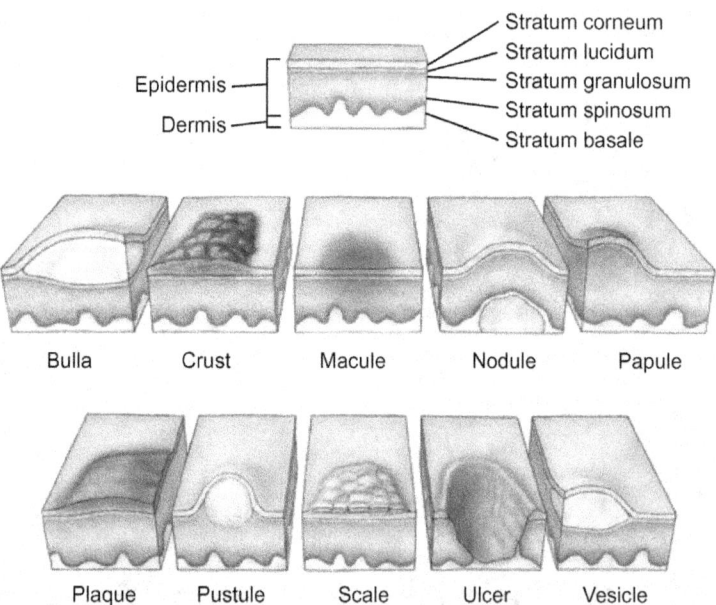

20-Lesions of the Skin

Appendices: Color Plates

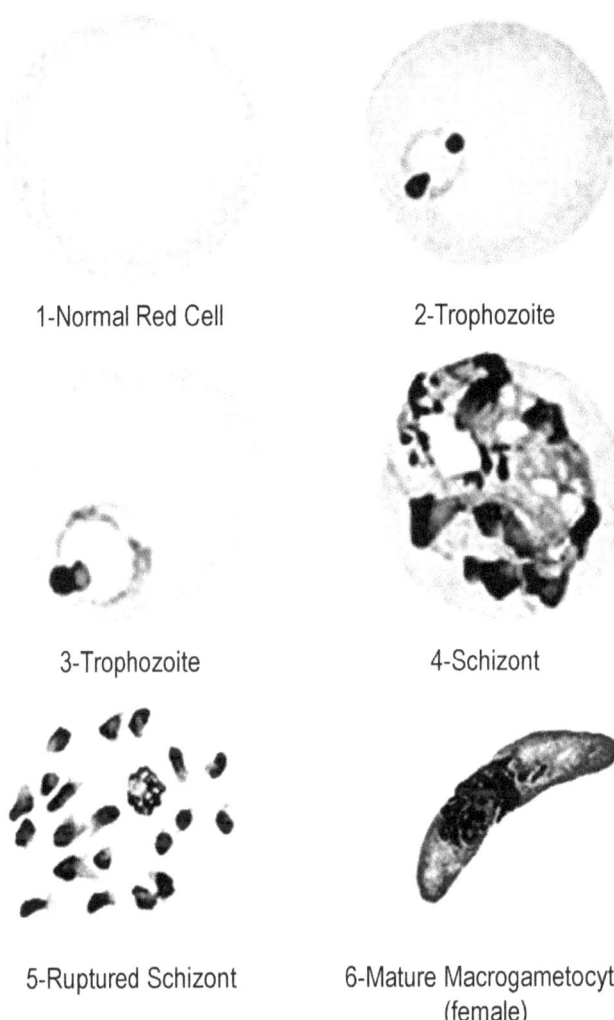

1-Normal Red Cell 2-Trophozoite

3-Trophozoite 4-Schizont

5-Ruptured Schizont 6-Mature Macrogametocyte (female)

0 ⎣⎯⎯⎯⎯⎯⎯⎯⎦ 10 u

Plasmodium falciparum

21-Courtesy of CDC Parasite Image Library

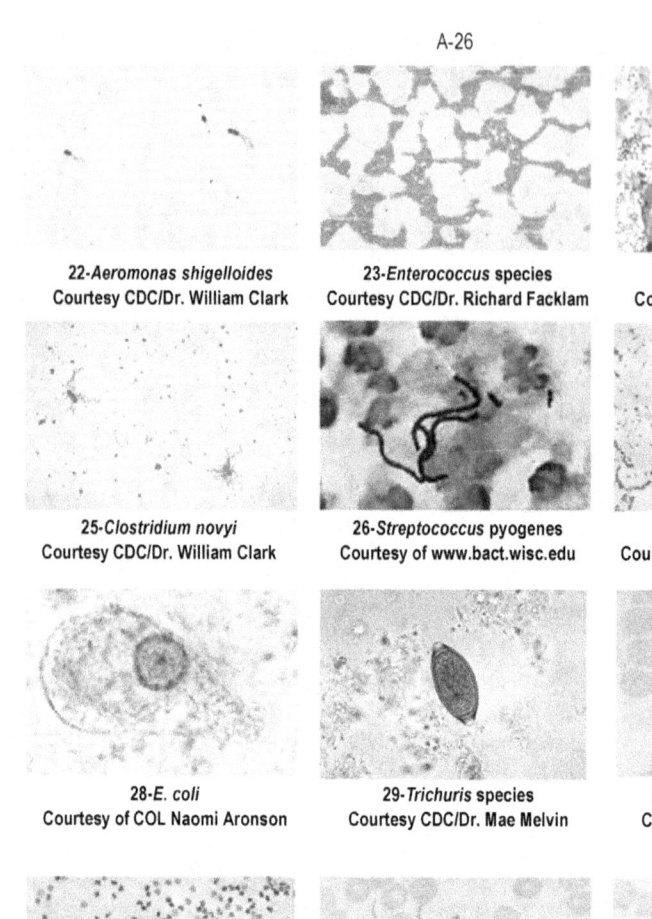
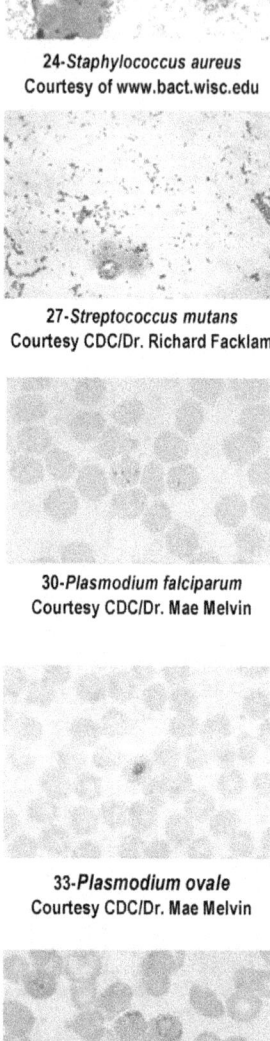
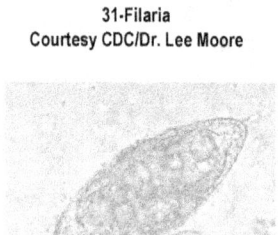
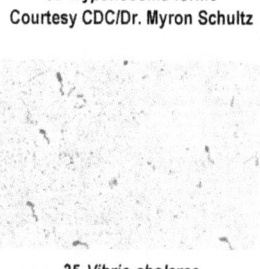
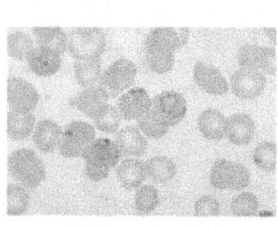

22-*Aeromonas shigelloides*
Courtesy CDC/Dr. William Clark

23-*Enterococcus* species
Courtesy CDC/Dr. Richard Facklam

24-*Staphylococcus aureus*
Courtesy of www.bact.wisc.edu

25-*Clostridium novyi*
Courtesy CDC/Dr. William Clark

26-*Streptococcus* pyogenes
Courtesy of www.bact.wisc.edu

27-*Streptococcus mutans*
Courtesy CDC/Dr. Richard Facklam

28-*E. coli*
Courtesy of COL Naomi Aronson

29-*Trichuris* species
Courtesy CDC/Dr. Mae Melvin

30-*Plasmodium falciparum*
Courtesy CDC/Dr. Mae Melvin

31-Filaria
Courtesy CDC/Dr. Lee Moore

32-*Trypanosoma* forms
Courtesy CDC/Dr. Myron Schultz

33-*Plasmodium ovale*
Courtesy CDC/Dr. Mae Melvin

34-*Schistosoma haematobium*
Courtesy of the CDC-PHIL

35-*Vibrio cholerae*
Courtesy CDC/Dr. William Clark

36-*Plasmodium vivax*
Courtesy of the CDC-PHIL

Appendices: Color Plates
Identify Cellular Blood Components
COL Richard Tenglin, MC, USA

When: To identify white cells, red cells and platelets in a properly prepared peripheral blood smear slide.
What You Need: Microscope, light source, peripheral blood smear slide.
What To Do: Use the following photos and descriptions to properly identify various cellular blood components.

1. White cell differential: ***Note:** Determine the total White Blood Cell Count as directed in the Laboratory Procedures Section under WBC Differential Count. Normal counts range from 4000- 12,000 white blood cells/ml3. To determine a white blood cell differential, a total of 100 cells should be counted with the number of each of the above noted as a percentage of the total.
    a. The polymorphonuclear leukocytes (PMNs or segs) (Figure 1) are white cells with the nucleus appearing as a string of 2-5 sausage links. Normal= 40-60% of the total white cell count.

    b. Lymphocytes (Figure 2) are round cells with a thin rim of deep blue cytoplasm surrounding a dark, condensed nucleus which takes up the majority of the cell. Normal = 20-40% of the total white cell count.

    c. Monocytes (Figure 3) are large cells with a bluish-gray cytoplasm and a nucleus often in an "M" shape. Normal = 5-10% of the total white cell count.

    d. Lesser numbers of white cells with large, bright red granules are called eosinophils (Figure 4). Normal = <5% of the total white cell count.

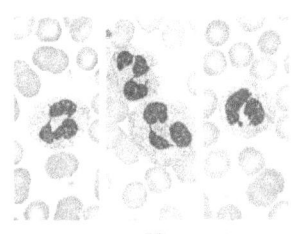

Figure 1
Normal neutrophils

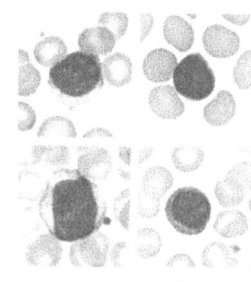

Figure 2
Normal lymphocytes
Note that the red cells are about the same size as the nucleus of lymphocyte in the upper left.

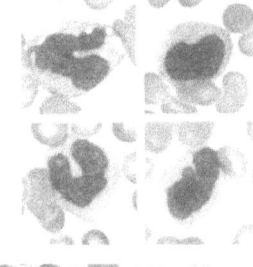

Figure 3
Normal monocytes
These are usually the largest white cells seen in normal blood.

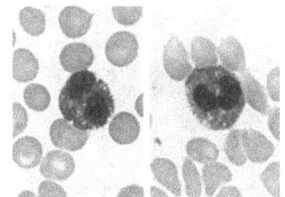

Figure 4
Eosinophils

Appendices: Color Plates

e. Cells with large dark purple granules are basophils (Figure 5). Normal = <1-2% of the total white cell count.

Figure 5
Basophils

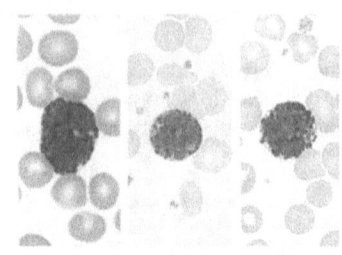

2. The size and shape of red cells should be noted. A normal red cell is about the size of the nucleus of a normal lymphocyte (see above Figure 2.), and extreme variations towards larger (macrocytes), smaller (microcytes), or variations between the size of individual red cells (anisocytosis) should be noted. Poikilocytosis refers to significant differences in the shape (normally a bi-concave disc) of individual red cells. Sickle cells have abnormal hemoglobin, which distorts the cell when deoxygenated.

Figure 6
Sickle cells and target cells.

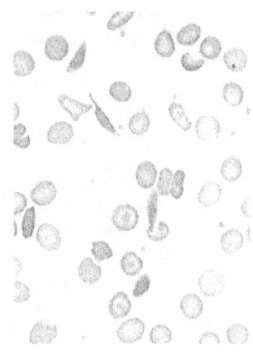

Cells in which the normally pale central area has a collection of hemoglobin surrounded by a pale rim are called Target Cells (Figure 6). Normal red cell counts are between 4-6 million/cubic microliter.

c. Platelets (Figure 7), cellular fragments (often with distinct granules) much smaller than the red or white cells, should be noted. On high power oil immersion (1000X) each platelet in a field represents a peripheral count of 20,000/cubic microliter. The platelets in several fields should be counted and averaged to get an idea of the peripheral count. Normal platelet numbers are between 150,000 to 450,000/cubic microliter.

Figure 7
Normal red cells with normal platelets (the small purple cells are platelets-there are 5)

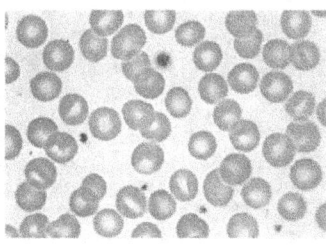

All pictures taken with permission from Hoffbrand AV, and Pettit JE. Clinical Hematology. Gower Medical Publishing, London and New York. 1988

What Not To Do: Do not break slide or microscope. Do not confuse the different types of WBCs.

Appendices: Organism/Antibiotic Chart

Developed by LTC Glenn Wortmann, MC, USA, LTC Duane Hospenthal, MC, USA & LCDR (sel) Edward Moldenhauer, MSC, USN

Infecting Organism	Medication of Choice	Alternatives
I. AEROBIC BACTERIA		
A. Gram-positive Cocci		
1. Staphylococi		
a. *S. aureus* (methicillin-sensitive)	PRP, Cephalosporin (1st generation)	Vanocomycin, Clindamycin, Sulfa-trimethaprim
b. *S. aureus* (methicillin-resistant)	Vancomycin	Clindamycin, Sulfa-trimethoprim, Linezolid, Quinupristin-dalfopristin
c. *S. epidermidis*	Vancomycin	Linezolid, Quinupristin-dalfopristin
2. Streptococci		
a. Streptoccocus, Group A, B, C, G, *S. viridans* and penicillin-sensitive *S. pneumoniae*	Penicillin G or V	Cephalosporin (1st generation), Vancomycin, Clindamycin, Erythromycin, Azithromycin
b. *S. pneumoniae* with high level penicillin-resistance	Vancomycin	Levofloxacin
c. Enterococci	Ampicillin and Gentamicin	Vancomycin and Gentamicin
B. Gram-positive Bacilli		
1. *Bacillus anthracis*	Penicillin or Ciprofloxacin	Doxycycline
2. *Corynebacterium diphtheriae*	Erythromycin and Antitoxin	Clindamycin, Penicillin G
3. *Listeria monocytogenes*	Ampicillin or Penicillin	Sulfa-trimethoprim
C. Gram-negative Cocci		
1. *Neisseria gonorrhoeae*	Ceftriaxone or Fluoroquinolone	Cefixime
2. *Neisseria meningitidis*	Penicillin G or V	Ceftriaxone
3. *Moraxella catarrhalis*	Sulfa-trimethoprim, Amoxicillin-clavulanate	Azithromycin, doxycycline
D. Gram-negative Bacilli		
1. *Escherichia coli*	Sulfa-trimethoprim (if sensitive)	Cephalosporin (3rd generation), fluoroquinolone
2. *Enterobacter* species	Imipenem	Fluoroquinolone, Cephalosporin (3rd generation)
3. *Klebsiella* species	Imipenem	Fluoroquinolone, Cephalosporin (3rd generation)
4. *Pseudomonas aeruginosa*	APP	Ceftazidime
5. *Proteus* species	APP	Cephalosporin (3rd generation), fluoroquinolone
6. *Serratia marcescens*	Cephalosporin (3rd generation)	Imipenem, Fluoroquinolone
7. *Salmonella typhi*	Ceftriaxone or fluoroquinolone	Sulfa-trimethoprim
8. *Haemophilus influenzae*	Ceftriaxone or fluoroquinolone	Azithromycin, doxycycline, Amoxacillin-clavulanic acid
9. *Haemophilus ducreyi*	Ceftriaxone or fluoroquinolone	Amoxicillin-clavulanic acid, Azithromycin, Erythromycin

Infecting Organism	Medication of Choice	Alternatives
10. *Brucella* species	Doxycycline and rifampin	Doxycycline and Gentamicin
11. *Francisella tularensis*	Streptomycin or gentamicin	Doxycycline
12. *Yersinia pestis*	Streptomycin	Gentamicin

II. ANAEROBIC BACTERIA

1. *Clostridium tetani*	Penicillin and tetanus toxoid and tetanus immune globulin	Imipenem, Clindamycin, Metronidazole
2. *Bacteroides* species	Metronidazole	Imipenem, Clindamycin, Betalactam-betalactamase inhibitor

III. ACID FAST BACILLI

1. *Mycobacterium tuberculosis*	Isoniazid + Rifampin + Pyrazinamide + Ethambutol	
2. *M. kansasii*	Isoniazid + Rifampin + Ethambutol	
3. *M. avium intracellulare* complex	Clarithromycin + Ethambutol + Rifabutin	Azithromycin + Ethambutol + Rifabutin
4. *M. leprae*	Rifampin + Dapsone (paucibacillary disease)	Rifampin + Dapsone + Clofazamine (multibacillary disease)

IV. ACTINOMYCETES

1. *Nocardia* species	Sulfa-trimethoprim	Minocycline + sulfonamide
2. *Actinomyces* species	Penicillin	Clindamycin

V. MISCELLANEOUS

1. *Chlamydia* species	Doxycycline	Erythromycin
2. *Rickettsia* species	Doxycycline	Chloramphenicol
3. *Treponema pallidum* (syphilis)	Penicillin	Ceftriaxone
4. *Mycoplasma pneumoniae*	Azithromycin	Doxycycline
5. *Pneumocystis carinii*	Sulfa-trimethoprim	Pentamidine

VI. FUNGAL

1. *Aspergillus* species	Amphotericin	Itraconazole
2. *Blastomyces dermatitidis*	Amphotericin	Itraconazole
3. *Coccidioides immitis*	Fluconazole	Amphotericin
4. *Cryptococcus neoformans*	Amphotericin	Fluconazole
5. *Histoplasma capsulatum*	Amphotericin	Itraconazole
6. *Candida* species	Amphotericin	Fluconazole
7. *Sporothrix schenckii*	Itraconazole	Saturated solution of potassium iodide
8. Tinea versicolor	Topical selenium sulfide	Ketoconazole

VII. VIRUSES

1. Herpes simplex	Acyclovir	Valacyclovir, Famciclovir
2. Herpes zoster	Acyclovir	Valacyclovir, Famciclovir
3. Influenza	Amantadine or rimantadine or zanamivir or oseltamivir	

Abbreviations/Classifications: APP (Anti-pseudomonal penicillin): Piperacillin, Mezlocillin, Ticarcillin
1st Gen Cephalosporin: Cefazolin, Cephalexin
2nd Gen Cephalosporin: Cefoxitin, Cefaclor, Cefuroxime axetil

3rd Gen Cephalosporin: Ceftazidime, Ceftriaxone, Cefixime Betalactam-Blactamase combination: Ampicillin-sulbactam, Piperacillin-tazobactam, Amoxacillin-clavulanic acid
PRP (Penicillinase-resistant penicillin): Nafcillin, Dicloxacillin Fluoroquinolone: Ciprofloxacin, Levofloxacin, Gatifloxacin, Moxifloxacin

The preceding chart does not account for the fact that these organisms may cause a range of disease from UTI or mild skin infections to life-threatening sepsis and meningitis. It also does not account for local resistance patterns, cost-effectiveness, or what drugs may be available. Microbiology support to identify organisms and produce susceptibility results greatly increases the success of antibiotic therapy

Appendices: Medications Causing Photosensitivity
Prepared by LCDR (sel) Edward Moldenhauer, MSC, USN

Alprazolam (Xanax)
Amiodarone (Cordarone)
Amitriptyline (Elavil)
Captopril (Capoten)
Chlordiazepoxide (Librium)
Chloroquine (Aralen)
Chlorpropamide (Diabinese)
Ciprofloxacin (Cipro)
Cyproheptadine (Periactin)
Dantrolene (Dantrium)
Dapsone
Diethylstilbestrol (DES)
Doxycycline (Vibramycin)
Erythromycin/Sulfisoxazole (Pediazole)
Griseofulvin (Gris-Peg)
Haloperidol (Haldol)
Hydrochlorothiazide (HCTZ)
Ibuprofen (Motrin)
Isotretinoin (Retin-A)
Ketoprofen (Orudis)

Methazolamide (Niaprazine)
Minocycline (Minocin)
Mupirocin (Bactroban)
Ofloxacin (Floxin)
Omeprazole (Prilosec)
Non-steroidal Anti-inflammatory Drugs (NSAIDs)
Norfloxacin (Noroxin)
Quinidine (Quinaglute)
Quinine
Risperidone (Risperdal)
Selegiline (Eldepryl)
Sotalol (Betapace)
Sulfa medications
Sulfamethoxazole (Gantanol)
Terfenadine (Seldane)
Tetracycline (Achromycin)
Tolazamide (Tolinase)
Triamterene w/ HCTZ (Maxzide, Dyazide)
Tripelennamine (PBZ)
Vinblastine (Velban)

Appendices: IV Drip Rates

CPT Joseph Fasano, AN, USA

Drip Chamber Drop Sizes (Select Your Drip Chamber Size)

* Hours To Deliver 1000 ml.	ml per hour	10 Drops per ml. Drops Per Minute (DPM)	15 Drops per ml. Drops Per Minute (DPM)	20 Drops per ml. Drops Per Minute (DPM)	60 Drops per ml. Drops Per Minute (DPM)
Fractions are rounded to closest number, except where shown.	10	1.7 DPM	2.5 DPM	3 DPM	10 DPM
	30	5 DPM	7.5 DPM	10 DPM	30 DPM
	50	8 DPM	12.5 DPM	17 DPM	50 DPM
	70	12 DPM	17.5 DPM	23 DPM	70 DPM
	90	15 DPM	22.5 DPM	30 DPM	90 DPM
* 8 Hours	125	20.8 DPM	31 DPM	42 DPM	
	175	29 DPM	43.8 DPM	58 DPM	
	250	41.6 DPM	62.5 DPM	83 DPM	

Appendices: Glasgow Coma Scale

Eye Opening Response	Spontaneous	4
	To Verbal Command	3
	With Painful Stimulus	2
	No Response	1
Best Verbal Response	Oriented	5
	Confused	4
	Inappropriate Words	3
	Incomprehensible Sounds	2
	None	1
Best Motor Response	Obeys Command	6
	Localizes Pain	5
	Withdraws (pain)	4
	Flexion (pain)	3
	Extension (pain)	2
	None	1
Total		3 to 15

Any patient with a GCS of 8 or less is considered to have a severe head injury. Those in the 9 to 12 range are considered moderate, but may require airway control. Any GCS of 13 to 15 is considered indicative of mild head injury, but even these patients can deteriorate and should be observed or evacuated per the guidelines in the handbook.

Appendices: Mini-Mental Status Exam

The Mini-Mental Status Exam (MMSE) is the most useful tool for assessing brain function in the field. Once a baseline score is established with an initial MMSE, improvement or worsening mental status can be observed with repeated testing.

Specific Test Function	Brain Area Tested	Points
What is the year/season/month/date/day of the week? (1 for each correct answer)	Orientation (Frontal)	5
What state/county/hospital/floor are you in? (Use other cues, such as military unit, etc., when appropriate).	Orientation (Frontal)	5
List three items. Have the patient repeat the list.	Registration (Frontal)	3
Serial sevens- ask the patient to count backwards by 7 from 100, or have them spell "WORLD" backwards.	Concentration (Frontal)	5
Name an object you point to, such as a wristwatch or pen.	Naming (Dominant Temporoparietal)	2
Have patient repeat, "No ifs, ands. or buts."	Expressive Speech (Dominant Frontal)	1
3-part command: Take this paper in your right hand, fold it in half & put it on the table.	Command (Frontal)	3
Read "close your eyes," on paper and do it.	Reading (Dominant Temporoparietal)	1
Recall the three items listed earlier.	Short-term Memory (Hippocampal)	3
Write a sentence.	Writing (Dominant Temporoparietal)	1
Copy intersecting pentagons	Construction (Nondominant Parietal)	1
TOTAL		**30**

A-34

NEUROLOGICAL EXAMINATION CHECKLIST

(Sheet 1 of 2)

Patient's Name: _____ Date/Time: _____
Describe pain/numbness: _____
_____

HISTORY

Type of dive last performed: _____ Depth: _____ How long: _____
Number of dives in last 24 hours: _____
Was symptom noticed before, during, or after the dive? _____
If during, was it while descending, on the bottom, or ascending? _____
Has symptom increased or decreased since it was first noticed? _____
Have any other symptoms occurred since the first one was noticed? _____
Describe: _____
Has patient ever had a similar symptom before? _____ When: _____
Has patient ever had decompression sickness or an air embolism before? _____ When: _____

MENTAL STATUS/STATE OF CONSCIOUSNESS

COORDINATION

Walk: _____
Heel-to-Toe: _____
Romberg: _____
Finger-to-Nose: _____
Heel-Shin Slide: _____
Rapid Movement: _____

CRANIAL NERVES

Sense of Smell: (I): _____
Vision/Visual Fld (II): _____
Eye Movements, Pupils (III, IV, VI): _____
Facial Sensation, Chewing (V): _____
Facial Expression Muscles (VII): _____
Hearing (VIII): _____
Upper Mouth, Throat Sensation (IX): _____
Gag & Voice (X): _____
Shoulder Shrug (XI): _____
Tongue (XII): _____

STRENGTH (Grade 0 to 5)

Upper Body

Deltoids	L _____	R _____
Latissimus	L _____	R _____
Biceps	L _____	R _____
Triceps	L _____	R _____
Forearms	L _____	R _____
Hands	L _____	R _____

Lower Body

Hips

Flexion	L _____	R _____
Extension	L _____	R _____
Abduction	L _____	R _____
Adduction	L _____	R _____

Knees

Flexion	L _____	R _____
Extension	L _____	R _____

NEUROLOGICAL EXAMINATION CHECKLIST

(Sheet 2 of 2)

REFLEXES
(Grade: Normal, Hypoactive, Hyperactive, Absent)

Biceps	L ___ R ___	**Ankles**	
Triceps	L ___ R ___	Dorsiflexion	L ___ R ___
Knees	L ___ R ___	Plantarflexion	L ___ R ___
Ankles	L ___ R ___	**Toes**	L ___ R ___

Sensory Examination for Skin Sensation
(Use diagram to record location of sensory abnormalities – numbness, tingling, etc.)

LOCATION

Indicate results as follows:

|||| Painful Area

= Decreased Sensation

COMMENTS

_____
_____
_____
_____
_____
_____
_____
_____

Examination Performed by: _____

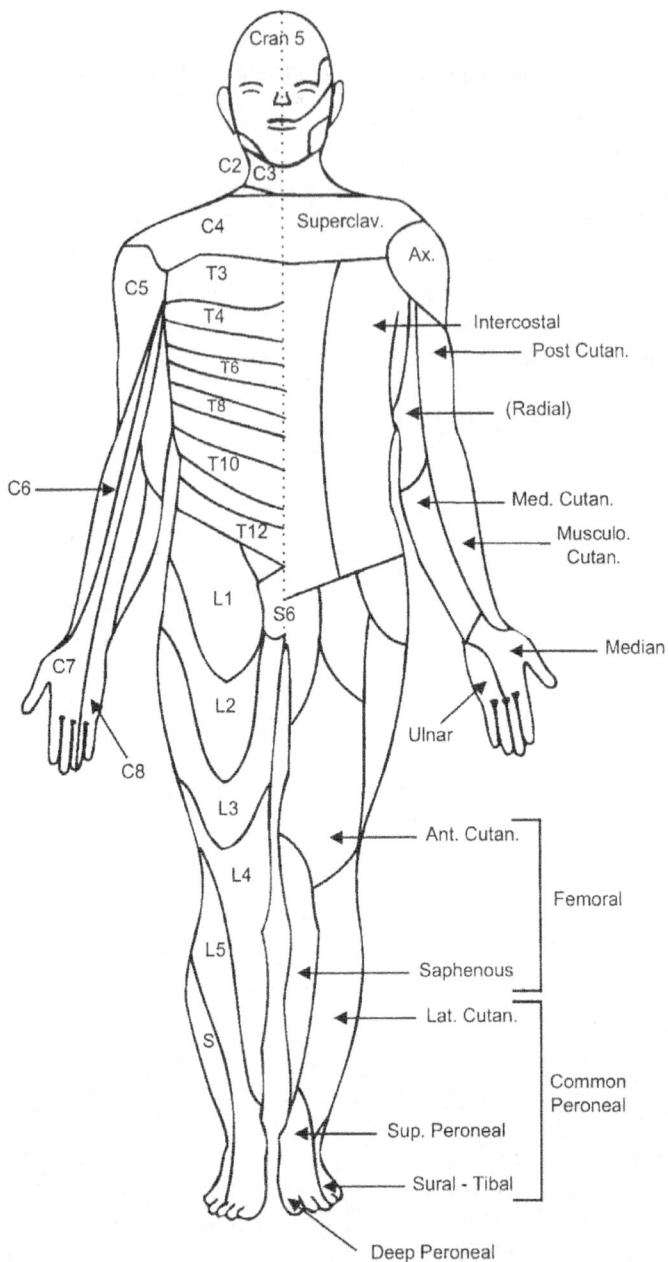

Dermatomes of Cutaneous Innervation, Anterior View
(United States Navy Dive Manual)

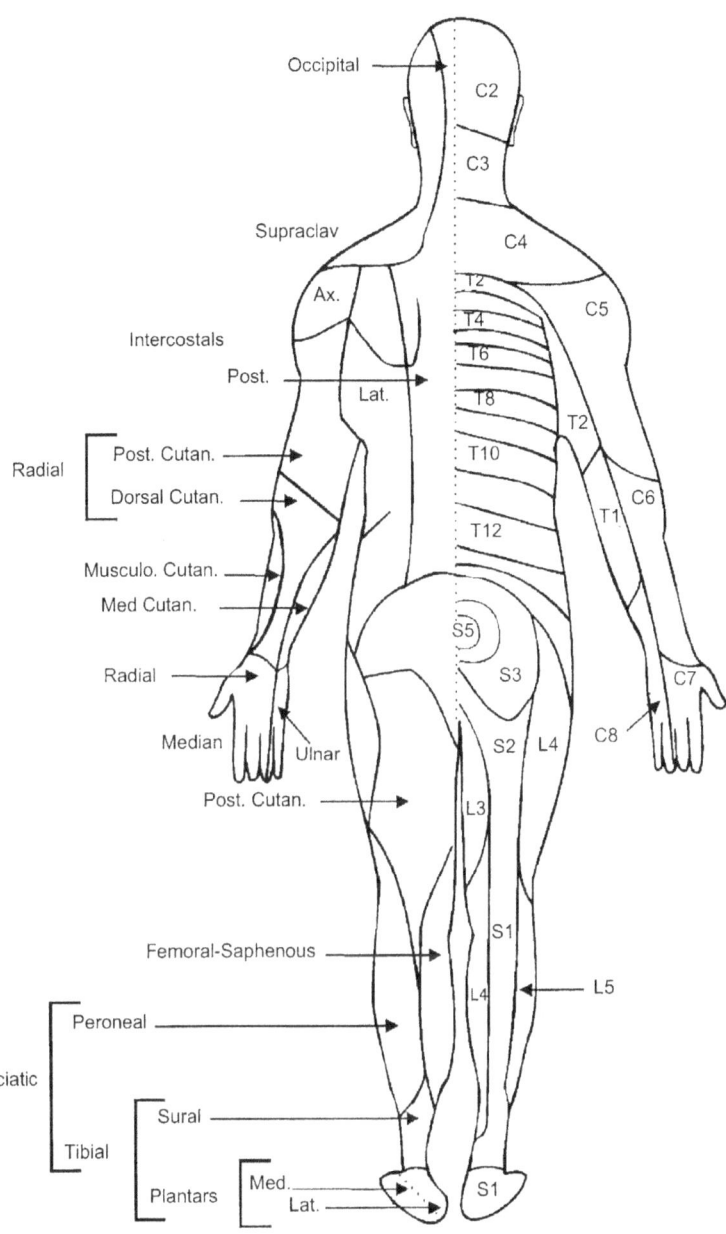

Dermatomes of Cutaneous Innervation, Posterior View
(United States Navy Dive Manual)

Appendices: Wind Chill Chart

Estimated wind speed (in MPH)	Actual Thermometer Reading (°F)										
	50	40	30	20	10	0	-10	-20	-30	-40	-50	-60
	EQUIVALENT CHILL TEMPERATURE (°F)											
Calm	50	40	30	20	10	0	-10	-20	-30	-40	-50	-60
5	48	37	27	16	6	-5	-15	-26	-36	-47	-57	-68
10	40	28	16	4	-9	-24	-33	-46	-58	-70	-83	-95
15	36	22	9	-5	-18	-32	-45	-58	-72	-85	-99	-112
20	32	18	4	-10	-25	-39	-53	-67	-82	-96	-110	-124
25	30	16	0	-15	-29	-44	-59	-74	-88	-104	-118	-133
30	28	13	-2	-18	-33	-48	-63	-79	-94	-109	-125	-140
35	27	11	-4	-21	-35	-51	-67	-82	-98	-113	-129	-145
40	26	10	-6	-21	-37	-53	-69	-85	-100	-116	-132	-148
Wind speeds greater than 40 MPH have little additional effect	LITTLE DANGER Under 5 hours with dry skin. Maximum danger of false sense of security.				INCREASING DANGER Flesh may freeze within one minute.				GREAT DANGER Flesh may freeze within 30 seconds.			
	Danger from freezing of exposed flesh.											
	Immersion foot (trench foot) may occur at any point on this chart.											

Appendices: Heat

Guidelines for Physical Activity

Category	WBGT Index	Nonacclimated Personnel	Acclimated Personnel
I	82°-84.9°F	Use discretion in planning intense physical activity. Limit intensity of work and exposure to sun. Provide constant supervision.	Normal duties
II	85°-87.9°F	Strenuous exercises will be cancelled. Outdoor classes in the sun will be cancelled.	Use discretion in planning intense physical activities. Limit intensity of work and exposure to sun. Provide constant supervision
III	88°-89.9°F	All physical training, strenuous activities, and parades will be cancelled.	Strenuous outdoor activities will be minimized for all personnel with less than 12 weeks of training in hot weather.
IV	≥90°F	Strenuous activities and non-essential duty will be cancelled.	Strenuous activities and non-essential duty will be cancelled.

Fluid Replacement Guidelines for Warm Weather Training

(Applies to average acclimated soldier wearing BDU, Hot Weather)

Heat Category	WBGT Index °F	Easy Work		Moderate Work		Hard Work	
		Work/Rest	Water Intake Q/hr	Work/Rest	Water Intake Q/hr	Work/Rest	Water Intake Q/hr
1	78-81.9	NL	1/2	NL	3/4	40/20 min	3/4
2 Green	82-84.9	NL	1/2	50/10 min	3/4	30/30 min	1
3 Yellow	84-87.9	NL	1/2	40/20 min	3/4	30/30 min	1
4 Red	88-89.9	NL	3/4	30/30 min	3/4	20/40 min	1
5 Black	≥90	50/10 min	1	20/40 min	1	10/50 min	1

- The work/rest times and fluid replacement volumes will sustain performance and hydration for at least 4 hours of work in the specified heat category. Individual water needs will vary ±1/4 qt/hr.
- NL= no limit to work time per hour
- Rest means minimal physical activity (sitting or standing), accomplished in shade if possible.
- **CAUTION: Hourly fluid intake should not exceed 1½ quarts.**
- **Daily fluid intake should not exceed 12 quarts.**
- Wearing body armor adds 5°F to WBGT Index.
- Wearing Mission Oriented Protective Posture (MOPP) overgarment adds 10°F to WBGT Index.

Wet Bulb Globe Temperature (WBGT)

Microclimates:
The WBGT reflects the effects of radiant (sun) energy and the effects of evaporation on effective temperatures. It should be noted that anything that changes these factors changes the WBGT. Standing in the shade on a sunny day is cooler, and being below the crest of a hill, protected from the breeze eliminates the cooling effects of the wind, so it is warmer. Being in a narrow valley by a stream (humid) will be hotter then a WBGT taken in a clearing. Local differences can be extremely important. Do not generalize from a single WBGT.

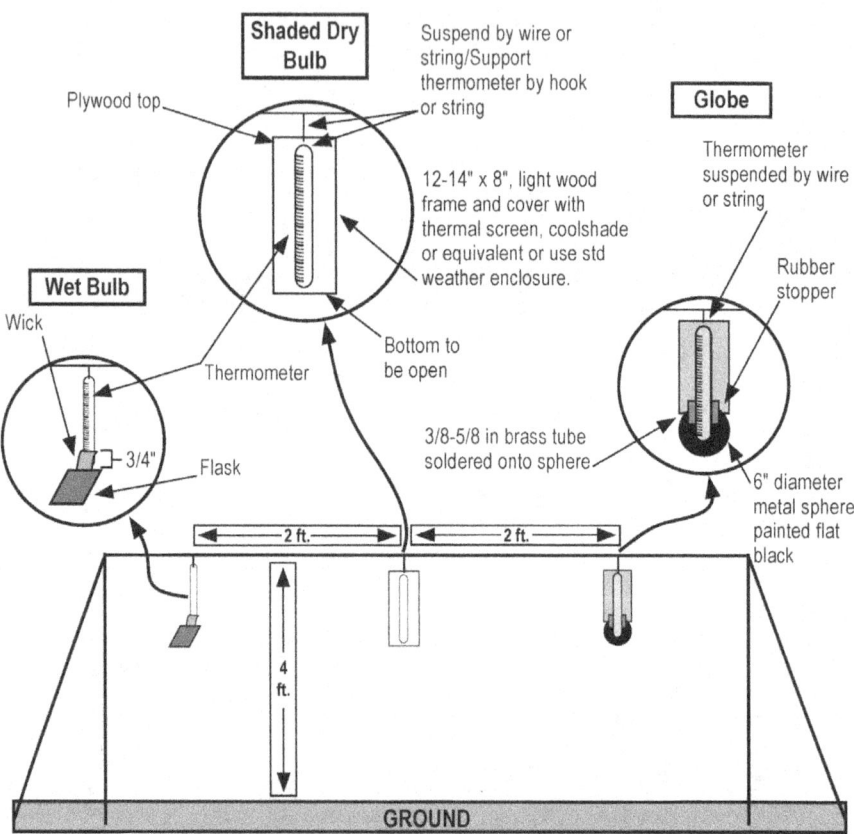

Appendices: Do Not Resuscitate Guidelines

Lt Col John McAtee, USAF, BSC

1. **Do NOT resuscitate** (DNR) casualties under the following circumstances:
   a. Combat (direct fire) situations:
      1) Patient with no pulse, regardless of cause.
      2) Patient with a pulse but no respirations, who cannot be resuscitated without endangering the lives of care givers and/or rescuers.
   b. **Non-combat situations**: Clear the decision not to resuscitate with medical consultants if possible. If it is not possible, do not resuscitate the following casualties:
      1) Victim is obviously dead, characterized by signs such as:
         i. Obvious decomposition (bloating, etc.)
         ii. Body partially consumed by scavengers
         iii. Skin over dependent areas is mottled (dependent lividity)
         iv. Rigor mortis (may resemble severe hypothermia, so check body core temperature)
      2) Victim is decapitated
      3) Victim partially decapitated with no pulse present
      4) Victim is dismembered, or body is fragmented
      5) Victim has an open head injury, with brain matter exposed and no pulse present
      6) Victim has an injury to the trunk with chest contents exposed, and no pulse present
      7) Hypothermia victim 'frozen', e.g., ice formation in the airway
      8) Hypothermia victim with frozen, incompressible chest
      9) Victim has total body burns or body carbonization, no pulse present
      10) Victim has suffered massive blunt trauma (e.g., fall of over 100 feet, etc) and has no pulse
2. Decisions to not initiate resuscitation must be completely documented, including time/date of decision, reason for decision, name of medical consultant (if able to contact), and location of victim (GPS coordinates if possible).
3. **NOTE: The decision not to initiate resuscitation IS NOT a legal declaration of death, unless a qualified physician declares the patient dead.**
4. In a **combat** situation, body recovery should be attempted unless the attempt exposes the rescue team to undue danger. If the body cannot be safely recovered, the location should be noted as accurately as possible (GPS coordinates preferred) for later recovery efforts. If the body has a set of ID tags that can be safely recovered, one should be left with the body, and one should be secured.
5. In **non-combat** situations, attempt body recovery only if it can be accomplished with a minimum of risk to the rescue team. If there is any suspicion of death as a result of foul play or other forensic circumstances (suicide, homicide, neglect, accident, etc.), the body and the area around it should be undisturbed until law enforcement authorities have had an opportunity to examine the scene.
6. In the event of a **military aircraft crash**, do not disturb the scene except to assess and resuscitate any casualties which are not dead (see above). If the casualties must be moved to perform medical treatment, every attempt should be made to record the exact location where the patient was found, and his/her exact position (photographs from multiple angles are helpful). Body recovery may be the responsibility of local law enforcement or military authority, depending on the circumstances and location of the mishap. In most circumstances, it is best to leave the body or bodies in position until investigating authorities arrive and survey the site.

Appendices: Normal Laboratory Values

TEST	NORMAL RANGE	
	Conventional Units	SI Units
Complete Blood Count (CBC)		
Hemoglobin (Hb)	Male: 13.8 - 17.2 g/dL Female: 12.0 - 15.6 g/dL	Male: 138 - 172 g/L Female: 120 - 156 g/L
Hematocrit (Hct)	Male: 41-50% Female: 35 - 46%	Male: 0.41 - 0.50 Female: 0.35 - 0.46
RBC Count	Male: $4.4 - 5.8 \times 10^6/\mu L$ Female: $3.9 - 5.2 \times 10^6/\mu L$	Male: $4.4 - 5.8 \times 10^{12}/L$ Female: $3.9 - 5.2 \times 10^{12}/L$
RBC Indices	Mean corpuscular vol: 78 - 102 fL Mean corpuscular Hb: 27 - 33 pg Mean corpuscular Hb concentration: 32 - 36 g/dL RBC distribution width: $\leq 15\%$	Mean corpuscular vol: 78 - 102 fL Mean corpuscular Hb: 27 - 33 pg Mean corpuscular Hb concentration: 320 - 360 g/L RBC distribution width: ≤ 0.15
WBC Count WBC Differential	$3.8 - 10.8 \times 10^3/\mu L$ Absolute neutrophils: 1500-7800 cells/μL Absolute eosinophils: 50-500 cells/μL Absolute basophils: 0 - 200 cells/μL Absolute lymphocytes: 850 - 4100 cells/μL Absolute monocytes: 200 - 1100 cells/μL	$3.8 - 10.8 \times 10^9/L$ Absolute neutrophils: $1.5 - 7.8 \times 10^9/L$ Absolute eosinophils: $0.05 - 0.55 \times 10^9/L$ Absolute basophils: $0 - 0.2 \times 10^9/L$ Absolute lymphocytes: $0.85 - 4.10 \times 10^9/L$ Absolute monocytes: $0.2 - 1.1 \times 10^9/L$
Platelet Count	$130 - 400 \times 10^3/\mu L$	$130 - 400 \times 10^9/L$
Chem 7		
Carbon dioxide	20 - 32 mmol/L	20 - 32 mmol/L
Chloride	95 - 108 mmol/L	95 - 108 mmol/L
Creatinine	≤ 1.2 mg/dL	≤ 106 μmol/L
Glucose Fasting Random	 < 110 mg/dL 70 - 125 mg/dL	 < 6.1 mmol/L 3.9 - 6.9 mmol/L
Potassium	3.5 - 5.3 mmol/L	3.5 - 5.3 mmol/L
Sodium	135 - 146 mmol/L	135 - 146 mmol/L
Urea Nitrogen, Blood (BUN)	7 - 30 mg/dL	2.5 - 10.7 mmol/L

Lipid Profile

Total Cholesterol, Total	Desirable: < 200mg/dL Borderline-high: 200 - 239mg/dL High: ≥ 240mg/dL	Desirable: < 5.17mmo/L Borderline-high: 5.17 - 6.18mmo/L High: ≥ 6.21mmo/L
High Density Lipo- protein (HDL) Cholesterol	≥ 35 mg/dL "Negative" risk factor: ≥ 60 mg/dL	≥ 0.9 mmol/L "Negative" risk factor: ≥ 1.55 mmol/L
Low density Lipo- protein (LDL) Cholesterol, direct	Desirable: < 130 mg/dL Borderline high: 130 - 159 mg/dL High: ≥ 160 mg/dL	Desirable < 3.36 mmol/L Borderline high: 3.36-4.11 mmol/L High: ≥ 4.14 mmol/L
Triglycerides	< 200 mg/dL	< 2.26 mmol/L
Urinalysis, Complete	Appearance: clear, yellow Specific gravity: 1.001 - 1.035 pH: 4.6 - 8.0 Protein: negative Glucose: negative Reducing substances: negative Ketones: negative Bilirubin: negative Occult blood: negative WBC esterase: negative Nitrite: negative WBC: ≤ 5/high-power field RBC: ≤ 3/high-power field Renal epithelial cells: ≤ 3/high-power field Squamous epithelial cells: none or few/high-power field Casts: none Bacteria: none Yeast: none	

All lab values are used by permission from The Merck Manual of Diagnosis and Therapy, Edition 17, edited by Mark H. Beers and Robert Berkow. Copyright 1999 by Merck & Co., Inc., Whitehouse Station, NJ.

SOF Medical Handbook Abbreviations

ABC- airway, breathing, circulation; atomic, biological, chemical
ABE - acute bacterial endocarditis
ABG - arterial blood gas
A/B ratio - acid/base ratio
ac - before meals
ACD - allergic contact dermatitis
ACL - anterior cruciate ligament
ACLS - advanced cardiac life support
ACTH - adrenocorticotropic hormone
ACVD - acute cardiovascular disease
ADH - antidiuretic hormone (vasopressin)
ADL - activities of daily living
ad lib - as desired
AE - above elbow
A/E - air evacuation
AED - automatic external defibrillator
AFB - acid-fast bacilli
AFib/AFlut - atrial fibrillation/atrial flutter
AFIP - Armed Forces Institute of Pathology
AFMIC - Armed Forces Medical Intelligence Center
AGA - appropriate for gestational age
AGE - arterial gas embolism
AGL - above ground level
A/G ratio - albumin/globin ratio
AHD - atherosclerotic heart disease
AIDS - acquired immune deficiency syndrome
AJ - ankle jerk
AK - above knee
AKA - above-the-knee amputation; also known as
ALL - acute lymphoblastic or lymphocytic leukemia
ALS - amyotrophic lateral sclerosis
AMI - acute myocardial infarction
AML - acute myelocytic/myeloblastic leukemia
AMPLE - allergies, medications, previous injury, last meal, and events preceding injury
AMS - altered mental status; acute mountain sickness
AN - Army Nurse
ANUG - acute necrotizing ulcerative gingivitis
ant - anterior
ante - before
AP - anterior-posterior
AP & Lat - anteroposterior and lateral
ARDS - acute respiratory distress syndrome
AROM - active range of motion

ASA - acetylsalicylic acid (aspirin)
ASAP - as soon as possible
ASD - atrial septal defect
ASHD - arteriosclerotic heart disease
ATA - atmosphere absolute
ATLS - advanced trauma life support
AUB - abnormal uterine bleeding
AV, A-V - arteriovenous; atrioventricular
AVPU - alert, verbal stimulus response, painful stimulus response, unresponsive
BAC - blood alcohol concentration
BAT - blood alcohol test
BBB - bundle branch block
BCG - Bacillus Calmette-Guerin (vaccine)
BCP - birth control pills
BE - barium enema
bicarb - bicarbonate
bid - twice a day
bil or bilat - bilateral
bili - bilirubin
BK - below knee
BKA - below-knee amputation
BLS - basic life support
BM - bowel movement
BMR - basal metabolic rate
BNTI - blind naso-tracheal intubation
BP - blood pressure
BPH - benign prostatic hypertrophy
BPM - beats per minute
BPV - benign positional vertigo
BR - bed rest
BSR - blood sedimentation rate
BSI - blood stream infection; body substance isolation
BUN - blood urea nitrogen
BVM - bag, valve, mask
BW - biological warfare
Bx - biopsy
C - Celsius or centigrade
C1 to C7 - cervical nerves or vertebrae 1 to 7
c - with
Ca - calcium; cancer; carcinoma
CABG - coronary artery bypass graft
CAD - coronary artery disease
CAM - chemical agent monitor
CANA - convulsant antidote for nerve agents

CAT - computerized axial tomography
cath - catheter
CASEVAC - casualty evacuation
CBC - complete blood count
CBR - chemical, biological, radiation
CC - chief or current complaint
cc - cubic centimeter
CDC - Centers for Disease Control
CF - cystic fibrosis
CHF - congestive heart failure
Cl - chloride
CLC - cutaneous larva currens
CLM - cutaneous larva migrans
cm - centimeter
CNS - central nervous system
CO - carbon monoxide
CO_2 - carbon dioxide
c/o - complains of
COPD - chronic obstructive pulmonary disease
CPAP - continuous positive airway pressure
CPD - cephalopelvic disproportion
CPK - creatine phosphokinase
CPR - cardiopulmonary resuscitation
CRAO - central retinal artery occlusion
CRF - chronic renal failure
CRNA - certified registered nurse anesthetist
CS - compartment syndrome
C & S - culture and sensitivity
C-section - cesarean section
CSF - cerebral spinal fluid
CT - clotting time
CVA - cerebrovascular accident; costovertebral angle
CVD - cardiovascular disease
CVP - central venous pressure
CXR - chest x-ray
d - day
DBP - diastolic blood pressure
DC - duty cycle; Dental Corps
D/C - discharge or discontinue
D & C - dilatation and curettage or curettement
DCS - decompression sickness
DIC - diffuse intravascular coagulation
DIP - distal interphalangeal
DJD - degenerative joint disease
DKA - diabetic ketoacidosis
DM - diabetes mellitus
DMARD - disease modifying antirheumatic drug
DMO - Diving Medical Officer
DNA - deoxyribonucleic acid
DNBI - disease/non-battle injuries
DNR - do not resuscitate
DO - Doctor of Osteopathy
DOA - dead on arrival

DOB - date of birth
DOE - dyspnea on exertion
DOT - directly observed therapy
DPN - drops per minute
DT - diphtheria toxoid and tetanus toxoid (for children under 7 years of age)
DTP - diphtheria toxoid, tetanus toxoid, pertussis vaccine
DTR - deep tendon reflexes
DTs - delirium tremens
DVT - deep vein thrombosis
Dx - diagnosis
EAC - external auditory meatus
EBA - emergency breathing apparatus
EBL - estimated blood loss
EBV - Epstein-Barr virus
E. coli - Escherichia coli
EDC - estimated date of confinement
EEG - electroencephalogram
e.g. - for example
EGA - estimated gestational age
EKG; ECG - electrocardiogram
EM - erythema migrans
EMD - electromechanical disassociation
EMG - electromyogram
EMS - emergency medical service
ENT - ear, nose, and throat
EOM - extraocular movement
ESR - erythrocyte sedimentation rate
ESRD - end-stage renal disease
ET - endotracheal tube, Eustachian tube
ETOH - ethyl alcohol
F - Fahrenheit
FB - foreign body
FBS - fasting blood sugar
FDA - Food and Drug Administration
Fe - iron
FEV - forced expiratory volume
FFP - fresh frozen plasma
FHR - fetal heart rate
FHT - fetal heart tone
F Hx - family history
fib - fibrillation
FITT - frequency, intensity, time, type
FP - family practice
fsw - feet of salt water
FT - full term
ft - foot; feet
F/U - follow-up
FUO - fever of unknown or undetermined origin
FVC - forced vital capacity
Fx - fracture
g - gram(s)
ga - gauge

GB - gallbladder
GBS - group B streptococcus
GC - gonococcus; gonococcal
GCA - giant cell arteritis
GCS - Glasgow coma scale
GERD - gastroesophageal reflux disease
GI - gastrointestinal
gm - gram
GPS - global positioning system
GSW - gunshot wound
gt; gtt - drop; drops
GTT - glucose tolerance test
GU - genitourinary
GYN; Gyn - gynecology
H - hydrogen
H_2O - water
H_2O_2 - hydrogen peroxide
HA or H/A - headache
HAA - hepatitis-associated antigen
HACE - high altitude cerebral edema
HAHO - high altitude, high opening (parachute)
HALO - high altitude, low opening (parachute)
HAPE - high altitude pulmonary edema
HAV - hepatitis A virus
Hb; hgb - hemoglobin
HBO - hyperbaric oxygen
HBP - high blood pressure
HBV - hepatitis B virus
HC - head circumference
HCC - hepatocellular carcinoma
HCG - human chorionic gonadotropin
HCl - hydrochloric acid
Hct or HCT - hematocrit
HCV - hepatitis C virus
HDL - high-density lipoprotein
HDV - hepatitis D virus
HE - high explosive
HEENT - head, eyes, ears, nose, throat
HEV - hepatitis E virus
HF - hydrofluoric
HFRS - hemorrhagic fever with renal syndrome
Hg - mercury
HIV - human immunodeficiency virus
HLZ - helicopter landing zone
HN - host nation
H/O - history of
H & P - history and physical
HPF - high power field
HPI - history of present illness
HPS - hantavirus pulmonary syndrome
hr - hour
HR - heart rate
HRO - human relief organization
hs - at bedtime

HSV - herpes simplex virus
ht - height
HTN - hypertension
HTS - hypertonic saline
Hx - history
IAPP - inspection, auscultation, palpation and percussion
IAW - in accordance with
IBS - irritable bowel syndrome
ICD - irritant contact dermatitis
ICP - intracranial pressure
ICRC - International Committee of the Red Cross
ID - infectious disease
I & D - incision and drainage
i.e. - that is
IM - intramuscular
in - inch
inf - inferior
INH - isonicotinic acid hydrazide; isoniazid; isonicotinoylhydrazide
I & O - intake and output
IOP - intraocular pressure
IPPB - intermittent positive pressure breathing
IPV - inactivated polio vaccine
IRM - intermediate restoration material
ITBS - iliotibial band syndrome
IU - international unit
IUCD; IUD - intrauterine contraceptive device
IV - intravenous
IVP - intravenous pyelogram
JOMAC - judgment, orientation, mentation, abstraction, calculation
JVD - jugular vein distention
K - potassium
kg - kilogram
KIA - killed in action
KJ - knee jerk
kL - kiloliter
km - kilometer
KOH - potassium hydroxide
KUB - kidney, ureter, and bladder
KVO - keep vein open
L - liter
lac - laceration
lap - laparotomy
laser; LASER - light amplification by stimulated emission of radiation
lat - lateral
lb - pound
L/B - live birth
LBBB - left bundle branch block
LBP - low back pain
LBW - low birth weight
L & D - labor and delivery

LCP disease - Legg-Calves-Perthes disease
LDL - low-density lipoprotein
LE - lower extremity
LGV - lymphogranuloma venereum
lig - ligament
LLE - left lower extremity
LLL - left lower lobe (of lung)
LLQ - left lower quadrant
LMA - laryngeal mask airway
LMP - left mentoposterior (position of fetus); last menstrual period
L/ M - liters per minute
LMWH - low molecular weight heparin
LOC - loss of consciousness; level of consciousness
LOD - line of duty
LOM - limitation of motion
LP - lumbar puncture
LQ - lower quadrant
LR - Lactated Ringer's
L-S - lumbosacral
LTBI - latent tuberculosis infection
LUL - left upper lobe (of lung)
LUQ - left upper quadrant
LV - left ventricular
lymphs - lymphocytes
m - meter
MA - mortuary affairs
ma - milliampere
maint - maintenance
mAs - milliamps
max - maximum
MC - Medical Corps
mc; mCi - millicurie
mcg - microgram
MCHC - mean corpuscular hemoglobin concentration or count
MCL - mid clavicular line
MCP - metacarpal phalangeal
MDI - multidirectional instability
MEA/DMSO - monoethylamine/dimethylsulfocxide
MEDCAP - medical civic action program
MEDEVAC - medical evacuation
MEB - medical evaluation board
med - medicine or medication
mEq - milli equivalent
MES - medical equipment set
MET - meteorological; mission essential task
MG - myasthenia gravis
Mg - magnesium
mg - milligram
MHz - megahertz
MI - myocardial infarction
MIA - missing in action

min - minute
misc - miscellaneous
MK1 - MARK 1
ml - milliliter
MMSE - mini mental status exam
mm - millimeter
mm Hg - millimeters of mercury
MOI - mechanism of injury
Mono - mononucleosis
monos - monocytes
MOPP - mission oriented protection posture
MOTT - mycobacteria other than tuberculosis
MRI - magnetic resonance imaging
MS - multiple sclerosis
MSL - mean sea level
MTF - medical treatment facility
Na - sodium
NA - not applicable
NAD - no acute distress
NAIAD - nerve agent immobilized enzyme alarm and detector
NaPent - sodium Pentothal
NBC - nuclear, biological, chemical
neg - negative
NG - nasogastric
NGU - non-gonococcal urethritis
NKA - no known allergies
NKDA - no known drug allergies
nl; norm - normal limits
NLT - not later than
No - number
NPA - nasopharyngeal airway
NPH insulin - neutral protamine Hagedorn insulin
NPO - nothing by mouth
NS - nervous system; normal saline
NSAID - non-steroidal anti-inflammatory drug
NSN - national stock number; nonstandard number
NTG - nitroglycerin
NTM - non-tuberculous mycobacterial
N & V - nausea and vomiting
NWB - non-weight bearing
NSS - normal saline solution
O_2 - oxygen
OB - obstetrics
OB-GYN - obstetrics and gynecology
OBS - organic brain syndrome
OC - oral contraceptive
OCOKA - Observation and fire, Concealment and cover, Obstacles, Key terrain, and Avenues of approach
OD - overdose; right eye
O & P - ova and parasites
OPA - oropharyngeal airway
OS - left eye

OPV - oral poliomyelitis vaccine
OPQRST - onset, provokes, quality, radiates, severity, time
OR - operating room
ORS - oral rehydration solution
os, per os - mouth; by mouth
OSA - obstructive sleep apnea
OTC - over the counter (drugs)
OU - each eye
oz - ounce
PAC - premature atrial contractions
PAM Cl - pralidoxime chloride
Pap test - Papanicolaou's test
PASG - pneumatic anti shock garment
pc - after meals
PE - physical examination; pulmonary embolus
PEA - pulseless electrical activity
PEEP - positive end-expiratory pressure
PERRLA - pupils equal, round, and react to light and accommodation
PE tubes - pressure-equalizing tubes
PFB - pseudofolliculitis barbae
PFS - patellofemoral syndrome
PFSH - past, family, social history
PFT - pulmonary function test
PH - past history
PHF - potentially hazardous foods
PHTLS - pre-hospital trauma life support
PI - present illness
PID - pelvic inflammatory disease
PIP - proximal interphalangeal
Pit - Pitocin
PKU - phenylketonuria
PM - preventive medicine
PMCC - patient movement control center
PMH - past medical history
PMI - point of maximum impulse
PNM - polymorphonuclear neutrophil leukocytes
PND - paroxysmal nocturnal dyspnea
PNS - peripheral nerve stimulator
PO - post-operative
po - by mouth; orally
pO2 - partial pressure oxygen
POD - post-operative day
POIS - pulmonary over inflation syndrome
pos - positive
postop - postoperative
POW - prisoner of war
PP - post partum; pulsus paradoxicus
PPB - positive pressure breathing
PPD - purified protein derivative
ppm - parts per million
PPV - positive-pressure ventilation
Pre med - premedication

pre-op - preoperative
prn - as needed
PROM - premature rupture of membranes; passive range of motion
PROMED - Program for Monitoring Emerging Diseases
PSI - pounds per square inch
PSVT - paroxysmal supraventricular tachycardia
PT - physical therapy; prothrombin time
PTL - preterm labor
PTSD - post-traumatic stress disorder
PTT -
PTB - primary tuberculosis
PUD - peptic ulcer disease
PULHES - physical profile factors: P--physical capacity or stamina; U--upper extremities; L--lower extremities; H--hearing and ears; E--eyes; S--psychiatric
PVC - premature ventricular contractions
q - every
QC - quality control
qd - every day
qh - every hour
q2h, q3h, and so on - every 2 hours, every 3 hours, and so on
qid - four times a day
qt - quart
qty - quantity
r - roentgen
R - right
RA - rheumatoid arthritis
Ra - radium
RBC - red blood cells or corpuscles
RDA - recommended dietary allowance
RDS - respiratory distress syndrome
REM - rapid eye movement
REF - reference
RPM - revolutions per minute
RPR - rapid plasma reagin
Rh factor - Rhesus blood factor
RICE - rest, ice, compression, elevation
RLL - right lower lobe (of lung)
RLQ - right lower quadrant
RML - right middle lobe (of lung)
RMSF - Rocky Mountain spotted fever
R/O - rule out
ROM - range of motion
ROS - review of systems
RPR - Reiter protein reagin
RR - respiratory rate
RUL - right upper lobe (of lung)
RUQ - right upper quadrant
Rx - prescription; treatment; take
S-A; SA node - sinoatrial node

SBE - subacute bacterial endocarditis
SBP - systolic blood pressure
SC - subcutaneous
SCI - spinal cord injury
SCFE - slipped capital femoral epiphysis
SCUBA - self contained underwater breathing apparatus
Sed rate - erythrocyte sedimentation rate
SEV - surface equivalent
SF - Special Forces; Standard Form
SG - specific gravity; Surgeon General
SGA - small for gestational age
SLE - systemic lupus erythematosus
SLR - short leg raise
SOAP - S-subject O-objective A-assessment P-plans (progress note format)
SOB - shortness of breath
SODARS - special operations debrief and retrieval system
SO - special operations
SOF - special operations forces
SOP - standing operating procedures
S/P - status post
sp. gr. - specific gravity
SPK - superficial punctate keratitis
spp. - species
SQ - subcutaneous
SSRI - selective serotonin reuptake inhibitor
S & S - signs and symptoms
staph - staphylococcus
STAT - immediately
STB - super tropical bleach
STD - sexually transmitted disease
strep - streptococcus
STS - serologic test for syphilis
Sx - signs; symptoms
T - temperature
T & A - tonsillectomy and adenoidectomy
tab - tablet
TAH - total abdominal hysterectomy
TB - tuberculosis
TBD - to be determined; to be developed
tbs; tbsp - tablespoon
TBSA - total body surface area
Td - tetanus toxoid and diphtheria toxoid (for older children and adults)
temp - temperature
TIA - transient ischemic attacks
tid - three times a day
TIG - tetanus immune globulin
TIVA - total intravenous anesthesia
TKO - to keep open
TM - tympanic membrane
TMJ - temporomandibular joint

TNTC - too numerous to count
TPR - temperature, pulse, and respiration
TSE - testicular self exam
TSH - thyroid-stimulating hormone
tsp - teaspoon
TURP - transurethral resection, prostate
TVH - total vaginal hysterectomy
Tx - treatment
U - unit
UA - urinalysis
UE - upper extremity
UGI - upper gastrointestinal
UNHCR - United Nations High Commissioner for Refugees
UQ - upper quadrant
URI - upper respiratory infection
URQ - upper right quadrant
US - ultrasound
USAID - United States Agency for International Development
UTI - urinary tract infection
UV - ultraviolet
UW - unconventional warfare
VC - Veterinary Corps
VDRL - venereal disease research laboratory test
V-Fib - ventricular fibrillation
VHF - viral hemorrhagic fevers
VIG - Vaccinia Immune Globulin
vit - vitamin
Vol - Volume
VS - vital sign(s)
vs - against
V-Tach - ventricular tachycardia
VZV - varicella-zoster virus
WBC - white blood cell
WBGT - wet bulb globe temperature
WHO - World Health Organization
WIA - wounded in action
wk - week
WMD - weapons of mass destruction
WNL - within normal limits
WP - white phosphorus
wt - weight
W/U - workup
X -times
y/o -year old
yr -year
xr - x-ray

Comments and suggestions about the Special Operations Forces Medical Handbook can be directed to:

Mr. Robert Clayton
Medical R&D
Bldg. 153
ATTN: SOCS-SG
USSOCOM
7701 Tampa Blvd
MacDill, AFB, FL 33621-5323

Or by email to:

MedTruth@socom.mil

INDEX

Index

A vitamin deficiency, 5–138*t*
ABC protocol mnemonic, 8–1
   with chemical burns, 7–18
   for electrical or lightning injury, 7–27
   for hypoxia, 6–9
   for marine animal bites, 6–17—6–18
   for trauma management, 7–1—7–3
Abdomen
   acute pain in
      assessment of, 3–1
      causes of, 3–3*t*
      field management goals with, 3–2
      follow-up, 3–2
      objective signs of, 3–4*t*
      patient education on, 3–2
      signs of, 3–1
      symptoms of, 3–1
      treatment plan, 3–2
   aortic aneurysm of, 3–6
   examination of
      in secondary trauma survey, 7–4
      for urinary tract problems, 4–87
   pain in
      in acute organic intestinal obstruction, 4–86
      in appendicitis, 4–71
      in bacterial food poisoning, 4–79
      non-food-poisoning etiologies in, 4–80
      in tapeworm infections, 5–48
   tenderness in, 5–35
   trauma to
      in adrenal insufficiency, 4–27
      in horses, 5–133
ABO grouping, 8–59
   in crossmatching, 8–60
Abrasions, debridement of, 8–26
Abscess
   Bartholin's gland, 3–52—3–53
      incision and drainage of, 3–53—3–55
   breast, 3–8
      drainage of, 3–9
      incision and drainage of, 3–9—3–10
   inflamed, removal of, 8–27
   periapical, 5–9, 5–13
      untreated, 5–13—5–14
   periodontal, 5–15
   removal of, 8–26
   tooth, draining of, 5–26
Acclimatization, 6–47—6–48
Acetaminophen
   for acute mountain sickness, 6–38
   for avulsed tooth, 5–14
   for breast engorgement, 3–9
   for centipede bites, 4–56
   for common cold and flu, 4–11
   in hepatitis A, 5–72
   for infectious mononucleosis, 5–80
   for joint pain, 3–63
   for mastitis, 3–8
   overdose of, 3–58
   for pain control, 8–38, 8–40
   replacing aspirin, 4–85
   for temporomandibular joint dislocation, 5–19
   for venomous snake bite, 5–145
   in yellow fever, 5–70
Acetazolamide
   for acute mountain sickness, 6–38
   for high altitude cerebral edema, 6–39
   for high altitude pulmonary edema, 6–41
   for red eye, 3–25
   side effects of, 6–38
   for vision loss, 3–23
Acetest, 8–42
Acetic acid, 6–15
Achalasia, 6–22
Achilles stretching, 5–1
Acid-alcohol reagent, 8–47
Acid burns, 7–19
Acid-fast bacilli, Ziehl-Neilson stain for, 8–47
Acid-fast bacteria, fecal, 8–51
Acid suppression, 4–85
ACLS. *See* Advanced cardiac life support
Acne
   differential diagnosis of, 4–65
   pustules in, 4–38
Acoustic neuroma, 3–20
Acquired immune deficiency syndrome, 5–77.
      See also Human immunodeficiency
      virus
   differential diagnosis of, 3–112*t*
Acral lentiginous melanoma, 4–68
Acromioclavicular joint, 3–70
   anesthetic injection into, 3–71
   capsulitis of, 3–71
Activated charcoal
   for anaphylactic shock, 7–11
   for poisoning, 5–142
Acute respiratory distress syndrome
   adenoviruses in, 5–64
   alleviating or aggravating factors, 4–24
   assessment of, 4–25

causes of, 4–24
differential diagnosis of, 5–100
follow-up, 4–25
in pancreatitis, 4–82
patient education for, 4–25
phases of, 4–24
signs of, 4–25
symptoms of, 4–24
treatment of, 4–25
Acute stress disorder, 3–5
Acyclovir
for Bell's palsy, 4–38
for herpes simplex, 5–29
for herpes simplex encephalitis, 3–86
for herpes zoster, 4–47
for herpetic oral lesions, 5–16
for meningitis, 4–36
Addison's disease, 4–27
Adenopathy
in animals, 5–126
suppurative, 5–93
Adenovirus infection
assessment of, 5–64
differential diagnosis of, 3–112t, 4–11
follow-up for, 5–65
patient education for, 5–65
risk factors for, 5–64
signs of, 5–64
symptoms of, 5–64
treatment for, 5–64
vaccination for, 4–13
Adrenal insufficiency
alleviating or aggravating factors in, 4–27
assessment of, 4–27—4–28
causes of, 4–27
follow-up, 4–28
patient education, 4–28
signs of, 4–27
sub-acute and chronic, 4–27
symptoms of, 4–27
treatment of, 4–28
Advanced cardiac life support
drugs
cardiac resuscitation, 4–7
against chemical blood agents, 6–52
protocol, for decompression sickness, 6–5
Advil, 4–84
ADVON (advanced party), 1–3
AEIOU-TIPS mnemonic, 5–140
Aeromonas food poisoning, 4–79
Aerospace medicine
barodontalgia in, 6–32—6–33
barosinusitis in, 6–33—6–34
barotitis in, 6–34—6–35
decompression sickness in, 6–35—6–37
hypoxia in, 6–31—6–32
Aerozoin, 5–8
AF Form 3899, 1–22
African trypanosomiasis
assessment of, 5–52

follow-up for, 5–53
patient education for, 5–52—5–53
signs of, 5–52
symptoms of, 5–52
transmission of, 5–52
treatment for, 5–52
zoonotic disease considerations in, 5–53
Afrin
for barosinusitis, 6–34
for barotitis, 6–35
for common cold and flu, 4–11
for ear barotrauma, 6–2
AGE. *See* Arterial gas embolism
AIDS. *See* Acquired immune deficiency syndrome; Human immunodeficiency virus
Air conditioner lung, 4–18
Air embolism
with blood transfusion, 8–19
treatment of, 8–20
Air evacuation
CASEVAC, marking and lighting of landing zone, 1–31, 1–32
patient loads in, 1–33
phone list for, 1–33
Air Force Patient Movement Precedence, 1–22
Airflow obstruction, 4–21
Airway
assessment of
during intubation, 8–4
for trauma, 7–5
compromise of
differential diagnosis of, 7–24
with massive hemoptysis, 7–24, 7–25f
management of
assessment for, 8–1
with chemical burns, 7–18
equipment for, 8–1
indications for, 8–1
Naval special warfare combat trauma AMAL items for, 1–21
precautions in, 8–4
procedure for, 8–1—8–4
for trauma, 7–1
obstruction of
dyspnea with, 3–116
treatment of, 3–117
patency of, 8–1
securing, for poisoning, 5–141
support of, for venomous snake bite, 5–146
Airway/breathing management, M5 item list, 1–18
Albendazole
for ascariasis, 5–34
for clonorchiasis, 5–36
for cutaneous larva migrans, 5–42
for enterobiasis, 5–37
for hookworm, 5–42
for tapeworm infections, 5–49
for trichuriasis, 5–51
Albumin, 7–22
Albuterol

for allergic pneumonitis, 4–19
for anaphylactic shock, 7–11
for asthma, 4–21
Alcohol
dependence on as disqualifying condition for diving, 6–21—6–22
in gastritis, 4–81
in pancreatitis, 4–82
withdrawal from, 5–150t
Alcohol-related fatigue, 3–29
Alcohol-related syncope, 3–118
Alcoholic stupor, 6–44
Aldomet, 4–82
Aleve, 4–84
Alkaline wash, 3–79
Allergic contact dermatitis, 4–69
Allergic pneumonitis
assessment of, 4–18
causes of, 4–18
follow-up, 4–19
patient education for, 4–19
signs of, 4–18
symptoms of, 4–18
treatment of, 4–19
Allergic reactions
to blood transfusion, 8–19
treatment of, 8–20
differential diagnosis of, 6–31, 6–36
generalized, 7–10—7–11
Allergic rhinitis
differential diagnosis of, 4–11
treatment of, 4–11
Allergy-related cough, 3–14
Allopurinol, 3–63
Aloe, 4–59
Alpha-blockers
drug interactions of, 4–91
for prostatitis, 3–82
side effects of, 3–82
for urinary incontinence, 4–91
Alphaviruses, 5–66
Altered mental status (AMS), 1–1
Altitude illness, 8–36. See also High altitude illness; Mountain sickness, acute
Alum, 5–8
Aluminum-containing antacids, 7–18
Aluminum crown, tooth, 5–12
Alupent (metaproterenol), 4–22
Alveolar nerve blockade, 5–23f
Amantadine, 4–12
Ambien (zolpidem), 3–30
Ambu bag, 8–7
Ambulance points, 6–55
Amebiasis, 5–33
Amebic colitis, 4–71
Amebic dysentery, 3–18
Amebic liver abscess, 5–104
Amelanotic melanoma, 4–68
differential diagnosis of, 4–66
American trypanosomiasis
assessment of, 5–53
follow-up for, 5–54
patient education for, 5–53
symptoms and signs of, 5–53
transmission of, 5–53
treatment for, 5–53
zoonotic disease considerations in, 5–54
Aminoglycoside, 7–19
5-Aminosalicyclic acid, 4–82
Amnestics, dosing guidelines for, 5–156t
Amoxicillin
for disseminated gonococcal infection, 4–40
for leptospirosis, 5–87
for peptic ulcer disease, 4–85
for rat bite fever, 5–97
Amoxicillin/clavulanate (Augmentin)
for breast abscess drainage, 3–10
for human and animal bites, 7–9
for mastitis, 3–8
for tympanic membrane rupture, 7–26
Amoxil, 3–21
Amphotericin B
for blastomycosis, 5–59
for candidiasis, 5–58
for coccidioidomycosis, 5–61
for histoplasmosis, 5–62
liposomal, for leishmaniasis, 5–43
for paracoccidioidomycosis, 5–63
side effects of, 5–58, 5–61
Ampicillin
for acute diarrhea, 3–19
for bacterial vaginosis, 3–48t, 3–49
for epididymitis, 3–84
for Fournier's gangrene, 3–78
for meningitis, 4–36
for oroya fever, 5–93
for preterm labor infection, 3–94
for prostatitis, 3–81
for urinary tract infections, 4–94
for urolithiasis, 4–92
Ampicillin/sulbactam
for abnormal uterine bleeding, 3–41
for electrical or lightning injury, 7–27
for human and animal bites, 7–9
Amputation, traumatic, 1–1
AMS. See Altered mental status (AMS)
Anal pruritus, 3–114
Analgesia, 3–62—3–63
Analgesics
for barodontalgia, 6–32
for barotrauma, 6–4
dosing guidelines for, 5–156t
for equine colic, 5–133
Anaphylactic shock, 7–10
assessment of, 7–10
differential diagnosis of, 7–13
follow-up for, 7–11
patient education for, 7–11
symptoms and signs of, 7–10
treatment for, 7–10—7–11

Ancef, 3-100
Ancylostoma
　brazilense, 5-41
　caninum, 5-41
　duodenale, 5-41
Anemia
　with abnormal uterine bleeding, 3-39
　acute and chronic, 4-8
　assessment of, 4-9
　causes of, 4-8—4-9
　with copper deficiency, 5-139t
　fatigue with, 3-30t
　follow-up, 4-10
　in hookworm, 5-42
　macrocytic, 5-137t
　signs of, 4-9
　symptoms of, 4-9
　treatment of, 4-9—4-10
Anesthesia
　in alveolar nerve blockade, 5-23
　for cesarean section, 3-100
　dental, 5-20—5-23
　facial injection of, 5-21f
　in lingual nerve blockade, 5-23
　long buccal injection of, 5-24f
　for mandibular extraction, 5-21—5-23
　for maxillary extraction, 5-20
　palatal injection of, 5-21, 5-22f
　regional, 5-159—5-175
　topical, for red eye assessment, 3-24
　total intravenous
　　complications of, 5-156—5-157
　　contraindications to, 5-156
　　dosing guidelines for, 5-156t
　　equipment for, 5-155—5-156
　　field techniques for, 5-157—5-158t
　　fundamentals of, 5-155
　　indications for, 5-155
　　pre- and intraoperative system checks
　　　for continuous infusion in,
　　　5-157t
　　precautions in, 5-156—5-157
　　procedures in, 5-156
Anesthetics
　for digital blocks, 5-163t
　local
　　local infiltration of area, 5-159
　　for pain control, 8-39
Aneurysm, ruptured, 3-3t
Angina pain, 4-1
Animal bites
　assessment of, 7-8
　follow-up for, 7-9
　incidence of, 7-7
　patient education for, 7-9
　in rabies, 5-81
　signs of, 7-8
　symptoms of, 7-7—7-8
　treatment of, 7-8—7-9
Animals

　care and management of, 5-130
　diseases in, 5-132—5-136
　normal physiologic values in, 5-130t
　obstetrics for, 5-130—5-131
　postmortem exam of, 5-126
　restraint of, 5-128—5-130
Anisocytosis, 8-55
Ankle
　nerve blockade of, 5-168—5-174
　pain in
　　assessment of, 3-76
　　causes of, 3-76
　　follow-up, 3-77
　　patient education for, 3-77
　　signs of, 3-76
　　symptoms of, 3-76
　　treatment of, 3-76—3-77
　sprains
　　follow-up, 3-77
　　treatment of, 3-76
Ankylosing spondylitis
　in low back pain, 3-6
　patient education for, 3-64
Anovulation, 3-41
Ant stings, 4-57—4-59
Antacids
　for chemical burns, 7-18
　for self-limiting abdominal pain, 3-1
Antemortem exam, veterinary, 5-123
Anthrax, 5-31
　assessment of, 5-91
　follow-up for, 5-92
　inhalational, 6-56—6-57
　patient education for, 5-91
　required vaccination for, 5-105t
　signs of, 5-90—5-91
　symptoms of, 5-90
　transmission of, 5-90
　treatment for, 5-91
　zoonotic disease considerations in, 5-92
Antibiotics, 6-15
　for abnormal uterine bleeding, 3-41
　for acute diarrhea, 3-19
　for acute mastitis in animals, 5-135
　for acute organic intestinal obstruction,
　　4-86—4-86
　for acute respiratory distress syndrome,
　　4-25
　for ankle pain, 3-77
　for appendicitis, 4-72
　for avulsed tooth, 5-15
　for biological warfare agents, 6-56
　in candida vaginitis, 3-49
　for cesarean section, 3-100, 3-102
　for chancroid, 5-29
　for cholecystitis, 4-78
　for congestive heart failure, 4-4
　for cough, 3-15
　dental, 5-19—5-20
　for empyema, 4-17

for febrile infections, 3–33f
for foot rot in caprines, 5–134
for Fournier's gangrene, 3–78
for herpetic oral lesions, 5–16
for human and animal bites, 7–9
for localized osteitis, 5–18
M5 item list for, 1–19
for mastitis, 3–8, 3–9
for meningitis, 4–36
ointment, for impetigo contagiosa, 4–44
for pancreatitis, 4–82
for pericoronitis, 5–17
for peritonitis, 4–84
for phimosis, 3–78
for pleural effusion, 4–15
for pneumonia, 4–12, 4–13
postthoracostomy, 8–9
prophylaxis
    with bladder catheterization, 8–34—8–35
    for chemical burns, 7–19
for prostatitis, 3–81—3–82
for septic joint, 3–63
for trauma patient, 7–5
for typhoid fever, 5–104
for underwater blast injury, 6–19
for untreated periapical abscess, 5–14
for urinary incontinence, 4–91
for urinary tract infections, 4–94
for urolithiasis, 4–92
for venomous fish bites, 6–15
for yaws, 4–63
Anticoagulants, 4–24
Antidepressants, 3–31
Antiemetics
  for acute organic intestinal obstruction, 4–86
  for bacterial food poisoning, 4–80
  for cholecystitis, 4–78
  for gastritis, 4–81
  for peritonitis, 4–84
  for urolithiasis, 4–92
Antifungals
  for candidiasis, 5–58
  M5 item list for, 1–19
Antihistamines
  for bed bug bites, 4–55
  for dizziness, 3–21
  for mites, 4–59
  for pediculosis, 4–63
  for pruritus, 3–115
  for sea urchin bites, 6–15
  for serous otitis media, 3–21
  for sleep disturbance with fatigue, 3–30
  for swimming dermatitis, 4–54
Antihypertensives, 4–5
Anti–inflammatory agents
  for heel spur syndrome, 5–2
  for low back pain, 3–7
  for shoulder pain, 3–71
Antimotility drugs, 3–19
Antimycobacterial drugs, 5–55

Antipyretics, 4–78
Antiseptics, 8–34
Antitussives, 6–53
Antivenin
  for venomous fish bites, 6–15
  for venomous snake bite, 5–145
Antivert, 3–21
Anuria, 4–88
Anxiety, 3–2
  assessment of, 3–5
  in chest pain, 3–11
  dyspnea with, 3–116
  follow-up, 3–5
  free-floating, 3–2
  memory loss with, 3–86
  patient education, 3–5
  in pruritus, 3–115
  signs of, 3–2—3–5
  symptoms of, 3–2
  treatment for, 3–5
Aortic dissection
  chest pain with, 3–10
    quality of, 3–10—3–11
  in hypertensive emergency, 4–5
  in pericarditis, 4–6
  treatment for, 3–12
  vital signs suggesting, 3–11
Aphthous ulcers, 5–16—5–17
Apnea
  aggravating factors in, 4–26
  assessment of, 4–26
  causes of, 4–25
  follow-up, 4–26—4–27
  patient education for, 4–26
  risk factors for, 4–25
  signs of, 4–26
  symptoms of, 4–26
  treatment of, 4–26
Appendectomy
  for appendicitis, 4–72
  emergency field, 4–72
    cautions in, 4–77—4–78
    equipment for, 4–72—4–73
    goal for, 4–72
    postoperative orders for, 4–77
    procedures for, 4–73—4–77
Appendicitis
  acute abdominal pain in, 3–3t
  acute pelvic pain with, 3–42
  assessment of, 3–42, 4–71
  follow-up for, 4–72
  incidence of, 4–70
  patient education for, 4–72
  in peritonitis, 4–83
  signs of, 4–71
  symptoms of, 4–70—4–71
  treatment for, 4–72
Apresoline, 3–109
Arboviral encephalitis
  assessment of, 5–67

causes of, 5-66
follow-up for, 5-67
patient education for, 5-67
seasonal variation in, 5-66
signs of, 5-67
symptoms of, 5-67
treatment for, 5-67
zoonotic disease considerations in, 5-68
Arch supports
 for heel spur syndrome, 5-1
 for stress fractures of foot, 5-7
ARDS. See Acute respiratory distress syndrome
Areola, hyperpigmented, 3-87
Argentinian hemorrhagic fever, 6-61
Armed Forces Medical Intelligence Center (AFMIC), fever diagnosis criteria, 3-31
Arrhythmias
 with blood transfusion, 8-19
 malignant
  with cardiac resuscitation, 4-8
  syncope with, 3-117
 pericardiocentesis for, 8-13
 treatment of, 8-20
 with venomous snake bite, 5-144—5-145
Arterial gas embolism
 assessment of under fire, 1-1
 with blast injury, 7-26
 in carbon dioxide poisoning, 6-12
 chest pain with, 3-10
 in decompression sickness, 6-36
 differential diagnosis of, 6-5, 6-9, 6-10, 6-36
 disqualifying condition for diving, 6-22
 portable hyperbaric chamber for, 8-36
 in pulmonary over inflation syndrome, 6-7, 6-8
 treatment of, 6-6f, 7-24
 with underwater blast injury, 6-19
Arthralgia, 3-62
Arthritis. See also Monoarthritis; Oligoarthritis; Polyarthritis
 of ankle, 3-76
 assessment of, 3-61—3-62
 gonococcal, 5-98
 heel, 5-1
 hip, 3-73—3-74
 in joint pain, 3-59
 non-infectious, 3-63
 septic, 3-74
 traumatic, 3-59
 treatment of, 3-62—3-63
Arthropod-borne viruses, 5-66—5-68
Artificial blood, 7-15f
Artificial tears, 4-38
ASA
 for chest pain, 3-12
 for pericarditis, 4-6
 side effects of, 4-7
Ascariasis, 5-34
 in acute organic intestinal obstruction, 4-86

differential diagnosis of, 5-49
Ascaris, 5-34
 differential diagnosis for, 4-71
 suum, 5-31
Ascites
 in hepatitis, 3-57—358
 in jaundice, 3-58
Ascorbic acid deficiency, 5-137t
Aseptic necrosis
 hip pain with, 3-73
 treatment of, 3-74
Asphalt burn, 7-19
Aspiration
 for joint pain, 3-63
 for knee pain, 3-75
Aspirin
 for acute mountain sickness, 6-38
 for acute rheumatic fever, 5-99
 in asthma, 4-21
 for avulsed tooth, 5-14
 for chest pain, 3-12, 3-13
 in children, 4-12
 in gastritis, 4-81
 for headache, 3-56
 in hemorrhagic fevers, 6-61
 for myocardial infarction, 4-3
 for pain control, 8-39
 in peptic ulcer disease, 4-84, 4-85
 in pulmonary embolus, 4-24
 for trench fever, 5-93
 in yellow fever, 5-70
Assessment under fire, 1-1
Asthma
 assessment of, 4-20
 causes of, 4-19
 cough with, 3-14, 3-15
 differential diagnosis for, 6-41
 dyspnea with, 3-116
 follow-up, 4-21
 medications for in palpitations, 3-110
 patient education for, 4-21
 signs of, 4-20
 symptoms of, 4-19
 treatment of, 3-117, 4-20—4-21
 triggers of, 4-19
Asthma attacks, 4-19
Atarax (hydroxyzine)
 for contact dermatitis, 4-70
 for pruritus, 3-115
 for psoriasis, 4-65
 for sleep disturbance with fatigue, 3-30
Ataxia, 3-20
Atelectasis
 differential diagnosis of, 4-13
 treatment of, 3-117
Athlete's foot, 4-49—4-51
Ativan, 5-148
ATLS protocols
 for electrical or lightning injury, 7-27
 for hip dislocation, 3-69

for patellar dislocation, 3–69
Atopic allergy, 3–113
Atopic dermatitis, 3–114
Atrial fibrillation
    differential diagnosis of, 3–111
    treatment of, 3–111
Atrophy, skin, 4–39
Atropine
    for chemical nerve agents, 6–52
    for chest pain, 3–12
    for heat stroke, 6–51
    for syncope, 3–118
Atrovent (ipratropium)
    for common cold and flu, 4–11
    for COPD, 4–22
AUB. *See* Uterine bleeding, abnormal
Augmentin (amoxicillin/clavulanate)
    for breast abscess drainage, 3–10
    for human and animal bites, 7–9
    for mastitis, 3–8
    for prostatitis, 3–81
    for purulent otitis media, 3–21
    for urinary tract infections, 4–93, 4–94
Auspitz sign, 4–64
Autoimmune disease
    in adrenal insufficiency, 4–27
    in aphthous ulcers, 5–16
    in diabetes mellitus, 4–28
Automatic external defibrillator, 4–7
Aveeno Colloidal Oatmeal bath, 3–114
AVPU scale, 7–3
Axillary blockade
    anesthetics for, 5–174*t*
    assessment of, 5–175
    contraindications and complications of, 5–175
    equipment for, 5–174
    indications for, 5–174
    procedure for, 5–174—5–175
Azathioprine, 4–82
Azithromycin
    for abnormal uterine bleeding, 3–41
    for acute rheumatic fever, 5–99
    for cat scratch disease, 5–93
    for chancroid, 5–29
    for fever, 3–34*f*
    for granuloma inguinale, 5–29
    for knee pain, 3–75
    for Lyme disease, 5–88
    for lymphogranuloma venereum, 5–29
    for mastitis, 3–9
    for meningitis, 4–37
    for pelvic inflammatory disease, 3–51
    for pneumonia, 4–13
    for typhoid fever, 5–104
    for typhus, 5–84
    for urethral discharges, 5–27
Azole antifungals, 5–61
Aztreonam
    for acute abdominal pain, 3–2

    for acute organic intestinal obstruction, 4–87
    for cholecystitis, 4–78
    for peritonitis, 4–84

B1 deficiency, 5–137*t*
B2 deficiency, 5–137*t*
B3 deficiency, 5–137*t*
B6 deficiency, 5–137*t*
B12 vitamin deficiency, 4–8
Babesia
    bovis, 5–35
    divergens, 5–35
    microti, 5–35
    protozoa, 5–34
Babesiosis
    assessment of, 5–35
    causes of, 5–34
    follow-up for, 5–35
    patient education for, 5–35
    signs of, 5–35
    symptoms of, 5–35
    treatment for, 5–35
    zoonotic disease considerations in, 5–35
Bacillus
    anthracis, 5–90
        identification of colonies of, 8–50
    cereus, in food poisoning, 4–79
    subtilis enzymes, 4–18
Bacitracin
    for chemical burns, 7–19
    for friction foot blisters, 5–8
    ointment
        for eye injury, 3–28
        for red eye, 3–25
    in suturing, 8–23
    for venomous fish bites, 6–15
Bacterial infections
    acute rheumatic fever, 5–98—5–99
    anthrax, 5–90—5–92
    bartonellosis, 5–92—5–93
    brucellosis, 5–94
    differential diagnosis of, 3–112*t*
    ehrlichiosis, 5–95
    food-borne, 5–108
    plague, 5–95—5–97
    rat bite fever, 5–97—5–98
    skin
        cutaneous tuberculosis, 4–44—4–45
        erysipelas, 4–41—4–42
        gonococcal, 4–40
        impetigo contagiosa, 4–43—4–44
        leprosy, 4–45
        meningococcemia, 4–41
        staphylococcal scalded skin syndrome, 4–42—4–43
    streptococcal, 5–99—5–100
    tetanus, 5–100—5–102
    tularemia, 5–102—5–103
    typhoid fever, 5–104—5–105

Bacterial vaginosis
    assessment of, 3–47—3–49
    causes of, 3–47
    diagnosis and treatment of, 3–48t
    follow-up, 3–49
    infections causing, 3–48t
    patient education for, 3–49
    signs of, 3–47
    symptoms of, 3–47
    treatment plan for, 3–49
Bactrim
    for acute diarrhea, 3–19
    in pancreatitis, 4–82
Baffle grease trap, 5–116f
Bag-valve-mask (BVM) system, 5–155, 8–1
Bagassosis, 4–18
Baker's cyst, 3–74
Balance, loss of, 6–1
Balanitis, 3–77—3–78
    candidal, 3–78
    patient education for, 3–79
    treatment of, 3–78
Banamine, 5–135
Bancroftian filariasis, 5–39
Barbiturates
    overdose of, 5–150, 5–150t
    withdrawal from, 5–151t
Barodontalgia, 5–18, 6–3—6–4
    assessment of, 6–32
    cause of, 6–32
    differential diagnosis of, 6–33
    follow-up for, 6–33
    patient education for, 6–33
    signs of, 6–32
    symptoms of, 6–32
    treatment for, 6–32—6–33
Barosinusitis
    assessment of, 6–33—6–34
    cause of, 6–33
    differential diagnosis of, 6–32
    follow-up for, 6–34
    patient education for, 6–34
    signs of, 6–33
    symptoms of, 6–33
    treatment for, 6–34
Barotitis
    assessment of, 6–35
    cause of, 6–34
    externa, 6–35
    follow-up for, 6–35
    media, 6–35
    patient education for, 6–35
    signs of, 6–35
    symptoms of, 6–34
    treatment for, 6–35
Barotrauma
    dental, 6–3—6–4
    differential diagnosis of, 6–35
    dizziness in, 3–20
    to ears, 6–1—6–2

    external, 6–1
    inner, 6–1
    middle, 6–1
    TEED classification of, 6–2
    GI, 6–3—6–4
    sinus, 6–3—6–4
    skin, 6–3—6–4
Barracudas, 6–17
Barrel filter grease trap, 5–115f
Barrel incinerator, 5–116f
Bartholin's gland, 3–52
    cyst/abscess of
        assessment of, 3–52
        causes of, 3–52
        follow-up, 3–53
        incision and drainage of, 3–53—3–55
        patient education for, 3–53
        signs of, 3–52
        symptoms of, 3–52
        treatment plan for, 3–52
    marsupialization of, 3–54f
Bartonella
    bacilliformis, 5–93
    henselae, 5–93
Bartonellosis
    assessment of, 5–93
    cause of, 5–92
    follow-up for, 5–93
    patient education for, 5–93
    signs of, 5–92—5–93
    symptoms of, 5–92
    treatment for, 5–93
    zoonotic disease considerations in, 5–93
Basal cell carcinoma
    assessment of, 4–66
    follow-up for, 4–67
    incidence of, 4–66
    metastasis of, 4–66
    prevention of, 4–67
    signs of, 4–66
    symptoms of, 4–66
    treatment of, 4–66—4–67
Basal cell skin cancer, 4–39
Base burns, 7–19
Basic medical skills, procedures in, 8–1—8–40
Basilar skull fracture, 7–4
Basophilic stippling, 8–56
Bather's eruption, 4–54
Battle fatigue, 5–147—5–148
Battle's sign, 7–4
BCG vaccine, 4–44
Beard, dandruff of, 4–49—4–51
Bechet's syndrome, 5–16
Beclomethasone dipropionate, 4–20
Bed bug bites, 4–55
Bedrest
    after cesarean section, 3–102
    for epididymitis, 3–84
    for low back pain, 3–7
    for meningitis, 4–37

Bee sting, 4-57—4-59
Beef tapeworm disease, 5-49
Bees, 4-57
    killer, 4-57, 4-59
Bell's palsy
    assessment of, 4-37
    course of, 4-37
    patient education for, 4-38
    return evaluation for, 4-38
    signs of, 4-37
    symptoms of, 4-37
    treatment for, 4-37—4-38
Benadryl
    for allergic reaction to blood transfusion, 8-20
    for anaphylactic shock, 7-11
    for bed bug bites, 4-55
    for contact dermatitis, 4-70
        of penis, 3-78
    for hemolytic reaction, 8-20
    for sleep disturbance with fatigue, 3-30
Bench press, 3-72
Bends. *See* Decompression sickness
Benign paroxysmal vertigo
    differential diagnosis of, 3-20—3-21
    dizziness in, 3-20
Benign positional vertigo, 3-20—3-21
Benzathine penicillin
    for acute rheumatic fever, 5-99
    for syphilis, 5-30
Benzidazole, 5-53
Benzocaine, 5-16
Benzocaine/menthol, 4-58
Benzodiazepam, 5-145
Benzodiazepines
    abuse of, 5-150
    for anxiety, 3-5
    for elbow dislocation, 3-67
    intoxication of, 5-150*t*
    for mania, 3-17
    for pain control, 8-38, 8-39
    for psychosis and delirium, 5-152
    withdrawal from, 5-151*t*
Benzoin, tincture of, 5-8
Benzyl benzoate, 4-60
Beriberi, dry, 5-137*t*
Beta-agonists, 4-20
Beta-blockers, 3-12
Beta-lactam, 4-13
Betadine
    after appendectomy, 4-77
    for cesarean section, 3-101
    in digital block of finger or toe, 5-159
    in dorsal slit procedure, 3-79
    for friction foot blisters, 5-8
    for ingrown toenail, 5-3
    in IV infusion for animals, 5-132
    in local infiltration of area, 5-159
    in needle thoracostomy, 8-8
    ointment, for chemical burns, 7-19
    in skin mass removal, 8-27

    in suturing, 8-25
    in tube thoracostomy, 8-8
Betamethasone, 4-65
Bicarbonate
    for blood transfusion reaction, 8-19
    for hemolytic reaction, 8-20
Biliary colic pain, 4-78
Biliary disease, pruritus of, 3-114
Biliary obstruction
    bilirubin metabolism in, 3-57
    in jaundice, 3-58
Bilirubin
    abnormal metabolism of, 3-57
    normal metabolism of, 3-57
Bilirubin-Icotest procedure, 8-42
Bilirubinuria, 8-43
Biological agents, 6-56—6-61
Biological warfare, 6-56
    biological agents in, 6-56—6-61
Bipolar disorder
    differential diagnosis of, 5-152
    symptoms of, 3-16—3-17
Bird breeder's lung, 4-18
Birth control pills. *See* Oral contraceptives
Bisacodyl suppositories, 3-13
Bithionol
    for fascioliasis, 5-39
    side effects of, 5-39
Black flies, 4-53
Black widow spider bite, 4-60
Blackheads, 4-38
Bladder
    antispasmodics for, 3-82
    catheterization of
        equipment for, 8-34
        indications for, 8-33—8-34
        precautions for, 8-35
        procedure for, 8-34—8-35
    infections of, 4-93
    lacerations of in cesarean section, 3-102
    sudden emptying of, 8-36
    suprapubic aspiration of, 8-35—8-36
Bladder tap. *See* Bladder, suprapubic aspiration of
Blast injuries, 7-23. *See also* Underwater blast
        injury
    assessment of, 7-24
    in congestive heart failure, 4-4
    under fire
        assessing, 1-1
        care of, 1-1
    signs of, 7-23—7-24
    symptoms of, 7-23
    treatment for, 7-24—7-26
Blastomyces dermatitidis, 5-59
Blastomycosis, 5-31
    assessment of, 5-59
    cause of, 5-59
    follow-up for, 5-60
    patient education for, 5-60
    signs of, 5-59

symptoms of, 5-59
treatment for, 5-59—5-60
zoonotic disease considerations in, 5-60
Bleached rubber syndrome, 4-70
Bleeding. *See* Hemorrhage
Bleeding disorder, snakebite-induced, 5-145—5-146
Bleeding ulcer, 4-85
Blepharitis
  differential diagnosis of, 3-25
  red eye with, 3-24
  treatment of, 3-25
Blister agents, 6-53—6-54
Blisters
  friction, 5-7—5-8
  genitourinary, 4-89
Blood
  components of, 4-8
  Giemsa stain for parasites of, 8-47—8-48
  glucose level of. *See* Hypoglycemia
  glucose testing of, 4-29
Blood agents, 6-52—6-53
Blood collection system, 8-18
Blood diseases, 4-8
  anemia, 4-8—4-10
  tests for, 4-8
Blood donor, 8-18
Blood fluke infection, 5-46—5-47
Blood plasma, 4-8
Blood pressure
  in abdominal pain, 3-1
  diastolic, 4-5
  systolic, 4-6
  in trauma patients, 7-3
Blood products
  handling of, 8-16
  safe, 4-9
Blood smear
  for African trypanosomiasis, 5-52
  Giemsa stain of, 8-47—8-48
  for trichenellosis, 5-50
  Wright's stain of, 8-46—8-47
Blood supply, crossmatching of, 8-60—8-62
Blood transfusion
  for acute diarrhea, 3-19
  for acute pelvic pain, 3-43
  in American trypansomiasis, 5-53
  for anemia, 4-9
  equipment for, 8-14
  in field, 8-18
  for hantavirus, 5-69
  indications for, 8-14
  monitoring during, 8-17
  precautions for, 8-18
  procedure for, 8-15—8-18
  reactions to, 8-19—8-21
    precautions with, 8-21
Blood types, 4-9
  type O, 8-19
Blood typing

  in blood transfusion, 8-15—8-16
  confirmation tests for, 8-59
BLS/ACLS protocol, electrical or lightning injury, 7-27
Blunt injuries, under fire, 1-1
Body heat, preserving, 7-13
Body lice, 4-62—4-63
  in relapsing fever, 5-89
Body surface area, 7-20f
Body temperature
  rapid reduction of, 6-50
  regulation of, 6-47
Boiling, water treatment, 5-119—5-120
Boils, 4-52
Bolivian hemorrhagic fever, 6-61
Bone marrow
  failure of, 4-8
  suppression of, 4-8—4-9
Borborygmi, 4-86
Borrelia
  burgdorferi, 5-88
  recurrentis, 5-89
Bottled water, 5-119
Botulism, 5-31
  in biological warfare, 6-57—6-58
  cause of, 6-57—6-58
  patient education for, 6-58
  symptoms and signs of, 6-58
Boutonneuse fever, 5-83
  differential diagnosis of, 3-112t
  treatment of, 5-84
Bowel obstruction, 4-71
  acute organic
    assessment of, 4-86
    causes of, 4-86
    follow-up for, 4-87
    patient education for, 4-87
    signs of, 4-86
    symptoms of, 4-86
    treatment for, 4-86—4-87
Bowel perforation, 4-83
Bowel sounds, 4-86
Bowel-wall stretching injuries, 7-26
Bowen's disease, 4-67
Box jellyfish, 6-13—6-14
Bradycardia
  differential diagnosis of, 3-111
  with myocardial infarction, 4-1
  syncope with, 3-117, 3-118
Brain
  abscess of, 4-36, 5-80
  memory loss with injury of, 3-86
  tumors of
    headache with, 3-55
    manic and depressive symptoms with, 3-16
Breakbone fever, 5-65
Breast
  abscess of
    assessment of, 3-8
    causes of, 3-9

drainage of, 3–9
  incision and drainage of
    equipment for, 3–9
    follow-up, 3–10
    procedure in, 3–9—3–10
  cancer of, 3–8
  engorgement of
    assessment of, 3–8
    treatment of, 3–9
  enlargement of in pregnancy, 3–87
  examination of, 2–4
  problems of
    abscess incision and drainage for, 3–9—3–10
    mastitis, 3–7—3–9
Breastfeeding
  bacterial vaginosis treatment during, 3–49
  for breast engorgement, 3–9
  with mastitis, 3–7, 3–8
Breathing
  assessment of, 8–1
    with chemical burns, 7–18
    for trauma, 7–5
  management of, 7–2
Breech delivery
  cesarean section in, 3–100
  precautions, 3–96
  presentation, 3–96, 3–97f
  procedures in, 3–96, 3–97—99f
  risks of, 3–96
  Wigand maneuver for, 3–99f
Brief psychotic disorder, 5–152
Bronchial spasm, 4–19
Bronchiectasis, 4–22
Bronchitis
  chest pain with, 3–10
  in COPD, 4–21
  cough with, 3–14, 3–15
  fatigue with, 3–30t
Bronchodilators
  for allergic pneumonitis, 4–19
  for asthma, 4–21
  for COPD, 4–22
Bronchogenic carcinoma, 4–22
Bronchospasm, 7–10
Brown recluse spider bite, 4–60
Brucella
  abortus, 5–94
  canis, 5–31, 5–94
  mellitensis, 5–94
  suis, 5–31, 5–94
Brucellosis, 5–31
  assessment of, 5–94
  diagnostic algorithm for, 3–34f
  differential diagnosis of, 5–104
  follow-up for, 5–94
  patient education for, 5–94
  signs of, 5–94
  symptoms of, 5–94
  test for, 8–45
  transmission of, 5–94

  treatment for, 5–94
  zoonotic disease consideration in, 5–94
Brugia
  malayi, 5–39
  timori, 5–39
Bruising, 5–138t
Bubonic plague, 5–96
Buddy watch, suicide prevention, 5–149
Bullae, 4–38
  in staphylococcal scalded skin syndrome, 4–42
BUMED-21, 6–22
Bunion, 5–4
  assessment of, 5–5
  follow-up for, 5–5
  patient education for, 5–5
  signs of, 5–4—5–5
  symptoms of, 5–4
  treatment for, 5–5
Bunyaviruses, California group of, 5–66
Bupivacaine
  in axillary blockade, 5–174t
  for digital block of finger or toe, 5–163t
  with epinephrine (Marcaine), 5–20
  local infiltration of, 5–159
  for pain control, 8–39
Burley restraint method, 5–129
Burn out latrine, 5–113f
Burns
  assessment of, 7–17
  from blister agents, 6–53
  caustic cocktail, 6–20—6–21
  classification of, 7–17
  compartment syndrome management in, 8–30
  with electrical or lightning injury, 7–26, 7–27
  follow-up for, 7–22
  partial-thickness, 7–17
  patient education for, 7–22
  symptoms and signs of, 7–17
  treatment for, 7–17—7–18
    for chemical injuries, 7–19
    for fluid resuscitation, 7–19—7–22
    primary, 7–18—7–19
Bursitis
  differential diagnosis of, 5–1
  hip, 3–73
Butorphanol
  for equine colic, 5–133
  for pain control, 8–39

C vitamin deficiency, 5–137t
Cabot's ring, 8–56
Café au lait spots, 4–38
Cafergot, 3–56
Caffeine
  for COPD, 4–22
  for headache, 3–56
  in palpitations, 3–110
Calabar swellings, 4–52

IN-11

Calamine lotion
  for blister agents, 6–53
  for mites, 4–59
Calciferol, 5–138t
Calcium
  deficiency of, 5–139t
  supplement of, 3–74
Calcium diethylenetriaminepentaacetic acid (DPTA), 6–62
Calcium gluconate
  for black widow spider bite, 4–60
  for hydrofluoric acid burn, 7–19
  for preeclampsia, 3–109
  for yellow fever, 5–70
Calcium hydroxide, 5–13
California Mastitis Test, 5–135
Callus
  foot, 5–5
    follow-up for, 5–6
    patient education for, 5–6
    signs of, 5–6
    treatment for, 5–6
    versus plantar warts, 5–4
Calymmatobacterium granulomatis, genital, 5–28
Cameco Quik Stain, 8–46—8–47
  for blood disorders, 4–8
Cameco Wright-Giemsa stain, 8–47
Campylobacter, food poisoning, 4–79, 4–80
Campylobacteriosis, 5–31
Cancer
  in acute organic intestinal obstruction, 4–86
  breast, 3–8
  pruritus with, 3–114
  of testis and scrotum, 3–79—3–80
Candida
  albicans, 5–57—5–58
  vaginitis, 3–47, 3–49, 5–58
    assessment of, 3–50
    diagnosis and treatment of, 3–48t
    follow-up for, 3–50
    patient education for, 3–50
    risk factors for, 3–49
    signs of, 3–49—3–50
    symptoms of, 3–49
    treatment plan for, 3–50
Candidal infection
  of male genitals, 3–78
  penile, 3–78
Candidiasis. *See also* Intertrigo; Thrush
  assessment of, 5–58
  cause of, 5–57—5–58
  differential diagnosis of, 4–65
  esophageal, 5–58
  follow-up for, 5–59
  oropharyngeal, 5–58
  patient education for, 5–58—5–59
  signs of, 5–58
  symptoms of, 5–58
  treatment for, 5–58
  vaginal, 5–58
  wet mount and KOH preparation for, 8–43—8–44
Canker sores, 5–16—5–17
Canning, meat, 5–128
Canteen, individual, 5–119
Caprine, foot rot in, 5–133—5–134
Carbon dioxide absorption device, 8–37
Carbon dioxide poisoning
  assessment of, 6–12
  in diving, 6–12
  follow-up for, 6–13
  patient education for, 6–13
  symptoms and signs of, 6–12
  treatment for, 6–12
Carbon monoxide poisoning
  assessment of, 6–11
  differential diagnosis of, 6–9, 6–37, 6–39
  in diving, 6–11
  follow-up for, 6–12
  patient education for, 6–11—6–12
  symptoms and signs of, 6–11
  treatment for, 6–11
Carboxyhemoglobin, 8–9
Cardiac arrest, 8–19
Cardiac monitoring
  for myocardial infarction, 4–3
  with venomous snake bite, 5–144—5–145
Cardiac pain, 4–78
Cardiac resuscitation
  equipment for, 4–7
  indications for, 4–7
  precautions, 4–8
  procedure for, 4–7
  for sudden cardiac death, 4–7
Cardiac risk factors, 4–1
Cardiac shock, 7–13
Cardiac symptoms
  cardiac resuscitation, 4–7—4–8
  congestive heart failure, 4–3—4–4
  hypertensive emergency, 4–5—4–6
  pericarditis, 4–6—4–7
Cardiac system, acute myocardial infarction of, 4–1—4–3
Cardiac tamponade
  dyspnea with, 3–116
  treatment of, 3–117
Cardiopulmonary arrest, 4–8
  with electrical injury, 7–26
  from hemorrhage, 7–3
Cardiopulmonary resuscitation (CPR), 4–7
  for hypothermia, 6–44
  precautions, 4–8
Cardiovascular examination, 2–3—2–4
Carditis, 5–99
Cardura (doxazosin)
  for prostatitis, 3–82
  for urinary incontinence, 4–91
Care under fire
  assessment in, 1–1
  signs, 1–1
  symptoms, 1–1
  treatment plan, 1–1—1–2

types of injuries, 1–1
Caries. See Dental caries
Carotenoids, 5–138t
Carotid massage, 3–111
Cartilage tear, knee, 3–59
CASEVAC, with fixed winged aircraft, 1–31
   marking and lighting of landing zone
      daytime, 1–31
      nighttime, 1–32
Casting, cow, 5–129
Casts
   for ankle injury, 3–77
   splitting of, 8–31
Casualty decontamination station set up, 6–54—6–55
Cat hole latrine, 5–111f
Cat scratch disease, 5–92—5–93
   diagnostic algorithm for, 3–34f
   differential diagnosis of, 3–112t, 5–103
Caterpillar sting, 4–57
Catheterization, bladder, 8–33—8–35
Cattle
   obstetrics for, 5–130—5–131
   restraint of, 5–129—5–130
Cauda equina syndrome
   in low back pain, 3–6
   treatment of, 3–7
Caustic cocktail chemical burn
   assessment of, 6–21
   with diving, 6–20
   follow-up for, 6–21
   patient education for, 6–21
   signs of, 6–21
   symptoms of, 6–20
   treatment for, 6–21
Cavity tooth varnish, 5–13
Cefadyl, 5–20
Cefazolin, 3–8
Cefixime
   for knee pain, 3–75
   for urethral discharges, 5–27
Cefmetazole
   for acute abdominal pain, 3–2
   for acute organic intestinal obstruction, 4–86
   for peritonitis, 4–84
Cefotaxime
   for airway obstruction, 3–117
   for disseminated gonococcal infection, 4–40
   for hip pain, 3–74
   for meningococcemia, 4–41
   for pancreatitis, 4–82
   for urinary tract infections, 4–94
Cefotetan
   for abnormal uterine bleeding, 3–41
   for acute abdominal pain, 3–2
   for acute organic intestinal obstruction, 4–86
   for appendicitis, 4–72
   for pelvic inflammatory disease, 3–51
   for peritonitis, 4–84
   for trauma patient, 7–5
Cefoxitin
   for acute abdominal pain, 3–2
   for acute organic intestinal obstruction, 4–86
   for blast injury, 7–26
   for pelvic inflammatory disease, 3–51
   for peritonitis, 4–84
Ceftizoxime
   for disseminated gonococcal infection, 4–40
   for hip pain, 3–74
Ceftriaxone (Rochephin)
   for abnormal uterine bleeding, 3–41
   after thoracostomy, 8–9
   for Bartholin's gland abscess or cyst, 3–52
   for blast injury, 7–26
   for chancroid, 5–29
   for disseminated gonococcal infection, 4–40
   for electrical or lightning injury, 7–27
   for empyema, 4–17
   for epididymitis, 3–84—3–85
   for febrile infections, 3–33f
   for granuloma inguinale, 5–29
   for hip pain, 3–74
   for knee pain, 3–75
   for Lyme disease, 5–88
   for lymphogranuloma venereum, 5–29
   for mastitis, 3–9
   for meningitis, 4–36, 4–37
   for meningococcemia, 4–41
   for pelvic inflammatory disease, 3–51
   for pneumonia, 4–13
   for septic joint, 3–63
   for shoulder pain, 3–71
   for streptococcal infections, 5–100
   for syphilis, 5–30
   for tularemia, 5–103
   for typhoid fever, 5–104
   for urethral discharges, 5–27
   for urinary tract infections, 4–94
Cefuroxime
   for airway obstruction, 3–117
   for empyema, 4–17
Cefuroxime axetil, 4–40
Celestone, 5–2
Cellulitis
   differential diagnosis of, 5–100
   in erysipelas, 4–41
   of male genitals, 3–78
      treatment of, 3–78
   in mastitis, 3–7
   periorbital, 3–26
Centipede bites, 4–55—4–56
   assessment of, 4–56
   follow-up for, 4–56
   patient education for, 4–56
   signs of, 4–56
   symptoms for, 4–56
   treatment for, 4–56
Central nervous system
   disease of, 3–116
   injury of with anxiety, 3–5
   trauma to, 3–55

Central nervous system stimulants, 4–26
Central retinal artery occlusion
 differential diagnosis of, 3–23
 treatment of, 3–23
Central retinal vein occlusion
 in acute vision loss, 3–22
 differential diagnosis of, 3–23
 treatment of, 3–23
Cephalexin, 3–8
Cephalosporin
 for chemical burns, 7–19
 for pelvic inflammatory disease, 3–51
 for pneumonia, 4–13
Cephalothin, 4–4
Cephazolin, 3–74
Cephotoxin, 7–9
Cercarial dermatitis, 4–54
Cerebellar function, 2–4
Cerebral edema, high altitude, 6–37, 6–38—6–40
Cerebrospinal fluid leak, 6–22
CERTEC bag
 for high altitude cerebral edema, 6–40
 for high altitude pulmonary edema, 6–42
CERTEC chamber, 8–36
Cervical burn injury, 7–18
Cervical carcinoma
 with abnormal uterine bleeding, 3–41
 pelvic examination for, 3–37
Cervical cultures, 3–38
 for Bartholin's gland abscess or cyst, 3–52
 for chronic pelvic pain, 3–46
 for pelvic inflammatory disease, 3–51
Cervical meningismus, 4–36
Cervical spine injury, primary survey for, 7–1
Cervicitis, 3–41
Cervix, bimanual examination of, 3–38—3–39
Cesarean section, 3–100
 for breech delivery, 3–96
 for eclampsia, 3–109
 equipment for, 3–100
 indications for, 3–100
 postoperative orders for, 3–102
 precautions, 3–102
 surgical procedure, 3–101—3–102
Cetaphil lotion
 for contact dermatitis, 4–70
 for pruritus, 3–114
Chagas' disease. See American trypanosomiasis
Chagoma, 5–53
Chancroid
 in genital ulcers, 5–28
 treatment of, 5–29
Charcoal, activated. See Activated charcoal
Chemical burns
 alternate treatment of, 7–19
 caustic cocktail, 6–20—6–21
 primary treatment of, 7–18—7–19
Chemical nerve agents, 1–1
Chemical weapons of mass destruction, 6–51
 blister agents, 6–53—6–54
 blood agents, 6–52—6–53
 nerve agents, 6–52
Chemosis, 3–28
Chest auscultation
 for common cold and flu, 4–10
 for pleural effusion, 4–14
Chest-eschar syndrome, 7–19
Chest examination, 2–4
 in secondary trauma survey, 7–4
Chest pain
 alleviating or aggravating factors in, 3–11
 assessment of, 3–12
 associated symptoms with, 3–11
 cardiac resuscitation and, 4–8
 causes of, 3–10
 coronary risk factors with, 3–11
 with dyspnea, 3–115
 follow-up, 3–12—3–13
 life-threatening causes of, 3–10
 with myocardial infarction, 4–1
 patient education for, 3–12
 in pleural effusion, 4–14
 in pulmonary embolus, 4–23
 signs of, 3–11—3–12
 symptoms of, 3–10—3–11
 treatment for, 3–12
Chest tube drainage, 4–17
Chest tube thoracostomy, 4–15
Chest wall splinting, 7–2
Chest wound, pericardiocentesis for, 8–12
Chest x-ray
 for allergic pneumonitis, 4–18
 for COPD, 4–22
 for inhalational anthrax, 6–57
 for pneumonic plague, 6–58
 for pulmonary embolus, 4–23
Cheyletiellosis, 5–31
Chickenpox, 6–59
Chief complaint, 2–2
Chig-a-rid, 4–59
Chiggers, 4–59
 in rickettsial infections, 5–83
 toxicants for, 4–60
Chilblains, 6–44
 in trenchfoot, 6–47
Chinese liver fluke. See Clonorchiasis
Chinococcosis, 5–50
Chlamydia
 cultures of, 3–37, 3–38
 pneumoniae, 4–14
 psittaci, 5–31
 trachomatis, urethral discharges of, 5–26
 in urethral discharges, 5–26
Chlor-Floc tablets, 5–120
Chloramphenicol
 for febrile infections, 3–33f
 for oroya fever, 5–93
 for plague, 5–96
 for Rocky Mountain spotted fever, 5–84

for scrub typhus, 5-84
Chlorhexidine gluconate, 4-43
Chlorhexidine (Peridex), 5-16
Chlorine water treatment, 5-119
Chloroprocaine axillary blockade, 5-174t
Chloroquine, 5-44, 5-45
Chlorpromazine
　for mania, 3-17
　for psychosis and delirium, 5-153
Chokes
　in aerospace medicine, 6-36
　differential diagnosis for, 6-10
Cholangitis
　ascending, 4-78
　treatment of, 3-58
Cholecystectomy, 4-78
Cholecystitis
　acute
　　abdominal pain in, 3-3t
　　assessment of, 4-78
　　follow-up for, 4-79
　　incidence of, 4-78
　　patient education for, 4-79
　　signs of, 4-78
　　symptoms of, 4-78
　　treatment for, 4-78
　in peritonitis, 4-83
Cholera, 3-18
Choline magnesium trisalicylate, 8-38
Chromium deficiency, 5-139t
Chronic obstructive pulmonary disease (COPD)
　with apnea, 4-25
　assessment of, 4-22
　causes of, 4-21
　dyspnea with, 3-115, 3-116
　fatigue with, 3-30t
　follow-up, 4-22
　patient education for, 4-22
　signs of, 4-22
　symptoms of, 4-21
　treatment of, 3-117, 4-22
Chronic pain syndromes, 8-37
Ciclopirox, 4-50
Cimetidine
　for anaphylactic shock, 7-11
　for chemical burns, 7-18
　for chest pain, 3-12
　for peptic ulcer disease, 4-85
　for verruca plana, 4-49
Cimex lectularius, 4-55
Cipro
　for epididymitis, 3-84, 3-85
　for prostatitis, 3-81
　in testis torsion treatment, 3-83
　for urinary incontinence, 4-91
　for urinary tract infections, 4-93, 4-94
Ciprofloxacin
　for acute diarrhea, 3-19
　for anthrax, 5-91
　for biological warfare agents, 6-56

　for blast injury, 7-26
　for chancroid, 5-29
　for disseminated gonococcal infection, 4-40
　for electrical or lightning injury, 7-27
　for eye injury, 3-28
　for febrile infections, 3-33f
　for fever, 3-34f
　for granuloma inguinale, 5-29
　for hip pain, 3-74
　for human and animal bites, 7-9
　for inhalational anthrax, 6-57
　for knee pain, 3-75
　for lymphogranuloma venereum, 5-29
　for meningitis, 4-37
　for meningococcemia, 4-41
　for plague, 5-96
　for pneumonic plague, 6-59
　for Q fever, 5-86
　for red eye, 3-25
　for streptococcal infections, 5-100
　for tularemia, 6-60
　for tympanic membrane rupture, 7-26
　for typhoid fever, 5-104
　for typhus, 5-84
　for underwater blast injury, 6-19
　for urethral discharges, 5-27
Ciprofloxin, 3-78
Circulation
　assessment of
　　with chemical burns, 7-18
　　for trauma, 7-5—7-6
　management of
　　M5 item list for, 1-18
　　in trauma, 7-2—7-3
Circulatory overload
　with blood transfusion, 8-19
　treatment of, 8-20
Circumcision
　for male genital inflammation, 3-79
　for paraphimosis, 3-78
　procedure for, 3-79
Cirrhosis
　in hepatitis B, 5-72
　jaundice in, 3-58
Clam digger's itch, 4-54
Clarithromycin
　for Lyme disease, 5-88
　for peptic ulcer disease, 4-85
Clavicle fractures, 3-71
Clavulanic acid, 4-40. *See also* Amoxicillin/
　　clavulanate; Ticarcillin/clavulanate
Cleaning procedures, 5-110—5-111
Clear Lice Egg remover, 4-63
Clenched fist injury, 7-7
　signs of, 7-8
Cleocin (clindamycin), 5-19, 5-20
Clindamycin (Cleocin)
　for abnormal uterine bleeding, 3-41
　for babesiosis, 5-35
　for bacterial vaginosis, 3-48t, 3-49

IN-15

for cholecystitis, 4–78
in dentistry, 5–19, 5–20
for empyema, 4–17
for human and animal bites, 7–9
for malaria, 5–45
for mastitis, 3–8, 3–9
for preterm labor infection, 3–94
for streptococcal infections, 5–100
Clinical process
chief complaint in, 2–2
history of present illness, 2–2
medical examination in, 2–2
medical/family history, 2–3
physical examination, 2–3—2–5
review of systems, 2–3
SOAP format in, 2–1—2–2
Clinitest, 8–42
Clinoril, 4–84
Clobestasol, 4–65
Clofazimine, 4–45
Clonidine, 4–5
Clonorchiasis, 5–35
assessment of, 5–36
follow-up for, 5–36
patient education for, 5–36
signs of, 5–36
symptoms of, 5–35
treatment for, 5–36
zoonotic disease considerations in, 5–36
Clonorchis sinensis, 5–35
Clostridium
botulinum, 6–57
in food poisoning, 4–79
perfringens
in food poisoning, 4–79
identification of colonies of, 8–50
tetani bacteria, 5–100
Clotrimazole
for candida vaginitis, 3–48t, 3–50
for candidiasis, 5–58
for dermatophyte infections, 4–50
for trichomonas, 3–48t
for vaginal trichomonas, 5–31
Cloxacillin, 3–74
Clue cells, wet mount, 8–44
Cluster headache, 3–55
Coagulation factors, 4–8
Coagulopathy, 3–41
Coban, 5–8
Cobras, 5–143
Coccidioides immitis, 5–60
Coccidioidomycosis
assessment of, 5–61
causes of, 5–60
follow-up for, 5–61
patient education for, 5–61
risk factors for, 5–60
signs of, 5–60—5–61
symptoms of, 5–60
treatment for, 5–61

Codeine, 5–150
for acute diarrhea, 3–19
for chest pain, 3–12
for common cold and flu, 4–11
for cough, 3–15
for joint pain, 3–63
for pain control, 8–39
for tetanus, 5–101
Coelenterates, 6–13—6–14
treatment for, 6–15
Colchicine, 3–63
Cold illness/injury, 6–42—6–47
non-freezing, 6–44—6–47
Cold sores, 5–16
Cold test, 5–20
Colic, equine, 5–132—5–133
Coliforms, 5–135
Colloids, 7–15t
Colonies, identification of, 8–49
Color vision testing, 3–23
Coma, 4–31
Combat/jungle litter, Naval special warfare combat trauma AMAL, 1–21
Combat resiliency, 5–148
Combat stress reaction, 5–147—5–148
Combivir (zidovudine/lamivudine), 5–78
Comedone, 4–38
Common cold
assessment of, 4–10—4–11
causes of, 4–10
follow-up, 4–12
patient education for, 4–12
signs of, 4–10
symptoms of, 4–10
treatment of, 4–11—4–12
Compartment syndrome, 8–30
causes of, 8–30
with electrical or lightning injury, 7–27
equipment for, 8–30
management of, 8–30—8–32
signs and symptoms of, 8–30—8–31
streptococcal infection in, 5–99
in trauma, 7–5
Compazine
for acute abdominal pain, 3–2
after appendectomy, 4–77
for bacterial food poisoning, 4–80
for cholecystitis, 4–78
for gastritis, 4–81
for nausea with headache, 3–56
for venomous snake bite, 5–145
Complete blood count (CBC)
RBC count in, 8–54—8–56
Wright's stain with, 8–46—8–47
Compression, 3–73—3–74
Compur M1100 method, 8–53—8–54
Con-Trate method, 8–52—8–53
Condoms, 3–82
Cone shells, 6–14
treatment for, 6–16

Congestive heart failure, 4–3
    assessment of, 4–4
    conditions causing, 4–3
    cough with, 3–14
    fatigue with, 3–30t
    follow-up, 4–4
    patient education for, 4–4
    pleural effusion with, 4–15
    signs of, 4–4
    symptoms of, 4–3
    treatment of, 4–4
Conjunctivitis
    in animals, 5–136
    differential diagnosis of, 3–25
    red eye with, 3–24
    treatment of, 3–25
Connective tissues disease, 3–60
Consciousness, loss of. *See also* Syncope; Unconsciousness
Constipation, 3–13
    acute pelvic pain with, 3–42
    assessment of, 3–13
    differential diagnosis for, 4–71
    follow-up, 3–13
    patient education for, 3–13
    self-limiting abdominal pain with, 3–1
    signs of, 3–13
    symptoms of, 3–13
    treatment for, 3–13
Constitution, physical examination, 2–3
Contact dermatitis
    allergic, 4–69
    assessment of, 4–70
    categories of, 4–69
    causes of, 4–69
    differential diagnosis of, 3–114
    irritant, 4–69
        differential diagnosis of, 4–65
    of male genitals, 3–78
    pruritus with, 3–113
    signs of, 4–69—4–70
    symptoms of, 4–69
    treatment for, 3–78, 4–70
Contact lens overwear syndrome
    differential diagnosis of, 3–25
    red eye with, 3–24
    treatment of, 3–25
Contagious diseases, pruritus with, 3–114
Contaminated dump, 6–55
Continuous infusion techniques
    general considerations in, 5–156
    pre- and intraoperative system checks for, 5–157t
    in total intravenous anesthesia, 5–155
Continuous positive airway pressure (CPAP)
    for apnea, 4–26
    devices for, 8–4
Continuous suture, 8–23
Conversion disorders, 5–148
Convulsive events, 4–34

Copalite, 5–13
COPD. *See* Chronic obstructive pulmonary disease (COPD)
Copper deficiency, 5–139t
Copper sulfate
    for equine lameness, 5–134
    in swimming dermatitis prevention, 4–54
Copperhead snakes, 5–143
    characteristics of, 5–146
Coral, 6–14
    treatment for, 6–15
Coral snakes, 5–143
    characteristics of, 5–146
    identification of, 5–147
Cornea
    abrasion of
        differential diagnosis of, 3–28
        treatment of, 3–28
    erosion of
        differential diagnosis of, 3–24
        red eye with, 3–24
        treatment of, 3–25
    laser injury of, 7–29
    ulcers of
        in animals, 5–136
        differential diagnosis of, 3–28
        with eye injury, 3–27
        treatment of, 3–28
Corns, 5–5
    follow-up for, 5–6
    patient education for, 5–6
    signs of, 5–6
    treatment for, 5–6
Coronary ischemic syndromes, 3–10
Coronary risk factors, 3–11
Corticosteroids
    for asthma, 4–20
    in pancreatitis, 4–82
    for psoriasis, 4–65
Cortisone, 5–2
Cortisporin optic drops, 6–2
Costovertebral angle tenderness
    in appendicitis, 4–71
    with urolithiasis, 4–91—4–92
Cotton worker's lung, 4–18
Cottonmouth snakes, 5–143
    characteristics of, 5–146
Coudé catheter, 8–34
Cough
    in adenoviruses, 5–64
    alleviating or aggravating factors in, 3–14
    assessment of, 3–15
    in coccidioidomycosis, 5–60
    in common cold and flu, 4–10
    high-altitude, 6–41
    mechanisms of, 3–13
    in paracoccidioidomycosis, 5–63
    in paragonimiasis, 5–45, 5–46
    patient education and follow-up, 3–15
    psychogenic, 3–15

quality of, 3-14
signs of, 3-14—3-15
symptoms of, 3-14
treatment for, 3-15, 4-11
in tuberculosis, 5-54
Coxiella burnetii, 5-85
Coxsackievirus, 3-112t
CPR. *See* Cardiopulmonary resuscitation
Crab lice, 4-62—4-63
Cranberry juice, 4-94
Creeps, 6-36
Cricothyroidotomy, 7-1
 in airway management, 8-3
 for airway obstruction, 3-117
 equipment for, 8-5—8-6
 indications for, 8-5
 needle technique, 8-6
 precautions in, 8-7
 surgical technique, 8-6—8-7
Crimean-Congo hemorrhagic fever, 6-61
Crocodiles, 6-17
Crohn's disease, aphthous ulcers of, 5-16
Crossmatching, 8-60—8-62
Crotalidae snakes, 5-143, 5-144
Crotamiton (Eurax), 4-61
Crush injury, 8-30
Crusts, skin, 4-39
Crutches
 for ankle injury, 3-77
 for hip pain, 3-74
Cryptosporidiosis, 5-32
Crystal arthritides, 3-59
Crystalloid fluid bolus IV, 7-11
Crystalloids, 7-15t
Cullen's sign, 4-82
Cultures, interpretation of, 8-49—8-50
Curing, 5-127
Cutaneous anthrax, 5-90, 5-91
Cutaneous larva currens rash, 5-47. *See also* Strongyloidiasis
Cutaneous larva migrans, 5-32
 cause of, 5-41
 rash of, 5-42
 treatment of, 5-42
Cyanide poisoning
 assessment of under fire, 1-1
 differential diagnosis of, 6-11
Cyanocobalamin deficiency, 5-137t
Cyanosis
 in pneumonia, 4-13
 in pulmonary embolus, 4-23
Cyclobenzaprine (Flexeril), 3-63
Cyclospora infections, 5-36
Cyclosporiasis
 assessment of, 5-36—5-37
 follow-up for, 5-37
 patient education for, 5-37
 signs of, 5-36
 symptoms of, 5-36
 transmission of, 5-36

treatment for, 5-37
Cystic hydrated disease, 5-50
Cystitis, urinary tract infection, 4-93
Cysts, 4-38—4-39
Cytomegalovirus, 5-87

D vitamin deficiency, 5-138t
Da Nang lung. *See* Acute respiratory distress syndrome
Dacryocystitis, 3-26
 differential diagnosis of, 3-26
 treatment of, 3-27
Dairy products, boiling of, 5-109
Dandruff, 4-49—4-51
Dapsone
 for brown recluse spider bite, 4-61
 for leprosy, 4-45
Darkfield microscopy, 4-64
Darling's disease. *See* Histoplasmosis
DD Form 600, 1-22
DD Form 602, 1-22
DD Form 1380, 1-22
Debridement
 of abrasions, 8-26
 equipment for, 8-25
 for frostbite, 6-43
 for human and animal bites, 7-8—7-9
 indications for, 8-25
 for marine animal bites, 6-18
 of minor wound not involving muscle, 8-25
 of penetrating wounds, 8-26
 for plantar warts, 5-4
 procedure in, 8-25
 for streptococcal infections, 5-100
Decadron, 4-36
Decadron rinse, 5-17
Decompression
 for compartment syndrome, 8-31
 in diving medicine, 6-29
Decompression sickness
 in aerospace medicine, 6-35—6-37
 assessment of, 6-5
 chest pain with, 3-10
 differential diagnosis of, 6-7, 6-9—6-10, 6-12
 disqualifying condition for diving, 6-22
 dizziness with, 3-20
 follow-up for, 6-7
 hypoxia in, 6-31
 mechanisms of, 6-4
 memory loss with, 3-86
 neurologic, 6-36
 patient education for, 6-5—6-7
 portable hyperbaric chamber for, 8-36
 signs of, 6-5
 symptoms of, 6-4—6-5
 treatment for, 6-5, 6-6f
 Type I, 6-4
 Type II, 6-4—6-5

with underwater blast injury, 6–19
Decongestant spray
  for barosinusitis, 6–34
  for barotitis, 6–35
Decongestants
  for serous otitis media, 3–21
  for sleep apnea with fatigue, 3–31
Decontamination procedures, poisoning, 5–141—5–142
Decontamination station
  clean side of, 6–55
  equipment for, 6–54
  hot-line in, 6–55
  primary and alternate sites for, 6–54
  setup of, 6–54—6–55
  upwind side of, 6–55
Deep pit latrine, 5–112f
Deep tendon reflexes, 3–6
Deer tick bites, 5–34—5–35
DEET, 4–52
  as mite repellent, 4–60
  in pest control, 5–121
Defibrillator, 4–7
Degenerative joint disease, shoulder, 3–71
Dehydration, 6–37, 6–39, 6–49
Delirium
  versus psychosis, 5–151—5–153
  symptoms of, 5–152
Delivery
  in animals, 5–131
  breech, 3–96—3–99
  postpartum complications, 3–93
  vaginal. See Vaginal delivery
Delusions, symptoms of, 5–152
Demerol
  after cesarean section, 3–102
  for urolithiasis, 4–92
Demodex folliculorum, 5–32
Demulcents, 6–53
Dengue fever
  assessment of, 5–65—5–66
  diagnostic algorithm for, 3–30—3–34f
  differential diagnosis of, 5–84, 5–87, 5–89
  follow-up for, 5–66
  patient education for, 5–66
  risk factors for, 5–65
  signs of, 5–65
  symptoms of, 5–65
  transmission of, 5–65
  treatment for, 5–66
Dengue hemorrhagic fever, 5–65
Dental barotrauma, 6–3—6–4
Dental caries, 5–9, 5–12
  in barodontalgia, 6–32
  with fluoride deficiency, 5–139t
  thermal test for, 5–20
Dental problems, 5–9—5–19
Dental procedures
  anesthesia, 5–20—5–23
  temporary restorations, 5–24
  thermal test for caries, 5–20

tooth extraction, 5–25—5–26
Dental restorations, temporary, 5–24
Dentin, 5–13
Dentistry
  anatomy in, 5–9
  antibiotics in, 5–19—5–20
  minimal dental field kit for, 5–9
  oral and dental problems of, 5–9—5–19
  procedures for, 5–20—5–26
Depakote
  cautions with, 3–56
  for headache, 3–56
Depo-Provera, 3–44t
Depression, 3–15
  assessment of, 3–17
  fatigue with, 3–29
    treatment of, 3–31
  follow up, 3–17—3–18
  memory loss with, 3–86
  patient education for, 3–17
  signs of, 3–16
  symptoms of, 3–16
  treatment for, 3–17
Depressive disorder, major, 3–17
Dermatitis
  in schistosomiasis, 5–46
  swimming, 4–54
Dermatologic disease
  assessment of, 4–39
  diagnosis and disposition of, 4–38—4–40
  signs of, 4–38—4–39
  skin lesions in, 4–38—4–39
  symptoms of, 4–38
  treatment of, 4–40
Dermatomes, of cutaneous innervation of hand, 5–168f
Dermatophyte infections
  assessment of, 4–50
  causes of, 4–49—4–50
  follow-up for, 4–50
  patient education for, 4–50
  signs of, 4–50
  symptoms of, 4–50
  treatment for, 4–50
  zoonotic disease considerations in, 4–50—4–51
Dermatophytosis, 4–50—4–51
Dermatosis, pediatric, 4–42—4–43
Dermoplast, 4–58
Desert rheumatism. *See* Coccidioidomycosis
Dexamethasone
  for acute mountain sickness, 6–38
  for adrenal insufficiency, 4–28
  for high altitude cerebral edema, 6–39
  for preeclampsia, 3–109
  for preterm neonate, 3–94
  for typhoid fever, 5–104
Dexamethasone acetate
  for bunion, 5–5
  for heel spur syndrome, 5–2
Dexamethasone sodium phosphate, 8–20

Dextran, 6–43
Dextromethorphan
  for common cold and flu, 4–11
  for cough, 3–15
Dextrose
  for drug-induced dyspnea, 3–117
  for hypoglycemia, 4–31
Diabetes
  in candida vaginitis, 3–49
  disqualifying condition for diving, 6–22
  hypoglycemia and, 4–30
  insipidus, 4–29
  mellitus
    assessment of, 4–29
    causes of, 4–28
    follow-up, 4–30
    gestational, 4–28
    patient education for, 4–30
    signs of, 4–29
    symptoms of, 4–29
    treatment of, 4–29
    type 1, 4–28
    type 2, 4–28—4–29
Diabetic ketoacidosis, 3–118
Diagnosis, SOAP format in, 2–1—2–2
Diarrhea
  acute
    assessment of, 3–18—3–19
    episodes of, 3–18
    follow-up, 3–19
    patient education for, 3–19
    signs of, 3–18
    symptoms of, 3–18
    treatment plan for, 3–19
  bloody, in trichuriasis, 5–51
  chronic, symptoms of, 3–18
  in cyclosporiasis, 5–36
  differential diagnosis of, 5–48
  in fasciolopsiasis, 5–38
  with fever, 3–34f
  in giardiasis, 5–40, 5–41
  parasitic infections causing, 5–33
  porcine, 5–135—5–136
  in strongyloides, 5–47
  with zinc deficiency, 5–139t
Diarrhea storms, 5–136
Diazepam (Valium)
  for anxiety, 3–5
  for chemical nerve agents, 6–52
  for chest pain, 3–12, 3–13
  for heat stroke, 6–51
  for hypertensive emergency, 4–5
  for mania, 3–17
  for pain control, 8–39
  for palpitations, 3–111
  for psychosis and delirium, 5–152
  for temporomandibular joint dislocation, 5–19
  for tetanus, 5–101
  for urinary incontinence, 4–91
  for venomous fish bites, 6–15

Dibenzyoline (phenoxybenzamine hydrochloride), 6–43
Diclofenac
  for elbow dislocation, 3–68
  for eye injury, 3–28
  for patellar dislocation, 3–69
Dicloxacillin
  for cellulitris of penis, 3–78
  for erysipelas, 4–42
  for febrile infection, 3–34f
  for impetigo contagiosa, 4–43
  for ingrown toenail, 5–3
  for mastitis, 3–8
  for staphylococcal scalded skin syndrome, 4–43
Didanosine, 4–82
Diet
  for acute diarrhea, 3–19
  for adrenal insufficiency, 4–28
  after cesarean section, 3–102
  for bacterial food poisoning, 4–80
  for burn injury, 7–22
  for candida vaginitis prevention, 3–50
  for chest pain, 3–12
  for cholecystitis, 4–79
  for chronic pelvic pain, 3–47
  for congestive heart failure prevention, 4–4
  for constipation prevention, 3–13
  for diabetes mellitus, 4–30
  for empyema, 4–17
  for heat cramps prevention, 6–48
  for hip stress fracture, 3–74
  for hypertensive emergency, 4–5
  for jaundice, 3–58
  for myocardial infarction prevention, 4–3
  for non-freezing cold injury of foot, 6–47
  for paragoniamiasis prevention, 5–46
  for pericarditis, 4–7
  for peritonitis, 4–84
  for pleural effusion, 4–15
  for seizure disorder, 4–35
  for tetanus, 5–102
  for thyroid disorders, 4–33
  for urolithiasis, 4–92
Dietary goitrogens, 4–33—4–34
Diethylcarbamazine
  citrate
    for filariasis, 5–40
    side effects of, 5–40
  for loiasis, 4–52
Digital block, 5–159—5–160
  anesthetics for, 5–163t
  classical approach in, 5–161f
  contraindications and complications of, 5–163
  metacarpal approach in, 5–161f
  metatarsal, 5–163f
  web space of finger, 5–160f
  web space of toe, 5–162f
Dihydrostreptomycin, 5–134
Dilantin, 4–35

Dimenhydrinate
   for dizziness, 3–21
   for yellow fever, 5–70
Dimercaprol, 6–54
Diphenhydramine (Benadryl)
   for allergic reaction to blood transfusion, 8–20
   for common cold and flu, 4–11
   for contact dermatitis of penis, 3–78
   for mania, 3–17
   for mites, 4–59
   for psychosis and delirium, 5–153
   for venomous snake bite, 5–145
Diphtheria, 4–11
Dipylidiasis, 5–50
Dirty Sock technique, 3–67
Disability
   assessment of with chemical burns, 7–18
   AVPU scale of, 7–3
   M5 item list for managing, 1–18
Disc herniation
   in low back pain, 3–6
   treatment of, 3–74
Disinfection, 5–110—5–111
Dislocations, 3–64—3–70. See also specific joints
Distributive shock, 7–13
Ditropan
   for prostatitis, 3–82
   side effects of, 4–91
   for urinary incontinence, 4–90
Diuresis, 4–5
Dive medicine
   barotrauma to ears, 6–1—6–2
   treatment tables for, 6–23—6–30
Diver's Alert Network (DAN), 6–22
Diverticular stricture, 4–86
Diverticulitis
   acute abdominal pain in, 3–3t
   in acute organic intestinal obstruction, 4–86
   acute pelvic pain with, 3–42
   in peritonitis, 4–83
Diving
   ascent rate, 6–23, 6–27
   depth/time profile for, 6–24f, 6–25, 6–26, 6–28f, 6–30f
   descent rate, 6–23, 6–27
   disqualifying conditions for, 6–21—6–22
   resources for, 6–22
   time at depth, 6–25f
Diving Medical Officer (DMO), 6–21
Diving reflex
   alleviating palpitations, 3–110
   for palpitations, 3–111
Diving rigs, closed-circuit and semi closed-circuit, 6–20
Dix-Hallpike Maneuver, 3–20
Dixie cup technique, 5–1
Dizziness, 3–20
   assessment of, 3–20—3–21
   follow-up, 3–22
   patient education for, 3–21—3–22
   signs of, 3–20

   symptoms of, 3–20
   treatment plan for, 3–21
DNBI (disease non-battle injury) report, 5–107—5–108
Documentation, patient, 1–22
DOD Arthropod Repellent System, 5–121
Dog tapeworm, 5–50
Domeboro, 4–70
Donor blood units, crossmatching of, 8–60—8–62
Dopamine
   for peritonitis, 4–84
   for venomous snake bite, 5–145
Dorsal slit procedure
   for male genital inflammation, 3–79
   for phimosis, 4–89—4–90
Dove Sensitive Skin soap, 4–70
Doxazosin (Cardura)
   for prostatitis, 3–82
   for urinary incontinence, 4–91
Doxepin, 3–115
Doxycycline
   for abnormal uterine bleeding, 3–41
   for anthrax, 5–91
   for Bartholin's gland abscess or cyst, 3–52
   for biological warfare agents, 6–56
   for brucellosis, 5–94
   cautions with, 3–53
   for chancroid, 5–29
   for ehrlichiosis, 5–95
   for epididymitis, 3–85
   for granuloma inguinale, 5–29
   for human and animal bites, 7–9
   for inhalational anthrax, 6–57
   for knee pain, 3–75
   for leptospirosis, 5–87
   for Lyme disease, 5–88
   for lymphogranuloma venereum, 5–29
   for malaria, 5–44, 5–45
      prophylaxis, 5–120
   for nontuberculous mycobacterial infections, 5–57
   for oroya fever, 5–93
   for pelvic inflammatory disease, 3–51
   for plague, 5–96
   for pneumonia, 4–13
   for pneumonic plague, 6–59
   precautions with, 3–80, 5–85
   for prostatitis, 3–81
   for Q fever, 5–86
   for rat bite fever, 5–97
   for relapsing fever, 5–90
   for Rocky Mountain spotted fever, 5–84
   side effects of, 5–45, 5–87, 5–95
   for syphilis, 5–30
   for testicular/scrotal masses, 3–80
   for tetanus, 5–101
   for trench fever, 5–93
   for typhus, 5–69, 5–84
   for urethral discharges, 5–27
   for urinary tract infections, 4–93
Dressings

for burn injury, 7–22
   volume-restricting, 8–31
   wet-to-dry, 8–28
Driving restrictions, seizure disorder, 4–35
Drug eruptions, 3–114
Drug hypersensitivity, 3–113t
Drug reactions
   differential diagnosis of, 4–65, 6–31
   fever with, 3–31
Drug-related fatigue, 3–29
Drug-related pruritus, 3–114
Drug-related syncope, 3–118
Dry eye
   differential diagnosis of, 3–25
   red eye with, 3–24
   treatment of, 3–25
Dry socket. See Osteitis, localized
Drysol, 5–8
Dumbbell shoulder flies, 3–72
   lateral, 3–72
Dumping syndrome, 6–22
Dwarf tapeworm disease, 5–49
Dycal, 5–13
Dysentery
   assessment of, 3–18
   symptoms of, 3–18
Dysmenorrhea
   diagnosis of, 3–45
   differential diagnosis of, 3–44t
   primary, 3–45
   symptoms of, 3–45
   treatment of, 3–45
Dyspareunia, 3–51
Dyspepsia
   alleviating or aggravating factors in, 3–11
   treatment for, 3–12
Dyspnea, 3–115
   assessment of, 3–116
   on exertion, with congestive heart failure, 4–3
   exposure history in, 3–115
   follow-up, 3–117
   medical history in, 3–115
   neurological exam for, 3–116
   patient education for, 3–117
   signs of, 3–115—3–116
   symptoms of, 3–115
   treatment of, 3–117
Dysrrhythmias, 8–10

E vitamin deficiency, 5–138t
Ears
   barotrauma to, 6–1—6–2
   dizziness in trauma to, 3–20
   middle, 6–34
Eastern equine encephalitis, 5–66
Ebola virus, 6–61
Echinococcosis, 5–32
Echovirus, 3–112t

Eclampsia, 3–104. See also Preeclampsia
   signs of, 3–105, 3–108
   treatment of, 3–109
Ecthyma contagiosum
   assessment of, 4–46
   cause of, 4–45—4–46
   follow-up for, 4–46
   patient education for, 4–46
   signs of, 4–46
   symptoms of, 4–46
   treatment for, 4–46
Ectopic pregnancy
   acute pelvic pain with, 3–41, 3–42
   ruptured, 3–3t
   treatment of, 3–43
Eczema, 3–113
Eczematous dermatitis, 4–61
Edema, generalized, 4–89
EEG. See Electroencephalography
Efavirenz, 5–78
Eflornithine, 5–52
Eggs, cooking, 5–109
Ehrlichia, 5–95
Ehrlichiosis, 5–95
   differential diagnosis of, 5–89
EKG. See Electrocardiography
Elapidae snakes, 5–143, 5–144
Elastoplast, 5–8
Elavil, 3–82
Elbow joint
   aspiration at, 8–29
   dislocation of
      anterior, 3–65
      assessment of, 3–67
      diagnostic tests for, 3–67
      differential diagnosis of, 3–67
      follow-up, 3–68
      patient education for, 3–68
      post treatment care for, 3–67—3–68
      posterior, 3–65
      procedure for, 3–67
Electrical injuries, 7–26
   assessment of, 7–27
   follow-up for, 7–28
   patient education for, 7–28
   signs of, 7–27
   symptoms of, 7–26—7–27
   treatment for, 7–27—7–28
Electrocardiogram rhythm, 3–118
Electrocardiography
   in congestive heart failure, 4–4
   for myocardial infarction, 4–1, 4–2f
   normal with chest pain, 4–8
   in pericarditis, 4–6
   for pulmonary embolus, 4–23
   three-lead, 8–10—8–12
      applying chest electrodes in, 8–11
      equipment for, 8–10
      interpreting, 8–11—8–12
      patient preparation for, 8–11

preparation for, 8–10—8–11
Electrocution, 7–26
Electroencephalography
　for memory loss, 3–87
　for seizures, 4–35
Elephantiasis. See Filariasis
Elimite, 4–61
Emaciation, veterinary exam for, 5–123
Embolism without infarction, 4–23
Emergency medications, M5 item list for, 1–19
Emergency treatment, casualty decontamination station, 6–54
Emphysema
　in COPD, 4–21
　cough with, 3–14, 3–15
　fatigue with, 3–30t
Empyema. See also Pleural effusion
　assessment of, 4–17
　causes of, 4–17
　follow-up, 4–18
　patient education for, 4–17
　pleural effusion with, 4–15
　in pneumonia, 4–13, 4–14
　signs of, 4–17
　symptoms of, 4–17
　treatment of, 4–17
Enalapril, 4–5
Encephalitis
　arboviral, 5–66—5–68
　differential diagnosis of, 5–82
　European tick-borne, 5–32
　in genital ulcers, 5–28
End-positive-airway-pressure mask, 6–41
Endocrine disorders
　with abnormal uterine bleeding, 3–41
　adrenal insufficiency, 4–27—4–28
　diabetes mellitus, 4–28—4–30
　hypoglycemia, 4–30—4–31
　thyroid, 4–31—4–34
Endometriosis
　in chronic pelvic pain, 3–45
　differential diagnosis of, 3–44t
　focused history for, 3–45
　pelvic examination for, 3–45
　symptoms of, 3–45
　treatment of, 3–45
Endometritis, 3–41
Endotracheal intubation
　for high altitude pulmonary edema, 6–41
　indications for, 8–4
　procedure for, 8–5
ENT examination, 2–3
ENT problems. See also Asthma; Common cold; Cough; Flu
Entamoeba histolytica, 5–33
Enteric fever, diagnostic algorithm for, 3–34f
Enterobiasis
　assessment of, 5–37
　differential diagnosis of, 3–114
　follow-up for, 5–38

patient education for, 5–37—5–38
　symptoms and signs of, 5–37
　transmission of, 5–37
　treatment for, 5–37
Enterobius vermicularis nematode, 5–37
Enteroviral exanthems, 6–59
Enterovirus infections, 3–112t
Eosin, 8–46
Eosinophilia, 5–41
Epidermal inclusion, 4–38—4–39
Epidermal inclusion cyst
　differential diagnosis of, 4–48
　removal of, 8–26, 8–27
Epidermophyton, 4–49
Epididymis torsion, 3–83
Epididymitis
　assessment of, 3–84
　follow-up, 3–85
　location of, 3–84
　patient education for, 3–85
　signs of, 3–84
　symptoms of, 3–84
　with testicular mass, 3–80
　treatment of, 3–84—3–85
Epididymo-orchitis, 3–83
Epidural hematoma, 4–36
Epilepsy
　assessment of, 4–34
　causes of, 4–34
　follow-up for, 4–35
　patient education for, 4–35
　signs of, 4–34
　symptoms of, 4–34
　treatment of, 4–34—4–35
Epinephrine
　for allergic reaction to blood transfusion, 8–20
　for anaphylactic shock, 7–11
　in axillary blockade, 5–174
　for cesarean section, 3–100
　for coelenterate bites, 6–15
　in dental procedures, 5–20
　for heat stroke, 6–51
　in local anesthesia infiltration, 5–159
　for sea urchin bites, 6–15
　in skin mass removal, 8–27
　in suturing, 8–22
　for venomous fish bites, 6–15
　for venomous snake bite, 5–145
Epinephrine (Marcaine), bupivacaine with, 5–20
Epinephrine (Xylocaine), lidocaine with, 5–20
Episcleritis
　differential diagnosis of, 3–25
　red eye with, 3–24
　treatment of, 3–25
Episiotomy, 3–102, 3–105f
　classification of, 3–103
　equipment for, 3–103
　first degree repair, 3–103
　indications for, 3–102—3–103
　management after, 3–104

IN-23

precautions, 3–104
second degree repair, 3–103, 3–106—107f
third and fourth degree repair, 3–104, 3–108f
Epley maneuver, 3–21
Epstein-Barr virus, mononucleosis in, 5–78—5–79
Equine colic, 5–132—5–133
Equine encephalitis, 5–66—5–68
Equine morbillivirus, 5–32
Erection, persistent, 4–89
Ergots, 3–56
Erosion, skin, 4–39
Erysipelas
  assessment of, 4–42
  cause of, 4–41—4–42
  differential diagnosis of, 3–112t
  follow-up for, 4–42
  patient education for, 4–42
  risk factors for, 4–42
  signs of, 4–42
  streptococcal infection in, 5–99
  symptoms of, 4–42
  treatment for, 4–42
Erysipelothrix rhusiopathiae, 5–32
Erythema infectiosum, 3–113t
Erythema migrans, 5–88
Erythema multiforma, 3–113t
Erythema nodosum, 3–60
Erythematous macules
  in bed bug bites, 4–55
  in disseminated gonococcal infection, 4–40
  in impetigo contagiosa, 4–43
  in pinta, 4–64
Erythematous papules
  in ecthyma contagiosum, 4–46
  in herpes zoster infection, 4–47
  in pediculosis, 4–62
Erythematous plaque, 4–60
Erythromycin
  for acute rheumatic fever, 5–99
  for chancroid, 5–29
  for chemical burns, 7–18
  in dentistry, 5–19
  for erysipelas, 4–42
  for granuloma inguinale, 5–29
  for impetigo contagiosa, 4–43
  for ingrown toenail, 5–3
  for lymphogranuloma venereum, 5–29
  for mastitis, 3–8
  for pelvic inflammatory disease, 3–51
  for pinta, 4–64
  for pneumonia, 4–13
  for preterm labor infection, 3–94
  for prostatitis, 3–82
  for relapsing fever, 5–90
  for urethral discharges, 5–27
  for yaws, 4–63
Eschar, 7–17
Escharotomy
  chest, 7–19, 7–21f

regions of, 7–21f
Escherichia coli, 5–32
  in food poisoning, 4–79, 4–80
  identification of colonies of, 8–50
  in porcine diarrhea, 5–135
Esophageal reflux disease, 4–78
Esophageal rupture
  alleviating or aggravating factors in, 3–11
  chest pain with, 3–10
  treatment for, 3–12
Estrogens, 4–82
Ethambutol, 5–55
Ethyl acetate, 8–52
Eugenol
  for dental caries, 5–12
  for localized osteitis, 5–18
  in temporary dental restorations, 5–24
  for tooth fracture, 5–13
Eurax, 4–61
Eustachian tube
  dysfunction of, 6–22
  swelling of, 6–34
Evacuation, trauma patient, 7–6
Excision
  in malignant melanoma biopsy, 4–68
  for nontuberculous mycobacterial infections, 5–57
Excoriation, 4–39
Exercise
  aerobic, for diabetes mellitus, 4–30
  for diabetes mellitus, 4–30
  for dysmenorrhea, 3–45
  for hypertensive emergency, 4–5
  for joint pain, 3–62
  for shoulder rehabilitation, 3–72
  for stress fractures of foot, 5–6
Expectorants, 3–15
Explosion, underwater. See Underwater blast injury
Exposure management, M5 item list for, 1–18
External auditory canal, blocking of, 6–34
Extraocular muscle
  derangement of, 3–27
  examination of for red eye, 3–24
Extremities, secondary trauma survey of, 7–4
Eye globe, ruptured, 3–27
  differential diagnosis of, 3–28
  treatment of, 3–28
Eye patch, 3–25
Eyelid laceration, 3–28
Eyes
  drainage of, 3–27
  injuries of
    assessment of, 3–27—3–28
    from blister agents, 6–53
    follow-up, 3–29
    with hymenoptera sting, 4–58
    laser, 7–28—7–29
    patient education for, 3–29
    signs of, 3–27

symptoms of, 3–27
treatment for, 3–28
pain in with red eye, 3–24
problems of
    acute red eye without trauma, 3–24—3–26
    acute vision loss without trauma, 3–22—3–24
    eye injury, 3–27—3–29
    orbital or periorbital inflammation, 3–26—3–27
protection of in Bell's palsy, 4–37—4–38

FABER maneuver, 3–60
Facemask squeeze, 6–3, 6–4
Facemasks, airway management, 8–4
Facial nerve palsy, idiopathic. *See* Bell's palsy
Fainting. *See* Syncope
Falling, dizziness with, 3–20
Fallopian tube, damage to, 3–50
Famciclovir, 5–29
Family support group, 5–154—5–155
Famotidine
    for chest pain, 3–12
    for peptic ulcer disease, 4–85
Farmer's lung, 4–18
Fasciola hepatica infections, 5–38—5–39
Fascioliasis, 5–38—5–39
Fasciolopsiasis, 5–38
Fasciolopsis buski, 5–38
Fasciotomy
    in compartment syndrome, 8–31
    precautions for, 8–32
    for venomous snake bite, 5–145—5–146
Fasinex (triclabendazole), 5–39
Fatigue, 3–29
    follow-up, 3–31
    patient education for, 3–31
    signs of, 3–30
    symptoms of, 3–29
    treatment for, 3–30—3–31
Febrile disease
    diagnostic algorithm for, 3–33—3–34f
    differential diagnosis of, 6–49
    incubation period of, 3–31
Febrile reactions, blood transfusion, 8–19
    treatment of, 8–20
Febrile syndromes, adenoviruses in, 5–64
Fecal contamination
    in cyclosporiasis, 5–36
    in giardiasis, 5–40
Fecal leukocytes, 3–34f
Feces
    Con-Trate method for ova and parasite detection in, 8–52—8–53
    examination of for ova and parasites, 8–51—8–52
    macroscopic examination of, 8–50—8–51
Feflex, 4–94
Feka Con-Trate System, 8–52

Feldene, 4–84
Femoral neck stress fracture, 3–73
Femoral shaft stress fracture, 3–74
Femur fractures, hemorrhage control in, 7–13
Fentanyl
    for chemical burns, 7–18
    continuous infusion of in field, 5–158t
    dosing guidelines for, 5–156t
    for pain control, 8–39
Fetal complications, cesarean section for, 3–100
Fetal head crowning, 3–88, 3–90f, 3–91f
Fetal heart tones, 3–87
Fever. *See also* Febrile disease
    with abdominal pain, 3–1
    in adenoviruses, 5–64
    in American trypansomiasis, 5–53
    assessment of, 3–32—3–35
    in coccidioidomycosis, 5–60
    in histoplasmosis, 5–61
    incubation period of, 3–31
    low-grade, 3–31
    in malaria, 5–44
    patient education for, 3–35
    principles of initial approach to, 3–31
    purpura of, 4–39
    rash with, 3–112
        differential diagnosis of, 3–112—3–113t
    in schistosomiasis, 5–46, 5–47
    signs of, 3–32
    symptoms of, 3–31—3–32
    treatment plan for, 3–35
    in trichenellosis, 5–50
    with urolithiasis, 4–91
Fever blisters, 5–16
Field dressing, game, 5–123—5–125
Field kits, dental, 5–9
Field sanitation
    cleaning and disinfecting in, 5–110—5–111
    food procurement in, 5–108—5–109
    food storage and preservation in, 5–109—5–110
    general procedures in, 5–108
    preparing and serving food in, 5–110
Field transfusion procedure, 8–18
Field water purification, 5–118—5–119
Filariasis
    assessment of, 5–40
    follow-up for, 5–40
    patient education for, 5–40
    signs of, 5–40
    symptoms of, 5–39—5–40
    transmission of, 5–39
    treatment for, 5–40
Fingers, human bite to, 7–7
    signs of, 7–8
Fioricet, 3–56
Fiorinal, 3–56
Fire ants, 4–58
Firewood, 5–127
Fish

highly toxic, 6-13
  treatment for, 6-14—6-15
  venomous, 6-13
Fish tapeworm disease, 5-49
Fistula
  test, for ear barotrauma, 6-1
  in urinary incontinence, 4-90
Flaccid limb paralysis, 5-80
Flagyl (metronidazole)
  for bacterial vaginosis, 3-48t
  for balanitis, 3-78
  for candida vaginitis, 3-48t
  for Fournier's gangrene, 3-78
  for jaundice, 3-58
  for prostatitis, 3-82
  for trichomonas, 3-48t
Flail chest, 7-2
Flamazine, 7-18
Flank pain, 3-1. See also Urolithiasis
  assessing, 4-87
  hematuria with, 4-88
Flashlight examination
  for orbital or periorbital inflammation, 3-26
  for red eye, 3-24
Flaviviruses, 5-66
Flavoxate
  for prostatitis, 3-82
  for urinary incontinence, 4-90
Flexeril (cyclobenzaprine)
  for joint pain, 3-63
  side effects of, 3-64
Flight operations, decompression sickness risk in, 6-37
Flomax (tamsulosin)
  for prostatitis, 3-82
  for urinary incontinence, 4-91
Flonase, 3-31
Floricet, withdrawal from, 5-150
Florinal, withdrawal from, 5-150
Floxin
  for epididymitis, 3-84, 3-85
  for urinary tract infections, 4-94
Flu. See also Common cold; Influenza
  assessment of, 4-10—4-11
  causes of, 4-10
  follow-up, 4-12
  patient education for, 4-12
  signs of, 4-10
  symptoms of, 4-10
  treatment of, 4-11—4-12
Flucinolone, 4-65
Fluconazole
  for blastomycosis, 5-60
  for candida vaginitis, 3-48t, 3-50
  for candidal penile infection, 3-78
  for candidiasis, 5-58
  for coccidioidomycosis, 5-61
  for paracoccidioidomycosis, 5-63
  for pityriasis versicolor, 4-51
  for urinary tract infections, 5-93
Fluid overload, 4-3

Fluid resuscitation
  for acute diarrhea, 3-19
  for acute pelvic pain, 3-43
  administration routes for, 7-16t
  for blister agents, 6-53
  in burn injury, 7-19—7-22
  with chemical burns, 7-18
  end points of, 7-14
  for hypovolemic shock, 7-13
  indications, benefits, and cautions for, 7-15t
  maintenance rate calculation in, 7-14
  for poisoning, 5-141
  for pulmonary embolus, 4-24
  in uncontrolled hemorrhage, 7-3
  for valvular malfunction, 3-117
  for venomous snake bites, 5-144
Fluid therapy
  for acute respiratory distress syndrome, 4-25
  for hypothermia, 6-44
Flumazenil, 8-40
Flunixin meglumine (banamine)
  for equine colic, 5-133
  for equine lameness, 5-135
Fluocinonide, 4-70
Fluorescein strip staining
  for acute vision loss without trauma, 3-23
  for eye injury, 3-27
  for orbital or periorbital inflammation, 3-26
  for red eye assessment, 3-24
Fluoride deficiency, 5-139t
Fluoroquinolones
  for biological warfare agents, 6-56
  for inhalational anthrax, 6-57
  for pneumonia, 4-13
  for urinary tract infections, 4-94
Fluoxetine (Prozac)
  for apnea, 4-26
  for depression, 3-17
Fly eggs, myiasis transmission, 4-52
Focal motor seizures, 4-34—4-35
Focal neurologic signs, 4-31
Folate deficiency, 5-137t
Foley catheterization
  in cesarean section, 3-102
  for prostatitis, 3-81
  for venomous snake bite, 5-145
Foley catheters
  for bladder catheterization, 8-34
  in episiotomy, 3-103
Folic acid deficiency, 5-137t
  in anemia, 4-8
Folliculitis, 3-114
Food
  acquisition of, 5-108
  contamination of
    categories of, 5-108
    in cyclosporiasis transmission, 5-36
    in leptospirosis, 5-86
  potentially hazardous, 5-108
  preparing and serving of, 5-110

procurement of, 5-108—5-109
   storage and preservation of, 5-109—5-110
     of meat and animal products, 5-126—5-128
Food-borne disease
   factors in, 5-108
   prevention of, 5-126
Food poisoning
   acute bacterial
     assessment of, 4-80
     causes of, 4-79
     follow-up for, 4-80
     patient education for, 4-80
     signs of, 4-80
     symptoms of, 4-79—4-80
     treatment for, 4-80
   diarrhea with, 3-19
   differential diagnosis for, 4-71
Foot
   cutaneous innervation of, 5-173f
   disorders of. See Podiatry; specific disorders
   friction blisters of, 5-7
     assessment of, 5-7
     follow-up for, 5-8
     patient education for, 5-8
     signs of, 5-7
     symptoms of, 5-7
     treatment for, 5-7—5-8
   nerve blockade of, 5-168—5-174
   non-freezing cold injury of, 6-44—6-47
   stress fractures of, 5-6
     assessment of, 5-6
     follow-up for, 5-7
     patient education for, 5-7
     signs of, 5-6
     symptoms of, 5-6
     treatment for, 5-6—5-7
Foot rot, caprine, 5-133—5-134
Footling breech, 3-96
Foreign body
   in airway obstruction, 3-117
   aspiration of
     in asthma, 4-20
     cough with, 3-15
     dyspnea with, 3-115, 3-116
   conjunctival
     differential diagnosis of, 3-25
     red eye with, 3-24
     treatment of, 3-25
   ocular
     differential diagnosis of, 3-28
     treatment of, 3-28
Foreign-body sensation, eye, 3-27
Formalin
   in Con-Trate method, 8-52
   for foot rot in caprines, 5-134
Fournier's gangrene, 3-77, 3-78
   consultation criteria for, 3-79
   treatment of, 3-78
Fractures
   of ankle, 3-76

in compartment syndrome, 8-31
compartment syndrome management in, 8-30
femoral shaft, 3-74
hip, 3-73
   assessment of, 3-73
   treatment of, 3-74
in hip dislocation, 3-68
in hypovolemic shock, 7-12
in joint pain, 3-59, 3-60
with low back pain, 3-6, 3-7
shoulder, 3-71
splinting for, 8-32
stress, foot, 5-6—5-7
of tooth or crown, 5-12—5-13
traction for, 8-33
with underwater blast injury, 6-19
Francisella tularensis, 5-102, 6-60
Frank breech, 3-96
Freckles, 4-38
Freezing injury. *See* Frostbite
Fresh frozen plasma, 8-15
Frostbite, 6-42
   assessment of, 6-42
   differential diagnosis of, 6-47
   follow-up for, 6-43
   patient education for, 6-43
   symptoms and signs of, 6-42
   treatment for, 6-42—6-43
Fruit, disinfection of, 5-109
Fruit juice, 4-31
Fulminant hepatic failure, 3-58
Fundus exam, 3-23
Fungal infections, 5-57
   blastomycosis, 5-59—5-60
   candidiasis, 5-57—5-59
   coccidioidomycosis, 5-60—5-61
   dermatophyte, 4-49—4-51
   histoplasmosis, 5-61—5-62
   paracoccidioidomycosis, 5-62—5-63
   pityriasis versicolor, 4-51
   pruritic, 3-114
Furosemide (Lasix)
   for chest pain, 3-12
   for congestive heart failure, 4-4
   for electrical or lightning injury, 7-27
   for hypertensive emergency, 4-5
   in pancreatitis, 4-82
   side effects of, 7-28
   for venomous snake bite, 5-145
Furunculosis, bacterial, 4-53

GA (tabun), 6-52
Gait
   with non-freezing cold injury of foot, 6-47
   testing in neurologic examination, 2-4
Gallstones. *See also* Cholecystitis, acute
   alleviating or aggravating factors in pain of, 3-11
   chest pain with, 3-10

in jaundice, 3-58
in pancreatitis, 4-82
Game, humane slaughter and field dressing of, 5-123—5-125
Gamma benzene hexachloride, 4-61
Gamow bag, 8-36
    for decompression sickness, 6-36
    for high altitude cerebral edema, 6-40
    for high altitude pulmonary edema, 6-42
Gangrene
    with frostbite, 6-42
    with peritonitis, 4-83
Garbage disposal, 5-118
Gardnerella infection, glans penis, 3-77—3-78
Gas, intestinal, 3-1
Gas gangrene, *Clostridium perfringens* colonies in, 8-50
Gastric aspiration, 3-15
Gastric lavage, 5-142
Gastritis
    acute
        assessment of, 4-81
        causes of, 4-81
        follow-up for, 4-81—4-82
        patient education for, 4-81
        signs and symptoms of, 4-81
        treatment for, 4-81
    self-limiting abdominal pain with, 3-1
Gastroenteritis
    acute abdominal pain in, 3-3t
    assessment of, 3-19
    differential diagnosis of, 4-71, 5-91
    viral. *See* Diarrhea
Gastroesophageal reflux
    cough with, 3-14, 3-15
    differential diagnosis of, 4-22
    disqualifying condition for diving, 6-22
    self-limiting abdominal pain with, 3-1
    treatment for, 3-12
Gastrointestinal anthrax, 5-91
Gastrointestinal barotrauma, 6-3—6-4
Gastrointestinal bleeding
    assessment of, 3-19
    with blast injuries, 7-23
    treatment for, 7-24—7-26
Gastrointestinal disorders
    acute bacterial food poisoning, 4-79—4-80
    acute cholecystitis, 4-78—4-79
    acute gastritis, 4-81—4-82
    acute organic intestinal obstruction, 4-86—4-87
    acute pancreatitis, 4-82—4-83
    acute peritonitis, 4-83—4-84
    appendicitis, 4-70—4-72
    disqualifying condition for diving, 6-22
    emergency field appendectomy, 4-72—4-78
    peptic ulcer disease, 4-84—4-86
Gaze, deviated, 3-27
GD (soman), 6-52
General medical site survey checklist, MedCAP,

1-3, 1-7—1-9
Generalized anxiety disorder, 3-5
Genital inflammation, male, 3-77—3-79
Genital skin lesions, 4-89
Genital ulcers
    assessment of, 5-29
    causes of, 5-28
    follow-up for, 5-30
    male, 3-77
    patient education for, 5-30
    signs of, 5-28—5-29
    symptoms of, 5-28
    treatment for, 5-29, 5-30
Genitourinary disorders
    urinary incontinence, 4-90—4-91
    urinary tract infection, 4-93—4-94
    urinary tract problems of, 4-87—4-90
    urolithiasis, 4-91—4-92
Genitourinary examination
    female, 2-4
    male, 2-4
Gentamicin
    for abnormal uterine bleeding, 3-41
    for chemical burns, 7-18
    for congestive heart failure, 4-4
    for conjunctivitis in animals, 5-136
    for epididymitis, 3-84
    for febrile infection, 3-34f
    for Fournier's gangrene, 3-78
    for pericarditis, 4-6
    for plague, 5-96
    for pneumonic plague, 6-59
    for prostatitis, 3-81
    for tularemia, 5-103, 6-60
    for tympanic membrane rupture, 7-26
    for urinary tract infections, 4-94
    for urolithiasis, 4-92
Gentian violet, 5-58
Gestational age, assessment of, 3-88
Giant cell arteritis
    in acute vision loss, 3-22
    differential diagnosis of, 3-23
    treatment of, 3-23
Giardia cysts, 5-40
Giardiasis, 5-32
    assessment of, 3-19, 5-41
    follow-up for, 5-41
    patient education for, 5-41
    signs of, 5-41
    symptoms of, 5-40
    transmission of, 5-40
    treatment for, 5-41
    zoonotic disease considerations in, 5-41
Giemsa stain
    for blood parasites, 8-47—8-48
    for malaria, 5-44
    for relapsing fever, 5-89
    in Tzanck stain, 8-48
Gingival swelling, 5-13
Gingivitis, acute necrotizing ulcerative,

5-15—5-16
Glanders, 5-32
Glans penis inflammation, 3-77—3-78
Glasgow Coma Scale, 2-4
    for arboviral encephalitis, 5-67
    for meningitis, 4-36
    for rabies, 5-82
    for syncope, 3-118
    for trauma patient, 7-4
    for underwater blast injury, 6-19
Glass ionomer cement, 5-12, 5-13
Glaucoma, acute angle-closure, 3-25
    red eye with, 3-24
Glenohumeral joint, 3-70
Glipizide, 4-29
Glucagon, 4-31
Glucose, metabolism of, 4-30
Glucose-Clinitest procedure, 8-42
Glyburide, 4-29
Glycerin suppository, 3-13
Glycosuria, 8-42
Gnathostomiasis, 4-52
Goats, foot rot in, 5-133—5-134
Goiter, 4-31
    diagnosis of, 4-32
    differential diagnosis of, 4-33
    with iodine deficiency, 5-139t
    symptoms of, 4-32
    treatment of, 4-33
Gonococcal arthritis
    differential diagnosis of, 5-98
    knee pain with, 3-75
Gonococcal infection
    in Bartholin's gland abscess or cyst, 3-52
    disseminated
        assessment of, 4-40
        cause of, 4-40
        follow-up for, 4-40
        patient education for, 4-40
        signs of, 4-40
        symptoms of, 4-40
        treatment of, 4-40
Gonococcal peritonitis, 4-71
Gonococcal pharyngitis, 4-11
Gonococcemia, 3-112t
Gonorrhea
    in Bartholin's gland abscess or cyst, 3-52
    cultures for, 3-37, 3-38
        for pelvic inflammatory disease, 3-51
    in urethral discharges, 5-26
Gout, 3-59—3-60
    natural history of, 3-63
    signs of, 3-60
    treatment of, 3-63
Gram-negative bacterial sepsis, 4-27
Gram stain
    for blastomycosis, 5-59
    for coccidioidomycosis, 5-61
    equipment for, 8-44
    for impetigo contagiosa, 4-43
    indications for, 8-44
    in joint pain assessment, 3-61
    for meningococcemia, 4-41
    procedures in, 8-44—8-45
    quality control testing in, 8-45
    for urethral discharges, 5-27
Granisetron (Kytril), 6-62
Granuloma
    inguinale
        in genital ulcers, 5-28
        treatment of, 5-29
    pleural effusion with, 4-15
Graves' disease
    in diabetes mellitus, 4-28
    in hyperthyroidism, 4-31—4-32
Grease traps, 5-115f, 5-116f
Grepafloxacin, 4-13
Grey Turner's sign, 4-82
Griseofulvin, 4-50
Gross motor strength testing, 7-4
Guanifenesin
    for common cold and flu, 4-11
    for cough, 3-15
Guillain-Barre syndrome, 4-37, 5-81
Gum boil, 5-13
Gurgling breath sounds, 6-40
Gynecologic symptoms
    abdominal pain in, 3-1
    abnormal uterine bleeding, 3-39—3-41
    bacterial vaginosis, 3-47—3-49
    Bartholin's gland cyst/abscess, 3-52—3-55
    candida vaginitis/vulvitis, 3-49—3-50
    female pelvic examination for, 3-37—3-39
    pelvic inflammatory disease, 3-50—3-52
    pelvic pain
        acute, 3-41—3-43
        chronic, 3-43—3-47
    vaginitis, 3-47

H-2 blockers
    for gastritis, 4-81
    for self-limiting abdominal pain, 3-1
HACE. *See* High altitude cerebral edema (HACE)
Haemophilus influenzae, 4-14
HAHO missions, hypoxia in, 6-32
Hallucinations
    in high altitude cerebral edema, 6-39
    symptoms of, 5-152
Hallux abductor valgus. *See* Bunion
HALO missions
    decompression sickness risk in, 6-37
    hypoxia in, 6-32
Haloperidol
    for mania, 3-17
    for psychosis and delirium, 5-152, 5-153
Halter, 5-128
Hammertoe, 5-5
Hand
    dermatomes of cutaneous innervation of, 5-168f

nerve blockade at, 5-164—5-168
PIPJ or DIPJ joint dislocations of, 3-65
Hand signals, in care under fire, 1-1
Handwashing station, improvised, 5-117f
Hansen's disease. *See* Leprosy
Hantaan virus, 5-68
Hantavirus, 6-61
   assessment of, 5-69
   follow-up, 5-69
   patient education for, 5-69
   signs of, 5-68—5-69
   symptoms of, 5-68
   transmission of, 5-68
   treatment for, 5-69
   zoonotic disease considerations in, 5-69
Hantavirus pulmonary syndrome, 5-68, 5-69
   differential diagnosis of, 5-87
HAPE. *See* High altitude pulmonary edema (HAPE)
Head
   bites about, 7-8
   puncture wounds from human bites, 7-7
   trauma to
      with blast injury, 7-26
      differential diagnosis of, 6-39, 6-44
      as disqualifying condition for diving, 6-22
      dyspnea with, 3-116, 3-117
      in hypertensive emergency, 4-5
      manic and depressive symptoms with, 3-16
Head lice, 4-62—4-63
Headache
   assessment of, 3-56
   cluster, 3-55, 3-56
   in common cold and flu, 4-10
   conditions causing, 3-55
   disqualifying condition for diving, 6-22
   follow-up, 3-57
   in high altitude cerebral edema, 6-39
   migraine, 3-55, 3-56
   patient education for, 3-56
   signs of, 3-55
   symptoms of, 3-55
   tension, 3-55, 3-56
   treatment of, 3-56
Hearing
   dysfunction of as disqualifying condition for diving, 6-22
   loss of with ear barotrauma, 6-1
Heart abnormalities, EKG, 8-12
Heart attacks
   chest pain with, 3-10
   referred pain to shoulder in, 3-70
Heart block, 3-111
Heart disease, dyspnea with, 3-116
Heart murmur, 3-116
Heart sounds
   with dyspnea, 3-116
   muffled, 8-12

Heat cramps, 6-48
Heat exhaustion, 6-49—6-50
   syncope with, 3-117, 3-118
Heat-related illness, 6-48—6-51
   acclimatization and, 6-47—6-48
   fever with, 3-31
   mechanisms of, 6-47
Heat stroke, 6-50
   assessment of, 6-50
   differential diagnosis of, 6-49
   follow-up for, 6-51
   patient education for, 6-51
   symptoms and signs of, 6-50
   treatment for, 6-50—6-51
Heat test, dental caries, 5-20
Heel spur syndrome
   assessment of, 5-1
   causes of, 5-1
   follow-up for, 5-2
   patient education for, 5-2
   signs of, 5-1
   symptoms of, 5-1
   treatment for, 5-1—5-2
HEENT
   examination in secondary trauma survey, 7-4
   referred pain from in headache, 3-55, 3-56
   for underwater blast injury, 6-19
Helicobacter pylori
   in peptic ulcer disease, 4-84
   recurrence of, 4-85—4-86
   triple therapy for, 4-85
Helicopter landing sites
   criteria for, 1-25
   field expedient landing zone
      daytime, *1-29*
      nighttime, *1-30*
   identifying, 1-25
   night operation considerations, 1-25—1-26
   semi-fixed base operations
      daytime, *1-27*
      nighttime, *1-28*
Hemacytometer, 8-55, 8-57
Hematemesis, 4-85
Hematocrit, 8-53
   in anemia, 4-8
   for blood disorders, 4-8
   normal values of, 8-54
Hematoma, subdural/epidural, 4-36
Hematospermia, 4-88
Hematuria, 4-88
   differential diagnosis of, 5-47
   urinalysis in, 8-43
Heme protein, 3-57
Hemoglobin
   low blood levels of, 4-8, 5-139t
   variations in, 8-56, 8-58
Hemoglobinuria, 8-43
Hemolysis
   bilirubin metabolism in, 3-57
   with vitamin E deficiency, 5-138t

Hemolytic reactions
   to blood transfusion, 8–19
   treatment for, 8–20
Hemophilus
   ducreyi, 5–28
   influenzae, colonies of, 8–50
Hemorrhage. *See also* Gastrointestinal bleeding
   with blast injuries
      gastrointestinal, 7–23
      pulmonary, 7–23
   cardiopulmonary arrest from, 7–3
   controlled
      fluid resuscitation in, 7–14
      in hypovolemic shock, 7–13
   in hypovolemic shock, 7–12
   identifying sites of under fire, 1–1
   internal sources of, 7–2—7–3
   tourniquet for under fire, 1–2
   uncontrolled, 7–2
      fluid resuscitation in, 7–14
   with vitamin K deficiency, 5–138t
Hemorrhagic fever
   diagnostic algorithm for, 3–33—3–34f
   differential diagnosis of, 5–69, 5–70
   with renal syndrome, 5–68, 5–69
   viral, 6–60—6–61
   with yellow fever virus, 5–69
Hemorrhagic lesions, skin, 4–39
Hemorrhagic shock, 7–12
Hemostasis, during appendectomy, 4–73—4–77
Hemothorax
   thoracostomy for, 8–7
   in trauma, 7–2
   treatment of, 3–117, 7–24
Heparin
   for chest pain, 3–10, 3–12
   for frostbite, 6–43
   for pulmonary embolus, 4–24
Hepatitis
   bilirubin metabolism in, 3–57
   chronic, 5–73—5–75
   differential diagnosis of, 5–86, 5–87, 5–104
   in jaundice, 3–58
   preventive measures for, 3–59
   treatment of, 3–58, 5–74t
Hepatitis A
   assessment of, 5–71
   cause of, 5–71
   differential diagnosis of, 5–79
   follow-up for, 5–72
   patient education for, 5–72
   signs of, 5–71
   symptoms of, 5–71
   vaccination against, 3–59
      required, 5–105t
Hepatitis B, 5–72—5–73
   differential diagnosis of, 5–71, 5–79
   vaccination against, 3–59
      required, 5–105t
Hepatitis C, 5–73—5–75

   differential diagnosis of, 5–71
Hepatitis D, 5–72—5–73
   differential diagnosis of, 5–71
Hepatitis E, 5–75—5–77
   differential diagnosis of, 5–71
Hepatocellular carcinoma, 5–72
Hepatomegaly, 5–79
Hernia, incarcerated, 3–83
Herpes simplex virus
   disseminated, 3–113t
   encephalitis
      differential diagnosis of, 5–67
      memory loss with, 3–86
      treatment of, 3–86
   in genital ulcers, 5–28
   keratitis
      differential diagnosis of, 3–24
      red eye with, 3–24
      treatment of, 3–25
   treatment of, 5–29
   Tzanck stain for, 8–48—8–49
   in urethral discharges, 5–26
Herpes zoster virus, 4–47
   chest pain with, 3–10
   differential diagnosis of, 3–113t, 4–37, 5–16
   Tzanck stain for, 8–48—8–49
Herpesvirus
   differential diagnosis of, 5–67
   genital blisters and nodules with, 4–89
Herpetic lesions, oral, 5–16
Hiatal hernia
   chest pain with, 3–10
   differential diagnosis of, 4–78
   disqualifying condition for diving, 6–22
Hibiclens soap, 4–43
High altitude cerebral edema (HACE), 6–37, 6–38
   assessment of, 6–39
   differential diagnosis for, 6–41
   follow-up for, 6–40
   in high altitude pulmonary edema, 6–40
   patient education for, 6–39—6–40
   signs of, 6–39
   symptoms of, 6–38—6–39
   treatment for, 6–39
High altitude illness, 6–37—6–42
High altitude pulmonary edema (HAPE), 6–37
   assessment of, 6–41
   cause of, 6–40
   follow-up for, 6–42
   high altitude cerebral edema and, 6–38, 6–39
   patient education for, 6–41—6–42
   portable hyperbaric chamber for, 8–36
   signs of, 6–40—6–41
   symptoms of, 6–40
   treatment for, 6–41
High-altitude retinal damage, 3–22
Hip
   dislocation of

assessment of, 3-68
diagnostic tests for, 3-69
differential diagnosis of, 3-69
post treatment care for, 3-69
posterior, 3-65
procedure for, 3-69
fractures of, 3-74
pain of
assessment of, 3-73
causes of, 3-72
follow-up, 3-74
patient education for, 3-74
referred, 3-73
risk factors for, 3-72
signs of, 3-73
symptoms of, 3-73
treatment of, 3-73—3-74
tendonitis of, 3-73
Histoplasma capsulatum, 5-61
Histoplasmosis
assessment of, 5-62
causes of, 5-61
disseminated, 5-78
follow-up for, 5-62
patient education for, 5-62
risk factors for, 5-61
signs of, 5-61—5-62
symptoms of, 5-61
treatment for, 5-62
zoonotic disease considerations in, 5-62
History of present illness, 2-2
History questions, in care under fire, 1-1
HIV. *See* Acquired immune deficiency syndrome; Human immunodeficiency virus
Hives, 4-39
in anaphylactic shock, 7-10
Hoof testers, 5-134
Hookworm. *See also* Cutaneous larva migrans
assessment of, 5-42
differential diagnosis of, 5-48
follow-up for, 5-42
incidence of, 5-41
patient education for, 5-42
signs of, 5-42
species of, 5-41
symptoms of, 5-41—5-42
treatment for, 5-42
zoonotic considerations in, 5-42
Hornet sting, 4-57—4-59
Horses
colic in, 5-132—5-133
lameness in, 5-134—5-135
restraint of, 5-128—5-129
Hospital survey checklist, in MedCAP guide, 1-4—1-7
Hot compresses, 6-15
Howell-Jolly bodies, 8-56
Human bites
assessment of, 7-8
follow-up for, 7-9
patient education for, 7-9

risk associated with, 7-7
signs of, 7-8
symptoms of, 7-7—7-8
treatment of, 7-8—7-9
Human diploid cell rabies vaccine, 5-82
for human and animal bites, 7-9
Human immunodeficiency virus. *See also* Acquired immune deficiency syndrome
in aphthous ulcers, 5-16
assessment of, 5-78
course of untreated, 5-76f
differential diagnosis of in anemia, 4-9
occupational exposure prophylaxis for, 5-78
patient education for, 5-78
signs of, 5-78
symptoms of, 5-77
transmission of, 5-77
treatment for, 5-78
Human papillomavirus
in plantar warts, 5-3
warts with, 4-48
Human rabies immune globulin
for human and animal bites, 7-9
for rabies, 5-82
Human remains, recovering
cautions for, 5-155
equipment for, 5-153
mission, 5-153
procedures for, 5-154—5-155
what to expect in, 5-153
Hydralazine (Apresoline), 3-109
Hydration
for adrenal insufficiency, 4-28
for decompression sickness, 6-5
for heat stroke, 6-51
for mastitis, 3-8
for pulmonary over inflation syndrome, 6-8
for syncope, 3-118
for urolithiasis, 4-92
Hydrocele, above testis, 3-80
Hydrocortisone
for adrenal insufficiency, 4-28
for bed bug bites, 4-55
for contact dermatitis of penis, 3-78
for millipede exposure, 4-57
Hydrofluoric acid burn, 7-19
Hydrogen peroxide, 6-15
Hydromorphone, 3-102
Hydropel, 5-8
Hydrophidae snakes, 5-143, 5-144
Hydrophobia. *See* Rabies
Hydroxyzine (Atarax)
for contact dermatitis, 4-70
for psoriasis, 4-65
Hygiene
for candida vaginitis prevention, 3-50
for diarrhea prevention, 3-19
for epididymitis prevention, 3-85
for jaundice, 3-59
for male genital inflammation, 3-79

Hymenoptera stings, 4–57—4–58
 assessment of, 4–58
 follow-up for, 4–59
 patient education for, 4–59
 signs of, 4–58
 symptoms of, 4–58
 treatment for, 4–58
Hyoscyamine
 for prostatitis, 3–82
 for urinary incontinence, 4–90
Hyperbaric chamber
 for high altitude cerebral edema, 6–40
 for high altitude pulmonary edema, 6–41, 6–42
 portable, 8–36
  equipment for use of, 8–36—8–37
  indications for, 8–36
  precautions with, 8–37
  treatment with, 8–37
Hyperbaric oxygen recompression therapy
 for carbon monoxide poisoning, 6–11
 for decompression sickness, 6–5, 6–36
 for pulmonary over inflation syndrome, 6–8
Hypercapnia. See Carbon dioxide poisoning
Hyperglycemia, 4–29
Hyperhidrosis, 5–7, 5–8
Hyperkeratosis, 4–67
 in corns and calluses, 5–5
Hyperlite transport chamber, 8–36
 for decompression sickness, 6–36
 for high altitude cerebral edema, 6–40
 for high altitude pulmonary edema, 6–42
Hypersensitivity pneumonitis, 4–18
Hypertension
 in eclampsia, 3–105
 headache in, 3–55
 in preeclampsia/eclampsia, 3–104
 with sleep apnea, 4–25
Hypertensive emergency
 assessment of, 4–5
 in congestive heart failure, 4–4
 defined, 4–5
 follow-up, 4–6
 patient education for, 4–5
 signs of, 4–5
 symptoms of, 4–5
 treatment of, 4–5
Hyperthermia, 6–37, 6–39
Hyperthyroidism, 4–31—4–32
 in diabetes mellitus, 4–28
 diagnosis of, 4–32—4–33
 differential diagnosis of, 4–33
 fever with, 3–31
 in mania, 3–16
 symptoms of, 4–32
 treatment of, 4–33
Hypertonic saline-colloid combinations, 7–15t
Hyphema
 differential diagnosis of, 3–27
 treatment of, 3–28

Hypocalcemic tetany, 5–101
Hypochromasia, hemoglobin variations in, 8–56
Hypoglycemia
 alleviating or aggravating factors of, 4–30
 assessment of, 4–31
 causes of, 4–30
 differential diagnosis of, 6–44
 follow-up, 4–31
 patient education for, 4–31
 signs of, 4–30
 symptoms of, 4–30
 treatment of, 4–31
Hypoglycemic agents, 4–29
Hypotension, 7–12
Hypotensive states, 4–27
Hypothermia, 6–46t
 assessment of, 6–44
 with chemical burns, 7–18
 classification of level of, 6–45t
 definition of, 6–43
 differential diagnosis of, 6–37, 6–39
 follow-up for, 6–44
 patient education for, 6–44
 symptoms and signs of, 6–43
 in trauma patients, 7–3
 treatment for, 6–44
Hypothyroidism, 4–32
 apnea with, 4–26
 in depression, 3–16
 in diabetes mellitus, 4–28
 diagnosis of, 4–33
 differential diagnosis of, 4–33
 fatigue with, 3–30t
 symptoms of, 4–32
 treatment of, 4–33
Hypoventilation, voluntary, 6–12
Hypovolemia, syncope with, 3–118
Hypovolemic shock
 assessment of, 7–12—7–13
 cause of, 7–11—7–12
 hemorrhage control in, 7–13
 signs of, 7–12
 symptoms of, 7–12
 treatment for, 7–13—7–14
Hypoxemia
 in apnea, 4–26
 fatigue with, 3–30t
Hypoxia
 in aerospace medicine, 6–31
 assessment of, 6–9, 6–31
 in carbon dioxide poisoning, 6–12
 differential diagnosis of, 6–11, 6–12, 6–36
 in diving operations, 6–8—6–9
 follow-up for, 6–9, 6–32
 memory loss with, 3–86
 patient education for, 6–9, 6–32
 signs of, 6–9, 6–31
 symptoms of, 6–9, 6–31
 treatment for, 6–9, 6–31
Hytrin (terazosin)

for prostatitis, 3-82
for urinary incontinence, 4-91

IAPP assessment technique, 7-4—7-5
Ibuprofen
  for abnormal uterine bleeding, 3-41
  for acute mountain sickness, 6-38
  for avulsed tooth, 5-14
  for Bartholin's gland abscess or cyst, 3-52
  for blood transfusion reaction, 8-19
  for breast engorgement, 3-9
  for chest pain, 3-12
  for elbow dislocation, 3-68
  for endometriosis, 3-45
  for epididymitis, 3-84
  for febrile reaction to blood transfusion, 8-20
  for joint pain, 3-63
  for low back pain, 3-6—3-7
  for mastitis, 3-8
  for mittleschmerz, 3-46
  for pain control, 8-39
  for patellar dislocation, 3-69
  for pericarditis, 4-6
  for temporomandibular joint dislocation, 5-19
  for testicular/scrotal masses, 3-80
  for thrombosis of penile vein and sclerosing lymphangitis, 3-78
  for urolithiasis, 4-92
Ice
  contamination of, 5-109
  for low back pain, 3-7
  for stress fractures of foot, 5-6
Ice massage, heel, 5-1
Icotest, 8-42
IgG immunoglobulins, 8-19
Illness surveillance, 5-107—5-108
Imipenem
  for jaundice, 3-58
  for underwater blast injury, 6-19
Imipramine, 4-90
Imiquimod
  for molluscum contagiosum, 4-48
  for verruca plana, 4-49
Imitrex (sumatriptan), 3-56
Immersion foot, 6-44—6-47
Immobilization, septic joint, 3-63
Immunizations
  required, 5-105—5-106t
  for tetanus, 5-101, 5-102
Immunoglobulin, 4-8
  for hepatitis A, 5-72
  for streptococcal infections, 5-100
Immunosuppression, 3-49
Impetigo
  contagiosa
    assessment of, 4-43
    cause of, 4-43
    follow-up for, 4-44
    patient education for, 4-44

  symptoms and signs of, 4-43
  treatment for, 4-43
  pustules in, 4-38
  streptococcal infection in, 5-99
Incinerator, barrel, 5-116f
Incontinence, urinary, 4-88
  assessment of, 4-90
  follow-up for, 4-91
  incidence of, 4-90
  patient education for, 4-91
  signs of, 4-90
  symptoms of, 4-90
  treatment for, 4-90—4-91
Inderal
  cautions with, 3-56
  for headache, 3-56
Indinavir, 5-78
Indomethacin (Indocin)
  for acute mountain sickness, 6-38
  for gout, 3-63
  for joint pain, 3-63
  for pericarditis, 4-6
  for preterm labor, 3-94
  for urolithiasis, 4-92
Infarction, pulmonary embolus, 4-23
Infectious diseases
  bacterial, 5-90—5-105
  fatigue with, 3-29
  in fever, 3-31
  fungal, 5-57—5-63
  mycobacterial, 5-54—5-57
  parasitic, 5-33—5-54
  rickettsia, 5-83—5-l86
  spirochetal, 5-86—5-90
  viral, 5-63—5-83
Inflammatory bowel disease, 3-19
Influenza
  differential diagnosis of, 5-64, 5-86, 5-87
  required vaccination for, 5-105t
  treatment of, 4-11—4-12
  vaccination for, 4-12
    for COPD prevention, 4-22
    in pneumonia prevention, 4-13
  viruses of, 4-10
    in pneumonia, 4-14
  zoonotic, 5-32
Injury
  caused by weapons, 1-1
  surveillance of, 5-107—5-108
Insect bites, 7-8
  bed bugs, 4-55
  centipede, 4-55—4-56
  differential diagnosis of, 4-53
  hymenoptera, 4-57—4-59
  millipede, 4-56—4-57
  mite, 4-59—4-60
  pediculosis, 4-62—4-63
  pruritus with, 3-114
  scabies, 4-61—4-62
  spider, 4-60—4-61

wheals in, 4–39
Insect envenomation
 periocular, 3–26, 3–27
 in periorbital inflammation, 3–26
Insect netting, 4–52
Insecticides, 4–55
Insulin
 deficiency of, 4–28
 for diabetes mellitus, 4–29
 resistance to, 4–28
Integument, physical examination of, 2–5
Interferon, 5–75
Intertrigo, 5–57—5–58
Intestinal flu. *See* Diarrhea; Flu; Influenza
Intestinal flukes, giant, 5–38
Intestinal ischemia/infarction, 3–3t
Intestinal obstruction. *See* Bowel obstruction
Intoxication, 6–37, 6–39
Intracranial hemorrhage, memory loss with, 3–86
Intracranial pressure elevation, 4–5
Intracranial swelling, 3–55
Intraosseous infusion, 8–21—8–22
Intravenous access, 7–3
 for blister agents, 6–53
Intravenous anesthetic medications, 5–155
 dosing guidelines for, 5–156t
Intravenous catheters, for animals, 5–131—5–132
Intravenous drug therapy, Naval special warfare combat trauma AMAL items for, 1–20—1–21
Intravenous fluids
 for acute organic intestinal obstruction, 4–86
 for electrical or lightning injury, 7–27
 for fever, 3–31
 for gastritis, 4–81
 for hypovolemic shock, 7–13
 for pancreatitis, 4–82
 for peritonitis, 4–84
 for poisoning, 5–141
 for streptococcal infections, 5–100
 for urinary tract infections, 4–94
Intravenous infusion, veterinary, 5–131—5–132
Intraventricular hemorrhage, 3–93
Intubation procedure, 8–4
 patient evaluation in, 8–4
 precautions in, 8–5
 technique in, 8–4—8–5
Intussusception
 in acute organic intestinal obstruction, 4–86
 differential diagnosis for, 4–71
Iodine
 deficiency of, 5–139t
 in goiter, 4–31
 for equine lameness, 5–134
Ipecac syrup, 5–141, 5–142
Ipratropium (Atrovent), 4–22
Ipratropium bromide
 for asthma, 4–20
 for common cold and flu, 4–11
Ipratropium nebulizer, 4–21

Iritis
 differential diagnosis of, 3–25
 traumatic
  differential diagnosis of, 3–28
  treatment of, 3–28
 treatment of, 3–25
Iron
 deficiency of, 5–139t
 supplementation of, 4–9
Iron deficiency anemia, 4–8
 differential diagnosis of, 4–9
Irritable bowel syndrome
 acute pelvic pain with, 3–43
 differential diagnosis of, 3–44t
 physical examination for, 3–46
 symptoms of, 3–46
 treatment of, 3–46
Irritant contact dermatitis, 4–69
Irritant rhinitis, 4–11
Irritated seborrheic keratosis, 4–66
Isocyanate lung, 4–18
Isoniazid
 for pulmonary disease, 6–21
 for tuberculosis, 5–55
Itching
 in aerospace medicine, 6–36
 with pruritus, 3–113—3–115
Itraconazole
 for blastomycosis, 5–59
 for candidiasis, 5–58
 for coccidioidomycosis, 5–61
 for dermatophyte infections, 4–50
 for histoplasmosis, 5–62
 for paracoccidioidomycosis, 5–63
IV. *See* Intravenous
Ivermectin
 for cutaneous larva migrans, 5–42
 for filariasis, 5–40
 for loiasis, 4–52
 for onchocerciasis, 4–53
 for scabies, 4–61
 for strongyloides, 5–48

J tube, 8–3
Japanese B encephalitis vaccination, 5–105t
Japanese encephalitis, 5–66, 5–67, 5–68
Jarisch-Herxheimer reaction, 5–28
Jaundice
 assessment of, 3–58
 characteristics and mechanisms of, 3–57
 common causes of, 3–58
 follow-up, 3–59
 in hepatitis A, 5–71
 in hepatitis C, 5–73
 in hepatitis E, 5–75
 patient education for, 3–59
 signs of, 3–57—3–58
 symptoms of, 3–57
 treatment of, 3–58

Jaw injuries, 5–18
Jellyfish stings, 6–13—6–14
Jellyfish toxin, 4–54
Jerky, 5–127
Jock itch, 4–49—4–51
Joint
  aspiration of, 8–28—8–30
    equipment for, 8–28
    indications for, 8–28
    precautions in, 8–30
    sites of, 8–29
  examination of
    palpation of, 3–60
    range of motion of, 3–60
    stability of, 3–60
  pain in
    activity with, 3–64
    arthritis and, 3–59
    assessment of, 3–61—3–62
    causes of, 3–59
    follow-up, 3–64
    hip, 3–72—3–74
    inflammatory conditions in, 3–59
    with joint dislocation, 3–64—3–70
    knee, 3–74—3–76
    mechanical processes in, 3–59
    patient education for, 3–63—3–64
    shoulder, 3–70—3–72
    signs of, 3–60—3–61
    symptoms of, 3–59—3–60
    treatment of, 3–62—3–63
Joint fluid analysis, 8–28
Jugular venous distention
  in congestive heart failure, 4–4
  in trauma, 7–2, 7–4

K factor, 8–57—8–58
K vitamin, 3–59
K vitamin deficiency, 5–138t
Kala azar, 5–42, 5–43
Kaposi's sarcoma, 5–93
Katayama fever
  in schistosomiasis, 5–46
  treatment of, 5–47
Kawasaki disease, 3–113t
Keflex
  for cellulitris of penis, 3–78
  for impetigo contagiosa, 4–43
  for ingrown toenail, 5–3
  for phimosis, 3–78
  for prostatitis, 3–81
  in testis torsion treatment, 3–83
  for urinary incontinence, 4–91
  for urinary tract infections, 4–93
Keflex/Ancef, 5–146
Kegel exercises, 4–90
Kenalog
  in Orabase gel, for aphthous ulcers, 5–17
  for trochanteric bursitis, 3–73

Keratitis
  prevention of in Bell's palsy, 4–37—4–38
  ultraviolet
    differential diagnosis of, 3–25
    red eye with, 3–24
Keratoacanthoma, 4–67
Keratoconjunctivitis, 5–28
Ketamine
  for chemical burns, 7–18
  continuous infusion of in field, 5–157—5–158t
  dosing guidelines for, 5–156t
  for pain control, 8–40
Ketoacidosis, 4–29
Ketoconazole
  for blastomycosis, 5–60
  for candidiasis, 5–58
  for histoplasmosis, 5–62
  for leishmaniasis, 5–43
  for paracoccidioidomycosis, 5–63
  for pityriasis versicolor, 4–51
  side effects of, 5–58
Ketones-Acetest procedure, 8–42
Ketonuria, 8–42—8–43
Ketorolac (Toradol)
  for pain control, 8–38
  for urolithiasis, 4–92
Kidney failure, pruritus with, 3–114
Kidney stones, 3–83. See also Urolithiasis
Killer whales, 6–17
Killip classification, myocardial infarction, 4–3
Kinyon's carbol-fuchsin stain, 8–47
Kissing bug bite, 5–53
Kissing disease. See Mononucleosis, infectious
Knees
  cartilage or ligament tear at, 3–59
  dislocation of, 3–65—3–66
    consultation for, 3–76
  joint aspiration at, 8–29
  pain in
    assessment of, 3–75
    causes of, 3–74
    follow-up, 3–76
    patient education for, 3–75—3–76
    risk factors for, 3–74
    signs of, 3–75
    symptoms of, 3–74—3–75
    treatment of, 3–75
  palpation of, 3–75
KOH wet mount
  for bacterial vaginosis, 3–47, 3–49
  for candida vaginitis, 3–49—3–50
  for candidiasis, 5–58
  for coccidioidomycosis, 5–61
  for female pelvic examination, 3–37
  procedures in, 8–43—8–44
  in vaginal examination, 3–38
Kotorolac, 8–39
Kraits, 5–143
Kunjin, 5–66

Kwell (lindane), 4-61
Kytril (granisetron), 6-62

La Cross virus, 5-66
Labetalol, 3-109
Labor, 3-87
   in animals, 5-130—5-131
   preterm, 3-93—3-95
   shoulder dystocia with, 3-95—3-96
   stages of, 3-87—3-88
   vaginal delivery during
      first stage, 3-88
      fourth stage, 3-93
      second stage, 3-88
      third stage, 3-88—3-93
Laboratory procedures
   ABO grouping and confirmation tests, 8-59
   brucellosis test, 8-45
   Con-Trate method, 8-52—8-53
   crossmatching, 8-60—8-62
   culture interpretation, 8-49—8-50
   for feces for ova and parasites, 8-51—8-52
   Giemsa stain, 8-47—8-48
   Gram stain, 8-44—8-45
   macroscopic examination of feces and occult blood test, 8-50—8-51
   microhematocrit determination, 8-53—8-54
   preparing specimens for transport, 8-40—8-41
   RBC count/morphology, 8-54—8-56
   Rh typing, 8-59—8-60
   Tzanck stain, 8-48—8-49
   urinalysis, 8-41—8-43
   WBC count on whole blood, 8-56—8-58
   WBC differential count, 8-58
   wet mount and KOH prep, 8-43—8-44
   Wright's stain, 8-46—8-47
   Ziehl-Neilson stain, 8-47
Laboratory tests
   for COPD, 4-22
   for fatigue, 3-30
   in joint fluid analysis, 8-29
   in joint pain assessment, 3-61
   for knee pain, 3-75
   for memory loss, 3-86
   for palpitations, 3-110
   for pruritus, 3-114
   for syncope, 3-118
Labyrinthitis
   differential diagnosis of, 3-20
   dizziness in, 3-20
   treatment of, 3-21
Lac-Hydrin, 4-50
Lacrimal gland inflammation, 3-26
Lactulose, 3-59
Lameness
   antemortem veterinary exam for, 5-123
   equine, 5-134—5-135
Laminitis, equine, 5-134

Lamivudine, 5-78
Landfill management, 5-121
Lansoprazole, 4-85
Laryngeal mask airway, 7-1
Laryngospasms, 6-21
Laser exposure evacuation criteria, 7-29
Laser eye injuries, 7-28—7-29
Lasix (furosemide)
   for congestive heart failure, 4-4
   for electrical or lightning injury, 7-27
   for hemolytic reaction, 8-20
   for pleural effusion, 4-15
   for venomous snake bite, 5-145
Lassa fever, 6-61
Latrines, 5-111—5-113f, 5-115
   closing, 5-118
   maintenance of, 5-115
Latrodectus spider, 4-60
Laxatives
   for equine colic, 5-133
   for self-limiting abdominal pain, 3-1
Lead poisoning, 4-71
Left bundle branch block, 4-1
Legionella species, 4-14
Legs, restraining of (cows), 5-129—5-130
Leishmaniasis
   assessment of, 5-43
   cutaneous type, 5-43
   follow-up for, 5-43
   patient education for, 5-43
   signs of, 5-43
   symptoms of, 5-43
   transmission of, 5-42
   treatment for, 5-43
   visceral, 5-43
      differential diagnosis of, 5-78
   zoonotic disease considerations in, 5-43—5-44
Lepidoptera sting, 4-57
Leprosy, 4-45
Leptospira, 5-86
Leptospirosis, 5-32
   assessment of, 5-87
   differential diagnosis of, 4-36, 5-69, 5-70, 5-71, 5-84, 5-89, 5-104
   follow-up for, 5-87
   patient education for, 5-87
   prevention of, 5-74t
   signs of, 5-86—5-87
   symptoms of, 5-86
   transmission of, 5-86
   treatment for, 5-74t, 5-87
   zoonotic disease considerations in, 5-87
Leukemia, 4-8
Levaquin
   for epididymitis, 3-84, 3-85
   for prostatitis, 3-81, 3-82
   for urinary tract infections, 4-93, 4-94
Levofloxacin
   for anthrax, 5-91
   for eye injury, 3-28

for orbital and periorbital inflammation, 3-27
for pneumonia, 4-13
for urolithiasis, 4-92
Lewisite, 6-53—6-54
Libido loss, 3-29
Lice, 4-62
　body, 4-62—4-63
　head, 4-62—4-63
Lichen simplex chronicus, 4-38
　differential diagnosis of, 3-114, 4-65
Lichenification, 4-39
Lichenoid keratosis, benign, 4-66
Lidex gel, 5-17
Lidocaine
　in appendectomy, 4-77
　in axillary blockade, 5-174t
　in Bartholin's gland abscess drainage, 3-53
　for bladder catheterization, 8-34
　for cesarean section, 3-100, 3-101
　for chest pain, 3-12
　for digital block of finger or toe, 5-163t
　in dorsal slit procedure, 3-79
　for elbow dislocation, 3-67
　with epinephrine, 5-20
　　for breast abscess drainage, 3-9
　in episiotomy, 3-103
　for eye injury, 3-28
　local infiltration of, 5-159
　for pain control, 8-39
　for patellar dislocation, 3-69
　for suprapubic bladder aspiration, 8-35
　in suturing, 8-22
　in testis torsion treatment, 3-83
　in thoracentesis, 4-16
　in thoracostomy, 8-7
　for trochanteric bursitis, 3-73
　in tube thoracostomy, 8-8
　for venomous fish bites, 6-15
Ligament tear, knee, 3-59
Lightning injuries, 7-26
　assessment of, 7-27
　follow-up for, 7-28
　patient education for, 7-28
　signs of, 7-27
　symptoms of, 7-26—7-27
　treatment for, 7-27—7-28
Lindane
　for scabies, 4-61
　shampoo, for pediculosis, 4-63
Lingual nerve blockade, 5-23f
Lipoma, removal of, 8-26
Liposomal amphotericin B, 5-43
Liquid waste disposal, 5-118
Lisinopril, 4-5
Listeria, food poisoning, 4-79
Lithotomy table, 3-38
Liver
　amebic abscess of, 5-104
　disease of, 3-113
　enlargement of

　　in clonorchiasis, 5-35
　　differential diagnosis of, 5-47
　　with jaundice, 3-57—3-58
　failure of
　　jaundice in, 3-57—3-58
　　pleural effusion with, 4-15
　masses in, 4-78
Liver fluke, Chinese. See Clonorchiasis
Loa loa. See Loiasis
Local infiltration, anesthesia, 5-159
Lodine, 4-33
Loiasis, 4-52
Loperamide
　for acute diarrhea, 3-19
　for fever, 3-34f
Lorazepam
　for anxiety, 3-5
　for mania, 3-17
　for pain control, 8-39
　for psychosis and delirium, 5-152
　for tetanus, 5-101
Lotrimin
　for balanitis, 3-78
　for candidal penile infection, 3-78
Low back pain, 3-6
　assessment of, 3-6
　follow-up, 3-7
　mechanical, 3-6
　　in joint pain, 3-62
　　treatment of, 3-6—3-7
　patient education for, 3-7
　prevention of, 3-7
　referred, 5-1
　signs of, 3-6
　symptoms of, 3-6
　treatment plan for, 3-6—3-7
Loxosceles spider, 4-60
Lumbar mobility assessment, 3-61
Lumbar puncture, 3-87
Lung fluke, 5-45
Lungs
　cancer of
　　cough with, 3-15
　　differential diagnosis of, 5-54
　　pleural effusion with, 4-15
　disease of, 3-14
　granulomas of in histoplasmosis, 5-62
　intubation of for airway compromise, 7-25f
Lupus
　erythematosus, 4-39
　fatigue with, 3-30t
　joint pain with, 3-60
　treatment of, 3-63
　vulgaris, in cutaneous tuberculosis, 4-44
Lyme disease, 5-32
　assessment of, 5-88
　differential diagnosis of, 3-113t, 4-37, 5-89, 5-98
　fatigue with, 3-30t
　follow-up for, 5-88

patient education for, 5-88
signs of, 5-88
symptoms of, 5-88
transmission of, 5-88
treatment for, 5-88
zoonotic disease considerations in, 5-88—5-89
Lymph nodes, tender or draining, 3-34f
Lymphadenitis
differential diagnosis of, 5-57
in nontuberculous mycobacterial infections, 5-56
treatment of, 5-57
Lymphadenopathy
in American trypansomiasis, 5-53
in animals, 5-126
differential diagnosis of, 5-79
in erysipelas, 4-42
pruritus with, 3-114
in schistosomiasis, 5-47
Lymphangitis
differential diagnosis of, 5-100
in ecthyma contagiosum, 4-46
Lymphatic system examination, 2-5
Lymphedema, 5-40
Lymphogranuloma venereum
differential diagnosis of, 5-103
in genital ulcers, 5-28
treatment of, 5-29
Lymphoma
differential diagnosis of, 5-78
fatigue with, 3-30t

M5 packing list, suggested, 1-18—1-20
MacRobert's position, 3-95
Macrobid
for cesarean section, 3-102
for urinary tract infections, 4-93
Macrodantin, 4-94
Macrolides
for allergic pneumonitis, 4-19
for pneumonia, 4-13
Maculae caeruleae, 4-62
Macule, 4-38
Maculotoxin, 6-14
Mad dog syndrome, 5-82—5-83
Mafenide acetate (Sulfamylon), 7-18
Magnesium citrate, 3-13
Magnesium-containing antacids, 7-18
Magnesium deficiency, 5-139f
Magnesium sulfate
for eclampsia, 3-109
for preeclampsia, 3-109
for preterm labor, 3-94
Major depressive disorder, 5-152
Malaria
assessment of, 5-44
diagnostic algorithm for, 3-33f, 3-34f
differential diagnosis of, 4-71, 5-66, 5-71,
5-84, 5-86, 5-87, 5-89, 5-104, 6-50
fever with, 3-32
follow-up for, 5-45
Giemsa stain for, 8-47—8-48
patient education for, 5-45
prevention and control of, 5-120
prophylaxis for, 5-45
signs of, 5-44
symptoms of, 5-44
transmission and incidence of, 5-44
treatment for, 5-44—5-45
Malarone, 5-45
Malayan filariasis, 5-39
Male genital problems. *See also* Sexually transmitted disease
epididymitis, 3-84—3-85
genital inflammation, 3-77—3-79
prostatitis, 3-80—3-82
testis/scrotal mass, 3-79—3-80
testis torsion, 3-82—3-84
Malignancy
fever with, 3-31
meningial signs in, 4-36
Mallory-Weiss tear
differential diagnosis of, 4-81
with syrup of Ipecac, 5-142
Mamba snake, 5-143
Mandibular extraction, 5-21—5-23
Mandibular hypermobility, 5-18
Mandibular Universal Forceps, 5-25
Mania, 3-15
assessment of, 3-17
follow-up, 3-17—3-18
patient education for, 3-17
signs of, 3-16
symptoms of, 3-16
treatment for, 3-17
Mannitol
for blood transfusion reaction, 8-19
for electrical or lightning injury, 7-27
for hemolytic reaction, 8-20
side effects of, 7-28
for venomous snake bite, 5-145
Marburg virus, 6-61
Marcaine (epinephrine)
for bunion, 5-5
bupivacaine with, 5-20
for heel spur syndrome, 5-2
for ingrown toenail, 5-3
for trochanteric bursitis, 3-73
Marine life
biting, 6-16—6-18
venomous, 6-13—6-16
Mark 1 auto-injector, 6-52
Mask
for blister agents, 6-53
for chemical blood agents, 6-52
for chemical nerve agents, 6-52
Massage, for milk duct obstruction, 3-9

MAST. *See* Military anti-shock trousers (MAST)
Mastitis, 3–7
  in animals, 5–135
  assessment of, 3–7—3–8
  chest pain with, 3–10
  follow-up, 3–8—3–9
  patient education for, 3–8
  signs of, 3–7
  symptoms of, 3–7
  treatment for, 3–8
  vital signs suggesting, 3–11, 3–12
Maxalt (rizatriptan), 3–56
Maxillary extraction, 5–20
Maxillary Universal Forceps, 5–25
Measles
  differential diagnosis of, 5–66
  required vaccination for, 5–106t
Meat
  cooking, 5–109
  salting of, 5–128
  storage and preservation of, 5–126—5–128
Mebendazole (Vermox)
  for ascariasis, 5–34
  for enterobiasis, 5–37
  for hookworm, 5–42
  for trichenellosis, 5–50
  for trichuriasis, 5–51
Meclizine (Antivert), 3–21
MedCAP guide, 1–2
  final planning conference (FPC), 1–3
  hospital survey checklist, 1–3
  initial planning conference (IPC), 1–2
  mid planning conference (MPC), 1–2—1–3
  planning checklist of, 1–2—1–3
  predeployment site survey (PDSS), 1–3
  site survey checklist, 1–3
  veterinary site survey checklist, 1–3
MedEvac
  in MedCAP guide, 1–3
  Request, 9 line, 1–23—1–24
Median nerve block, at wrist, 5–165—5–165
  contraindications to, 5–166
Mediastinitis, 5–91
MEDIC network, fever diagnosis, 3–31
Medical documentation, 1–22
Medical examination, 2–2
Medical history, 2–3
Medications
  M5 item list for, 1–19
  Naval special warfare combat trauma AMAL items for, 1–20
Mefloquine, 5–45
Melanoma
  amelanotic, 4–68, 6–66
  malignant, 4–68
Melarsoprol, 5–52
Melatonin, 3–31
Melena, 3–1
Memory function assessment, 3–85—3–86
Memory loss
  assessment of, 3–86
  etiology of, 3–85
  follow-up, 3–87
  patient education for, 3–86
  signs of, 3–85—3–86
  symptoms of, 3–85
  treatment for, 3–86
Meniere's disease
  differential diagnosis of, 3–20, 6–2
  dizziness in, 3–20
  treatment of, 3–21
Meningitis
  assessment of, 4–37
  bacterial, 4–35
    treatment of, 4–36
  causes of, 4–35
  in coccidioidomycosis, 5–60
  diagnostic algorithm for, 3–33—3–34f
  differential diagnosis of, 5–61
  follow-up for, 4–38
  headache in, 3–55
  patient education, 4–38
  signs of, 4–37
  symptoms of, 4–35—4–36
  treatment of, 3–56, 4–37
Meningococcal fever, 5–66
Meningococcal infection
  diagnostic algorithm for, 3–33—3–34f
  differential diagnosis of, 5–89, 5–103
  required vaccination for, 5–106t
Meningococcemia, 4–41
  differential diagnosis of, 3–112t, 4–40, 5–84, 6–50
Meningoencephalitis
  differential diagnosis of, 5–101
  flaviviruses in, 5–66
Meniscal tear, 3–74
Menstrual flow, changes in interval and duration of, 3–39
Mental disorders
  associated with anxiety, 3–5
  differential diagnosis of, 5–152
  operational stress in, 5–148
Mental health
  mood disorders. *See* Depression
  operational stress, 5–147—5–148
  psychosis versus delirium, 5–151—5–153
  recovering human remains, 5–153—5–155
  substance abuse, 5–150—5–151
  suicide prevention, 5–149—5–150
Mental status assessment
  for African trypanosomiasis, 5–52
  for anxiety, 3–2—3–5
  for delirium and psychosis, 5–152
  for depression and mania, 3–16
  for headache, 3–55
  in hypoglycemia, 4–30
  with hypoglycemia, 4–31
  in hypovolemic shock, 7–12
  for meningitis, 4–36

with poisoning, 5-140
Mepivacaine
　in axillary blockade, 5-174t
　in dental procedures, 5-20
　for pain control, 8-39
6-Mercaptopurine, 4-82
Mesenteric lymphadenitis, 4-71
Mess wash setup, 5-110f
Metabolic acidosis, 4-29
Metamucil
　for chronic pelvic pain, 3-44t
　for irritable bowel syndrome, 3-46
Metaproterenol (Alupent), 4-22
Metastatic cancer, adrenal, 4-27
Metatarsal pad, 5-7
Metatarsal stress fractures, 5-6
　treatment of, 5-7
Metered dose inhaler, 4-20
Methadone, 7-18
Methemoglobin, 8-10
Methylcellulose, 3-13
Methylene blue
　in Wright's stain, 8-46
　in Ziehl-Neilson stain, 8-47
Methylprednisolone (Solu-Medrol)
　for anaphylactic shock, 7-11
　side effects of, 6-20
　for underwater blast injury, 6-19
Metrogel, 5-31
Metronidazole (Flagyl)
　for acute abdominal pain, 3-2
　for acute diarrhea, 3-19
　for acute organic intestinal obstruction, 4-87
　for amebiasis, 5-33
　for bacterial vaginosis, 3-49
　for blast injury, 7-26
　for cholecystitis, 4-78
　for fever, 3-34f
　for giardiasis, 5-41
　in pancreatitis, 4-82
　for peritonitis, 4-84
　precautions with use of, 5-41
　for prostatitis, 3-82
　for tetanus, 5-101
　for urethral discharges, 5-27
　for vaginal trichomonas, 5-31
"Michelin man" look, 5-100
Miconazole
　for candida vaginitis, 3-48t, 3-50
　for candidiasis, 5-58
Microhematocrit
　centrifuge/reader method, 8-53
　determination of, 8-53—8-54
Microhematuria, 4-93
Microorganisms, identification of, 8-49—8-50
Microsporum, 4-49
Midazolam (Versed)
　continuous infusion of in field, 5-157—5-158t
　dosing guidelines for, 5-156t
　for pain control, 8-39

for temporomandibular joint dislocation, 5-19
　for tetanus, 5-101
Midrin, 3-56
Migraine headache
　memory loss with, 3-86
　symptoms of, 3-55
Military anti-shock trousers (MAST), 7-2
　for anaphylactic shock, 7-11
　for hemorrhage control in hypovolemic shock, 7-13
Military diving, disqualifying conditions for, 6-21—6-22
Milk duct obstruction, 3-7
　assessment of, 3-7—3-8
　treatment of, 3-9
Milker's nodules, 4-45—4-46
Millaria rubra, 4-53
Millipede exposure, 4-56
　assessment of, 4-57
　follow-up for, 4-57
　patient education for, 4-57
　signs of, 4-56
　symptoms of, 4-56
　treatment for, 4-57
Mineral deficiencies, 5-137—5-139t
Mini-Mental State Examination (MMSE), 3-85—3-86
Mini-mental status exam
　for carbon dioxide poisoning, 6-12
　for carbon monoxide poisoning, 6-11
Minipress
　for prostatitis, 3-82
　for urinary incontinence, 4-91
Minocycline, 4-66
Mites, 4-59
　assessment of, 4-59
　follow-up for, 4-60
　patient education for, 4-60
　signs of, 4-59
　symptoms of, 4-59
　treatment for, 4-59
Mitral valve prolapse, 3-10
Mittelschmerz
　in chronic pelvic pain, 3-46
　diagnosis of, 3-46
　differential diagnosis of, 3-44t
　pelvic examination for, 3-46
　symptoms of, 3-46
　treatment of, 3-46
Moles, removal of, 8-26
Molluscum contagiosum
　assessment of, 4-48
　causes of, 4-47
　follow-up for, 4-48
　papules of, 4-38
　patient education for, 4-48
　signs of, 4-48
　symptoms of, 4-48
　treatment for, 4-48
Monkeypox, 6-59
Monoarthritis, 3-61—3-62

Monochloroacetic acid, 5-4
Mononucleosis, infectious
    assessment of, 5-79
    cause of, 5-78—5-79
    differential diagnosis of, 4-11, 5-64, 5-71
    fatigue with, 3-30t
    follow-up for, 5-80
    patient education for, 5-79—5-80
    signs of, 5-79
    symptoms of, 5-79
    treatment for, 5-79
Mood disorders. See Depression
Moraxella
    bovis, 5-136
    catarrhalis, 4-14
Moray eels, 6-17
Morphine
    for chemical burns, 7-18
    for heat stroke, 6-51
    for hip dislocation, 3-69
    for marine animal bites, 6-18
    for myocardial infarction, 4-3
    for pain control, 8-38
    for pericarditis, 4-6
    for trauma pain management, 7-4
Morphine sulfate
    for chest pain, 3-12
    for urolithiasis, 4-92
Mosaic wart, 5-3
Mosquitos
    in arboviral encephalitis transmission, 5-68
    in dengue fever, 5-65
    in myiasis transmission, 4-52
Motrin
    after cesarean section, 3-102
    for bunion, 5-5
    for chronic pelvic pain, 3-44t
    for headache, 3-56
    for heel spur syndrome, 5-2
    in peptic ulcer disease, 4-84
Mountain sickness, acute, 6-37
    assessment of, 6-37
    in congestive heart failure, 4-4
    differential diagnosis for, 6-41
    follow-up for, 6-38
    high altitude cerebral edema and, 6-38
    patient education for, 6-38
    signs of, 6-37
    symptoms of, 6-37
    treatment for, 6-38
Mouth, inflammation of, 5-137t
Mucocutaneous disease, 5-63
Multi-organ failure, 5-100
Multiple sclerosis, 4-90
Mumps
    differential diagnosis of, 4-11, 5-80
    required vaccination for, 5-106t
Mupirocin ointment, 4-43
Murine typhus, 5-83
    treatment of, 5-84—5-85

Murphy's sign, 4-78
Murray Valley encephalitis, 5-66
Muscle relaxants
    dosing guidelines for, 5-156t
    for joint pain, 3-63
Muscles
    heat-related cramping of, 6-48
    pain in with magnesium deficiency, 5-139t
    weakness of in trichenellosis, 5-50
Musculoskeletal chest wall pain, 3-10
    alleviating or aggravating factors in, 3-11
    quality of, 3-11
Musculoskeletal system examination, 2-4
Mustard gas, 6-53—6-54
Myalgia
    assessment of in joint pain, 3-62
    in trichenellosis, 5-50
Myasthenia gravis, 4-37
Mycobacterial infections
    atypical, 4-53
    nontuberculosis, 5-56—5-57
    tuberculosis, 5-54—5-56
Mycobacterium
    abscessus, 5-56—5-57
    avium complex, 5-56—5-57
    bovis, 5-32, 5-54, 5-56
    fortuitum, 5-56—5-57
    kansasii, 5-56—5-57
    leprae, 4-45
    marinum, 5-56—5-57
    tuberculosis, 5-54
        in cutaneous tuberculosis, 4-44
    ulcerans, 5-56—5-57
Mycolog
    for balanitis, 3-78
    for candidal penile infection, 3-78
Mycoplasma
    pneumoniae, 4-14
        differential diagnosis of, 4-13
    in urethral discharges, 5-26
Mycoses, 5-57—5-63
Myiasis
    assessment of, 4-52—4-53
    cause of, 4-52
    follow-up for, 4-53
    patient education for, 4-53
    signs of, 4-52
    symptoms of, 4-52
    treatment of, 4-53
Myocardial infarction, acute
    assessment of, 4-1
    cardiac resuscitation in, 4-8
    in congestive heart failure, 4-4
    differential diagnosis of, 6-5
    dyspnea with, 3-116
    EKG showing, 4-2f
    follow-up, 4-3
    incidence of, 4-1
    Killip classification of, 4-3
    patient education for, 4-3

in pericarditis, 4-6
quality of pain in, 3-11
risk factors for, 4-1
signs of, 4-1
symptoms of, 4-1
syncope with, 3-118
treatment for, 4-3
Myocarditis, 5-99
Myoglobinuria
   with electrical or lightning injury, 7-27
   urinalysis in, 8-43
Myositis
   signs of, 5-100
   streptococcal infection in, 5-99
Myringotomy, 6-35

N-acetyl cysteine, 3-58
Nafcillin
   for erysipelas, 4-42
   for hip pain, 3-74
   for mastitis, 3-8
   for pericarditis, 4-6
   for septic joint, 3-63
Nalbuphine, 8-39
Naloxone
   for opiate side effects, 8-40
   for trauma pain management, 7-4
Naprosyn
   for abnormal uterine bleeding, 3-41
   for chronic pelvic pain, 3-44t
   for endometriosis, 3-45
   for headache, 3-56
   for testicular/scrotal masses, 3-80
Naproxen
   for acute mountain sickness, 6-38
   for joint pain, 3-63
Narcan
   for chest pain, 3-12
   for drug-induced dyspnea, 3-117
   for syncope, 3-118
Narcotics
   for ankle pain, 3-77
   for Bartholin's gland abscess or cyst, 3-52
   for chronic pelvic pain, 3-44t
   for joint pain, 3-63
   for urolithiasis, 4-92
Nasal cannula, 8-4
Nasal cellulitis, 3-26
Nasal discharge, 6-33
Nasal polyps, 4-20
Nasogastric intubation
   for acute abdominal pain, 3-2
   after appendectomy, 4-77
Nasogastric tube decompression, 4-84
Nasotracheal intubation, 8-5
   blind, 7-1
Naval Experimental Diving Unit (NEDU), 6-22
Naval special warfare combat trauma AMAL, 1-20—1-21

Nebulizer, 4-20, 4-21
Necator americanus, 5-41
Neck
   examination of in secondary trauma survey, 7-4
   referred pain from, 3-70
Necrotic gingival ulcers, 5-15—5-16
Necrotic ulceration, 4-60
Necrotizing enterocolitis, 3-93
Necrotizing fasciitis
   differential diagnosis of, 5-100
   of male genitals, 3-77
Needle aspiration
   for breast abscess drainage, 3-9
   in thoracentesis, 4-15, 4-16
Needle cricothyroidotomy, 8-6
Needle decompression, 6-19
Needle thoracostomy
   indications for, 8-7
   procedures in, 8-8
Needle thoracotomy, 3-117
Needlestick injury prophylaxis, 5-78
Neisseria, 5-32
   gonorrhoeae, 4-40
      identification of colonies of, 8-50
      in urethral discharges, 5-26
   meningitides, 4-41
Nelfinavir, 5-78
Nematodes
   Enterobius vermicularis, 5-37
   lymphatic-dwelling, 5-39—5-40
Neocobefrin, 5-20
Neomycin
   in contact dermatitis, 4-69
   for venomous fish bites, 6-15
Neoplasms, in intestinal obstruction, 4-86
Neosporin
   for bed bug bites, 4-55
   for friction foot blisters, 5-8
Nephrotic syndrome, 4-15
Nerve agents, 6-52
Nerve blockade
   for equine lameness, 5-135
   of foot and ankle, 5-168—5-174
      indications for, 5-168
      standard equipment for, 5-168
   of hand and wrist, 5-164—5-168
      median nerve, 5-165—5-166
      radial nerve, 5-166—5-168
      standard approach to, 5-163—5-164
Nerve entrapment, 5-1
Nerve impingement syndrome
   rotator cuff tendonitis in, 3-70
      treatment of, 3-71
   shoulder pain in, 3-70
Neural tissue disease, 4-90
Neurofibroma, 4-66
Neurogenic shock, 7-13
Neuroimaging, 3-87
Neuroleptics
   for mania, 3-17

for psychosis and delirium, 5-152
Neurologic disorders
   Bell's palsy, 4-37—4-38
   disqualifying condition for diving, 6-22
   meningitis, 4-35—4-37
   pruritus with, 3-114
   seizures and epilepsy, 4-34—4-35
Neurological deficits, focal, 7-24
Neurological examination, 2-4
   for arboviral encephalitis, 5-67
   for carbon dioxide poisoning, 6-12
   for decompression sickness, 6-5
   for dizziness, 3-20
   for dyspnea, 3-116
   for headache, 3-55
   in low back pain, 3-6
   for memory loss, 3-85—3-86
   for meningitis, 4-36
   for pulmonary over inflation syndrome, 6-7
   for rabies, 5-82
   in secondary trauma survey, 7-4
   for trauma, 7-3
   for underwater blast injury, 6-19
Neuromuscular disorders, 3-30*t*
Neurontin
   cautions with, 3-56
   for headache, 3-56
Neuropathy, peripheral, 5-139*t*
Neurosyphilis, 5-30
Nevirapine, 5-78
Nevus, intradermal, 4-66
Newcastle disease virus, 5-32
Niacin deficiency, 5-137*t*
Nickel contact dermatitis, 4-69
Niclosamide, 5-49
Nifedipine
   for high altitude pulmonary edema, 6-41
   side effects of, 6-41
Nifurtimox, 5-53
Night blindness, 5-138*t*
Night sweats, 3-29
Nightsoil, 5-109
Nikolsky's sign, 4-42
Nipples, hyperpigmentation of, 3-87
Nitrite poisoning, 8-10
Nitrofurantoin
   in pancreatitis, 4-82
   for prostatitis, 3-81
   for urinary incontinence, 4-91
   for urinary tract infections, 4-93, 4-94
Nitrofurazone, 5-136
Nitrogen
   in atmosphere, 6-35—6-36
   liquid
      for plantar warts, 5-4
      for warts, 4-49
Nitroglycerin
   for chest pain, 3-11, 3-12
   for congestive heart failure, 4-4
   for myocardial infarction, 4-3

Nodules, 4-38
   deep subcutaneous, 4-53
   genitourinary, 4-89
   weeping, in ecthyma contagiosum, 4-46
Nonsteroidal anti-inflammatory drugs
   for acute rheumatic fever, 5-99
   for ankle pain, 3-77
   for aseptic necrosis of hip, 3-74
   for bunion, 5-5
   cautions with, 3-56
   for dysmenorrhea, 3-45
   for elbow dislocation, 3-68
   for endometriosis, 3-45
   for epididymitis, 3-84
   for fire ant stings, 4-58
   for frostbite, 6-43
   in gastritis, 4-81
   for herniated disc, 3-74
   for joint pain, 3-63
   for knee pain, 3-75
   for localized osteitis, 5-18
   for mittleschmerz, 3-46
   for non-freezing cold injury of foot, 6-47
   for pain control, 8-38, 8-39
   for patellar dislocation, 3-69
   in peptic ulcer disease, 4-84
   in pulmonary embolus, 4-24
   for sickle cell crisis, 4-9
   side effects of, 3-64, 3-72, 3-75, 3-77
   for stress fractures of foot, 5-6
   for testicular/scrotal masses, 3-80
   for thrombosis of penile vein and sclerosing lymphangitis, 3-78
Norfloxacin, 3-19
NPH insulin, 4-29
NSAIDs. *See* Nonsteroidal anti-inflammatory drugs
Nuchal cord, 3-95
Nursing techniques, 3-9
Nutritional deficiencies, vitamin and mineral, 5-137—5-139*t*
Nystatin
   for balanitis, 3-78
   for candidal penile infection, 3-78
   for candidiasis, 5-58
   for dermatophyte infections, 4-50

Oatmeal bath, 3-114
Obesity
   in apnea, 4-25, 4-26
   in diabetes mellitus, 4-29, 4-30
Obstetric symptoms
   breech delivery, 3-96—3-99
   cesarean section, 3-100—3-102
   episiotomy and repair, 3-102—3-104
   preeclampsia/eclampsia, 3-104—3-109
   pregnancy, 3-87
   preterm labor, 3-93—3-95
   shoulder dystocia relief, 3-95—3-96

vaginal delivery, 3-87—3-93
Obstetrics, large animal, 5-130—5-131
Obstructive biliary disease, 3-114
Obstructive sleep apnea, 4-25
Occult blood, feces examination for, 8-50—8-51
Octopi, 6-14
  treatment for, 6-15
Ocular conditions, disqualifying for diving, 6-21
Ocular larva migrans, 4-52
Ocular massage, 3-23
Ofloxacin
  for eye injury, 3-28
  for red eye, 3-25
Oligoarthritis, 3-62
Omeprazole, 4-85
Omnifloxacin, 4-13
Onchocerca volvulus, 4-53
Onchocerciasis
  assessment of, 4-53
  cause of, 4-53
  differential diagnosis of, 4-52
  follow-up for, 4-54
  patient education for, 4-53
  signs of, 4-53
  symptoms of, 4-53
  treatment for, 4-53
Ondansetron (Zofran), 6-62
Onycholysis, 4-64
Onychomycosis, 4-50
Open pneumothorax, 7-2
Operational environments
  aerospace medicine, 6-31—6-37
  biological warfare, 6-56—6-61
  chemical weapons of mass destruction, 6-51—6-55
  cold illnesses and injuries, 6-42—6-47
  dive medicine, 6-1—6-30
  heat-related illnesses, 6-47—6-51
  high altitude illnesses, 6-37—6-42
  radiation injury, 6-61—6-62
Operational issues
  air evacuation phone list, 1-33
  aircraft patient loads, 1-33
  care under fire
    assessment, 1-1—1-2
    signs, 1-1
    symptoms, 1-1
  CASEVAC with fixed winged aircraft, 1-31
  field expedient landing zone (day), 1-29f
  field expedient landing zone (night), 1-30f
  general medical site survey checklist, 1-7—1-9
  helicopter landing sites, 1-25—1-26
  hospital survey, 1-4—1-7
  marking and lighting of airplane landing zone (night), 1-32f
  MedCAP planning checklist for, 1-2—1-4
  naval special warfare combat trauma AMAL, 1-20—1-21
  9 line MEDEVAC request, 1-23—1-24

  pararescue primary medical kit packing list, 1-12—1-14
  patient considerations, 1-22
  semi-fixed base operations (day), 1-27f
  semi-fixed base operations (night), 1-28f
  site survey, veterinary annex, 1-10—1-11
  suggested M5 packing list, 1-18—1-20
  USAF SOF trauma ruck pack list, 1-15—1-16
  USAF SOF trauma vest pack list, 1-17
Ophthalmic ointments, 5-136
Ophthalmic zoster, 4-47
Ophthalmoscopy
  for acute vision loss without trauma, 3-23
  for blast injuries, 7-24
  for eye injury, 3-27
  for laser injury, 7-29
  for orbital or periorbital inflammation, 3-26
Opiates, 5-150
  for elbow dislocation, 3-67
  intoxication, 5-150t
  withdrawal from, 5-151t
Opioids, 8-38—8-39
Opium, tincture of, 3-19
OPQRST pain assessment mnemonic, 8-38
Optic neuritis, 3-22
Optic neuropathy, anterior ischemic
  in acute vision loss, 3-22
  differential diagnosis of, 3-23
Oral azoles, 5-59—5-60
Oral contraceptives
  for abnormal uterine bleeding, 3-41
  for chronic pelvic pain, 3-44t, 3-47
  for dysmenorrhea, 3-45
  for endometriosis, 3-45
Oral disorders, 5-9—5-19
Oral hygiene, 5-15—5-16
Oral rehydration solution (ORS)
  for electrical or lightning injury, 7-28
  indications, benefits, and cautions for, 7-15t
  recipe for, 7-28
Orbital cellulitis
  differential diagnosis of, 3-26—3-27
  in periorbital inflammation, 3-26
  treatment of, 3-27
Orbital fracture, 3-27
Orbital/periorbital inflammation, 3-26
  assessment of, 3-26—3-27
  follow-up, 3-27
  patient education for, 3-27
  signs of, 3-26
  symptoms of, 3-26
  treatment plan for, 3-27
Orcas, 6-17
Orchitis, 3-83
Orf virus, 4-46, 5-32
  differential diagnosis of, 5-91
Organ systems. *See also specific organs*
  veterinary, 5-123
Oropharyngeal obstruction, 3-115

Orotracheal intubation, 8–5
Oroya fever, 5–92—5–93
Orthopedic disorders, 6–22
Orthopoxviruses, 6–59
Orthostatic hypotension, 7–12
Osteitis, localized, 5–17—5–18
Osteoarthritis
  assessment of, 3–73
  hip pain with, 3–72
  shoulder, 3–71
  traumatic, 5–5
Osteomalacia, 5–138*t*
Osteoporosis, 5–139*t*
Otitis media
  differential diagnosis of, 6–35
  dizziness in, 3–20
  serous, 3–21
  treatment of, 3–21
Otoscope, with insufflation bulb, 6–1
Otoscopy
  for barotitis, 6–35
  for blast injuries, 7–24
  for common cold and flu, 4–10
  for dizziness, 3–20
Ottawa Ankle Rules, 3–76
Ova
  Con-Trate method of detection for, 8–52—8–53
  in feces, 8–51—8–52
Ovarian torsion
  acute pelvic pain with, 3–41, 3–42
  treatment of, 3–43
Ovaries
  bimanual examination of, 3–39
  ruptured cyst of
    acute pelvic pain with, 3–41, 3–42
    treatment of, 3–43
Overflow incontinence, 4–90
  treatment of, 4–90—4–91
Overhead cover, in decontamination area, 6–55
Overuse injuries
  in hip pain, 3–72
  knee, 3–76
  shoulder pain with, 3–71
  treatment of, 3–71
Overuse syndromes, patellar, 3–74
Oxacillin
  for cellulitis of penis, 3–78
  for erysipelas, 4–42
  for hip pain, 3–74
  for mastitis, 3–8
  for staphylococcal scalded skin syndrome, 4–43
Oxycodone
  for joint pain, 3–63
  for pain control, 8–39
  preparations of, 5–150
Oxygen
  for acute respiratory distress syndrome, 4–25
  for anaphylactic shock, 7–11
  for atelectasis, 3–117
  in cardiac resuscitation, 4–7
  for congestive heart failure, 4–4
  for COPD, 4–22
  for decompression sickness, 6–5, 6–36
  for high altitude cerebral edema, 6–39
  for high altitude pulmonary edema, 6–41
  for hypoxia, 6–31
  partial pressure of, 6–9, 6–31
  for pulmonary embolus, 4–24
  for pulmonary over inflation syndrome, 6–8
  RBC transport of, 4–8
  saturation, 8–9—8–10
  single depth exposure limits of, 6–10
  for streptococcal infections, 5–100
  supplementation of for vision loss, 3–23
  toxicity of
    assessment of, 6–10
    differential diagnosis of, 6–12
    in diving, 6–9
    follow-up for, 6–10
    patient education for, 6–10
    single depth oxygen exposure limits and, 6–10
    symptoms and signs of, 6–10
    treatment for, 6–10
Oxygen treatment gas, 6–23
Oxygenation, 6–53
Oxyhemoglobin, 8–9
Oxymetazoline, 4–11
Oxytocin, 3–100, 3–101

P-wave analysis, 8–12
Packed red blood cells, 8–15
Pail latrine, 5–113*f*
Pain
  acute abdominal. *See* Abdomen, acute pain
  assessment of, 8–38
  control of, 8–37
    analgesics for, 8–38—8–40
    for cholecystitis, 4–78
    equipment for, 8–38
    indications for, 8–38
    OPQRST mnemonic for, 8–38
    precautions in, 8–40
  low back, 3–6—3–7
  management of in secondary trauma survey, 7–4—7–5
Pain diary, 3–46
Palpations, 3–110
  assessment of, 3–111
  follow-up, 3–111
  patient education for, 3–111
  signs of, 3–110
  symptoms of, 3–110
  treatment of, 3–111
Palpitations, 3–110
Pamelor
  cautions with, 3–56
  for headache, 3–56

Pancreatitis
　acute abdominal pain in, 3–3f
　assessment of, 4–82
　causes of, 4–82
　chest pain with, 3–10
　differential diagnosis of, 4–78
　follow-up for, 4–82—4–83
　in low back pain, 3–6
　patient education for, 4–82
　in peritonitis, 4–83
　Ranson's criteria for severity of, 4–82, 4–83
　signs of, 4–82
　symptoms of, 4–82
　treatment for, 4–82
Panic attack, 3–2
Panic disorder, 3–5
Pap smear, 3–37
Pap stain, 8–48
Papules, 4–38
　dome-shaped with umbilication, 4–48
　pinpoint, 4–48
　of psoriasis, 4–64
　red, scaly, 4–67
　with scabies, 4–61
　of verruca vulgaris, 4–48—4–49
　waxy or pearly, 4–66
Paracoccidioides brasiliensis, 5–62
Paracoccidioidomycosis
　assessment of, 5–63
　causes of, 5–62
　follow-up for, 5–63
　patient education for, 5–63
　symptoms and signs of, 5–63
　treatment for, 5–63
Paraesophageal hiatal hernia, 6–22
Paraesthesia axillary blockade techniques, 5–174—5–175
Paragonimiasis, 5–45–46
Paragonimus, 5–45
Paralytic poliomyelitis, 5–80—5–81
Paranoia, 5–152
Paraphimosis, 3–77, 4–89
　consultation criteria for, 3–79
　patient education for, 3–79
　treatment of, 3–78
Pararescue primary medical kit packing list, 1–12—1–14
Parasitic infections
　African trypanosomiasis, 5–52—5–53
　amebiasis, 5–33
　American trypanosomiasis, 5–53—5–54
　ascariasis, 5–34
　babesiosis, 5–34—5–35
　blood-borne, 8–47—8–48
　clonorchiasis, 5–35—5–36
　Con-Trate method of detection for, 8–52—8–53
　cyclosporiasis, 5–36—5–37
　enterobiasis, 5–37—5–38
　fascioliasis, 5–38—5–39
　fasciolopsiasis, 5–38
　in fecal specimen, 8–50—8–52
　filariasis, 5–39—5–40
　giardiasis, 5–40—5–41
　hookworm and cutaneous larva migrans, 5–41—5–42
　leishmaniasis, 5–42—5–44
　malaria, 5–44—5–45
　paragonimiasis, 5–45—5–46
　schistosomiasis, 5–46—5–47
　of skin
　　loiasis, 4–52
　　myiasis, 4–52—4–53
　　onchocerciasis, 4–53—4–54
　　swimming dermatitis, 4–54
　strongyloidiasis, 5–47—5–48
　tapeworm infections, 5–48—5–50
　trichinellosis, 5–50—5–51
　trichuriasis, 5–51—5–52
Paregoric, 3–19
Paromomycin, 5–33
Paroxetine (Paxil), 3–17
Paroxysmal atrial tachycardia, 3–110
Paroxysmal nocturnal dyspnea, 4–3
Paroxysmal supraventricular tachycardia
　differential diagnosis of, 3–111
　palpitations with, 3–110
PASG. *See* Pneumatic anti-shock garment
Pastereulla multocida, 5–32
Patch, 4–38
Patella
　dislocations of
　　assessment of, 3–69
　　diagnostic tests for, 3–69
　　differential diagnosis of, 3–69
　　follow-up, 3–69—3–70
　　patient education for, 3–69
　　post treatment care for, 3–69
　　procedures for, 3–69
　　spontaneous, 3–65
　　traumatic, 3–65
　tendinitis of, 3–74
Patellofemoral syndrome
　assessment of, 3–75
　knee pain with, 3–74
Patient examination, Naval special warfare combat trauma AMAL items for, 1–20
Patient holding area, decontamination, 6–54
Patient loads, aircraft, 1–33
Patients
　classifications/categories of, 1–22
　information on, 1–22
　preparation/documentation of, 1–22
Paxil, 3–17, 3–31
Pediatric dermatosis, 4–42—4–43
Pediculosis, 4–62
　assessment of, 4–62
　follow-up for, 4–63
　patient education for, 4–63
　signs of, 4–62
　symptoms of, 4–62

treatment for, 4-62—4-63
Pediculus humanus
  capitus, 4-62
  corporis, 4-62
Pellagra, 5-137t
Pelvic compression, 3-61
Pelvic examination
  in abdominal pain, 3-1
  for bacterial vaginosis, 3-47
  bimanual, 3-40f
  for endometriosis, 3-45
  female
    bimanual, 3-38—3-39
    equipment for, 3-37
    external genitalia, 3-38
    in field, 3-37
    indications for, 3-37
    procedures for, 3-37—3-39
    rectovaginal, 3-39
    unprofessional, 3-39
    vaginal, 3-38
  for mittleschmerz, 3-46
  in secondary trauma survey, 7-4
Pelvic inflammatory disease
  with abnormal uterine bleeding, 3-41
  acute abdominal pain in, 3-3t
  acute pelvic pain with, 3-41—3-42
  assessment of, 3-51
  causes of, 3-50
  diagnostic criteria for, 3-51
  differential diagnosis for, 4-71
  follow-up, 3-52
  patient education for, 3-52
  with pelvic pain, 3-47
  risk factors for, 3-50
  signs of, 3-51
  symptoms of, 3-51
  treatment for, 3-43
    inpatient, 3-51
    outpatient, 3-51
Pelvic masses, 3-43—3-44
Pelvic pain, 3-1
  acute
    assessment of, 3-42—3-43
    follow-up, 3-43
    gynecologic causes of, 3-41—3-42
    patient education for, 3-43
    signs of, 3-42
    symptoms of, 3-42
    treatment for, 3-43
  chronic
    assessment of, 3-44t
    with dysmenorrhea, 3-45
    with endometriosis, 3-45
    gynecologic causes of, 3-43—3-44
      diagnosis and treatment of, 3-46—3-47
      follow-up, 3-47
      patient education for, 3-47
    with irritable bowel syndrome, 3-46
    with mittelschmerz, 3-46

Pemmican, 5-127—5-128
Penetrating wounds
  care of under fire, 1-1
  chest, pericardiocentesis for, 8-12
  debridement of, 8-26
  vascular, 1-1
Penicillin
  for acute rheumatic fever, 5-99
  allergic reaction to, 4-37
  for chemical burns, 7-19
  for conjunctivitis in animals, 5-136
  for foot rot in caprines, 5-134
  for human and animal bites, 7-9
  for leptospirosis, 5-87
  for Lyme disease, 5-88
  for mastitis, 3-8
  for meningitis, 4-36
  for pinta, 4-64
  for preterm labor infection, 3-94
  for rat bite fever, 5-97
  for streptococcal infections, 5-100
  for yaws, 4-63
Penicillin G
  for anthrax, 5-91
  aqueous, 5-19—5-20
  for human and animal bites, 7-9
  for inhalational anthrax, 6-57
  for syphilis, 5-30
  for tetanus, 5-101
Penicillin V
  for anthrax, 5-91
  for inhalational anthrax, 6-57
Penicillin VK
  in dentistry, 5-19
  for human and animal bites, 7-9
Penile vein thrombosis, 3-78
  treatment of, 3-78
Penis
  cellulitis of, 3-78
  discharge from. *See under* Sexually transmitted disease
  inflammation of foreskin of, 3-77—3-79
Penlac, 4-50
Penrose drain, 3-10
Pentamidine, 4-82
Pentavalent antimony, 5-43
Pentostam (sodium stibogluconate), 5-43
Pepcid
  for acute abdominal pain, 3-2
  for gastritis, 4-81
Peptic ulcer disease
  assessment of, 4-85
  causes of, 4-84
  chest pain with, 3-11
  complications of, 4-84
  differential diagnosis of, 4-78
  follow-up for, 4-85—4-86
  patient education for, 4-85
  signs of, 4-85
  symptoms of, 4-84—4-85

treatment for, 4–85
Peptobismol, 3–34f
Peracute, 5–135
Periapical abscess, 5–9, 5–13
   differential diagnosis of, 6–32, 6–33
   untreated, 5–13—5–14
Pericardial fluid, removal of, 8–13
Pericardial friction rub, 4–6
Pericardial tamponade
   chest pain with, 3–10
   in pericarditis, 4–6
   treatment for, 3–12
Pericardiocentesis, 8–12—8–13
   equipment for, 8–12
   indications for, 8–12
   for pericarditis, 4–6
   preparation for, 8–12
   for tamponade, 4–7
Pericarditis
   assessment of, 4–6
   causes of, 4–6
   differential diagnosis of, 5–99
   follow-up, 4–7
   patient education for, 4–6—4–7
   referred pain to shoulder in, 3–70
   signs of, 4–6
   symptoms of, 4–6
   treatment for, 3–12, 4–6
   viral, 8–12
   vital signs suggesting, 3–11, 3–12
Pericoronitis, 5–17
Peridex (chlorhexidine), 5–16
Perineum, incision of. See Episiotomy
Periodontal abscess, 5–15
Periorbital inflammation. See Orbital/periorbital inflammation
Periosteal Elevator, 5–25
Peripheral nerve system lesions, 4–37
Peritonitis
   assessment of, 4–84
   causes of, 4–83
   differential diagnosis of, 4–71, 4–86
   follow-up for, 4–84
   mortality with, 4–83
   patient education for, 4–84
   signs of, 4–83—4–83
   symptoms of, 4–83
   treatment for, 4–84
Permethrin, 4–52
   for pediculosis, 4–62
   in pest control, 5–121
   for scabies, 4–61
Pernio, 6–47
Peroneal nerve block
   deep, 5–170, 5–171f
   superficial, 5–171f
Pertussis, 4–11
Pest control, 5–120—5–121
Pesticides, 5–120—5–121
Petrolatum, 4–44

Pharyngitis
   differential diagnosis of, 4–11, 5–64, 5–79
   streptococcal, 5–98, 5–99
Phenergan
   for acute organic intestinal obstruction, 4–86
   after cesarean section, 3–102
   for chemical burns, 7–18
   for peritonitis, 4–84
Phenobarbital, 4–35
Phenothiazine toxicity, 5–101
Phenoxybenzamine hydrochloride (Dibenzyline), 6–43
Phenylbutazone, 5–134—5–135
Phimosis, 3–77, 4–89
   consultation criteria for, 3–79
   patient education for, 3–79
   signs of, 3–78
   treatment of, 3–78, 4–89—4–90
Phlebotomy
   for congestive heart failure, 4–4
   for hypertensive emergency, 4–5
Phobias, 3–5
Phosgene, 6–53—6–54
Phosphorus deficiency, 5–139t
Photophobia, 3–27
Physical examination
   animal restraint for, 5–128—5–130
   of animals, 5–130
   of breasts (chest), 2–4
   cardiovascular, 2–3—2–4
   constitutional, 2–3
   ear, nose, mouth, and throat, 2–3
   gastrointestinal, 2–4
   genitourinary, 2–4
   integumentary, 2–5
   lymphatic, 2–5
   musculoskeletal, 2–4
   neurologic, 2–4
   psychiatric, 2–4—2–5
   respiratory, 2–3
   of trauma patient, 7–7
Physiologic values, animal, 5–130t
Phytonadione, 5–138t
Pickling, 5–128
Pickwickian syndrome, 4–25
PID. See Pelvic inflammatory disease
PIES regimen, 5–148
Pigs
   diarrhea in, 5–135—5–136
   humane slaughter and field dressing of, 5–125
Pilar cysts, 4–38—4–39
Pinkeye. See Conjunctivitis
Pinta, 4–64
Pinworm. See Enterobiasis
Pipe urinal, 5–114f
Piperacillin
   for abnormal uterine bleeding, 3–41
   for chemical burns, 7–19
   for jaundice, 3–58

Piperacillin/tazobactam, 7–9
Piroxicam, 3–63
Pit vipers, 5–143
　characteristics of, 5–146
　identification of, 5–147
Pityriasis rosea
　differential diagnosis of, 3–114
　pruritus with, 3–113
Pityriasis versicolor
　assessment of, 4–51
　causes of, 4–51
　follow-up for, 4–51
　patient education for, 4–51
　signs of, 4–51
　symptoms of, 4–51
　treatment for, 4–51
Pityrosporum orbicularis, 4–51
Placenta previa, 3–100
Placental abruption, 3–100
Placental delivery, 3–88, 3–92f, 3–93
Plague, 5–32
　assessment of, 5–96
　cause and transmission of, 5–95—5–96
　follow-up for, 5–97
　patient education for, 5–96
　pneumonic, 6–58—6–59
　required vaccination for, 5–106t
　signs of, 5–96
　symptoms of, 5–96
　treatment for, 5–96
　zoonotic disease considerations in, 5–97
Plantar warts
　assessment of, 5–4
　cause of, 5–3
　follow-up for, 5–4
　patient education for, 5–4
　signs of, 5–3
　symptoms of, 5–3
　treatment for, 5–4
Plaque
　in erysipelas, 4–42
　in herpes zoster infection, 4–47
　of psoriasis, 4–64
　skin, 4–38
Plasmodium
　malariae, 5–44
　vivax, 5–44
Plaster casts, volume-restricting, 8–31
Platelets
　count of, 5–65
　morphological variations of, 8–58
Pleural effusion
　assessment of, 4–15
　exudative, 4–14, 4–15
　follow-up, 4–15
　patient education for, 4–15
　of pus, 4–17
　signs of, 4–14—4–15
　symptoms of, 4–14
　transudative, 4–14, 4–15

　treatment of, 3–117, 4–15
　types of, 4–14
Pleural friction rub, 4–17
Pleurectomy, 4–18
Pleurisy
　differential diagnosis of, 4–78
　in pericarditis, 4–6
Pneumatic anti-shock garment, 7–2
　for anaphylactic shock, 7–11
　for hemorrhage control in hypovolemic shock, 7–13
　indications for, 8–13
　premature removal of, 8–14
　procedure for, 8–13—8–14
Pneumococcal peritonitis, 4–71
Pneumococcal vaccine, 4–22
Pneumococcus colonies, identification of, 8–50
Pneumocystis carinii pneumonia, 4–18
Pneumonia
　alleviating or aggravating factors in, 4–12—4–13
　assessment of, 4–13
　atypical, 4–11, 4–18, 5–86
　chest pain with, 3–10
　common causes of, 4–14
　in congestive heart failure, 4–4
　cough with, 3–15
　differential diagnosis of, 4–78, 5–64, 6–41, 6–57, 6–59, 6–60
　dyspnea with, 3–115, 3–116
　　treatment of, 3–117
　empyema and, 4–17
　follow-up, 4–14
　fungal, 5–54
　lobar, 4–13
　mycoplasma, 4–13
　patient education for, 4–13
　pleural effusion with, 4–15
　signs of, 4–13
　symptoms of, 4–12—4–13
　treatment for, 4–13
　　principles of, 4–12
　viral, 4–13
　vital signs suggesting, 3–11
Pneumonic plague, 6–58—6–59
Pneumopericardium, 8–12
Pneumothorax
　with blast injury, 7–26
　chest pain with, 3–10
　　quality of, 3–10
　as disqualifying condition for diving, 6–21
　pericardiocentesis for, 8–13
　in pericarditis, 4–6
　referred pain to shoulder in, 3–70
　with thoracentesis, 4–15—4–16
　thoracostomy for, 8–7
　in trauma, 7–2
　treatment for, 3–12, 3–117, 7–24
　with underwater blast injury, 6–19
Podiatry
　bunions, 5–4—5–5

corns and calluses, 5-5—5-6
friction foot blisters, 5-7—5-8
heel spur syndrome, 5-1—5-2
ingrown toenails, 5-2—5-3
plantar warts, 5-3—5-4
stress fractures of foot, 5-6—5-7
Podophyllin, 4-49
Poikilocytosis, RBC abnormality in, 8-55—8-56
POIS. *See* Pulmonary over inflation syndrome
Poison oak, 4-69
Poisoning
  assessment of, 5-141
  follow-up for, 5-142
  patient education for, 5-142
  routes of entry in, 5-140
  signs of, 5-141
  symptoms of, 5-140—5-141
  toxins in, 5-140
  treatment for, 5-141—5-142
Polio vaccination, 5-80
  required, 5-106*t*
Poliovirus
  assessment of, 5-80—5-81
  follow-up for, 5-81
  patient education for, 5-81
  risk factors for, 5-80
  signs of, 5-80
  symptoms of, 5-80
  transmissions of, 5-80
  treatment for, 5-81
Politzer bag, 6-35
Polyarthritis
  assessment of, 3-62
  differential diagnosis of, 5-98
Polydipsia, 4-29
Polymorphonuclear leukocytes, 4-9
Pork tapeworm disease, 5-49—5-50
Porphyria, acute, 4-71
Port-wine stains, 4-38
Portable altitude chamber (PAC)
  for high altitude cerebral edema, 6-40
  for high altitude pulmonary edema, 6-42
Portable pressure chamber. *See* Hyperbaric chamber, portable
Portuguese man-of-war, 6-13—6-14
Positive end-expiratory pressure (PEEP), 7-24
Positive-pressure ventilation (PPV)
  for blast injury, 7-24
  devices, 5-155
Post-traumatic stress disorder, 3-2, 5-148
  differential diagnosis of, 3-5
Postmortem exam, veterinary, 5-126
Postnasal drip, 3-14
Postpartum complications, 3-93
Postural drainage, 4-13
Potassium hydroxide, 4-48
Potassium iodide, 6-62
Pott's disease, 4-71
Poultry, humane slaughter and field dressing of, 5-125

Povidone-iodine
  for coral sting, 6-15
  in water treatment, 5-120
PR depression, 4-6
PrameGel
  for pediculosis, 4-63
  for scabies, 4-61
Praziquantel
  for clonorchiasis, 5-36
  for fasciolopsiasis, 5-38
  for paragonimiasis, 5-46
  for schistosomiasis, 5-47
  for tapeworm infections, 5-49
Predeployment site survey, MedCAP guide, 1-3
Prednisolone
  for eye injury, 3-28
  for red eye, 3-25
Prednisone
  for adrenal insufficiency, 4-28
  for allergic pneumonitis, 4-19
  for Bell's palsy, 4-38
  for chest pain, 3-12
  for contact dermatitis of penis, 3-78
  for COPD, 4-22
  for infectious mononucleosis, 5-79
  for inflammatory arthritis, 3-63
  for pericarditis, 4-6
  precautions with, 3-64
  for red eye, 3-25
  for vision loss, 3-23
Preeclampsia, 3-104
  assessment of, 3-108
  follow-up, 3-109
  patient education for, 3-109
  signs of
    mild, 3-104
    severe, 3-104—3-105
  stabilization and evacuation for, 3-108
  symptoms of, 3-104
  treatment of
    mild, 3-108—3-109
    severe, 3-109
Pregnancy
  with abnormal uterine bleeding, 3-41
  assessment of, 3-87
  bacterial vaginosis treatment during, 3-49
  in candida vaginitis, 3-49
  carbon monoxide poisoning during, 6-11
  disqualifying condition for diving, 6-22
  in horses, 5-133
  pruritus with, 3-114
  signs of, 3-87
  symptoms of, 3-87
  test for, 3-87
  thyroid disorders during, 4-34
  treatment of.*See under specific conditions*
Premature atrial contractions
  differential diagnosis of, 3-111
  palpitations with, 3-110
Premature rupture of membranes (PROM)

IN-51

differential diagnosis of, 3-94
  with preterm birth, 3-93
Premature ventricular contractions
  differential diagnosis of, 3-111
  palpitations with, 3-110
Prepatellar bursitis, 3-75
Preseptal cellulitis
  differential diagnosis of, 3-26
  treatment of, 3-27
Preservation records, meat, 5-128
Pressure bandage
  for cone shell injury, 6-16
  for octopi bites, 6-15
  for sea snake bite, 6-16
Preterm birth, 3-93
  with syphilis, 5-30
Preterm labor, 3-93
  follow-up, 3-95
  patient education on, 3-94
  signs of, 3-94
  symptoms of, 3-93—3-94
  treatment of, 3-94
Preventive medicine, 5-105
  DNBI report in, 5-107—5-108
  field sanitation in, 5-108—5-111
  field water purification in, 5-118—5-120
  landfill management in, 5-121
  for malaria, 5-120
  pest control in, 5-120—5-121
  rabies control in, 5-121
  required immunizations in, 5-105—5-106t
  surveillance in, 5-107
  waste disposal in, 5-111—5-118
Priapism, 4-89
Prilocaine
  for digital block of finger or toe, 5-163t
  local infiltration of, 5-159
Primaquine
  for malaria, 5-45
  side effects of, 5-45
Primaquine phosphate, 5-44—5-45
Probenecid
  for gout, 3-63
  for pelvic inflammatory disease, 3-51
Procaine penicillin G
  for acute mastitis in animals, 5-135
  for rat bite fever, 5-97
Procedures
  airway management, 8-1—8-4
  bladder catheterization, 8-33—8-35
  blood transfusion, 8-14—8-18
  blood transfusion reaction, 8-19—8-21
  common external traction devices in, 8-33
  compartment syndrome management, 8-30—8-32
  cricothyroidotomy, needle and surgical, 8-5—8-7
  field transfusion, 8-18
  intraosseous infusion, 8-21—8-22
  intubation, 8-4—8-5

joint aspiration, 8-28—8-30
laboratory, 8-40—8-62
pain assessment and control, 8-37—8-40
pericardiocentesis, 8-12—8-13
pneumatic anti-shock garment, 8-13—8-14
portable pressure chamber, 8-36—8-37
pulse oximetry monitoring, 8-9—8-10
skin mass removal, 8-26—8-28
splint application, 8-32
suprapubic bladder aspiration, 8-35—8-36
suturing, 8-22—8-25
thoracostomy, needle and chest tube, 8-7—8-9
three-lead electrocardiography, 8-10—8-12
wound debridement, 8-25—8-26
Prochlorperazine
  for acute mountain sickness, 6-38
  for blast injury, 7-26
  for dizziness, 3-21
  for yellow fever, 5-70
Progressive muscle relaxation, 3-5
Prolacin secretion disorders, 3-41
PROMED network, 3-31
Promethazine, 7-26
Proparacaine, 3-28
Propofol
  continuous infusion of in field, 5-158t
  dosing guidelines for, 5-156t
Propranolol
  for chest pain, 3-12
  for hyperthyroidism, 4-33
  for palpitations, 3-111
  side effects of, 3-111, 3-118
  for syncope, 3-118
Propulsid, 3-110
Propylene glycol, 4-51
Prostate, painful, 3-81
Prostatitis, 3-80—3-81
  assessment of, 3-81
  follow-up, 3-8
  patient education for, 3-82
  signs of, 3-81
  symptoms of, 3-80
  treatment of
    alternative, 3-81
    with no infection, 3-81—3-82
    primary, 3-81
  in urinary tract infection, 4-93
Protein-sulfosalicylic acid test procedure, 8-41—8-42
Proteinuria, 8-42
Protriptyline, 4-26
Proventil, 4-22
Prozac, 3-17, 3-31
Pruritus, 3-113
  assessment of, 3-114
  follow-up, 3-115
  with mite infestation, 4-59
  patient education for, 3-115
  with scabies, 4-61—4-62
  in schistosomiasis, 5-46

signs of, 3-113—3-114
in swimming dermatitis, 4-54
symptoms of, 3-113
treatment of, 3-114—3-115
without rash, 3-114
Pseudoephedrine
for barosinusitis, 6-34
for barotitis, 6-35
for common cold and flu, 4-11
Pseudofolliculitis barbae
assessment of, 4-65
causes of, 4-65
follow-up for, 4-66
patient education for, 4-66
signs of, 4-65
symptoms of, 4-65
treatment for, 4-66
Pseudomembranous colitis, 3-18
Pseudomonas aeruginosa, 3-112t
Pseudopregnancy, in animals, 5-131
Pseudotumor, periorbital, 3-26
Psoriasis
assessment of, 4-65
causes of, 4-64
differential diagnosis of, 3-113t
follow-up for, 4-65
patient education for, 4-65
plaque of, 4-38
pustular, 4-38, 4-64
signs of, 4-64—4-65
symptoms of, 4-64
treatment for, 4-65
Psychiatric disorders
as disqualifying condition for diving, 6-21—6-22
fatigue in, 3-29
memory loss with, 3-86
syncope with, 3-118
Psychiatric examination, 2-4—2-5
Psychogenic dyspnea
differential diagnosis of, 3-116
treatment of, 3-117
Psychosis, 5-151—5-153
Psychotic disorders, operational stress in, 5-148
Psyllium, 3-13
Pthirus pubis, 4-62
PTSD. *See* Post-traumatic stress disorder
Pulmonary anthrax, 5-90, 5-91
Pulmonary contusions
with blast injuries, 7-23
treatment of, 7-24
Pulmonary disease
chronic, progressive, 5-62—5-63
differential diagnosis of, 5-57, 5-61, 5-63
in histoplasmosis, 5-61—5-62
in hypoxia, 6-31
Pulmonary edema
high altitude, 6-37, 6-38, 6-39, 6-40, 6-42
in hypertensive emergency, 4-5
non-cardiac. *See* Acute respiratory distress syndrome

Pulmonary embolism
chest pain with, 3-10
alleviating or aggravating factors in, 3-11
in pericarditis, 4-6
treatment for, 3-12
vital signs suggesting, 3-12
Pulmonary embolus
assessment of, 4-23
causes of, 4-23
differential diagnosis of, 6-5, 6-41
follow-up, 4-24
massive, 4-23
differential diagnosis of, 4-23
patient education for, 4-24
pleural effusion with, 4-15
risk factors for, 4-23
signs of, 4-23
symptoms of, 4-23
treatment of, 4-24
Pulmonary function tests
for allergic pneumonitis, 4-18
for COPD, 4-22
Pulmonary hemorrhage, 7-23
Pulmonary infection
blastomycosis, 5-59
chronic, 5-54—5-56
Pulmonary over inflation syndrome
assessment of, 6-7
cause of, 6-7
differential diagnosis of, 6-10, 6-12
follow-up for, 6-8
patient education for, 6-8
portable hyperbaric chamber for, 8-36
signs of, 6-7
symptoms of, 6-7
treatment for, 6-8
Pulmonary status monitoring, 7-26
Pulmonary syndromes, nontuberculous mycobacterial, 5-56
Pulmonary tuberculosis, 3-32
Pulp, tooth fractures involving, 5-13
Pulpitis, 6-33
Pulse
with dyspnea, 3-115
in trauma patients, 7-3
Pulse oximeter, 8-9
Pulse oximetry
for acute respiratory distress syndrome, 4-25
for anaphylactic shock, 7-10
for blast injuries, 7-23—7-24
for COPD, 4-22
for hantavirus, 5-69
for high altitude pulmonary edema, 6-41
monitoring with, 8-9—8-10
in pneumonia, 4-13
for poisoning, 5-141
for venomous snake bite, 5-144
Pulse pressure, in hypovolemic shock, 7-12
Puncture wounds, 8-32
Pupil

irregular, 3–27
in red eye, 3–24
Purified protein derivative (PPD) skin testing, 5–54
  for cutaneous tuberculosis, 4–45
  for nontuberculous mycobacterial infections, 5–56
  procedure and interpretation of, 5–56
Purpura, 4–39
Pustule, 4–38
Pyariformis syndrome, 3–73
Pyelonephritis
  treatment of, 4–94
  in urinary tract infection, 4–93
Pyogenic granuloma, 5–4
Pyrantel pamoate, 5–37
Pyrazinamide, 5–55
Pyrethrins shampoo, 4–63
Pyridostigmine, 6–52
Pyridoxine deficiency, 5–137t

Q fever, 5–32
  assessment of, 5–86
  cause of, 5–85
  patient education for, 5–86
  signs of, 5–85—5–86
  symptoms of, 5–85
  treatment for, 5–86
  zoonotic disease considerations in, 5–86
QRS complex, 3–111
Quinidine, 5–44, 5–45
Quinine
  for babesiosis, 5–35
  for malaria, 5–45
Quinine sulfate, 5–44
Quinolones, 4–94

Rabies, 5–32
  from animal bites, 7–8
  assessment of, 5–82
  cause and transmission of, 5–81
  control of, 5–121
  follow-up for, 5–82
  patient education for, 5–82
  required vaccination for, 5–106t
  signs of, 5–82
  symptoms of, 5–81
  treatment for, 5–82
  vaccine for, 7–9
  zoonotic disease considerations in, 5–82—5–83
RADIACS meter, 6–62
Radial head subluxation
  assessment of, 3–68
  in children, 3–65
  diagnostic tests for, 3–68
  differential diagnosis of, 3–68
  follow-up, 3–68
  patient education for, 3–68

post treatment care for, 3–68
procedure for, 3–68
Radial nerve block, at wrist, 5–166—5–168
Radiation injury, 6–61—6–62
  follow-up for, 6–62
  patient education for, 6–62
  symptoms and signs of, 6–62
  treatment of, 6–62
Radiation monitoring, 6–62
Radiograph, pelvic, 3–46
Rales
  with dyspnea, 3–116
  in pneumonia, 4–13
Ramsay Hunt syndrome, 4–47
Range of motion exercise
  for joint pain, 3–62
  for knee injury, 3–75
Ranitidine, 4–85
Rash
  in coccidioidomycosis, 5–60
  cutaneous larvae currens, 5–47
  with fever, 3–33f, 3–112
    differential diagnosis of, 3–112—3–113t
  with zinc deficiency, 5–139t
Rat bite fever
  assessment of, 5–97
  cause of, 5–97
  follow-up for, 5–98
  patient education for, 5–97
  symptoms and signs of, 5–97
  treatment for, 5–97
  zoonotic disease considerations in, 5–98
Rat snakes, 5–146
Rattlesnakes, 5–143
  identification of, 5–147
Razor bump fighter, 4–66
RBCs. See Red blood cells
Rebound tenderness, abdominal, 3–1
Recompression therapy
  for decompression sickness, 6–5, 6–36
  for pulmonary over inflation syndrome, 6–8
Rectal bleeding, 3–18
Rectal examination
  for abdominal pain, 3–1
  for female pelvic complaints, 3–37, 3–39
  for urinary tract problems, 4–87
Rectovaginal examination, 3–39, 3–40f
  pelvic examination for, 3–37
Red blood cells, 4–8
  bizarre forms of, 8–56
  coloration differences in, 8–56
  count/morphology of, 8–54—8–56
  crenated, 8–56
  inclusions in, 8–56
  indications, benefits, and cautions for, 7–15t
  morphological variations of, 8–58
  nucleated, 8–58
  as percentage of total blood volume, 8–53—8–54
  transfusion of in field, 8–18

Red eye, non-traumatic, 3–24
  assessment of, 3–24—3–25
  follow-up, 3–26
  patient education for, 3–25
  signs of, 3–24
  symptoms of, 3–24
  treatment for, 3–25
Red rubber Robinson, 8–34
Reduction
  for elbow joint dislocation, 3–67
  of hip dislocation, 3–69
  of patellar dislocation, 3–69
  of radial head subluxation, 3–68
  for shoulder dislocation, 3–66—3–67
  of shoulder fracture, 3–71
  of temporomandibular joint dislocation, 5–18—5–19
Referred pain, abdominal, 3–1
Refractometer method, 8–41
Regional blocks, 5–163—5–164
Reglan, 3–56
Rehabilitation, shoulder joint, 3–72
Rehydration
  for bacterial food poisoning, 4–80
  for electrical or lightning injury, 7–28
  for fever, 3–34f
  for heat cramps, 6–48
  for hypovolemic shock, 7–14
  indications, benefits, and cautions for, 7–15t
Reiter's syndrome
  differential diagnosis of, 5–98
  joint pain with, 3–60
  patient education for, 3–64
Relapsing fever
  assessment of, 5–89
  cause and transmission of, 5–89
  follow-up for, 5–90
  patient education for, 5–90
  signs of, 5–89
  symptoms of, 5–89
  treatment for, 5–90
Relaxation exercises
  for anxiety, 3–5
  for chronic pelvic pain, 3–44t, 3–47
Relenza (zanamivir), 4–11—4–12
REM sleep apnea, 4–25
Renal failure, hypertensive, 4–5
Respiratory disorders
  acute respiratory distress syndrome, 4–24—4–25
  allergic pneumonitis, 4–18—4–19
  apnea, 4–25—4–27
  asthma, 4–19—4–21
  chronic obstructive pulmonary disease, 4–21—4–22
  common cold and flu, 4–10—4–12
  empyema, 4–17—4–18
  pleural effusion, 4–14—4–15
  pneumonia, 4–12—4–14
  pulmonary embolus, 4–23—4–24
  thoracentesis for, 4–15—4–16

Respiratory distress
  assessment of under fire, 1–1
  in trauma, 7–2
Respiratory distress syndrome, 3–93
Respiratory examination, 2–3
Respiratory obstruction, 3–14
Respiratory rate, 7–3
Rest, for operational stress, 5–148
Restraint, of animals, 5–128—5–130
Resuscitation
  algorithms for, 4–8
  for blast injury, 7–24
Retinal detachment
  in acute vision loss, 3–22
  differential diagnosis of, 3–23
Retinoic acid, 4–49
Retinoid gel, 4–48
Retinoids, 5–138t
Reverse osmosis water purification unit (ROWPU), 5–118
Reversing agents, pain control, 8–40
Review of systems (ROS), 2–3
Reye's syndrome, 5–80
Rh factor, 8–15
Rh typing, 8–59—8–60
  in crossmatching, 8–60—8–61
Rheumatic fever, acute
  assessment of, 5–98—5–99
  cause of, 5–98
  follow-up for, 5–99
  Jones diagnostic criteria of, 5–98
  patient education for, 5–99
  signs of, 5–98
  symptoms of, 5–98
  treatment for, 5–99
Rheumatoid arthritis
  in diabetes mellitus, 4–28
  fatigue with, 3–30t
  treatment of, 3–63
Rheumatologic disorders, 3–31
Rhinitis
  allergic, 4–11
  irritant, 4–11
Rhonchi, 4–22
Rhythm disturbance, syncope with, 3–118
Ribavirin, 5–69
Riboflavin deficiency, 5–137t
RICE regimen, 3–75
Rickets, 5–138t
Rickettsia
  akari, 5–83
  austrtalis, 5–83
  conorii, 5–83
  rickettsia, 5–83
  typhi, 5–83
Rickettsial fever, 5–66, 5–70
Rickettsial infections, 5–83—5–85
  assessment of, 5–84
  causes of, 5–83
  diagnostic algorithm for, 3–34f

differential diagnosis of, 3–112t, 5–103
follow-up for, 5–85
in meningitis, 4–36
patient education for, 5–85
Q fever, 5–85—5–86
signs of, 5–83
symptoms of, 5–83
treatment for, 5–84
zoonotic disease considerations in, 5–84—5–85
Rickettsialpox
differential diagnosis of, 3–112t, 5–83
treatment of, 5–85
Rifampin
for brucellosis, 5–94
for hip pain, 3–74
for leprosy, 4–45
for meningitis, 4–37
for Q fever, 5–86
for tuberculosis, 5–55
for verruga peruana, 5–93
Rimantadine, 4–12
Ring block, 5–175
Ringer's lactate
for blister agents, 6–53
for electrical or lightning injury, 7–27
saline, 3–43
Ringworm, 4–49—4–51
River blindness. *See* Onchocerciasis
Rizatriptan (Maxalt), 3–56
Rocephin (ceftriaxone)
after thoracostomy, 8–9
for cesarean section, 3–100
for epididymitis, 3–84
for preterm labor infection, 3–94
with splinting, 8–32
Rocky Mountain spotted fever, 5–32, 5–83
differential diagnosis of, 3–112t
joint pain with, 3–60
treatment of, 5–84
Rodent mite, 5–83
Rodents
in hantavirus transmission, 5–68, 5–69
in plague transmission, 5–95—5–96, 5–97
Romana's sign, 5–53
Rosacea, 4–39
Rosy anemone, 6–13—6–14
Rotator cuff injury, 3–70
Rotator cuff tendonitis
assessment of, 3–70
treatment of, 3–71
Rotavirus, porcine, 5–135
Rouleaux formations, 8–56, 8–58
RPR, 3–61
Rubbish disposal, 5–118
Rubella
differential diagnosis of, 3–113t, 5–66, 5–79
required vaccination for, 5–106t
Rubeola, 4–11
Rule of nines, 7–20f, 7–22

Sacroiliac joint dysfunction, 3–6
Salicylates, 5–99
Salicylic acids, 4–49
paste, 5–4
Saline
for blister agents, 6–53
boluses, for heat exhaustion, 6–49
for diabetes mellitus, 4–29
in fluid resuscitation for burns, 7–22
indications, benefits, and cautions for, 7–15t
Salmeterol (Serevent)
for asthma, 4–20
for COPD, 4–22
Salmonella
differential diagnosis of, 5–86
in food poisoning, 4–79, 4–80
nontyphoidal, 5–104
in porcine diarrhea, 5–135
typhi, 5–104
vaccination against, 5–105
Salmonellosis, 5–32
assessment of, 3–18
Salsalate, 8–38
Salt restriction
for congestive heart failure prevention, 4–4
for hypertensive emergency, 4–5
for pericarditis, 4–7
Sandfly bite, 5–42
Sanitation, in field. *See* Field sanitation
Saphenous nerve block, 5–168—5–169
Saran lotion, 4–61
Sarcoidosis
differential diagnosis of, 4–37
as disqualifying condition for diving, 6–21
Sarcoptes scabiei, 4–59, 4–61—4–62
Sarcosporidiosis, 5–32
Sarna lotion, 3–115
Scabies, 4–59, 5–32
assessment of, 4–61
cause of, 4–61
differential diagnosis of, 3–114, 4–53
follow-up for, 4–62
patient education for, 4–61—4–62
signs of, 4–61
symptoms of, 4–61
treatment for, 4–61
zoonotic disease considerations in, 4–62
Scabs, 4–39
Scalded skin syndrome, 3–112t
Scales, 4–39
Scarlet fever, 5–99
diagnostic algorithm for, 3–30—3–34f
differential diagnosis of, 3–112t
Scars, 4–39
Schistosoma
haemotobium, 5–47
japonica, 5–47
mansoni, 5–47

Schistosomiasis
　assessment of, 5–47
　diagnostic algorithm for, 3–30—3–34f
　follow-up for, 5–47
　patient education for, 5–47
　signs of, 5–46
　symptoms of, 5–46
　transmission of, 5–46
　treatment for, 5–47
　zoonotic disease considerations in, 5–47
Schizophrenia, 5–152
Schizophreniform disorder, 5–152
Schober maneuver, 3–61
Sciatica
　assessment of, 3–73
　in low back pain, 3–6
Scleral icterus, jaundice of, 3–57
Scleritis
　differential diagnosis of, 3–25
　red eye with, 3–24
　treatment of, 3–25
Sclerosing lymphangitis, penile, 3–78
　treatment of, 3–78
Scoliosis, 3–6
Scopolamine
　for eye injury, 3–28
　for red eye, 3–25
　side effects of, 3–118
Scorpionfish, 6–13
Scotch tape test, 8–51—8–52
Scours, 5–135—5–136
Scrofuloderma, 4–44
Scrotal support
　for epididymitis, 3–84
　for testis torsion prevention, 3–83
Scrotum
　cellulitis of, 3–78
　examination of for urinary tract problems, 4–87
　mass in, 3–79—3–80
　pain in, 4–88
Scrub typhus, 5–83
　differential diagnosis of, 5–91
　treatment of, 5–84
SCUBA tanks, carbon monoxide in, 6–11
Scurvy, 5–137t
　differential diagnosis of, 5–15
Sea anemone, 6–13—6–14
Sea blubber, 6–13—6–14
Sea lions, 6–17
Sea nettle, 6–13—6–14
Sea snakes, 5–143, 6–14
　treatment for, 6–16
Sea urchins, 6–14
　treatment for, 6–15—6–16
Sea wasps, 6–13—6–14
Seafood
　cooking, 5–109
　in paragonimiasis, 5–45—5–46
Sebaceous gland hyperplasia, 4–48

Seborrheic dermatitis, 3–114, 4–65
Seborrheic keratosis, 4–68
　assessment of, 4–69
　follow-up for, 4–69
　patient education for, 4–69
　signs of, 4–69
　symptoms of, 4–68
Seizure disorders
　assessment of, 4–34
　causes of, 4–34
　follow-up for, 4–35
　memory loss with, 3–86
　patient education for, 4–35
　signs of, 4–34
　symptoms of, 4–34
　treatment of, 4–34—4–35
Seizures
　assessing under fire, 1–1
　atypical, 6–31
　differential diagnosis of, 4–31, 5–49, 5–101, 6–36
　in eclampsia, 3–105
　with preterm birth, 3–93
　syncope with, 3–118
Selective serotonin reuptake inhibitors (SSRIs)
　for anxiety, 3–5
　for depression, 3–17, 3–31
Selenium sulfide, 4–51
Self-reduction, shoulder, 3–66
Semen, blood in, 4–88
Senna bisacodyl, 3–13
Seoul virus, 5–68
Sepsis
　in adrenal insufficiency, 4–27
　fever with, 3–32
　with preterm birth, 3–93
Septic arthritis
　diagnostic tests for, 3–71
　knee pain with, 3–75
　treatment of, 3–74
Septic joint, 3–59, 3–60
　treatment of, 3–63
Septic shock, 7–13
Septra
　for urinary incontinence, 4–91
　for urinary tract infections, 4–93, 4–94
Septra DS, 3–84
　for prostatitis, 3–81
　in testis torsion treatment, 3–83
Serevent (salmeterol)
　for asthma, 4–20
　for COPD, 4–22
Serotonin specific reuptake inhibitors (SSRIs). *See* Selective serotonin reuptake inhibitors (SSRIs)
Sertraline (Zoloft), 3–17
Serum sickness, 3–113t
Sexually transmitted diseases
　genital ulcers, 5–28—5–30
　screening for, 3–49

IN-57

urethral discharges, 5-26—5-27
vaginal trichomonas, 5-30—5-31
Shallow water blackout, 6-9
Sharks, 6-17
Sheep
  foot rot in, 5-133—5-134
  obstetrics for, 5-131
Shigella
  in food poisoning, 4-79, 4-80
  identification of colonies of, 8-50
Shigellosis dysentery, 3-18
Shingles, 4-47
  chest pain with, 3-10, 3-11
  treatment for, 3-12
Shock. *See also* Anaphylactic shock
  anaphylactic, 7-10—7-11
  differential diagnosis of, 6-9, 7-24
  fluid resuscitation for, 7-14—7-17
  with hemorrhage, 7-3
  hypovolemic, 7-11—7-14
  treatment of, 7-24
Shock waves, underwater, 6-18
Shortness of breath. *See also* Dyspnea
  with congestive heart failure, 4-3
  in COPD, 4-21
Shoulder
  aspiration at, 8-29
  auscultation of, 3-70
  dislocation of
    anterior, 3-64
    assessment of, 3-66
    diagnostic tests for, 3-66
    differential diagnosis of, 3-66
    follow-up, 3-67
    patient education for, 3-67
    post treatment care for, 3-67
    posterior, 3-65
    procedures for, 3-66—3-67
    treatment of, 3-71
  dystocia of, 3-95—3-96
  fractures of, 3-71
  multidirectional instability of, 3-70
  pain in
    assessment of, 3-71
    diagnostic tests for, 3-71
    follow-up, 3-72
    location of, 3-70
    patient education for, 3-71—3-72
    procedures for, 3-71
    referred, 3-70
    rehabilitation guide for, 3-72
    risk factors for, 3-70
    signs of, 3-70
    symptoms of, 3-70
    treatment of, 3-71
Shower, solar heated, 5-117f
Shy-Drager syndrome, 3-117
Sickle cell anemia, 4-9, 4-71
Sickle cell crisis
  acute abdominal pain in, 3-3t

follow-up for, 4-10
treatment of, 4-9
Silo filler's lung, 4-18
Silvadene
  for blister agents, 6-53
  for chemical burns, 7-18
  for staphylococcal scalded skin syndrome, 4-43
Silver nitrate, 7-19
Silver sulfadiazine
  for chemical burns, 7-18
  for staphylococcal scalded skin syndrome, 4-43
Simple interrupted suture, 8-23
Sinus barotrauma, 6-3—6-4
Sinus squeeze, 6-33
  prevention of, 6-34
Sinusitis
  cough with, 3-15
  differential diagnosis of, 4-11
  infectious, 6-32, 6-34
  treatment of, 3-56
Site survey checklist, MedCAP guide, 1-3, 1-7—1-9
Sitz bath, 3-52
Skin
  barotrauma of, 6-3—6-4
  cancer of
    basal cell, 4-39, 4-48, 4-66—4-67
    malignant, 4-68
    squamous cell, 4-48, 4-67
  disorders of
    bacterial infections, 4-40—4-45
    bug bites and stings, 4-55—4-63
    cancer, 4-66—4-68
    contact dermatitis, 4-69—4-70
    dermatology and, 4-38—4-40
    fungal infections, 4-49—4-51
    parasitic infections, 4-52—4-54
    pseudofolliculitis barbae, 4-65—4-66
    psoriasis, 4-64—4-65
    seborrheic keratosis, 4-68—4-69
    spirochetal, 4-63—4-64
    viral infections, 4-45—4-49
  full-thickness injury of in frostbite, 6-42
  lesions of
    arrangement of, 4-39
    distribution of, 4-39
    primary, 4-38—4-39
    secondary, 4-39
    shape of, 4-39
  mass removal from
    equipment for, 8-27
    indications for, 8-26—8-27
    precautions for, 8-28
    procedure for, 8-27—8-28
  rashes of in joint pain, 3-60
  tenting of in diabetes mellitus, 4-29
Skin decontamination area, 6-54
Skinning, 5-125
Skip breathing, 6-12

Skull fracture
    basilar, 7–4
    disqualifying condition for diving, 6–22
Slaughter, humane procedure for, 5–123—5–125
Sleep
    altered patterns of, 3–30—3–31
    disturbances of
        in depression, 3–16
        fatigue and, 3–29
        memory loss with deprivation of, 3–86
Sleep apnea, 4–25
    fatigue with, 3–29
    treatment of, 3–31
Sleep wedge, 3–31
Sleeping sickness. *See* African trypanosomiasis
Sling, shoulder, 3–67
Slnuyrtol, 4–20
Slow deep breathing, 3–5
Small bowel obstruction, 3–3f
Smallpox
    in biological warfare, 6–59—6–60
    vaccine for, 6–60
Smoke inhalation, 4–4
Smokehouse, 5–127
Smoking
    in COPD, 4–21
    meat, 5–127
Snake bites, venomous, 5–142—5–143
    assessment of, 5–144
    follow-up for, 5–147
    patient education for, 5–146
    signs of, 5–143—5–144
    snake identification in, 5–147
    symptoms of, 5–143
    treatment for, 5–144—5–146
Snakes
    characteristics of, 5–146
    identification of, 5–147
    poisonous, 5–142—5–143
    sea, 6–14
        treatment for, 6–16
Snellen chart
    for acute vision loss without trauma, 3–23
    for red eye, 3–24
Snoring
    with apnea, 4–26
    treatment of, 3–31
SOAP format
    condensed, 2–1—2–2
    value of, 2–1
SODARS reports, 5–107
Sodium amytal, 6–51
Sodium bicarbonate
    in fluid resuscitation, 7–19—7–22
    for venomous snake bite, 5–145
Sodium iodide, 6–62
Sodium pentochlorphenate, 4–54
Sodium stibogluconate (Pentostam), 5–43
Solar heated shower, 5–117f
Solu-Medrol (methylprednisolone), 7–11

Somatoparietal pain, 3–1
Sonata (zaleplon), 3–30
Sound injury, with underwater blast, 6–18—6–20
South American bartonella. *See* Oroya fever
Specimen transport, preparation for, 8–40—8–41
Spectinomycin
    for disseminated gonococcal infection, 4–40
    for epididymitis, 3–85
Speech, slurred, 4–37
Spermatocele, 3–80
Spider angiomata, 3–87
Spider bites, 4–60
    assessment of, 4–60
    differential diagnosis of, 5–91
    follow-up for, 4–61
    patient education for, 4–61
    signs of, 4–60
    symptoms of, 4–60
    treatment for, 4–60—4–61
Spinal cord compression, 4–90
Spinal examination
    for joint pain, 3–60—3–61
    range of motion of, 3–61
Spinal mass, 3–6
Spinal needle, 8–35
Spine
    cervical injury of, 7–1
    stenosis of, 3–6
Spine board, 3–7
Spinning, 3–20
Spiramycin, 4–41
Spirillum minor, 5–97
Spirochetal infections
    leptospirosis, 5–86—5–87
    Lyme disease, 5–88—5–5–89
    relapsing fever, 5–89—5–90
Spirochetal skin disease
    pinta, 4–64
    yaws, 4–63
Splenic pain, referred to chest, 3–11
Splinting
    for avulsed tooth, 5–14
    for luxated tooth, 5–14
Splints
    for ankle injury, 3–76—3–77
    application of, 8–32
    for elbow dislocation, 3–67
    indications for, 8–32
    for marine animal bites, 6–18
    precautions with, 8–32
Spondylolysis, 3–6
Sporotrichosis, 5–103
Sports drinks, 4–31
Sprains, ankle, 3–76
Sputum
    Gram-stain for pneumonic plague, 6–58
    Ziehl-Neilson stain of, 8–47
Squamous cell carcinoma, 4–67
St. Louis encephalitis, 5–66
ST segment

analysis of, 8–12
elevated in pericarditis, 4–6
in myocardial infarction, 4–1
Staphylococcal boil, 5–91
Staphylococcal scalded skin syndrome
  assessment for, 4–43
  cause of, 4–42
  follow-up for, 4–43
  signs of, 4–42—4–43
  symptoms of, 4–42
  treatment for, 4–43
Staphylococcus
  in acute mastitis in animals, 5–135
  aureus
    in erysipelas, 4–41
    in food poisoning, 4–79
    identification of colonies of, 8–49
    in impetigo contagiosa, 4–43
    in pneumonia, 4–14
    in staphylococcal scalded skin syndrome, 4–42
  differential diagnosis of, 3–112*t*
  intermedius, 5–32
  of male genitals, 3–77
Status epilepticus, 4–35
STDs. See Sexually transmitted diseases
Steroid cream
  for bed bug bites, 4–55
  for millipede exposure, 4–57
Steroids
  for adrenal insufficiency, 4–28
  for aphthous ulcers, 5–17
  for contact dermatitis, 4–70
  inhaled, 4–20
  for low back pain, 3–7
  for meningitis, 4–36
  for mites, 4–59
  for pediculosis, 4–63
  in pruritus, 3–115
  for psoriasis, 4–65
  side effects of, 3–64
  for swimming dermatitis, 4–54
  topical, for blister agents, 6–53
Stevens-Johnson syndrome, 3–113*t*, 5–16
Stimson maneuver, 3–66—3–67
Stingrays, 6–13
  treatment for, 6–14—6–15
Stonefish, 6–13
Stool guaiac, 7–24
Stool sample
  for occult blood, 3–37
  for strongyloides, 5–48
  for tapeworm infections, 5–48
  for trichuriasis, 5–51
Stool softener, postepisiotomy, 3–104
Straddle trench latrine, 5–112*f*
Stratum corneum, hyperproliferative layers of, 4–39
Street vendors, 5–109
Strengthening exercises, shoulder, 3–72

Streptobacillus moniliformis, 5–97
Streptococcal pharyngitis
  differential diagnosis of, 4–11
  untreated, 5–98
Streptococcus, 5–32
  assessment of, 5–100
  differential diagnosis of, 3–112*t*
  follow-up for, 5–100
  group A, 5–100
    beta-hemolytic, in impetigo contagiosa, 4–43
  group B, 5–99
  in early neonatal infection, 3–93
  identifying colonies of, 8–50
  of male genitals, 3–77
  pathogens in, 5–99
  patient education for, 5–100
  pneumoniae, 4–14
  pyogenes, group A beta-hemolytic, 4–41
  signs of, 5–99—5–100
  symptoms of, 5–99
  treatment for, 5–100
  zoonotic disease considerations in, 5–100
Streptomycin
  for pneumonic plague, 6–59
  for tuberculosis, 5–55
  for tularemia, 6–60
Stress
  memory loss with, 3–86
  operational
    assessment of, 5–147—5–148
    causes of, 5–147
    follow-up for, 5–148
    signs of, 5–147
    symptoms of, 5–147
    treatment for, 5–148
  in peptic ulcer disease, 4–84
  in pruritus, 3–115
Stress fractures
  foot, 5–6—5–7
  heel, 5–1
Stress incontinence, 4–90
Stretching exercise
  for hip strain or arthritis, 3–73—3–74
  to prevent knee injury, 3–76
Striae gravidarum, 3–87
Stridor, with dyspnea, 3–116
Stroke
  dyspnea with, 3–116
  in hypertensive emergency, 4–5
  memory loss with, 3–86
  meningeal signs in, 4–36
  treatment of dyspnea in, 3–117
Strongyloides
  hyperinfection syndrome, 5–47
  stercoralis, 5–32, 5–47
Strongyloidiasis
  assessment of, 5–48
  follow-up for, 5–48
  patient education for, 5–48
  signs of, 5–48

symptoms of, 5–47
transmission of, 5–47
treatment for, 5–48
zoonotic disease considerations in, 5–48
Strychnine poisoning, 5–101
Subacromial bursitis, 3–10
Subconjunctival hemorrhage
   differential diagnosis of, 3–25, 3–28
   treatment of, 3–26, 3–28
Subcutaneous mass
   inflamed, 8–28
   non-inflamed, 8–27—8–28
Subdural hematoma, 4–36, 5–67
Subperiosteal abscess, 5–13
Substance abuse, 5–150
   differential diagnosis of, 5–150—5–151t, 5–152, 6–31
Subtalar joint dislocation
   lateral, 3–66
   medial, 3–66
Subungal hyperkeratosis, 4–64
Subungual exostosis, 5–2
Succinylcholine, dosing guidelines for, 5–156t
Sudden cardiac death, resuscitation of, 4–7
Sufentanil, dosing guidelines for, 5–156t
Suicidal ideation, 5–149
Suicide
   assessment for, 5–149
   follow-up for, 5–150
   incidence of, 5–149
   prevention of, 5–149
   risk of with depression, 3–17
   symptoms and signs of, 5–149
Suicide attempts, 5–149
Suicide watch, for HIV patients, 5–78
Sulfa in Vaseline, 4–61
Sulfadiazine, 5–63
Sulfamylon, 7–18
Sulfasalazine, 4–82
Sulfonamides, 4–82
Sumatriptan (Imitrex), 3–56
Sun exposure
   in basal cell carcinoma, 4–66
   in squamous cell carcinoma, 4–67
Sunburn
   differential diagnosis of, 3–114
   pruritus with, 3–113
Superficial nerve block, foot, 5–168—5–169, 5–171f
Supertropical bleach, 6–54, 6–55
Suppositories, 3–13
Suprapubic bladder aspiration, 8–35—8–36
Sural nerve block, 5–169—5–170
Surgery
   for cesarean section, 3–101—3–102
      instruments in, 3–100
   M5 item list for, 1–19
   for peritonitis, 4–83, 4–84
   for sea urchin bites, 6–15
Surgical cricothyroidotomy, 8–6—8–7
Surgical excision
   for myiasis, 4–53
   for onchocerciasis, 4–53
   for plantar warts, 5–4
Surgilube, 7–19
Sutures
   selection of, 8–22—8–23
   types of, 8–23
Suturing
   equipment for, 8–22
   hints for, 8–23
   precautions in, 8–22—8–24
   procedure for, 8–22
   techniques for, 8–23, 8–24f
Sweating, 6–47
   absence of, 6–50
Swimmer's ear, 6–2
Swimmer's itch, 5–46
Swimming dermatitis
   assessment of, 4–54
   cause of, 4–54
   follow-up for, 4–54
   patient education for, 4–54
   signs of, 4–54
   symptoms of, 4–54
   treatment for, 4–54
Symptoms
   acute abdominal pain, 3–1—3–2
      causes of, 3–3t
      objective signs of, 3–4t
   anxiety, 3–2—3–5
   Bartholin's gland cyst/abscess, 3–52—3–55
   breast problems
      abscess incision and drainage, 3–9—3–10
      mastitis, 3–7—3–9
   chest pain, 3–10—3–13
   constipation, 3–13
   cough, 3–13—3–15
   depression and mania, 3–15—3–18
   diarrhea, acute, 3–18—3–19
   dizziness, 3–20—3–22
   eye problems, 3–22—3–29
   fatigue, 3–29—3–31
   fever, 3–31—3–35
   gynecologic, 3–37—3–55
   headache, 3–55—3–57
   itching, 3–113—3–115
   jaundice, 3–57—3–59
   joint pain, 3–59—3–64
      hip, 3–72—3–74
      with joint dislocations, 3–64—3–70
      knee, 3–74—3–76
      shoulder, 3–70—3–72
   low back pain, 3–6—3–7
   male genital problems, 3–77—3–85
   memory loss, 3–85—3–87
   obstetric, 3–87—3–109
   palpitations, 3–110—3–111
   rash with fever, 3–112—3–113
   shortness of breath, 3–115—3–117
   syncope, 3–117—3–119

Syncope, 3-117
  assessment of, 3-118
  follow-up, 3-119
  patient education for, 3-118
  signs of, 3-118
  symptoms of, 3-117
  treatment of, 3-118
Synovial fluid
  analysis of, 8-29
  cell count in joint pain assessment, 3-61
Syphilis
  diagnostic algorithm for, 3-30—3-34f
  differential diagnosis of, 5-103
  in genital ulcers, 5-28
  joint pain with, 3-60
  secondary, 3-113t
  signs of, 5-29
  treatment of, 5-30
Systemic lupus erythematosus, 3-113t

Tabanid flies, 4-52
Tache noire lesions, 4-36
Tachycardia
  with blast injuries, 7-23
  in congestive heart failure, 4-4
  differential diagnosis of, 3-111
  headache in, 3-55
  in high altitude pulmonary edema, 6-40
  in hypovolemic shock, 7-12
  with myocardial infarction, 4-1
  syncope with, 3-117
Tachypnea, high altitude, 6-40
Taenia
  saginata, 5-32
  solium, 5-32
Taeniasis. *See* Tapeworm infections
Tail restraint, cattle, 5-130
Tamponade
  after pericarditis treatment, 4-7
  chest pain with, 3-11
  vital signs suggesting, 3-11
Tamsulosin (Flomax)
  for prostatitis, 3-82
  for urinary incontinence, 4-91
Tanning chemicals, 4-69
Tapeworm infections
  assessment of, 5-48—5-49
  follow-up for, 5-49
  patient education for, 5-49
  signs of, 5-48
  symptoms of, 5-48
  transmission of, 5-48
  treatment for, 5-49
  zoonotic disease considerations in, 5-49—5-50
Tar burn, 7-19
Target cells, hemoglobin variations of, 8-56
Tarsal tunnel syndrome, 5-1
TEED classification, 6-2

Telangiectasia, 4-39
Temperature monitoring, 7-3
Temporomandibular joint dislocation, 5-18
  follow-up for, 5-19
  patient education for, 5-19
  treatment for, 5-18—5-19
Tendonitis, 3-73—3-74
Tension headache, 3-55
Tension pneumoperitoneum, 7-26
Tension pneumothorax
  in trauma, 7-2
  treatment of, 3-117, 7-24
Terazosin (Hytrin)
  for prostatitis, 3-82
  for urinary incontinence, 4-91
Terbinafine, 4-50
Terbutaline, 3-94
Terconazole, 4-93
Testis
  loss of, 3-84
  mass in, 3-79—3-80
  ruptured, 3-83
  torsion of
    assessment of, 3-83
    follow-up, 3-84
    patient education for, 3-83
    salvage of, 3-82
    signs of, 3-83
    symptoms of, 3-82—3-83
    with testicular mass, 3-80
    treatment of, 3-83
  tumor of, 3-83
Tetanospasmin, 5-100
Tetanus
  antitoxin, 7-9
  assessment of, 5-101
  cause and incidence of, 5-100—5-101
  follow-up for, 5-102
  immunization chart for, 5-102
  patient education for, 5-101—5-102
  prophylaxis, 6-15
    for brown recluse spider bite, 4-60
    for chemical burns, 7-18
    for cone shell sting, 6-16
    for electrical or lightning injury, 7-27
    for marine animal bites, 6-18
    for octopi bites, 6-15
    for sea snake bite, 6-16
    for sea urchin bites, 6-16
    for venomous fish bites, 6-15
  signs of, 5-101
  symptoms of, 5-101
  treatment for, 5-101
Tetanus-diphtheria toxoid, 5-106t
Tetanus immune globulin, 5-101
Tetany, 6-48
Tetracaine
  in axillary blockade, 5-174t
  for eye injury, 3-28
Tetracycline

for biological warfare agents, 6–56
for conjunctivitis in animals, 5–136
for inhalational anthrax, 6–57
for malaria, 5–45
in pancreatitis, 4–82
for pelvic inflammatory disease, 3–51
for pinta, 4–64
for pneumonic plague, 6–59
for rat bite fever, 5–97
for rickettsialpox, 5–85
for tularemia, 5–103, 6–60
for typhus, 5–84
for venomous fish bites, 6–15
for yaws, 4–63
Thalassemia, 4–9
Theophylline
for asthma, 4–20
for COPD, 4–22
Thermal injuries, care of under fire, 1–1
Thermal test, dental caries, 5–20
Thermophilic actinomycetes, 4–18
Thermoregulatory status, 6–45*t*
Thiabendazole
for cutaneous larva migrans, 5–42
for strongyloides, 5–48
Thiamine deficiency, 5–137*t*
memory loss with, 3–86
Thoracentesis
for empyema, 4–17
equipment for, 4–16
indications for, 4–15
for pleural effusion, 4–15
precautions, 4–16
procedure in, 4–16
risks of, 4–15—4–16
Thoracostomy
equipment for, 8–7
indications for, 8–7
needle, 8–8
precautions for, 8–9
procedures in, 8–7—8–8
tube, 8–8—8–9
Thrush, 5–57—5–58
equine, 5–134
esophageal, 5–58
oral, 5–58
vaginal, 5–58
Thyroid disorders
alleviating and aggravating factors in, 4–32
assessment of, 4–33
follow-up for, 4–33—4–34
patient education for, 4–33
during pregnancy, 4–34
pruritus with, 3–114
signs of, 4–32—4–33
symptoms of, 4–32
treatment for, 4–33
types and causes of, 4–31—4–32
Thyroid dysfunction, 6–22
Thyroid peroxidase, 4–33—4–34

Thyroid stimulating hormone, overproduction of, 4–31—4–32
Thyroid storm, 4–34
Tibial nerve block, 5–170—5–174
classic approach in, 5–172*f*
sustenaculum tali approach in, 5–172*f*
Tibial stress fracture, 5–6
Ticarcillin
for chemical burns, 7–19
for cholecystitis, 4–78
for jaundice, 3–58
Ticarcillin/clavulanate
for abnormal uterine bleeding, 3–41
for human and animal bites, 7–9
Tick-borne disease
ehrlichiosis, 5–95
Lyme disease, 5–88—5–89
relapsing fever, 5–89
in rickettsial infections, 5–83—5–85
tularemia, 5–102
Tick-borne encephalitis, 5–66
Tilt test, 3–118
Timorean filariasis, 5–39
Tinea
barbae, 4–50
barbae/faciale, 4–49—4–51
capitis, 4–49—4–51
signs of, 4–50
corporis, 4–49—4–51
signs of, 4–50
cruris, 4–49—4–51
signs of, 4–50
manuum, 4–49—4–51
signs of, 4–50
pedis, 4–49—4–51
signs of, 4–50
unguium, 4–49—4–51
signs of, 4–50
versicolor. *See* Pityriasis versicolor
Tinel's sign, 5–1
Tinnitus
dizziness with, 3–20
with ear barotrauma, 6–1
TIVA. *See* Anesthesia, total intravenous
Tocolytics, 3–94
Tocopherol, 5–138*t*
Todd's paralysis, 4–34—4–35
Toenail
fungal infection of, 5–3
ingrown
assessment of, 5–2—5–3
causes of, 5–2
follow-up for, 5–3
patient education for, 5–3
signs of, 5–2
symptoms of, 5–2
treatment for, 5–3
Tolectin DS, 3–63
Tooth
anatomy of, 5–9, 5–10*f*

avulsed, 5-14—5-15
   preserving/transporting, 5-26
crown surfaces of, 5-9, 5-11f
draining abscess of, 5-26
extraction of, 5-25—5-26
   for untreated periapical abscess, 5-14
fractures of, 5-9, 5-12—5-13
   involving dentin and pulp, 5-13
luxated, 5-14
severe pain in, 5-13
temporary restorations of, 5-24
Toothaches, 5-9
Toradol (ketorolac)
   for chronic pelvic pain, 3-44t
   for headache, 3-56
   for urolithiasis, 4-92
Total body surface area (TBSA), 7-20f
   of burn injury, 7-22
Total intravenous anesthesia. *See* Anesthesia, total intravenous
Tourniquet test, 5-65
Tourniquets, 7-2
   of external hemorrhage under fire, 1-2
Toxic epidermal necrosis, 3-113t
Toxic shock syndrome, 3-112t
Toxicology
   poisoning, 5-140—5-142
   venomous snake bite, 5-142—5-147
Toxocariasis, 4-52
Toxoplasma, 5-79
Traction devices, external, 8-33
Traction splint, hip, 3-74
Transient global amnesia, 3-86
Trauma
   ankle, 3-76
   in arthritis, 3-59
   dizziness with, 3-20
   in hip dislocation, 3-68
   hip fractures with, 3-73
   hip pain with, 3-72
   human and animal bites, 7-7—7-9
   knee pain with, 3-74—3-75
   low back pain with, 3-6
   Naval special warfare combat trauma AMAL items for, 1-21
   in patellar dislocation, 3-69
   primary survey for, 7-1—7-3
   red eye with.*See* Eye injury
   secondary survey for, 7-3—7-5
      checklist for, 7-5—7-7
   in urinary incontinence, 4-90
Trauma ruck pack list, USAF SOF, 1-15—1-16
Trauma vest pack list, USAF SOF, 1-17
Traveler's diarrhea, 3-19
Treatment plan, under fire, 1-1—1-2
Trench fever, 5-92—5-93
   differential diagnosis of, 3-112t
Trench mouth, 5-15—5-16
Trenchfoot, 6-44—6-47
Treponema

hyodysenteriae, in porcine diarrhea, 5-135
pallidum
   carateum, 4-64
   in genital ulcers, 5-28
   pertenue, 4-63
Triage
   in casualty decontamination station, 6-54
   in female pelvic examination, 3-37—3-38
   reverse, 7-26
Triamcinolone ointment
   for contact dermatitis, 4-70
   for psoriasis, 4-65
Trichenellosis
   assessment of, 5-50
   cause of, 5-50
   follow-up for, 5-51
   patient education for, 5-50—5-51
   symptoms and signs of, 5-50
   treatment for, 5-50
   zoonotic disease considerations in, 5-51
Trichinella spiralis, 5-32, 5-50
Trichinosis, 5-32. *See also* Trichenellosis
Trichloroacetic acid, 5-4
Tricholemmoma, 4-66
Trichomonas, 3-47
   diagnosis and treatment of, 3-48t
   in urethral discharges, 5-26
   vaginalis, 5-30—5-31
   in wet mount and KOH prep, 8-44
Trichophyton, 4-49
Trichuriasis
   assessment of, 5-51
   cause of, 5-51
   patient education for, 5-51
   symptoms and signs of, 5-51
   treatment for, 5-51
   zoonotic disease considerations in, 5-51—5-52
Trichuris trichiura, 5-51
Triclabendazole (Fasinex), 5-39
Tricyclic antidepressants, 3-31
Trimethoprim-sulfamethoxazole
   for brucellosis, 5-94
   for cyclosporiasis, 5-37
   for human and animal bites, 7-9
   for nontuberculous mycobacterial infections, 5-57
Triptans, cautions with, 3-56
Trismus, 5-101
Trochanteric bursitis, 3-73
Tropical pulmonary eosinophilia, 5-40
Trough urinal, 5-114f
Trypanosoma brucei, 5-52
   gambiense, 5-52
   rhodesiense, 5-52
Trypanosomiasis
   African, 5-52—5-53
   American, 5-53—5-54
Trypansoma cruzi, 5-53
Tsetse fly bites, 5-52

Tube thoracostomy
  indications for, 8–7
  procedures in, 8–8—8–9
Tuberculosis
  assessment of, 5–54
  cause and transmission of, 5–54
  cough with, 3–15
  cutaneous, 5–54
    assessment of, 4–44
    cause of, 4–44
    follow-up for, 4–45
    patient education for, 4–44—4–45
    signs of, 4–44
    symptoms of, 4–44
    treatment for, 4–44
  diagnostic algorithm for, 3–34f
  differential diagnosis of, 4–13, 5–78, 5–104
  extrapulmonary, 5–54
  follow-up for, 5–55—5–56
  latent, 5–54
  mastitis, 3–7
  patient education for, 5–55
  pleural effusion with, 4–15
  pulmonary, 4–44
  signs of, 5–54
  symptoms of, 5–54
  treatment for, 5–55
  zoonotic disease considerations in, 5–56
Tuberculous meningitis, 5–80
Tularemia
  assessment of, 5–103
  in biological warfare, 6–60
  cause of, 5–102
  differential diagnosis of, 3–112t
  follow-up for, 5–103
  patient education for, 5–103
  signs of, 5–102—5–103
  symptoms of, 5–102
  treatment for, 5–103
  zoonotic disease considerations in, 5–103
Tumors
  low back pain with, 3–6
  of testis and scrotum, 3–79—3–80
Tuning fork, 6–1
Turtle sign, 3–95
Twitch, 5–129
Tylenol
  for barotrauma, 6–4
  for blood transfusion reaction, 8–19
  for cholecystitis, 4–78
  for chronic pelvic pain, 3–44t
  for episiotomy pain, 3–104
  for febrile reaction to blood transfusion, 8–20
  for fever, 3–35
  for headache, 3–56
  for leptospirosis, 5–87
  for meningitis, 4–36
Tylox, 4–92
Tympanic membrane rupture, 6–35
  with blast injuries, 7–23, 7–24
  treatment for, 7–26
Tympanoplasty, 6–22
Typhoid
  diagnostic algorithm for, 3–30—3–34f
  required vaccination for, 5–106t
Typhoid fever
  assessment of, 5–104
  cause of, 5–104
  differential diagnosis of, 3–112t, 4–71, 5–84
  follow-up for, 5–105
  patient education for, 5–105
  signs of, 5–104
  symptoms of, 5–104
  treatment for, 5–104
Typhoid polysaccharide Vi vaccine, 5–105
Typhus, 5–83
  differential diagnosis of, 3–112t, 5–69
  treatment of, 5–84
Tzanck smear, 4–47
Tzanck stain, 8–48—8–49

Ulcerated skin lesions, yaws, 4–63
Ulcers
  aphthous, 5–16—5–17
  genital. See Genital ulcers
  herpetic, 5–16
  perforated
    acute abdominal pain in, 3–3t
    chest pain with, 3–10
  skin, 4–39
  urinary tract, 4–89
Ulnar nerve block
  contraindications to, 5–165
  medial approach in, 5–165f
  ventral approach in, 5–164f
  at wrist, 5–164—5–165
Ultraviolet exposure, 4–65
Ultraviolet keratitis
  differential diagnosis of, 3–25
  treatment of, 3–25
Ultraviolet light exam, for vision loss, 3–23
Ultraviolet sunlight, 3–115
Unconsciousness. See also Consciousness, loss of
  assessment of under fire, 1–1
  with blast injuries, 7–24
  with electrical injury, 7–26
  with heat stroke, 6–50
  with hypoxia, 6–8—6–9
Underwater blast injury, 6–18
  assessment of, 6–19
  follow-up for, 6–20
  patient education for, 6–20
  prevention of, 6–20
  signs of, 6–19
  symptoms of, 6–18
  treatment for, 6–19—6–20
Underwater detonation, calculating safe distance from, 6–20

Unopette system, 8-54
Upper respiratory tract infections
  in barosinusitis, 6-33
  in barotitis, 6-34
Ureaplasma, 5-26
Uremia, pruritus with, 3-114
Ureteral stone, 4-91
Urethral discharges
  assessment of, 5-27
  causes of, 5-26
  follow-up for, 5-27
  male, 3-77
  patient education for, 5-27
  signs of, 5-27
  symptoms of, 5-26
  treatment for, 5-27
Urge incontinence, 4-90
Urinals, 5-114f, 5-115
  construction of, 5-115—5-118
Urinalysis
  confirmation tests in, 8-41—8-42
  for epididymitis, 3-84
  equipment for, 8-41
  indications for, 8-41
  in joint pain assessment, 3-61
  for male genital inflammation, 3-78
  microscopic examination of, 8-42
  for poisoning, 5-141
  procedure in, 8-41—8-43
  for prostatitis, 3-81
  for testis torsion, 3-83
  for urinary incontinence, 4-90
  for urinary tract problems, 4-87—4-88
Urinary catheterization, 4-90, 4-91
Urinary incontinence. See Incontinence, urinary
Urinary retention, 4-90
  treatment of, 4-90—4-91
Urinary tract disorders
  with anuria, 4-88
  blisters and nodules in, 4-89
  edema in, 4-89
  examination tips for, 4-87
  with hematospermia, 4-88
  with hematuria, 4-88
  with normal voiding, 4-88
  with pain at urination, 4-89
  with pain in scrotum, 4-88
  phimosis/paraphimosis, 4-89—4-90
  with priapism, 4-89
  skin lesions in, 4-89
  urinalysis for, 4-87—4-88
Urinary tract infections
  assessment of, 4-93
  causes of, 4-93
  follow-up for, 4-94
  patient education for, 4-94
  signs of, 4-93
  symptoms of, 4-93
  treatment for, 4-93—4-94
Urination, pain with, 4-89

Urine
  appearance of, 8-41
  blood in, 4-88
  centrifuging, 8-42
  cultures of for chronic pelvic pain, 3-46
  nitrites in, 8-43
  output of after bladder decompression, 8-36
Urine dipstick test
  for bacterial vaginosis, 3-47
  for candida vaginitis, 3-50
Urine specific gravity-refractometer method, 8-41
Urised, 4-90
Urobilinogen, 8-43
Urochrome, 8-41
Urolithiasis, 4-91
  assessment of, 4-92
  follow-up for, 4-92
  patient education for, 4-92
  signs of, 4-91—4-92
  symptoms of, 4-91
  treatment for, 4-92
Urticaria, 4-39
  in anaphylactic shock, 7-10
  differential diagnosis of, 3-114
USAF SOF trauma ruck pack list, 1-15—1-16
USAF SOF trauma vest pack list, 1-17
Uterine bleeding, abnormal, 3-39
  assessment of, 3-39—3-41
  follow-up, 3-41
  signs of, 3-39
  symptoms of, 3-39
  treatment for, 3-41
Uterine fibroids
  abnormal uterine bleeding with, 3-39, 3-41
  pelvic examination for, 3-37
Uterus
  bimanual examination of, 3-38—3-39
  complications with cesarean section, 3-100
UV. See under Ultraviolet

Vaccinations
  rabies, 5-82
  required, 5-105—5-106t
  typhoid, 5-105
  vaccinia, 6-60
Vaccinia immune globulin, 6-60
Vaginal delivery, 3-87
  complications of, 3-100
  equipment for, 3-88
  normal, 3-89—3-92f
  precautions, 3-93
  procedure for
    during first stage, 3-88
    during fourth stage, 3-93
    during second stage, 3-88
    during third stage, 3-88—3-93
  stages of, 3-87—3-88
Vaginal discharge, 3-38
Vaginal examination, 3-38

bimanual, 3-38—3-39, 3-40f
Vaginal overgrowth, bacterial, 3-47—3-49
Vaginal speculum, 3-37f
   insertion of, 3-38
Vaginal trichomonas, 5-30
   assessment of, 5-31
   follow-up for, 5-31
   patient education for, 5-31
   symptoms and signs of, 5-30
   treatment for, 5-31
Vaginitis, 3-47
Valacyclovir, 5-29
Valium (diazepam)
   for dizziness, 3-21
   for hip dislocation, 3-69
   for operational stress, 5-148
   for psychogenic dyspnea, 3-117
   for seizures, 4-35
   for temporomandibular joint dislocation, 5-19
   for urinary incontinence, 4-91
   for venomous snake bite, 5-145
Valley fever. *See* Coccidioidomycosis
Valproate, 4-82
Valsalva maneuver
   for barosinusitis, 6-34
   for barotitis, 6-35
   for dizziness, 3-21
   for palpitations, 3-111
Valvular malfunctions, 3-117
Vancomycin
   for cellulitris of penis, 3-78
   for chemical burns, 7-19
   for meningitis, 4-36
   for meningococcemia, 4-41
   for preterm labor infection, 3-94
   for prostatitis, 3-81
   for urinary tract infections, 4-94
Varicella
   differential diagnosis of, 3-113t
   required vaccination for, 5-106t
Varicella zoster virus
   in herpes zoster infection, 4-47
   Tzanck stain for, 8-48
Varicoceles, 3-80
Variola virus, 6-59—6-60
Vaseline
   for blister agents, 6-53
   for tar and asphalt burns, 7-19
Vasovagal faint, 7-12
Vasovagal reaction
   in anaphylactic shock, 7-10
   with sudden emptying of bladder, 8-36
Vasovagal syncope, 1-1
Vecuronium
   continuous infusion of in field, 5-157—5-158t
   dosing guidelines for, 5-156t
Vegetables, disinfection of, 5-109
Venereal warts, 4-48
Venezuelan equine encephalitis, 5-66
Venomous fish, 6-14—6-15

Venomous marine life, 6-13—6-16
Venous thromboembolism, 4-23
VENTID-C mnemonic, 6-10
Ventilation
   for hypothermia, 6-44
   for massive hemothorax, 7-2
Ventilation pump, 8-37
Ventilator support, 4-25
Ventilatory insufficiency, 7-24
Ventolin, 4-22
Ventricular arrhythmia, 6-44
Ventricular fibrillation
   with hypothermia, 6-43
   palpitations with, 3-110
Ventricular tachycardia
   palpitations with, 3-110
   syncope with, 3-118
   treatment of, 3-118
Verapamil
   for chest pain, 3-12
   for palpitations, 3-111
   side effects of, 3-111
Vermox (mebendazole), 5-51
Verruca
   papules of, 4-38
   plana, 4-49
      differential diagnosis of, 4-48
   plantaris, 4-49
   vulgaris, 4-48—4-49
Verruga peruana, 5-93
Versed (midazolam), 5-19
Vertebral arthritis, 3-60
Vertical mattress suture, 8-23
Vesicle, 4-38
Vestibular sedatives, 3-21
Veterinary medicine
   animal disease in, 5-132—5-136
   food storage and preservation in, 5-126—5-128
   IV infusion in, 5-131—5-132
   large animal obstetrics in, 5-130—5-131
   procedures in, 5-123—5-126
   restraint and physical exam in, 5-128—5-130
Veterinary site survey checklist, MedCAP guide, 1-3, 1-10—1-11
Vibramycin
   for allergic pneumonitis, 4-19
   for prostatitis, 3-81
Vibrio
   cholera
      in food poisoning, 4-79
      identification of colonies of, 8-50
   parahaemolyticus, in food poisoning, 4-79
Vincent's infection, 5-15—5-16
Viral infections, 5-63—5-64
   adenoviral, 5-64—5-65
   arboviral encephalitis, 5-66—5-68
   in common cold and flu, 4-10
   dengue fever, 5-65—5-66

hantavirus, 5-68—5-69
hepatitis A, 5-71—5-72
hepatitis B, 5-72—5-73
hepatitis C, 5-73—5-75
hepatitis E, 5-75—5-77
HIV, 5-77—5-78
mononucleosis, 5-78—5-80
poliovirus, 5-80—5-81
prevention and treatment of, 5-74t
rabies, 5-81—5-83
with rash and fever, 3-112—113t
of skin
    ecthyma contagiosum, 4-45—4-46
    herpes zoster, 4-47
    molluscum contagiosum, 4-47—4-48
    warts, 4-48—4-49
yellow fever, 5-69—5-71
Viruses, 5-63—5-64
    culture of, 5-64
Visceral larva migrans, 5-32
    differential diagnosis of, 4-52
Visceral pain, 3-1
Vision loss
    alleviating or aggravating factors in, 3-22
    duration of, 3-22
    with laser injury, 7-28—7-29
    quality of, 3-22
    without trauma
        assessment of, 3-23
        disorders causing, 3-22
        follow-up, 3-24
        patient education for, 3-23
        signs of, 3-22—3-23
        symptoms of, 3-22
        treatment for, 3-23
Visual acuity
    assessing for laser injury, 7-29
    correction of, 6-21
    with eye injury, 3-27
Visual field testing, 3-23
Visualization therapy, 3-5
Vital signs, 2-3
    with acute abdominal pain, 3-4t
    in chest pain, 3-11
    in dyspnea, 3-115—3-116
    in hypovolemic shock, 7-12
    M5 item list for assessing, 1-18
    with palpitations, 3-110
    in trauma patients, 7-3
Vitamin deficiencies, 5-137—5-139t
Vitreoretinal injury, laser, 7-29
Vitreous hemorrhage
    in acute vision loss, 3-22
    differential diagnosis of, 3-23
    treatment of, 3-23
Voiding
    bladder catheterization for problems with, 8-33—8-34
    normal, 4-88
Volvulus, 4-86

Vomiting, radiogenic versus psychogenic, 6-62
Vulvar pruritus, 3-114
Vulvitis
    assessment of, 3-50
    follow-up for, 3-50
    patient education for, 3-50
    signs of, 3-49—3-50
    symptoms of, 3-49
    treatment for, 3-50

Warm bath, 3-47
Warm compress
    for Bartholin's gland abscess or cyst, 3-52
    for chronic pelvic pain, 3-47
Warts. *See also* Plantar warts; Venereal warts
    assessment of, 4-49
    causes of, 4-48
    follow-up for, 4-49
    signs of, 4-48—4-49
    symptoms of, 4-48
    treatment for, 4-49
Washed red blood cells, 8-15—8-16
Washing powder lung, 4-18
Wasp sting, 4-57—4-59
Waste disposal, 5-111
    garbage and rubbish, 5-118
    human, 5-111—5-118
    liquid, 5-118
Water
    contamination of
        in cyclosporiasis transmission, 5-36
        in *Fasciola hepatica* infections, 5-39
        in giardiasis, 5-40
        in leptospirosis, 5-86
    emergency, 5-119
    field purification of, 5-118—5-119
    quality of, 5-119
    quantity of, 5-119
    treatment of, 5-119—5-120
Water-borne disease, prevention of, 5-118—5-119
Water immersion
    for frostbite, 6-42
    hypothermia with, 6-43
Water production points, 5-118
Water ski technique
    for shoulder dislocation, 3-67
    for shoulder pain, 3-71
WBC Unopette, 8-57
WBCs. *See* White blood cells
Weapon injuries, non-lethal, 7-28—7-29
Weapons of mass destruction
    chemical, 6-51—6-54
    symptoms of, 6-51
Web space digital block
    of finger, 5-160f
    of toe, 5-162f
Weight
    control of for hypertensive emergency, 4-5
    loss of, 3-29

Weil's disease, 5-86
West Nile encephalitis, 5-66
Western equine encephalitis, 5-66
Wet lung syndrome. *See* Acute respiratory distress syndrome
Wet mount
    for candida vaginitis, 3-50
    procedures in, 8-43—8-44
Wet-to-dry dressings, 8-28
Wheals, 4-39
Wheezing, 4-19, 4-20
Whipworm. *See* Trichuriasis
White blood cells, 4-8
    count of
        differential, 8-58
        in joint pain assessment, 3-61
        on whole blood, 8-56—8-58
    elevated levels in anemia, 4-9
    morphological variations of, 8-58
White phosphorus burns, 7-19
Whiteheads, 4-38
Wigand maneuver, 3-99f
Woodson Plastic Instrument, 5-25
Worm infection, horse, 5-133
Wound care
    after appendectomy, 4-77
    in Bartholin's gland cyst or abscess, 3-53
    for burn injury, 7-22
    in shoulder injury, 3-72
    for venomous snake bite, 5-146
Wound debridement, 8-25—8-26
Wounded in action (WIA) casualties, 7-3—7-4
Wright stained peripheral smear, 4-8
Wright's stain
    for relapsing fever, 5-89
    in Tzanck stain, 8-48
    using Cameco Quik Stain, 8-46—8-47
Wrist
    joint aspiration at, 8-29
    nerve blockade at, 5-164—5-168
Wuchereria bancrofti, 5-39

X-ray studies
    for ankle pain, 3-76
    for shoulder joint pain, 3-71
Xerophthalmia, 5-138t
Xerosis, 3-114
Xylazine, 5-133
Xylocaine (epinephrine)
    for ingrown toenail, 5-3
    lidocaine with, 5-20

Yaws, 4-63
Yeast infections, glans penis, 3-78
Yellow fever, 5-69—5-70
    assessment of, 5-70
    in biological warfare, 6-61
    differential diagnosis of, 5-71, 5-87
    follow-up for, 5-71
    patient education for, 5-70—5-71
    prevention and treatment of, 5-74t
    required vaccination for, 5-106t
    signs of, 5-70
    symptoms of, 5-70
    treatment for, 5-70
Yellowjacket sting, 4-57—4-59
Yersinia
    enterocolitica, 5-32
    in food poisoning, 4-79
    pestis, 5-95, 6-58
    pseudotuberculosis, 5-32
Yersiniosis, 5-32

Zaleplon (Sonata), 3-30
Zanamivir (Relenza), 4-11—4-12
Zebrafish, 6-13
Zidovudine/lamivudine (Combivir), 5-78
Ziehl-Neilson stain, 8-47
Zinc deficiency, 5-139t
Zinc oxide
    for dental caries, 5-12
    powder, 5-24
    for tooth fracture, 5-13
Zinc oxide-eugenol paste, 5-12
Zithromax, 3-102
Zofran (ondansetron), 6-62
Zoloft, 3-17, 3-31
Zolpidem (Ambien), 3-30
Zoonotic diseases, 5-31—5-32
    African trypanosomiasis, 5-53
    American trypanosomiasis, 5-54
    anthrax, 5-92
    arboviral encephalitis, 5-68
    Babesia in, 5-35
    bartonellosis, 5-93
    blastomycosis, 5-60
    brucellosis, 5-94
    clonorchiasis, 5-36
    dermatophyte infections, 4-50—4-51
    Fasciola hepatica infections, 5-39
    giardiasis, 5-41
    hantavirus, 5-69
    histoplasmosis, 5-62
    hookworm and cutaneous larva migrans, 5-42
    leishmaniasis, 5-43—5-44
    leptospirosis, 5-87
    Lyme disease, 5-88—5-89
    Mycobacterium bovis tuberculosis, 5-56
    paragonimiasis, 5-46
    plague, 5-97
    Q fever, 5-86
    rabies, 5-82—5-83
    rat bite fever, 5-98
    rickettsial fever, 5-84—5-85
    scabies, 4-62
    schistosomiasis, 5-47

streptococcal infections, 5–100
strongyloidiasis, 5–48
tapeworm infections, 5–49—5–50
trichenellosis, 5–51
trichuriasis, 5–51—5–52
tularemia, 5–103

www.ingramcontent.com/pod-product-compliance
Lightning Source LLC
Chambersburg PA
CBHW071204240526
45470CB00018B/1398